MARINE MICROBIOLOGY

The 3rd Edition of this book captures all the recent amazing advances in our understanding of the marine microbiology world but still manages to present the concepts in a an easy, informative and entertaining way that will engage the novice to the expert. What a great book and a fun read.

- David Bourne, James Cook University and The Australian Institute of Marine Science

It is great to see another edition of the book given that marine microbiology is such a fast moving and scientifically diverse field. Munn's new edition will be a great resource for new students and advanced scientists alike.

- Greta Reintjes, Max Planck Institute for Marine Microbiology, Bremen, Germany

Reading this textbook has made me realise how much the field of marine microbiology has progressed in recent decades. I recommend this book also to biogeochemists and ecologists in search of the Big Picture of ocean functioning. The many details of interactions emerging from the microbial world are amazing and shed light on the factors driving evolution of these ancient ecosystems.

- Victor Smetacek, Alfred Wegener Institute for Polar and Marine Research, Bremerhaven, Germany

It has been astonishing to see the evolution of this book over the years. With its many 'RESEARCH FOCUS' boxes and 'SIDEBARS', this is more than your usual textbook. It is written in an enthusiastic, thought-provoking manner, encompassing the most up-to-date concepts in marine microbiology. From planktonic tunicates involved in carbon cycling, to viruses infecting other viruses, this essential read has it all!

- Jozef I. Nissimov, Lecturer in Microbial Oceanography, SAMS, Scottish Marine Institute, UK

Colin B. Munn obtained a BSc (Hons) degree from University College London and a PhD from the University of Birmingham. He is an Honorary Fellow of the Marine Institute of University of Plymouth, England and was Associate Professor of Microbiology and Admissions Tutor for Marine Biology programs until 2017. He has a long experience of research in various aspects of microbiology, with particular interests in the interactions between symbiotic and pathogenic microorganisms and their hosts. He has held former positions as Visiting Professor at the University of Victoria, Central University of Venezuela and St Georges University, Grenada and a visiting researcher (Leverhulme Fellow) at James Cook University and the Australian Institute of Marine Science. He has served as external examiner for Bachelor's, Master's, and PhD degrees in many countries and as a special assessor for molecular and organismal biology for the UK Quality Assurance Agency for Higher Education. He has extensive research experience in a range of microbiological topics, with special interest in interactions between pathogenic and symbiotic microbes and their hosts.

MARINE MICROBIOLOGY

ECOLOGY & APPLICATIONS

Third Edition

Colin B. Munn

CRC Press
Taylor & Francis Group
Boca Raton London New York

CRC Press is an imprint of the
Taylor & Francis Group, an **informa** business

CRC Press
Taylor & Francis Group
6000 Broken Sound Parkway NW, Suite 300
Boca Raton, FL 33487-2742

Printed on acid-free paper

International Standard Book Number-13: 978-0-367-18356-1 (Paperback)
International Standard Book Number-13: 978-0-367-18359-2 (Hardback)

Visit the Taylor & Francis Web site at
http://www.taylorandfrancis.com

and the CRC Press Web site at
http://www.crcpress.com

Contents

Chapter 4 Diversity of Marine Bacteria   113

Chapter 5 Marine Archaea   149

Chapter 6 Marine Eukaryotic Microbes   165

Chapter 7 Marine Viruses   195

Chapter 8 Microbes in Ocean Processes—Carbon Cycling   219

Chapter 12 Marine Microbes as Agents of Human Disease   331

Chapter 13 Microbial Aspects of Marine Biofouling, Biodeterioration, and Pollution   355

Chapter 14 Marine Microbial Biotechnology   387

List of Research Focus Boxes

Preface

In the Foreword to the Second Edition of this textbook published in 2011, Farooq Azam, Distinguished Professor of Scripps Institution of Oceanography, wrote:

> When we think of the ocean we might think of whales, waves, dolphins, fish, the smell of the sea, its blue color, and its vastness; most of us would not look out at the sea and think of marine microbes, nor did marine scientists for over a century. They sailed the seas and strenuously dragged plankton nets through the ocean's pelagic zone to capture what they judged would represent the marine biota. But they were unaware that the great majority of the biota, perhaps 99 percent, easily streamed through the holes of their nets; the holes were simply too big to capture these microbes. Even when membrane filters and microscopy were used, and they revealed great diversity of microplankton and nanoplankton, most microbes evaded detection. The view of the pelagic web of life that emerged, and became entrenched for a century, was then based on a tiny fraction of marine biota. As a result, fisheries scientists used models that did not include marine microbes, as did marine chemists and geochemists who studied how biological forces influenced the grand cycles of elements in the ocean. Much had to be revised as the major roles of the microbes were discovered, following the development of new concepts, incisive imaging, and molecular methods to observe and study marine plankton.

In this Third Edition, it will become clear that astonishing advances in marine microbiology have continued to propel the discipline to be one of the most important areas of modern science. I hope that readers will share a sense of my excitement of learning about new discoveries in this fascinating and fast-moving subject.

This book is intended primarily for upper-level undergraduates and graduate students, but I anticipate that it will also prove useful to researchers who are interested in some of the broader aspects and applications outside of their specific area of investigation. University courses often include some element of microbiology as a specialist option for oceanography or marine biology majors who have little previous knowledge of microbiology, but marine microbiology is lightly covered in most textbooks. I hope that this book may play some part in rectifying this deficiency. I also hope that the book will be useful to microbiology majors studying courses in environmental microbiology, who may have little knowledge about ocean processes or the applications of the study of marine microorganisms. I have attempted to make the book sufficiently self-contained to satisfy all of the various potential audiences. Above all, I wanted to create a book that is enjoyable to read, with the overall aim of bringing together an understanding of microbial biology and ecology with consideration of the applications for environmental management, human welfare, health, and economic activity.

As will become evident, many common themes and recurring concepts link the activities, diversity, ecology, and applications of marine microbes. In each chapter, I have attempted to summarize the current state of knowledge about each aspect, with extensive cross-linking to other sections. To improve readability, I rarely cite specific references in the main text, but each chapter contains special interest boxes, which contain references to recent research. Short boxes marked with the symbols ⑦ or ⓘ highlight important questions or interesting facts that supplement the main text. The choice of these topics is entirely my personal whim—they represent subjects that I think are particularly intriguing, exciting, controversial, or sometimes fun. Each chapter contains one to three *Research Focus Boxes*—these are intended to be relatively self-contained "mini-essays," which explore in more detail some hot topics of investigation. Examples include the impacts of rising CO_2 levels on microbial community structure and ocean processes, interactions of microbes with plastic pollution, symbiotic interactions, and emerging diseases of marine life. They are not intended to be exhaustive reviews, but I hope that they serve as a stimulus for students to follow up some of the original research papers suggested and use these as a starting point for further inquiry or discussion in seminars. The reference list for each chapter contain numerous suggestions for further reading.

The great advances in marine microbiology have occurred because of the development and application of innovative new techniques. Therefore, an understanding of the main principles of these methods is essential if the student is to make sense of research findings. *Chapter 2* is written in a style that concentrates on the principles and avoids too much technical detail, so I implore you to read this chapter right through at an early stage to gain an overview of methods, referring back whenever a particular technique is mentioned later.

What's new in this edition? The general aims and structure of the book are similar, but all material has been updated and most sections have been completely rewritten and expanded to take account of the many new discoveries in this field since the second edition was published in 2011, especially the astonishing advances due to extensive application of high-through-put sequencing, single cell genomics, and analysis of large datasets. Significant advances in understanding the diversity and evolution of bacteria, archaea, fungi, protists, and viruses are discussed and their importance in marine processes is explored in detail. There are many new color diagrams, illustrations, and boxes to aid students' interest and understanding. I have also tried to incorporate the numerous helpful comments received from students, course leaders, and reviewers about previous editions and early drafts of this one. In addition, the book has a companion website, which provides additional online resources for instructors and students, including a summary of key concepts and terminology for each chapter, links to further resources, artwork, videos, and more. My personal blog at www.marinemicro.org also contains news and longer research articles.

Despite an exhaustive review process, astute readers will undoubtedly spot some errors and omissions, or have suggestions for improvement. I welcome your comments—please e-mail me at c.munn@plymouth.ac.uk or via the publishers.

Acknowledgments

This book would not have been possible without the continued stimulation of ideas through teaching my university courses and mentoring research students. It has been a source of immense pleasure and pride to see my efforts come full circle through the authoritative input of several of my former students who have gone on to establish successful research careers in marine microbiology. I am indebted to them and to the many other scientific colleagues listed below, who gave up their time to comment on the original proposal, review drafts, and provide valuable suggestions to improve the text. I also thank all those colleagues who provided images, especially Davis Laundon for his spectacular cover design. I am particularly grateful to Alice Oven, Senior Editor at CRC Press, for her enthusiasm and encouragement to produce a Third Edition and for developing the new improved design for the book. I also thank Sadé Lee, Damanpreet Kaur, Andrew Corrigan, and other members of the editorial and design team at CRC Press for their expertise and help. Finally, to Sheila—thank you for your constant love, support, and patience during this venture.

Colin B. Munn,
Marine Institute,
University of Plymouth, UK.

Reviewers

Mike Allen, University of Exeter and Plymouth Marine Laboratory; Rudolf Amann, Max Planck Institute for Marine Microbiology; Craig Baker-Austin, CEFAS Laboratory; David Bourne, AIMS and James Cook University; Mya Breitbart, University of South Florida; Craig Carlson, University of California, Santa Barbara; Ian Cooper, University of Brighton; Michael Cunliffe, Marine Biological Association of the UK; Jesse Dillon, California State University, Long Beach; Stuart Donachie, University of Hawaii, Manoa; Alexander Gruhl, Max Planck Institute for Marine Microbiology; Marcelo Gutiérrez, Universidad de Concepción, Chile; Oliver Jäckle, Max Planck Institute for Marine Microbiology; Christina Kellogg, US Geological Survey, St. Petersburg Coastal and Marine Science Center; Anne Leonard, University of Exeter; Davis Laundon; Marine Biological Association of the UK; Sophie Leterme, Flinders University; Lauren Messer, University of Queensland; Jozef Nissimov, Scottish Association for Marine Science; Mircea Podar, Oak Ridge National Laboratory; Elizabeth Robertson, Gothenburg University; Emma Ransome, Imperial College London; Greta Reintjes, Max Planck Institute for Marine Microbiology; Wolfgang Sand, University of Duisberg-Essen; Val Smith, University of St Andrews; Victor Smetacek, Alfred Wegener Institute Helmholtz Centre for Polar and Marine Research; Roman Stocker, Environmental Microfluidics Group, ETH Zürich; Ben Temperton, Exeter University; Jack Thomson, University of Liverpool; Malin Tietjen, Max Planck Institute for Marine Microbiology; Richard Thompson, University of Plymouth; Rebecca Vega-Thurber, Oregon State University; Jack Wang, University of Queensland; Joanna Warwick-Dugdale, Plymouth Marine Laboratory and University of Exeter; Robyn Wright, University of Warwick; Willie Wilson, Marine Biological Association of the UK; Erik Zettler, NIOZ Royal Netherlands Institute for Sea Research and Utrecht University; Xiao-Hua Zhang, Ocean University of China.

Chapter 1

Microbes in the Marine Environment

Viewed from space, it is clear why our planet would be better named "Ocean" than "Earth." More than 70% of the planet's surface is covered by interconnected bodies of water. Life originated in the oceans about 3.5 billion years ago and microbes were the only form of life for two-thirds of the planet's existence. The development and maintenance of all other forms of life depend absolutely on the past and present activities of marine microbes. Yet the vast majority of humans—including many marine scientists—live their lives completely unaware of the diversity and importance of marine microbes. Such understanding is vital, as we now live in a period of rapid global change. This chapter introduces the scope of marine microbiology, the different types of marine microbe (viruses, bacteria, archaea, fungi, and protists), and their place in the living world. The activities of microbes in the many diverse habitats found in the marine environment are introduced to provide the background for more detailed consideration in later chapters.

Key Concepts

- Modern methods have led to new ideas about the diversity and evolution of microbial life.

- Marine microbes are highly diverse and exist in huge numbers, forming a major component of biomass on Earth.

- The most abundant marine microbes are exceptionally small.

- The oceans provide diverse specialized habitats, in which physical and chemical conditions determine microbial activities.

- Planktonic microbes are responsible for primary productivity and recycling of organic compounds in a continuum of dissolved and particulate matter.

- Microbes are important in the formation and fate of sediments and there is abundant life below the seafloor.

- Microbes colonize the surfaces of inanimate objects and other living organisms by the formation of biofilms and microbial mats.

ORIGINS AND SCOPE OF MARINE MICROBIOLOGY

Marine microbiology has developed into one of the most important areas of modern science

Ever since a detailed study of the microbial world began in the late nineteenth century, scientists have asked questions about the diversity of microbial life in the sea, its role in ocean processes, its interactions with other marine life, and its importance to humans. However, despite excellent work by pioneering scientists, progress in understanding accumulated gradually and some aspects were poorly understood until recently. However, toward the end of the twentieth century, several factors conspired to propel marine microbiology to the forefront of "mainstream" science. The involvement of more investigators and the subsequent application of new technology mean that it is now one of the most exciting and fast-moving areas of investigation. Today, our subject is characterized by multidisciplinary investigations and widespread application of powerful new tools in molecular biology, information technology, remote sensing, and deep-sea exploration, leading to astonishing discoveries of the abundance, diversity, and interactions of marine microbial life and its role in global ecology. These continuing new discoveries necessitate radical rethinking of our understanding of ocean processes. We now realize the vital role that marine microbes play in the maintenance of our planet, a fact that will have great bearing on our ability to respond to problems such as the increase in human population, overexploitation of fisheries, climate change, ocean acidification, and marine pollution. Studies of the interactions of marine microbes with other organisms are providing intriguing insights into the phenomena of food webs, symbiosis, pathogenicity, and the important role microbiomes play in metazoan biology. Since some marine microbes produce disease or damage, we need to study these processes and develop ways to control them. Finally, marine microbes have beneficial properties such as the manufacture of new drugs and materials, and the development of new processes in the growing field of marine biotechnology. This chapter sets the scene for the discussion of all these topics in this book.

Microbes include microscopic cellular organisms and non-cellular viruses

Defining the terms "microbiology" and "microorganism" is surprisingly difficult! Microbiology is the study of very small organisms that are too small to be seen clearly with the naked eye (i.e. less than about 1 mm diameter), but most microbiologists are concerned with the activities or molecular properties of microbial communities rather than viewing individual cells with a microscope. The term "microorganism" simply refers to a form of life that falls within the microscopic size range, but there is a huge spectrum of diversity concealed by this all-encompassing term. Indeed, some "microorganisms" are large enough to see without using a microscope, so this is not entirely satisfactory either. Some scientists would argue that the distinguishing features of microorganisms are small size, unicellular organization, and osmotrophy (feeding by absorption of nutrients). The osmotrophic characteristic is important because diffusion processes are a major limitation to cell size, as discussed in the next section. However, this characteristic would exclude many microscopic unicellular eukaryotes, many of which feed by phagotrophy (engulfment of particles). For many years, these microorganisms were studied by specialists who had a traditional background in botany or zoology and classified into "plant-like" (algae) or "animal-like" (protozoa) groups. However, many of these organisms are mixotrophic and can switch from photosynthesis to phagotrophic feeding, so the "plant" or "animal" similarity is meaningless. This loose grouping of organisms is therefore called "protists," a diverse category encompassing most of the diversity within the domain Bacteria. Depending on their size (see below), they may also be referred to as microeukaryotes or picoeukaryotes. The study of marine protists and recognition that they are microbes with a major role in ocean processes has lagged behind the study of bacteria until recently. Where do viruses fit? Viruses are obviously microscopic, so I consider them to be microbes. However, they are not cellular, so cannot be described as micro*organisms* and many would argue that they are not living (this question is explored in depth in *Chapter 7*). In summary, in this book I use the term "microbe" as a generic descriptor for microscopic cellular organisms including bacteria, archaea, fungi, and protists, together with the non-cellular viruses.

 TINY MICROBES... HUGE NUMBERS

Whitman et al. (1998) estimated the total number of bacterial and archaeal cells in the oceans to be about 10^{29}. This figure was confirmed by Bar-On et al. (2018) in a recent recalculation based on analysis of many new datasets; they also estimated the biomass of marine bacteria and archaea at 1.3 and 0.3 gigatons (Gt) of carbon, respectively. Suttle (2005) calculated the number of viruses to be about 10^{30}—again, this was confirmed by Bar-On et al. (2018). This is an unimaginably huge number—1 million, million, million, million, million, million. If all the marine virus particles were placed end to end, they would span about 10 million light years (100 times the distance across our own galaxy).

Marine microorganisms are found in all three domains of cellular life

Biologists usually rely on the study of morphology and physiological properties to classify living organisms, but these characteristics have always proved frustratingly unhelpful when dealing with microbes. Modern methods of classification group organisms by attempting to determine the evolutionary relationships. Such phylogenetic systems of classification depend on comparisons of the genetic information contained in their macromolecules, especially nucleic acids and proteins. If two organisms are very closely related, we expect the sequence of the individual units (nucleotides or amino acids) in a macromolecule to be more similar than they would be in two unrelated organisms. In the 1970s, Carl Woese and colleagues pioneered the use of ribosomal RNA (rRNA) sequencing in order to develop a better view of microbial diversity. Our view of the living world has since been revolutionized by advances in this approach, made possible because of the parallel advances in molecular biological techniques and computer processing of the large amounts of information generated. Because the secondary structure of rRNA is so important in the ribosome and the vital cell function of protein synthesis, base sequence changes in the rRNA molecule occur quite slowly in evolution. In fact, some parts of rRNA are highly conserved and sequence comparisons can be used to ascertain the similarity of organisms on a broad basis. The methods and applications of this major technique are described in *Chapter 2*.

In 1990, Woese identified three distinct lineages of cellular life, referred to as the domains Archaea, Bacteria, and Eukarya. A phylogenetic "tree of life" based on rRNA sequences envisaged divergence of these three domains from an original "universal ancestor" (*Figure 1.1A*). A phylogenetic approach to classification is now widely accepted, although some biologists prefer other systems. Microbiologists like it because we can say that we study two entire domains of life and a significant proportion of the third! Traditionally, members of the domains Bacteria and Archaea have been grouped together as "the prokaryotes," because they share a simple internal cellular structure without a nucleus. However, the most important consequence of the three-domain tree of life is that we now realize that the Bacteria and Archaea are completely different phylogenetic groups. Archaea are not a peculiar, specialized

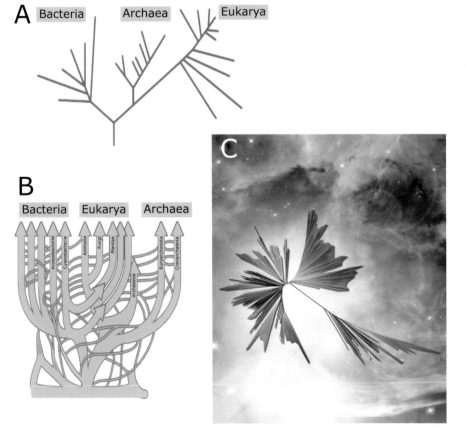

Figure 1.1 Representations of the three domains of life. A. Simple tree based on early interpretation of ribosomal RNA sequencing. In this model, the root of the tree is envisaged as a hypothetical universal ancestor from which all life evolved. B. A three-domain tree based on evidence of extensive lateral gene transfer, revealed by studies of other genes. (Drawn before discovery of other archaeal branches; see *Box 5.1*). C. An artistic representation of major divisions of the tree of life by Hug et al. (2016). The numerous known groups of the Bacteria are shown on the left, with the large group of currently uncultivable Bacteria termed the Candidate Phyla Radiation at upper right. The Archaea are shown at the left of the lower branch, with the Eukarya at the lower right. (See *Figure 4.1* for an updated detailed version of the tree). Credits: B. Gary J. Olsen, University of Illinois, based on concept of W. Ford Doolittle. C. Zosia Rostomian, Berkeley Lab.

**TWO DOMAINS
OR THREE?**

There are many differences of opinion about the relationships between organisms, especially when we try to explain deep evolutionary branches. Some authors believe that new evidence about the evolution of rRNA and the use of new models to compute phylogenetic trees calls the three-domain tree concept into question. Embley and Williams (2016) argue that the discovery of new archaeal groups (see *Chapter 4*) supports the idea of a two-domain "eocyte" tree to explain the origin of the eukaryotes from within the Archaea. The "archaeal ancestor hypothesis" for the origin of eukaryotes is gaining acceptance, although Forterre (2015) and Nasir et al. (2016) provide alternative explanations.

group of bacteria as originally thought (for many years they were called the archaebacteria) but are in fact a completely separate group that actually has closer phylogenetic relationships to the Eukarya than to the Bacteria. This concept has proved to be very influential in shaping our thinking about the evolution of organisms. As new methods and knowledge about genomes has developed, the simple three-domain tree has changed, as illustrated in *Figures 1.1B* and *1.1C*. These developments are discussed in detail in *Chapters 4* and *5*.

The members of the Eukarya domain are the protists, fungi, plants, and animals. Their cells are distinguished by a membrane-bound nucleus and organelles with specific functions. Mitochondria occur in all eukaryotic cells, with the exception of a few anaerobic protozoa, and carry out the processes of respiratory electron transport. In photosynthetic eukaryotes, chloroplasts carry out reactions for the transfer of light energy for cellular metabolism. Various lines of evidence (especially the molecular analysis of the nucleic acids and proteins) support the "endosymbiosis theory" (originally developed by Lynn Margulis in the 1960s) that the organelles of eukaryotic cells have evolved by a process of endosymbiosis, in which one cell lives inside another cell for mutual benefit. This theory proposes that the original source of mitochondria in eukaryotic cells occurred when primitive cells acquired respiratory bacteria (most closely related to proteobacteria) and that the chloroplasts evolved from endosymbiosis with cyanobacteria. Such interactions between different types of cell have continued throughout evolution, and *Chapter 10* contains many examples of endosymbiosis involving microbes.

Horizontal gene transfer confounds our understanding of evolution

The use of rRNA sequences as a basis for phylogenetic classification has revolutionized our understanding of microbial diversity and phylogeny. However, since advances in DNA sequencing (see *Chapter 2*) made it possible to study the sequences of many other genes, we have found increasing evidence of extensive horizontal gene transfer (HGT, also known as lateral gene transfer, LGT) between microbes. Such transfers occur most commonly between related organisms but transfers across bigger genetic distances also occur—even between domains. Members of the Bacteria and Archaea contain some genes with very similar sequences, and members of the Eukarya contain genes from both of the other domains. Some members of the domain Bacteria have even been shown to contain eukaryotic genes. Previously, evolution was explained only by the processes of mutation and sexual recombination, but we now know that the pace of evolution is accelerated by the transfer and acquisition of modules of genetic information. This phenomenon is widespread in modern members of the Bacteria and Archaea and can occur via three processes. During the process known as transformation, cells may take up and express naked DNA from the environment, while conjugation relies on cell–cell contact mediated by pili. The most important source of HGT is the process of transduction by phages (viruses infecting bacterial or archaeal cells and introducing "foreign" DNA); this is explored in detail in *Chapter 7*. The enormous diversity of marine viruses and the identification of a viral origin of genes in many marine organisms indicate how important this process has been throughout evolution.

Viruses are non-cellular entities with great importance in marine ecosystems

Virus particles (virions) consist of a protein capsid containing the viral genome composed of either RNA or DNA. Because they only contain one type of nucleic acid, viruses must infect living cells and take over the host's cellular machinery in order to replicate. It is often thought that viruses could have evolved (perhaps from bacteria) as obligate parasites that have progressively lost genetic information until they consist of only a few genes, or that they represent fragments of host-cell RNA or DNA that have gained independence from cellular control. New ideas about the evolution of viruses are discussed in *Chapter 7*. The genome of viruses often contains sequences that are equivalent to specific sequences in the host cell. Viruses exist for every major group of cellular organisms (Bacteria, Archaea, Fungi, protists, plants, and animals), but at present we have knowledge of only a tiny proportion of the viruses infecting marine life. As discussed in *Chapter 7*, recognition of the abundance and diversity of marine viruses, and the role that they play in biogeochemical cycles and the control of diversity in marine microbial communities, has been one of the most important discoveries of recent years.

Microbial processes shape the living world

Probably the most important overriding features of microbes are their exceptional diversity and ability to occupy every conceivable habitat for life. Indeed, what we consider "conceivable" is challenged constantly by the discovery of new microbial communities in habitats previously thought of as inhospitable or carrying out processes that we had no idea were microbial in nature. Bacteria and archaea have shaped the subsequent development of life on Earth ever since their first appearance—the metabolic processes that they carry out in the transformation of elements, degradation of organic matter, and recycling of nutrients play a central role in innumerable activities that affect the support and maintenance of all other forms of life. Microbial life and the Earth have evolved together, and the activities of microbes have affected the physical and geochemical properties of the planet. Indeed, they are actually the driving forces responsible for major planetary processes like changes in the composition of the atmosphere, oceans, soil, and rocks. This is especially relevant to our consideration of the marine environment, in view of the huge proportion of the biosphere that this constitutes. Despite the preponderance of microbes and the importance of their activities, they are unseen in everyday human experience. Microbes live and grow almost everywhere, using a huge range of resources, whereas plants and animals occupy only a small subset of possible environments and have a comparatively narrow range of lifestyles.

Marine microbes show great variation in size

Table 1.1 shows the range of dimensions and volumes of some representative marine microbes. Even by the usual standards of microbiology, the most abundant microbes found in seawater are *exceptionally* small—*much* smaller than implied by the common textbook or internet statement that "bacteria are typically a few micrometers in length." Their very small size is the main reason that appreciation of their abundance eluded us for so long. As described in *Chapter 2*, recognition of the abundance of marine microbes depended on the development of fine-pore filters and direct counting methods using epifluorescence microscopy and flow cytometry. Small cell size has great significance in terms of the physical processes that affect life. At the microscale, the rate of molecular diffusion becomes the most important mechanism for transport of substances into and out of the cell. Small cells feeding by absorption (osmotrophy) can take up nutrients more efficiently than larger cells. The surface area to volume ratio (SA/V) is the critical factor because as cell size increases, volume (V) increases more quickly than surface area (SA), as shown in *Figure 1.2A*.

The most abundant ocean bacteria and archaea have very small cell volumes and large SA/V ratios. The majority are smaller than about 0.6 μm in their largest dimension, and many are less than 0.3 μm, with cell volumes as low as 0.003 μm³. Indeed, the most abundant type of organism in the oceans (the SAR11 clade, *Figure 1.3A*) has some of the smallest known cells. If nutrients are severely limiting, as they are in most of the oceans, selection will favor small cells. Since the first description of such small cells, termed "ultramicrobacteria," their size has provoked considerable controversy. Such extremely small cells could result from a genetically fixed phenotype maintained throughout the cell cycle or because of physiological changes associated with starvation. The latter explanation is supported by the fact that some cultured bacteria become much smaller when starved. Most naturally occurring bacteria (identified only by their genetic "signature") have been impossible to grow in culture—this is a central problem in marine microbiology, which we shall return to on several occasions in future chapters. Because of this, it has been difficult to determine whether small size is a genotypically determined condition for marine bacteria. However, studies with some recently cultured marine bacteria from low-nutrient (oligotrophic) ocean environments show that addition of nutrients does *not* cause an increase in cell size. Cells use various strategies to increase the SA/V ratio and thus improve efficiency of diffusion and transport. In fact, spherical cells are the least efficient shape for nutrient uptake, and many marine bacteria and archaea are thin rods or filaments or may have appendages such as stalks or buds. *Figure 1.2B* shows examples of the diverse morphology of marine bacteria in a sample of ocean water. Many of

TIME TO CHOP DOWN THE "TREE OF LIFE"?

The idea that relationships between all living organisms can be represented as a tree of life helped to shape Darwin's theory of evolution by natural selection and has been deeply embedded in the philosophy of biology for more than 150 years. As the importance of endosymbiosis and HGT became better understood, some evolutionary scientists began to question the validity of the "tree of life" concept. A seminal paper by Doolittle (1999) argued that "Molecular phylogeneticists will have failed to find the 'true tree', not because their methods are inadequate or because they have chosen the wrong genes, but because the history of life cannot properly be represented as a tree." Relationships are now envisaged as complex intertwined branches, more like a web (*Figure 1.1B*) or network of genomes (Dagan and Martin, 2009). However, this remains a controversial topic, and some have argued that analysis of genome sequences for "core genes" still supports the idea of a common ancestor and branching tree (Ciccarelli et al., 2006)—an approach that worked successfully for Hug et al. (2016) to develop a new tree of all major groups of life (*Figure 1.1C*).

Table 1.1 *Size range of some representative marine microbes*

Organism	Characteristics	Size (μm)[a]	Volume (μm³)[b]
Brevidensovirus	Icosahedral DNA virus infecting shrimp	0.02	0.000004
Coccolithovirus	Icosahedral DNA virus infecting *Emiliania huxleyi*	0.17	0.003
Thermodiscus	Disc-shaped. Hyperthermophilic archaeon	0.2×0.08	0.003
"*Ca.* Pelagibacter ubique"[c]	Crescent-shaped bacterium ubiquitous in ocean plankton (cultured example of SAR11 clade)	0.1×0.9	0.01
Megavirus chilensis	Giant virus infecting marine amoebae	0.44	0.045
Prochlorococcus	Cocci. Dominant photosynthetic ocean bacterium	0.6	0.1
Ostreococcus	Cocci. Prasinophyte alga. Smallest known eukaryote	0.8	0.3
Vibrio	Curved rods. Bacteria common in coastal environments and associated with animals and human diseases	1×2	2
Pelagomonas calceolata	Photosynthetic flagellate adapted to low light	2	24
Pseudo-nitzschia	Pennate diatom which produces toxic domoic acid	5×80	1600
Staphylothermus marinus	Cocci. Hyperthermophilic archaeon	15	1800
Thioploca auraucae	Filamentous. Sulfur bacterium	30×43	30000
Lingulodinium polyedra	Bioluminescent bloom-forming dinoflagellate	50	65000
Beggiatoa	Filamentous. Sulfur bacterium	50×160	314000
Epulopiscium fishelsoni	Rods. Bacteria symbiotic in fish gut	80×600	3000000
Thiomargarita namibiensis	Cocci. Sulfur bacterium	300[d]	14137100

[a]Approximate diameter×length; where one value is given, this is the diameter of spherical virus particles or cells. [b]Approximate values calculated assuming spherical or cylindrical shapes. [c]*Candidatus*, provisional taxonomic name—see *Chapter 4*. [d]Cells up to 750 μm have been recorded.

the larger organisms overcome the problems of diffusion by having extensive invaginations of the cytoplasmic membrane or large intracellular vacuoles, increasing the SA. Small cell size also has important implications for mechanisms of active motility and chemotaxis, because of the microscale effects of Brownian movement (bombardment by water molecules) and shear forces. Small marine bacteria have mechanisms of motility and chemotaxis quite unlike those with which we are familiar from laboratory studies of organisms such as *Escherichia coli*.

As shown in *Table 1.1*, marine eukaryotic microbes also show a considerable variation in size. Many protists have cellular dimensions that are more typical of the familiar bacteria. The smallest known eukaryote is *Ostreococcus tauri*, which is only about 0.8 μm in diameter (*Figure 1.3B*). Again, the realization that such small cells (now referred to as "picoeukaryotes") play a key role in ocean processes escaped attention until quite recently. Many small protists seem capable of engulfing bacteria of almost the same size as themselves or can prey on much larger organisms. Many groups of the flagellates, ciliates, diatoms, and dinoflagellates are somewhat larger, reaching sizes up to 200 μm, and amoeboid types (radiolarians and foraminifera) can be millimeters in diameter. Finally, a few types of bacteria can be bigger than many protists. The largest of these is *Thiomargarita namibiensis* (*Figure 1.3C*). Further discussion of the size range of microbes is given in the section on plankton below.

OCEAN HABITATS

The world's oceans and seas form an interconnected water system

The oceans cover 3.6×10^8 km² (71% of the Earth's surface) and contain 1.4×10^{21} liters of water (97% of the total on Earth). The average depth of the oceans is 3.8 km, with a number of deep-sea trenches, the deepest of which is the Marianas Trench in the Pacific (11 km). The

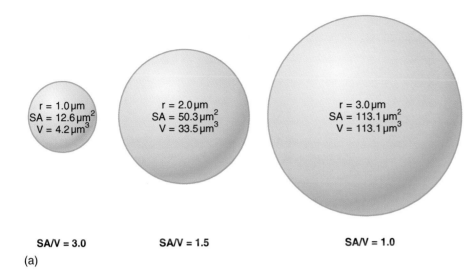

r = 1.0 μm
SA = 12.6 μm^2
V = 4.2 μm^3

r = 2.0 μm
SA = 50.3 μm^2
V = 33.5 μm^3

r = 3.0 μm
SA = 113.1 μm^2
V = 113.1 μm^3

SA/V = 3.0   **SA/V = 1.5**   **SA/V = 1.0**

(a)

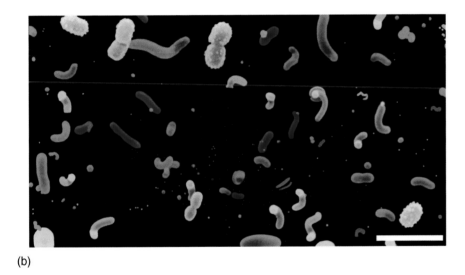

(b)

Figure 1.2 (a) Diagrammatic representation of three spherical cells showing a reduction in the ratio of surface area (SA) to volume (V) as size increases. V is a function of the cube of the radius (V = (4/3)πr^3) whereas SA is a function of the square of the radius (SA=4πr^2). Cells with large SA/V ratios are more efficient at obtaining scarce nutrients by absorption across the membrane. (b) Scanning electron micrograph of picoplankton, showing various cell morphologies of marine bacteria (cells are artificially colorized for effect). Bar represents ~ 1 μm. Credit: Ed DeLong, Massachusetts Institute of Technology.

ocean floor contains large mountain ranges and is the site of almost all the volcanic activity on Earth. More than 80% of the area and 90% of the volume of the oceans occurs beyond the continental shelf. Most of the deep-sea remains unexplored. It is usual to recognize five major ocean basins, although they actually form one interconnected water system.

The Pacific is the deepest and largest ocean, almost as large as all the others combined. This single body of water has an area of 1.6×10^8 km^2. The ocean floor in the eastern Pacific is dominated by the East Pacific Rise, while the western Pacific is dissected by deep trenches. The Atlantic Ocean is the second largest with an area of 7.7×10^7 km^2 lying between Africa, Europe, the Southern Ocean, and the Americas. The mid-Atlantic Ridge is an underwater mountain range stretching down the entire Atlantic basin and the deepest point is the Puerto Rico Trench (8.1 km). The Indian Ocean has an area of 6.9×10^7 km^2 and lies between Africa, the Southern Ocean, Asia, and Australia. A series of ocean ridges cross the basin, and the deepest point is the Java Trench (7.3 km). The Southern Ocean is the body of water between 60°S and Antarctica. It covers 2.0×10^7 km^2 and has a fairly uniform depth of 4–5 km, with a continual eastward water movement called the Atlantic Circumpolar Current. The Arctic Ocean, lying north of the Arctic Circle, is the smallest ocean, with an area of 1.4×10^7 km^2. As well as the major oceans, there are marginal seas, including the Mediterranean, Caribbean, Baltic, Bering, South China Seas, and many others.

At the margins of major landmasses, the ocean is shallow and lies over an extension of the land called the continental shelf. This extends offshore for a distance ranging from a few kilometers to several hundred kilometers and slopes gently to a depth of about 100–200 m,

Figure 1.3 Extremes of size in marine microbes (note different scale bars). A. Cryo-electron tomography of "*Candidatus* Pelagibacter ubique" cells, one of the smallest bacteria known (a cultivated representative of the abundant SAR11 clade). Left: a tomographic slice of a typical log-phase cell. Right: the 3D isosurface-rendered model of the same cell reveals internal cellular organization. Model coloring: outer membrane (blue), inner membrane (cyan), peptidoglycan (white), cytoplasm (orange), nucleoid (red), ribosome-like particles are represented by yellow spheres. B. Transmission electron micrograph (TEM) of section of *Ostreococcus tauri*, the smallest known eukaryote. N = nucleus, m = mitochondrion, c = chloroplast. C. Light micrograph of *Thiomargarita namibiensis*, the largest known bacterium, showing sulfur granules. Credits: A. Xiaowei Zhao and Daniela Nicastro, University of Texas SW Medical Center, Stephen Giovannoni, Oregon State University, and J. Richard McIntosh, University of Colorado. B. Reprinted from Henderson et al. (2007), CC-BY-2.0. C. Heide Schulz-Vogt, Max Planck Institute for Marine Microbiology, Bremen.

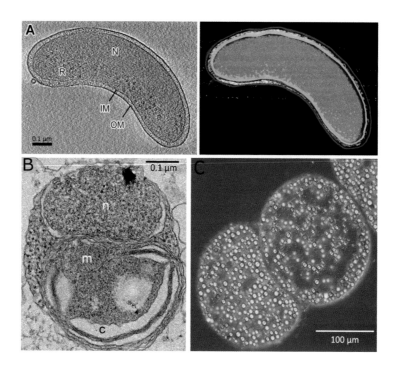

before there is a steeper drop-off to become the continental slope. The abyssal plain covers much of the ocean floor; this is a mostly flat surface with few features, but is broken in various places by ocean ridges, deep-sea trenches, undersea mountains, and volcanic sites.

The upper surface of the ocean is in constant motion owing to winds

Rotation of the Earth deflects moving air and water in a phenomenon known as the Coriolis Effect. Wind belts created by differential heating of air masses move the surface water, and in combination with the Coriolis Effect they generate major surface current systems. This leads to large circular gyres that move clockwise in the northern hemisphere and anticlockwise in the southern hemisphere. Such gyres and currents affect the distribution of nutrients and marine organisms. On the basis of surface ocean temperatures, the marine ecosystem can be divided into four major biogeographical zones, namely polar, cold temperate, warm temperate (subtropical), and tropical (equatorial). The boundaries between these zones are roughly defined by latitude but are strongly affected by surface currents moving heat away from the equator, as well as varying with the season.

Deep-water circulation systems transport water between the ocean basins

Below a depth of about 200 m, ocean water is less affected by mixing and wind-generated currents. However, a system of vast undersea rivers transports water around the globe and has a major influence on the distribution of nutrients (*Figures 1.4 and 8.1*). This thermohaline circulation system—often referred to as the "global ocean conveyor belt"—is formed by the effects of temperature and salinity causing differences in the density of water. Surface water in the North Atlantic flows toward the pole as the Gulf Stream. Water is removed to the atmosphere by evaporation and during the formation of sea ice in high latitudes, resulting in higher salinity. The cold, salty water becomes denser and sinks to form a deep pool, which then flows south toward Antarctica, where more cold, dense water is added. The current then splits, with one branch going toward the Indian Ocean and the other to the Pacific Ocean. As the current nears the equator it warms and becomes less dense, so upwelling occurs. The warmer waters loop back to the Atlantic Ocean, where they start the cycle again.

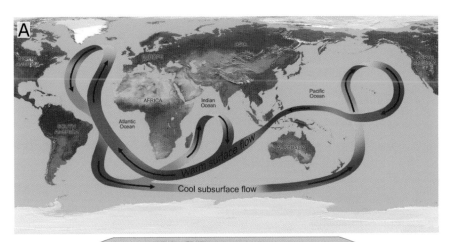

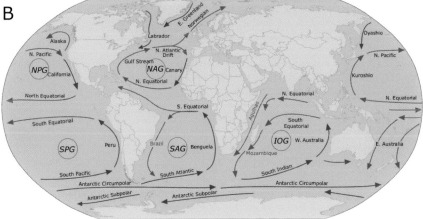

Figure 1.4 A. The thermohaline circulation system (global ocean "conveyor belt"). (Credit: NASA/JPL.) B. Approximate locations of the major warm (red) and cold (blue) ocean currents and the gyres in the South Pacific (SPG), North Pacific (NPG), North Atlantic (NAG), South Atlantic (SAG), and Indian Ocean (IOG).

Light and temperature have important effects on microbial processes

Light is of fundamental importance in the ecology of microbes that use light energy for photosynthesis and other functions, thus affecting primary productivity. The extent to which light of different wavelengths penetrates seawater depends on a number of factors, notably cloud cover, the polar ice caps, dust in the atmosphere, and variation of the incident angle of solar radiation according to season and location on the Earth's surface. Light is absorbed or scattered by organisms and suspended particles. Even in the clearest ocean water, photosynthesis is restricted by the availability of light of sufficient intensity to the upper 150–200 m (*Figure 1.5*). This is termed the photic or euphotic zone (from the Greek for "well lit"). Blue light has the deepest penetration, and photosynthetic microbes at the lower part of the photic zone have light-harvesting systems that are tuned to collect blue light most efficiently (p.134). In turbid coastal waters, during seasonal plankton blooms, the euphotic zone may be only a few meters deep, and green and yellow light have the deepest penetration. Solar radiation also heats surface waters and leads to thermal stratification of seawater. In tropical seas, the continual input of energy from sunlight leads to warming of the surface waters to 25–30°C, causing a considerable difference in density from that of deeper waters. Thus, throughout the year, there is a marked thermocline at about 100–150 m, below which there is a sudden reduction in temperature to 10°C or less. Little mixing occurs between these layers. In polar seas, the water is permanently cold except for a brief period in the summer, when a slight thermocline develops. Apart from this period, turbulence created by surface winds generates mixing of the water to considerable depths. Temperate seas show the greatest seasonal variation in the thermocline, with strong winds and low temperatures leading to extensive mixing in the winter. The thermocline develops in the spring, leading to a marked shallow surface layer of warmer water in summer. As the sea cools and wind increases, the thermocline breaks down again in the autumn. Combined with seasonal variations in light intensity, these effects of temperature stratification and vertical mixing have a great impact on rates of photosynthesis and other microbial activities that affect the entire trophic system.

Figure 1.5 Penetration of light of different wavelengths in seawater. Credit: Kyle Carothers, NOAA-OE.

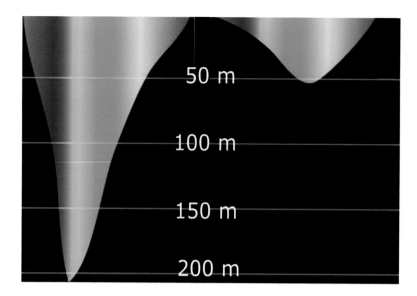

Microbes occur in all the varied habitats found in the oceans

Various ecological zones can be recognized in the marine environment, as shown in *Figure 1.6*. Microbes are found everywhere—as members of the plankton, associated with suspended particles and colloidal materials, attached to surfaces like rocks and submerged structures, in association with plants and animals, or in sediments.

Plankton is a general term in marine biology referring to organisms suspended in the water column that do not have sufficient power of locomotion to resist large-scale water currents (in contrast to the nekton, which are strong-swimming animals). Traditionally, biologists refer to phytoplankton (plants) and zooplankton (animals). Using this approach, we can add the terms bacterioplankton for bacteria, virioplankton for viruses, and mycoplankton for fungi. Traditional concepts of "plant" and "animal" are now unsatisfactory, and the term phytoplankton therefore refers to all photosynthetic microbes, including cyanobacteria, as well as algae and other eukaryotic protists. Of course, phytoplankton is only active in the photic zone, but heterotrophic protists are found at all depths, where active bacterial and archaeal production provides their food source. Another approach to classifying the plankton is in terms of size classes, for which a logarithmic scale ranging from megaplankton (>20 mm) to femtoplankton (<0.2 µm) has been devised. *Table 1.2* shows the size classes that encompass marine microbes. Thus, the viruses constitute the femtoplankton, and bacteria and archaea mainly occur in the picoplankton. The recently discovered giant viruses (see *Chapter 7*) may also be considered to be part of the picoplankton size range. While protists (eukaryotic microbes) have a wide size range and occur in the picoplankton, nanoplankton, and microplankton, we now know that in most marine samples, the majority of protistan cells are in the picoplankton range. This system of tenfold progression is not rigorously adhered to, and many investigators define picoplankton as organisms ≤3 µm. (This cut-off value provides a more coherent pattern in surveys to estimate seasonal or geographic changes in abundance of small eukaryotes measured by filtration). Because individual taxa of photosynthetic and heterotrophic protists can span these three size ranges, this system is of little use in identification and classification. However, it is useful for specifying the size ranges that are likely to be collected using filters or meshes with different cut-off values. It is also useful when considering the feeding of heterotrophic protists, which generally graze by phagocytosis of organisms in the next size class down. Although there are important exceptions, the predator–prey size ratio is typically about 10:1.

Seawater is a complex mixture of inorganic and organic compounds, colloids, and gels

Seawater is a slightly alkaline (pH 7.5–8.4) aqueous solution—a complex mixture of more than 80 solid elements, gases, and dissolved organic substances. The concentration of these

---

? WHY IS THE SEA SALTY?

The constant percolation of rainwater through soil and rocks leads to weathering, in which some of the minerals are dissolved. Ground water has very low levels of salts and we cannot taste it in the water we drink. The addition of salts to the oceans from rivers is thus a very slow process, but evaporation of water from the oceans to form clouds means that the salt concentration has increased to its present level over hundreds of millions of years. Seawater also percolates into the ocean crust where it becomes heated and dissolves minerals, emerging at hydrothermal vents (*Figure 1.13*). Submarine volcanoes also result in reactions between seawater and hot rock, resulting in the release of salts. The salt concentration in the oceans appears to be stable, with deposition of salts in sediments balancing the inputs from weathering, hydrothermal vents, and volcanic activity.

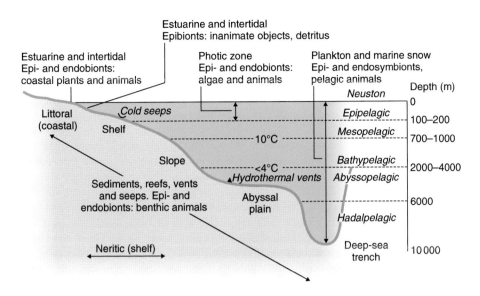

Figure 1.6 The major ecological zones of the oceans and marine microbial habitats (not to scale).

varies considerably according to geographic and physical factors, and it is customary to refer to the salinity of seawater in parts per thousand (‰) to indicate the concentration of dissolved substances. The open ocean has a salinity in the range 34–37‰, with differences in different regions due to dilution by rainfall and evaporation. Oceans in subtropical latitudes have the highest salinity as a result of higher temperatures, while temperate oceans have lower salinity as a result of less evaporation. In coastal regions, seawater is diluted considerably by fresh-water from rivers and terrestrial runoff and is in the range 10–32‰. Conversely, in enclosed areas such as the Red Sea and Arabian Gulf, the salinity may be as high as 44‰. In polar areas, the removal of freshwater by the formation of ice also leads to increased salinity. The major ionic components of seawater are sodium (55% w/v), chloride (31%), sulfate (8%), magnesium (4%), calcium (1%), and potassium (1%). Together, these constitute more than 99% of the weight of salts. There are four minor ions—namely bicarbonate, bromide, borate, and strontium—which together make up just less than 1% of seawater. Many other elements are present in trace amounts (<0.01%), including key nutrients such as nitrate, phosphate, silicate, and iron. The concentration of these is crucial in determining the growth of marine microbes and the net productivity of marine systems, as discussed in *Chapter 9*.

Table 1.2 Classification of plankton by size

Size category		Size range (μm)	Examples of microbial groups
SIZE ↓	Femtoplankton	0.01–0.2	Viruses[a]
	Picoplankton	0.2–2 (3)[b]	Bacteria[c], archaea, prasinophytes, haptophytes, some flagellates
	Nanoplankton	2–20	Coccolithophores, diatoms, dinoflagellates, flagellates
	Microplankton	20–200	Ciliates, diatoms, dinoflagellates, foraminifera, yeasts

[a]Some giant viruses are >1 μm long. [b]A value of ≤3 μm is most commonly used, see text. [c]Some filamentous cyanobacteria and sulfur-oxidizing bacteria occur in larger size classes (see *Table 1.1*).

The concentration of salts has a marked effect on the physical properties of seawater. The freezing point of seawater at 35‰ is −1.9°C, and seawater increases in density up to this point. As noted above, this results in the formation of masses of cold, dense water in polar regions, which sink to the bottom of the ocean basins and are dispersed by deep-water circulation currents. Differences in the density of seawater create a discontinuity called the pycnocline, which separates the top few hundred meters of the water column from deeper water. This has great significance, because the gases oxygen and carbon dioxide are more soluble in cold water.

Oxygen is at its highest concentrations in the top 10–20 m of water, owing to exchange with the atmosphere and production of oxygen by photosynthetic plankton. Concentration decreases with distance from the surface until it reaches a minimum between 200 and 1000 m, and bacterial decomposition of organic matter may create conditions that are almost anoxic. Below this, the oxygen content increases again as a result of the presence of dense water (with increased solubility at lower temperature) that has sunk from polar regions and been transported on the thermohaline circulation system. This oxygen gradient varies greatly in different regions, and there are several regions where large bodies of hypoxic water occur at relatively shallow depths—these are the oxygen minimum zones (*Figure 9.7*).

The solubility of carbon dioxide is an important factor in controlling the exchange of carbon between the atmosphere and the oceans and therefore is of huge significance in understanding climatic processes, as discussed in *Chapter 8*. Only a very small proportion of dissolved inorganic carbon (DIC) is present in the form of dissolved CO_2 gas. Carbon dioxide reacts with water to form carbonic acid, which rapidly dissociates to form bicarbonate, hydrogen ions, and carbonate in the reactions:

$$CO_2\left(gas\right)+H_2O \leftrightharpoons H_2CO_3 \leftrightharpoons H^+ + HCO_3^- \leftrightharpoons 2H^+ + CO_3^{2-}$$

These reactions tend to stay in equilibrium, buffering the pH of seawater within a narrow range and constraining the amount of CO_2 taken up from the atmosphere. However, the increasing atmospheric levels of CO_2 since the industrial revolution are shifting the equilibrium and causing the pH to fall because of increased levels of H^+ ions, leading to the phenomenon known as ocean acidification, which may have major consequences for ocean life.

It is tempting to think of seawater as a homogeneous fluid, with planktonic microbes and nutrients evenly distributed within it. However, a growing body of evidence indicates that there is microscale heterogeneity in the distribution of nutrients around organisms and particles of organic matter. Large-scale processes like productivity, nutrient recycling, and geochemical cycles are the result of microbial activity. In turn, physical processes like turbulence, photon flux, and gas exchange are translated down to the microscale level, affecting microbial behavior and metabolism. Physical factors such as diffusion, shear forces, and viscosity must be considered in this context. A pool of small, soluble organic molecules provides the starting material for bacterial productivity and recycling or carbon compounds that drives ocean food webs (see *Chapter 8*). Only molecules of less than about 600 Da can be assimilated across the membrane—any larger molecules must be broken down by extracellular enzymes. Seawater contains vast amounts of polymers in the form of proteins, carbohydrates, and nucleic acids, resulting from excretion by organisms or the release of their cellular material by lysis. Free dissolved organic matter (DOM) in the form of polymeric macromolecules can spontaneously assemble to form gels in surface waters, which coalesce to form larger aggregates that diffuse into the bulk seawater. It is estimated that at least 10% of organic carbon in the oceans exists in this form. We can envisage seawater as a complex three-dimensional gel-like network with a continuum between dissolved, colloidal, and particulate material (*Figure 1.7*). As a result, at a micrometer scale—the realm of the microbes—there is great patchiness in the distribution of nutrients and the physical properties of microenvironments.

These microgels can further coalesce into larger structures termed transparent exopolymeric particles (TEP). These are especially important in the carbon cycle, because they are critical in promoting the formation of particles that are sufficiently dense to sink through the water column, depositing carbon in the depths of the ocean and its sediments. This continuous shower of clumps and strings of material which falls through the water column is termed "marine snow" because of its resemblance to falling snowflakes when illuminated

IS "DISSOLVED ORGANIC MATTER" REALLY DISSOLVED?

Oceanographers traditionally describe organic matter as "dissolved" and "particulate' (DOM and POM, respectively). Measurements of concentrations and fluxes of DOM and its constituent elements carbon (DOC), nitrogen (DON), and phosphorus (DOP) are among the most important factors in the study of ocean processes. It is important to remember that the difference between "dissolved" and "particulate" is a purely empirical distinction, reflecting the size of filters used in sample preparation. There is no absolute definition, but most filters used in studies of DOM and POM have pore sizes from about 0.45 to 1.0 μm. Many bacteria and almost all viruses would pass through such filters and therefore appear in the DOM fraction! Colloidal material and polymers aggregate to form particles, and it is only low-molecular-weight compounds like sugars and amino acids that are truly dissolved. Thus, DOM and POM form a continuum, with microbes spanning both fractions.

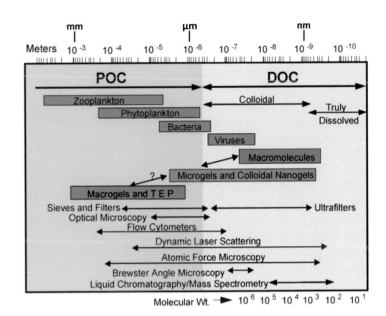

Figure 1.7 Representation of the size continuum of marine particles, indicating the size range of planktonic microbes and methods used to study the different fractions. Reprinted from Verdugo et al. (2004) with permission from Elsevier.

underwater. Marine snow consists of aggregates of plankton cells, detritus from dead or dying plankton, zooplankton fecal material, and inorganic particles, glued together by the matrix of TEP released from plankton (*Figure 1.8*). Most particles are 0.5 to a few micrometers in diameter, but they can grow to several centimeters in calm waters. Dissolved polymers aggregate to form nanogels, stabilized by calcium binding. Larger aggregates form microgels as a result of collision and coagulation of primary particles, and they increase in size as they acquire more material through these physical processes. The nucleus for snow formation is often the mucus-based feeding structures used by salps and larvaceans in the zooplankton. Dying diatoms, at the end of a bloom, often precipitate large-scale snow formation owing to the production of large amounts of mucopolysaccharides in their cell walls. The generation of water currents during feeding by flagellates and ciliates colonizing the aggregate also collects particles from the surrounding water and leads to growth of the snow particle.

Marine snow is mainly produced in the upper 100–200 m of the water column, and large particles can sink up to 100 m per day, allowing them to travel from the surface to the ocean deep within a matter of days. This is the main mechanism by which a proportion of the photosynthetically fixed carbon is transported from the surface layers of the ocean to deeper waters and the seafloor. However, aggregates also contain active complex assemblages of bacteria and protists that graze on them. Levels of microorganisms in marine snow are typically 10^8–10^9 mL^{-1}, which are about 100–10000-fold higher than in the bulk water column. As particles sink, organic material is degraded by extracellular enzymes produced by the resident microbial population. Microbial respiration creates anoxic conditions, so that diverse aerobic and anaerobic microbes colonize different niches within the snow particle. The rate of solubilization exceeds the rate of remineralization, so dissolved material leaks from snow particles, leaving a plume of nutrients in its wake as it spreads by diffusion and advection. This may send chemical signals that attract small zooplankton to consume the particle as food. The trailing plume also provides a concentrated nutrient source for suspended planktonic bacteria, which may show chemotactic behavior in order to remain within favorable concentrations. Thus, much of the carbon is recycled during its descent, but some material reaches the ocean bottom, where it is consumed by benthic organisms or leads to the formation of sediments. Photosynthesis by algae and bacteria leads to the formation of organic material through CO_2 fixation, but viruses, heterotrophic bacteria, and protists all play a part in the fate of this fixed carbon. The balance of their activities throughout the water column determines the proportions of fixed carbon that are remineralized to CO_2, transferred to higher trophic levels, or reach the seafloor. The discovery of this mechanism, termed the microbial loop, was one of the most important conceptual advances in biological oceanography, and its significance is considered further in *Chapter 8*. The sinking rate of marine snow particles depends greatly in their composition—one area of considerable current interest and importance is the extent to which microplastics and microbes can affect the settlement of particles. This is explored in *Box 13.1*.

Figure 1.8 Schematic diagram showing the microbial processes occurring in the formation and fate of a marine snow particle as it falls through the water column. The action of extracellular enzymes and viral lysis leads to the release of dissolved organic material (DOM).

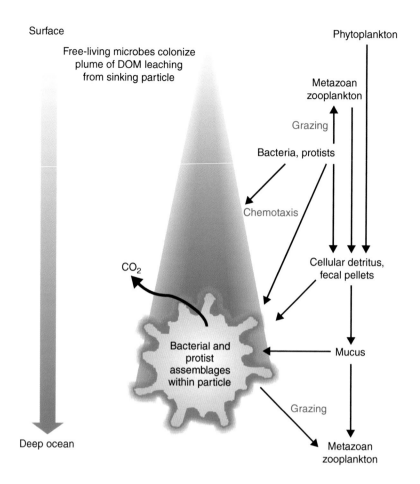

HOW DID CHERNOBYL FALLOUT HELP THE STUDY OF MARINE SNOW?

In 1986, a major environmental catastrophe occurred when the nuclear reactor in Chernobyl, Ukraine exploded, releasing hundreds of tons of radioactive particles. These were carried by wind to many distant regions and settled with rain on land and sea. Fowler et al. (1987) were able to extract some benefit from this tragedy. They had set up sediment traps to measure vertical transport in the Mediterranean Sea. Following the Chernobyl fallout, they found that the pulse of radioactivity—especially the rare nuclides ^{141}Ce and ^{144}Ce—was transported to depths of over 200 m within a few days at a rate of 29 m per day. Physical processes alone could not account for such rapid settlement. Fowler and colleagues concluded that zooplankton were ingesting radioactive particles adsorbed to their food source and repackaging them as larger, denser particles in fecal pellets that aggregated with marine snow to sink at a high rate. Subsequent studies have shown that the fecal pellets of different zooplankton species vary greatly in their density and sinking rate.

The sea surface is covered by a gelatinous biofilm

The interface between the sea surface and the atmosphere is the site of the exchange of gases, aerosols and trace elements, in both directions *Figure 1.9*). We know now that the sea surface microlayer (SML, typically up to 1 mm thick) has very different physicochemical and microbial composition from the underlying seawater. It contains high concentrations of lipids, proteins, and polysaccharides, much of it aggregated into TEPs, formed mainly from phytoplankton as described above. The organisms associated with this surface layer are known as the neuston. Analysis of the microbial community of the SML shows that the bacterial community composition contains distinct taxa, but these are interconnected to those in bulk seawater below. Surface wind, exposure to UV irradiation, deposition of air-borne dust, and other factors are important in determining the composition of the bacterioneuston, and these vary greatly according to location. The composition of the phytoneuston—microalgae found in the SML—is also very different from the phytoplankton, especially in the diversity of diatoms. Recently, fungi have been shown to be a significant component of the SML (p.187). The physical properties of the sea surface are altered by the presence of the SML through the production of surfactants, which modify phenomena such as turbulence and the formation of bubbles and micro-waves. This is often visible in the form of surface slicks of calmer water. This affects the transfer of microbes and organic compounds into the atmosphere via aerosols produced by wind action, which is linked to the formation of clouds and ice, affecting local and global climatic conditions (*Figure 1.9*). This topic is explored further in *Chapter 9*.

SEDIMENTS AND SURFACES

Microbes play a major role in marine sediments

More than 70% of the Earth's surface is covered by marine sediments. On the continental shelf, sediments are formed due to the accumulation of eroded materials transported into the ocean as particles of mud, sand, or gravel, together with copious particulate organic matter,

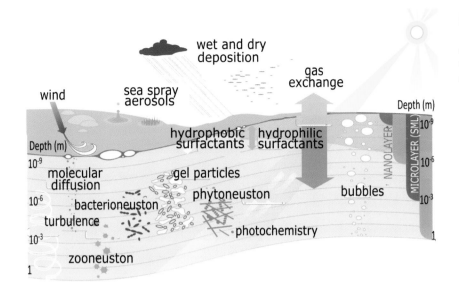

carbonate- and silica-rich compounds produced by biological activity. The mineral composition reflects the nature of the rocks and the type of weathering. Large rivers such as the Amazon, Orinoco, or Ganges transfer millions of tons of fine sediments to the ocean each year. Most of this mud settles along the continental margins or is funneled as dilute suspensions by submarine canyons. Here, the sediments may be more than 10 km thick. By contrast, in much of the Atlantic and Pacific Oceans underneath the ocean gyres, the sediments may only be 100–1000 m thick. Here, far from the continental land masses, abyssal clays are formed by the deposition of fine sediments from the continents mixed with wind-blown dust, volcanic ash, and cosmogenic dust from meteor impact. These accumulate very slowly—less than 1 mm per 1000 years—while biogenous oozes accumulate at up to 4 cm per 1000 years. Biogenous oozes contain over 30% of material of biological origin, mainly shells of protistan plankton, mixed with clay. Oozes are usually insignificant in the shallow waters near continents. Calcareous oozes or muds cover nearly 50% of the ocean floor, especially in the Indian and Atlantic Oceans. They are formed by the deposition of the calcium carbonate shells (tests) of two main types of protist: the coccolithophorids and the foraminifera (see *Chapter 6*). Siliceous oozes are formed from the shells (frustules) of diatoms and radiolarians, which are composed of opaline silica ($SiO_2.nH_2O$). The rate of accumulation of biogenous oozes depends on the rate of production of organisms in the plankton, the rate of destruction during descent to the seafloor, and the extent to which they are diluted by mixing with other sediments.

In the case of coccolithophorids and foraminifera, depth has an important effect on dissolution of the calcified scales or shells. At relatively high temperatures near the surface, seawater is saturated with $CaCO_3$. As calcareous shells sink, $CaCO_3$ becomes more soluble as a result of the increased content of CO_2 in water at lower temperatures and higher pressures. The carbonate compensation depth is the depth at which carbonate input from the surface waters is balanced by dissolution in deep waters; this varies between 3000 m in polar waters and 5000 m in tropical waters. For this reason, calcareous oozes tend not to form in waters more than 5000 m deep. Similarly, not all of the silica in the frustules of diatoms reaches the ocean floor because bacterial action has been shown to play a large part in the dissolution of diatom shells during their descent (p.182). The rate of deposition of protist remains to the seabed is much more rapid than would be assumed from their small size. This is because they are aggregated into larger particles through egestion as fecal pellets after grazing by zooplankton and through the formation of marine snow as described above. In shallower waters near the continental shelf, the high input of terrigenous sediments mixes with and dilutes sediments of biogenous origin.

There is increasing recognition of the importance of microbial activities in the sediment–water interface (SWI) and deep-sea benthic boundary layer (BBL), which is a layer of homogeneous water 10 m or more thick, adjacent to the sediment surface. The SWI includes high

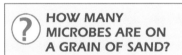

? HOW MANY MICROBES ARE ON A GRAIN OF SAND?

To answer this question, Probandt et al. (2018) developed a method to examine individual grains of sand from a coastal sediment using fluorescent *in situ* hybridization (FISH) and confocal microscopy, together with polymerase chain reaction (PCR) amplification of rRNA genes (see *Chapter 2* for methods). They found that individual grains (just 202–635 μm diameter) were colonized by a highly diverse community of 10^4–10^5 microbes (mostly bacteria, with smaller numbers of archaea and a few eukaryotes), densely packed in a thin film in the indentations on the sand grain. The wide range of taxa and functional groups found shows that the variable physical conditions and nutrient availability in these grains of surface sediment clearly provide the right microenvironments for many of the major metabolic transformations carried out by different microbial groups—a microbial zoo on a tiny grain of sand!

concentrations of particulate organic debris and dissolved organic compounds that become adsorbed onto mineral particles.

In nutrient-rich areas with high rates of microbial activity, oxygen may be present in only the top few millimeters or few centimeters of the sediment, but the structure and composition of the microbial habitat is modified by benthic "storms" and the action of animals such as worms and burrowing shrimps, which move and resuspend sediments, transporting oxygen into deeper layers (bioturbation).

As well as the constant "snowfall" of plankton-derived material, concentrated nutrient inputs reach the seabed in the form of large animal carcasses. For example, time-lapse photography has shown how quickly fallen whale carcasses attract colonies of animals, and microbiological studies accompanying these investigations have yielded novel bacteria, some with biotechnological applications (p.390). The microbial communities and symbioses that develop are similar to those found at hydrothermal vents and cold seeps. Other types of sediment that provide special habitats for microbes include those in salt marshes, mangroves, and coral reefs.

Studies of the extent to which carbon fixed in the photic zone finds its way to the seabed, and its fate in sediments, are important in understanding the role of the oceans in the planetary carbon cycle. Microbial processes such as production and oxidation of methane and oxidation and reduction of sulfur compounds are of special interest. Studies of the diversity and activity of microbial life in the various types of sediment are yielding many new insights, mainly because of the application of molecular techniques, and these are described in subsequent chapters.

Deep marine sediments contain a vast reservoir of ancient microbes

The microbiology of deep marine sediments and subsurface rocks is an area of current active investigation using deep-core drilling and novel coring devices, and microbes have been detected to a depth of 1.6 km in porous layers that were laid down as sediments tens or hundreds of million years ago (MYA). Current estimates of the global distribution of bacteria and archaea in deep sediments, obtained using a variety of techniques, give a consensus value of about 3×10^{29} bacterial and archaeal cells. The deep biosphere remains one of the most inaccessible ecosystems on Earth and investigation is expensive and technologically challenging, but recent research is providing new insights leading to the conclusion that most organisms are "barely alive" descendants of cells buried over millions of years, as discussed in *Box 1.1)*

Microbes colonize surfaces through formation of biofilms and mats

In the last few decades, the special phenomena that govern the colonization of surfaces by microbes have come under intense scrutiny, with the growing recognition that such biofilm formation involves complex physicochemical processes and community interactions. Biofilms consist of a collection of microbes bound to a solid surface by their extracellular products, which trap organic and inorganic components. In the marine environment, all kinds of surfaces including other microbes, plants, animals, sediment particles, rocks, and fabricated structures may become colonized by biofilms. Biofilm formation is considered in more detail in *Chapter 3*, and its economic importance in biofouling is discussed in *Chapter 13*.

As a result of metabolic processes, ecological succession can result in the development of multi-layered structures known as microbial mats. Mats can be several millimeters to a few centimeters thick. Depending on the nutritional and environmental conditions, mats may contain multiple types of bacteria, archaea, protists, and fungi (and their viruses) in combination with microbial polymers and sedimentary materials. These are particularly important in shallow and intertidal waters, but they are also found in deeper nutrient-rich water. The composition of microbial mats is affected by physical factors such as light, temperature, water

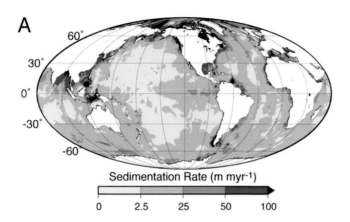

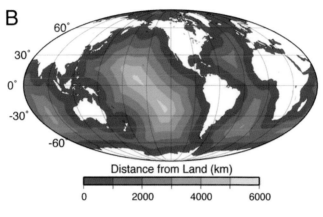

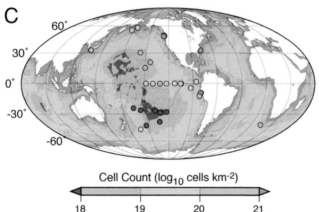

Figure 1.10 Global distribution of subseafloor sedimentary cell abundance. A. Sedimentation rate. B. Distance from shore. C. Integrated number of cells. Reprinted from Kallmeyer et al. (2012) with permission from National Academy of Sciences.

content, and flow rate; and by chemical factors such as pH, redox potential, the concentration of molecular oxygen, sulfide, nitrate, iron, and dissolved organic compounds. Phototrophic bacteria and diatoms are major components of stratified microbial mats in the photic zone, and the species composition and zonation are determined by the intensity and wavelength of light penetration into the mat. Light normally only penetrates about 1 mm into the mat and anoxic conditions develop below this. The formation and diurnal variations of physico-chemical gradients (especially of oxygen and sulfide) have a major effect on the distribution of organisms in the mat. Microbial mats formed by chemosynthetic bacteria are common at hydrothermal vents (*Figure 1.12A*).

Stromatolites are formed by the trapping of sedimentary particles, cemented together by microbial exudates to form reefs or pillar-like structures. Fossils of stromatolites are common in ancient rocks (*Figure 1.12B*). Cyanobacteria are active on the surface of the structures and the underlying material is formed by a slow build-up of lithified remains of the former growth (<1 mm per year). They occur today in a few shallow marine lagoons, such as Shark Bay in Western Australia (*Figure 1.12C*).

BOX 1.1   RESEARCH FOCUS

Deep subsurface microbes—are they dead, dormant, or just staying alive?

Studying deep subsurface sediments. Ensuring that samples are recovered without contamination during drilling is a major challenge, depending on multi-national consortia of microbiologists, geologists, and chemists, such as the Integrated Ocean Discovery Programme (IODP) (Hoehler and Jørgensen, 2013) operating specialized drilling ships (*Figure 1.11*). Based on analysis of various studies using epifluorescence counts, Whitman et al. (1998) estimated of the number of bacteria and archaeal cells in the deep marine subsurface sediments at 3.5×10^{30} bacterial and archaeal cells. More recent calculations put the figure about ten times lower, with an estimated biomass of 10 Gt (Pg) carbon (Bar-On et al., 2018). What are all these organisms doing—are they dead, dormant, or alive? Different investigators have produced conflicting opinions on this issue (reviewed by Orcutt et al., 2013).

In ocean regions with high productivity, oxygen becomes quickly depleted below the seafloor and the anaerobic reduction of sulfate coupled to the oxidation of organic matter is the dominant metabolic process in sediments. However, under the extremely oligotrophic subtropical ocean gyres, the rate of deposition of organic material reaching the seabed is so low (a few millimeters every 1000 years) that oxygen can penetrate far below the seafloor (Røy et al., 2012). The amount of energy available to sustain life under these conditions is very limited. Bradley et al. (2019) developed a mathematical model

Figure 1.11 A. The ocean drilling vessel *JOIDES Resolution*. B. Sampling a sediment core for microbial analysis. Credit: William Crawford, Integrated Ocean Drilling Program, US Implementing Organization.

to explain the long-term maintenance of microbes in the sediments over millions of years. They concluded that the organisms in the deep oligotrophic sediments survive by using the low available energy to just "staying alive," rather than growing and reproducing. The rate of respiration in these deep sediments was found to be about 10^4 times lower than at the seafloor and the density of organisms is only about 100–1000 cells cm^{-3}. Jørgensen (2011) estimates that cells with such a low metabolic rate might reproduce only once every several thousand years. One possibility is that cells in this nutrient-deprived, semi-solid environment are not subject to predation by protists or attack by viruses and this enables a stable population to exist, with cells devoting all their metabolic efforts to maintenance (turnover and repair of essential cell components like DNA, proteins, and membrane lipids) rather than growth and reproduction.

Could such cells be stimulated into a more active lifestyle? In experiments conducted by Morono et al. (2011), deep sediments from beneath the Japan sea were incubated with isotope-labeled nutrients such as glucose or amino acids. They found that most of the cells incorporated these nutrients, increasing their metabolic rate 1000 times. This could be explained if most of the cells are in a truly dormant state, such as endospores that become activated into vegetative growth by the addition of nutrients. This is known to occur in laboratory studies of endospores. Although endospores are found in deep sediments and could survive long periods of environmental stress such as starvation, Jørgensen (2011) argues that the amount of energy needed to germinate and return to an active growth state far exceeds that found in these deeply buried, ancient sediments. Trembath-Reichert et al. (2017) analyzed material recovered from a coal seam 2.5 km below the seabed off the coast of Japan. In this deep sample, the tiny microbial cells were very scarce (10–100 cm^{-3}) and DNA sequence analysis showed that the bacterial community composition was more closely related to that found in forest soil than in marine sediments. This indicates that they originate from rich organic material from primeval forests that was buried by subsidence into the ocean over 20 million years ago and subsequently overlaid by 2 km thick shales. Studies of the diversity of deep subsurface communities are still rather limited, although metagenomic analyses are beginning to yield valuable information. Based on studies of distinctive membrane lipids, supported by nucleic acid-based techniques, in sediments more than 1 m deep, it appears that archaea constitute a much greater proportion of the biomass than bacteria (Lipp et al. 2008). Archaea may have a selective advantage due to the nature of their cell membranes.

Do populations of these buried microbes adapt to the new low-nutrient conditions by evolving? To answer this question, Starnawski et al. (2017) investigated the assembly and evolution of microbial communities in 8700-year-old 10 m thick sediments in Arhus Bay, Denmark. They used DNA sequencing techniques to show that a small number of specific bacterial types were present throughout all sample depths. This subset represented a small proportion of the diversity near the surface but made up 40–50% of the total microbial community in the deeper layers. The authors concluded that rare members of surface sediment microbial communities become predominant as they become buried over thousands of years. By comparing the whole genome sequences obtained from single cells from different depths, they showed that their genetic diversification is minimal. Since the stable structure of the sediments prevents exchange of cells or genes, Starnawski et al. (2017) concluded that as populations were separated over time (thousands) of years and space (tens of meters) during burial, there was hardly any accumulation of mutations. Given the extremely low generation times of cells due to low-nutrient availability, they suggested that their findings can be generalized to the deepest sediments, which are millions of years old.

SOME EXAMPLES OF SPECIAL HABITATS—THE DEEP SEA, POLAR OCEANS, CORAL REEFS, AND LIVING ORGANISMS

Microbial activity at hydrothermal vents fuels an oasis of life in the deep sea

Hydrothermal vents form a specialized and highly significant habitat for microbes. They occur mainly at the mid-ocean ridges at the boundary of the Earth's tectonic plates, where seafloor spreading and formation of new ocean crust is occurring (*Figure 1.13A*). Over 300 such sites have been studied in the Pacific and Atlantic Oceans and many others are predicted on the basis of geological surveys. Seawater permeates through cracks and fissures in the crust and interacts with the heated underlying rocks, thereby changing the chemical and physical characteristics of both the seawater and the rock. The permeability structure of the ocean crust and the location of the heat source determine the circulation patterns of hydrothermal fluids. As cold seawater penetrates into the ocean crust, it is gradually heated along its flow path, leading to the removal of magnesium from the fluid into the rock, with production of acid during the process. This leads to the leaching of other major elements and transition metals from the rock into the hydrothermal fluid, and sulfate in the seawater is removed by precipitation and reduction to hydrogen sulfide. As the percolating fluids reach the proximity of the magma heat source, extensive chemical reactions occur within the rock and the pressurized fluids are heated to over 350°C, becoming buoyant and rising toward the ocean floor. As they rise, decompression causes the fluids to cool slightly, and precipitation of metal sulfides and other compounds occurs en route. The hydrothermal fluid is injected

(i) MICROBIAL MATS AND EVOLUTION

Some of Earth's oldest sedimentary rocks in Western Australia contain evidence of complex microbial communities that existed 3.48 billion years ago (BYA). These fossilized microbially induced sedimentary structures (MISS) and stromatolites are believed to have formed from mats colonizing an ancient shoreline or lagoon and consist of layers of sediment and organic material (Noffke et al., 2013). These ancient MISS probably contained anaerobic phototrophic bacteria. Cyanobacteria evolved subsequently (~3 BYA), and this led to the development of an oxygen-containing atmosphere. This heralded the evolution of eukaryotic organisms (~2 BYA) and eventually, multicellular life (0.6 BYA). Until the start of the Cambrian period (~0.54 BYA), microbial mats are believed to have dominated the surface of the ocean floor, but this is thought to have ended with the diversification of animals that began burrowing into the seabed.

Figure 1.12 A. Chemosynthetic microbial mats covering red algae and coral in an area where hydrothermal vent and coral reef communities overlap at 190 m depth. B. Cross section of stromatolite fossil (Eocene period, 56–34 MYA) from Fort Laclede Bed, Wyoming showing layered structure due to microbial growth and sediment accumulation. C. Stromatolites at the hypersaline Hamelin Pool, Shark Bay, W. Australia. Credits: A. Submarine Ring of Fire 2004 Exploration, NOAA Vents Program; B. James St. John, CC-BY-SA-2.0 via Wikimedia Commons; C. Bryn Pnzauger. CC BY 2.0 via Wikimedia Commons.

Figure 1.13 A. Schematic diagram of processes occurring at hydro-thermal vent systems. Gradients of temperature and chemical elements create a variety of habitats for diverse microbial and animal communities. B. A venting black smoker emit-ting jets of particles of iron sulfide. Image shows a dense colony of *Riftia pachyptila* giant tubeworms. Credit: NOAA Pacific Marine Environmental Laboratory.

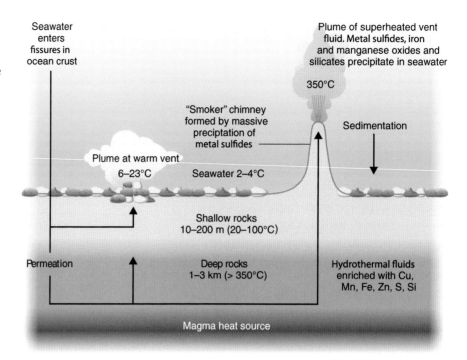

A.

B.

into the ocean as plumes of mineral-rich superheated water. The hottest plumes (up to 350°C) are generally black, because of the high content of metal sulfide and sulfate particles, and precipitation occurs as the hot plume mixes with the cold seawater. Some of these precipitates form chimney structures called "black smokers" (*Figure 1.13B*) while others are dispersed through the water and form sediments in the vicinity. In other parts of the vent field, the cir-culation of hydrothermal fluid may be shallower, leading to diffuse plumes of water heated to 6–23°C. The gradients of temperature and nutrients that exist at hydrothermal systems provide a great diversity of habitats for microbes suspended in the surrounding heated waters, in sediments, and attached to surfaces of the chimneys. Many of these are hyperthermo-philic bacteria and archaea, which can grow at temperatures up to 121°C, while others grow at lower temperatures further from the fluid emissions. Molecular studies are revealing an astonishing diversity of such organisms, many of which have biotechnological applications. The microbiology of the deep subsurface rocks beneath vents is also now under investigation, and many novel microbes and metabolic processes are being discovered. Microbial activity in the deep subsurface contributes to the chemical changes in composition during circulation of the hydrothermal fluids.

Hydrothermal vent systems were first described in 1977, when scientists aboard the submersible *Alvin*, from Woods Hole Oceanographic Institution, were exploring the seabed about 2500 m deep near the Galapagos Islands. The discovery of life around the vents was totally unexpected. The *Alvin* scientists observed dense communities of previously unknown animals, including giant tubeworms, clams, anemones, crabs, and many others. Subsequent research showed that the warm waters near hydrothermal vents contain large populations of chemosynthetic bacteria and archaea, which fix CO_2 using energy from the oxidation of sulfides in the vent fluids. This metabolism supports a food chain with many trophic levels that is independent of photosynthesis. We now know that many of the animals at vent sites contain chemosynthetic bacteria as symbionts within their tissues or on their surfaces—these relationships are discussed in *Chapter 10*. In addition, bacterial populations directly support the growth of filter-feeding animals, such as clams and mussels, or shrimp that graze on microbial mats. Previously, animal life was thought always to rely ultimately on the fixation of CO_2 by photosynthesis, but the vent communities function without the input of material derived from the use of light energy, although the sulfide oxidation depends on dissolved oxygen in the water, and the origin of this is photosynthetic.

Cold seeps also support diverse life based on chemosynthesis

Cold seeps are abundant along the continental shelf and slope, where the upwards percolation of fluids through fissures in the sediments is caused by plate tectonic activity and other geological processes, allowing high concentrations of hydrocarbons to seep into the water column. The temperature of emissions is the same as ambient seawater, or just a little higher, but they were described as "cold" seeps to distinguish them from the warm and superheated hydrothermal vents. Cold seeps were first discovered in 1983 in the Gulf of Mexico at a depth of 3.2 km and have subsequently been found in many other regions. They are most common along the continental margins, and although they usually occur in deep water (the deepest is at 7.3 km in the Japan Trench), there are some shallow seeps off the coasts of Chile, northern California, Oregon, and Denmark.

Methane is produced in anoxic sediments by a range of methanogenic archaea (p.87), and in very deep sediments it is formed by through thermogenic chemical transformation of organic matter at high pressure and temperature. Methane has a low density and rises to stability zones with a specific combination of temperature and pressure under the seabed, where it combines with water molecules combine to form a stable crystalline structure (clathrate). Seeps are formed when fissures occur, destabilizing the clathrate due to changes in temperature or pressure. Some sites are associated with seeps of hypersaline brines or leakage of oil or gas from hydrocarbons reservoirs. The methane supports prolific chemosynthetic communities consisting of free-living methanotrophic bacteria and archaea, as well as those living symbiotically with invertebrates, including bivalve mollusks from five different families and siboglinid tubeworms (*Chapter 10*). Reef-like structures are created by the deposition of calcium carbonate as a product of the anaerobic oxidation of methane (p.89) by consortia of methanotrophic archaea and sulfate-reducing and bacteria. A wide range of other animals including polychaete worms, sea stars, urchins, echinoderms, gastropods, crabs and lobsters, and shrimp are sustained by this chemosynthetic community (*Figure 1.14*).

Microbes inhabit the interface of brine pools in the deep sea

Distinct pools of hypersaline water occur in the Gulf of Mexico, Red Sea, and Eastern Mediterranean Sea, created by movement of the Earth's crust. For example, those in the Mediterranean Sea are thought to have been formed when a large area of the sea became

HOW VULNERABLE ARE VENT COMMUNITIES?

Hydrothermal vent communities are among the most productive ecosystems on Earth, but they cover a relatively small area and are very ephemeral, lasting only months or a few years. The flow of vent fluids may gradually decline, or volcanic eruptions may destroy the site completely. The key animals inhabiting these sites depend on production of large quantities of larvae with sufficiently long lifespans for them to be dispersed via currents until a new vent site is located. Goffredi et al. (2017) showed that differences in the overlying sediment and the chemistry of vent fluids can result in colonization of neighboring vents by very different animal species. Vent ecosystems clearly show resilience to natural disturbance but may be threatened by undersea mining for rare minerals deposited around vents unless effective conservation measures are put in place (Van Dover, 2014).

isolated from the Atlantic Ocean about 250 MYA. Regions of the sea evaporated, leaving a layer of rock salt, which later became covered by sediments after the Mediterranean basin reflooded. Movement of the tectonic plates has exposed the salt deposits in a few areas. Here, the salts have dissolved to form dense pockets of highly concentrated solutions of different salts, separated from the overlying seawater by extremely sharp density gradients (pycnoclines) and environmental interfaces (chemoclines). These undersea brine pools have extreme physical and chemical conditions, notably the complete lack of oxygen and near-saturated solutions of salts (up to ten times the salinity of seawater). They are therefore termed Deep Sea, Hypersaline, Anoxic Basins (DHABs).

The interface is just a few meters thick, and accurate sampling within this chemocline at such a great depth requires great ingenuity. Conventional remotely operated vehicles (ROVs) and manned submersibles cannot be used because of the damaging properties of the brines. Because of the extreme conditions, organisms growing in such habitats require special adaptations, and it might be expected that the diversity of microbial types would be low. However, through the use of DNA-based identification methods, many previously unknown taxa of bacteria have been identified, occupying narrow niches in the highly stratified interface. Archaea appear to be less prevalent than bacteria. The low availability of organic compounds at such depths suggests that chemoautotrophic metabolism sustains these communities and there is evidence for production of methane and sulfur cycling. Eukaryotic microbes (heterotrophic protists) are also abundant at the interface, and there is extensive grazing of the bacteria. It appears that some protists may be protected from the extreme anoxic and sulfidic chemical conditions by symbiotic interactions or partnerships with bacteria. Each basin seems to be inhabited by a distinctive set of microbial species, which have adapted through evolution to the unique chemical conditions of each DHAB.

At the edge of some brine pools, there are communities of mussels containing chemosynthetic bacteria and associated invertebrates, similar to those found at methane seeps.

Microbes in sea ice form an important part of the food web in polar regions

At the poles, the temperature is so low during the winter that large areas of seawater freeze to form sea ice, some of which forms adjacent to the coastal shoreline and some of which forms floating masses of pack ice. Sea ice forms when the temperature is less than −1.9°C, the freezing point of water at 35‰ salinity. The first stage in sea-ice formation is the accumulation of minute crystals of frazil ice on the surface, which are driven by wind and wave action into aggregated clumps called grease ice. These turn into pancake-shaped ice floes that freeze together and form a solid ice cover. At the winter maxima, the combined coverage by sea ice at the north and south polar regions is almost 10% of the Earth's surface (1.8×10^7 km^2 in the Antarctic and 1.5×10^7 km^2 in the Arctic). During the formation of frazil ice, planktonic microbes become trapped between the ice crystals and wave motion transports more organisms into the grease ice during its formation. Near the ice-air interface, temperatures may be

Figure 1.14 Mussels and shrimp at a chemosynthetic cold-seep community, Gulf of Mexico. Credit: NOAA-OE, Expedition to the Deep Slope 2007.

as low as −20°C during the depths of the polar winter, while the temperature at the ice-water interface remains fairly constant at about −2°C. When seawater freezes, it forms a crystalline lattice of pure water, excluding salts from the crystal structure. The salinity of the liquid phase increases, and its freezing point drops still further. This very cold, high-density, high-salinity (up to 150‰) water forms brine pockets or channels within the ice, which can remain liquid to −35°C. The ice becomes less dense than seawater and rises above sea level, with the channels draining brine through the ice to the underlying seawater. Thus, sea ice is very different from freshwater glacial ice. Loss of sea ice cover is a topic of great concern (Box 1.2).

The structure of sea ice provides a labyrinth of different microhabitats for microbes, with variations in temperature, salinity, nutrient concentration, and light penetration. This enables colonization and active metabolism by distinctive mixed communities of cold-adapted (psychrophilic) photosynthetic and heterotrophic protists and bacteria, as well as viruses. Microbial activities also alter the physicochemical conditions, mainly owing to the production of large amounts of cryoprotectant compounds and extracellular polymers, leading to the creation of additional microenvironments. The dominant photosynthetic organisms near the ice–sea interface are pennate diatoms and small dinoflagellates (*Figure 1.15A*). The density of diatoms in sea ice may be up to 1000 times that in surface waters. Through photosynthesis, the microalgae make a small, but significant, contribution to primary productivity in the polar regions. For example, the contribution of sea ice to primary productivity in the Southern Ocean is only about 5% of the total, but it extends the short summer period of primary production and provides a concentrated food source that sustains the food web during the winter. During the Antarctic winter, microalgae on the undersurface of sea ice are the main source of food for grazing krill, shrimp-like crustaceans that are the main diet of fish, birds, and mammals in the Southern Ocean. A wide range of protists and heterotrophic bacteria have been found in sea ice, including new species with biotechnological potential. Some microbes remain active—albeit at a much-reduced metabolic rate—even in the coldest parts of the ice and in "frost flowers" formed on the surface of the ice, where they are trapped in pockets of very low temperatures and high-salinity (*Figure 1.15B*).

(a)

(b)

Figure 1.15 (a) Light microscopy image of diatoms living in annual sea ice, McMurdo Sound, Antarctica. (b) Frost flowers over young sea ice in the central Arctic Ocean. Credits: (a) Gordon T. Taylor, NOAA Corps Collection. (b) Matthias Wietz, MPI Marine Microbiology, Bremen.

BOX 1.2   RESEARCH FOCUS

What will happen when the sea-ice melts?

Global warming is affecting polar regions more rapidly than any other parts of the planet. Although there is a large natural annual variation in the thickness and extent of summer sea ice in the Arctic Ocean, these have shown a dramatic decline since the 1970s. In 2018, some of the thickest and oldest areas of ice to the north of Greenland started to break up (*Figure 1.16*). Notz and Stroeve (2016) provide compelling evidence that this is directly linked to CO_2 emissions as a result of human activity, and their models predict that the Arctic Ocean will be free of summer ice by the end of this century. The loss of ice cover results in changes to wind patterns and warmer and less saline waters in the upper ocean. Also, we now know that sea ice absorbs large amounts of CO_2 from the atmosphere (Søgaard et al., 2013). Most significantly, loss of summer ice also reduces the albedo effect resulting in a positive feedback, which increases the absorption of solar energy by seawater rather than reflecting sunlight. This further accelerates the rise in temperature. Many scientific programmes are investigating the impact of these changes and microbiological research is especially important. We urgently need to know how changes in microbial communities and their activities are altering the cycling of carbon and other elements.

Changes in Arctic picophytoplankton. One example of change has been observed in apparent shifts of picophytoplankton diversity in the Arctic Ocean. Paulsen et al, (2016) found high numbers of the cyanobacterium *Synechococcus* at high latitudes. This was surprising, as Arctic water masses are usually thought to be dominated by larger picoeukaryotes adapted to low temperatures, such as *Micromonas* spp. This could be attributed in part to a change in the flow of water from the Atlantic into the Arctic Ocean due to global warming, but analysis of the distribution of different genetic clades of the cyanobacteria suggests that some of them have adapted to Arctic conditions. Because populations of *Synechococcus* are controlled by grazing by small heterotrophic protists, Paulsen and colleagues concluded that much of their biomass would be recycled within the microbial loop (p.220), rather than transferring to higher

trophic levels. Thus, an increase in small *Synechococcus* in place of larger phytoplankton in a warmer Arctic Ocean could have major effects on transfer of productivity to other organisms.

Benthic microbes are also affected. Antje Boetius and colleagues at the Alfred Wegener Institute and Max Planck Institute of Marine Microbiology are undertaking an intensive research program to investigate the microbiology of current and archived samples of Arctic deep-sea sediments, revealing important information about changes in microbial diversity and how the flux of particulate organic matter is affected by the loss of permanent ice cover. Rapp et al. (2018) report the analysis of microbial communities in areas affected by the summer ice melt in 2012, which was the largest decline to date. Vast quantities of ice algae associated with the underside of the sea ice became detached and sank to the seafloor at over 4000 m in depth. Using high-throughput DNA sequencing, Rapp and colleagues studied the diversity of different groups of microbes in the sea ice, in melt water and in the surface sediment beneath the melting ice. The sea-ice algae aggregates were shown to be mostly composed of diatoms, especially *Melosira arctica*. The large filaments of this diatom sank rapidly to the seabed. The sinking aggregates were shown to transport bacteria such as Flavobacteriia and Gammaproteobacteria, including *Glaciecola* and *Paraglaciecola*. These are known to be actively involved in the breakdown of algal and other organic material via the production of extracellular enzymes. Labyrinthulids were also detected—these protists are known to be active in degradation of algae (p.183). Rapp et al. found that the algal deposits changed the community composition on the seafloor by introducing bacteria from the sea ice, although these appeared to be overgrown by specific groups of indigenous deep-sea bacteria within a few months. If strong summer ice melting becomes a regular occurrence as predicted, the abrupt export of large amounts of sea-ice algae and associated microbiota will become more frequent. The authors suggest that this could lead to permanent changes in bacterial community composition, which may lead to significant changes in the cycling of nutrients in the surface sediments.

Changes in Antarctic food webs. Warming is also occurring in the Antarctic, where increased melting of the ice-sheets increases the input of fresh water, altering the salinity and structure of the surrounding water column. Consequent changes in ocean food web structures will also affect benthic communities, as seasonal shifts in different types of phytoplankton in the water column or underside of sea ice will affect the flow of fixed carbon to the seafloor. Based on a large-scale study of microbial diversity in Antarctic sediments, Learman et al. (2016) found that community composition varied greatly with geographic location and was strongly influenced by the nature of organic matter reaching the seafloor. Warming could lead to an increase in meltwater, stimulating increased phytoplankton blooms transporting organic matter to the seabed. Learman and colleagues suggest that this might cause a shift from communities dominated by lithotrophic organisms (such as some types of archaea) to a greater proportion of chemoheterotrophs. This could cause significant changes to biogeochemical cycling.

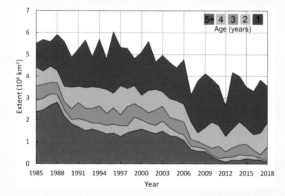

Figure 1.16 Time series showing the Arctic ice remaining at the end each summer melt season since 1985. The area of the oldest ice has declined 16-fold. Credit: M. Tschudi, S. Stewart, University of Colorado, and W. Meier, J. Stroeve, NSIDC.

Microbial activity underpins productive food webs in coral reefs

The most familiar coral reefs are those found in the shallow photic zone, especially in tropical and subtropical areas. These present specialized habitats, in which symbiotic interactions between corals and microbes are responsible for primary production. Symbiotic dinoflagellates (zooxanthellae) are responsible for the fixation of CO_2, which promotes growth of the coral animal tissue and skeleton, but also leads to large excesses of fixed carbon that fuels microbial processes in the seawater and sediments surrounding the reef. Corals produce large amounts of mucus, much of which forms TEPs and gels which provide a major source of nutrients sustaining the growth of planktonic bacteria or aggregate to form sinking particles which fuel benthic activities. Thus, although coral reefs only occupy about 0.1% of the sea-floor area and occur in ocean areas of generally low productivity (due to lack of nutrient input), they are as productive as tropical rain forests and are estimated to contain a third of all marine species. Recent studies have shown that many other microbes associated with corals and other reef organisms are involved in the primary production and recycling of carbon, nitrogen, and other key elements. We now recognize that major phase shifts in microbial activity are responsible for the degradation of coral reefs, with profound implications for marine life and the destabilization of safety and livelihood of millions of people. Climate change is a major threat to the world's reefs, with severe mass bleaching affecting many tropical reefs (most notably, Australia's Great Barrier Reef) in years when average sea surface temperatures have been just 1°C or so above normal. As sea levels rise, reefs may not be able to grow quickly enough to maintain optimum light levels, while ocean acidification will reduce their ability to calcify. These issues are discussed in detail in *Chapter 10*.

Living organisms are the habitats of many microbes

Microbial biofilms also form on the surfaces of all kinds of animals, seaweeds, and coastal plants; these provide a highly nutritive environment through secretion or leaching of organic compounds. Many organisms seem selectively to enhance surface colonization by certain microbes and discourage colonization by others. This may occur by the production of specific compounds that inhibit attachment or growth of certain microbes. Once established, particular microbes may themselves influence colonization by other types. These processes offer obvious applications in the control of biofouling (p.356). As well as surface (epibiotic) associations, microorganisms can form endosymbiotic associations within the body cavities, tissue, or cells of living organisms.

Many microalgae (such as diatoms, dinoflagellates, and prymnesiophytes) and other protists (such as ciliates) harbor bacteria on their surfaces, or as endosymbionts within their cells (*Figure 4.11*). Intimate associations between bacterial and archaeal cells are also being revealed by new imaging techniques (*Figure 5.5*). Seaweeds and seagrasses have dense populations of bacteria (up to 10^6 per cm^2) on their surfaces, although this varies considerably with species, geographic location, and climatic conditions.

The external surfaces and intestinal content of animals provide a variety of habitats to a wide diversity of microbes. Such associations commonly lead to some mutual benefit for host and microbe—examples of symbiotic interactions between animals and microbes are considered in *Chapter 10*. Pathogenic interactions may also result—microbial diseases of marine organisms are discussed in *Chapter 11*.

Conclusions

This chapter has introduced the various types of marine microbes and some of the wide range of habitats that they occupy. The adaptations of marine microbes to the different physical, chemical, and biological conditions encountered have led to the evolution of highly diverse microbes. The discovery that these tiny microbes are present in such large numbers and biomass—and that they are responsible for the biogeochemical processes that shape our planet—can be viewed as one of the most important advances of modern science. Subsequent chapters build on this introduction by exploring the mechanisms underlying this diversity of form and function.

References and further reading

Origins and scope of marine microbiology

Bar-On, Y. M., Phillips, R., and Milo, R. (2018). The biomass distribution on Earth. *Proc. Nat. Acad. Sci.* **115**: 6506–6511.

Brown, J. W. (2015). *Principles of Microbial Diversity*. ASM Press.

Caron, D. A., Worden, A. Z., Countway, P. D., et al. (2008). Protists are microbes too: A perspective. *ISME J.* **3**: 4–12.

Ciccarelli, F. D., Doerks, T., von Mering, C., et al. (2006). Toward automatic reconstruction of a highly resolved tree of life. *Science* **312**: 697.

Dagan, T. and Martin, W. (2009). Getting a better picture of microbial evolution *en route* to a network of genomes. *Phil. Trans. Roy. Soc. B Biol. Sci.* **364**: 2187–2196.

Doolittle, W. F. (1999). Phylogenetic classification and the universal tree. *Science* **286**: 1433.

Embley, M. and Williams T. (2016). Only two domains, not three: Changing views on the tree of life. *Microbiol. Today* **2016**: 70–73.

Forterre, P. (2015). The universal tree of life: An update. *Front. Microbiol.* **6**: 717.

Gasol, J. M. and Kirchman, D. (2018). Introduction: The evolution of microbial ecology of the ocean. In *Microbial Ecology of the Oceans*, ed. Gasol, J. M. and Kirchman, D. L., pp. 1–46, Wiley Blackwell.

Grossart, H.-P. and Rojas-Jiminez, K. (2016). Aquatic fungi: Targeting the forgotten in microbial ecology. *Curr. Opin. Microbiol.* **31**: 140–145.

Henderson, G. P., Gan, L., and Jensen, G. J. (2007). 3-D ultrastructure of *O. tauri*: Electron cryotomography of an entire eukaryotic cell. *PLoS ONE* **2**: e749.

Hug, L. A., Baker, B. J., Anantharaman, K., et al. (2016). A new view of the tree of life. *Nat. Microbiol.* **1**: 16048.

Karl, D. M. and Proctor, L. M. (2008). Foundations of microbial oceanography. *Oceanography* **20**: 16–27.

Keeling, P. J. and de Campo, J. (2017). Marine protists are not just big bacteria. *Curr. Biol.* **27**: R451–R459.

Kolter, R. and Maloy, S. (eds.) (2012). *Microbes and Evolution: The World That Darwin Never Saw*. ASM Press.

Lawton, G. (2009). Why Darwin was wrong about the tree of life. *New Sci.* **2692**: 34–39.

Martin, W. and Embley, T. M. (2004). Evolutionary biology: Early evolution comes full circle. *Nature* **431**: 134–137.

Nasir, A., Kim, K. M., Da Cunha, V., and Caetano-Anollés, G. (2016). Arguments reinforcing the three-domain view of diversified cellular life. *Archaea* **2016**: 1851865.

Schulz, H. N. and Jørgensen, B. B. (2001). Big bacteria. *Annu. Rev. Microbiol.* **55**: 105–137.

Sherr, B. F., Sherr, E. B., Caron, D. A., et al. (2007). Oceanic protists. *Oceanog.* **20**: 130–134.

Suttle, C. (2005). The viriosphere: The greatest biological diversity on Earth and driver of global processes. *Environ. Microbiol.* **7**: 481–482.

Whitman, W. B., Coleman, D. C., Wiebe, W. J. (1998). Prokaryotes: The unseen majority. *Proc. Nat. Acad. Sci. USA* **95**: 6578–6583.

Yayanos, A. A. (2003). Marine microbiology at Scripps. *Oceanography* **16**: 63–75.

Zhao, X., Schwartz, C. L., Pierson, J., et al. (2017). Three-dimensional structure of the ultraoligotrophic marine bacterium "*Candidatus* Pelagibacter ubique". *Appl. Environ. Microbiol.* **83**: e02807–16.

Ocean habitats

Azam, F. and Long, R. A. (2001). Sea snow microcosms. *Nature* **414**: 495–498.

Azam, F. and Malfatti, F. (2007). Microbial structuring of marine ecosystems. *Nat. Rev. Microbiol.* **5**: 782–791.

Costello, M. J. and Breyer, S. (2017). Ocean depths: The mesopelagic and implications for global warming. *Curr. Biol.* **27**: R19–R41.

Cunliffe, M., Engel, A., Frka, S., et al. (2013). Sea surface microlayers: A unified physicochemical and biological perspective of the air–ocean interface. *Prog. Oceanog.* **109**: 104–116.

Cunliffe, M. and Murrell, J. C. (2009). The sea-surface microlayer is a gelatinous biofilm. *ISME J.* **3**: 1001–1003.

Engel, A., Bange, H. W., Cunliffe, M., et al. (2017). The ocean's vital skin: Toward an integrated understanding of the sea surface microlayer. *Front. Mar. Sci.* **4**: 165.

Fowler, S. W., Buat-Menard, P., Yokoyama, et al. (1987). Rapid removal of Chernobyl fallout from Mediterranean surface waters by biological activity. *Nature* **329**: 56–58.

Grossart, H. P., Kiorboe, T., Tang, K. W., et al. (2006). Interactions between marine snow and heterotrophic bacteria: Aggregate formation and microbial dynamics. *Aquat. Microb. Ecol.* **42**: 19–26.

Kjørboe, T. (2001). Formation and fate of marine snow: Small-scale processes with large-scale implications. *Sci. Mar.* **65**: 57–71.

Mühlenbruch, M., Grossart, H. P., Eigemann, F., and Voss, M. (2018). Phytoplankton-derived polysaccharides in the marine environment and their interactions with heterotrophic bacteria. *Environ. Microbiol.* **20**: 2671–2685.

Nissimov, J. I. and Bidle, K. D. (2017). Stress, death, and the biological glue of sinking matter. *J. Phycol.* **53**: 241–244.

Seymour, J. R. and Stocker, R. (2018). The ocean's microscale: A microbe's view of the sea. In *Microbial Ecology of the Oceans*, ed. Gasol, J. M. and Kirchman, D. L., pp. 289–466, Wiley Blackwell.

Verde, C., Giordano, D., Bellas, C. M., et al. (2016). Polar marine microorganisms and climate change. *Adv. Microb. Physiol.* **69**: 187–215.

Verdugo, P., Alldredge, A. L., Azam, F., et al. (2004). The oceanic gel phase: A bridge in the DOM-POM continuum. *Mar. Chem.* **92**: 67–85.

Sediments and surfaces

Betts, H. C., Puttick, M. N., Clark, J. W., et al. (2018). Integrated genomic and fossil evidence illuminates life's early evolution and eukaryote origin. *Nat. Ecol. Evol.* **2**: 1556–1562.

Bolhuis, H., Cretoiu, M. S. and Stal, L. J. (2014). Molecular ecology of microbial mats. *FEMS Microbiol. Ecol.* **90**: 335–350.

Bradley, J. A., Amend, J. P., and LaRowe, D. E. (2019). Survival of the fewest: Microbial dormancy and maintenance in marine sediments through deep time. *Geobiol.* **17**: 43–59.

Corinaldesi, C. (2015). New perspectives in benthic deep-sea microbial ecology. *Front. Mar. Sci.* **2**: 5–17.

Danovaro, R., Corinaldesi, C., Dell'Anno, A., and Rastelli, E. (2017). Potential impact of global climate change on benthic deep-sea microbes. *FEMS Microbiol. Lett.* **364**: 23, fnx214.

Hoehler, T. M. and Jørgensen, B. B. (2013). Microbial life under extreme energy limitation. *Nat. Rev. Microbiol.* **11**: 83–94.

Jørgensen, B. B. (2011). Deep subseafloor microbial cells on physiological standby. *Proc. Natl. Acad. Sci. USA* **108**: 18193–18194.

Jørgensen, B. B. and Boetius, A. (2007). Feast and famine—Microbial life in the deep-sea bed. *Nat. Rev. Microbiol.* **5**: 770–781.

Kallmeyer, J., Pockalny, R., Adhikari, R. R., et al. (2012). Global distribution of microbial abundance and biomass in subseafloor sediment. *Proc. Natl. Acad. Sci. USA* **109**: 16213–16216.

Learman, D. R., Henson, M. W., Thrash, J. C., et al. (2016). Biogeochemical and microbial variation across 5500 km of Antarctic surface sediment implicates organic matter as a driver of benthic community structure. *Front. Microbiol.* **7**: 284.

Lipp, J. S., Morono, Y., Inagaki, F., and Hinrichs, K. U. (2008). Significant contribution of Archaea to extant biomass in marine subsurface sediments. *Nature* **454**: 991–994.

Morono, Y., Terada, T., Nishizawa, M., et al. (2011). Carbon and nitrogen assimilation in deep subseafloor microbial cells. *Proc. Natl. Acad. Sci. USA* **108**: 18295–18300.

Noffke, N., Christian, D., Wacey, D., and Hazen, R. M. (2013). Microbially induced sedimentary structures recording an ancient ecosystem in the ca. 3.48 billion-year-old Dresser Formation, Pilbara, Western Australia. *Astrobiology* **13**: 1103–1124.

Orsi, W. D. (2018). Ecology and evolution of seafloor and subseafloor microbial communities. *Nat. Rev. Microbiol.* **16**: 671–683.

Prieto-Barajas, C., Valencia-Cantero, E. and Santoyo, G. (2018). Microbial mat ecosystems: Structure types, functional diversity, and biotechnological application. *Electron. J. Biotechnol.* **31**: 48–56.

Probandt, D., Eickhorst, T., Ellrott, A., et al. (2018). Microbial life on a sand grain: From bulk sediment to single grains. *ISME J.* **12**: 623–633.

Røy, H., Kallmeyer, J., Adhikari, R. R., et al. (2012). Aerobic microbial respiration in 86-million-year-old deep-sea red clay. *Science* **336**: 922–925.

Schrenk, M. O., Huber, J. A., and Edwards, K. J. (2010). Microbial provinces in the subseafloor. *Ann. Rev. Mar. Sci.* **2**: 279–304.

Starnawski, P., Bataillon, T., Ettema, T. J. G., et al. (2017). Microbial community assembly and evolution in subseafloor sediment. *Proc. Natl. Acad. Sci. USA* **114**: 2940–2945.

Trembath-Reichert, E., Morono, Y., Ijiri, A., et al. (2017). Methyl-compound use and slow growth characterize microbial life in 2-km-deep subseafloor coal and shale beds. *Proc. Natl. Acad. Sci. USA* **114**: E9206–E9215.

Some special habitats—the deep sea, polar oceans, coral reefs, and living organisms

Arrigo, K. R. and Thomas, D. N. (2004). Large scale importance of sea ice biology in the Southern Ocean. *Antarctic Sci.* **16**: 471–486.

Caron, D. A., Gast, R. J. and Garneau, M. (2017). Sea ice as a habitat for micrograzers. In *Sea Ice*, ed. Thomas, D. N. pp. 370–393. John Wiley & Sons Ltd.

Cavicchioli, R. (2015). Microbial ecology of Antarctic aquatic systems. *Nat. Rev. Microbiol.* **13**: 691–706.

Deming, J. W. and Collins, R. E. (2017). Sea ice as a habitat for Bacteria, Archaea and viruses. In *Sea Ice*, ed. Thomas, D. N., pp. 370–393. John Wiley & Sons Ltd.

Dick, G. J., Anantharaman, K., Baker, B. J., et al. (2013). The microbiology of deep-sea hydrothermal vent plumes: Ecological and biogeographic linkages to seafloor and water column habitats. *Front. Microbiol.* **4**: 124.

Dziak, R. P., Matsumoto, H., Embley, R. W., et al. (2018). Passive acoustic records of seafloor methane bubble streams on the Oregon continental margin. *Deep Sea Res. II: Top. Stud. Oceanogr.* **150**: 210–217.

Ewert, M. and Deming, J. W. (2013). Sea ice microorganisms: Environmental constraints and extracellular responses. *Biology* **2**: 603–628.

Garren, M. and Azam, F. (2012). New directions in coral reef microbial ecology. *Environ. Microbiol.* **14**: 833–844.

Goffredi, S. K., Johnson, S., Tunnicliffe, V., et al. (2017). Hydrothermal vent fields discovered in the southern Gulf of California clarify role of habitat in augmenting regional diversity. *Proc. Roy Soc. B: Biol. Sci.* **284**: 20170817.

Hoegh-Guldberg, O., Hughes, T., Anthony, K., et al. (2009). Coral reefs and rapid climate change: Impacts, risks and implications for tropical societies. *IOP Conf. Ser.:Earth Environ. Sci.* **6**: 302004.

Kallmeyer, J. ed. (2017). *Life at Vents and Seeps.* Walter de Gruyter, GmbH, Berlin/Boston.

Kallmeyer, J., Pockalny, R., Adhikari, R. R., et al. (2012). Global distribution of microbial abundance and biomass in subseafloor sediment. *Proc. Natl. Acad. Sci. USA* **109**: 16213–16216.

Kirchman, D. L., Moran, X. A. G. and Ducklow, H. (2009). Microbial growth in the polar oceans—Role of temperature and potential impact of climate change. *Nat. Rev. Microbiol.* **7**: 451–459.

Levin, L. A., Baco, A. R., Bowden, D. A., et al. (2016). Hydrothermal vents and methane Seeps: Rethinking the sphere of influence. *Front. Mar. Sci.* **3**: 72.

Martin, W., Baross, J., Kelley, D. and Russell, M. J. (2008). Hydrothermal vents and the origin of life. *Nat. Rev. Microbiol.* **6**: 805–814.

Mock, T. and Thomas, D. N. (2005). Recent advances in sea-ice microbiology. *Environ. Microbiol.* **7**: 605–619.

Notz, D. and Stroeve, J. (2016). Observed Arctic sea-ice loss directly follows anthropogenic CO_2 emission. *Science* **354**: 747–750.

Orcutt, B. N., LaRowe, D. E., Biddle, J. F., et al. (2013). Microbial activity in the marine deep biosphere: Progress and prospects. *Front. Microbiol.* **4**: 189.

Paulsen, M. L., Doré, H., Garczarek, L., et al. (2016). *Synechococcus* in the Atlantic Gateway to the Arctic Ocean. *Front. Mar. Sci.* **3**: 191.

Rapp, J. Z., Fernández-Méndez, M., Bienhold, C., and Boetius, A. (2018). Effects of ice-algal aggregate export on the connectivity of bacterial communities in the central Arctic Ocean. *Front. Microbiol.* **9**: 1035.

Reysenbach, A.-L. and Cady, S. L. (2001). Microbiology of ancient and modern hydrothermal systems. *Trends Microbiol.* **9**: 79–86.

Rowher, F. and Youle, M. (2010). *Coral Reefs in the Microbial Seas.* Plaid Press, San Diego, CA.

Silveira, C. B., Cavalcanti, G. S., Walter, J. M., et al. (2017). Microbial processes driving coral reef organic carbon flow. *FEMS Microbiol. Rev.* **41**: 575–595.

Søgaard, D. H., Thomas, D. N., Rysgaard, S., et al. (2013). The relative contributions of biological and abiotic processes to carbon dynamics in subarctic sea ice. *Polar Biol.* **36**, 1761–1777.

Torda, G., Donelson, J. M., Aranda, M., et al. (2017). Rapid adaptive responses to climate change in corals. *Nat. Clim. Change* **7**: 627–636.

Van Dover, C. L. (2014). Impacts of anthropogenic disturbances at deep-sea hydrothermal vent ecosystems: A review. *Mar. Environ. Res.* **102**: 59–72.

Verde, C., Giordano, D., Bellas, C. M., et al. (2016). Polar marine microorganisms and climate change. *Adv. Microb. Physiol.* **69**: 187–215.

Chapter 2

Methods in Marine Microbiology

The continuing development and application of new techniques has resulted in a huge recent expansion of our knowledge of the diversity of marine microbes, their structure and functions, and their roles in ecology and ocean processes. It is essential for the student to appreciate the scope of available methods and their underlying principles, in order to understand how they are applied in research. A key ambition for this book is to encourage you to read the primary literature and explore the latest developments. This can be daunting, because it is easy to be put off by a plethora of technical details, which can sometimes appear quite impenetrable. This chapter doesn't aim to describe in detail every possible technique you might encounter, but instead aims to provide you with an introductory overview that will help you to understand research described in later chapters and in your own discovery of scientific papers. The chapter is divided into sections covering: (1) sampling methods; (2) imaging techniques; (3) cultivation methods; (4) RNA- and DNA-based techniques; and (5) approaches to studying the collective *in situ* activities and interactions of microbial communities in the environment. Today, we have methods that enable us to study microbial ecology at multiple levels of scale—from the function of single cells, to community activities, to global processes.

Key Concepts

- Sampling and experimental observation of marine microbes requires special techniques.

- Epifluorescence microscopy and flow cytometry led to recognition of the abundance and diversity of marine microbes in the oceans.

- Most marine microorganisms cannot currently be cultivated.

- The most significant recent advances in marine microbiology have occurred as a result of the development of methods for the study of nucleic acids collected directly from the environment.

- New sequencing technologies and bioinformatics tools are resulting in massive expansion of knowledge of microbial ecology through metagenomics, metatranscriptomics, metaproteomics, and metabolomics.

- Progress towards full understanding of marine microbial diversity, physiology, and ecology requires the integration of field observations and measurements, imaging, culture-based methods, and molecular biological approaches.

SAMPLING METHODS

Sampling the marine environment requires special techniques

The overall goal of microbial ecology is to understand how cells, populations, and communities of microbes interact with each other and with their environment. A wide range of methods in microbial ecology are designed to answer three fundamental questions about microbes in the environment: "who's there?" (biodiversity), "what are they doing?" and "how are they doing it?" (metabolic functions and activities). The importance of scale cannot be overemphasized: *micro*organisms carry out their activities in *micro*environments. Experimental approaches and measurement techniques must be designed to cause minimum disturbance of the microenvironment that they are probing, because microbial communities may exist in a delicate three-dimensional structured organization, which can be easily disrupted. At the micrometer scale, physicochemical gradients are very steep. For example, anaerobic, microaerophilic, and aerobic conditions may occur in a sediment particle or a microbial mat just a few hundred micrometers thick. Other conditions such as boundary layer effects, diffusion, turbulence, and flow patterns will affect the microenvironment in the vicinity of a microbe—a microbe swimming for a few seconds through a patch of seawater might encounter huge differences in physical conditions and concentrations of different nutrients. Most important of all, the microbes carry out biochemical reactions that alter the physical and chemical conditions in their immediate vicinity. Bulk measurements of substrates, metabolites, pH, or oxygen in marine samples may therefore not reflect the heterogeneity of the natural environment. Increasingly, we recognize the importance of gradients in the marine environment.

The traditional method of collecting plankton at sea is to tow a funnel-shaped net behind a boat. The smallest nylon mesh used is about 60 μm, which is clearly too large to capture most microbes. However, microbes will be attached to larger phytoplankton and zooplankton, and some aggregates of free-living microbes, such as clumps of filamentous cyanobacteria or diatoms, are also large enough to be collected in this way.

Obtaining samples of seawater and sediments is straightforward in shallow coastal waters, with simple collection from beaches, piers, harbors, or small boats. Samples of sediments and benthic organisms are obtained using grabs and corers from a boat, or by scuba divers who can find specific locations for collection. For most microbiological work, samples can be collected in sterile plastic bags or bottles, taking great care to prevent contamination. Sampling in the open ocean requires the use of research vessels equipped with suitable sampling gear and onboard laboratory facilities. Research investigations are facilitated through routine monitoring of oceanographic and atmospheric data and research cruises coordinated through international collaboration. Well-known examples of established long-term sampling programs include the Pacific Ocean time series site known as station ALOHA (A Long-term Oligotrophic Habitat Assessment) off Hawaii, the Bermuda Atlantic Time Series (BATS) site, and the Western Channel Observatory in the English Channel off Plymouth, UK.

It is important to collect water samples in a form suitable for subsequent analysis together with appropriate environmental data about the sampling site. Because ocean water is highly heterogeneous, proper attention must be paid to the replication, frequency, and location of sampling. For sampling from specific depths within the water column, specialized hydrological bottles are used. Niskin bottles are plastic cylinders that are open at both ends; they are lowered into the sea on a wire until they reach the required depth; they are then closed by a weighted trigger that is sent down the cable from the surface or by an electrical signal or automatic pressure sensor. One of the most widely used systems is a circular rosette system of up to 24 Niskin bottles holding 10 or 20 L, attached to a steel frame equipped with sensors for conductivity, temperature, and depth (CTD), and sometimes other parameters such as turbidity, dissolved oxygen, fluorescence, and photosynthetically active radiation, so that the water properties can be monitored in real time at the sampling locations. It is lowered by a winch from the research vessel and closed selectively at specific depths of up to 6000 m (*Figure 2.1A*). Careful attention is needed in the choice of construction materials for frames and sampling containers because many microbial processes are affected

(a)   (b)

(c)   (d)

Figure 2.1 Procedures for sampling water and sediments. (a) CTD frame and rosette full of water sampling containers being recovered from Antarctic waters. (b) Water filtration system on board RRS *James Cook*. Samples of up to 200 L are sequentially passed through filters of different pore sizes to trap different fractions of the microbial plankton. (c) Biomass collected by filtration of Arctic seawater. (d) Box corer (50×50×50 cm) for sampling of sediments, being deployed in the Southern Ocean from RV *Polarstern*. Credits: (a)-(c) Jozef I. Nissimov, Scottish Association of Marine Science, Oban. (d) Haanes Grobe, Alfred Wegener Institute, via Wikimedia Commons, CC-BY-2.5.

by the presence of trace metals or organic materials in hoses and stoppers; specialized systems are needed for some applications. For some investigations in biologically productive waters, it may be necessary to collect only a few liters of water (*Figures 2.1B, 2.1C*), whereas in low productivity waters, hundreds of liters may be required. Samples are sieved through a mesh to remove large aggregates and micro- or mesoplankton. Microbes and suspended particles are usually obtained by filtration though a series of filters of varying pore sizes depending on the size fractions of interest, from glass fiber prefilters of ~30 μm down to 0.1 μm polycarbonate filters or 0.02 anodisc filters for viruses. Such selective filtration is frequently used to denote divisions between particle-associated and free-living microbes or to discriminate plankton size classes (pico-, nano-, microplankton, see *Table 1.2*). For concentration of small microorganisms and viruses, tangential flow filtration is commonly used, which allows for the filtration of thousands of liters without clogging of membranes. Viruses can also be concentrated by flocculation with iron salts followed by large-pore-size filtration.

Several methods can be used for sampling the sea surface microlayer (SML, p.14). A common approach is to dip a glass plate into the water or to float a hydrophobic membrane filter or fine nylon mesh screen on the surface of the water.

For collecting sediment samples, various types of grabs and corers are used. Small samples of coastal mud can be collected simply with a cutoff syringe, but special designs are needed for operation in deep waters and for collecting sands or loose sediments. Tubular corers consist of weighted sections of steel tube that bore into the sediment and are sealed with caps to keep the sample intact. Specialized drilling rigs allow the recovery of deep sediment cores from hundreds of meters below the seabed (*Figure 1.10*). Box corers are designed to collect undisturbed sediment samples, typically ~0.1–0.2 m³. A weighted box is lowered into the sediment and spring-loaded flaps close the box to prevent disturbance during recovery (*Figure 2.1D*).

A device called the Continuous Plankton Recorder (CPR) invented by Alister Hardy in 1931 is designed to be towed by ships at a depth of 5–10 m. Its robust casing and mechanical structure mean that it is highly reliable and can be towed from merchant ships on regular routes. The CPR filters plankton from the water over long distances on continuously moving bands of fine silk, which is wound through the CPR on rollers turned by gears, powered by a propeller. On return to the laboratory, the silk is removed from the mechanism and divided into samples representing ten nautical miles of towing and the plankton trapped on the silk is identified and enumerated by microscopy. Since 1931, recorders have been towed over 5 million miles, producing more than a quarter-million analyses, with the CPR survey producing one of the longest archived biological datasets. Methods to extract and amplify DNA from preserved samples mean that retrospective analyses of bacteria (*Box 12.1*) and viruses associated with host cells or attached to particles have been possible.

Investigations in deep water are increasingly carried out using remotely operated vehicles (ROVs), which are usually equipped with cameras, lights, and robotic sampling arms and linked by an umbilical cable to a surface vessel for control by a pilot (*Figure 2.2A*). Improvements in battery technology and electronic systems are leading to the development of fully autonomous underwater vehicles (AUVs, gliders, or rovers, *Figure 2.2B*) that can travel for several months along preplanned routes either in the water column or on the seabed. These can be equipped with a range of measuring instruments to record environmental data and devices to collect samples of water, sediment, or benthic animals.

One example especially suited to plankton microbiology is the Environmental Sample Processor, which uses an electromechanical fluidic system to collect water samples that are filtered, preserved, and stored. The Environmental Sample Processor is carried aboard the long-range AUV developed by Monterey Bay Aquarium Research Institute (MBARI), designed to take multiple samples over long periods from eddies of the ocean gyres that are otherwise hard to investigate. Some processes such as automated microarrays or polymerase chain reaction (PCR; see below) can be carried out on board. The AUV *Clio* developed by Woods Hole Oceanographic Institute (WHOI) can dive to a depth of 6000 m, collecting multiple water samples which are immediately filtered and sampled before the vessel returns to its mother ship. Free-falling lander systems may be deposited on the seabed; some of these have mini-laboratories that collect samples for *in situ* molecular analysis and analysis of microbial activity. Purpose-specific equipment has also been used in the vicinity of hydrothermal vents.

Robotics and artificial intelligence systems are increasingly used in oceanographic research, but there is no replacement for a human presence to explore the deep sea. The most famous of the human-operated submersible vehicles is DSV *Alvin*, owned by the US Navy and operated by WHOI (*Figure 2.2C, D*). *Alvin* has been rebuilt numerous times and has made >4800 deep-sea dives since its first launch in 1964. A typical dive can take two scientists and a pilot down to 4500 m, taking about two hours for *Alvin* to reach the seafloor and another two to return to the surface, with four hours of carefully planned recording and sampling work on the seafloor. Viewing ports and video cameras allow direct observation so that specific samples can be taken. Several other agencies operate research programs with human-operated submersibles capable of diving to greater depths; these include China National Deep Sea Center (*Jialolong*, 7000 m), Japan Agency for Marine-Earth Science & Technology (*Shinkai*, 6500 m), Shirshov Institute of Oceanology (Russia, *MIR-1* and *MIR-2*, 6000 m), *and IFREMER (France, Nautile*, 6000 m). Special technology is needed to collect samples from abyssal depths at great pressure, and from high-temperature environments near hydrothermal vents.

IMAGING TECHNIQUES

Light and electron microscopy are used to study morphology and structure of microbes

The study of microbes began in the mid-17th century with Antonie van Leeuwenhoek's observations of "animalcules" using a simple handheld microscope and development of the light microscope by Robert Hooke. Microscopy remains an important method for the initial examination of plankton samples and cultured microbes. Eukaryotic marine microbes like diatoms, dinoflagellates, ciliates, and fungi are large enough for microscopy to be useful in distinguishing morphological and structural features—in these groups, microscopic appearance is usually the main criterion for classification (for examples, see figures in *Chapter 6*). Direct light microscopy is also used for enumeration of eukaryotic plankton in seawater samples. A device called the FlowCam is used for rapid, high-throughput analysis of plankton samples; it uses a microsyringe to draw a sample through a cuvette, and a high-resolution camera photographs each cell as it flows, recording size, shape, and concentration of plankton samples. However, with bacteria and archaea, light microscopy reveals little more than the general shape and morphology, although the use of special stains and illumination techniques can improve the amount of information revealed. The wavelength of visible light limits the effective magnification of the light microscope to ~1000–1500 times and it is not possible to resolve objects or structures smaller than ~200 nm. Nevertheless, high-resolution phase-contrast light microscopy linked to digital recording equipment provides the best method

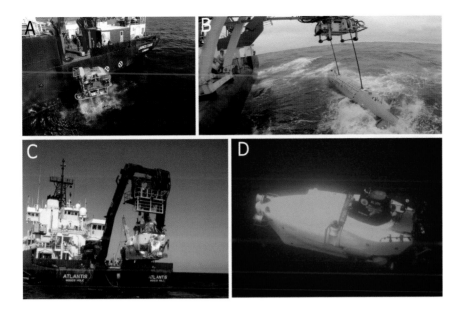

Figure 2.2 Recovery of the AUV AUTOSUB-3 following a seabed mapping mission in 2000 m deep water. B. Recovery of the ROV ISIS following a seabed sampling mission on 1500 m deep seamounts for collection of targeted benthic invertebrates for genetic and microbiological analysis. C. DSV *Alvin* being deployed from the RVV *Atlantis*. D. *Alvin* beginning its descent. Credit: A, B: Alex Nimmo-Smith, University of Plymouth, NERC DEEPLINKS project. C. Mountains in the Sea Research Team; the IFE Crew; and NOAA/OAR/OER. D. Gavin Eppard, NOAA Photo Library, via Flickr CC-BY-2.0.

for observing microbial behavior, such as bacterial movement (*Figure 3.14*) and predation of bacteria by protists.

The development of the electron microscope enabled the study of the ultrastructure of cells and viruses. In the transmission electron microscope (TEM), a beam of electrons is focused onto an ultrathin section of the specimen in a vacuum, usually after staining with lead or uranium salts. Electrons are scattered as they pass through the specimen according to different densities of material in the cell. Because the wavelength of an electron beam is much smaller than that of light, the TEM has an effective magnification up to 1 million times and objects as small as 0.5 nm can be resolved (e.g. *Figure 4.4*). Various techniques such as shadowing, freeze-etching, and negative staining are used to visualize the membranes, internal structures, and surface appendages of cells. However, it must always be borne in mind that TEM images are only obtained after staining and fixing the cells and observing them in a vacuum. Therefore, the appearance of structures in sections may not reflect their actual organization in the living cell. Refinement of techniques for preparing ultrathin sections allows visualization of the 3D ultrastructure of microbial cells (e.g. *Figure 6.6*). The recent development of cryo-electron tomography involves tilting the specimen in an electron beam to create a series of 2D projection images, which are then used to reconstruct a 3D image of the object. This reveals a life-like, frozen hydrated state with sufficient resolution to visualize the macromolecular organization of intact cells (e.g. *Figure 1.3A*). In scanning electron microscopy (SEM), a very fine electron beam scans the surface of the object, generating a 3D image. Therefore, SEM is particularly useful for studying the structure of cell surfaces (e.g. *Figure 6.7*).

Other exciting modern developments of electron microscopy include atomic force microscopy (AFM), in which a tiny probe is held in place very close to the surface of an object using weak atomic repulsion forces. Effective magnifications up to 100 million are possible and the atomic structure of molecules such as DNA or proteins can be visualized. A major advantage of this technique is that specimens do not need to be fixed or stained and can be examined in the living state. This technique has been used to reveal complex surface architectures at the nanometer to micrometer scale, including structures which connect planktonic microbes in microscale networks (*Figure 2.3*).

Epifluorescence light microscopy enables enumeration of marine microbes

Fluorescence occurs when material absorbs light at one wavelength (the excitation or absorption spectrum) and then re-emits it at a different wavelength (the emission spectrum). The specimen is illuminated with a tungsten-halogen or mercury vapor lamp. In

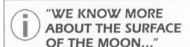

"WE KNOW MORE ABOUT THE SURFACE OF THE MOON…"

While 12 humans have walked on the Moon, only four have been to the deepest part of the ocean. In 1960, Jaques Piccard and Don Walsh descended to the Challenger Deep at the bottom of the Marianas Trench almost 11 km deep in the Pacific Ocean in a bathyscape (*Trieste*) designed to withstand the enormous pressures. The mission took 8 hours, and the men spent only 20 minutes on the seabed. *Trieste* was expensive to maintain and had almost no scientific value, so further missions never took place and the excitement of space exploration attracted greater public interest and geopolitical impetus than exploring the ocean floor. In 2012 and 2019 respectively, film director James Cameron and investor/explorer Victor Vescovo made privately funded solo trips to the Deep. While the statement that we know more about the surface of the Moon than that of our own oceans has become a cliché and is hard to justify objectively, it does highlight the difference in public expenditure. For example, in 2013, the USA spent $23.7 million on NOAA marine exploration research, but $3.8 billion on NASA space exploration research (Conathan, 2013).

Figure 2.3 Atomic force microscopy. (a) Interactions between pelagic bacteria; a heterotrophic bacterium (red) is closely associated with a larger *Synechococcus* cell (yellow). (b) Bacteria surrounded by an apparent network of gel and nanometer-sized particles. Scale bars indicates cell height. Credit: Francesca Malfatti and Farooq Azam, Scripps Institution of Oceanography (see Malfatti & Azam, 2009).

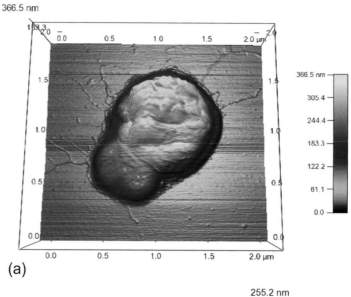

(a)

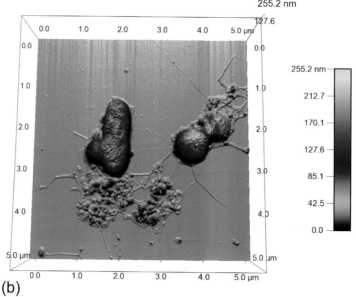

(b)

FLUORESCENCE MICROSCOPY AND THE "GREAT PLATE COUNT ANOMALY"

The development in 1977 of epifluorescence microscopy for examination of seawater by John Hobbie and colleagues at the Woods Hole Marine Biological Laboratory led to a major upheaval in microbial oceanography. Routine measurements became possible because of improvements in the type of filter used, revealing that bacterial numbers in plankton were a thousand or more times higher than previously realized. This prompted a major reappraisal of their role in marine processes. The term "great plate count anomaly"—first used by Staley and Konopka (1985)—became embedded in microbial ecology as a concept to explain the large difference between the total number of bacteria revealed in direct cell counts with those recovered by culture techniques such as viable counts on agar plates.

direct fluorescence microscopy, a filter is placed between the light source and the specimen, allowing only light of the desired excitation wavelength to be transmitted, while a barrier filter placed between the specimen and the eyepiece transmits the emitted fluorescence and absorbs longer wavelengths. Epifluorescence microscopy depends on the use of dichroic mirrors as interference filters that transmit one set of wavelengths and reflect the others. Water samples are usually fixed with paraformaldehyde or glutaraldehyde and filtered through membrane filters of a pore size appropriate to the microbial group under study—0.22 µm filters are most commonly used for bacteria. For enumeration of viruses, larger particles are removed by pre-filtration before trapping the viruses on 0.02 µm aluminum oxide filters. This method can therefore be used for observation and enumeration of all groups of marine microbes. The original stain (fluorochrome) used in plankton studies was acridine orange (AO, 3,6-bis[dimethylamino]acridinium chloride), which binds to DNA and RNA. Problems arise with AO owing to background fluorescence and difficulties in distinguishing microbes from inanimate particles, so the most widely used stain today is DAPI (4′,6′-diamidino-2-phenylindole), which binds to DNA and largely overcomes these problems. DAPI is excited by ultraviolet light and emits bright blue light. The use of incident light also permits observation of microbes attached to suspended particles. Although viruses are below the limits of resolution of the light microscope, they may be visualized sufficiently for enumeration if stained with SYBR Green, which emits a very bright fluorescence (*Figure 7.1*). This method only works for DNA viruses.

Some staining methods can indicate the physiological state of individual cells within a population, especially to distinguish between living and dead cells based on differences in membrane permeability. CTC (5-cyano-2,3-dilotyl tetrazolium chloride) is a fluorogenic redox dye that detects an active electron transport chain. Although there have been some criticisms of the technique, the CTC assay has been used to estimate the distribution of cell-specific metabolic activity in natural assemblages of marine bacteria. The Live–Dead kit is a widely used test for bacterial viability; a mixture of SYTO 9 and propidium iodide stains bacteria fluorescent green if they have intact cell membranes, whereas bacteria with damaged membranes appear fluorescent red. This is widely used to assess the viability of bacterial populations following environmental changes or chemical treatments.

Confocal laser scanning microscopy enables recognition of living microbes within their habitat

Confocal microscopy uses a laser light source coupled to an optical microscope and computer-aided digital imaging system. The beam is directed through a scanner and then through an aperture that adjusts the focal plane of the beam, meaning that it is possible to zoom in on particular vertical layers in the specimen. The advantage of confocal microscopy lies in its ability to be used on living specimens and its generation of a 3D image via the assembly of digital outputs from each layer. Because the magnification is much less than that of electron microscopy, it is not as useful for revealing ultrastructural detail. However, it is proving enormously useful in ecological studies of microbial communities, especially when combined with fluorescent *in situ* hybridization (FISH) techniques (p.36, below), which enables specific microbes to be located on surfaces or within other organisms. In particular, understanding of the structure of biofilms and microbial mats has advanced considerably using this technique.

Flow cytometry measures the number and size of particles

Flow cytometry was originally developed for biomedical uses and was first applied to marine studies in the late 1970s. Since then, use of this technique has produced spectacular advances in the enumeration and characterization of ocean microbes. A flow cytometer is an instrument that can be used to identify cells according to a specific fluorescence signature, leading to a wide range of physiological and ecological information. Flow cytometry simultaneously measures and analyzes multiple physical characteristics of single particles, ranging in size from 0.2 to 150 μm, as they flow in a fluid stream through a beam of light. The coupling of the optical detection system to an electronic analyzer records how the particle scatters incident laser light and emits fluorescence. A flow cytometer is made up of three main components (*Figure 2.4A*).

The fluidics system injects particles into the instrument in a stream of fluid so that they pass in single file through a laser beam. The optics system consists of lasers—there may be four or more different wavelengths—to illuminate the particles in the sample stream and optical filters to direct the resulting light signals to the appropriate detectors. Each particle scatters light at different angles and may also emit fluorescence. The scattered and emitted light signals are converted to electronic pulses that can be processed by the electronics system. In a fluorescent-activated cell sorter (FACS), the electronics system can also transfer a charge and deflect particles with certain characteristics, so that individual cells or different cell populations can be separated and collected. Flow cytometry is used for a range of specific measurements in marine microbiology, including quantifying different microbial groups and their diversity, estimation of cell dimensions and volume, determination of DNA content, and assessment of viability and growth rates. In marine studies, flow cytometry was first used for the analysis of phytoplankton cells, as they are naturally labeled with photosynthetic pigments such as chlorophyll and phycobilins, which are autofluorescent. However, most microbes can be detected in flow cytometry by using the same principles described for epifluorescence microscopy—labeling cells with fluorochromes that emit light of various wavelengths when illuminated by the appropriate laser. Flow cytometry has thus been particularly valuable in the analysis of bacteria and small eukaryotes in the picoplankton, especially following the development of portable instruments with stable

THE "JELLYFISH GENE" IS A MAJOR TOOL IN MOLECULAR BIOLOGY

Green fluorescent protein (GFP) is produced in the jellyfish *Aequorea victoria* and has very wide applications in all areas of biology. Its discovery and application by Tsien, Chalfie, and Shimomura led to the award of the Nobel Prize in 2008. The *gfp* gene can be easily inserted into the genome of many organisms and used as a marker for the expression of specific genes, with many applications in cell biology and genetic modification. Some examples of its use in marine microbiology include following the fate of *gfp*-marked bacteria after their ingestion by grazing protists or the process of infection of animal hosts by symbiotic or pathogenic bacteria. If specific genes are tagged, it is possible to visualize when they are "switched on" during the development of a microbial process.

optical and electronic systems that can be deployed on ships for direct examination of water samples. Improvements in the sensitivity of flow cytometry instruments and the introduction of new dyes mean that flow cytometry can now be used for detection and quantification of marine viruses (see p.200 for discussion of precautions needed in the interpretation of such counts). *Table 2.1* lists examples of fluorochromes used in flow cytometry and epifluorescence microscopy. Using an appropriate mixture of different fluorochromes, lasers, and light of different wavelengths, different populations of microbes in aquatic ecosystems can be analyzed based on size, abundance, and specific properties. Exciting new advances in flow cytometry have been possible owing to a fusion with FISH technology (*Figure 2.5*), in which fluorescently labeled oligonucleotides can be used to discriminate specific taxa in heterogeneous natural marine microbial communities.

Fluorescent *in situ* hybridization (FISH) allows visualization and quantification of specific microbes

The acronym FISH is no accident. This technique permits an investigator to "fish" for a specific nucleic acid sequence in a "pool" of unrelated sequences. It is widely used in biomedical imaging and its use in marine microbiology was pioneered by Rudolf Amann and Bernard Fuchs of the Max Plank Institute (MPI) for Marine Microbiology, Bremen. It has proved to be one of the most useful culture-independent techniques for identification of particular organisms or groups of organisms in the marine environment. Another major advantage is the ability to quantify different members of a microbial community and their dynamics because oligonucleotide probes can be designed to recognize

Figure 2.4 Flow cytometry. (a) Schematic diagram of the components of a flow cytometer. Particles pass single file through a narrow channel and into the laser beam, producing a single wavelength of light at a specific frequency to activate the fluorescence signal. Detectors measure forward and side scattering of light from fluorochrome-stained cells. Different light signals are split into different wavelengths by a series of filters and mirrors for detection by specific photomultiplying tubes. Signals are converted by the electronic system for visualization and interpretation. (b), (c). Examples of typical flow cytometry plots obtained during a coastal mesocosm experiment conducted at the Espegrend Marine Biological Station, Bergen, Norway in 2017. (b) The 585 nm (phycoerythrin) fluorescence vs. 692 nm (chlorophyll) fluorescence of particles excited with a 488 nm laser. Distinct phytoplankton groups can be distinguished based on the types and amounts of pigments present within cells. (c) The nanophytoplankton group is further distinguished by side scatter (SSC) vs. forward scatter (FSC). Elevated SSC is indicative of cell surface roughness, granularity, or light-scattering properties, which can be caused by the presence of calcifying cell wall features (e.g. coccoliths on calcifying coccolithophores). FSC is a proxy for cell size. Using cultures of known calcified coccolithophores a threshold gate can be set to delineate calcifying and non-calcifying nanophytoplankton. Here, calcified nanophytoplankton were enumerated by setting a gate using calcified *Emiliania huxleyi* strain DHB607 (Johns et al., 2019). Credits: (a) Boster Biological Technology, Pleasanton, Ca.; (b), (c) Brittany Schieler and Kay Bidle, Rutgers University, NJ.

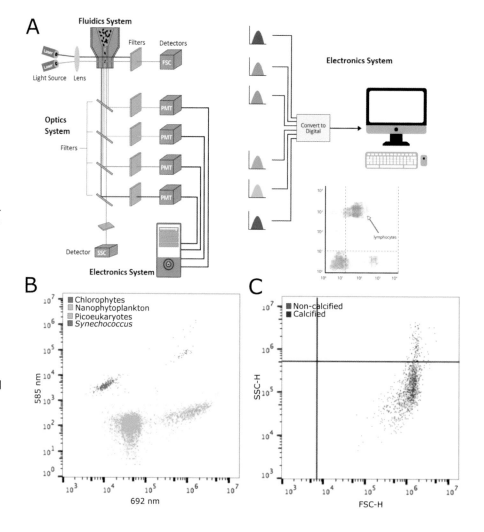

Table 2.1 Some representative fluorochromes used in epifluorescence microscopy and flow cytometry

Fluorochrome	Excitation/emission (nm)	Target
Acridine orange	500/526; 460/650	DNA, RNA
DAPI	358/461	DNA
Ethidium bromide	518/605	Double-stranded DNA, RNA
FITC	495/520	General fluorescent
Hoechst 33342	350/461	AT-rich DNA
Mithramycin	425/550	GC-rich DNA
Propidium iodide	535/617	Double-stranded DNA, RNA
Rhodamine 123	480/540	Membrane potential
SYBR green	494/521	DNA, can be used for viruses
SYTOX Green	504/523	Cell viability

AT, adenine and thymine; DAPI, 4′,6′-diamidino-2-phenylindole; FITC, fluorescein isothiocyanate; GC, guanine and cytosine.

and hybridize to complementary sequences that are unique to specific microbial groups (*Figure 2.8A*). The oligonucleotide probe is covalently linked to a fluorescent dye such as fluorescein, Texas Red, or indocarbocyanines. Hybridization of the probe with the target sequence within a morphologically intact microbial cell can then be visualized by fluorescence or confocal microscopy. The signal that is produced after hybridization and washing away the excess unbound probe is proportional to the amount of target nucleic acid present in the sample. This allows an investigator to visualize and quantify the microbes in marine samples and to determine cell morphology and spatial distributions of specific taxa *in situ*. FISH has been particularly useful for examining microbes attached to particles and spatial localization of microbes in biofilms and sediments, in symbiotic associations, and in infected tissues. Probes for different taxa can be linked to alternative fluorochromes and used together, enabling two or more dissimilar organisms to be located within a sample. Examples of micrographs using this technique can be seen in *Figures 3.13* and *9.6*; these show the importance of revealing associations between different microbes in significant biogeochemical processes like anoxic methane oxidation and nitrogen fixation. It can also be used to assess diversity of planktonic microbes collected on filters. *Figure 5.3* shows an important example, in which the relative abundance of archaea and bacteria in the water column was assessed by FISH microscopy. Small subunit (SSU) rRNA genes are most commonly used for diversity studies. Cells usually contain thousands of copies of rRNA molecules, which are relatively stable. However, many marine microbes have very small cells and may be slow growing or

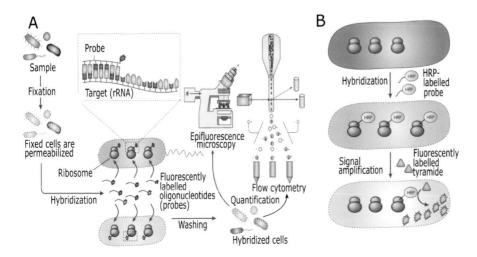

Figure 2.5 Fluorescent *in situ* hybridization (FISH). A. Principles of the basic method. The sample is fixed to stabilize cells and make the membrane permeable to the oligonucleotide probe, which attaches to its intracellular targets. After washing, single cells can be identified by epifluorescence microscopy or flow cytometry. B. Principles of catalyzed reporter deposition in CARD-FISH. Hybridization involves a single oligonucleotide that is covalently linked to horseradish peroxidase (HRP). The signal is amplified because multiple tyramide molecules are radicalized by a single HRP enzyme. Reprinted from Amann and Fuchs (2008) with permission of Springer Nature.

dormant and this can result in variable results because they may have low numbers of ribosomes or have membrane modifications that lower the accessibility of the probe. This places some limits on the uses of FISH in investigation of oligotrophic ocean environments, but modifications have improved the efficiency of hybridization; multiple probes, different sequences, and different fluorochromes can be applied to single samples to improve sensitivity and to assess genetic diversity. Enzyme-linked methods such as catalyzed reporter deposition (CARD) FISH (*Figure 2.8B*) can also provide greater signal intensity. Here, hybridization involves an oligonucleotide that has been covalently cross-linked to the enzyme horseradish peroxidase (HRP). Because the HRP molecule is much larger than the fluorochromes, permeabilization requires harsher enzymatic or chemical treatments, conducted in agarose gels to avoid loss of cell shape. Fluorescently labeled tyramide (a derivative of the amino acid tyrosine) is then added, and multiple molecules are activated by a single HRP enzyme which binds permanently to the cells, resulting in amplification to give a strong fluorescent signal. Besides detection of rRNA, the FISH technique can be extended to search for a wide range of specific gene sequences characteristic of particular structures or metabolic functions, assuming that sufficient copies of the targeted nucleic acid are present in the sample.

CULTIVATION OF MICROORGANISMS

Different microorganisms require specific culture media and conditions for growth

As noted above, there is a large discrepancy between the number of microbes that can be observed in direct counts of marine samples and those that can be cultured in the laboratory. Even those organisms that can be successfully cultured differ enormously in their physiological properties (see *Chapter 3*), so the composition of culture media and environmental conditions (temperature, pH, atmospheric conditions, and pressure) must be chosen with great care.

Many autotrophic bacteria and archaea can be grown in a simple defined synthetic medium containing bicarbonate, nitrate or ammonium salts, sulfates, phosphates, and trace metals. Appropriate light conditions for photolithotrophs, and the requisite substrates as energy sources and electron donors for chemolithotrophs must be provided. Chemoorganotrophic (heterotrophic) bacteria, archaea, protists, and fungi can be grown on a similarly defined medium with the addition of appropriate organic substrates such as sugars, organic acids, or amino acids—organisms differ greatly in their preference for particular nutrients. Initial isolation and routine culture are usually carried out using complex media, made from semi-defined ingredients such as peptone (a proteolytic digest of meat or vegetable protein, which provides amino acids) and yeast extract (which provides amino acids, purines, pyrimidines, and vitamins). Many heterotrophs are more fastidious and require the addition of specific growth substances. An example of a widely used medium for routine culture of many marine bacteria is Zobell's 2216E medium, which contains low concentrations of peptone and yeast extract plus a mixture of various mineral salts at concentrations similar to those found in seawater. It must be stressed that there is no single medium suitable for all microbial types, and microbiologists employ a very wide range of recipes for different purposes. Despite this, only a tiny proportion of the bacteria and archaea identified in gene surveys have been cultured, but there is a renewed interest in finding appropriate media and incubation conditions to rectify this (*Box 2.1*).

Microorganisms may be grown in liquid media (broth) or on media solidified with agar. Cultivation on agar plates (Petri dishes) permits samples to be streaked out to obtain single colonies, which can then be re-streaked to form pure cultures. In the shake tube and roll tube techniques, samples are mixed with molten agar, which allows culture of some anaerobic bacteria. Agar is a polysaccharide obtained from red algae such as *Gracilaria* and it has several advantages as a gelling agent for microbiological work. It is colorless, transparent, and remains solid at normal incubation temperatures, but can be held in a molten state above 42°C. Its main advantage in microbiology is that it does not normally affect the nutritional status of the medium because most bacteria and fungi do not possess extracellular enzymes for its degradation. However, not surprisingly, many marine microorganisms can degrade this algal compound by production of agarase enzymes. When marine samples are plated onto agar, craters are frequently observed surrounding the colonies as a result of enzymatic digestion of the agar.

The concentration of cultivable microorganisms in a sample can be determined by plating appropriate tenfold dilutions on agar plates, either by mixing with molten agar (pour plate), spreading evenly over the surface of the agar (spread plate), or by pipetting small drops onto the surface (drop count or Miles–Misra technique). Colonies are counted after incubation at an appropriate temperature and the viable count is expressed as colony-forming units per milliliter (CFU mL^{-1}). When plating samples from natural sources, it is important to check plates at regular intervals, as slow-growing colonies may take days or weeks to develop and may be obscured by rapidly growing types. Even if the right culture conditions can be found, some marine bacteria and archaea grow very slowly. To produce a visible colony on agar after 12–24 h incubation, bacteria must divide about once per hour, whereas bacteria with a doubling time of 48 h would take several weeks to produce a visible colony, assuming they are capable of growth under the specific conditions. Organisms in low abundance in the natural environment are often out-competed by those which grow more rapidly or produce inhibitory substances. Selective media containing antibiotics or other chemicals that inhibit the growth of certain microorganisms can be used, thus allowing only selected groups of interest to multiply. They may also include dyes or chromogenic substances that give a differential reaction when metabolized by bacteria belonging to particular groups. These media are used as an enrichment method in liquid culture or to select colonies of the desired type on agar plates. In marine microbiology, selective media are used mainly for the isolation and growth of animal and human pathogens and allochthonous indicators of pollution. Selective media frequently contain bile or synthetic detergents, which select for enteric bacteria and related groups by their effects on membrane function. For example, the medium thiosulfate-citrate-bile-sucrose (TCBS) agar is widely used for isolation of vibrios, while lauryl sulfate broth is used in the enumeration of coliform bacteria in assessing fecal pollution of seawater or shellfish (*Figures 13.6 and 13.9*). Selective antibiotics are often added to suppress the growth of bacteria on fungal isolation plates.

As well as plating, successive dilution in liquid media can be used to separate individual bacteria to obtain pure cultures. This method has been used successfully to grow previously nonculturable bacteria in seawater without the addition of nutrients (see *Box 2.1*). The dilution-to-extinction method is also used in the most probable number (MPN) technique for quantification of microbial populations. One widely used application of this is in the determination of coliforms as indicators of fecal pollution in seawater or shellfish (*Figure 13.9*).

Pure culture methods for the cultivation of protists and fungi follow the same basic principles described above. Many photosynthetic microalgae can be maintained in culture, requiring mineral salts solution and appropriate conditions of temperature and light/dark cycles, and a method of keeping cells in suspension through shaking or aerating. Large-scale cultures may be maintained with bioreactors (*Figure 14.1*). However, algal cultures can be very difficult to maintain and there are numerous recipes for special media for growing and maintaining different species. Culture media can be made of defined chemicals or from natural seawater enriched with extracts from various sources. It is relatively easy to isolate single microalgae by streaking or making serial dilutions to obtain a clonal population, or individual cells of larger species can be isolated using a fine pipette. A recent advance in micromanipulation allows the isolation of individual cells, which can be trapped using a highly focused laser beam which provides a force, depending on the relative refractive index between the cell and the surrounding medium. The cell can be separated off for subsequent pure culture. Such "optical tweezers" provide a very powerful technique when used in conjunction with gene probes or antibody staining. Phagotrophic flagellates, dinoflagellates, and other heterotrophic protists are much more difficult to culture, and established cultures of only a few types are available. Because the nutritional requirements can vary so widely, considerable experimentation is needed to optimize conditions for isolation and maintenance of different types. For many protists, especially diatoms and dinoflagellates, it is very difficult to obtain pure axenic cultures because bacteria and fungi often occur in close association with their cells. For loose associations, axenic cultures can sometimes be obtained by extensive rinsing of the isolated protists, or treatment with antibiotics, but some bacteria or fungi are tightly associated (for example,

"BUT THEY ALL LOOK THE SAME!"

Many culture-based investigations in marine microbiology begin with the collection of seawater samples, sediments, material scraped from surfaces, or extracts of animal or seaweed tissues. These are plated on agar media and, after incubation, colonies are examined, and different types can be distinguished by noting the colony size, color, form, elevation, texture, and margin. Different types are then picked off, streaked onto fresh plates and re-incubated to obtain pure colonies for further characterization. While some isolates have bright colors or other distinctive features (*Figure 2.6*), many marine bacteria have very similar-looking colonies. Students starting an isolation project are often frustrated that so many colonies appear as off-white, circular, with entire margins. Patience, careful observation of slight differences, and methodical recording after different periods of time are all essential. The chance of obtaining a good representative collection of isolates is further enhanced by using different culture media and conditions, but the results will always be only a partial reflection of the true diversity in a natural microbial community.

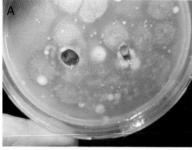

Figure 2.6 Iridescent colonies of *Cellulophaga lytica* isolated from the anemone *Actinia equine* on marine agar at 25°C. A. The initial isolation plate shows colored *C. lytica* colonies together with agarolytic and white bacterial colonies. B. A pure culture of *C. lytica* observed under direct epi-illumination shows the intense structural "glitter-like" or "pointillist" iridescence. Subsequent studies revealed that the iridescence is caused by self-organization of the bacterial cells within the colony biofilm, creating hexagonal photonic crystals (Kientz et al., 2013, 2016). Credit: courtesy of Eric Rosenfeld, Université de La Rochelle.

within the frustules of diatoms) or occur as intracellular symbionts or parasites. Such extracellular and intracellular associations are often symbiotic and necessary for successful growth of the algae, but they add major complications to the analysis of the genomes of protists.

Enrichment culture selects for microbes with specific growth requirements

Enrichment culture depends on the provision of particular nutrients and incubation conditions to select for a certain group. For example, the addition of cellulose or chitin plus ammonium salts to a marine sample under aerobic conditions will enrich for *Cytophaga*-like bacteria, while sulfates with acetic or propionic acid under strictly anaerobic conditions will favor growth of *Desulfovibrio* and related types. The Winogradsky column is especially useful in investigating communities in marine sediments. A tall glass tube is filled with mud and seawater and various substrates added prior to incubation under appropriate conditions. A gradient of anaerobic to aerobic conditions will develop across the sediment and a succession of microbial types will be enriched. In the light, microalgae and cyanobacteria will grow in the upper parts of the column and will generate aerobic conditions by the production of oxygen. Depending on the substrate added, the activity of various groups will lead to the production of organic acids, alcohols, and hydrogen, which favor the growth of sulfate-reducing bacteria. In turn, the sulfide produced leads to the growth of anaerobic phototrophs that use hydrogen sulfide as an electron acceptor. A wide range of enrichment conditions may be adapted for specific groups, and the resulting communities may then be subject to further enrichment or selective plating to obtain pure cultures. The results of enrichment cultures must be interpreted carefully because they often lead to overrepresentation of groups that may not be the dominant members of the natural community.

Phenotypic testing is used for characterization of many cultured bacteria

After isolation, the initial characterization of bacterial cultures usually begins with preparation of a Gram stain and observation of cell morphology and grouping under the microscope, but this does not provide sufficient information for identification and taxonomy. Cultivated bacteria can be identified and characterized using a combination of growth characteristics and tests for the production of particular enzyme activities, particularly those involved in carbohydrate and amino acid metabolism. *Table 2.2* shows a summary of the methods used for bacterial identification and classification. Once it is certain that a pure culture has been obtained, it is usually easier to use 16S rRNA gene sequencing (p.43) to place a new isolate into a particular phylogenetic group, before deciding which phenotypic tests should be performed to provide more detailed information. Of course, phenotypic characterization based on biochemical test methods is suitable only for the small proportion of heterotrophic marine bacteria that can be easily cultured. Diagnostic tables and keys aid in the identification of unknown bacteria, although the inherent strain-to-strain variation in individual characteristics often causes problems for accurate identification and taxonomy. Commercial kits employing a battery of tests (e.g. API®, Micro-Bact®) offer advantages for standardization of methods and processing of results with the aid of databases developed using numerical taxonomic principles. However, most biochemical methods were developed for the identification of bacteria of medical importance and require adaptation for examination of marine environmental samples. These methods are most useful in areas such as identification of pathogens or fish spoilage organisms. For example, there are extensive databases and established diagnostic keys based on biochemical tests for the identification of *Vibrio* and related pathogens in aquaculture. Some tests are very characteristic of particular groups. For example, the detection of the enzymes β-galactosidase and β-glucuronidase using fluorogenic substrates is important in the identification of coliforms and *E. coli* in polluted waters (*Figure 13.4*). The BIOLOG® system differentiates microbial taxa based on their nutritional requirements and incorporates up to 95 different substrates as a sole source of carbon or nitrogen, together with a tetrazolium salt. Different microbial groups use a range of substrates at characteristic levels and rates. Positive reactions are tested via a color reaction based on reduction of the tetrazolium salt. BIOLOG systems have been developed specifically for Gram-positive bacteria, Gram-negative bacteria, and yeasts. Again, this method is not suitable for detection of

BOX 2.1   RESEARCH FOCUS

Cultivating the uncultured

Many major divisions of the Bacteria and Archaea contain no known cultured species. As discussed in *Chapters 4* and *5*, our knowledge of bacterial and archaeal diversity in the oceans has been revolutionized by the application of molecular biology techniques, especially sequencing of 16S rRNA genes, which allow recognition and relatedness of groups of organisms based solely on their genetic sequence. In the 1990s, as the number of previously unknown taxa recognizable by direct sequencing of genes in environmental samples increased, many microbial ecologists assumed that these organisms are inherently unculturable and that it was futile to attempt to grow them. In the excitement generated by the rapidly accumulating results produced by new techniques, many molecular microbiologists became vociferous in their dismissal of the traditional methods applied to laboratory cultivation, considering them old-fashioned, distinctly "uncool", and of little further use. The phrase "more than 99% of bacteria are unculturable" became dogma. As pointed out by Donachie et al. (2007), this attitude ignores important lessons from the history of microbiology; Beijerinck, Winogradsky and other pioneers of general microbiology used numerous approaches to overcome the difficulties of microbial cultivation more than 100 years ago. Provided sufficient effort and ingenuity are applied, there is every reason to be optimistic that we will be able to culture many more marine microbes. As Carini (2019) reminds us: "unculturable" is a frame of mind, not a state of microbiology—we should regard these organisms as *not yet* cultured" instead.

Overcoming the problem of extreme oligotrophy. The most likely reason that so many marine bacteria are so difficult to grow is that they are extreme oligotrophs, adapted to growing very slowly at low cell densities with very low-nutrient concentrations (p.92). It is very difficult to reproduce these conditions in the laboratory, especially if we try to make organisms grow quickly at thousands of times the density at which they exist in nature. Common laboratory media used for the culture of marine heterotrophic bacteria typically contain nutrients at levels 10^6–10^9 times higher than those found in seawater. Button et al. (1993) developed techniques for culturing slow-growing marine bacteria by making successive dilutions of natural seawater in filtered, sterilized seawater, so that only one or two bacteria per tube are obtained. This leads to a successful enrichment of the dominant cell types, which can grow to densities of about 10^5–10^6 mL^{-1}, similar to those occurring in seawater. Schut et al. (1997) used this technique to isolate an oligotrophic *Sphingomonas* sp. This approach was developed in order to culture representatives of the SAR11 clade, which was discovered using 16S rRNA gene sequencing in the Sargasso Sea in 1990 (Giovannoni et al., 1990) and later recognized as the most abundant ocean bacteria (Morris et al., 2002; Giovannoni, 2017). Rappé et al. (2002) developed a high-throughput method in microtiter trays containing sterile seawater supplemented with phosphate, ammonia, or low levels of organic compounds. Bringing SAR11 into culture allowed the

physiological properties of this ubiquitous organism to be investigated for the first time and enabled determination of the genome sequence of the cultured isolate, which was named "*Candidatus* Pelagibacter ubique" (Giovannoni et al., 2005). Even with the genome sequences, identifying the nutrients required for growth proved difficult and cultivation of more strains of SAR11 and other Alphaproteobacteria required further modification of the method by using ultraclean techniques and Teflon culture vessels (Stingl et al., 2007). Zengler et al. (2002) used encapsulation of single bacteria in microdroplets of gel to allow large-scale culture of individual cells. Concentrated seawater was emulsified in molten agarose to form very small droplets, and flow cytometry was used to sort droplets containing single bacteria. Various media and growth conditions were evaluated and microbial diversity in the microcolonies within the droplets was examined using 16S rRNA gene typing. When unsupplemented filtered seawater was used as a growth substrate, a broad range of bacteria, including previously uncultured types, were isolated and grown.

Why is cultivation so important in the age of metagenomics? Zengler (2009) emphasized the importance of developing new cultivation strategies to enable detailed studies of cell physiology and genetics, essential to understand community-level processes. Clearly, cultivation-independent and cultivation-dependent approaches to microbial community assessment are complementary. But as Carini (2019) points out, it is not always possible to predict the activity of microbes and their ecosystem function from genome sequences alone. He gives the analogy "… one cannot understand the experience of driving a Ferrari from the list of its components; a parts list does not convey the handling, the sound, or the driver's connection with the machine." Carini emphasizes the great value obtained from current cultivation-independent genomic methods but reminds us that "… genomes are a parts list: we seek an understanding of the emergent principles of the cells themselves and the communities that they constitute." Furthermore, metagenomic analysis of microbial communities relies heavily on data obtained from the sequencing of genomes of cultivated species. To interpret the likely activities of new genes identified in metagenomes, we need genes from cultivated organisms to compare them with. We can use molecular data to identify which uncultured taxa we should target for cultivation. How abundant are they? Do they diverge substantially from cultured taxa? Might they have biotechnological potential? Do they have a key role in ecology or biogeochemical processes? Only by studying the growth, physiology, and metabolism of live bacteria and archaea in culture can we determine the properties of the whole organism and relate them to its ecosystem function. One of the most compelling examples of the benefits of ingenuity, patience, and persistence is the successful cultivation —after painstaking attempts spanning 12 years—of an extremely slow-growing Asgard archaeon found in sediments; an organism that that could hold the key to the origin of eukaryotic cells, as described in *Box 5.1*.

nonculturable microorganisms, but it provides a less stressful environment than the surface of solid growth media and some investigators have used BIOLOG systems for community characterization of metabolic activities in sediments or surface layers by direct inoculation without an intervening culture step.

Analysis of microbial cell components can be used for bacterial classification and identification

Analysis of the composition of bacterial membranes via gas chromatography of fatty acid methyl esters (FAME) is a powerful technique that can detect very small differences between species and strains. Individual taxa have a distinct fatty acid fingerprint. However, careful standardization of growth conditions is needed because membrane composition is strongly affected by environmental conditions and the nature of the culture medium.

Analysis of the protein profile of microbes can also be used for comparison of microbial species. Extracted proteins are dissociated with the detergent sodium dodecyl sulfate and separated by polyacrylamide gel electrophoresis (SDS-PAGE). However, gene expression and the resulting protein pattern produced are extremely dependent on environmental conditions and the technique needs to be used with caution in identification and taxonomy. Indeed, its main use is in assessing the effect of various factors on the synthesis of particular proteins, such as bacterial colonization or virulence factors during infection of a host, or responses to temperature or nutritional shifts.

Pyrolysis mass spectrometry (PyMS) generates a chemical fingerprint of the whole microbe by thermal degradation of complex organic material in a vacuum to generate a mixture of low-molecular-weight organic compounds that are separated by mass spectrometry. PyMS is easily automated and a high throughput of samples can be processed, but the equipment is expensive and is found only in a few specialized laboratories.

METHODS BASED ON DNA AND RNA ANALYSIS

Nucleic acid-based methods have transformed understanding of marine microbial diversity and ecology

Marine microbiologists use a wide range of techniques based on the isolation and analysis of nucleic acids. Many of these can be applied to the characterization of microbes in culture, for example, for accurate identification and taxonomy or for diagnosis of disease, as shown in *Table 2.2*. The complete genome sequence of many cultured microbes has also been determined, as shown in *Table 3.1*. However, the most dramatic advances in marine microbial ecology have occurred as a result of the development of "environmental genomics"—methods for the detection and identification of microbes based on direct extraction of genetic material from the environment, without the need for culture. First introduced in the mid-1980s, these culture-independent molecular techniques are very sensitive and can allow the recognition of very small numbers of specific organisms or viruses among many thousands of others. Direct analysis of the genes that are present in an environmental sample or comparison of community profiles using "genetic fingerprinting" techniques allows us to make inferences about the diversity, abundance, and activity of microbes in the oceans. Since 2000, there have been rapid developments in techniques for isolation, amplification, and sequencing of DNA, together with advances in bioinformatics—the mathematical and computing approaches needed to analyze, interpret, and manipulate the genomic data generated. It has become possible to assemble and analyze the genetic information obtained from microbial communities in the environment in a way comparable to the analysis of single genomes, leading to what we now term "metagenomics."

Amplification and sequencing of ribosomal RNA genes is widely used in microbial systematics and diversity studies

Chapter 1 introduced the pioneering work of Woese and colleagues based on sequencing of ribosomal RNA, which led to revision of ideas about evolution of the major domains of life.

BREAKTHROUGH—DIRECT ISOLATION OF NUCLEIC ACIDS FROM THE SEA

Following Woese's use of rRNA for studies of relationships between organisms, parallel advances in nucleic acid sequencing technologies and computing led to the rapid generation of large databases of information linking sequences from known organisms to their taxonomic position. Because it was clear that many marine microbes could not be cultured, microbial ecologists asked whether it would be possible to obtain sequence information directly from environmental samples. The first application of direct sequencing of nucleic acids from the marine environment was in 1986 by Norman Pace and colleagues at the University of Colorado. Development of the PCR technique in 1988 enabled rapid application in many different diversity studies cataloging ocean microbes. A number of research teams, most notably including those led by Stephen Giovannoni (Oregon State University), Ed DeLong (then at Monterey Bay Aquarium Research Institute), and Jed Fuhrman (University of Southern California), pioneered research in this field in the 1990s, providing the breakthrough that paved the way for the current importance of marine microbiology.

Table 2.2 *Summary of techniques used for the identification and classification of cultured bacteria*[a]

Technique	Usefulness at different taxonomic levels
Genome properties	
DNA sequencing	All levels, for phylogenetic analysis
DNA–DNA hybridization	All levels above species, 70% hybridization level is used for species definition
GC ratios	Comparison of species known to be closely related
PCR-based fingerprinting (e.g. RAPD, AFLP, rep-PCR)	Good discrimination at strain level
RFLP, ribotyping	Strain differentiation, but prone to variability
MLST	Reliable and robust; good discrimination of species and strains
16S rRNA gene sequencing	Higher taxonomic levels, for phylogenetic analysis
ITS region sequencing	Strain differentiation
Phenotypic characteristics	
Bacteriophage typing	Strain differentiation, highly specific
Biochemical tests (e.g. API® system, BIOLOG Phenotypic Microarrays™)	Species and strain differentiation, routine identification, community analysis
Morphology	Limited information except in some groups
Plasmid and protein profiles	Strain differentiation but highly variable
Serology	Strain (serotype) differentiation
Chemotaxonomic markers	
FAME	Rapid typing to genus level; further discrimination may be possible with careful standardization
PyMS	Rapid typing to genus level; further discrimination may be possible with careful standardization
Quinones and other biomarkers	Good differentiation above species level

[a]Many of these methods can also be used for archaea, fungi, and protists. AFLP, amplified fragment length polymorphism; FAME, fatty acid methyl ester analysis; GC ratio, guanine–cytosine base pair ratio; ITS, internal transcribed spacer; MLST, multilocus sequence typing; PCR, polymerase chain reaction; PyMS, pyrolysis mass spectroscopy; RAPD, random amplification of polymorphic DNA; RFLP, restriction fragment length polymorphism.

The ribosome is composed of a large number of proteins and rRNA molecules, as shown in *Figure 2.7*. The rRNA molecules (16S for members of the Bacteria and Archaea; 18S for Eurkarya) in the small ribosomal subunit (SSU) quickly became the first choice for microbial community analysis and diversity studies for several reasons. First, the rRNAs are universally present in all organisms, because of the role of the ribosome in the essential function of protein synthesis. Second, the rRNA molecule has a complex secondary structure and mutations in parts of the gene that affect critical aspects of structure and function in the ribosome are often lethal; therefore, changes in rRNA occur slowly over evolutionary time. Some parts of the rRNA molecules—and consequently the genes that encode them—are highly conserved, while others have a high degree of variability. These genetic differences act as a proxy for differences between genomes and form the basis of modern phylogenetic classification systems.

Figure 2.7 Structure of the ribosomes in the three domains of life. The subunits, proteins, and rRNA molecules are classified by the Svedberg unit (S), a measure of the rate of sedimentation during centrifugation. Sedimentation rate depends on the mass and the shape of the particles; hence when the large (LSU) and small (SSU) subunits are measured separately, they do not add up to the S value of the intact ribosome. The 16S (Bacteria and Archaea) and 18S rRNA (Eukarya) molecules in the SSU are most commonly analyzed. Molecular masses of the ribosomes and their subunits are shown in daltons (Da).

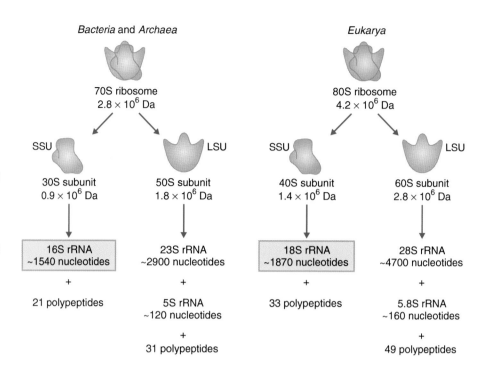

Another advantage is that growing microbes often contain multiple copies of the rRNA genes, so use of the PCR (see below) allows them to be amplified from very small amounts of DNA.

PCR amplification and Sanger sequencing (described below) still remains a reliable and inexpensive method in identification and classification of cultivated bacteria and archaea. Because PCR is carried out with primers directed against specific sequences in the SSU rRNA genes, it can also be used to selectively amplify gene fragments from different organisms in an environmental sample. Most environmental studies use broad-spectrum ("universal") primers directed against highly conserved regions in the SSU rRNA genes while amplifying the more variable sequences between the primer annealing sites. Specific primers can be designed to target particular taxonomic groups that share a known sequence. The PCR-amplification products can be cloned in *E. coli* to create a gene library. Proprietary cloning vector kits are widely used; these ensure high efficiency of cloning and allow detection of transformant colonies containing DNA inserts, which are purified by re-amplification before sequencing. After determination of DNA sequences, data are compared to generate phylogenetic trees. A brief description of each of the main stages in this process is given in the following sections. In large-scale environmental surveys, these methods have been replaced by high-throughput sequencing (HTS), but they still remain an important part of many research projects.

Although tremendous advances have been made using SSU rRNA gene sequencing—this method dominates much of the discussion about diversity in this book—it should be remembered that it is not the only approach and it has some inherent limitations. The highly conserved nature and the small size of rRNA genes (1500–1800 bases) means that their useful information content is relatively low. SSU rRNA gene sequencing is best suited to comparison of higher-level microbial taxa, and there can be drawbacks when it is used to delineate genera and species. In theory, genes encoding proteins contain more information because the genetic code, based on four nucleotides, generates a protein sequence based on 20 amino acids. Increasingly, the sequences of genes encoding key structural proteins (e.g. protein synthesis elongation factors) or enzymes (e.g. DNA polymerase or ATPase) are determined for detailed phylogenetic comparisons. In addition, the use of probes directed against specific metabolic genes can provide information about the importance of microbial activities in the environment. For very fine discrimination between closely related lower taxonomic levels (such as between species, or between strains within a species) it is often preferable to use noncoding regions between genes, such as the internal transcribed spacer (ITS) regions of nonfunctional RNA that occurs between the structural rRNA genes. These regions of

DNA are not under the same constraints as functional genes and are often hypervariable. Multilocus sequence typing (MLST) is a technique for the typing of multiple loci, using 7–10 "housekeeping" genes for major metabolic processes. Internal 450–500-bp fragments are sequenced on both strands and separated by electrophoresis, providing allelic profiles that allow high discriminatory power. MLST remains useful, especially for epidemiological studies of pathogenic bacteria, enabling characterization of isolates from different hosts and geographic regions, but it is less commonly used since the advent of HTS. Examples of recent research findings using HTS in studies of microbial diversity are discussed in *Box 4.1*.

Isolation of genomic DNA or RNA is the first step in all nucleic acid-based investigations

Protocols for DNA or RNA extraction from cells in culture are usually straightforward and reproducible. By contrast, isolations from environmental samples are often problematic and require considerable optimization. Samples must usually be processed rapidly or preserved in such a way that the nucleic acids are protected from degradation (e.g. immediate freezing in liquid nitrogen or transfer to a −80°C freezer). RNA is particularly prone to degradation by nucleases, and special solutions for sample preservation (e.g. RNA*later*) and avoidance of contamination in the laboratory are needed. For extraction from planktonic microbes, it may be necessary to concentrate biomass by centrifugation or tangential flow filtration of large volumes of water. Cells are lysed using an ionic detergent (SDS) and chelating agent (EDTA) plus proteinase K to break the membranes and dissociate proteins. Gram-positive bacteria require treatment with lysozyme to disrupt the cell wall. RNA is removed with ribonuclease. With a few techniques such as rep-PCR, preparation may involve simple procedures such as boiling a sample of culture, but for most methods, good-quality purified RNA or DNA is needed. The traditional method is based on phenol-chloroform extraction, which results in separation of DNA from proteins, based on their relative polarity. However, most studies employ commercially available kits in which DNA is collected by ethanol precipitation onto silica or anion exchange resin inserts in spin column microcentrifuge tubes and washed with buffers containing chaotropic agents and high salt concentration, before eluting in pure water or low salt buffer. These provide advantages in speed and reproducibility. All nucleic-acid-based methods for analysis of diversity in the environment depend on the assumption that DNA is extracted in a more or less pure state equally from the different members of the community, but there are several reasons why this may not be the case. Microbes differ greatly in their sensitivity to the treatments used to break open the cells. In particular, the nature of the cell envelope varies greatly, and cells may be protected by association with organic material or because they are inside the cells of other organisms as symbionts or pathogens, meaning that DNA may be extracted preferentially from some types according to the treatment used. Techniques such as freeze-thawing, bead beating (rapid shaking of the mixture with tiny glass beads), ultrasonic disintegration, or use of powerful lytic chemicals enhance the recovery of DNA but can result in extensive degradation. Problems often arise when trying to amplify DNA from samples such as sediments, microbial mats, or animal tissue due to the presence of PCR inhibitors. Again, specialized proprietary kits may overcome many of these problems.

When investigating microbial symbionts or parasites, extra steps such as selective lysis or differential centrifugation may be needed to separate microbial cells from host material. Special treatments may be needed to remove inhibitors of DNA extraction, amplification, and sequencing. Filtration or fluorescence-activated cell sorting (FACS) can be used if only part of the community is under investigation; for example, viruses or a particular size class of plankton.

The polymerase chain reaction (PCR) forms the basis of many techniques

PCR is a method of amplifying specific regions of DNA, depending on the hybridization of specific DNA primers to complementary sequences in the target organism's DNA. It is a routine process in many laboratories, but a methodical approach and avoidance of contamination are essential. The target DNA is mixed with a DNA polymerase, a pair

SECRETS FOR SUCCESS WITH THE PCR

An exhibition called *Chain Reaction* held in Canterbury UK in 2013 celebrated 30 years since the invention of the PCR. The co-curator of the exhibition, Charlotte Sleigh (Reader in the History of Science at the University of Kent) has commented on how the exhibition was used to show how the PCR has led to huge advances in genetics and biological understanding, despite its simple process and relatively humble equipment—"it's smaller than a microwave, and even less exciting to look at" (Sleigh, 2013). The article also highlights the rituals and superstitions of laboratory work with the PCR. Sometimes it doesn't work, and a tedious and frustrating cycle of troubleshooting will ensue. Sleigh reports how many scientists devise comic routines and superstitions to "ensure" success. She reports of a scientist who had to see a "lucky cloning rabbit" from the lab window before starting the thermocycler. In my lab, there were students who had to wear a "lucky lab coat" or turned the PCR hood into a mock altar, where they made offerings of reagents to the "PCR god" and carried out ritual cleaning with bleach wipes. And, woe betide anyone who touched a lab colleague's micropipettes!

Figure 2.8 The polymerase chain reaction (PCR). The initial stages are shown. (1) The reaction is initiated by heating (usually 94–98°C for 1 min), so that the double-stranded DNA dissociates into single strands. (2) The temperature is then lowered (usually 50–60°C for 0.5–1 min), so that the primers can attach to the complementary sequences on the DNA (the annealing temperature is calculated to be close to the Tm to ensure efficient but specific binding). (3) The temperature is raised to 72°C (the optimum for *Taq* polymerase), and primer extension begins. After the first few cycles, the number of short duplexes containing the amplified sequence will increase exponentially; one target molecule will become 2^n molecules after *n* cycles (~1 billion after 30 cycles). Credit: modified from Enzoklop (CC-BY-SA 3.0) via Wikimedia Commons.

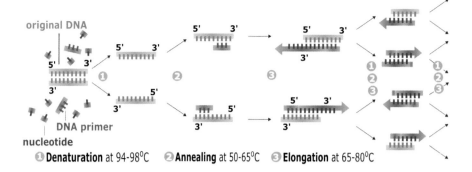

❶ **Denaturation** at 94-98°C   ❷ **Annealing** at 50-65°C   ❸ **Elongation** at 65-80°C

of oligonucleotide primers and a mixture of the four deoxyribonucleoside triphosphates (dNTPs). The PCR depends on the use of a thermostable DNA polymerase which is able to function during the repeated cycles of heating and cooling in the reaction, as illustrated in *Figure 2.8*. The original and most widely used enzyme, *Taq* polymerase, is a thermostable enzyme isolated from the hot-spring thermophilic bacterium *Thermus aquaticus*. Alternative thermostable polymerases, including some from deep-sea thermophilic bacteria, are now also used (see *Table 14.3*). The primers are designed to anneal to opposite strands of the target DNA so that the polymerase extends the sequences toward each other by addition of nucleotides from the 3′ ends, by base pairing using the target DNA sequence as a template for synthesis. The heating and cooling cycles of the PCR are carried out in specially designed electronically controlled thermal cyclers, in which the reaction tubes are placed in blocks designed to ensure rapid and precise heating and cooling. Usually, ~30 cycles are used, resulting in amplification of a single target sequence to over 250 million PCR products (usually slightly less than the theoretical maximum). The outcome of the reaction can be checked by agarose gel electrophoresis, which separates the DNA molecules according to size; bands are then visualized under ultraviolet light after staining the gel with ethidium bromide, Safe-Red, or SYBR Green dye. Because of the high degree of conservation in the SSU gene (or other conserved genes used as targets), the PCR products will all have approximately the same size and should migrate as a single band. The molecular weight of the amplicon can be assessed by comparison with a "ladder" of DNA fragments that is separated in the same gel. Before submitting it for sequencing, further purification steps are usually carried out, before checking the quantity and purity of the DNA by a drop-based spectrophotometer that measures microliter volumes.

Choosing a primer for use in the PCR obviously depends on knowing the sequence of the target DNA, but several factors need to be considered in order to ensure a successful PCR. For frequently studied genes, such as those for rRNA, there are standard primers and PCR conditions that are widely used. For studying novel genes, it is necessary to compare sequences in the databases to find a consensus sequence that recognizes conserved regions of a gene. An oligonucleotide primer, usually about 17–28 bases in length, is made using a DNA synthesizer—this is often carried out by commercial services. The ratio of G+C bases should usually be about 50–60%, depending on the gene, and primers should end in G or C to increase efficiency. Care needs to be taken to avoid complementary sequences within the primer, and between the pair of primers, as this can lead to secondary structures such as "hairpins" or preferential synthesis of primer dimers. Alternative nucleotide bases can be placed at specific positions, giving a set of degenerate primers that will allow for strain variation in DNA sequence for a particular gene. Computer programs are used to design the appropriate sequence of bases to give good results and to calculate the temperatures to be used in the thermal cycler program.

Suitable positive and negative controls are always included in the PCR run and each application of the PCR must be carefully optimized. This is especially important when trying to amplify sequences from a large mixed population of DNA, such as direct examination of environmental samples. Conditions such as the annealing temperature, amount of template DNA, and concentration of Mg^{2+} are critical. For example, *Taq* polymerase is inactive in the absence of Mg^{2+}, but at excess concentration the fidelity of the polymerase is impaired and nonspecific amplification can result. Designing a PCR protocol to amplify a new gene requires extensive optimization.

There are many variations of the basic PCR technique. In nested PCR, the level of specificity and efficiency of amplification is improved by carrying out a PCR for 15–30 cycles with one primer set, and then an additional 15–30 cycles with a second set of primers that anneal within the region of DNA amplified by the first primer set. Multiplex PCR involves the use of multiple sets of primers and results in the production of multiple products. This is often used as a quick screening method for the presence of certain organisms within water or infected tissue. Reverse transcriptase PCR (RT-PCR) detects gene expression by quantifying messenger RNA (mRNA). A DNA copy of mRNA (cDNA) is made with reverse transcriptase, before running the PCR. This technique requires great care in sample preparation and handling, since mRNA is very short-lived, and the technique is prone to contamination from genomic DNA and nucleases. In quantitative real-time PCR (Q-PCR), the formation of a fluorescent reporter is measured during the reaction process. By recording the amount of fluorescence emission at each cycle and incorporation of appropriate standards, it is possible to monitor the PCR reaction during the exponential phase in which the first significant increase in the amount of PCR product correlates to the initial amount of target template. Q-PCR offers significant advantages for the diagnostic detection of microbes in the environment and in infected tissue samples, and handheld instruments for use in the field are now available.

Genomic fingerprinting can be used to assess diversity of cultured isolates

Investigation of diversity among closely related individual species, or strains within a species, of bacterial or archaeal cultures can be achieved by a variety of genomic fingerprinting techniques. The most widely used methods make use of differences in genetic sequences revealed by the action of restriction enzymes, which cut DNA at specific sites determined by short sequences of nucleotide bases. All of these fingerprinting techniques produce band patterns that lend themselves to computer-assisted pattern analysis, leading to the generation of databases and phylogenetic trees useful for studying diversity among cultured isolates.

Ribotyping is a technique used for bacterial identification in which the PCR products from rRNA gene amplification are cut with restriction endonucleases. These enzymes recognize short sequences in the DNA and cut the molecule wherever those sequences occur. The DNA fragments are separated by agarose electrophoresis and the resulting band pattern gives a rapid indication of similarities and differences between strains. While this method obviously lacks the resolution of sequence analysis, it is quick and inexpensive.

Restriction fragment length polymorphism (RFLP) analysis involves cutting DNA with a combination of restriction endonucleases and separating the fragments according to size by gel electrophoresis. Fragments containing a specific base sequence can be identified by Southern blotting, which involves transfer to a sheet of nitrocellulose and addition of a radioactively labeled complementary DNA probe. Hybridization of the probe is detected using autoradiography or with an enzyme-linked color reaction. This enables identification of the restriction fragment with a sequence complementary to that of the DNA probe. Fragments of different length caused by mutations in duplicate copies of the gene can be detected and used for strain differentiation. The RFLP method can be combined with PCR amplification of specific genes, such as SSU rRNA genes, followed by restriction endonuclease digestion and electrophoresis. This avoids the need for blotting and radiography but is not always successful in providing useful markers for strain differentiation.

The RAPD (random amplified polymorphic DNA) technique is a more commonly used PCR marker technique for strain differentiation. It uses a single short primer in a PCR reaction, leading to a fingerprint of multiple bands generated by single nucleotide differences between individuals that prevent or allow primer binding. The method is quick and easy to perform but does not always give reliable results for detecting strain differences. AFLP (amplified fragment length polymorphism) gives better resolution than either RFLP or RAPD analysis. Genomic DNA is digested using a pair of restriction endonucleases, one of which cuts at common sites and the other at rare sites. Adapter sequences are ligated to the resulting fragments and a PCR is performed with primers homologous to the adapters plus selected additional bases. A subset of the restriction fragments is amplified, resulting in a large but distinct set of bands on a polyacrylamide gel. Band patterns are compared using imaging systems,

THE PCR CAN HAVE INHERENT BIAS AND LIMITATIONS

A considerable amount of prior investigation is needed to ensure efficient extraction and amplification of DNA from samples. So-called "universal" primers targeted against specific groups can hybridize more favorably with some templates; for example, DNA templates from thermophilic organisms with a high content of G+C base pairs may denature less rapidly than those from mesophiles. An additional problem arises from the production of chimeric PCR products. These are artifacts caused by the joining together of the amplification products of two separate sequences and, if not detected, such sequences could suggest the presence of a novel nonexistent organism. This is a particular pitfall when multi-template PCR is carried out using DNA derived from environmental samples. Various techniques are available to limit the formation of chimeras, and software to detect chimeric sequences should be used. Analysis of rRNA databases such as GENBANK has revealed that numerous chimeric sequences have been deposited in the past, leading to confusion for phylogenetic analysis.

and similarities and differences between strains can be computed to generate phylogenetic trees. Despite its advantage, AFLP analysis is a technique requiring highly purified DNA and its use is generally restricted to laboratories specializing in taxonomy.

Microsatellite markers are repeat regions of short nucleotide sequences that can be analyzed using PCR amplification of unique flanking sequences and separation of the resulting bands by gel electrophoresis. Microsatellite markers are used extensively in the population analysis of eukaryotic organisms such as microalgae.

The genomic fingerprinting method for bacteria known as rep-PCR is a simple and reproducible method for distinguishing closely related strains and deducing phylogenetic relationships. The method is based on the fact that bacterial DNA naturally contains interspersed repetitive elements, such as the REP, ERIC, and BOX sequences, for which standard PCR primers are available. The method requires very simple preparation of DNA samples without the need for extensive purification and is especially useful for the initial screening of cultures collected from marine samples.

Pulsed-field gel electrophoresis (PFGE) is a method in which the orientation and duration of an electric field is periodically changed, allowing large-molecular-weight fragments of DNA (up to 1000 kb) to migrate through the gel. Bacteria can be embedded in the gel and lysed *in situ* into large fragments, using a restriction enzyme that recognizes rare restriction sites. PFGE typing detects small changes in the genome resulting from the insertion or deletion of gene sequences or mutations that alter the restriction enzyme sites. PFGE is also very useful for analysis of virus diversity.

Determination of DNA properties is used in bacterial and archaeal taxonomy

Determination of the ratio of nucleotide base pairs has been used in taxonomy since the 1970s, and currently still forms part of the formal description of species, although this may soon change due to the development of improved methods (see *Chapter 4*). The principle of this method is that two organisms with very similar DNA sequences are likely to have the same ratio of guanine (G)+cytosine (C) to the total bases. DNA is extracted from the bacterium or archaeon and the concentration of the bases is determined by high-performance liquid chromatography (HPLC) or by determining the "melting point" (T_m) of DNA. This is the temperature at which double-stranded DNA dissociates; because the hydrogen bonds connecting G–C pairs are stronger than those connecting adenine (A) and thymine (T), the T_m increases with higher ratios of G+C. Closely related organisms have very similar GC ratios, and this can be used to compare species within a genus. However, it is important to remember that the converse is not true: bacteria and archaea possess a wide range of GC ratios and completely unrelated organisms can share a similar GC ratio.

A more useful measure of relatedness for the definition of species is DNA–DNA hybridization. Purified DNA from two organisms to be compared is denatured by melting and mixing their DNA. When cooled, DNA with a large number of homologous sequences will reanneal to form duplexes. The amount of hybridization can be measured if the DNA from one of the organisms is labeled by prior incorporation of bases containing radioactive isotopes of ^{14}C or ^{32}P. The amount of radioactivity in the reannealed DNA collected on a membrane filter is measured and the percentage hybridization is calculated by comparison with suitable controls. Taxonomists often use the benchmark of 70% or greater hybridization (under carefully standardized conditions) as the definition of a bacterial species. The difficulties associated with the concept of bacterial and archaeal species, and the use of improved genomic methods to overcome them, are discussed in *Chapter 4*.

DNA sequence data are used for identification and phylogenetic analysis

Sequence data derived both from microbes cultured in the laboratory and from direct community analysis of the environment are used to determine the degree of phylogenetic relatedness between organisms. Data processing depends on the availability of publicly accessible databases (e.g. GENBANK, Ribosome Database Project, and EMBL) and specialized computer software. The first step is usually to use BLASTn, which compares a newly generated nucleotide sequence to previously described sequences deposited in databases. Various mathematical methods and computer algorithms are used for the construction of phylogenetic trees, but details of the theory of each technique and its merits are beyond the scope of this book. In one common method (neighbor-joining tree), the different sequences are compared in order to determine the evolutionary distance between all the permutated pairs of sequences in the dataset by calculating the percentage of non-identical sequences. Statistical corrections to allow for the possibility of multiple mutations at a given site are included in the calculation. A distance matrix is then constructed, and a phylogenetic tree is drawn by grouping the most similar sequences and then adding less similar groups of sequences. The lengths of the lines in the tree are in proportion to their evolutionary distances apart (e.g. *Figure 4.1*). In another approach, called parsimony, the tree is constructed after calculating the minimum number of mutational events needed for each pair to transform one sequence into the other. The tree is then drawn using the minimum number of such parsimonious steps. This method relies on the inherent assumption that evolution proceeds in one direction by the fastest route; this is undoubtedly an oversimplification. The validity of trees can be tested by "bootstrapping"—a computational technique in which multiple iterations are made of trees based on subsamples of the sites in an alignment.

Sequence analysis of rRNA molecules shows that certain nucleotide sequences are highly conserved in certain regions. The nucleotide bases are numbered in a standard notation and signature sequences are derived which are characteristic of groups of organisms at different taxonomic levels. For example, the sequence CACYYG (where Y is any pyrimidine) occurs at approximate position 315 in the 16S rRNA molecule in almost all Bacteria but does not occur in Archaea or Eukarya. Sequence CACACACCG at position 1400 is distinctive of Archaea but does not occur in Bacteria or Eukarya. Signature sequences can therefore be used to design domain-specific oligonucleotide probes for identification of any member of the respective group in a marine sample. Other sequences may be diagnostic for phyla, families, or even genera. An essential feature of modern taxonomy is to target these regions of the DNA by PCR amplification to produce multiple copies that can be sequenced. If the complete sequences are obtained from cultured microbes, the approved process to ascribe a taxonomic name requires additional information about the physiological properties and phylogenetic relationships of the organism. Therefore, when deposited in the databases, these sequences serve as a reference source against which sequences from the environment can be compared. This is a very powerful technique for culture-independent community analysis, especially in combination with metagenomics and FISH.

DGGE and TRFLP can be used to assess composition of microbial communities

Different strategies for microbial community analysis are shown in *Figure 2.9*. Denaturing gradient gel electrophoresis (DGGE) and terminal restriction fragment length polymorphism (TRFLP) are methods for separating PCR amplicons. They were the mainstay of community analysis studies throughout the 1990s–2000s but they have been now largely replaced by HTS. They are still used by some laboratories, and many important research papers refer to these methods, so a basic understanding of how they work is important.

In DGGE, PCR products are run in a polyacrylamide gel under a linear gradient of denaturing conditions—typically, 7 M urea and 40% formamide at a temperature of 50–65°C. Small sequence variations result in different migration of DNA fragments. Double-stranded DNA migrates through the gel until it reaches denaturing conditions that cause separation of the double helix ("melting"). DGGE primers must contain a guanine- and cytosine-rich sequence (GC clamp), which is resistant to complete denaturation under the conditions used, so that the band is "locked" in place according to the T_m. In analysis of community DNA, a set of bands will be generated representing variations in the sequence of the target gene (*Figure 2.10*). A typical analysis of an environmental sample will generate more than 50 bands, and many different samples can be compared on a single gel. This provides a rapid and efficient method of "community fingerprinting" for monitoring qualitative changes in community composition at

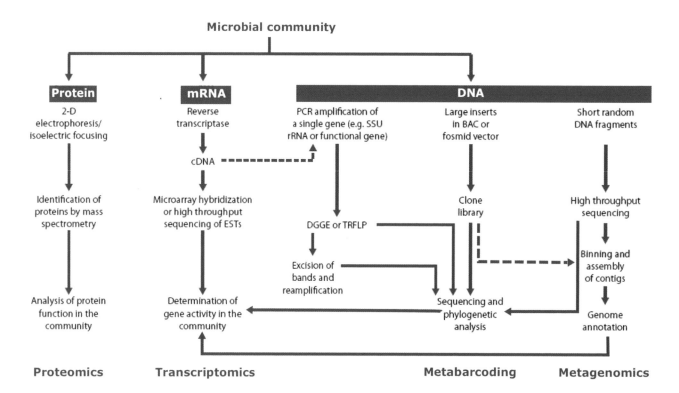

Figure 2.9 Strategies for cultivation-independent analysis of microbial communities using molecular biological techniques. BAC, bacterial artificial chromosome; DGGE, denaturing gradient gel electrophoresis; EST, expressed sequence tag; SSU, small subunit; TRFLP, terminal restriction fragment length polymorphism.

different locations or under different conditions. Sometimes, the results of DGGE community analysis are interpreted in a semi-quantitative fashion, using the assumption that the intensity of bands indicates the relative abundance of the species producing them. However, great caution is needed in drawing such conclusions because the techniques used for DNA extraction, the presence of inhibitors, the nature of the primers, and the PCR reaction conditions can all have significant effects on the results obtained. Temperature gradient gel electrophoresis (TGGE) works using similar principles, but the concentration of denaturing chemicals remains the same while the temperature of the gel is increased gradually and uniformly.

In TRFLP analysis, PCR is performed on DNA extracted from the mixed community with one of the primers labeled with a fluorescent probe. After PCR, the products are cut using specific restriction enzymes. Each PCR product produces a fluorescent molecule with the probe at the end. The size of the fragments is determined by differences in the presence or absence of restriction sites in a particular region of the molecule or by deletions and insertions in the region. Like DGGE, TRFLP can provide a quick semi-quantitative picture of community diversity but, again, results must be interpreted carefully.

Advances in DNA sequencing enable improved microbial community analysis

In the well-established technique of chain termination during synthesis of a DNA copy developed by Frederick Sanger, a sequencing primer is added to a single-stranded DNA (ssDNA) molecule in four separate reaction mixtures and chain extension begins using the normal nucleotides (dNTPs). The different reactions contain a small amount of dideoxy nucleotides (ddNTPs), which are synthetic analogs of the nucleotides lacking an −OH group on the sugar. Because of this, the DNA chain is extended only up to the point at which one of the ddNTPs is incorporated. The products from the four reactions are compared by electrophoretic separation, and the sequence of the DNA is determined by reading the sequence of nucleotides at which the chain extension is terminated. The original technique was modified to allow direct sequencing of PCR products as double-stranded DNA (dsDNA) using dNTPs labeled with different fluorescent dyes, which can be read by lasers. Automated DNA sequencers allow robotic handling and high throughput of multiple samples, generating color-coded printouts of the sequence and data suitable for direct analysis by computer. For community analysis, it is often necessary to obtain only a partial sequence. A recent improvement of the Sanger

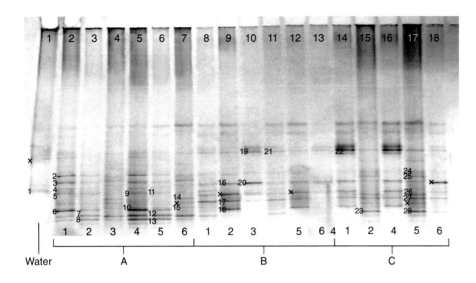

Figure 2.10 Example of a DGGE gel for microbial community analysis. The gels show 16S rRNA gene fragments from water, and replicate tissue slurry samples from three colonies of the coral *Pocillopora damicornis*. Numbers indicate bands of interest that were excised and sequenced. Reprinted from Bourne and Munn (2005) with permission of John Wiley & Sons.

method involves the application of microfluidic separation devices so that all steps in the process are fabricated onto an electronic wafer to produce a "lab on a chip."

Even with advances in automation, a Sanger sequencing run for community analysis typically produces only ~100 sequences with a length of ~650 base pairs (bp). In the mid-1980s, intensive research by biotechnology companies led to new methods that permit rapid and inexpensive DNA sequencing. These were called "next-generation sequencing" methods to emphasize their novelty. This term persists, although it is clearly a misnomer once a technique has passed the design and development stage; it is more appropriate to use the term HTS, emphasizing the length of sequence reads provided.

This development continues to occur at a very rapid rate and only a brief summary of some of the main methods is given here—images and animations of the methods are available on the company websites. The method of pyrosequencing was developed in 2005 by the company 454 Life Sciences and quickly became the main method for metagenomic studies in marine microbiology. The process depends on PCR amplification in minute individual reactions with a one-by-one nucleotide addition cycle, where the pyrophosphate (PPi) released from the DNA polymerization reaction is transformed into chemiluminescent signals. This method is now obsolete.

The Applied Biosystems SOLiD sequencer is based on sequencing catalyzed by DNA ligase. DNA fragments are linked by adaptor sequences to complementary oligonucleotides on 1 mm magnetic beads. After amplification by emulsion PCR, the beads are attached to the surface of a specially treated slide inside a fluidics cassette. A universal sequencing primer that is complementary to the adapters on the DNA fragments is added, with sequential addition of a set of four fluorescently labeled oligonucleotides that hybridize to the DNA fragment sequence adjacent to the universal primer, so that DNA ligase can join the nucleotides. Multiple cycles of ligation, detection, and cleavage are performed, with the number of cycles determining the eventual read length.

The Ion Torrent system contains electronic systems combining integrated circuits with metal-oxide semiconductors and transistors that detect the change in potential each time a proton is released after a nucleotide is added during DNA synthesis. It is very cost-effective, yielding ~500 million reads with a length of ~400 bp.

The method currently favored by most microbial ecologists is the Illumina sequencing technology. This involves the attachment of genomic DNA fragments to a transparent surface of a flow cell, where they are amplified to create millions of clusters, each containing ~1000 copies of the same template. These DNA templates are sequenced by synthesis with DNA polymerase, employing reversible terminators with fluorescent dyes that can be removed after each round of synthesis. The instrument scans the incorporation of each nucleotide using a

HIGH THROUGHPUT, LOW COST, HIGH IMPACT

Much of the commercial impetus for the development of new HTS technologies is driven by biomedical companies seeking to develop personalized medicine based on individual human genome sequencing. Sequencing the first human genome in 2003 involved multinational consortia and cost $3 billion. By 2006, it had decreased to ~$300,000 and today some companies offer a service costing less than $1000. Widespread genome screening offers many potential benefits for advances in healthcare (although there are significant ethical and data privacy concerns). The spin-off for microbial ecologists is that it is now possible to explore biodiversity at an unprecedented scale. Individual laboratories can conduct massive metagenomic analyses, or sequence multiple strains of cultured microbes at relatively low cost within a short period (although the cost of processing the enormous amounts of data generated can be considerable). Common bioinformatics protocols and public databases facilitate a rapid rate of discovery, although improvements in standardization are needed to improve interpretation (Thompson et al., 2017). Pedros-Alio et al. (2018) review the dramatic impact of HTS in marine microbial ecology but remind us of the challenge that still remains to describe the diversity of the millions of microbial species that exist in the oceans (Salazar and Sunagawa, 2017).

laser and an algorithm assigns the base sequence, using a reference sequence for internal quality control. Although the read length of sequences (~300 bp) is much lower than other technologies, this is compensated by very high output (millions of sequences per run at very low cost) and a very low error rate.

The most recent phase in the evolution of HTS is the ability to sequence single DNA molecules without any necessity for an amplification stage. One of the most successful developments is the PacBio SMRT (single molecule, real time) platform manufactured by the company Pacific Biosciences. DNA polymerization occurs in arrays of fabricated nanostructures—tiny holes in a metallic film covering a chip. Because these apertures are smaller than the wavelength of laser excitation light, only a tiny area at the bottom of the well is illuminated. DNA polymerase molecules are attached to the bottom of each well—the incorporation of nucleotides being polymerized produces real-time bursts of fluorescent signal within the illuminated region and the extension of DNA chains by single fluorescent nucleotides can be monitored in real time. This process can sequence single DNA molecules rapidly, producing long sequence reads of over 10^4 bases, although it currently has high error rates. It can also detect modification of DNA bases by methylation, which is an important phenomenon leading to alterations in the control and activity of DNA-mediated processes like transcription, or defense against phage infection.

Another exciting development is the MinION sequencer produced by Oxford Nanopore Technologies. This is a USB device the size of a desk stapler, which can provide rapid results, even in the field, without the need for highly expensive laboratory sequencers. The dsDNA is denatured by a processive enzyme which ratchets one of the single strands (ssDNA) through a biological nanopore embedded in a synthetic membrane. The technology has also been used successfully to directly sequence mRNA and ss RNA viruses. A voltage is applied across the membrane and the different nucleotide bases restrict the flow of ions in a distinctive way as the ssDNA passes through the nanopore. Machine learning is then used to translate current fluctuations into nucleotide base calls, resulting in a real-time sequence analysis for each channel. The length of Oxford Nanopore reads is determined by the integrity of the input DNA and single reads >1 million bases in length have been successfully sequenced. The DNA sequence can be inferred by monitoring the current at each channel. Another promising technology has been developed by the BGI company; their sequencer employs combinatorial chemistry to incorporate a fluorescent probe onto a DNA anchor on a "nanoball" followed by high-resolution digital mapping.

Each method has advantages for different applications and costs per gigabase vary more than 100-fold (excluding the cost of the infrastructure). The long and accurate read lengths obtained with Sanger methods mean that they continue to be used for sequencing PCR amplicons, genomes from cultured organisms, or sequences with repetitive regions. The current generation of long-read sequencers generate data quickly but have error rates that are rather high for community analysis. However, with sufficient coverage it is possible to correct read errors sufficiently to partially predict protein functions. Hybrid approaches are used increasingly, combining the advantages of long reads offered by some systems with the low error rates of other technologies such as Illumina. Future technological developments will undoubtedly lead to further capabilities.

Elucidating the full genome sequence of microbes provides insights into their functional roles

The whole genome shotgun sequencing method was pioneered in the 1990s by J. Craig Venter and colleagues to accelerate progress in the Human Genome Project. It has since been applied to many thousands of marine bacteria and archaea and a few species of eukaryotic microbes. *Tables 3.1* and *6.1* list some important examples illustrating how genomics has provided insights into the physiology, adaptations, and ecology of marine microbes. DNA is fragmented into small segments up to 1000 bp and sequenced from both ends. Typically, when using Sanger or long-read HTS methods, ~10 replicate sequences will be obtained for different sections of the genome, but 50–100 runs are needed. Overlap consensus algorithms often fail at these levels due to high memory requirements, and a network analysis method (de Bruijn graph) is used. Computer algorithms are used to check the sequences for overlapping, contiguous segments (contigs) and assemble them into larger pieces (scaffolds), which are then compiled into a draft genome. In the early days, this process often involved multi-partner teams and took many months. Today, robotic handling of samples and automated HTS methods mean that it is possible for an individual laboratory to compile a draft genome within hours or days. Nevertheless, alignment of contiguous nucleotide sequences, genome mapping, and annotation of the sequences—the prediction of gene sequences and attachment of predicted protein-coding sequences and other biological information—often relies on rigorous collaborative efforts between different laboratories and is most useful when different strains of the microbe are sequenced. For this reason, most genome sequences published in databases are at the draft stage. Most bacterial and archaeal species possess single circular chromosomes of a manageable size, but there are many exceptions to this rule which can lead to complications in the interpretation of genomic data. The technical difficulties of sequencing the much larger and more complex genomes of eukaryotic microbes are discussed in *Chapter 6*.

Once sequences are complete and gaps are closed, auto-annotation of the genome sequence is carried out by computer software that predicts the open reading frames (ORFs); these are sequences of codons beginning with a start codon (usually AUG) and ending with a stop codon (UAA, UAG, or UGA). These are recognized by searching for characteristic conserved sequences at the promoter region (where DNA polymerase binds), ribosomal binding sites, and start and stop codons on the transcribed mRNA. DNA sequences—and the predicted amino acid sequences of the proteins that they would encode—are compared with those in DNA and protein databases. More detailed examination using other evidence is used to refine the annotation of the genome. Often, many of the predicted ORFs cannot be reliably linked to a function using current knowledge of bioinformatics. Therefore, the next stages in molecular analysis are functional genomics and proteomics. This can involve inactivation of specific regions of the genome and high-resolution separation and identification of proteins, although for many organisms (especially those with streamlined genomes), this can be problematic and transcriptomics under different conditions will prove more useful. Such post-genomic analyses of marine microbes are revealing many previously unknown properties and activities. It is important to bear in mind that each genome sequence is only a "snapshot" of one individual microbial strain. As reduced sequencing costs and enhanced bioinformatics enable analysis of genomes from multiple isolates of the same species, very large variations in genome properties can be revealed. Indeed, such comparative genomics can yield important information about adaptations to specific environments.

Metabarcoding and metagenomics have led to major advances in microbial community analysis

In the past two few decades, techniques based on analysis of SSU rRNA gene sequences have underpinned microbiome research in the marine environment, as well as in plants, soils, and the human body. Progress has been slower for eukaryotic microbes, but discoveries in this field are now accelerating, and use of other genomic sequences have been useful, such as internal transcribed spacer (ITS) regions of the genome encoding the rRNAs or the *rbc* gene encoding RubisCO in phototrophic protists (see *Chapter 6*). The study of viral diversity in the environment (viromes) remains difficult, because there is no equivalent universal marker gene that can be identified, even for the most abundant viruses (see *Chapter 7*).

The concept of metagenomics developed in the late 1990s as a method of analyzing collections of genes sequenced from microbial communities in the environment (*Figure 2.9*). In the first studies, methods for handling large DNA fragments up to 300 kb were developed, so that pieces of DNA could be isolated and recognized by detection of their signature sequence of SSU rRNA (or other marker gene) using Southern blotting (a process by which DNA fragments separated by electrophoresis are transferred to a membrane and identified by hybridization with a probe sequence). The large fragments were cloned directly into the F plasmids of *E. coli* to create bacterial artificial chromosomes (BACs), which have the additional advantage that expression of some of the inserted genes may occur in the *E. coli* host harboring the vector.

This is a slow and difficult process, but fortunately it was quickly replaced by the application of the shotgun sequencing method described above. This allowed large numbers of random fragments of DNA to be extracted from environmental samples. Mass sequencing was made possible by the deployment of multiple automated DNA sequencers and the development of new bioinformatics software to reconstruct the overlapping DNA sequences (contigs) arising from many small fragments. This led to the first major marine environmental survey by Venter and colleagues in 2004–2007, (GOS, see p.55). The frequency of specific DNA sequences gives an indication of the abundance of a particular organism, and even very rare organisms should be represented if enough sequences are analyzed. Where this approach is used for community analysis based on a barcode such as the 16S rRNA gene, it is often referred to as metabarcoding. The advent of the 454 HTS platform enabled further exploration of marine microbial diversity by the International Census of Marine Microbes (ICoMM, see *Box 4.1*). Despite its success, it became increasingly obvious that the PCR-amplification stage used in early sequencing methods led to biases that distorted estimations of richness and diversity in microbial communities. To avoid these biases, it is preferable to directly identify DNA fragments containing the code for rRNA. This became possible with the advent of Illumina sequencing, which facilitated an enormous increase in throughput at comparatively low cost. An Illumina sequencing run can produce tens of millions of metagenomic reads, which are enough to recover the thousands of rRNA-encoding DNA fragments obtained from shotgun sequencing of all the DNA present in a sample. This can provide enough information to reliably capture the structure of the microbial community. These fragments are named metagenomic Illumina tags (miTags). Even though Illumina reads are very short, the multiple overlapping fragments produced can provide accurate estimates of entire rRNA gene sequences to be used for taxonomic characterization.

Advances in DNA sequencing have only been possible because of parallel advances in computing and bioinformatics to interpret the huge datasets produced by the community analysis sequencing runs. Many algorithms have been developed to filter, cluster, and annotate the sequences. Only a few of the main approaches are mentioned here. The first step is to check the quality of the sequencer output. In all platforms, there is a certain probability that the sequencer software will wrongly identify a base. Various software tools are available to remove artefacts such as low-quality base calls, primer and adapter sequences, reads that are too short, chimeras, or obvious contaminating sequences. If we are interested in knowing what kinds of organisms are present in a sample, the next step in a community analysis based on SSU rRNA sequencing will be to cluster similar sequences to identify the operational taxonomic units (OTUs). Because of the problems of defining bacterial and archaeal species, OTUs are used as a proxy, by grouping organisms using an arbitrary sequence similarity cutoff. The choice of similarity thresholds for determining similarity depends very much on the research question being investigated. In earlier studies, sequences which only occur once in the dataset (singletons) were removed to simplify statistical comparison between samples, but this is unnecessary with modern algorithms. Recently, the increased quality of DNA sequence results has led to reporting community analysis data as amplicon sequence variants (ASVs) rather than OTUs (these issues are discussed in more detail in *Chapter 4*). Taxon assignment of OTUs or ASVs depends on comparing sequences with those of reference organisms in databases such as SILVA, RDP, or GreenGenes. Algorithms such as QIIME2, MOTHUR, or DADA2 are used to sort and cluster the data to simplify the computational processing. These calculate a distance matrix set at various threshold levels, leading to a biome table, which clusters them according to the thresholds selected. The taxonomy assignment of taxonomic affiliations (genus, family, order, etc.) can then be added to the records of the OTUs or ASVs.

Further downstream analysis of the results of the survey can then be carried out. Data is often presented in a variety of different graphical formats to facilitate interpretation. For example, phylogenetic trees accompanied by a stacked bar chart, pie charts, or heat maps using different colors for different taxa such as orders or classes are often used (e.g. *Figure 10.9*). Diversity within a sample (alpha-diversity) is measured using various indices such as Chao1, Simpson's index, and Shannon–Wiener index (p.119). For diversity between samples (beta-diversity) correlation coefficients known as the Bray–Curtis dissimilarity index or Spearman rank correlation are commonly reported. Principal components analysis (PCA) and multidimensional scaling (MDS) plots are often used to highlight groupings within a community, for example by linking to metadata such as physical conditions, nutrient availability, or other variables.

It is important to recognize that any community analysis will have unavoidable biases, depending on the combination of sampling and sequencing methods and the algorithms used to analyze the results. Careful design of experiments and selection of data processing methods can minimize these biases, but it will always be impossible to be certain that we have found the "holy grail" of a description of the true community structure in complex marine communities.

Omics technologies provide information about the functional gene composition of a microbial community

As well as the narrower metabarcoding approach based on diversity of a subsection of a single gene, the full power of metagenomics is its provision of information about the range of genes within a microbiome. This provides insight into linkages between function and phylogeny for uncultured organisms and enables the discovery of genes encoding novel enzymes or other proteins. The endpoints of metagenomic studies get us closer to answering the key questions in microbial ecology mentioned earlier—"who's there, what are they doing, and how are they doing it?" Advances in DNA isolation methods and sequence analysis have enabled the full, or almost complete, genome sequence of many microbes to be determined without the need for isolation and culture. This has led to spectacular advances in our understanding of the potential functions of microbes in marine environments, including symbiotic and pathogenic interactions. There are many examples of ecologically important processes discovered through metagenomics throughout the book. For example, one of the first successes of this approach was the identification of proteorhodopsin genes revealing a new kind of photoheterotrophy in marine bacteria, discussed in *Box 3.1*. Other notable discoveries include chemosynthetic symbionts of invertebrates (*Chapter 10*), nitrogen-fixing bacteria, and ammonia-oxidizing archaea (*Chapter 9*), to name just a few. Assembling genomes from uncultivated organisms follows the same basic principles as that for cultivated organisms, with the important additional steps of binning the sequences and annotation of gene function (*Figures 2.9, 2.11* and *2.12*).

A range of software programs are available for assembling the sequences into closed partial or complete genomes via "binning." The key step of binning involves sorting the assembled contigs into the best estimate of groups that might belong together. If microdiversity is not too high, this gives a starting point for deriving the genome of a single population or group of closely related populations. A common approach is to assess the GC content or the

SAILING TOWARDS A BETTER UNDERSTANDING OF OCEAN MICROBIAL DIVERSITY

Several sampling expeditions using sailing vessels have been instrumental in vastly expanding our knowledge of microbial diversity in the oceans (*Figure 2.12*). The first major metagenomic dataset of 6.3×10^9 bases was published by J. Craig Venter in 2004. Further surveys conducted during the Global Ocean Sampling (GOS) expeditions with his yacht *Sorcerer II* resulted in the discovery of six million gene families in 2007. Since then, individual laboratories and international consortia have conducted many metagenomic surveys. The international Malaspina circumnavigation expedition (2010–2011) added many microbial sequences from the deep ocean. Perhaps the most productive of these coordinated research endeavors is the international *Tara Oceans* project from 2009–2012, which collected over 3.5×10^4 samples for diversity studies from surface and mesopelagic waters at 210 globally distributed sites. An initial analysis of 7.2×10^{12} trillion bases of genomic data were assembled into an inventory of 40 million genes—of which over 80% were novel. Hundreds of major research papers based on these samples have already been published and these are having major influences on our understanding of microbial ecology and the effects of climate change.

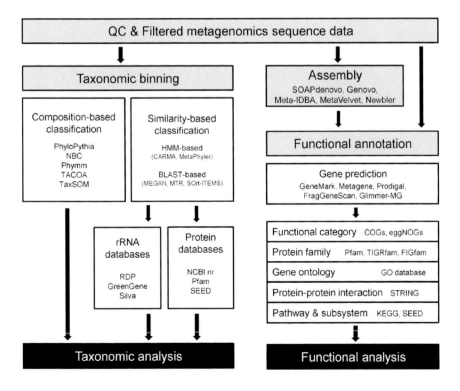

Figure 2.11 Overall workflow and bioinformatics tools for shotgun metagenomic analysis. HMM, hidden Markov model; KEGG, Kyoto Encyclopedia of Genes and Genomes; STRING, Search Tool for the Retrieval of Interacting Genes/Protein. Reprinted from Kim et al. (2013), CC-BY-3.0.

Figure 2.12 A selection of the scientific and technological advances in the recent history of marine microbial ecology are represented through time in the upper panel. Reprinted from Salazar and Sunagawa (2017) with permission of Elsevier.

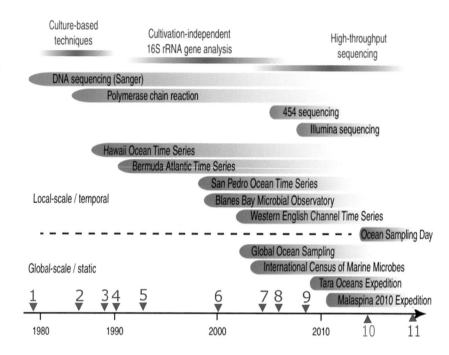

abundance distribution of k-mers (all the possible subsequences of a specific length, which will be characteristic of particular organisms. Many different systems ("pipelines") are available for sequence assembly, binning, and downstream analysis. The MEtaGenome ANalyzer (MEGAN) software is a well-established tool for analyzing HTS output, using BLAST analysis for taxonomic and functional binning. The faster methods work by co-assembling sequences with guidance from reference genomes though BLAST analysis for similarity, but larger computational resources and several days processing time on specialized servers are needed for *de novo* assembly. The expansion of metagenomics has resulted in accumulation of massive amounts of data. Archiving and curating sequences, together with the associated metadata such as descriptions of the sample source and details of methodology, is carried out by specialized databases such the Genomes OnLine Database (GOLD). This is essential to ensure conformity, reproducibility, and as a reference for other studies. MEGAN and other pipelines are used for comparative analysis of multiple datasets, using databases such as SEED, which organizes categories of gene functions into a hierarchical classification, and the Kyoto Encyclopedia for Genes and Genomes (KEGG), a collection of pathways integrating genes, proteins, RNAs, and metabolic pathways. A flowchart of the major steps in analysis of metagenomic data is shown in *Figure 2.11*.

Genomes can now be obtained from single cells in environmental samples

It is important to recognize that a genome constructed from metagenomic sequences extracted from the environment (a Metagenome Assembled Genome, MAG) is not the same as a genome constructed from a clonal population of a cultured organism. In view of the microdiversity within natural populations of bacteria and archaea, even the highest quality MAGs will likely represent an agglomeration of very closely related individuals. However, it is now possible to isolate a single microbial cell from the environment, amplify the whole genome, sequence it with HTS, and reconstruct the genome. This is known as a Single (cell) Amplified Genome or SAG. Whole genome amplification (WGA) is carried out using a thermostable mutant form of the high-fidelity DNA polymerase obtained from φ29 phage. The process is known as multiple displacement amplification—unlike PCR, it uses random primers—resulting in large fragments with low frequency of errors. Femtogram amounts of DNA from a single cell can be amplified over 1000-fold to provide the amount needed for HTS. There are several sources of error during the process, such as uneven genome coverage and chimera formation, which must be accounted for and overcome. Many other factors, such as nature of the sample, DNA extraction and preservation, nature of the cell wall, and packaging of DNA, can also affect the quality of the results. Appropriate methods for lysis of the cells under investigation must be used, with precautions needed to prevent DNA damage, contamination, or residues of chemicals that will interfere with amplification. Despite these difficulties, single-cell genomics has been successfully used in numerous studies of bacteria and archaea in the marine environment. It has proved especially important in the discovery of new processes in the carbon and nitrogen cycles (*Chapters 8* and *9*). It is now being used to obtain whole genome sequences from protists directly isolated from the environment, which has previously been very difficult (*Box 6.1*). FACS is the most commonly used method for isolating single cells; picoliter-sized droplets can be sorted into tubes or microwell plates. Another method is optofluidics, in which cells are captured by using a microfluidic chip and examined microscopically to select those of interest; these are moved into another compartment of the microfluidic device for WGA cells. This is particularly valuable for samples with a low biomass. The gel microencapsulation technique described on p.41 has also been adapted for single-cell genomics.

IN SITU ACTIVITY OF MICROBIAL COMMUNITIES

Microelectrodes and biosensors measure microbial processes at the microhabitat scale

Special electrochemical microelectrodes can be constructed with tip diameters of 10 μm or less that are inserted using micromanipulators into habitats to measure environmental and chemical changes. They are made of glass and metal with an ion-selective membrane that generates voltage by charge separation. Some have liquid ion-exchange membranes incorporating a synthetic ionophore, which is specific for target compounds. These probes can be left in place for long periods, and when linked to recorders, they provide continuous data about environmental changes and the rates of microbial and chemical processes in microhabitats. For example, the development and use of oxygen, sulfide, and pH microsensors made it possible to analyze the gradients of these factors in sediments and biofilms, which were used to show how sulfur bacteria (p.80) live in opposing gradients of sulfide and oxygen with only very small concentrations of both chemical species being present in the oxidation zone. Microsensor techniques can also be used to investigate oxygenic photosynthesis and respiration at a micrometer resolution in biofilms and sediments. Optical microsensors that detect small changes in light are also used. Recent innovations have included the development of microscale biosensors suitable for use in seawater and marine sediments to detect a range of compounds. Here, biological components such as immobilized enzymes or antibodies are incorporated into the electrodes and coupled to a signal converter so that the results of a reaction are amplified to produce an electronic reading. The development of such biosensors for various inorganic nitrogen species has been a major advance, permitting detailed study of the processes of nitrification and denitrification. Biosensors for methane and volatile fatty acids have also been used to probe microbial transformations in anaerobic sediments and planktonic aggregates such as marine snow particles. Studies using oxygen microelectrodes have shown that even the tiniest aggregates (*Figure 1.7*) contain anaerobic niches.

Radioisotopes can be used to detect metabolic activity in a community

Specific metabolic transformations can be studied by measuring the rate of incorporation of a radioactively labeled precursor of a key substrate. This method is widely used in biochemistry to follow the pathways of metabolic reactions—if one or more atom(s) of a compound is replaced with a radioisotope, we can track its progress as it is metabolized. One of the earliest applications of radioactive isotopes in marine microbial ecology was the introduction of ^{14}C-labeled bicarbonate ($H^{14}CO_3^-$) into bottle experiments to measure rates of photosynthesis (primary production) by phytoplankton. ^{14}C is a relatively safe isotope, because it has little energy and a long half-life. Bottles containing phytoplankton can be positioned on a mooring line at various depths to assess the effects of irradiance and temperature on productivity, with appropriate control bottles from which light is excluded. Phytoplankton cells can be collected by filtration after incubation, and the amount of labeled carbon fixed can be detected by measuring radioactive decay in a scintillation counter, or by autoradiography. Bacterial production in seawater can be measured by the incorporation of [^{3}H]-adenine or [^{3}H]-thymidine into RNA or DNA (respectively), or

VIRAL METAGENOMICS: THE LONG AND THE SHORT OF IT

The great advances in understanding of marine bacterial ecology through HTS and metagenomics have been possible because of the universal rRNA marker genes, but there is no equivalent in viruses. HTS with Illumina sequencing technology produces short sequences, which are very useful for predicting viral gene functions, but metagenome assembly is difficult because of nucleotide level diversity among viral strains. The advent of long-read HTS technology such as PacBio and MinION offers the potential to directly capture complete genomes of many marine viruses, but these both require large amounts of input DNA, which is challenging for natural viral community samples. Long-read technologies also suffer from high error rates, which are particularly problematic in viral metagenomics due to the shorter viral gene length compared to host genes, leading to frameshift errors. Warwick-Dugdale et al. (2019) of Plymouth Marine Laboratory and the University of Exeter adapted a method to amplify viral DNA from seawater and obtained long reads with MinION. A hybrid approach that employed short reads for long-read error correction, which they call VirION, improved the recovery of complete viral genomes and highly microdiverse virus populations from seawater from the Western Channel Observatory.

by incorporation of labeled amino acids such as [^{3}H]-leucine into proteins. The flux of organic material in seawater can be measured using [^{14}C]-glucose. Many other microbial processes can be measured using appropriately labeled compounds. Other commonly used radioisotopes for this type of study are ^{32}P, ^{18}O, and ^{35}S. Appropriate correction factors for the discrimination of enzyme processes against the heavier radioisotopes must be included in calculations.

Incorporation of the labeled precursor into individual cells can also be detected by microautoradiography (MAR), allowing us to assess the metabolic activities of different cells within a community. Radiolabeled substrate is added to a sample of a natural community and incubated—it could be a water sample, sediment, or animal tissue, for example—so that cells that incorporate the labeled compound become radioactive after incubation. Samples are then placed onto microscope slides or membranes and overlaid with a liquid photographic emulsion containing silver nitrate. After a few days, the film is examined microscopically. Radioactive decay will expose the film by deposition of silver salts, so in cells where the radioisotope has been incorporated, silver grains will become visible as dark spots under the light microscope. MAR is a long-established technique that has found many new applications in marine ecology in combination with the gene-based technique of FISH, discussed earlier. This combination of methods, called MAR-FISH, is extremely sensitive and is a powerful tool for revealing which organisms in a community are carrying out particular metabolic processes. Unlike some other methods, it is not subject to the biases that result from DNA extraction or PCR methods. In MAR-FISH, the incubated sample is treated with a permeabilizing agent and FISH oligonucleotides probes are added before adding the MAR photographic emulsion. Duplicate images of the resulting exposed specimen are examined. Under the light microscope, the isotope-labeled cells can be seen while the FISH-labeled cells are visible under the fluorescence microscope. The images can be overlaid digitally so that we can identify the organisms carrying out the metabolic reaction under study. In most cases, the FISH probe will be an oligonucleotide recognizing an rRNA target, which means that we can assign a taxonomic affiliation to active cells. An important example of the use of MAR-FISH is the discovery of chemolithotrophic fixation of CO_2 by deep-sea archaea, described in *Box 8.1*.

Stable-isotope probing (SIP) tracks fluxes of nutrients in communities

To follow the flux of major nutrients, particularly carbon and nitrogen, another widely used approach is to supply the community with a substrate enriched with a heavier stable isotope—^{13}C and ^{15}N are most commonly used—and to couple this with cultivation-independent methods to identify the organism. SIP provides direct evidence linking substrate assimilation to specific taxa, thus indicating their functional roles in the community. Usually, rRNA or DNA is used as the biomarker. For example, to find out which organisms in a mixed community are methanotrophs (consuming compounds with a single C atom [C1] such as methane or methanol, see p.90) we can add C1 compounds containing a high proportion of ^{13}C atoms. Since ^{13}C has a higher mass than the commoner ^{12}C, those organisms that are incorporating C1 compounds will contain DNA with a higher density than others. The heavy and light fractions of DNA can be separated by isopycnic centrifugation at high speed in a cesium chloride gradient. Ethidium bromide or another DNA stain is included so that the bands can be visualized and removed from the centrifuge tube with a needle. It can then be used for community analysis using PCR with universal primers or subjected to metagenomic sequencing by HTS. If the taxonomic identity of active groups is the goal, tracking incorporation of the isotope into rRNA has some advantages, because rates of rRNA synthesis are usually higher than DNA and are independent of replication. However, an additional step of reverse transcription to make cDNA will be needed before performing PCR. Isotope incorporation into proteins, lipid fatty acids, and metabolites can also be used to indicate a specific metabolic activity.

NanoSIMS allows metabolic transfers to be measured at subcellular levels

Although the SIP methods described above allow us to determine groups of organisms responsible for metabolic processes, they do not resolve the activities of individual cells, or compartments within cells. A technique called nanoscale secondary ion mass spectrometry (nanoSIMS) is a type of imaging system that enables elemental transformations to be monitored at subcellular resolution. This technique has proved particularly valuable in combination with SIP for determining interactions and nutrient flow between associated microbes, or between symbiotic (or parasitic) microbes and their hosts. Several important discoveries using this technique are described in later chapters, including the assimilation and transfer of nitrogen and carbon ($^{15}NH_4^+$, $^{15}N_2$, $^{13}CH_4$) in syntrophic consortia of methane-oxidizing archaea and sulfate-reducing bacteria in sediments (*Figure 3.13C*), $^{15}N_2$ fixation by intracellular cyanobacterial endosymbionts of diatoms (*Figure 4.11*), and chemolithotrophic symbionts of animals (*Chapter 10*).

In nanoSIMS, a beam of Cs^+ or O^- (the primary ion) is focused onto the surface of a sample with a resolution of 50 nm or less. The ions cause "sputtering" of a few atomic layers from the surface, and clusters of atoms are ejected; some of them become spontaneously ionized. The composition of these secondary ions is characteristic of the composition of the analyzed area. The secondary ion beam is mass filtered by a magnetic device, resulting in separation of up to seven different types of ion, each of which can be measured independently. The nanoSIMS-SIP technique can be made even more powerful by combining it with FISH or CARD-FISH, allowing phylogenetic identification of cells and subcellular structures containing the stable isotope. In a variant called HISH-nanoSIMS (halogen *in situ* hybridization), a rare element such as ^{19}F attached to oligonucleotide probes is introduced directly into the cells. Fluorescent antibodies can also be introduced to detect the location of specific proteins.

Microarrays enable assessment of gene activity in the environment

One of the major methods used in functional genomics is DNA microarray technology. DNA oligonucleotide sequences identified from microbial genomes are attached using photolithography or ink-jet technology on the surface of a silicon chip in a highly ordered array to produce a device that can act as a probe for hundreds or thousands of genes. For some bacteria, chips with probes for every expressed gene in the genome have been designed. The target nucleic acids (mRNA or cDNA) are labeled with fluorescent reporter groups and lasers detect hybridization of complementary sequences to the oligonucleotide on the microarray chip. Microarrays are particularly useful for determining the effects of changing conditions on the patterns of gene expression, for example, following nutrient changes in mesocosm experiments or during different phases of interactions with a host. The results of such experiments are often presented as color-coded heat maps indicating which genes are upregulated, downregulated, or unaltered during the transition. They have also been used to follow community changes by phylogenetic identification. The PhyloChip assay contains ~1.2×10^6 probes for variable regions of the 16S rRNA gene, enabling identification and measurement of relative abundance of over 5×10^4 microbial taxa; notable examples of its use include the assessment of healthy and disease states in corals (*Chapter 11*) and the rapid assessment of community changes during pollution incidents (*Chapter 13*). Microarrays have also been developed for high-throughput analysis of functional genes in community analysis. The GeoChip 5.0 microarray contains ~1.7×10^5 probes detecting sequences from >1500 gene families related to carbon, nitrogen, sulfur and phosphorus cycling, energy metabolism, antibiotic resistance, metal resistance/reduction, organic remediation, stress responses, bacteriophage, and virulence.

Metatranscriptomics, metaproteomics, and metabolomics reveal microbial activities in the environment

While knowledge of the metagenome indicates the potential activities within a microbial community, it does not reveal the expression of genes and hence their actual involvement in particular processes. Metatranscriptomics allow the analysis of expressed genes by measuring production of messenger RNA (mRNA). Because mRNA has a short half-life, it provides a near real-time picture of microbial activity. Samples can be frozen or preserved rapidly, so that they can be easily collected over time and analyzed later. As shown in *Figure 2.9*, metatranscriptomics involves the isolation of mRNA and the production of a complementary DNA (cDNA) copy using the enzyme reverse transcriptase. Specific gene transcripts can be identified using microarray hybridization. A more commonly applied technique is to sequence short segments (500–800 bp) of the transcribed cDNA; these are known as expressed sequence tags (ESTs). Databases are built up to link ESTs with specific gene functions. Metatranscriptomes may be mapped onto metagenomic bins to provide functional information on metagenome assembled genomes (MAGs).

Even knowing which genes are expressed is one step short of knowing which enzymes and other proteins are present in a community, because these may all be modified after translation owing to complex regulatory processes. Proteomics enables separation of individual proteins by two-dimensional separation of the proteins present in a mixed sample using a combination of SDS-PAGE and isoelectric focusing. The protein spots are digested in the electrophoresis gel and analyzed by mass spectrometry. Peptide patterns can be compared with those in databases, or full protein sequences can be obtained and homology with well-characterized proteins can be determined. Metabolomics aims to assess the total metabolic profile of a cell, organism, or community. This usually involves chromatographic separation of different classes of chemical compounds (amino acids, carbohydrates, lipids, etc.) and identification by mass spectrometry. New developments allow direct identification without the need for separation.

Microfluidics enables study of microscale processes

Microfluidics is a technology for handling and controlling fluids in microliter or picoliter quantities. It was developed in the 1990s by combining microanalytical and microelectronic circuits and has expanded due to the ability to fabricate microdevices using soft lithography. By combining it with microscopic examination, microfluidics allows us to study the behavior of individual microbial cells or groups of cells, and to manipulate their environment at a scale appropriate to their size. Nano-sized channels and porous membranes can be used to control the flow of nutrients and excretion products from cells while keeping the microbial cells in a fixed space. Application of this technology in microbial ecology has been championed by Roman Stocker and colleagues at MIT and ETH Zurich, resulting in many important findings related to the behavior of microbes in a "sea of gradients" and the effects of surface and boundary effects. Notable examples considered in detail elsewhere in the book involve the motility and chemotaxis of bacteria in response to nutrient gradients (*Box 3.1*) and its role in infection by coral pathogenic bacteria (*Box 11.1*).

Figure 2.13 An example of a mesocosm enclosure in a fjord at the Espegrend Marine Research Field Station at the University of Bergen, Norway, during an experiment conducted in the spring of 2017 led by Kay D. Bidle and Kimberlee Thamatrakoln (Rutgers University, NJ and Elizabeth Harvey (University of Georgia). In this study, 12 mesocosms consisted of plastic floating frames (~2 m diameter) attached to bags constructed of polyethylene, extending 8 m below the surface; each contained ~23,000 L of fjord water. Each bag tapered into a settling cone attached to a bottle to collect sinking particles. After the addition of nutrients to the enclosures, they were mixed with air bubblers for several days and subsequent sampling from various depths was done using pumps. Bottles containing sinking material were collected and replaced by scuba divers throughout the experiment. Credit: Kay D. Bidle (Rutgers) and Jozef I. Nissimov (Scottish Institute of Marine Science, Oban).

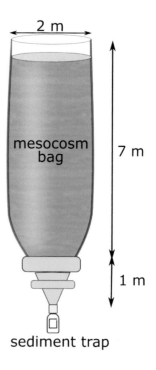

Mesocosm experiments attempt to simulate natural conditions

Carrying out controlled experiments is essential to understand the effects of factors such as nutrient additions, light, and temperature on community dynamics and microbial processes in the marine environment. Laboratory investigation of microbial processes in water uses bottle experiments, in which volumes of water up to a few liters can be handled. Experimental bottles can also be attached to CTD frames or mooring lines for incubation *in situ* at different depths. Use of these techniques has led to great advances in our understanding of phototrophic and heterotrophic processes, for example, by comparison of metabolic activities of samples incubated under light and dark conditions or the effects of temperature and nutrient additions. Such small-scale experiments are a vital first stage in the study of microbial processes in marine samples but extending studies to the natural environment is difficult. Great care is needed to ensure that vessels used in these experiments are sterile and free of chemical contaminants. Adsorption effects on the walls of containers can result in the local concentration of nutrients, which affects the composition and metabolic properties of the community. In general, the smaller the sample, the more likely it is that significant distortions due to these "bottle effects" will occur.

One way to overcome these problems is to use mesocosms—enclosures holding thousands of liters of seawater under semi-controlled conditions in an attempt to simulate natural sea environments or conduct experimental manipulations. Large tanks containing coastal seawater pumped in at controlled rates can be used in shore-based experimental stations. However, a more successful approach is the use of large enclosures constructed of polythene reinforced with vinyl and nylon—essentially, these are very big and very strong plastic bags immersed in the sea. The materials must be chosen carefully to simulate natural conditions as closely as possible. One of the longest-running facilities is based at the University of Bergen, Norway, which has maintained mesocosm systems available for international experiments since 1978 (*Figure 2.13*). Mesocosm bags are suspended in the relatively calm, deep waters of a fjord. The bags are filled by pumping in unfiltered fjord water so that natural communities of phytoplankton, zooplankton, bacteria, and viruses are introduced. Some success has been achieved in the development of very large mesocosm containers (up to 35 m³) for use in open ocean conditions, although there are major technical difficulties in the construction of enclosures strong enough to withstand waves and currents.

Generally, it is impractical to conduct experiments in the open sea because of the continuous turbulence and dispersion of water and the organisms it contains. There are also considerable logistical problems in conducting research at sites sufficiently far from the continental shelf to avoid land effects. However, a notable exception is the series of experiments carried out from 1993–2009 to investigate the effect of iron additions on phytoplankton composition and productivity. These experiments were possible because of the deployment of an inert tracer compound (sulfur hexafluoride, SF_6). The dispersion of SF_6 can be measured analytically, thus establishing the degree of dilution of iron in the water mass under study. The significance of these experiments and plans for further large-scale investigations involving measurement of microbial parameters are discussed in *Box 8.1*.

Remote sensing permits global analysis of microbial activities

At the opposite extreme from techniques to measure microenvironmental changes, the use of remote sensing by orbiting satellites has provided valuable information about the activities of planktonic organisms in the surface layers of the oceans. While the ocean appears blue to the human eye, sensitive instruments detect subtle changes in ocean color due to the presence of plankton, suspended sediments, photosynthetic pigments, and dissolved organic chemicals. The first such satellite was the Coastal Zone Color Scanner (CZCS) launched in 1978, which collected light reflected from the sea surface at different wavelengths, including the absorption maximum for chlorophyll. Scanning has been continued using the Sea-viewing Wide Field-of-View Sensor (SeaWiFS) system and other satellites. *Figure 8.2* shows a typical image from SeaWiFS, revealing chlorophyll levels

 CAN BACTERIAL LIGHT BE SEEN FROM SPACE?

Miller et al. (2005) asked this question after looking at reports of the "milky sea" phenomenon well known to sailors in the Arabian Sea and featured in the classic Jules Verne novel *Twenty Thousand Leagues under the Sea*, published in 1870. Miller and colleagues obtained the location and time of a milky sea occurrence reported by a ship's captain some years earlier and noted this record in the ship's log: "The bioluminescence appeared to cover the entire sea area, from horizon to horizon … and it appeared as though the ship was sailing over a field of snow or gliding over the clouds" (Nealson and Hastings, 2006). Miller et al. then checked the output from the meteorological satellites that monitor global cloud coverage under daytime and nighttime conditions. After adjusting spectral frequencies of the output, they produced images (*Figure 4.6*) of the milky sea, showing that it covered >15000 km²; they deduced that this was bacterial bioluminescence, probably due to *Vibrio harveyi* in association with massive blooms of *Phaeocystis* algae. They estimated the total bioluminescent bacterial population of this milky sea to be about 4×10^{22} cells.

for the monitoring of phytoplankton productivity. Satellite-based remote sensing is now used to measure a wide range of chemical and physical properties of the global ocean and can distinguish functional groups of microorganisms. Interpretation of the sensor data is a specialist field, with many different algorithms devised to interpret and display results. Different spatial resolution can be used to zoom in on particular areas, to identify processes and hazards influencing pelagic and coastal waters. For example, it can be used to monitor natural phytoplankton blooms (*Figure 6.9*) or track the development of harmful algal blooms, as discussed in *Chapter 11*.

Conclusions

This chapter has shown that marine microbiologists use a very wide range of methods to study the diversity and activity of microbes, and examples of their application in research will be found throughout the book. We now have techniques enabling us to explore the properties and activities of marine microbes at every level from single cells to global assemblages. The most significant advances in marine microbiology have occurred as a result of the development of new molecular methods and exciting technological developments, but we are also seeing a renaissance of longer-established techniques such as culturing and microscopy that provide essential insights that complement the molecular revolution. The next few years will undoubtedly see further technological progress and will be dominated by increasing application of metagenomics, transcriptomics, and proteomics, plus major advances in single-cell biology. The challenge for microbiologists will be to manage and interpret the huge datasets that are now accumulating and to translate them into real understanding of the ecology of marine microbes and their role in diverse activities.

References and further reading

Sampling

Conathon, M. (2013) Rockets top submarines: Space exploration dollars dwarf ocean spending. Center for American Progress. Online at https://www.americanprogress.org/issues/green/news/2013/06/18/66956/rockets-top-submarines-space-exploration-dollars-dwarf-ocean-spending/ (accessed 6 June 2019).

Moniz, R. & Coley, K. (2017) Challenges in the deep: The role of manned submersibles in an autonomous age. Online at https://www.oceannews.com/featured-stories/challenges-in-the-deep-the-role-of-manned-submersibles-in-an-autonomous-age (accessed 6 June 2019).

Rahlff, J., Stolle, C., Wurl, O., *et al.* (2017) SISI: A new device for in situ incubations at the ocean surface. *J. Mar. Sci. Eng.* **5**: 46.

Imaging and flow cytometry

Amann, R. & Fuchs, B.M. (2008) Single-cell identification in microbial communities by improved fluorescence *in situ* hybridization techniques. *Nat. Rev. Microbiol.* **6**: 339–348.

Chisholm, S.W., Olson, R.J., Zettler, E.R., *et al.* (1988) A novel free-living prochlorophyte abundant in the oceanic euphotic zone. *Nature* **334**: 340.

Johns, C.T., Nissimov, J.I., Grubb, A., *et al.* (2019) The mutual interplay between calcification and coccolithovirus infection. *Environ. Microbiol.* **21**: 1896–1915.

Malfatti, F. & Azam, F. (2009) Atomic force microscopy reveals microscale networks and possible symbioses among pelagic marine bacteria. *Aquat. Microb. Ecol.* **58**: 1–14.

Partensky, F. & Garczarek, L. (2010) Prochlorococcus: Advantages and limits of minimalism. *Ann. Rev. Mar. Sci.* **2**: 305–331.

Pennisi, E. (2017) Meet the obscure microbe that influences climate, ocean ecosystems, and perhaps even evolution. *Science.* doi: 10.1126/science.aal0873.

Cultivation

Button, D.K., Schut, F., Quang, P., *et al.* (1993) Viability and isolation of marine bacteria by dilution culture: Theory, procedures and initial results. *Appl. Environ. Microbiol.* **59**: 881–891.

Carini, P. (2019) A "cultural" renaissance: Genomics breathes new life into an old craft. *mSystems* **4**: e00092-19.

Connon, S.A. & Giovannoni, S.J. (2002) High-throughput methods for culturing microorganisms in very low-nutrient media yield diverse new marine isolates. *Appl. Environ. Microbiol.* **68**: 3878–3885.

Donachie, S.P., Foster, J.S., & Brown, M. V (2007) Culture clash: Challenging the dogma of microbial diversity—Commentaries. *ISME J.* **1**: 97–99.

Giovannoni, S.J., Foster, R.A., Rappé, M.S. & Epstein, S. (2007) New cultivation strategies bring more microbial plankton species into the laboratory. *Oceanography* **20**: 62–69.

Giovannoni, S. & Stingl, U. (2007) The importance of culturing bacterioplankton in the 'omics' age. *Nat. Rev. Microbiol.* **5**: 820–825.

Giovannoni, S.J., Tripp, H.J., Givan, S., *et al.* (2005) Genetics: Genome streamlining in a cosmopolitan oceanic bacterium. *Science* **309**: 1242–1245.

Morris, R.M., Rappe, M.S., Connon, S.A., *et al.* (2002) SAR11 clade dominates ocean surface bacterioplankton communities. *Nature* **420**: 806–810.

Rappé, M.S., Connon, S.A., Vergin, K.L. & Giovannoni, S.J. (2002) Cultivation of the ubiquitous SAR11 marine bacterioplankton clade. *Nature* **418**: 630–632.

Schut, F., Gottschal, J.C. & Prins, R.A. (1997) Isolation and characterisation of the marine ultramicrobacterium *Sphingomonas* sp. strain RB2256. *FEMS Microbiol. Rev.* **20**: 363–369.

Stingl, U., Tripp, H.J., & Giovannoni, S.J. (2007) Improvements of high-throughput culturing yielded novel SAR11 strains and other abundant marine bacteria from the Oregon coast and the Bermuda Atlantic Time Series study site. *ISME J.* **1**: 361–371.

Zengler, K. (2009) Central role of the cell in microbial ecology. *Microbiol. Mol. Biol. Rev.* **73**: 712–729.

Zengler, K., Toledo, G., Rappé, M., *et al.* (2002) Cultivating the uncultured. *Proc. Natl. Acad. Sci. USA* **99**: 15681–15686.

RNA- and DNA-based methods

Bourne, D.G. & Munn, C.B. (2005) Diversity of bacteria associated with the coral *Pocillopora damicornis* from the Great Barrier Reef. *Environ. Microbiol.* **7**: 1162–1174.

Compeau, P.E., Pevzner, P.A., & Tesler, G. (2011) Why are de Bruijn graphs useful for genome assembly?. *Nat. Biotechnol.* **29**: 987.

Coutinho, F.H., Gregoracci, G.B., Walter, J.M., *et al.* (2018) Metagenomics sheds light on the ecology of marine microbes and their viruses. *Trends Microbiol.* **26**: 955–965.

Escobar-Zepeda, A., Vera-Ponce de León, A., & Sanchez-Flores, A. (2015) The road to metagenomics: From microbiology to DNA sequencing technologies and bioinformatics. *Front. Genet.* **6**: 348.

Giovannoni, S.J., Britschgi, T.B., Moyer, C.L., & Field, K.G. (1990) Genetic diversity in Sargasso Sea bacterioplankton. *Nature* **345**: 60–63.

Heather, J.M. & Chain, B. (2016) The sequence of sequencers: The history of sequencing DNA. *Genomics* **107**: 1–8.

Hiraoka, S., Yang, C. & Iwasaki, W. (2016) Metagenomics and bioinformatics in microbial ecology: Current status and beyond. *Microbes Environ.* **31**: 204.

Hongoh, Y. & Toyoda, A. (2011) Whole-genome sequencing of unculturable bacterium using whole-genome amplification. *Methods Mol. Biol.* **733**: 25–33.

Hugerth, L.W. & Andersson, A.F. (2017) Analysing microbial community composition through amplicon sequencing: From sampling to hypothesis testing. Front. Microbiol. **8**: 1561.

Kim, M., Lee, K.-H., Yoon, S.-W., *et al.* (2013) Analytical tools and databases for metagenomics in the next-generation sequencing era. *Genom. Inform.* **11**: 102.

Kozińska, A., Seweryn, P., & Sitkiewicz, I. (2019) A crash course in sequencing for a microbiologist. *J. Appl. Genet.* **60**: 103–111.

Land, M., Hauser, L., Jun, S.-R., *et al.* (2015) Insights from 20 years of bacterial genome sequencing. *Funct. Integr. Genom.* **15**: 141–61.

Logares, R., Sunagawa, S., Salazar, G., *et al.* (2014) Metagenomic 16S rDNA Illumina tags are a powerful alternative to amplicon sequencing to explore diversity and structure of microbial communities. *Environ. Microbiol.* **16**: 2659–71.

Pedros-Alio, C., Acinas, S.G., Logares, R., & Massana, R. (2018) Marine microbial diversity as seen by high-throughput sequencing. In, *Microbial Ecology of the Oceans, Third edition.*, J.M. Gasol & D.L. Kirchman, eds., pp. 47–97. John Wiley & Sons.

Rhoads, A. & Au, K.F. (2015) PacBio sequencing and its applications. *Genom. Proteom. Bioinf.* **13**: 278–289.

Roh, S.W., Abell, G.C.J., Kim, K.-H., *et al.* (2010) Comparing microarrays and next-generation sequencing technologies for microbial ecology research. *Trends Biotechnol.* **28**: 291–9.

Salazar, G. & Sunagawa, S. (2017) Marine microbial diversity. *Curr. Biol.* **27**: R489–R494.

Singer, E., Wagner, M., & Woyke, T. (2017) Capturing the genetic makeup of the active microbiome in situ. *ISME J.* **11**: 1949–1963.

Sleigh, C. (2013) The secret life of the laboratory: PCR at 30. *The Guardian*. Online at https://www.theguardian.com/science/2013/nov/26/the-secret-life-of-the-laboratory-pcr-at-30 (accessed 7 June 2019).

Stepanauskas, R. (2012) Single cell genomics: An individual look at microbes. *Curr. Opin. Microbiol.* **15**: 613–620.

Thomas, T., Gilbert, J., & Meyer, F. (2012) Metagenomics—A guide from sampling to data analysis. *Microb. Inform. Exp.* **2**: 3.

Thompson, L.R., Sanders, J.G., McDonald, D., *et al.* (2017) A communal catalogue reveals Earth's multiscale microbial diversity. *Nature*, 551(7681): 457.

Warwick-Dugdale, J., Solonenko, N., Moore, K., *et al.* (2019) Long-read viral metagenomics captures abundant and microdiverse viral populations and their niche-defining genomic islands. *PeerJ* **7**: e6800.

In situ activity

Boughner, L.A. & Singh, P. (2016) Microbial ecology: Where are we now? *Postdoc. J.* **4**: 3–17.

Morando, M. & Capone, D.G. (2016) Intraclade heterogeneity in nitrogen utilization by marine prokaryotes revealed using stable isotope probing coupled with tag sequencing (Tag-SIP). *Front. Microbiol.* **7**: 1932.

Musat, N., Musat, F., Weber, P.K., & Pett-Ridge, J. (2016) Tracking microbial interactions with NanoSIMS. *Curr. Opin. Biotechnol.* **41**: 114–121.

Nagy, K., Ábrahám, Á., Keymer, J.E., & Galajda, P. (2018) Application of microfluidics in experimental ecology: The importance of being spatial. *Front. Microbiol.* **9**: 496.

Radajewski, S., Ineson, P., Parekh, N.R., & Murrell, J.C. (2000) Stable-isotope probing as a tool in microbial ecology. *Nature* **403**: 646.

Rusconi, R., Garren, M., & Stocker, R. (2014) Microfluidics expanding the frontiers of microbial ecology. *Annu. Rev. Biophys.* **43**: 65–91.

Rusconi, R. & Stocker, R. (2015) Microbes in flow. *Curr. Opin. Microbiol.* **25**: 1–8.

Staley, J.T. & Konopka, A. (1985) Measurement of in Situ activities of nonphotosynthetic microorganisms in aquatic and terrestrial habitats. *Annu. Rev. Microbiol.* **39**: 321–346.

Taylor, G.T. (2019) Windows into microbial seascapes: Advances in nanoscale imaging and application to marine sciences. *Ann. Rev. Mar. Sci.* **11**: 465–490.

Tu, Q., Yu, H., He, Z., *et al.* (2014) GeoChip 4: A functional gene-array-based high-throughput environmental technology for microbial community analysis. *Mol. Ecol. Resour.* **14**: 914–928.

Zark, M., Broda, N.K., Hornick, T., *et al.* (2017) Ocean acidification experiments in large-scale mesocosms reveal similar dynamics of dissolved organic matter production and biotransformation. *Front. Mar. Sci.* **4**: 271.

Remote sensing

Miller, S.D., Haddock, S.H.D., Elvidge, C.D. & Lee, T.F. (2005) Detection of a bioluminescent milky sea from space. *Proc. Natl Acad. Sci. USA* **102**: 14181–14184.

Nealson, K.H. & Hastings, J.W. (2006) Quorum sensing on a global scale: Massive numbers of bioluminescent bacteria make milky seas. *Appl. Environ. Microbiol.* **72**: 2295–2297.

Chapter 3

Metabolic Diversity and Ecophysiology

Microbes have been evolving for between three and four billion years, resulting in a huge diversity of metabolic types. Organisms have evolved to obtain energy from light or various inorganic substances, or by breaking down different organic compounds to their basic constituents. Some species obtain key elements for building cellular material directly from inorganic minerals, while others require complex "ready-made" organic compounds. The selective advantage of the ability to utilize particular nutrients under a set of physical conditions has led to the evolution of the enormous metabolic diversity that we see today. Such diversity results in the occurrence of microbes in every conceivable habitat for life. Microbial metabolism fuels biogeochemical cycles, transformation of nutrients, and other processes that maintain the web of life in the ocean, underpinning all aspects of marine ecology. The first part of this chapter provides a brief overview of cell structure and genome organization. This is followed by discussion of the diverse metabolic pathways by which microbes obtain and transform energy and fix carbon and nitrogen for cellular growth. Next, the strategies for nutrient acquisition and uptake are considered and the physiological factors affecting microbial growth in plankton and on surfaces are discussed. The main focus of this chapter is on the metabolism and ecophysiology of the Bacteria and Archaea, although some of the biochemical pathways and mechanisms (especially photosynthesis and transformations of organic molecules) are relevant to the eukaryotic fungi and protists, discussed in *Chapter 6*.

Key concepts

- Members of the Bacteria and Archaea share similar physiology and central metabolic processes, although there are substantial differences in cell structure.

- Microbes demonstrate huge diversity in the routes by which they obtain nutrients, extract energy, and produce the precursors needed for synthesis of macromolecules.

- All life depends on the activity of autotrophic organisms using light or chemical energy to fix CO_2 into cellular material.

- Bacteria, archaea, and protists in the photic zone harness energy from light by a range of mechanisms to support autotrophic and heterotrophic metabolism.

- Chemolithotrophic activity of bacteria and archaea is widespread in the water column, as well as in sediments, vents, and seeps.

- Diverse mechanisms are responsible for microbial transformations of carbon, sulfur, and nitrogen.

- Most planktonic microbes are genetically adapted to slow growth with minimum nutrients, while others can exploit patches of high nutrient concentration.
- Microscale bacterial activities such as chemotaxis result in global scale effects on biogeochemical cycles.
- Intercellular communication and antagonism are major features of biofilm communities.
- The most abundant marine microbes are adapted to growth under conditions of low temperature and high pressure.

A BRIEF OVERVIEW OF CELL STRUCTURE AND FUNCTION

Bacteria and archaea show a variety of cell forms and structural features

As introduced in *Chapter 1*, life may be organized into three phylogenetic domains, the Bacteria, Archaea, and Eukarya. Members of the Bacteria and Archaea are usually considered to be unicellular organisms with a simple cell structure. The basic cell forms are spherical or ovoid cells (cocci), rods (bacilli), and spiral cells (spirilla). Some groups have distinctive morphologies such as filaments, tightly coiled spirals, or cells with buds, stalks, or hyphae. As evident from *Table 1.2*, marine bacteria and archaea show enormous diversity in cell size, ranging from about 0.1 μm to 750 μm in diameter. Cell size has a great effect on metabolic and physiological processes in the cell.

The cytoplasmic membrane controls cell processes via transport of ions and molecules

All cells possess a 6–8 nm-wide cytoplasmic membrane, illustrated in *Figure 3.1*. In bacterial and eukaryotic cells, this is composed of a phospholipid bilayer containing hydrophobic fatty acids linked to glycerol moieties via ester linkages, plus a wide range of proteins. Archaeal membranes have a different structure and are more variable. The hydrophobic sidechains are 5-carbon hydrocarbons units (isoprenoids) linked to glycerol via ether linkages. The lipids can be either 20-carbon units (phytanyl), forming a bilayer, or 40-carbon diglycerol tetraethers in which the phytanyl sidechains are covalently linked to form a monolayer. The monolayer structure is more stable at high temperatures and is particularly prevalent in thermophilic archaea, such as those living at hydrothermal vents. Hydrocarbon ring structures also occur in different groups of Archaea.

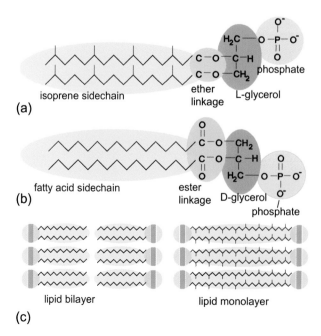

Figure 3.1 Structure of microbial membranes. (a) Lipid structure in Archaea, showing sidechains of isoprene units with ether link to glycerol phosphate. (b) Lipid structure in Bacteria and Eukarya, showing sidechains of fatty-acid units with ester link to glycerol phosphate. (c) Lipid bilayer of Bacteria, Eukarya, and some Archaea (left), and the monolayer structure found in most Archaea (right).

Bacterial membranes are stabilized by Mg^{2+} and Ca^{2+} and by pentacyclic compounds known as hopanoids. These are resistant to degradation and large quantities have accumulated in the environment over the many millennia of bacterial evolution. The total global mass of carbon deposited in hopanoids is estimated at about 10^3 Gt (Pg), approximately the same as the mass of organic carbon in all organisms living today. A very large fraction of fossil fuels such as petroleum and coal are composed of hopanoids, much of which has accumulated from the settlement of bacteria in the oceans. Because different bacteria produce characteristic hopanoid structures, these compounds are frequently used as markers for the analysis of microbial communities in marine sediments. Hopanoids have not been detected in archaea.

Despite these differences, the function of the cytoplasmic membrane is the same—acting as a selectively permeable barrier controlling movement of ions and molecules in and out of the cell. Transport mechanisms are needed for nutrient uptake, excretion, and secretion of extracellular products. As well as controlling the entry/exit of substances to/from the cell by its selective permeability and its specific transport proteins, the cytoplasmic membrane is the most important site of energy generation and conservation. During metabolism, electron carriers associated with the membrane bring about a charge separation across the membrane. H^+ ions (protons) become concentrated on the exterior surface of the membrane and OH^- ions concentrate on the inner surface. This charge separation creates an electrochemical potential that can be converted to chemical energy. Bacteria and archaea show enormous variation in the oxidation-reduction reactions used to produce this gradient, but the common "currency" of energy exchange is always the molecule ATP used for synthesis and mechanical work by the cell. When protons cross the membrane through a specific port, the contained energy is captured in the conversion of ADP to ATP.

Cells may contain organelles, microcompartments, and inclusion bodies

One of the original defining features of bacteria and archaea as "prokaryotic" cells (p.3) was that they have a simple cellular organization with no membrane-bound nucleus or organelles such as mitochondria or chloroplasts. However, many bacterial cells show extensive internal compartmentalization. Members of the phylum Planctomycetes clearly belong to the Bacteria, but have a membrane-bound nucleus and some have a membrane-bound organelle (the anammoxosome). Extensive invaginations or aggregates of membrane vesicles are especially important in phototrophs and chemolithotrophs with very high respiratory activity. Phototrophs grown in low light intensity will produce more internal membranes and pigments than those grown under high illumination, in order to harvest available light with the greatest efficiency.

Gas vesicles are also important in the ecology of marine phototrophs such as Cyanobacteria. These have a 2 nm thick, single-layered, gas-permeable wall composed of two hydrophobic proteins surrounding a central space. These form a very strong structure into which gas diffuses from the cytoplasm, resulting in increased buoyancy of the cell. This enables phototrophic bacteria to maintain themselves in the desired zone of light intensity in the water column. Organisms from deep habitats have narrow vesicles as they are more resistant to hydrostatic pressure.

Many types of autotrophic bacteria contain icosahedral structures known as carboxysomes. These are about 120 nm diameter and surrounded by a thin, selectively permeable envelope, architecturally similar to the protein coat of viruses. They contain RuBisCO, the key enzyme used by many autotrophs for the fixation of CO_2 (p.83) and serve as concentration mechanisms to enhance the efficiency of this process. Other similar structures act as centers for certain catabolic reactions, such as aldehyde oxidation.

Many bacteria also contain granules of organic or inorganic material used as a store of energy or structural components. Sometimes they occur as simple concentrations of material free within the cytoplasm, but they are often enclosed in a thin, single-layered lipid membrane. For example, a wide range of marine bacteria produce polyphosphate granules and carbon and energy storage polymers such as polyhydroxyalkanoates. Large inclusions of elemental liquid sulfur up to 1.0 μm in diameter occur in sulfur-oxidizing bacteria and can constitute up

CAN BACTERIA BE MULTICELLULAR?

Our blinkered view of bacteria as simple, independent unicellular organisms grew from the classical emphasis on isolating pure cultures. But bacteria show many properties of differentiation and coordinated behavior more commonly associated with eukaryotic organisms. For example, cells within some filamentous cyanobacteria communicate to coordinate formation of heterocysts for segregation of nitrogen fixation and photosynthesis (p.137). Myxobacteria form fruiting bodies consisting of hundreds of cells, some with different functions. Some magnetotactic bacteria aggregate together in clusters of up to 40 cells that swim in coordinated fashion (*Figure 4.5*). Cable bacteria exist as multicellular filaments for long-distance electron transport (*Box 4.1*). Many bacterial signaling molecules have been identified, with networks for the integration of signals between cells in order for the community to make "decisions" about gene expression and cellular differentiation. Whether in pure colony or a mixed community, bacteria communicate via these signals. Dunny et al. (2008) argue that "the sociobiology of bacteria, largely unappreciated and ignored by the microbiology research community two decades ago is now a major research area … increasingly considered in the context of evolutionary biology."

to 30% of the cell weight (see *Figure 1.4C*). Cyanobacteria also contain granules composed of large amounts of α-1,4-linked glucan, which is the polymerized carbohydrate product of photosynthesis. They also contain granules composed of polypeptides rich in the amino acids arginine and aspartic acid, which appear to be produced as a reserve of nitrogen.

The nature of the cell envelope has a major effect on physiology

Almost all bacteria and archaea possess some type of cell wall external to the cytoplasmic membrane. The cell wall is responsible for the shape of the cell and provides protection in an unfavorable osmotic environment. The key component of the cell wall in Bacteria is peptidoglycan, a mixed polymer composed of alternating residues of two amino sugars (*N*-acetyl glucosamine and *N*-acetyl muramic acid) with a tetrapeptide sidechain composed of a small range of amino acids. The great mechanical strength of peptidoglycan lies in the formation of cross-links between the peptide sidechains, forming a mesh-like molecule. Although variation in the types of amino acids and the nature of cross-linking occurs, peptidoglycan with essentially the same structure occurs in all Bacteria with the exception of the Planctomycetes.

It is traditional to differentiate bacteria as either Gram-positive or Gram-negative types, based on their staining reaction in a method devised by Gram in 1884. Although this basic distinction remains a useful first stage in the examination of unknown bacteria in culture and underpins some important physiological differences, it has no phylogenetic significance. As shown in *Figure 3.2*, Gram-positive bacteria have a relatively thick, simple wall composed of peptidoglycan, together with ribitol or phosphate polymers known as teichoic acids. Gram-negative bacteria have a thinner but more complex wall with a very thin peptidoglycan layer that lacks teichoic or teichuronic acids, together with an additional outer membrane (OM), anchored tightly to the peptidoglycan layer via a lipoprotein. The OM is a lipid-protein bilayer complex but is very different from the cytoplasmic membrane, containing unique outer membrane proteins (OMPs) and large amounts of lipopolysaccharide (LPS), both of which have important functions affecting cell physiology and interactions with other organisms. LPS is composed of lipid A covalently bound to a core polysaccharide and strain specific sidechains, which are immunogenic and known as the O-antigen. The OMPs have very distinctive properties. The major proteins, called porins, are trimeric structures that act as channels for the diffusion of low-molecular-weight solutes across the outer membrane. Other proteins act as specific receptors for key nutrients before transport into the cell. An important feature of many of these proteins is that they are not produced constitutively. Their synthesis can either be repressed or induced, according to the concentration of the specific substances. An especially important example of this in marine bacteria is in the acquisition of the essential nutrient iron, which may be imported into the cell via secreted compounds known as siderophores (p.92). Expression of genes encoding siderophores and OMP receptors for their entry into the cell (as well as many other genes) may only occur when iron is in short supply. As a consequence, bacteria grown in laboratory culture—which provides excess iron unless special precautions are taken—may have very different OM structure and physiological properties from those in natural environments, which are usually very iron-depleted. Other OMPs are responsible for the detection of environmental changes such as pressure, temperature, and pH, and communication of this information via signaling systems. Because the OM is an additional exterior permeability barrier with different selectivity to the cytoplasmic membrane, some substances cross one membrane, but not the other. This means that the periplasm—the space between the membranes—is an important feature of Gram-negative bacteria. Specific periplasmic binding proteins act as a shuttle to translocate a substrate from an OM receptor to a transport protein in the cytoplasmic membrane. Proteins such as enzymes and toxins may be transported across the cytoplasmic membrane but remain in the periplasm until the cell is lysed, unless there are specific mechanisms to excrete them across the OM. As cells grow, parts of the OM may balloon out to form small vesicles that are released from the cell, taking with them some of the periplasmic contents. These can include nucleic acids, enzymes, and toxins. Vesicles can act as a delivery system for these contents when they fuse with other membranes, which is especially important in pathogenic and symbiotic interactions of bacteria with their hosts. The vast majority of marine bacteria have the Gram-negative type of cell envelope. Recent work shows that vesicles carrying DNA from diverse bacteria are abundant in the oceans, providing discrete packages that may be involved in exchange of genetic information and energy sources among marine plankton.

The cell walls of Archaea are very variable in composition. They do not contain peptidoglycan, but a few types contain a related compound called pseudomurein or thick layers of other polysaccharides that provide the same osmoprotective function. The main cell walls in most Archaea are known as S-layers, which are composed of paracrystalline arrays of a single type of protein or glycoprotein arranged in a tetragonal or hexagonal lattice-like structure. These structures are built via self-assembly of identical subunits. Many Bacteria—both Gram-positive and Gram-negative—also possess an S-layer as an additional component as the outermost part of the envelope (*Figure 11.13*). Interestingly, S-layers are found in almost all phylogenetic groups of both Bacteria and Archaea and seem to represent one of the simplest biological membranes, which probably developed very early in evolution. Where the S-layer is the only component of the cell wall, it provides mechanical strength and osmotic stabilization, but when it occurs as an additional external surface layer it may have diverse functions such as protection against phages, predators, phagocytosis, enzymes, and toxins.

In addition to the cell envelope, many organisms secrete a slimy or sticky extracellular matrix. It is usually composed of polysaccharide, in which case it may form a network extending into the environment, sometimes termed the glycocalyx. In other cases, there may be a more distinct rigid layer (capsule), which may contain protein. The main importance of the glycocalyx in marine systems is in the attachment of bacteria to plant, animal, and inanimate surfaces, leading to the formation of biofilms. The bacterial glycocalyx is critical for the settlement of algal spores and invertebrate larvae; the significance of these processes in biofouling of surfaces in the marine environment is considered in *Chapter 13*. The release of these sticky exopolymers is also very important in the aggregation of bacteria and organic detritus in the aggregation of TEPs and marine snow particles discussed in *Chapter 1*. Furthermore, a large proportion of the dissolved organic carbon in ocean waters probably derives from the glycocalyx of bacteria. Slime layers can act as a protective layer, preventing the attachment of phages and penetration of some toxic chemicals. In pathogens, the presence of a capsule can inhibit engulfment by host phagocytes, and in free-living forms it is commonly assumed to inhibit ingestion by protists. However, many grazing flagellates seem to feed voraciously on capsulated planktonic bacteria, probably because the capsule contains additional nutrients.

Genome size and organization determines bacterial and archaeal lifestyles

Bacterial and archaeal genomes comprise the entire genetic information of the cell, containing the DNA code specifying all of the structural proteins, enzymes, regulatory proteins, and RNA molecules needed for growth and replication. As a general rule, bacterial and archaeal genomes contain very little non-coding DNA (unlike eukaryotes), so the size of the genome is directly proportional to the number of genes and hence the lifestyle of the organism. The number of genes for the core functions of the cell (energy generation, replication, and protein synthesis) is quite constant and therefore forms a higher proportion of small genomes. Organisms with larger genomes are likely to have a higher proportion of genes for functions such as nutrient transport, signaling, regulation of transcription, DNA repair, motility, and chemotaxis. This provides them with the greater ability to adapt to factors such as different nutrient sources and environmental conditions. Organisms with large genomes have the capacity to encode numerous regulatory pathways and specialized metabolic functions. *Table 3.1* illustrates this principle, showing the range of genome sizes in representative examples of Bacteria and Archaea.

As seen in the table, some bacterioplankton show very reduced genomes. The genomes found in cultured members of the OM43 and SAR11 clades—the most abundant groups of organisms in the oceans—are the smallest known for free-living organisms. A theory known as genome streamlining was proposed to explain this phenomenon, in which organisms gain a reproductive benefit by evolving to eliminate non-essential genes. As noted above, all bacterial and archaeal genomes can be considered to be streamlined by the lack of non-coding DNA, but some organisms such as the OM43 and SAR11 representatives have gone further by

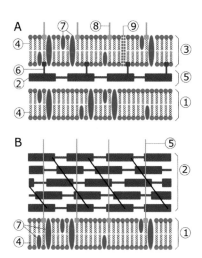

Figure 3.2 Schematic representation of the cell envelopes of Gram-negative (A) and Gram-positive (B) bacteria. Key: 1 cell membrane, 2 peptidoglycan, 3 outer membrane, 4 phospholipid, 5 periplasm, 6 lipoprotein, 7 protein, 8 LPS, 9 porin. Credit: Francisco2, CC BY 3.0 via Wikimedia Commons.

(i) PROPERTIES OF LPS

Serological typing based on differences in the O-antigen is important in identification and variation in immunological specificity is important in the epidemiology of bacterial diseases (discussed in *Chapter 12*). LPS also stimulates the activation of complement and related host responses, which is very important in the interactions of pathogenic or symbiotic bacteria with their animal hosts. Lipid A is also an endotoxin, causing hemorrhage, fever, shock, and other symptoms caused by Gram-negative pathogens of vertebrates. Blood collected from horseshoe crabs is used to screen for endotoxins in the pharmaceutical industry because it is sensitive to picomolar concentrations of LPS. Fortunately, this use, which threatens *Limulus* populations on the US East Coast, may soon be over, because a recombinant protein has gained regulatory acceptance as a replacement.

Table 3.1 Examples of genome sizes of selected marine members of the Bacteria and Archaea. (Data from NCBI Genome Database).

Microbe[a]	Size (Mb)[b,c]	ORFs[d]	Comments
Bacteria			
"*Ca.* Ruthia magnifica"	1.16	1099	Obligate intracellular sulfur-oxidizing chemolithoautotrophic symbiont of giant clam *Calyptogena magnifica* found at hydrothermal vents.
"*Ca.* Riegeria santandreae"	1.34	1344	Obligate intracellular sulfur-oxidizing chemolithoautotrophic symbiont of flatworm *Paracatenula.*
"*Ca.* Atelocyanobacterium thallassa" ALOHA	1.44	1148	Nitrogen-fixing cyanobacterium lacking many components of photosystems and CO_2 fixation. Symbiotic, dependent on algal host for nutrition.
Strain HTCC218 (OM43 clade)	1.30	1354	Methylotroph, metabolizes C1 compounds (but not CH_4). Cultured representative of OM43 clade. The smallest (just!) genome of a free-living organism.
"*Ca.* Pelagibacter ubique" HTCC1062 (SAR11 clade)	1.31	1354	Chemoorganotrophic heterotroph. Cultured representative of SAR11 clade. One of the smallest genomes of a free-living organism.
Aquifex aeolicus VF5	1.59	1526	Hyperthermophilic chemoautotroph. Despite its reduced genome, it possesses flagella and is motile.
Prochlorococcus marinus MED4	1.66	1716	Cyanobacterium. High-light adapted strain of the most abundant group of oxygenic photosynthetic organisms.
Thermotoga maritima MSB8	1.86	1858	Anaerobic, hyperthermophilic, fermentative, chemoorganotroph.
Thiomicrospira crunogena XCL-2	2.43	2244	Free-living sulfur-oxidizing chemolithoautotroph symbiont.
Prochlorococcus marinus MIT9313	2.14	2275	Strain of *P. marinus* found at intermediate depths, adapted to exploit more variable nutrient conditions.
"*Ca.* Endoriftia persephone"	3.20	n.d.	Intracellular symbiont of hydrothermal vent tubeworms but retains genes for free-living stage. Sulfur-oxidizing chemolithoautotroph.
Nitrosococcus oceani ATCC19707	3.53	3095	Aerobic, ammonia-oxidizing chemolithoautotroph (*Gammaproteobacteria*).
Synechocystis sp. PCC6803	3.57	3168	Oxygenic photosynthetic cyanobacterium, abundant in upper photic zone.
Bdellovibrio bacteriovorus HD100	3.78	3541	Bacterial predator that penetrates other bacteria, replicates, and causes cell lysis.
Pseudoalteromonas haloplanktis TAC125	3.97 (Chr*2)	3494	Psychrophilic chemoorganotrophic heterotroph. Adaptations for growth at low temperatures and reactive oxygen species.
Aliivibrio fischeri ES114	4.28 (Chr*2)	3814	Chemoorganotrophic heterotroph possessing two chromosomes; free-living and in bioluminescent symbiosis of squid.
Caulobacter crescentus NA1000	4.04	3886	Chemoorganotrophic heterotroph with a complex lifecycle of planktonic and stalked sessile stages.
Ruegeria pomeroyi DSS-3	4.60	4306	Member of the Roseobacter clade with a major role in the sulfur cycle via DMSP metabolism. Many amino acids and carboxylic acid transporters.
Trichodesmium erythraeum IMS101	7.75	4549	Filamentous cyanobacterium producing differentiated cells (diazocytes) for temporal regulation of photosynthesis and nitrogen fixation.
Vibrio parahaemolyticus RMID2210633	5.12 (Chr*2)	4692	Chemoorganotrophic heterotroph. Abundant in marine and estuarine inhabitants. Associated with chitinous surfaces of molluscs and crustacean shellfish. High replication rate at 37°C; human pathogen.
Rhodopirellula islandica K833	7.43	6851	Chemoorganotrophic, heterotrophic planctomycete. Complex cell structure with internal cell compartments.

(Continued)

Table 3.1 (Continued)

Microbe[a]	Size (Mb)[b,c]	ORFs[d]	Comments
Archaea			
Nanoarchaeum equitans Kin4-M	0.49	536	Hyperthermophilic, obligate symbiont attached to cells of *Igniococcus hospitalis*. One of the smallest known cells, minimal genome lacking genes for most metabolic pathways.
Igniococcus hospitalis KIN4/1	1.30	1442	Hyperthermophilic chemolithoautotroph from undersea volcano, reducing sulfur with H_2. Host for *N. equitans*.
Methanocaldococcus jannaschii DSM2611	1.74	1762	Hyperthermophilic obligate anaerobe from hydrothermal vents. Produces methane form $H_2 + CO_2$.
"*Ca.* Nitrosopumilus maritimus" SCM1	1.65	1795	Chemolithoautotroph which oxidizes very low concentrations of NH_3 in the deep ocean.
Pyrococcus furiosus DSM3638	1.91	1990	Hyperthermophilic obligate anaerobe; chemoorganotrophic heterotroph from hydrothermal vents. Highly motile.
Cenarchaeum symbiosum A	2.05	2017	Psychrophilic ammonia-oxidizing chemolithoautotroph associated with the sponge *Axinella*.
Archaeoglobus fulgidus DSM4304	2.04	2413	Hyperthermophilic obligate anaerobe from hydrothermal vents. Chemolithoautotroph via sulfate reduction.
Methanosarcina acetivorans C2A	5.75	4542	Metabolically and versatile archaeon from marine and coastal sediments capable of producing methane from multiple substrates, including acetate. The largest archaeal genome known.

[a] *Ca.* = *Candidatus* (interim name for uncultured organism); [b] Mb = megabases; [c] Chr*2 indicates possession of two chromosomes; [d] ORFs = open reading frames (predicted proteins).

losing the ability for metabolic flexibility and specializing in the use of low concentrations of a few carbon compounds that characterize ocean environments. One explanation for this is that competition between species will favor organisms with efficient and economic replication (by requiring less DNA). Large population sizes will increase this selective pressure, which is why genome streamlining is prevalent in bacteria and archaea, and particularly so in bacterioplankton. Extreme genome streamlining also minimizes overall size of the cell, increasing the surface area to volume ratio (see *Figure 1.2*), leading to more efficient uptake of scarce nutrients, in proportion to the cell's needs. The genome sizes of *Prochlorococcus marinus* are also extremely streamlined. Again, these have massive population sizes in surface waters, being the most abundant phototrophs. In this case, although classified as the same species, we can see large differences in genome size and gene content between different strains (ecotypes) adapted to different levels of light and the availability of nitrogen sources at different depths. In summary, genome streamlining increases the success of organisms in nutrient-poor environments, either by winning the competition to acquire their share of scarce resources, or by using them more efficiently. This is evident by the high gene frequencies of such organisms, as indicators of their abundance and diversity.

As discussed in detail later in this chapter, we can describe organisms adapted to these low levels of nutrients as oligotrophs, which have a "slow and steady" lifestyle in stable niches that show little variation in composition over time. The low average size of genomes in marine metagenomes is evidence that organisms with this lifestyle are most abundant. In contrast, other organisms described as copiotrophs—often associated with particles—have a "feast or famine" lifestyle and are able to grow rapidly when exposed to high substrate levels and respond to higher concentrations of nutrients. This alternative strategy is adopted by many marine bacteria with higher numbers of genes, conferring the ability to carry out a wider range of functions according to shifts in prevailing conditions. Variable factors such as availability of different nutrients, oxygen, light, temperature, pressure, surface attachment, or circadian rhythms require organisms to have a wider repertoire of enzymes and transport systems plus efficient systems for the regulation of gene expression.

It is also evident from *Table 3.1* that symbiotic and parasitic microbes often possess the smallest genomes of all. In this case, a different evolutionary mechanism is thought to operate, in which the organisms become dependent on their host for many of their metabolic functions. During evolution, reduction in genome size is thought to occur by genetic drift due to reduction in effective population sizes, with reduced opportunities for recombination and horizontal gene transfer. In addition, positive selection pressure results in loss of genes for metabolic functions that are provided by the host. The microbe may retain only the essential core genes for cell structure and central metabolism. The genome reduction of symbionts and pathogens of animals is explored in detail in *Chapters 10* and *11*, but one salient point will be discussed here to illustrate the principles by comparing the genome sizes of free-living and symbiotic sulfur-oxidizing bacteria (SOB). Note from *Table 3.1* that the bacterial symbiont of the clam *Calyptogena* has a highly reduced genome in comparison to free-living SOB, whereas the symbiont of the tubeworm *Riftia* has a genome of similar size to its free-living relatives. Both are intracellular symbionts of their host, but the clam symbiont is obligately dependent on its host and is passed from one generation to the next via the eggs, whereas the tubeworm symbiont has a free-living stage and is acquired at each host generation.

The table also shows examples of bacteria and archaea with minimal genomes that are symbionts of parasites of other microbes with which they are intimately associated. The archaeon *Nanoarchaeum* depends on its archaeon host *Igniococcus hospitalis* for most nucleotides, amino acids, and lipids and has a highly reduced genome. "*Ca.* Atelocyanobacterium thalassa" (UCYN-A) is a cyanobacterium that lacks many components of systems for photosynthesis and CO_2 fixation and depends on its unicellular algal hosts for energy and carbon compounds. In return, it fixes nitrogen for its host.

The most common form of bacterial genome organization is a singular circular chromosome. The length of the DNA molecule can be over 1000 times the length of the cell, and cells achieve the remarkable feat of coiling, folding, and packing the DNA in such a way that the processes of replication, transcription, and translation can occur efficiently. The circular chromosome exists as a compact structure organized into supercoiled domains by DNA gyrase. Replication of the DNA involves a number of polymerase and ligase enzymes and proceeds bidirectionally from an origin on the chromosome and is coordinated with cell division processes to ensure that each daughter cell receives a complete chromosome. In rapidly growing cells, the DNA may be replicating at more than one point on the chromosome. Some bacteria have linear chromosomes or possess more than one chromosome. Notably, members of the Vibrionaceae have two chromosomes, which confer important properties for their evolution by gene acquisition, as discussed in *Chapter 12*.

Although archaea possess small compact genomes like those of bacteria, organization of the nuclear material is more complex and variable in archaea. In some species, packaging and stabilization of the DNA structure is achieved by supercoiling, but other species possess positively charged proteins known as histones, which have amino acid sequences homologous to those found in eukaryotes. Short sequences of DNA are wound around clusters of histones to form tetrameric clusters called nucleosomes. Some extremely thermophilic archaea have an additional type of DNA gyrase that induces positive supercoiling in the DNA to protect it against denaturation. The process of DNA replication is also different from that in bacteria. Although most archaea possess single circular chromosomes, there are often multiple origins of replication, and the structure of the replication complex and polymerase enzymes again resembles that found in eukaryotes.

In addition to the chromosome(s), many bacteria and archaea contain small, circular, extrachromosomal elements of genetic information called plasmids. These replicate independently of the main chromosome, although they rely on enzymes encoded on it. Plasmids have great biological significance, because they encode genes that are useful to the cell under certain conditions, rather than those connected with mainstream metabolic processes. Often, plasmids can be transferred between cells and rapidly spread through a population via conjugation. This increases the adaptability of a population in various ways. The most important plasmids are those containing genes for antibiotic resistance, pathogenicity (e.g. toxins or colonization factors), or the ability to degrade recalcitrant organic compounds (e.g. naturally occurring or xenobiotic hydrocarbons).

Microbes use a variety of mechanisms to regulate cellular activities

Even the simplest microbes contain hundreds of genes encoding enzymes and other macromolecules required for the hundreds of biochemical reactions in metabolism. To ensure efficient use of resources and to respond to changes in environmental conditions, bacterial and archaeal cells need to regulate both the activity of enzymes and the level of macromolecular synthesis. Regulation of the activity of enzymes is a very rapid process, occurring by various mechanisms such as feedback allosteric inhibition, in which an excess of the end product of a biosynthetic pathway often inhibits the first enzyme of the pathway. Some other enzymes are modified by covalent addition of groups, which inhibits catalytic activity. Although some proteins are required under all conditions and are said to be constitutive, cells employ many mechanisms of controlling the level of gene expression to regulate the production of enzymes or other proteins that are only required under certain conditions. Because messenger RNA (mRNA) is very short-lived, switching transcription on or off provides a very rapid response to changing circumstances. Also, bacterial and archaeal genes are often grouped together in operons, so that expression of related genes (e.g. several enzymes in a biosynthetic pathway) is regulated together under the control of a single promoter and operator region, upstream of the structural genes. As noted above, copiotrophic planktonic microbes possess many regulatory systems, and the same applies to microbes that grow on surfaces or colonize animals or plants. The expression of a gene involves the binding of RNA polymerase to a specific region of the DNA (the promoter). Many genes are subject to negative control, when regulatory compounds bind to specific DNA-binding proteins. This causes an altered conformation of the protein, allowing it to bind to the DNA at the operator region (between the promoter and the gene) leading to the blocking of transcription. Genes can also be regulated positively—in this case, an activator protein acts to help the RNA polymerase bind efficiently to the promoter region of the DNA. Often, several operons will be regulated by the same repressor or activator, so they are expressed coordinately. Sometimes, regulation of a master gene results in a cascade of subsequent regulatory steps, so that expression of hundreds of genes is up- or down-regulated in response to a single stimulus. A set of genes like this is known as a regulon.

One of the most common methods used by cells to respond to external stimuli is via two-component regulatory systems. In this case, the signal does not act directly to control transcription, but is detected by a sensor kinase protein in the cell membrane, which then transduces the signal to a response regulator protein in the cytoplasm. The kinase enzyme is composed of two domains, one of which is exposed to the exterior and detects an environmental or chemical signal. When this occurs, the cytoplasmic domain of the protein becomes phosphorylated. The phosphoryl group is then transferred to a second specific protein called the response regulator. When phosphorylated, the response regulator interacts with specific regions of the DNA and can either activate or repress transcription of specific genes, depending on the system. *Chapters 11* and *12* describe important examples of such signaling and global control networks when considering infection of animals by symbiotic or pathogenic bacteria from the aquatic environment.

Translation of mRNA and protein synthesis in the Archaea and Bacteria are similar, employing the 70S ribosome (*Figure 2.7*). However, many of the component ribosomal proteins of the Archaea and Eukarya show closer homology than they do with those from the Bacteria.

SOURCES OF ENERGY AND CARBON

Microbes obtain energy from light or oxidation of compounds

Metabolism, the sum of all the biochemical reactions within a cell, consists of catabolism and anabolism. In catabolism, cells break down and oxidize larger molecules yielding energy via exergonic reactions, which is used for the synthesis of macromolecules by endergonic (energy-requiring) reactions. Adenosine triphosphate (ATP) is regarded as the chemical currency of the cell, mediating the transfer of energy during these processes (*Figure 3.3*). It is convenient to classify microbes in terms of their energy source as either chemotrophs (which

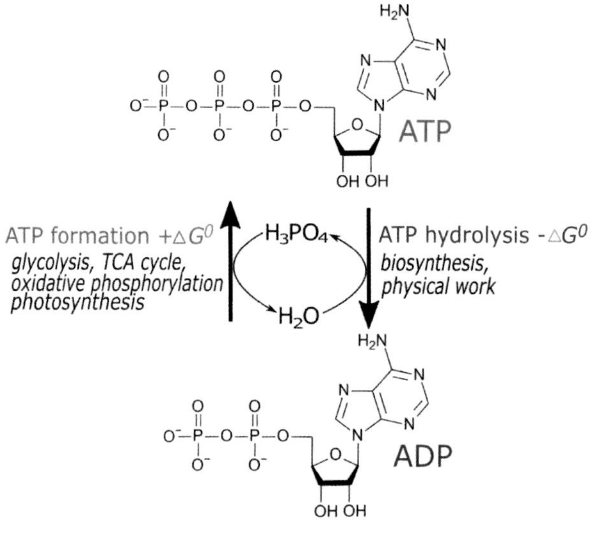

Figure 3.3 The energy metabolism of all cells depends on hydrolysis of adenosine triphosphate (ATP) to adenosine diphosphate (ADP). This is highly exergonic (negative Gibbs free energy, $-\Delta G$) and is used to drive endergonic ($+\Delta G$) biochemical processes such as biosynthesis, active transport, or movement.

ATP formation $+\Delta G^0$
glycolysis, TCA cycle,
oxidative phosphorylation
photosynthesis

ATP hydrolysis $-\Delta G^0$
biosynthesis,
physical work

produce ATP by the oxidation of inorganic or organic compounds) or phototrophs (which use light as a source of energy to produce ATP). Chemotrophs may be further divided into those that can obtain energy solely from inorganic compounds (chemolithotrophs) and those that use organic compounds (chemoorganotrophs). Chemolithotrophy is unique to members of the Bacteria and Archaea, while chemoorganotrophy is found in all groups of cellular microbes.

Microbes differ in their source of carbon to make cellular material

Those organisms that can use CO_2 as the sole source of carbon are termed autotrophs, while those that require preformed organic compounds as a source of carbon are termed heterotrophs. Autotrophs are primary producers that fix carbon from CO_2 into cellular organic compounds, which are then available to be used by heterotrophs. Most chemolithotrophs and phototrophs are also autotrophic, but some (e.g. purple non-sulfur photosynthetic bacteria and some sulfur-oxidizing chemolithotrophs) can switch between heterotrophic and autotrophic metabolism. The term mixotroph is used to describe organisms with a mixed mode of nutrition. For example, some sulfur-oxidizing bacteria are not fully lithotrophic, as they require certain organic compounds because they lack key biosynthetic enzymes. Many protists, including numerous types of flagellates and dinoflagellates, are mixotrophic because they derive nutrition both from photosynthesis and from the engulfment of prey such as bacteria. Indeed, among the protists there is a spectrum from absolute heterotrophy to absolute autotrophy. *Table 3.2* shows a summary of these nutritional categories.

PHOTOTROPHY AND CHEMOTROPHY

Phototrophy involves conversion of light energy to chemical energy

Virtually all life on Earth depends directly or indirectly on solar energy, and it is likely that simple mechanisms for harvesting light energy developed very early in the evolution of life. Phototrophs contain light-sensitive pigments, which use energy from sunlight to oxidize electron donors. Plants, algae, other phototrophic protists, and cyanobacteria use H_2O as the electron donor, resulting in the production of O_2 (oxygenic photosynthesis). Other types of phototrophic bacteria use H_2S, S^0, $H_2SO_3^{2-}$, or H_2, meaning that no O_2 is generated (anoxygenic photosynthesis). In the process of photophosphorylation, light energy creates a charge separation across membranes and "traps" energy from the excited electrons into a chemically stable molecule (ATP) and also generates NADPH (nicotinamide adenine dinucleotide phosphate [reduced]), as shown in *Figure 3.4A*. Many phototrophs are also autotrophic and use ATP and NADPH for the fixation of CO_2 into cellular material, in which case the process meets the strict definition of photosynthesis.

Table 3.2 Nutritional categories of microorganisms

Energy source	Carbon source	Hydrogen or electron source	Representative examples
Photolithoautotrophy			
Light	CO_2	Inorganic	Cyanobacteria Purple sulfur bacteria Phototrophic protists
Photoorganoheterotrophy			
Light	Organic compounds	Organic compounds or H_2	Purple non-sulfur bacteria Aerobic anoxygenic bacteria Proteorhodopsin-containing bacteria and archaea[a]
Chemolithoautotrophy			
Inorganic	CO_2	Inorganic	Sulfur-oxidizing bacteria Hydrogen bacteria Methanogens Nitrifying bacteria and archaea
Chemoorganoheterotrophy			
Organic compounds	Organic compounds	Organic compounds	Wide range of bacteria and archaea Fungi Phagotrophic protists
Mixotrophy (combination of lithoautotrophy and organoheterotrophy)			
Inorganic	Organic compounds	Inorganic	Some sulfur-oxidizing bacteria, e.g. *Beggiatoa*
Mixotrophy (combination of photoautotrophy and organoheterotrophy)			
Light + organic compounds	CO_2 + organic compounds	Inorganic or organic	Phagotrophic photosynthetic protists (some flagellates and dinoflagellates)

[a]The role of proteorhodopsin in nutrition is uncertain (see *Box 3.1*).

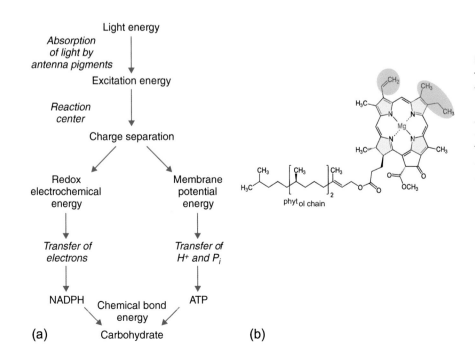

(a)   (b)

Figure 3.4 (a) Transformation of light energy to chemical energy in the light reactions of photosynthesis. (b) Structure of chlorophyll *a*, showing four pyrrole rings (green) containing magnesium at the center (yellow). Bacteriochlorophyll *a* has an identical structure, except for changes in the shaded regions (purple). Other chlorophylls and bacteriochlorophylls have other substitutions at various parts of the molecule, resulting in different light absorption properties.

Although many microbes possess photochemical mechanisms, true photosynthesis occurs only in cells containing the magnesium-containing pigments known as chlorophylls (found in oxygenic cyanobacteria and chloroplasts) or bacteriochlorophylls (found in anoxygenic phototrophic bacteria). Because of their efficiency in both transforming light energy into ATP through electron flow, the provision of reducing power (NADPH, NADH), and the use of water as reductant, chlorophyll-based photosynthesis has evolved to become the most prevalent photosystem. Chlorophyll and bacteriochlorophyll molecules are closely related, apart from a small region of the porphyrin molecule (*Figure 3.4B*). There are also small differences in other parts of the molecule, resulting in structures that absorb different wavelengths of light. A variety of accessory pigments are also present, notably carotenoids, which function primarily to protect cells from harmful photooxidation reactions that can occur in bright light, such as the generation of toxic forms of oxygen. Carotenoids such as fucoxanthin may be important in light harvesting for photosynthesis in eukaryotic algae. Antenna pigments surround complexes of 50–300 molecules of chlorophyll or bacteriochlorophyll combined with proteins (reaction centers). These arrays occupy a large surface area to trap the maximum amount of light of different wavelengths and transfer its energy to the reaction center. In cyanobacteria, the most important antenna pigments are the phycobilins phycoerythrin and phycocyanin. The diversity of different pigments with different light-absorbing properties has great ecological significance, because it determines the optimum niche within the water column and allows organisms with different combinations of pigments to coexist in illuminated habitats by absorbing a particular fraction of the light energy that other members of the community are not absorbing.

The photosynthetic pigments are contained within special membrane systems that create the structure necessary for generation of a proton-motive force. In photosynthetic bacteria, there are usually extensive invaginations of the cytoplasmic membrane; these can be complex and multilayered in cyanobacteria. In chloroplasts, the photosynthetic pigments are attached in stacks of sheet-like membranes called thylakoids.

Oxygenic photosynthesis involves two distinct but coupled photosystems

Oxygenic photosynthesis, in which H_2O is the electron donor leading to production of O_2, is the most important contributor to primary productivity accounting for the major roles played by cyanobacteria, diatoms, and dinoflagellates and other protists in ocean processes. Of course, it is also familiar as the mechanism used by seaweeds and plants. Photosynthesis is characterized by the presence of two coupled photosystems operating in series. As shown in *Figure 3.5*, photosystem I (PS I) absorbs light at long wavelengths (>680 nm) and transfers the energy to a specialized reaction-center chlorophyll (P700). Photosystem II (PS II) traps light at shorter wavelengths and transfers it to chlorophyll P680. Absorption of light energy by PS I leads to a very excited state, which then donates a high-energy electron via chlorophyll *a* and iron-sulfur proteins to ferredoxin. The electron can then pass through an electron transport chain, returning to P700 and leading to ATP synthesis via generation of a proton-motive force (cyclic photophosphorylation). ATP formation and reduction of $NADP^+$ to NADPH also occurs via a noncyclic route involving both photosystems. These processes are known as the light reaction of photosynthesis. In the light-independent (dark) reaction, ATP and NADPH molecules are used to reduce CO_2, leading to its fixation into carbohydrate. The reaction may be represented as follows, where (CH_2O) represents the basic unit of carbohydrates:

$$CO_2 + 3ATP + 2NADPH + 2H^+ + H_2O \rightarrow (CH_2O) + 3ADP + 3Pi + 2NADP^+$$

There are some important differences between the photosynthetic machinery of *Prochlorococcus* and other cyanobacteria, which are discussed in *Chapter 4*.

Anaerobic anoxygenic photosynthesis uses only one type of reaction center

A group known as the purple phototrophic bacteria contain bacteriochlorophyll *a*, but lack PS II and therefore cannot use H_2O as an electron donor in noncyclic electron transport. The

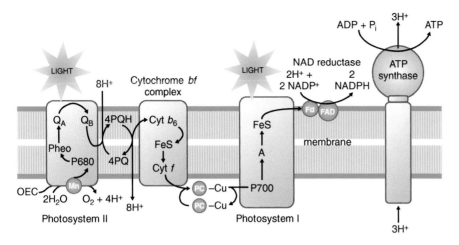

Figure 3.5 Electron flow in oxygenic photosynthesis. The light-driven electron flow generates an electrochemical gradient across the membrane, leading to ATP synthesis. Electrons flow through two photosystems, PS1 and PS2. Electrons are provided by water and the oxygen-evolving complex (OEC) generates oxygen. Key: Pheo, pheophytin; Q, quinone; Cyt, cytochrome; PC, plastocyanin; A, chlorophyll a; FeS, non-heme iron-sulfur protein; Fd, ferredoxin; FAD, flavin adenine dinucleotide; P680 and P700, the reaction centers of PSI and PSII, respectively.

light reaction generates insufficient reducing power to produce NADPH, and these bacteria therefore use H_2S, H_2 or S^0 because they are better electron donors. For this reason, they do not generate O_2 during photosynthesis and most are strict anaerobes found in shallow sediments and microbial mats. This mode of photosynthesis was almost certainly the first to evolve, before the development of PS II by cyanobacteria enabled water to be used as electron donor. Interestingly, some cyanobacteria growing in sulfide-rich habitats can carry out anoxygenic photosynthesis because they rely only on PS I.

Aerobic anoxygenic phototrophy is widespread in planktonic bacteria

Another form of phototrophy that does not result in generation of O_2 has recently been discovered to be widespread and important. Up to 20% of the bacteria in the photic zone may be aerobic anoxygenic phototrophs (AAnP). Screening of metagenomic datasets for *puf* genes, which code for the subunits of the light harvesting and reaction-center complexes, reveals their presence in a range of globally distributed bacterial groups in several phyla—they also occur in freshwater and the surface of soils. Like the purple bacteria, these bacteria do not generate O_2 during photosynthesis. However, unlike them, they do grow in aerobic conditions and sufficient transfer of electrons to drive ATP production by cyclic photophosphorylation only occurs in the presence of O_2. These organisms are generally photoheterotrophs and lack the enzymes for fixation of CO_2 into cell material. They use O_2 for respiratory metabolism of organic carbon and for the synthesis of bacteriochlorophyll a. However, AAnP bacteria metabolize more efficiently when light is available, suggesting that in oligotrophic waters they can use energy from both light and scarce nutrients simultaneously. In these bacteria, light inhibits bacteriochlorophyll a synthesis, so its activity diminishes during the day. Marine AAnP populations are complex and dynamic, forming a significant component of bacterioplankton in certain oceanic areas. Their high metabolic activity suggests that AAnP bacteria may contribute to ocean energy budgets and carbon cycling, but the extent of this is currently unclear.

Some phototrophs use rhodopsins as light-harvesting pigments

True photosynthesis is unknown in the Archaea, although some extremely halophilic types contain a specialized protein called bacteriorhodopsin, which is conjugated to the carotenoid pigment retinal. This appears as purple patches in the membrane of cells grown under low oxygen concentration and high-light intensity. Light energy is used to form a proton pump, which generates ATP to provide sufficient energy for slow growth when other energy-yielding reactions are not possible. Bacteriorhodopsin also functions to maintain the Na^+/K^+ balance in the cell in highly saline conditions. A similar light-mediated ATP synthesis system based on a molecule termed proteorhodopsin is now known to be very widespread amongst marine members of the Bacteria and Archaea in the photic zone. The discovery of proteorhodopsin and its ecological significance is discussed in *Box 3.1*.

> **? CAN AANP BACTERIA FIX CO₂?**
>
> Graham et al. (2018) recently screened genomes assembled from metagenomic databases obtained from the *Tara* Oceans dataset and, surprisingly, discovered draft genomes of nine bacteria that possess the potential for both AAnP and CO_2 fixation via the CBB cycle. These bacterial genomes contained the genes encoding two subunits (PufML) of the type II photochemical reaction center and the large and small subunits (RbcLS) of RuBisCO. These organisms form a clade with no cultured representatives within the alphaproteobacterial family Rhodobacteriaceae, for which Graham et al. propose the name "*Ca. Luxescamonaceae*." This group is globally distributed but constitutes only a minor component (0.1–1.0%) of free-living plankton in the photic zone. These bacteria also possess genes for the synthesis of bacteriochlorophyll and the oxidation of thiosulfate or sulfite, suggesting that they can grow photolithoautotrophically. Graham *et al.* note that their metagenomic assembly method has some limitations, and confirming that these bacteria do indeed possess a new method of photosynthesis will require proof of gene expression.

BOX 3.1   RESEARCH FOCUS

Microbes use solar power to top up energy levels

Discovery of a novel light-harvesting mechanism. The discovery of proteorhodopsin (PR) was one of the first demonstrations of the power of metagenomics to reveal previously unknown types of metabolism in ocean processes. As stated by Pinhassi et al. (2016): "With most techniques and research approaches, what one finds in nature is what one is looking for; the rest remains invisible." Rhodopsin is a membrane protein that contains a light-absorbing pigment (retinal) found in the rods of the retina in animal eyes. It has been known for some time that membranes of the archaean genus *Halobacterium* contain an analog of this protein—given the name bacteriorhodopsin—which use light energy to synthesize ATP (p.109). This was thought to be a process unique to these extremely halophilic Archaea. However, while carrying out sequence analysis of DNA from an uncultivated group of bacteria known as SAR86, Béjà et al. (2000) found evidence of a genetic sequence with strong sequence homology to the archaeal rhodopsins. Using bioinformatic tools, Béjà and colleagues made a structural model of the protein believed to be encoded by the gene and identified the transmembrane domains that are a critical feature associated with the rhodopsin function. Phylogenetic analysis showed that the gene for the proteobacterial protein (which they called proteorhodopsin, PR) is phylogenetically distinct from the archaeal protein bacteriorhodopsin (see p. 155 for comments on the naming of rhodopsins). The gene for the PR protein was then cloned in *Escherichia coli*, in which it was expressed as an active protein and acted as a proton pump employed for the synthesis of ATP. Béjà et al. (2001) then analyzed the properties of membranes isolated from marine bacterioplankton concentrated from seawater by filtration, showing that the functional reaction cycle is expressed in the natural environment. Subsequently, genes encoding PR have been shown to be present in a very diverse range of marine microbes, including the first cultured strain ("*Ca.* Pelagibacter ubique") of the highly abundant SAR11 clade (Giovannoni et al., 2005); see p.41. As more surveys were conducted, it became clear that proteorhodopsin is not restricted to the Proteobacteria, or even to the Bacteria. Frigaard et al. (2006) demonstrated PR genes in DNA of Archaea extracted from various depths in the Pacific Ocean and the GOS metagenomic datasets (see p.55) provided further evidence of the wide distribution of PRs in diverse microbes. Phylogenetic analysis of the genes required for synthesis of the PR protein and the associated retinal pigment provided strong evidence of horizontal gene transfer (HGT) between organisms. McCarren and DeLong (2007) showed that acquisition of a cassette of just six genes confers the ability to harvest biochemical energy from light on a microbe and suggested that this could happen easily in a single genetic transfer. Further studies showed that there is sequence variation in PR from bacteria isolated from different regions and depths, leading to altered biophysical properties. Spectral tuning of PR—due to small changes in amino acid composition at a critical region of the molecule—provides an ecophysiological adaptation to ensure maximum absorption of different wavelengths of light, which penetrate water to different depths (*Figure 1.5*). Over the next few years, evidence for PR-driven transduction of energy from light

was demonstrated in many of the major bacterial groups and in archaea, diatoms and dinoflagellates, other protists, and viruses (Ernst et al., 2014; Sharma et al., 2006). PR is now realized to be ubiquitous and present in over half of all heterotrophic bacterioplankton in the surface ocean (*Figure 3.6*).

PR-based phototrophy is dominant in picoplankton. So, we now know that life on Earth has evolved two efficient—but very different—light-harvesting strategies for phototrophy. Finkel et al. (2013) hypothesized that the simple, PR-based proton pump mechanisms should be more commonly found in organisms than the complex photochemical reaction centers (RCs) based on chlorophylls or bacteriochlorophylls. Although PRs appear not to provide enough energy for CO_2 fixation, they have the advantage of requiring the expression of only a single membrane protein and are easily spread by HGT. Photochemical RCs, on the other hand, are highly complex structures of numerous proteins and pigments and are "costly" for the cell to maintain. Finkel et al. used a systematic approach to examine 115 metagenomic datasets for the abundance of genes from oxygenic photosystems I and II and the anoxygenic photosystems RC1 and RC2 and compared these with rhodopsin homologs. The metagenomes had been prepared based on pre-filtration of different size fractions (see p.31) and samples filtered with a pore size greater than 0.8 μm—which would include most protists—were dominated by RCs. By contrast, in samples of the <0.8 μm fraction—containing most bacteria and archaea—48% of the cells contained a rhodopsin gene and only 18% had RC genes of any kind. Some organisms such as cyanobacteria, dinoflagellates, and diatoms can contain both phototrophic mechanisms. Finkel and colleagues conclude that the majority of bacterial and archaeal cells in the photic zone contain phototrophic potential. They point out that abundance of genes does not necessarily indicate the expression or function

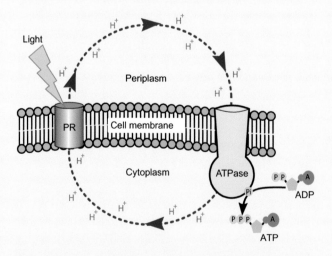

Figure 3.6 Proteorhodopsin (PR) uses light energy to translocate protons (H⁺) across the cell membrane which create a proton-motive force that drives ATP synthesis via a multi subunit membrane-bound ATPase.

BOX 3.1   RESEARCH FOCUS

of these genes, but numerous transcriptomic and proteomic studies have shown that PR is highly expressed in various ocean regions and depths (Pinhassi et al., 2016). Since many of these studies are based on single time points, Sieradzki et al. (2018) sought to determine whether PR metagenomes and metatranscriptomes varied according to the season at three contrasting locations, with major differences in nutrient concentration, chlorophyll levels, bacterial and viral counts, and heterotrophic production. PR dominated gene abundance and expression of phototrophy in this dynamic environment throughout the year, with oxygenic photosynthesis and AAnP at much lower levels. PR expression was dominated by the SAR11 cluster, followed by other Alpha- and Gammproteobacteria. In areas with low chlorophyll *a* concentrations, genes for the blue-light absorbing PR dominated, as found in other oligotrophic open ocean waters (Dubinsky et al., 2017).

What benefits does PR provide to cells? Given that PRs are so widespread in diverse microbes and that there may be over 20000 copies of the protein, occupying a considerable fraction of the cell membrane (Béjà et al., 2000), it seems likely that PR must provide a significant ecological advantage to planktonic bacteria and archaea that possess it. However, experimental proof of the physiological benefits that PR provides have proved elusive. Giovannoni et al. (2005) found that "*Ca.* Pelagibacter ubique" (SAR11) grew equally well in seawater cultures in dark and light conditions. However, measurement of oxygen consumption and gene expression under conditions of carbon starvation—leading to reduced respiration and oxidative phosphorylation—showed that, in the light, the bacterium switches to use of PR phototrophy to provide ATP (Steindler et al., 2011). Gómez-Consarnau et al. (2007) showed that *Dokdonia* sp. (a PR-containing member of the Flavobacteriia) showed a six-fold increase in growth yield when grown in light of the appropriate wavelength. The growth-enhancing effect of light was most effective at low concentrations of organic matter. Interestingly. PR genes

have also been acquired via HGT by some "classic" organohet-erotrophs like *Vibrio* spp. Gómez-Consarnau et al. (2010) showed increased long-term survival of a starved *Vibrio* strain in seawater exposed to light, rather than held in darkness. By contrast, no differences in survival were observed in a PR-deficient mutant strain. Wang et al. (2012) showed that cell viability of cells in nutrient-poor environments was due to light-induced proton pumping via PR, leading to increased ATP production. Use of bonus "solar panels" certainly seems to be an effective strategy of bacteria to supplement energy when organic nutrients are scarce, conferring a fitness advantage for bacteria to survive periods of resource deprivation. But not all bacteria possess PR—is it possible to detect genomic differences between related organisms with and without this mechanism? Kumagai et al. (2018) compared the genomes of 41 PR- and 35 PR+ strains of Flavobacteriia. Differences appear to be related to evolution of fundamentally different lifestyles and ecophysiological strategies, which they dubbed the "solar panel or parasol" hypothesis. Kumagai et al. conclude that PR- Flavobacteriia possess genes that encode synthesis of UV-absorbing pigments in their membranes ("parasols") that shield the cells from UV damage to their cells. This means that they cannot take advantage of light energy, whereas PR+ cells exploit the light for phototrophy (via PR "solar panels") and accept the cost of repairing the damage to their DNA (via synthesis of a photolyase enzyme).

In view of their presence in such diverse organisms, PR may have various physiological roles ranging from promoting long-term starvation survival to short-term adaptation to low-nutrient levels; also the presence of PR genes may have hitherto unidentified interactions with other gene systems that provide more subtle ecological or physiological benefits (Pinhassi et al., 2016). Further research should reveal how much contribution PR makes to the nutrition of the picoplankton microbes and to carbon cycling and ecosystem energy fluxes in the ocean.

Chemolithotrophs use inorganic electron donors as a source of energy and reducing power

Chemolithotrophs derive energy for the synthesis of ATP using an electron transport chain for the oxidation of inorganic molecules, and these bacteria play a major role in biogeochemical cycling in marine habitats. Until quite recently, it was thought that chemolithotrophy was only of major significance in sediments and around hydrothermal vents, but genomic data have shown that some of these processes are also highly important in the water column. The energetics of chemolithotrophy depend on the free-energy yield of coupled reactions between an electron donor and electron acceptor. Electron donors with sufficient reducing power to drive ATP synthesis include sulfur, sulfide, hydrogen, ammonium, nitrite, and iron (Fe^{2+}) and these have many geological or biological origins. A general feature is that the yield of ATP from oxidation of inorganic substances is very low; hence, the ecological and biogeochemical importance of chemolithotrophs is magnified because they need to oxidize large quantities of material in order to grow. Many chemolithotrophs are obligate aerobes, because oxygen is the most energetically favorable electron acceptor. However, anaerobic processes using sulfate, nitrate, and nitrite as electron acceptor are very important in anoxic sediments and deep ocean waters.

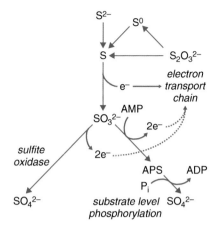

Figure 3.7 Oxidation of reduced sulfur compounds by chemolithotrophs. Sulfite is most commonly oxidized with the enzyme sulfite oxidase (left side of diagram) resulting in the direct generation of ATP via the electron transport chain and a proton-motive force generated across the membrane that leads to ATP synthesis by ATPase. Some sulfur-oxidizing bacteria use the enzyme adenosine phosphosulfate (APS) reductase (right side).

Many bacteria oxidize sulfur compounds

Many types of bacteria use sulfur in elemental (S^0) or reduced (S^{2-}, $S_2O_3^{2-}$) forms as electron donor for chemolithotrophic metabolism, using O_2 as electron acceptor. Most are also obligate autotrophs (fixing CO_2 via production of NADH) or mixotrophs (using organic compounds as a source of carbon). Sulfur-oxidizing bacteria (SOB) include genera such as *Thiobacillus*, *Thiothrix*, and *Thiovulum*—these are strict aerobes found in the top few millimeters of marine sediments that are rich in sulfur. SOB are also very prominent at hydrothermal vents and cold seeps, both as free-living forms and as symbionts of invertebrates (*Chapter 10*), where autotrophic SOB can form the base of food webs in the absence of input from photosynthesis. A major challenge facing SOB is the opposing gradients of reduced sulfur compounds they need for energy and the oxidants (O_2 or NO_3^-) needed as electron acceptors. SOB frequently show chemotactic or aerotactic motility in order to seek out the oxic-anoxic interface where O_2 and reduced sulfur compounds can exist together. Other SOB have other physiological or behavioral adaptations to overcome this challenge and symbionts of invertebrates often depend on their host for a specialized blood supply or movement through sediments to provide the correct balance of HS^- and O_2. Large sulfur bacteria such as the family Beggiatoaceae are very common in microbial mats and surface sediments. In order to bridge the gap between reductant and oxidant, they store large amounts of elemental S and NO_3^- as an alternative electron acceptor. Different strains are very diverse in their metabolism, varying from chemorganotrophy to chemolithoautotrophy. There are numerous variations in the pathways used for sulfur oxidation; an outline of the main biochemical reactions is shown in *Figure 3.7*. The complete oxidation of sulfide to sulfate generates eight electrons, and the intermediate oxidation states provide substrates for numerous other organisms and syntrophic interactions (*Figure 3.8*). In particular, the production of sulfate in sediments plays a key role in the breakdown of organic material in sediments, by providing nutrition for sulfate-reducing bacteria (SRB) as discussed below. *Thioploca* and *Thiomargarita* obtain energy by using sulfide as an electron donor, coupled with NO_3^- as electron acceptor and can grow autotrophically or mixotrophically using organic molecules as carbon source. They grow in nitrate-rich waters above anoxic bottom waters with high levels of organic matter. Each cell contains a very thin layer of cytoplasm around the periphery and a liquid vacuole that constitutes 80% of the cell volume and stores very high concentrations of NO_3^-. Some archaea also oxidize sulfur; these include the hyperthermophilic and acidophilic genus *Sulfolobus* found at hydrothermal vents. Some phototrophic bacteria (members of the Chromatiaceae, Chlorobiaceae, Rhodospirillaceae, and Cyanobacteria) found in microbial mats can grow anaerobically, using reduced sulfur compounds as electron donors for photosynthesis.

Figure 3.8 The sulfur cycle in marine sediments. Abbreviations: OSM = organo-sulfur molecules, ANME = anaerobic methane oxidation, SCI = sulfur cycle intermediates, C_{org} = organic matter, DIET = direct interspecies electron transport, LDET = long distance electron transport. Bacterial and archaeal groups are shown in green; note that the affiliations of the Deltaproteobacteria and Epsilonproteobacteria have recently been amended (see Chapter 4), Redrawn from Wasmund et al. (2017), CC BY 4.0.

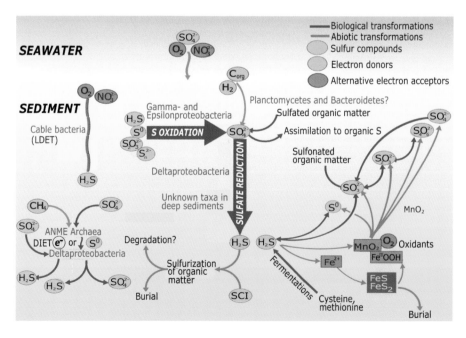

Mechanisms of sulfur oxidation are clearly very diverse, and SOB have evolved many interesting mechanisms for solving the challenge of obtaining their energy source and reducing power. Recently, an exciting new mechanism was discovered whereby some members of the family Desulfobulbaceae form very long, multicellular filaments that transfer electric currents over centimeter-long distances in sediments, remotely coupling sulfide oxidation in anoxic sediment layers with O_2 reduction at the surface. The properties of these "cable bacteria" and their ecological role are discussed in *Box 4.2*.

Many chemolithotrophs use hydrogen as an electron donor

Hydrogen is a common product of the breakdown of organic compounds and many bacteria can use it as the electron donor. In aerobic hydrogen bacteria, electrons from H_2 are transferred to quinone and proceed via a cytochrome chain, ending with the reduction of O_2 to H_2O. The enzyme hydrogenase catalyzes the reaction $H_2 + \frac{1}{2}O_2 \rightarrow H_2O$. The reaction has a high yield of energy, supporting the synthesis of ATP. Cultured examples found in marine habitats include *Alcaligenes*, *Pseudomonas*, and *Ralstonia*, most of which can fix CO_2 to grow autotrophically, although when organic compounds are available the synthesis of hydrogenase and CBB cycle enzymes is usually repressed. Hydrogen-oxidizing bacteria are typically associated with sediments and suspended particles where an oxic-anoxic interface provides the optimum conditions for growth. Some chemosynthetic endosymbionts of invertebrates in vents, seeps, and sediments can use H_2 for chemosynthesis. Some thermophilic chemolithotrophs found at hydrothermal vents use NO_3^- to oxidize H_2, S^0, or $S_2O_3^{2-}$.

Bacterial and archaeal nitrification is a major process in the marine nitrogen cycle

The inorganic compounds ammonia (NH_3), ammonium ion (NH_4^+), and nitrite (NO_2^-) serve as substrates for chemolithotrophy in the process known as nitrification. The process occurs in two steps; there is some variation in the enzymes and intermediates, but the overall reactions are:

1. Ammonium oxidation: $NH_4^+ + 2H_2O \rightarrow NO_2^- + 8H^+ + 6e^-$
2. Nitrite oxidation: $NO_2^- + 2H_2O \rightarrow NO_3^- + 2H^+ + 2e^-$

One group of microbes carry out oxidation of NH_3 or NH_4^+ using the enzyme ammonia monooxygenase (AMO, an integral membrane protein), which produces the intermediate hydroxylamine (NH_2OH); this is then oxidized to nitrite (NO_2^-) by hydroxylamine oxidoreductase in the periplasm. A different group of microbes oxidize NO_2^- to nitrate (NO_3^-) using the enzyme nitrite oxidoreductase (NXR). Nitrifying bacteria are widely distributed and active in suspended particles and in the upper, oxic layers of sediments. The ammonia oxidizers *Nitrosomonas* and *Nitrosococcus* fix CO_2 and are obligate chemolithoautotrophs, while the nitrite oxidizers *Nitrobacter*, *Nitrococcus*, and *Nitrospina* can be mixotrophic, using simple organic compounds as a source of carbon. These reactions yield little energy, and oxidation of 35 ammonia molecules or 15 nitrite molecules is required to produce fixation of one molecule of CO_2 resulting in low growth yields.

Through these activities, nitrifying bacteria play a major role in nitrogen cycling in the oceans by degrading organic matter to create nitrate, which is a major source of nitrogen for primary production by phytoplankton, algae, and plants; it is often the limiting nutrient (see *Chapter 9*). Nitrification is especially important in shallow coastal sediments and beneath upwelling areas such as the Peruvian coast and the Arabian Sea. Like the phototrophs, nitrifying bacteria have extensive internal structures to increase the surface area of the membrane. It is difficult to obtain estimates of the abundance and community structure of nitrifying bacteria. Although most can be cultivated in the laboratory, the energetics of this mode of chemolithotrophy mean that the bacteria grow slowly and are difficult to work with. Immunofluorescence and genomic methods reveal that *Nitrosococcus oceani* and similar strains are widespread in many marine environments, with worldwide distribution at concentrations between 10^3

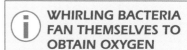

WHIRLING BACTERIA FAN THEMSELVES TO OBTAIN OXYGEN

Bacteria that oxidize sulfides show special adaptations to cope with the varying gradient of O_2 and S at the oxic-anoxic interface. Thar and Kuhil (2002) discovered unusual microaerophilic bacteria growing in highly sulfidic marine sediments that developed in Nivå Bay in Denmark, owing to the burial of large amounts of decaying seaweed in the sediment during winter storms. SRB generate large quantities of S^{2-}, and the whole area can become anoxic and colonized by other bacteria. One type uses an unusual sideways rotary movement to migrate toward O_2 and attaches to solid surfaces via stalks, forming conspicuous veils of mucus-like polymer. Stalks of different bacterial cells stick to each other to build up the veil, and holes in the structure are quickly repaired. The bacteria position themselves at the preferred concentration of 2 µM O_2, either by chemotaxis to O_2 or by increasing their stalk lengths. The attached cells keep rotating to enhance the rate of O_2 uptake through advection of a flow of water. Although the bacterium can be grown in enrichment cultures in the laboratory, it has not been isolated, so classification has been difficult. Muyzer et al. (2005) used molecular analysis of the enrichment cultures to show that the bacterium is a member of the Epsilonproteobacteria, which they named "*Ca.* Thioturbo danicus" ("the sulfur whirl of Denmark").

and 10^4 cells mL^{-1}. This organism is thought to be responsible for significant oxidation of ammonia in the open ocean. *Nitrospira* also seems to be distributed worldwide. Study of the activities of nitrifiers and their contribution to nitrogen cycling is usually carried out using isotopic methods with $^{15}NO_2^-$ or $^{15}NH_4^+$ or by using various inhibitors of nitrification enzymes (e.g. nitrapyrin inhibits AMO). Nitrification is a strictly aerobic process and sufficient O_2 usually only penetrates a few millimeters into sediments. However, the activity of burrowing worms, molluscs, and other animals (bioturbation) increases O_2 availability to deeper levels of sediments.

Until quite recently, ammonia oxidation was believed to be carried out exclusively by the specialized aerobic genera of the Bacteria named above. We now know that highly efficient nitrification is also carried out by the Thaumarchaeota phylum of the Archaea. This is of particular importance because this group is so abundant, especially in the deep ocean. This is explored in *Box 5.1*.

Ammonia can also support anaerobic chemolithoautotrophy

Anaerobic ammonia oxidation (anammox), using nitrite as the oxidant and yielding nitrogen as the end product, is now known to be an important component of the marine nitrogen cycle, with the overall reaction $NH_4^+ + NO_2^- \rightarrow N_2 + 2H_2O$. Anammox is only carried out by members of the Planctomycetes phylum, which possess an unusual cell structure including a membrane-bound nucleus and organelle-like structures (*Figure 4.13*). Those species that carry out anammox possess an organelle known as the anammoxosome, which can occupy up to half of the cell and has a membrane composed of unusual lipids known as ladderanes, densely packed in a rigid ladder-like array that is impervious to many agents. As shown in *Figure 3.9*, the enzyme nitrite reductase reduces NO_2^- to NO, which then reacts with NH_4^+ via hydrazine hydrolase yielding hydrazine (N_2H_4)—a highly reductive compound used as rocket fuel—so this special membrane is needed to prevent it leaking into the cytoplasm. Finally, N_2H_4 is oxidized to N_2 by the enzyme hydrazine dehydrogenase, yielding electrons, some of which are used by the anammoxosome electron transport chain to generate a proton-motive force for the synthesis of ATP. Like aerobic ammonia-oxidizing bacteria, anammox bacteria are autotrophic, using CO_2 as their sole carbon source and NO_2^- as electron donor in the overall reaction:

$$NH_4^+ + 1.32NO_2^- + 0.066HCO_3^- + 0.13H^+$$

$$\rightarrow 1.02N_2 + 0.256NO_3^- + 0.066(CH_2O_{0.5}H_{0.15}) + 2.0H_2O$$

The ecological importance of anammox is discussed in *Chapter 9*.

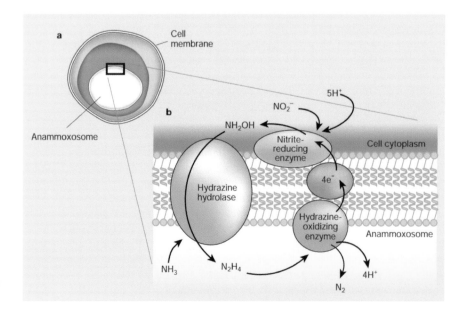

Figure 3.9 Anaerobic ammonia oxidation (anammox). a. A simplified depiction of a planctomycete anammox cell, showing the anammoxosome organelle, which has a dense impermeable membrane composes ladderane lipids. b. The pathway for the anammox reaction. Reprinted from DeLong (2002) with permission of Springer Nature.

CARBON AND NITROGEN FIXATION

The Calvin–Benson–Bassham (CBB) cycle is the main method of carbon fixation in autotrophs

In autotrophs, the generation of ATP by the mechanisms described in the previous sections is used to drive the incorporation of carbon into organic compounds for biosynthesis by oxidation of CO_2. The great majority of photoautotrophs and chemolithoautotrophs do this via the cycle of reactions first discovered in plants by Calvin, Benson, and Bassham in the 1950s. The key enzyme in this pathway is ribulose bisphosphate carboxylase/oxygenase (RuBisCO), which is unique to autotrophs. RuBisCO is a relatively inefficient enzyme, and cyanobacteria and some chemolithoautotrophs have evolved CO_2 concentrating mechanisms (CCMs) that increase CO_2 levels around the active site of the enzyme up to 1000 times. Bicarbonate (HCO_3^-) accumulates in the cytosol via the action of several high- and low-affinity transporters in both the plasma membrane and the thylakoid membrane (in cyanobacteria); synthesis of these is regulated in accordance with external levels of CO_2 and HCO_3^-. Large amounts of RuBisCO are contained in inclusion bodies called carboxysomes, about 100 nm in diameter and surrounded by a thin protein shell or envelope. They contain a crystal-like array of up to 250 RuBisCO molecules and another enzyme, carbonic anhydrase (CA). HCO_3^- diffuses into the carboxysome and CA converts it to CO_2. These mechanisms greatly increase the efficiency of carbon fixation, and if genetic mutations are generated resulting in structural changes or loss of the carboxysomes or CA, cells require much higher concentrations of CO_2 than normal for growth. Diversity in the components and activity of the CCMs contributes to the competitive ability of cyanobacteria to occupy different habitats.

As shown in *Figure 3.10A*, each turn of the CBB cycle incorporates one molecule of CO_2 by carboxylation of ribulose bisphosphate (RUBP). The intermediate molecule is unstable and immediately dissociates to two molecules of 3-phosphoglycerate (PGA). This is then phosphorylated to form glyceraldehyde-3-phosphate, which is also a key intermediate of the glycolytic pathway. Therefore, monomers of hexose sugars are synthesized by a reversal of glycolysis. The cycle requires 12 molecules of NADPH and ATP to reduce 12 PGAs to 12 glyceraldehyde-3-phosphates, and a further six ATPs to convert six ribulose phosphates to six ribulose bisphosphates. Thus, the overall reaction mechanism requires 12 NADPH plus 18 ATP, for the incorporation of six molecules of CO_2 to generate one molecule of C_6 monomer (hexose sugar). Hexose monomers are converted by various reactions to other metabolites and the building blocks of macromolecules. If energy and reducing power are in excess supply, hexoses are converted to storage polymers such as starch, glycogen, or polyhydroxybutyrate, and deposited as cellular inclusions. The final step in the CBB cycle is the regeneration of one molecule of RUBP by another important enzyme, phosphoribulokinase.

Some Archaea and Bacteria use alternative pathways to fix CO_2

Some Archaea and Bacteria in the deep sea, sediments, and vents, as well as some symbionts of marine invertebrates (see *Chapter 10*), can fix CO_2 via at least two other possible routes. The reverse or reductive tricarboxylic acid (TCA) cycle (*Figure 3.10B*) uses ferredoxin-linked enzymes leading to the formation of acetate. It was first described in *Aquifex* and *Chlorobium*, but sequence analysis of microbial communities from hydrothermal vents reveals that genes encoding the key enzyme of this pathway (ATP citrate lyase) are more common than genes for RuBisCO. This suggests that the reverse TCA cycle may be the main mechanism of CO_2 fixation in chemolithoautotrophs in these habitats.

Anammox bacteria use the acetyl-coA pathway, in which acetate is formed by combining the methyl group formed from reduction of one molecule of CO_2 with the carbonyl group from reduction of another CO_2 molecule with nitrite as the electron acceptor:

$$CO_2 + 2NO_3^- + H_2O \rightarrow CH_2O + 2NO_3^-$$

MICROBIOLOGISTS DOUBTED EXISTENCE OF ANAMMOX

Anammox was discovered in the late 1980s by the Delft microbiologist Gijs Kuenen, who was working with a yeast company to help them to reduce pollution from the effluent of their plant. When nitrate was added to an anaerobic digester, the concentration of toxic ammonia fell. Fellow microbiologists were very skeptical that a microorganism was responsible for this process—a reaction predicted on thermodynamic grounds by Broda (1977)— but Kuenen and colleagues persisted and eventually obtained highly enriched cultures of the bacterium responsible, which they named "*Ca.* Brocadia anammoxidans." Ten other *Candidatus* species have since been described in wastewater treatment plants, all members of the Planctomycetes. Biotechnological applications are being developed to enhance their use for treatment of wastewater and industrial effluent at much lower costs and reduced output of greenhouse gases than conventional methods (Kuenen, 2008; Pereira et al., 2017).

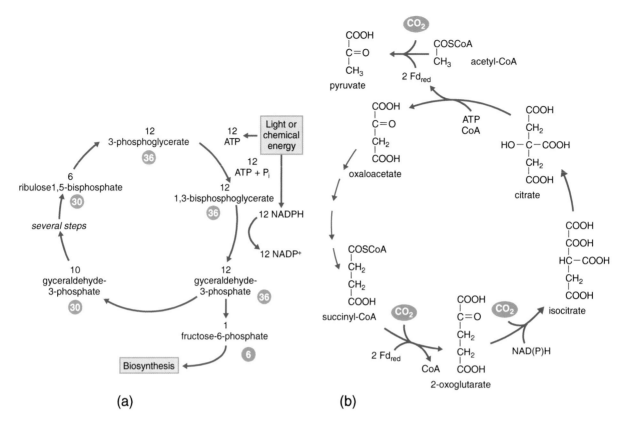

(a)   (b)

Figure 3.10 Major pathways of carbon fixation in autotrophs. (a) The CBB cycle. One molecule of hexose sugar is produced from each six molecules of CO_2 incorporated; figures in circles show the total number of carbon atoms involved. (c) The reverse (reductive) tricarboxylic acid cycle used in some chemolithotrophic autotrophs. Each cycle results in the fixation of one molecule of CO_2 and production of one pyruvate molecule, which is subsequently phosphorylated using ATP, leading to the biosynthesis of cellular material. (Intermediate steps are omitted for clarity).

The acetyl-coA pathway for reduction of CO_2 is also used by methanogens and by acetate-producing chemoorganotrophs growing in anaerobic sediments. In this case, H_2 serves as the electron donor in the reaction:

$$4H_2 + H^+ + 2HCO_3^- \rightarrow CH_3COO^- + 4H_2O.$$

Fixation of nitrogen makes this essential element available for building cellular material in all life

Although 80% of the Earth's atmosphere is dinitrogen gas (N_2), most organisms cannot metabolize it directly. Only about 5–8% of naturally occurring fixation of this N_2 occurs abiotically (as a result of lightning activity), so life on Earth depends almost entirely on diazotrophy, the reduction of N_2 into NH_3, which is then assimilated into organic compounds for biosynthesis of amino acids, nucleotides, and other essential compounds. Diazotrophy is a highly energy-demanding process that occurs only in members of the Bacteria and Archaea. The necessary genes and proteins are highly conserved, but there is great physiological and phylogenetic diversity among nitrogen-fixing species, which include photolithotrophs, chemolithotrophs, and chemoorganotrophs. The reduction of N_2 to NH_3 is extremely energy demanding, because N_2 is a very unreactive molecule due to the stability of the dinitrogen triple bond. At least 16 ATPs are required for the reaction: $N_2 + 8H^+ + 8e^- \rightarrow 2\,NH_3 + H_2$.

As shown in *Figure 3.11*, incorporation of N_2 depends on the nitrogenase complex, consisting of the enzymes dinitrogenase (which is composed of the two copies of each of the NifD and NifK subunits and contains Fe + Mo) and dinitrogenase reductase (which is composed of two copies of the NifH subunit and contains Fe). For reduction of each molecule of N_2, four pairs of electrons are sequentially passed from ferredoxin or flavodoxin to the dinitrogen reductase subunits (NifH). For each electron transferred, two ATP molecules are required to provide the energy for a change in conformation of the NifH subunits. This allow them to interact with the dinitrogenase component and transfer the electrons before dissociating. In the first step, a pair of electrons combine with two protons to form H_2, then three further pairs of electrons are sequentially routed through the reductase enzyme to complete the next steps in

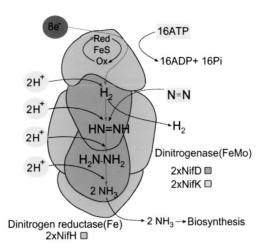

Figure 3.11 Nitrogen fixation. The nitrogenase complex consists of subunits of dinitrogenase reductase (NifH, orange) and dinitrogenase, composed of two copies of each of the NifD (blue) and NifK (green) subunits.

reduction of N_2, which occurs at the Fe-Mo cofactor of the dinitrogenase enzyme. The cycle of electron transfer and ATP-mediated association and disassociation of the nitrogenase complex is repeated at each step.

There are about 20 *nif* genes involved in synthesis and regulation of the nitrogen-fixing system. Although organisms show variation in some of the *nif* genes, the key genes *nifHDK* that encode the nitrogenase complex are highly conserved, enabling the use of probes and sequence analysis to detect nitrogen-fixing microbes in environmental samples, as discussed later. The *nif* genes are tightly regulated by different processes and occur in operons under common control (the *nif* regulon). Transcription of the *nif* structural genes encoding these enzymes is prevented by the presence of O_2 and fixed nitrogen such as NH_3 and NO_3^-. In addition, dinitrogenase is very labile in the presence of O_2. This poses no problem under anaerobic conditions, but aerobic nitrogen fixers such as photosynthetic cyanobacteria must employ special strategies for protecting the enzyme from O_2, discussed in *Chapter 4*. Furthermore, bacteria and archaea have a rapid and reversible system for shutting down nitrogen-fixing activity in the presence of excess NH_3 in the environment by limiting the flow of electrons to the nitrogenase complex. The aerobic cyanobacteria and the anaerobic archaeal methanogens play a particularly significant role in nitrogen cycling in the oceans, as discussed in *Chapter 9*.

HETEROTROPHIC METABOLISM

Many marine microbes obtain energy by the fermentation of organic compounds

All energy-yielding metabolic processes depend on coupled reduction-oxidation (redox) reactions, with conservation of the energy in the molecule ATP using the reaction:

$$ADP + Pi + energy \rightleftharpoons ATP + H_2O$$

Fermentation occurs only under anaerobic conditions and is the process by which substrates are transformed by sequential redox reactions without the involvement of an external electron acceptor. Fermentative microbes can be either strict or facultative anaerobes; the latter can often grow using respiration when O_2 is present. Fermentation yields much less ATP than aerobic respiration for each molecule of substrate, so growth of facultative anaerobes is better in the presence of O_2. A very wide range of fermentation pathways exist in marine bacteria and archaea, especially those associated with degradation of organic material in sediments, animal guts, and microbial mats. Carbohydrates, amino acids, purines, and pyrimidines can all serve as substrates. Some individual species are adapted to use only a limited range of substrates, while others can carry out fermentations with various starting materials.

 WHAT'S IN A NAME? GENES AND PROTEINS

Distinguishing between proteins and the genes that encode them is sometimes confusing, so it is important to become familiar with the conventions for abbreviations. Microbial proteins are typically represented as a three- or four-letter abbreviation, in normal type with the first letter capitalized, while the genes that encode them are written in lower case italic type. For example, Nif includes all members of the group of proteins involved in nitrogen fixation; the individual structural and regulatory proteins are called NifA, NifB, NifC, etc. The *genes* are represented by *nif, nifA, nifB,* etc. An operon contains several genes that are transcribed into a single mRNA under the control of a single regulatory gene, so we often need to describe these collectively. For example, *nifHDK* represents the three genes *nifH, nifD, nifK* in the order that they occur in the operon. Finally, different operons may be under the control of a single regulatory protein; this is called a regulon. In this example, we can say that all the Nif proteins are encoded by the *nif* regulon.

? DO MARINE PLANTS
AND ANIMALS
DEPEND ON
SYMBIOTIC NITROGEN
FIXATION?

In terrestrial biology, the symbiotic
partnership of leguminous plants
such as peas, beans, and alfalfa
with symbiotic bacteria belonging
to the genus *Rhizobium* has been
extensively investigated, because
of its importance in soil fertility.
Does something similar happen in
the marine environment? The high
productivity of seagrass beds and
mangroves is linked to close coupling with nitrogen-fixing bacteria
in a biofilm of bacteria growing on
the roots (the rhizosphere), making fixed nitrogen available to the
plants. In recent years, many examples of nitrogen-fixing symbionts
have also been discovered in corals,
sponges, molluscs, crustacean
larvae, and other invertebrates, as
well as inter-microbial symbioses in
diatoms and other microalgae. The
importance of these associations is
discussed in Chapters 9 and 10.

Anaerobic respiration has major importance in marine processes

In respiration, electrons are transferred via a sequence of redox reactions through an electron transport chain located in the cell membrane, which leads to the transfer of protons to the exterior of the membrane and generates a proton-motive force. The commonest and most familiar type of respiration is aerobic, in which O_2 is the terminal electron acceptor, but many microbes carry out anaerobic respiration using other terminal electron acceptors. Anaerobic respiration uses electron transport chains containing cytochromes, quinones, and iron-sulfur proteins like those in aerobic metabolism. Some organisms can carry out both aerobic and anaerobic respiration, using alternate electropositive electron acceptors such as Mn_4^+, Fe_3^+, NO_3^-, and NO_2^- that yield sufficient energy for anaerobic growth. But because O_2 is the most efficient electron acceptor and produces the most energy for oxidation of an electron donor, such facultative organisms will always use O_2 preferentially if it is available. The respiration of more electronegative electron acceptors like SO_4^{2-}, S_0, and CO_2 is only possible anaerobi-cally. Several types of anaerobic respiration have major importance in marine biogeochemi-cal processes; especially denitrification, sulfate reduction, methanogenesis, and methane oxidation, which are considered below.

Nitrate reduction and denitrification release nitrogen and other gases

Many organisms use nitrate (NO_3^-) as an electron acceptor in anaerobic respiration. The complete process of reduction is known as dissimilative reduction of nitrate. The sequence of transformations is $NO_3^- \rightarrow NO_2^- \rightarrow NO \rightarrow N_2O \rightarrow N_2$ and the overall net redox reaction is:

$$2NO_3^- \rightarrow 10e^- + 12H^+ \rightarrow N_2 + 6H_2O$$

This can be broken down into the following sequential reactions, catalyzed by the enzymes shown below:

(1) $NO_3^- + 2H^+ + 2e^- \rightarrow NO_3^- + H_2O$ (nitrate reductase)
(2) $NO_3^- + 2H^+ + e^- \rightarrow NO + H_2O$ (nitrite reductase)
(3) $2NO + 2H^+ + 2e^- \rightarrow N_2O + H_2O$ (nitric oxide reductase)
(4) $N_2O + 2H^+ + 2e^- \rightarrow N_2 + H_2O$ (nitrous oxide reductase)

Some organisms only carry out the first reaction, but others carry out the subsequent steps and this results in the formation of the gases nitric oxide, nitrous oxide, and nitrogen. Because these gases are released and may return to the atmosphere, resulting in the loss of biologically useful nitrogen, the process is known as denitrification. Furthermore, the intermediate gases have environmental consequences in climate warming and production of acid rain, so these processes are of critical importance in marine nitrogen cycling, especially in anoxic environ-ments and oxygen minimum zones (*Chapter 9*). The synthesis of the enzymes involved in denitrification is repressed by molecular O_2 and induced by high levels of NO_3^-. However, denitrification is an important beneficial process in the treatment of organic-rich wastewater because it lowers the levels of oxygen-demanding microbial growth. Most denitrifiers are phylogenetically and metabolically diverse members of the Proteobacteria, which are facul-tative anaerobes and can carry out aerobic respiration of many different organic compounds.

Sulfate reduction is a major process in marine sediments

Sulfate is abundant in seawater and sulfate-reducing bacteria (SRB) are very widely distrib-uted in anoxic marine sediments, as members of microbial mat communities, and at hydro-thermal vents. Although most marine members of the Bacteria and Archaea can carry out assimilative sulfate reduction for biosynthesis of cellular compounds, those known as SRB carry out dissimilative sulfate reduction. The SRB can link the oxidation of substrates to ATP generation, using SO_4^{2-} as electron acceptor and leading to the production of H_2S. Although SRB are found in several phyla of the Bacteria, the most important types are currently assigned to the class Deltaproteobacteria (see *Chapter 4* for discussion of their taxonomic

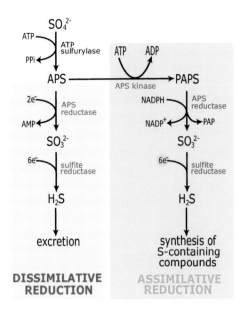

Figure 3.12 Pathways of assimilative and dissimilative reduction of sulfate.

position). Only one member of the Archaea is known to reduce sulfate: the hydrothermal vent genus *Archaeoglobus*. A wide range of organic compounds generated by many different members of the microbial community can be used by SRB as electron donors (including H_2, organic acids, and alcohols produced by fermentation). Many SRB oxidize acetate in order to provide electrons for SO^{2-} reduction using the acetyl-coA pathway. The reversible reactions of this pathway can also be employed for acetate synthesis, enabling the bacteria to grow autotrophically by incorporation of CO_2. A few SRB can also grow autotrophically using H_2.

As shown in *Figure 3.12*, the first step in sulfate reduction is activation of sulfate by ATP to adenosine-5′-phosphosulfate (APS). In assimilative sulfate reduction, another P is added to APS from ATP to form phosphoadenosine-5′phosphosulfate, which is then reduced to SO_3^{2-} and then to HS^- for the synthesis of the amino acids cysteine and methionine, coenzyme A, and other cell factors that contain sulfur. By contrast, in dissimilative reduction, APS is reduced directly to SO_3^{2-} and then H_2S. The dissimilative metabolism of SRB results in the production of large amounts of H_2S which is excreted from the cell, where it can serve as a substrate for SOB and react with metals to form metal sulfides that characterize marine sediments. Sulfate reduction is usually considered a strictly anaerobic process, but some types are tolerant of O_2. Some SRB can also reduce other oxidized sulfur compounds including sulfite (SO_3^{2-}), dithionite ($S_2O_4^{2-}$), thiosulfate ($S_2O_3^{2-}$), trithionate ($S_3O_6^{2-}$), tetrathionate ($S_2O_6^{2-}$), and elemental sulfur (S_0). This enables SRB to utilize intermediates produced by SOB as well as those produced from their own metabolism.

Because they are ubiquitous in anoxic sediments, SRB play a major role in the sulfur and carbon cycles through the oxidation of organic matter reaching the sea floor. An important aspect is the interaction of these activities with the production and oxidation of methane, discussed below. Their activities also cause problems of corrosion and toxicity in the oil industry and they influence the mobilization of toxic mercury, discussed in *Chapter 13*. SRB are also implicated in black band disease of corals, discussed in *Chapter 11*.

MICROBIAL PRODUCTION AND OXIDATION OF METHANE

Methanogenesis is unique to the Archaea

Many members of the Euryarchaeota produce methane as the final step in the anaerobic biodegradation of organic material. Bacterial fermentation produces H_2, CO_2, and acetate as end products, which are then used by methanogens. Most methanogens use H_2, with CO_2 serving as both the oxidant for energy generation and for incorporation into cellular material. The overall reaction is: $CO_2 + 4H_2 \rightarrow CH_4 + 2H_2O$. This reaction depends on the input of eight electrons, added two at a time, which leads to four steps with intermediate oxidation states. This depends on coenzymes, unique to this group of Archaea, that transfer electrons from hydrogen (or other donors, if used) and act as carriers for the C_1 unit from the substrate (CO_2) to the end product (CH_4). Besides $H_2 + CO_2$, other compounds methanogens form methane from CH_3-containing compounds such as methanol, acetate, and dimethyl sulfide. In all the possible pathways, the final step of reduction leads to the formation of a single molecule of ATP via generation of a proton-motive force. Sugars and fatty acids are not directly used for methane generation, but because methanogens exist in syntrophic communities with fermentative bacteria, virtually any organic compound can eventually be converted to methane. In the presence of high levels of SO_4^{2-}, methanogens are in competition with SRB because both groups use H_2 or acetate. Therefore, methanogens are usually found in deeper sediments where sulfate levels are lower.

Methane produced by archaeal activity in sediments has a number of fates. Much of the methane produced in anoxic sediments diffuses into adjacent oxic regions, where it sustains the growth of methanotrophic bacteria—thus recycling the gaseous end product of degradation of organic matter—and a large fraction is oxidized anaerobically, as discussed below. Methane is also trapped in a crystal lattice of water molecules, forming solid methane hydrate (clathrate). This only forms under appropriate conditions of geology, temperature, and pressure—the hydrate stability zone—forming widespread deposits on the continental shelf, typically in sediments between 250 and 550 m deep but sometimes mixed with other hydrocarbons nearer the surface. The production of methane in sediments and deep-sea seeps is the basis of many symbioses with invertebrates, as discussed in *Chapter 10*.

Besides anoxic sediments, methane is found in the oxygenated seawater of oceans. Some of this is due to the activity of methanogens as members of the intestinal microbiota of copepods, fish, and other marine animals or their activity in anoxic niches inside marine snow particles. Many protists, especially ciliates, also contain methanogens as endosymbionts.

Methane is produced in the surface ocean by bacterial cleavage of phosphonates

Oceanographers have long puzzled over the fact that large amounts of methane flux to the atmosphere occurs in well-oxygenated ocean gyres, whereas methanogenesis is a strictly anaerobic process. However, it has recently been found that many bacteria are capable of producing methane in the oxygenated conditions of the surface ocean and that this accounts for about 4% of the methane reaching the atmosphere. This occurs via the action of C-P lyase enzymes that, under phosphate-limited conditions, cleave methylphosphonate (MPn) compounds. These are simple, single-carbon compounds with a stable phosphonate (C-P) bond. Gene sequences for both the biosynthesis of MPn and for the C-P lyase enzyme are present in marine metagenomes. The reaction creates ethylene (which is consumed by heterotrophs) and methane, most of which escapes to the atmosphere; so the main benefit to the degrading organism appears to be the release of phosphate needed for biosynthesis of nucleotides, lipids, and other key cellular components. MPn-polysaccharides appear to be mostly synthesized by cyanobacteria such as *Prochlorococcus*, and experimental studies have shown DOM contains large amounts of these compounds that are degraded aerobically by bacteria in seawater in sufficient quantity to explain the daily flux of methane. Some cultured strains of SAR11, the most abundant clade of oceanic microbes, have been shown to cleave MPn when phosphate limited, although others do not. Intriguingly, some SAR11 have been shown to possess MPN synthase, suggesting that members of this clade may be either a source or sink of methane in the ocean.

Anaerobic oxidation of methane (AOM) in sediments is coupled to sulfate reduction

As described above, methane produced by the anaerobic process in deep sediments diffuses upwards to be consumed by aerobic methanotrophs in the oxic zone. However, for many years, geochemists had been puzzled by observations that as much as 80% of this disappeared before it reached zones where contact with oxygen could occur. Sulfate is the only possible electron acceptor that could account for anaerobic methane oxidation in these anoxic sediments. Stable isotope methods combined with 16S rRNA gene sequencing and FISH (*Figure 3.13B*) led to the discovery of a syntrophic consortium of *Archaea* and SRB carrying out AOM via the reaction:

$$CH_4 + SO_4^{2-} \rightarrow HCO_3^- + HS^- + H_2O$$

A model of these interactions is shown in *Figure 3.13A*. The anaerobic methanotrophic Archaea (ANME) are closely related to the methanogens and it is thought that the reaction functions as a reversal of methanogenesis. The syntrophy with SRB may be obligate, because attempts to culture them have been unsuccessful. The archaeal partners have been shown to fix nitrogen and transfer it to the SRB (*Figure 3.13C*). Nitrogen fixation is very expensive in terms of energy consumption and the growth of these consortia is very slow—the doubling time is 3–6 months—but it seems to be advantageous for the archaea to transfer fixed nitrogen to the SRB, which they need to utilize methane. Different microbial consortia have subsequently been discovered to carry out ANME via coupling to reduction of NO_3^-, NO_2^- Fe_3^+ and Mn_4^+ and it is likely that organisms capable of using other electron acceptors remain to be discovered.

The sulfide and bicarbonate products of the reaction are very important. These consortia contribute to the construction of reefs of precipitated carbonate at methane seeps, especially in the Black Sea. The sulfide can be used in chemosynthetic symbioses involving tubeworms, clams, or giant SOB, which form thick carpets on the seabed above the methanotrophic consortia. Elsewhere, extensive microbial growth above methane seeps may sustain deep-sea coral communities. This type of methanotrophic metabolism may have evolved very early on Earth, possibly predating the development of O_2 in the atmosphere from oxygenic photosynthesis (about 2.2 BYA).

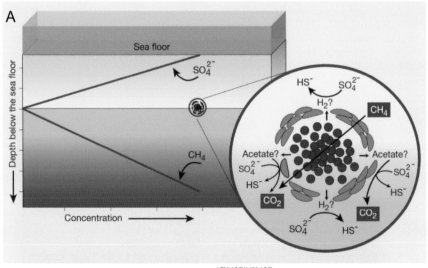

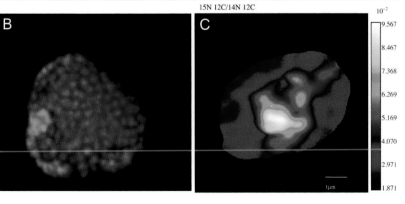

Figure 3.13 A. Model showing anaerobic oxidation of methane (ANME) in marine sediments, based on the findings of Boetius et al. (2000). The metabolic consortium contains methane-oxidizing archaea surrounded by sulfate-reducing bacteria (SRB). Other reaction pathways using alternative electron acceptors are now known. B. Confocal micrograph visualized by FISH with probes directed against the archaeal (red) and bacterial (green) partners. C. NanoSIMS image showing ^{15}N incorporation, indicating transfer of fixed nitrogen from the archaeal cells to SRB. Credits: A. Reprinted from DeLong (2000) with permission of Springer Nature; B, C. Anne Dekas and Victoria Orphan, California Institute of Technology.

Many marine microbes oxidize methane and other C_1 compounds

Chemoorganotrophs catabolize a wide range of organic compounds to obtain energy and a carbon source for biosynthesis. The simplest of these is methane (CH_4), and many organisms known as methylotrophs can utilize this or various other methyl- or one-carbon (C_1) compounds. A very wide range of bacteria in different phylogenetic groups can carry out this process and it is of great important in carbon and sulfur cycling in ocean processes. In particular, the compound dimethylsulfoniopropionate (DMSP) is produced in great amounts by phytoplankton and supplies much of the carbon and almost all of the sulfur for the growth of heterotrophic bacteria in ocean food webs, including the dominant SAR11 and *Roseobacter* clades (see *Chapter 9*). Many methylotrophs can also use more complex organic compounds, although some bacteria in the Alphaproteobacteria and Gammaproteobacteria are obligate methylotrophs and can use only C_1 compounds in their metabolism.

A subset known as the methanotrophs can grow using methane. These possess a unique enzyme system, methane monooxygenase (MMO), which occurs in membrane-bound and cytoplasmic forms whose synthesis depends on the availability of copper. MMO obtains electrons from NADH and incorporates O_2 leading to the formation of methanol, with subsequent conversion to formaldehyde by methanol dehydrogenase in the reaction:

$$CH_4 \rightarrow CH_3COOH \rightarrow CH_2O \rightarrow HCOO^- \rightarrow CO_2$$

The coenzymes tetrahydrofolate or methanopterin enable conversion of the CH_2O intermediate, producing electrons that enter the electron transport chain to produce a proton-motive force for ATP synthesis.

Different phylogenetic groups of methanotrophs vary in the route of assimilation of C_1 units into cell material. Some use the unique ribulose monophosphate pathway, which requires one molecule of ATP and no reducing power to assimilate three molecules of CH_2O into one molecule of glyceraldehyde-3-phosphate. Other types use the serine pathway for carbon assimilation into acetyl CoA, which is less efficient due to the requirement for two molecules of both NADH and ATP:

$$CH_2O + CO_2 + 2NADH + 2ATP + CoA \rightarrow CH_3\text{-}CO\text{~}S\text{-}CoA$$

Apart from the importance of free-living forms in methane oxidation in sediments and ocean processes, methanotrophs also occur as symbionts of mussels found near cold seeps of methane-rich material on the ocean floor, providing the animals with a direct source of nutrition (see *Chapter 10*). Methanotrophs are also important in bioremediation of low-molecular-weight halogenated compounds, an important process in contaminated marine sediments

NUTRIENT ACQUISITION AND MICROBIAL GROWTH

Microbial metabolism depends on nutrient uptake

The remainder of this chapter is mainly concerned with the growth of heterotrophic bacteria and archaea and the environmental factors that affect them. A huge diversity of biochemical pathways enables heterotrophs to obtain their carbon and energy from the breakdown of preformed organic compounds, including carbohydrates, organic acids, amino acids, and peptides by respiration and fermentation. Besides C and N—the major components of cellular material, which heterotrophs acquire from preformed organic compounds—microbes need a supply of the macronutrients P, S, K, Fe, Mg, Ca, and Na These are needed as key constituents of macromolecules like proteins, nucleotides, and lipids or in smaller quantities as cofactors for enzyme function or stabilization of structures such as nucleic acids, membranes, and ribosomes. The importance of these elements in ocean processes is discussed in *Chapters 8* and *9*. In addition, microbes often depend on the presence of very small amounts of metals such as Mn, Mg, Co, Zn, and Cu as enzyme cofactors. There is also growing recognition of

MOST MARINE BACTERIA REQUIRE SODIUM

A particular feature of the majority of marine bacteria is their absolute requirement for Na^+ (usually in the range 0.5–5.0%) and they fail to grow when K^+ is substituted in the culture medium. Sodium is required by transport proteins (permeases), which bring substrates into the cell via Na^+-substrate symport across the membrane and also help to stabilize the outer membrane found in Gram-negative bacteria. This distinguishes them from closely related terrestrial and freshwater species. As might be expected, most grow optimally at a concentration of NaCl similar to that found in seawater (about 3.0–3.5% NaCl), although they differ greatly in the lower and upper limits for growth. Gram-positive bacteria (which are much less common in seawater) do not seem to have this requirement.

the importance for marine microbial growth of growth factors such as vitamins (especially the B vitamins cobalamin, thiamin, and biotin), or specific amino acids, purines, or pyrimidines. Individual microorganisms are often auxotrophic—lacking the biosynthetic pathways needed for specific compounds—and they must obtain these compounds from other organisms within the community. This applies particularly to members of the picoplankton, which have often undergone genome reduction resulting in loss of genes for certain pathways and therefore depend on exogenous sources. Despite the perception that microalgae like diatoms and dinoflagellates just need light, CO_2, and a few inorganic nutrients, many are auxotrophic and do not synthesize particular essential vitamins. For example, half of surveyed species of algae require vitamin B12, cobalamin. The availability of vitamins depends on overall microbial community composition or tight associations between algae and bacteria such as *Roseobacter*, *Marinobacter*, *Sulfitobacter*, and their relatives. This has a major effect on the regional and seasonal distribution and activity of phytoplankton.

Apart from a few very small hydrophobic molecules that can diffuse into the cell, all nutrients are transported into the cell via transport proteins in the cytoplasmic membrane. All transport proteins depend on an energy-dependent conformational change when a substance binds to it, forming a channel that carries the substance across the membrane. Transporters may be specific for particular compounds or sometimes transport groups of related compounds. Some molecules like sugars and amino acids can cross the membrane using simple channels using energy via a proton pump and others are transported by a phosphotransferase system. The presence of the outer membrane (OM) in Gram-negative bacteria (the dominant form in bacterioplankton) requires them to use additional OM receptors and periplasmic binding proteins to bridge this additional permeability barrier. ATP-binding cassette (ABC) transporters consist of a periplasmic binding protein (PBP) that is highly efficient at binding low concentrations of substrates that are then transferred to a membrane-spanning transporter that is linked to a cytoplasmic ATP hydrolase, supplying energy. Many hundreds of high-affinity, multisubstrate ABC transporters and even greater numbers of the highly-expressed PBPs have been identified in metagenomic and metaproteomic analyses of low-nutrient ocean water. For example, the genome of SAR11 bacteria encodes individual transporters that import various organic compounds (including amino acids, DMSP, taurine, polyamines, simple sugars, and phosphonates) and inorganic compounds (e.g. iron and complexed phosphate). Proteins of another system, tripartite ATP-independent transporters (TRAP), which uses an electrochemical gradient to drive nutrient uptake, are also common. Another system of TonB-dependent transporters (TBDTs) in the OM is also widespread and particularly important for the transport of iron complexes, vitamins, and aromatic compounds. The presence of particular transporters can indicate which nutrients are limiting.

As noted in *Chapter 1*, apart from phagotrophic protists, microbes can only feed on relatively small molecules (usually <600 mDa) which are transported into the cell. The origin of these compounds in the marine environment as a component of dissolved organic matter (DOM) is discussed in *Chapter 8*. When cells die or are lysed by viruses, or when fecal pellets from zooplankton are released, large amounts of organic material are made available to heterotrophic microbes. But most of this will be in form of complex macromolecules like nucleic acids and cytoplasmic and membrane proteins and lipids, which cannot be assimilated by bacterial or archaeal cells. Therefore, many produce extracellular enzymes (exoenzymes), including chitinases, amylases, nucleases, and proteases for degradation of biopolymers into monomers. Chitinase is especially significant in marine bacteria, since chitin (a polymer of N-acetyl glucosamine, NAG) is such an abundant compound in the sea, due to its presence as a major structural component of the exoskeleton of many invertebrates (especially crustaceans), and the cell walls of fungi. Many marine bacteria possess surface-associated chitinases and the released NAG is an excellent source of both carbon and nitrogen. As in all cases in which bacteria degrade macromolecules in their environment, mechanisms must exist for transporting the monomers produced into the cell. In a natural bacterial assemblage, degradation of complex macromolecules by one type of microbe makes nutrients available to many others. Such processes are particularly important in particles of marine snow and may lead to a plume of DOM as the particle falls through the water column, because colonizing bacteria produce DOM more quickly than they can use it (*Figure 1.11*). Bacteria inhabiting this plume grow at much faster rates than those in bulk seawater.

SELFISH BACTERIA GOBBLE LARGER OLIGOSACCHARIDES

The notion that ocean bacteria always use exoenzymes to hydrolyze high molecular weight DOM to molecules <600 Da has been challenged in a discovery by Greta Reintjes and colleagues from MPI Marine Microbiology, Bremen. Reintjes et al. (2017) incubated natural ocean microbial communities with the fluorescently-labeled polysaccharides (FL-PS) laminarin and xylan (produced by algae and present in large amounts in seawater) and chondroitin sulfate (present in animal cartilage and known to be degraded by marine bacteria). Using fluorescence microscopy, she observed that 26% of cells took up FL-PS. To follow this up, Reintjes et al. combined FL-PS staining with single-cell identification by FISH and super-resolution structured illumination microscopy. FL-PS molecules were seen to bind rapidly to cells of the phyla Bacteroidetes and Planctomycetes and the gammaproteobacterial genus *Catenovulum*. FL-PS appear to be taken into the periplasm by a TonB-dependent OM transporter, where they are degraded into mono-, di-, and trisaccharides for transport into the cytoplasm. This "selfish" uptake prevents the polysaccharide breakdown products from being used by other bacteria.

There are two strategies for the deployment of exoenzymes. Some remain closely associated with the cell membrane, which maximizes the opportunity for the cell to benefit from the breakdown of the molecule. Others are excreted from the cell and diffuse freely into the surrounding water. Production of extracellular hydrolytic enzymes by a cell requires an "expensive" commitment because of the energetic costs of synthesizing and exporting the enzyme and of importing the degradation products. The cell-associated and free exoenzymes have different cost–benefit ratios—the cell-associated enzymes "pay off" at nanomolar concentrations of substrate, while the free enzymes are only profitable at micromolar concentrations. However, bacteria producing free exoenzymes in dense populations can benefit from the enzymes released by neighboring cells. In some species, exoenzyme production is coordinated via quorum sensing, discussed below.

Acquisition of iron is a major challenge for marine microbes

Iron is an essential constituent of cytochromes and iron-sulfur proteins, which have crucial roles in electron transport processes, especially photosynthesis and nitrogen fixation. Although iron is one of the most common elements in the Earth's crust, the amount of free iron in the open ocean can be virtually undetectable (picograms per liter) without ultraclean assay techniques. In oxygenated seawater, almost all the iron occurs in the ferric Fe(III) state bound tightly to organic ligands, so it is a major challenge for marine microbes to obtain enough iron for growth. Many bacteria, both autotrophs and heterotrophs, secrete chelating agents known as siderophores that bind to iron, preventing its oxidation and allowing active transport into the cell via specific receptors on the cell surface. In culture, these compounds are produced under conditions of iron restriction and some structures have been characterized. Over 500 chemical structures have been described; these include phenolics and derivatives of amino acids or hydroxamic acid. The first siderophores to be characterized from oceanic bacteria are structurally quite different from those previously described and behave as self-assembling amphiphilic molecules. Aquachelins from *Halomonas aquamarina* and marinobactins from *Marinobacter* both have a water-insoluble fatty-acid region and a water-soluble region of amino acids. In laboratory studies in the absence of iron, these compounds form clusters of molecules attached together by their fatty-acid tails. Upon binding to Fe(III), the micelles come together to form vesicles. It is not yet known whether this phenomenon of aggregation occurs in the natural environment, but it might be important for bacteria associated with particles containing local high concentrations of organic matter. The mechanism by which the cell takes up these siderophores is also unknown. As discussed in *Chapter 9*, iron cycling has a major role in the primary productivity of oceans and artificial iron fertilization of the oceans has been considered a potential method to mitigate the effects of elevated levels of atmospheric CO_2.

Iron deprivation is also a problem for pathogens of vertebrate animals, which produce iron-binding proteins as a defense mechanism. For example, *Vibrio vulnificus* and *V. anguillarum* produce siderophores, known as vulnibactin and anguibactin respectively, and these are important factors in virulence (*Figure 11.13*).

Siderophores are increasingly recognized as having additional functions besides iron transport, including the ability to transport a variety of other metals, boron, to act as signaling molecules and antibiotics, and to regulate oxidative stress. Some also sequester toxic heavy metals to prevent uptake by cells.

Marine bacterioplankton use two trophic strategies

Marine planktonic bacteria show two patterns of growth and multiplication, depending on their acquisition of organic matter. These lifestyle capabilities can be readily assessed from metagenome sequences, by identifying clusters of gene groups that characterize each type.

Some bacteria, known as copiotrophs, can grow rapidly when exposed to high substrate levels. This trophic type is associated with surfaces and particles or nutrient "hotspots" surrounding phytoplankton or marine snow particles (*Figure 1.11*). They show a "feast or famine" lifestyle, in which rapid growth rates and increased cell size can be induced in response to increased levels of nutrients. They often produce free exoenzymes and highly diverse nutrient transporters and have relatively large cells (>1 μm^3) that often show motility and chemotaxis in response to nutrient gradients (*Box 3.2*). Estimating the growth rate of bacteria in natural seawater is difficult because of the complicating effects of viral lysis and protistan grazing, but there is no doubt that growth rates are nothing like the rapid doubling times observed with those bacteria that can be easily grown in laboratory culture. Perhaps, in nature, copiotrophs have doubling times of a few hours under optimal conditions. Conversely, "famine" results in reduced synthesis of macromolecules, size reduction during cell division, and other adaptations to starvation, discussed below.

In contrast, most free-living planktonic bacteria can be described as oligotrophs, with less adaptability and regulatory flexibility, a smaller number of specific transporters and less reliance on exoenzymes for nutrient acquisition. Their small size—often associated with streamlined genomes—and slow growth rates (doubling once per day or longer) are an adaptation to the chronic low levels of nutrients.

Growth rate and turnover of organic material depend on nutrient concentrations

Bacterial growth efficiency (BGE) can be defined as the ratio of biomass or ATP produced to the amount of substrate utilized. The yield is always lower than predicted by theoretical calculations based on the energetics of biochemical pathways because organisms require a certain amount of nutrient for "housekeeping" functions such as transport, maintenance of internal pH and a proton gradient across the membrane, DNA repair, and motility (if present). These functions take priority over biosynthesis for growth or reproduction. At low growth rates due to nutrient limitation, the cell will use proportionately more of the available substrate for maintenance energy. Below a certain threshold, cells will divert all the substrate to just "staying alive" rather than growing. Uncoupling of catabolic and anabolic processes often occurs at low growth rates when growth is limited by a substrate other than the energy source. Maintaining the maximum possible energy state of the cell enables it to resume growth rapidly when conditions change. These effects of nutrient levels and BGE are important concepts in assessing the role of microbes in carbon flux and productivity in marine systems. Because nutrient concentrations in the sea are generally low, growth rates of oligotrophs are necessarily slow and a large proportion of energy intake is used for maintenance energy, especially active transport and synthesis of extracellular enzymes for acquiring nutrients. Therefore, net BGE will be low, despite considerable turnover of organic material. This is particularly true when N and P are in limiting concentrations. However, as noted in *Chapter 1*, there is an increasing recognition that the ocean environment is very heterogeneous with respect to nutrients and that many marine microbes are associated with local concentrations of organic matter.

Copiotrophic marine bacteria may show rapid growth in culture

Although the vast majority of marine bacteria have not yet been cultured, many copiotrophic bacteria from ocean water, as well as those associated with surfaces and sediments, can be isolated and grown in the laboratory. During growth and reproduction, a bacterial cell must synthesize additional cell membranes, cell wall components, hundreds of different RNA molecules, thousands of proteins, and a complete copy of its genome. The coordination of these activities is subject to complex regulation of gene expression. With a few exceptions (such as the budding and stalked bacteria), almost all bacteria reproduce by binary fission into two equal-sized cells. When growing under optimal or near-optimal conditions, the cell doubles in mass before division and regulation of DNA replication and cell septation are largely controlled by the rate of increase in cell mass. The growth cycle of culturable heterotrophs in the laboratory is well known. Upon introduction to the growth medium, unbalanced growth (without cell division) occurs during a lag phase, during which the cell synthesizes metabolites, enzymes, coenzymes, ribosomes, and transport proteins needed for growth under the prevailing conditions. Under ideal conditions, bacteria then enter a logarithmic phase and grow exponentially. Different species vary enormously in the time taken for the population

WHY IS IRON SO SCARCE, YET SO IMPORTANT?

Most parts of the open ocean contain very little iron to support marine microbial life, yet it is an essential element required for many key biological processes, including photosynthesis and nitrogen fixation. Why does life depend on such a scarce nutrient? The iron-containing proteins that are critical to these reactions are highly conserved across diverse phylogenetic lineages, suggesting that they evolved very early in the history of Earth. The early evolution of life occurred under anoxic conditions, but the development of oxygenic photosynthesis about 2.4 BYA resulted in a sudden transition of the atmosphere and oceans to an oxidizing state. Iron precipitated in large quantities and the amount of soluble iron in seawater today is lower than the concentration needed to sustain microbial life. Aquatic microbes, including the cyanobacterial ancestors that caused the change in oxygen content of the atmosphere, faced an iron-shortage crisis and this necessitated the development of efficient mechanisms for scavenging the increasingly low levels in the environment. Other transitional metals are occasionally found to replace iron, but they are equally scarce in the oceans, and iron is uniquely suited to its function.

to double (generation time). Environmental factors such as temperature, pH, and O_2 availability (for aerobes) have a major influence on the rate of growth. Growth rate increases with temperature. Many marine bacteria isolated from water at 10°C can be grown in the laboratory with doubling times of about 7–9 h, while those isolated at 25°C typically double every 0.7–1.5 h. Of course, laboratory culture will select for fast-growing copiotrophic bacteria. All organisms have a minimum, optimum, and maximum temperature for growth, and this often—but not always—reflects their evolution in a particular habitat. Above the minimum temperature, metabolic reactions increase in rate as temperature increases, and so the doubling time reduces. The maximum temperature is usually only a few degrees above the optimum because of the thermostability of enzymes and membrane systems. Some marine bacteria have exceptionally low generation times. For example, *Vibrio parahaemolyticus* can divide every 10–12 min at 37°C (although this temperature is not encountered in its normal estuarine habitat) and the hyperthermophile *Pyrococcus furiosus* doubles every 37 min at 100°C. At the other extreme, oligotrophic psychrophiles growing just above 0°C may have generation times of several days or weeks.

Bacteria adapt to starvation by coordinated changes to cell metabolism

In culture, bacteria continue exponential growth until they enter the stationary phase because a limiting nutrient is exhausted, toxic by-products of growth accumulate to become self-limiting, or some density-dependent signal limits growth. Cells entering the stationary phase do not simply "shut down" but undergo genetically programmed changes that involve the synthesis of new starvation-specific proteins induced via the RpoS global regulatory system. When cells of laboratory-cultured copiotrophic marine bacteria are starved, they become smaller and spherical. In the 1980s, Morita coined the term "ultramicrobacteria" to describe the very small cells (0.3 µm diameter or less) that develop in this way. To achieve reduction in cell size, some proteins and ribosomes are degraded in a controlled fashion, but 30–50 new proteins are induced in response to starvation. These include proteins that are essential for survival with limited sources of energy and nutrients as well as proteins that enable cells to acquire scarce substrates at very low concentrations. Cell surfaces become more hydrophobic, which increases the adhesion of bacteria to surfaces, and surface association in biofilms appears to be a major adaptation to nutrient limitation. Changes in the fatty-acid content of membranes promote better nutrient uptake and provide increased stability, while cell walls become more resistant to autolysis. Starved cells have a much-reduced ATP content, but maintain a proton-motive force across the membrane. Starved cells are more resistant to a variety of stress conditions, and mutations in any of the genes in the survival induction pathway may lead to death in the stationary phase. Most oligotrophic pelagic bacteria are also very small and have similar morphology to the ultramicrobacteria formed by starvation of cultured marine isolates in the laboratory. For some time, the prevailing paradigm was that marine bacteria are "normal" cells that have undergone reduction in size and are in a state of perpetual near-starvation. This led to the erroneous view that bacterioplankton are virtually inactive in the natural environment. In fact, we now know that the naturally occurring small cells found in the sea are *more* active than larger bacteria on a per volume basis and a large fraction of supposedly inactive cells can be induced to high metabolic activity by addition of substrates. However, determination of *in situ* growth rates of marine heterotrophic microbes remains one of the most difficult aspects. Recent techniques have relied on the use of different fluorochromes in flow cytometry (*Table 2.1*) to determine cell biomass. Some protocols attempt to distinguish dead and live bacteria by employing fluorochromes directed against nucleic acids, while others use fluorochromes that detect respiratory activity. One method combines nucleic acid staining with a permeable dye and an impermeable dye both attached to gene probes to measure membrane integrity and DNA content. Cell viability can also be assessed by measuring the uptake of radiolabeled compounds (such as amino acids) by microautoradiography (p.58) or the use of fluorescent electron acceptors to monitor the presence of an active electron transport chain. Contrary to earlier ideas, these methods reveal that planktonic microbes in the open sea are metabolically active and their small size is an evolutionary (genotypic) adaptation to ensure efficient use of scarce nutrients in their oligotrophic environment.

Some bacteria enter a "viable but nonculturable" state in the environment

The terms "viable but nonculturable" (VBNC) and "somnicell" ("sleeping" cells) were origi- nally introduced in the 1980s by Rita Colwell and coworkers, who observed that *Vibrio chol- erae* does not die off when introduced into the environment. Since then, the VBNC state has been studied extensively and shown to occur in a wide range of Gram-negative (and a few species of Gram-positive) bacteria, including pathogens of humans, fish, and invertebrates. The phenomenon is thus especially important in studies of the ecology of autochthonous pathogenic marine bacteria and the survival characteristics of introduced pathogens (e.g. from wastewater) and indicator organisms. This term has been used by some microbiologists to explain the discrepancy between the number of bacterial cells that can be visualized by direct observation (e.g. epifluorescence or immunological detection) and those that can be enumerated using viable plate counts. The VBNC state could constitute a dangerous reser- voir of pathogens, which cannot easily be detected, and it is also relevant to evaluation of the survival of introduced genetically modified organisms into the environment. A variety of fac- tors, especially nutrient deprivation and changes in pH, temperature, pressure, or salinity can initiate the cascade of cellular events leading to the VBNC state. The VBNC state remains a matter of considerable controversy among microbiologists. While many microbiologists believe these events to be an inducible response to promote survival under adverse condi- tions, many remain skeptical about the validity of the VBNC state as an adaptive stage. One of the most controversial aspects of the VBNC state is whether cells can be resuscitated to a form that grows on agar plates in the laboratory. Many studies have found that addition of nutrients or temperature changes can cause VBNC bacteria to revert, provided that the cells have not been in the VBNC state for too long. Others refute these findings and state that the effect is due to the regrowth of a few remaining culturable cells in the population. An alter- native explanation of the VBNC state view is that when exponentially growing bacteria face a sudden insult, such as the deprivation of nutrients, it results in the decoupling of growth from metabolism. Consequently, cells may suffer a burst of oxidative metabolism, resulting in lethal free radicals. Bacteria can avoid this if they can induce changes to protect DNA, proteins, and cell membranes. The shock of sudden transfer of cells into a rich medium when they are still in the process of adaptation to life in the aquatic environment could result in rapid death. The inability of cells in the VBNC state to detoxify peroxides and other free radicals commonly present in culture media or induced within the cells themselves during exposure to rich media seems to be a key explanation for the phenomenon of non-culturabil- ity. This may be due to repression of the gene-encoding periplasmic catalase, which breaks down toxic peroxide.

It is very important to realize the distinction between the VBNC state from the "unculturabil- ity" of most marine bacteria. As discussed in *Box 2.1*, these are better described as "not yet cultured," since, although many cannot be cultured using conventional methods, considerable progress is being made using dilution methods and other new approaches. This is completely different from the VBNC state, which is a transition shown by bacteria that grow readily in laboratory culture to a starvation or other shock.

Many bacteria use motility to search for nutrients and optimal conditions

The ability to swim toward high concentrations of favorable nutrients (chemotaxis) is an important property of many marine bacteria, which swim due to the presence of one or more flagella. These are long helical filaments attached to the cell by means of a hook-like structure and basal body embedded in the membrane. The flagellar filament in members of the Bacteria is about 20 nm in diameter and is made of many subunits of a single protein (flagellin) synthesized in the cytoplasm and passed up from the basal body through the cen- tral channel of the filament, to be added at the tip. Much less is known about flagella in the Archaea; although the basic architecture is similar to that in the Bacteria, the flagellum is made of several different proteins and is much thinner (about 13 nm in diameter); possibly too thin for the subunits to pass up the central channel. Genomic analysis shows little homology between bacterial and archaeal flagellum proteins.

Figure 3.14 Schematic drawing of the architecture of the basal body and hook region of bacterial flagella based on electron microscopy and identification the constituent proteins. The left side shows the complex of the *Salmonella* H+-driven motor and the right side is that of the *Vibrio* Na+-driven motor. The flagellin subunits are secreted from the cell and assemble at the hook to form a helical filament (not shown) that rotates to propel the cell. IM inner membrane; OM outer membrane; PG peptidoglycan layer. Credit: Reprinted from Minamino and Imada (2015) with permission from Elsevier.

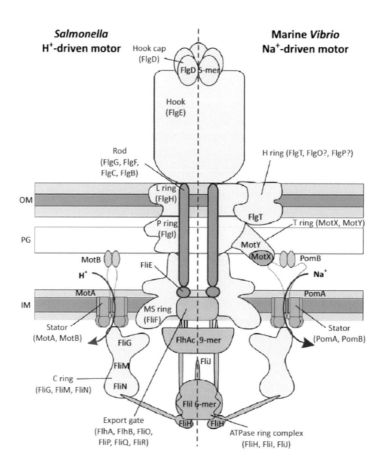

The basal body of the bacterial flagellum consists of a central rod and a series of rings embedded in the cell wall and membrane, composed of nearly 30 different proteins, as shown in *Figure 3.14*. Movement is caused by the rotation of the basal body acting as a tiny rotary motor, which couples the flow of ions to generation of force. The flagellum is rigid and behaves like a propeller. For many years, the remarkable mechanics of the motor and the process of self-assembly of the fine flagella tubules have attracted the attention of engineers for possible exploitation in nanotechnology. The mechanism has been extensively studied in thr human-associated bacteria *Escherichia coli* and *Salmonella*. Rotation is caused by a proton-motive force as H+ ions flow from the outside to the inside of the cell through the Mot protein subunits surrounding the C and MS rings in the membrane. In motion reminiscent of a turbine, the protons exert electrostatic forces on charges on the C and MS rings as they move through a channel in the Mot proteins, causing a rotatory movement. About 1000 protons are needed for each revolution. Genomic analysis shows that the model of flagellar assembly and function derived from studies of *E. coli* and *Salmonella* applies remarkably well to many bacteria in diverse phyla. Many genes are highly conserved—especially those connected with the MS ring, C ring, rod, and flagellin export process—although there are small but important differences between different systems. Most significantly for our discussion here is the finding that most motile marine bacteria that have been investigated seem able to switch between the use of either H+ or Na+ ions to create the ion motive force. This is an advantage in alkaline (~pH 8) seawater. When the flagellar motor of a marine bacterium like *Vibrio* is being driven by Na+, it can rotate at up to 1700 hz for short periods, about six times as fast as the H+-driven motor of *Salmonella*. In *Vibrio* spp. the Mot proteins that act as the ion channel for H+ are replaced by Pom proteins that act as a Na+ channel and additional H and T rings, as seen in *Figure 3.14*. Analysis of the polar flagellar motor shows that the Na+-driven motor has additional components that are responsible for detecting the speed of rotation or the flux of Na+ ions, discussed in *Box 3.2*.

The swimming behavior of many marine bacteria is very different to the "classical" system studied extensively in *E. coli* and *Salmonella*, which usually have multiple (4–10) flagella. When *E. coli* cells are observed in a neutral environment under the microscope, they move in an apparently random fashion. Counterclockwise (CCW) rotation of all motors causes the flagella to form a bundle that propels the bacterium on a relatively straight path ("run") at 10–30 μm s^{-1} before executing a brief "tumble" and swimming off in another direction. The tumble is caused by a brief change to clockwise (CW) rotation of the flagella, disrupting the bundle leading to random reorientation of the cell. The presence of attractants or repellents modifies the frequency of tumbling. Cells tumble less frequently and therefore spend longer swimming up a gradient of increasing concentration of an attractant and show a net movement toward the source of the substance. The bacterium senses minute changes in concentration with time, through a series of chemoreceptors on the cell surface and a chemical signaling system involving changes in the methylation state of chemotaxis (Che) proteins that relays information to a tumble generator at the base of the flagellum. This complex process involves many different proteins encoded by numerous genes. This method of movement is termed a "biased random walk" and biophysical considerations show that it is an energetically efficient mechanism for moving in high nutrient concentrations, but it depends on relatively low uniform speed in straight runs, with a random turn angle.

It is estimated that 40–70% of marine bacteria may be motile, although the observation of motility shows considerable diurnal and seasonal variations and depends strongly on the water composition, and the smallest cells (<0.4 μm) have the lowest motile fraction. Obviously, chemotaxis is only relevant for copiotrophic bacteria that have the genomic potential to respond to localized concentrations of organic carbon, although not all copiotrophs are motile. Most high-speed bacterial community members use both H$^+$ and Na$^+$ ion motors simultaneously. Marine bacteria almost always possess a single polar flagellum for swimming (although cells may become peritrichously flagellated in contact with surfaces, as discussed below).

Although there are far fewer studies of chemotaxis in marine bacteria, we now know that they behave quite differently, showing higher chemotactic speed (chemokinesis) and precision than *E. coli*. Very high speeds enable small marine bacteria to swim in a relatively straight line before a 180° reversal of direction or a "flick" mechanism of ~90° (*Figure 3.15*). Although energetically more expensive than the random walk, it is necessary if the bacteria are able to detect and respond to the small point sources of nutrients that occur in the sea due to turbulence. A few marine bacteria have been observed swimming in short bursts at up to 450 μm s^{-1}—equivalent to several hundred body lengths per second. The significance of these responses of marine bacteria to gradients in the sea is discussed in *Box 3.2*.

Flagella also have a mechanosensory function

When *Vibrio parahaemolyticus* is transferred from suspension to solid surfaces, the bacterial cells undergo a remarkable change in morphology. They stop dividing normally and elongate to about 30 μm. At the same time, they start synthesizing large numbers of lateral or peritrichous flagella as well as the normal polar flagellum (*Figure 3.16*). The Na$^+$-driven polar flagellum is effective when the cell is in the free-living, planktonic state, while the H$^+$-driven lateral flagella enable cells to move better through viscous environments or over surfaces. This can be observed in the laboratory by the phenomenon of swarming on the surface of agar plates. In the natural environment, differentiation to the swarmer form enables vibrios to adopt a sessile lifestyle, which is significant in the formation of biofilms on surfaces or the colonization of host tissues. This "swim-or-stick" switch is determined by the action of the polar flagellum acting as a sensor. Increased viscosity as the bacterium nears a surface slows down flagellar rotation, leading to induction of the lateral phenotype. Experimentally induced interference of the Na$^+$ ion flux with chemical agents also induces differentiation to the swarmer state. Genetic analysis of the Na+ motor is leading to an understanding of this mechanism. However, little is known at present about how the signal from the flagellum induces changes in the expression of the large number of genes required for altered cell division and synthesis of different flagella. It is likely that there is a master regulatory switch. In vibrios, the polar flagellum is enclosed in a sheath formed from the outer membrane. It is not known whether the flagellum rotates within the sheath or whether the filament and sheath rotate as a single unit. The lateral (H$^+$) flagella do not appear to possess a sheath. Motility and biofilm formation are also critical factors in infection by the cholera pathogen, *V. cholerae* (p.334) and in the life cycle of *Caulobacter crescentus* (p.123).

BOX 3.2   RESEARCH FOCUS

How can microscale bacterial swimming affect global processes?

High-speed swimming—"run and reverse" motility. The idea that seawater is a non-uniform environment with many small ephemeral hotspots of nutrients within a dynamic matrix of organic polymers was introduced in *Chapter 1*. Bacteria move over tiny, micrometer distances, whereas energy flow into the ocean through convection, wind, and planetary rotation creates turbulence on a scale many orders of magnitude greater. However, the effects of turbulence translate down to powerful micrometer-scale shear forces. Therefore, if motility and chemotaxis are to be effective, marine bacteria need to swim very fast and show rapid responses to altered conditions. Major advances in our understanding of marine bacterial motility and chemotaxis emerged from the work of Jim Mitchell and colleagues at Flinders University, who used video microscopy to record the behavior of natural assemblages of marine bacteria in enriched seawater samples. They observed tight microswarms of bacteria with cells swimming at a mean speed of 230 µm s^{-1} and exhibiting rapid accelerations (Mitchell et al., 1995). Further studies observed the clustering of bacteria around air bubbles and staying in these clusters by an almost instantaneous reversal of direction at the end of each run (Mitchell et al., 1996). The authors observed different speeds, which they attributed to the different flagellar motors driven by H$^+$ or Na$^+$. By observing the rotation of flagella of *Vibrio alginolyticus* that had stuck to glass slides, Muramoto et al. (1995) had previously shown that the sodium motor rotates rapidly at a stable speed. Mitchell and Barbara (1999) tested the effects of compounds that inhibit either H$^+$ or Na$^+$ ion gradients on an enriched seawater community, as well as isolated cultures of *E. coli* and the marine bacteria *Shewanella putrefaciens* and *Pseusoalteromonas haloplanktis*. Results indicated that marine bacteria use both kinds of motor simultaneously, with Na$^+$ motors accounting for ~60% of the speed. Other experiments showed that clusters of bacteria occur around particles of organic matter, or even track the release of nutrients as motile algal cells move through the water, with up to 12 consecutive correct turn turns enabling bacteria to remain in the wake of a swimming algal cell (Barbara and Mitchell, 2003a; Barbara and Mitchell, 2003b). This run and reverse mechanism was shown to be present in 70% of phylogenetically diverse motile marine bacteria examined (Johansen et al., 2002). In a generally low-nutrient environment such as the open ocean, it is energetically impossible for bacteria to seek out nutrients by biased random movement. Therefore, at this stage of the research, marine bacterial chemotaxis was best explained by bacteria relying on turbulent flow bringing them within a patch of increased nutrient concentration, where they would use the run and reverse movement strategy to maximize the opportunity that they remain associated with it.

Studying motility with microfluidics. Further major advances in the study of motility and chemotaxis have been made using microfluidic devices and the fusion of observations of microbial behavior with biophysical principles, pioneered by Roman Stocker and colleagues (MIT and ETH Zürich). Microfluidics devices enable study of very small volumes of fluid (µL to pL) in networks of micro channels mounted on a microscope slide. Video microscopy coupled with computational systems allows tracking and analysis of the movement of individual bacteria in real time (Rusconi et al., 2014). Using this approach, Stocker et al. (2008) measured the response of *Pseudoalteromonas haloplanktis* to nutrient patches mimicking those produced by a point source such as a lysed algal cell or plumes generated by a falling marine snow particle. The bacteria concentrated in a nutrient "hot spot" within a few seconds and the fastest 20% of the population gained a tenfold higher nutrient concentration than nonmotile cells. In response to a nutrient plume, motile bacteria gained a fourfold increase in nutrient exposure.

Flagella as a propeller and rudder—"run-reverse-flick" motility. This backwards and forwards movement alone could mean that the bacterium retraces its steps endlessly, which would not lead to the high-performance chemotaxis observed. This suggested that another aspect to the movement occurs. Xie et al. (2011) tracked the trajectories of over 800 individual cells of *V. alginolyticus* using fast video imaging and fluorescence microscopy. The strain used has a left-handed flagellar filament, meaning that forward direction results from CCW rotation of the motor, while reversal of the motor pulls the cell backward. They observed that the bacterium alternates between forward and backward runs and reorients by reversing (180° turn) or making a "flick" with an average angle of ~90° to the previous run. Close inspection of the video images shows that the transition from run to reverse occurs within ~1/30 s, but transitions in the opposite direction take longer, ~ 1/10 s. When cells were stained with a fluorescent dye, it could be seen that during the switch from forwards to backward swimming, a small kink forms between the cell body and the flagellum and this is amplified by the CCW rotation, pushing the cell body at an angle. After ~0.1 s, the flagellum realigns with the body axis of the cell and forward swimming resumes in a new direction. Xie et al. showed that the frequency of flicking correlates with swimming speed and flicking diminished when Na$^+$ ions were reduced, or the viscosity of the medium was increased. By introducing micropipette tips filled with different concentrations of a chemoattractant (the amino acid serine) at the edge of the microslide, Xie and colleagues demonstrated that *V. alginolyticus* uses the same three-step motility pattern to efficiently randomize their swimming directions, moving towards a point source of nutrient and remaining localized near it. The flick allows swift randomization of swimming direction, whereas the backtracking allows the bacterium to explore environments that are structured rather than spatially homogenous. If necessary, reversal of the motor can quickly bring the cell back to the high concentration of a nutrient. Stocker (2011) considers that the hybrid locomotion employing flicks could be commonplace among marine bacteria, and that earlier descriptions of run and reverse motility (e.g. in *P. haloplanktis*) may have overlooked flicks. To investigate the mechanical origin of the seemingly impossible off-axis flick, Son et al. (2013) used ultra-high-speed video microscopy to show that the flick occurs 10 ms after the cell reverses direction from a backward run and begins a forward run, revealed by the alignment between the direction of swimming and the cell head orientation (*Figure 3.15*). The flick takes ~60 ms. During backward swimming, the cell is under tension, because the thrust from the flagellum and the drag on the head are equal and directed away from each other, whereas it is under compression during forward swimming. This led Son et al. to reason that the reason that flicks only occur during forward swimming is due to a material instability of the hook region of the flagellum and they explain the mechanical basis by which this buckling occurs during a flick. They then showed experimentally that the probability of flicking during a cycle varies with swimming speed, as predicted.

BOX 3.2   RESEARCH FOCUS

The hook stiffens by twisting during swimming, preventing buckling until the motor reverses again at the end of a run. Further experiments showed that *P. haloplanktis* and a mixed seawater community show the same mechanism of buckling, as does the coral pathogen

Vibrio coralliilyticus which is chemotactic towards DMSP produced by coral (see *Box 11.1*).

Marine bacterial chemotaxis is both fast and precise. In follow-up experiments using microfluidics to track responses of thousands of individual *V. alginolyticus* cells to serine gradients, Son et al. (2016) investigated how swimming speed affects chemotaxis. By varying the Na⁺ concentration, they showed that the bacteria display chemokinesis—speeding up as they move up the concentration gradient, getting closer to the source. Faster cells also show greater chemotactic precision—the strength of accumulation of cells at the peak of the gradient. In other words, the bacteria reach a nutrient source more quickly and, once there, remain more closely associated with the hotspot. Higher speeds alone promote the frequency of flicks for the reason discussed in the previous section. The chemokinetic speed enhancement of a population occurs rapidly and is therefore an advantageous adaptation to exploit transient patches of resources in the sea.

The ecosystem scale effects of microscale processes. The experimental studies described above show how we have obtained considerable understanding of the microscale processes by which patchiness of nutrients leads to adaptations in bacterial chemotaxis, which affects nutrient acquisition and utilization. How microbial life in this "sea of gradients" is translated into a quantitative framework for foraging behavior of microbes and its consequences in the ecosystem is the subject of a major review by Stocker (2012). Adaptations such as rapid swimming and hybrid chemotactic responses are clearly successful adaptations whose benefits outweigh the major energetic costs of the mechanisms. They enable motile marine bacteria to exploit ephemeral nutrient patches such as sinking particles or DOM released from phytoplankton blooms. Mathematical models of the effect of chemotaxis in "chasing" a plume of marine snow indicate that growth rates of bacteria swimming into plumes increase 10-fold (Kiørboe and Jackson, 2001) and these estimates are consistent with microfluidics experiments by Stocker et al. (2008). Thus, motile copiotrophs might channel more DOM than oligotrophs and the remineralization of organic matter within sinking marine snow particles will stimulate primary production. Motile bacteria that cluster near phytoplankton enhance their productivity by feedback of inorganic nutrients, as well as accelerating the remineralization of phytoplankton DOM. Atomic force microscopy reveals that a large proportion of algae or cyanobacteria are intimately connected with heterotrophic bacteria (*Figure 2.3*), increasing direct nutrient flux between autotrophs and heterotrophs. Stocker (2012) concludes his review by discussing new tools for exploring microbial behavior at the single-cell level and the integration of these ideas into ecological frameworks. In an attempt to provide a unified view of the diverse microbial adaptations and their effects on the resource landscape, these ideas have recently been developed into a "foraging mandala" (Fernandez et al., 2019). This is a graphical representation that maps microbial behavior onto a space in which a variety of physical and biological factors are distilled into two fundamental parameters of gathering resources—the time taken to find new resources and the yield obtained from one resource patch. The biophysical processes included in the model encompass autotrophic and heterotrophic bacteria and algae as well as the impacts of virus predation and mortality. Fernandez et al. conclude that their model provides "a blueprint for quantitative understanding of microorganism–resource interactions at the level of single cells, from which the vast global flow of carbon and other elements starts."

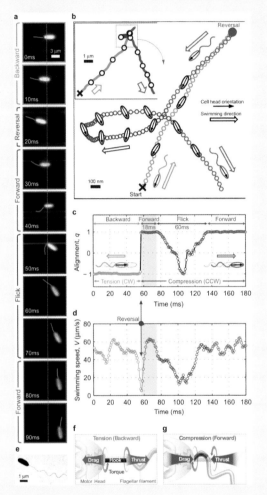

Figure 3.15 High-speed video microscopy of *V. alginolyticus*. a: Image sequence captured with high-intensity dark-field microscopy at 420 frames s⁻¹, showing the kinematics of the cell head and polar flagellum just before and during a flick. b: Schematic of the head orientation at selected times is overlaid on the cell trajectory captured at 1,000 frames s⁻¹. Cell head positions are shown by circles at 1 ms intervals. The inset shows the entire trajectory subsampled at 30 frames s⁻¹ (open circles). c: Alignment, q (directional cosine) between cell head and swimming direction, for the trajectory in b, reveals the elements of the swimming cycle, particularly the short forward swimming segment (here, 18 ms long; red) before the 60-ms-long flick (blue). d: Swimming speed during a flick. e: TEM showing the single polar flagellum (mean head length 3.2 μm, mean flagellar contour length 4.6 μm. f, g: Schematics (not to scale) of the flagellar filament, hook and rotary motor during backward swimming (f), when the hook is in tension, and during forward swimming (g), when the hook is in compression. Reprinted from Son et al. (2013) with permission of Springer Nature.

Figure 3.16 Transmission electron micrographs of *Vibrio parahaemolyticus*. (A) Grown in liquid, showing the swimmer cells with single, sheathed polar flagellum. (B) Grown on a surface showing the elongated swarmer cell with numerous lateral flagella. Bars represent ~1 μm. Image courtesy of Linda McCarter, University of Iowa; see McCarter (1999).

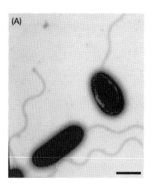
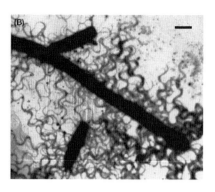

Microbes also respond to light, magnetic fields, and other stimuli

Movement toward light (phototaxis) is common in a diverse range of phototrophic bacteria and the genes controlling the response have probably been acquired in diverse phyla by horizontal gene transfer. Responses often depend on the ability to integrate information about changes in the intensity of light of different wavelengths and the physiological status of the cell. Negative phototaxis is also important, because many microbes need to move away from high-intensity light (especially in the UV range) that causes damage to DNA and proteins. Depending on the organism, sudden changes in motility in response to light intensity can result from biased run and tumble or run and reverse mechanisms. Scotophobotaxis can be observed under the microscope when a bacterium moves out of a light source and reverses flagellar rotation to reenter the light. Positive and negative photokinesis, i.e. greater swimming speed causing cells to accelerate towards or away from sources of illumination, is also common in cyanobacteria and purple photosynthetic bacteria. True phototaxis involves movement towards or away from a light source through direct sensing of the direction of illumination, rather than just detection of a spatial gradient of intensity. This is widespread in eukaryotic algae, but less so in bacteria. This mechanism occurs via photosensory proteins containing chromophores that change conformation when they absorb photons. Various types of photosensory proteins have been described. The archaeon *Halobacterium salinarum* possesses rhodopsin molecules, which act as light sensors influencing rotation of the flagella. Transduction of signals for the control of motility occurs via various mechanisms such as light-powered proton gradients or electron transport chains; in many cases, these are initiated by two-component signal transduction systems and systems homologous to the bacterial chemotaxis cascade Che proteins. It is unclear how small unicellular organisms can detect the direction of light. Pigments that shade part of the cell may be involved, or the cell may act as a lens that focuses light on the edge of the cell furthest from the light source.

Other tactic movements include those towards or away from high concentrations of oxygen (aerotaxis) and ionic substances (osmotaxis). Magnetotaxis is a specialized response seen in certain species of bacteria found in marine and freshwater muds which contain intracellular inclusion bodies called magnetosomes (*Figure 4.4*). These bacteria orient themselves in the Earth's magnetic field, although their movements are primarily in response to gradients of chemicals and oxygen.

Gliding and twitching motility occur on surfaces

Gliding motility does not involve flagella and only occurs when bacteria are in contact with an aqueous film on a solid. It is especially important in the responses of cyanobacteria and members of the *Cytophaga* and *Flavobacterium* genera to gradients of oxygen and nutrients in microbial mats. It is likely that there are several different mechanisms employed for gliding motility and these are not yet fully understood. The most common mechanism used by gliding cyanobacteria appears to be a kind of jet propulsion through the exclusion of slime via minute pores in the cell surface. However, in *Flavobacterium* it is thought that rotatory motors, similar to flagellar motors, transmit force to special outer membrane proteins, which results in a ratchet-like movement of the cell envelope, analogous to the movement of a tank on caterpillar tracks.

Many bacteria also show twitching motility, seen as jerky and irregular movements under the microscope. It occurs via the action of short filaments called type IV pili that occur in tufts or all over the cell surface. These are made of multiple subunits of pilin protein arranged helically. The pili bind to the surface and the pilin subunits rearrange, causing the filament to be under tension. This leads to retraction and pulls the cell forwards in a motion similar to the use of a grappling hook. Twitching motility is especially important in the formation and development of biofilms.

Microbes colonize surfaces via formation of biofilms

The importance of surface colonization by the formation of biofilms was introduced in *Chapter 1*. There are significant differences in the physiological properties of free planktonic cells and those that are in biofilms or associated with particles. Although biofilm formation has been recognized for more than a century, it is only in recent years that significant advances in the study of biofilm physiology have been achieved, largely because of the application of confocal microscopy and FISH techniques. Bacteria and diatoms both initiate biofilm formation as a result of environmental cues, particularly nutrient availability. Bacteria may undergo considerable developmental changes during transition from the suspended, planktonic form to the sessile, attached form. The events occurring in biofilm formation by single species of bacteria have been studied extensively in laboratory investigations, and the process can be regarded as a form of cellular development toward a multicellular lifestyle—some researchers liken it to a tissue. Organic molecules (largely polysaccharides and proteins) will coat any surface with a conditioning film within minutes of placing it into seawater. The initial stage in colonization by bacteria usually involves motility toward the surface, and changes in microviscosity as the cell approaches the surface may cause flagellar rotation and motility to slow, as discussed above. This causes a transient, reversible adsorption to the surface, mediated by electrostatic attraction and van der Waals forces. Genetic studies involving the creation of biofilm-deficient mutants have shown the importance of motility due to peritrichous flagella, gliding, and type IV pili in the movement of bacteria across the surface in order to contact other cells to form microcolonies.

During the process of biofilm formation, expression of specific genes is induced—this depends very much on the nature of the substrate. In experimental studies, bacteria will quickly colonize inert surfaces such as pieces of plastic, glass, or stainless steel immersed in seawater, but the genes expressed are quite different from those expressed when they colonize organic substrates such as chitin. Some marine heterotrophic bacteria express chitin binding and chitinase genes selectively when they encounter chitin, so that they can begin to utilize it as a nutrient source. This is especially important in the colonization of the exoskeletons of crustaceans, as discussed in *Chapters 11* and *12*. The most important developmental change following attachment is the expression of copious quantities of exopolymeric substances (EPS). These provide a strong and sticky framework that cements the cells together. The chemical and physical properties of EPS are very variable, and the nature and amount produced depends on the bacterial species, concentration of specific substrates, and environmental conditions. The majority are polyanionic because of the presence of acids (e.g. D-glucuronic, D-galacturonic, or D-mannuronic acids), ketal-linked pyruvate, or the presence of phosphate or sulfate residues. The EPS often form a complex network of long, interlinked strands surrounding the cell (glycocalyx). Both rigid and flexible properties can be conferred on the biofilm, depending on the secondary structure of the EPS, and interactions with other molecules such as proteins and lipids may produce a gel-like structure. The mature biofilm often has a complex architecture composed of pillars and channels. Stalked bacteria and diatoms may increase the length of their stalk when growing in dense biofilms. Once established, the dense packing of bacteria and the diversion of metabolism to the production of EPS may mean that the cells are metabolically active but divide at very slow rates.

In nature, biofilms are rarely composed of single species. Bacteria, algae, flagellates, ciliates, other protists, and viruses will all interact in the mature biofilm, and there is growing evidence that gene transfer between different microbial species is greatly enhanced within biofilms. This is of great significance in the evolution of organisms with altered characteristics.

BACTERIA STICK TOGETHER TO WARD OFF PREDATORS

One of the advantages of the biofilm mode of life for bacteria seems to be protection against predation (grazing) by phagotrophic protists. Matz et al. (2008) compared the grazing efficiency of two flagellate protists common in coastal waters—the surface-feeding *Rhynchomonas nasuta* and the suspension feeding *Cafeteria roenbergensis*—against a large number of bacteria isolated from the surface of a seaweed. *C. roenbergensis* increased in density in response to planktonic cells of the bacteria and predated them with high efficiency. By contrast, when grown as biofilms, most of the bacteria were resistant or toxic to the surface-feeding flagellate *R. nasuta*. One of the most effective anti-grazing compounds produced by biofilm cells was identified as the alkaloid violacein, which inhibits protozoan feeding at nanomolar concentrations by inducing a conserved eukaryotic cell death program. The authors concluded that biofilm-specific resistance against predation contributes to the successful persistence of biofilm bacteria in various environments. Unraveling the interactions between bacterial biofilms and eukaryotic cells may reveal important clues to the evolution of compounds that affect cell function.

Pili are important for bacterial attachment to surfaces and genetic exchange

Pili (also known as fimbriae) are fine hair-like protein filaments on the surface of many bacteria. Pili are typically about 3–5 nm in diameter and 1 μm long. They are composed of a single protein, although a cell may have different types and the amino acid sequence of each type can show significant strain-to-strain variation. The most common pili are those involved in the attachment of bacteria to surfaces (some microbiologists restrict the use of the term fimbriae to these adhesive structures). Their function has been particularly well studied in the interaction of pathogenic bacteria with mucous membranes of animals. There is often a specific receptor recognition site on the pili, which can explain why certain strains of bacteria attach specifically to particular hosts or tissues. Host immune responses to bacterial attachment are often important in resisting infection. Adhesive pili are often encoded by phages or plasmids and may be exchanged by transduction or conjugation, respectively. The human pathogen *V. cholerae* provides a good example: here, pili play a crucial role in infection, and the acquisition and expression of genes are important factors in explaining the transition of this bacterium from its estuarine or marine habitat to the human gut. Other examples discussed later in the book include the attachment of *Roseobacter* to invertebrate larvae and the attachment of swimming cells of *Caulobacter* to surfaces, where pili develop into a form of holdfast. More generally, pili play a key role in the attachment of bacteria to inanimate surfaces as the first stage in biofilm formation. As well as the more common adhesive pili, some cells possess the type IV pili responsible for a twitching motility observed on surfaces, discussed above.

Sex pili are quite different structures, often several micrometers long, formed only by donor ("male") bacteria involved in the process of conjugation. Sex pili and the machinery needed for replication and transfer of DNA from donor to recipient cells are encoded by conjugative plasmids. Sex pili are important in the transfer of genes, such as those encoding antibiotic resistance or degradative ability, between cells of bacteria in the marine environment. In addition, sex pili are the receptors for certain phages, and an application of this is the use of F^+ RNA phages as an indicator of fecal pollution in seawater.

Antagonistic interactions between microbes occur on particles or surfaces

As noted above, the activity of extracellular enzymes leads to the dissolution of organic material from marine snow particles as they sink through the water column, and much of this DOM becomes available to other members of the plankton. Extracellular enzyme activity is likely to provide little "return" for free-living bacteria at low density. Composition of the bacterial community inhabiting the particle and the plume of DOM that follows it could therefore have significant effects on the release and subsequent utilization of nutrients. We know that there are extensive antagonistic interactions between microbes in dense communities inhabiting the soil or the gut of animals, owing to the production of antibiotics that inhibit growth of organisms unrelated to the producer strain, but there have been few such investigations of antagonistic interactions between marine bacteria. However, studies indicate that a large proportion of marine-particle-associated bacterial isolates do possess antibiotic activity against other pelagic bacteria. Such antibiotic interactions are also likely to be widespread in sediments, microbial mats, and biofilms on the surface of algae, plants, and animals

Quorum sensing is an intercellular communication system for regulation of gene expression

Quorum sensing (QS) is a mechanism of intercellular communication through the controlled production, release, and detection of threshold concentrations of signal molecules. QS is widely used by a variety of bacteria to coordinate communal behavior. Usually, this involves the coordinated expression of certain genes in response to population density. Study of this phenomenon is leading to new insights into the physiology and ecology of attached marine microbial communities. The signal molecules used in QS are chemically diverse, but the commonest and best studied are N-acyl homoserine lactones (AHLs), which are widespread in diverse bacteria.

WHAT'S IN A NAME? QUORUM SENSING

The definition of quorum is the minimum number of members of a decision-making body (such as a committee or council) required to conduct the business of that group and make decision. Thus, the use of the term quorum sensing (QS) to describe density-dependent gene regulation implies a cooperative behavior, which is challenged by ecologists based on evolutionary theory. Redfield (2002) commented that "The appeal of the idea that bacteria act cooperatively has caused the postulated benefits of quorum sensing to be accepted uncritically as the explanation for the role of autoinducers in gene regulation." Alternative terms have been proposed to avoid this controversy: these include "positional" (Alberghini et al., 2009), "diffusion" (Redfield, 2002), and "efficiency" (Hense et al., 2007) sensing. The term "autoinduction" (used in the original description of LuxI/LuxR regulation) is the most neutral because it does not imply a general ecological function. However, Platt and Fuqua (2010) argue that these terms generate a "semantic quagmire" and that the ecological context of QS regulation is complex and affected by multiple aspects of natural environments.

AHLs have different chain lengths, and presence or absence of oxo- and hydroxyl groups. In a mixed bacterial population, there may be many different structures present.

QS was first studied in detail in the bacterium *Vibrio fischeri* (since renamed *Aliivibrio fischeri*), in which it is used to control bioluminescence. When grown in laboratory broth cultures, these bacteria emit no light until the bacteria enter the late logarithmic or stationary phase and the population reaches a certain critical density (typically about 10^7 cells mL^{-1}). This is because the bacteria synthesize a freely diffusible autoinducer molecule and release it into the medium. Low-density cultures can be induced to show bioluminescence by the addition of supernatants from high-density cultures. The autoinducer is an AHL synthesized from S-adenosylmethionine by the protein LuxI. As illustrated in *Figure 3.17*, AHL is produced in the cytoplasm and passively diffuses through the bacterial membrane. Accumulated AHL diffuses back into the cell, binding to the protein LuxR, a polypeptide of about 250 amino acids comprising two domains; the N-terminal domain binds the AHL and the C-terminal domain binds to a palindromic sequence (*lux* box) upstream of the *lux* operon promoter. Thus, when a certain threshold concentration is reached, the bioluminescence genes encoding the various proteins required for the bioluminescence system are expressed by activation of transcription of the operon (*Figure 3.17C*).

This AHL-mediated process QS known as the LuxI/LuxR system is now known to be very widespread and used by many Gram-negative bacteria to regulate the activity of many genes, including colonization and virulence factors. This description of the LuxI/LuxR QS mechanism is somewhat simplified, as bacteria differ considerably in the details of the regulatory circuits and other factors are often involved. The *luxI* and *luxR* genes are often located adjacent to one another on the chromosome. In the *A. fischeri* bioluminescence system, the transcription of *luxI* is controlled by the LuxR-AHL complex, which results in autoinduction of further AHL synthesis and amplification of the response. AHL-QS occurs only in Gram-negative bacteria and homologs of *luxI* are widespread in proteobacteria. Structural variations in the AHL sidechains, which bind to the cognate LuxR homolog, leads to high specificity of signaling in different bacterial taxa.

Elucidation of the regulation of bioluminescence in another marine vibrio, *V. harveyi*, has revealed three separate systems. Like *A. fischeri*, *V. harveyi* synthesizes and responds to the AHL molecule, in this case termed AI-1 synthesized by LuxLM, and a separate boron-containing furanosyl diester autoinducer, AI-2 synthesized by LuxS. In this case, the

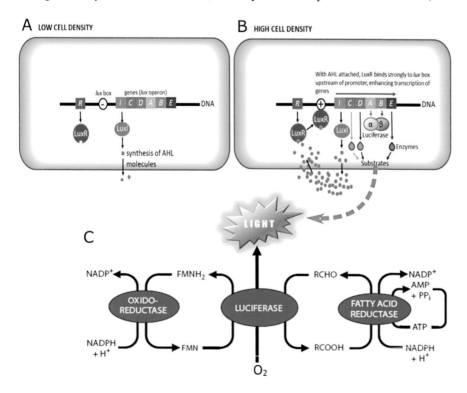

Figure 3.17 A. Quorum sensing control of bioluminescence in *Aliivibrio fischeri*. The process is controlled by two regulatory genes *luxI* and *luxR* in two different operons. The *luxI* gene encodes the enzyme responsible for synthesis of the AHL 3-oxo-C6-homoserine lactone. A. Representation of a cell at low in a low-density population, in which there is a low constitutive transcription of *luxI* and *luxR*. LuxI catalyzes synthesis of AHLs (green dots), which diffuse from the cell. B. As cell density increases, AHLs produced by other *A. fischeri* accumulate in the area surrounding the cells and diffuse into the cell at high concentration. LuxR is a repressor of the *lux* operon promoter (−). When LuxR binds the AHL autoinducer I, it binds strongly to the *lux* box upstream of the promoter and transcription of the right operon is enhanced (+). Production of AHL increases exponentially, giving an autocatalytic feedback loop as well as initiation of bioluminescence. Genes *luxA* and *luxB* encode the α and β subunits of the enzyme luciferase. Genes *luxC*, *luxD*, and *luxE* encode enzymes for the synthesis of the aldehyde substrates from fatty acids. The LuxR-AHL complex also binds at the *luxR* promoter, but in this case, it represses transcription, resulting in a compensatory negative feedback. C. Overview of reactions in bacterial bioluminescence.

DO ALGAE AND BARNACLES INTERPRET BACTERIAL QS SIGNALS?

Many mechanisms for enhancing and inhibiting QS have been identified, suggesting that manipulation of QS-controlled processes is important in many bacterial-bacterial and bacterial-eukaryotic associations. Karen Tait and colleagues at Plymouth Marine Laboratory found that biofilms composed of wildtype *Vibrio anguillarum* strongly enhanced settlement of zoospores of the alga *Enteromorpha*, but *vanM* mutants (defective in AHL synthesis) did not (Tait et al., 2005). Surprisingly, this is not due to chemotactic attraction up concentration gradients of the bacterial AHL, but due to negative chemokinesis—swimming speed of zoospores decreases as they near the wildtype biofilm (Wheeler et al., 2006). A similar role for AHLs was shown in the attraction of larvae of the barnacle *Balanus improvisus* to biofilms of *V. anguillarum*, *Aeromonas hydrophila*, and *Sulfitobacter* sp. (Tait and Havenhand, 2013). These studies provided the first evidence of such communication across the bacterial and eukaryotic domains, but it is a *cue* rather than a *signal*, because "signal" has a specific meaning in evolutionary biology—indicating something produced explicitly to invoke a response from other organisms to coordinate activities between the signal producer and the responder.

autoinducer molecules do not diffuse back into the cell and interact with a regulatory protein, but are recognized by sensor kinase proteins LuxN and LuxQ, which have a histidine kinase domain and a response regulator domain. The signal is then transduced by a phosphorylation mechanism to a cascade of other proteins that determines the expression of the structural genes for bioluminescence. Highly conserved *luxS* homologs occur in many Gram-negative and Gram-positive bacteria. These AI-2 synthase genes have been found in about half of all sequenced bacterial genomes and their wide distribution suggested that AI-2 might be a universal signaling molecule, which bacteria use for communication both within and between species. However, this idea has lost favor because many bacteria that produce AI-2 do not possess a cognate receptor for it to function in signaling. Also, recent work indicates that in many bacteria, LuxS may have a primary role in methyl metabolism rather than QS, which may explain its wide distribution.

Gram-positive bacteria do not use AHLs but rely on peptide signals via two-component response regulator proteins resembling those of *V. harveyi*. In this case, the autoinducers are usually short peptides and are highly species specific. They are actively exported from the cells by ATP-binding transporters.

Following its discovery and detailed investigation in marine vibrios, QS has emerged as one of the most important mechanisms of gene regulation in bacteria, and application of this knowledge extends across the whole field of microbiology. QS is especially important in surface-associated communities, with many properties such as motility, adhesion, EPS synthesis, production of allelopathic chemicals, and other aspects of biofilm formation being subject to density-dependent gene regulation. Experimental studies of single-species biofilms have shown that the production of AHLs is important in determining the three-dimensional structure of mature biofilms, and AHL mutants produce densely packed biofilms that are more easily dislodged from the surface. In multispecies marine biofilms bacteria may utilize signaling molecules produced by other species or actively inhibit or degrade them by production of specific enzymes (a phenomenon known as quorum quenching).

The critical role of QS in symbiotic or pathogenic interactions with plant and animal hosts has been particularly well studied. In the marine context, the regulation of bioluminescence in the symbiosis of *A. fischeri* with *Euprymna scolopes* squid (*Chapter 10*) is an important example. Besides the Lux/LuxR system. an additional QS system (AinS/AinR) is also important; this system regulates early colonization of the squid light organ using a different autoinducer. Recent studies indicate that QS may be even more complex, with a high diversity of AHLs being recognized. Other examples include the production of disease by *Vibrio* spp. in humans, fish and corals (*Chapters 11* and *12*). Bacteria associated with other organisms including corals, sponges, and algae—as well as marine snow particles—have been shown to produce AHLs, suggesting that QS is important in the production of extracellular enzymes and interspecies interactions in these densely populated habitats. The role of QS in bacterial associations with corals is discussed in *Box 11.2*.

Despite strong evidence from cultured isolates and gene surveys that many marine bacteria can produce the components of QS systems, there is limited insight into the role of QS *in situ*. Sensitive detection methods for AHLs and other autoinducers have shown them to be present in marine snow, algae, and a range of invertebrates, where they are likely involved in regulation of synthesis of exoenzymes involved in the cycling of carbon, phosphorus, nitrogen, and other key nutrients. The ocean bioluminescence due to association of *Vibrio* bacteria with algal blooms is a demonstration that QS is active in the ocean on a massive scale (*Figure 4.6*). The settlement and metamorphosis of algal spores and invertebrate larvae is strongly dependent on bacterial biofilms and QS may be directly involved in these processes via recognition of the signaling molecules. Most QS signals break down rapidly due to basic hydrolysis in seawater. In addition, many bacteria and algae have been shown to interfere with QS by the production of inhibitors of AHL synthesis or receptor binding and diverse bacteria produce acylase or lactonase enzymes that degrade AHLs. This "quorum quenching" (QQ) has the obvious consequence of conferring a competitive advantage to the organisms producing QQ molecules because of the removal of the signals from the QS producers and the degraded AHLs also provide valuable nutrition. Thus, in multispecies community, the competitive interactions are very complex. QQ is a promising method for the control of biofouling.

PHYSICAL EFFECTS ON MICROBIAL GROWTH AND SURVIVAL

Most marine microbes grow at low temperatures

Over 90% of the ocean has a temperature of 5°C or colder. The temperature of seawater in the deep sea and in polar regions ranges from −1 to 4°C, while internal fluids in sea ice can be as low as −35°C in winter. Except for sea ice, these temperatures are very stable and little affected by seasonal changes. The overwhelming majority of marine microorganisms are adapted to this cold environment and conditions can only be considered "extreme" from a human perspective. Psychrophilic (cold-loving) microbes are defined as those with an optimum growth temperature of less than 15°C, a maximum growth temperature of 20°C, and a minimum growth temperature of 0°C or less. In fact, many deep sea and polar bacteria have quite a narrow temperature growth range and may lose viability after brief exposure to typical laboratory temperatures. Therefore, special precautions are needed in their collection, transport, and culture. Psychrotolerant bacteria are those that can grow at temperatures as low as 0°C but have optima of 20–35°C; many organisms from shallow seawater or coastal temperate regions fall into this category. Apart from understanding their considerable importance in nutrient cycling and ocean processes, there has also been recent interest by astrobiologists in psychrophiles because of the planned future exploration and search for possible life on Europa, the ice-covered moon of Jupiter. Biotechnologists also study psychrophiles because of the industrial potential of their fatty acids and extracellular polymer-degrading enzymes such as chitinase, chitobiose, and xylanase (*Chapter 14*).

Low temperatures particularly affect the folding capacity of structural proteins, enzymes, and flexibility of membranes, inhibiting catalysis and transport functions. RNA and DNA become more stable, inhibiting the processes of replication, translation, and RNA. Psychrophiles have many adaptations to overcome these effects. Comparison of the genomes of diverse bacterial and archaeal psychrophiles with those of mesophiles confirms the importance of amino acid composition and the prevalence of certain residues at critical sites for protein function. Proteins from psychrophiles are more flexible at low temperatures because they have greater amounts of α-helix and lesser amounts of β-sheet than those from other organisms. Higher protein flexibility around the active site of enzymes enhances catalysis by reducing the activation energy needed for substrate binding. Adaptations of some DNA-associated proteins are critical to relaxation of DNA under cold stress, and genetic differences leading to altered secondary structure permits replication, transcription, and translation to occur effectively. Adaptations to ensure efficient active transport across membranes at low temperatures include the incorporation of large amounts of unsaturated fatty acids into the membrane, which helps to maintain membrane fluidity. Omega-3-polyunsaturated fatty acids (PUFAs), once thought to be nonexistent in bacteria, have been found in Antarctic and deep-sea isolates and have considerable biotechnological potential (see *Chapter 14*). Transcriptomic studies show that cells adapt to low temperatures by upregulating genes for the synthesis of transport proteins to overcome the problem of lower diffusion rates across the membrane at low temperatures, as well as enzymes for the synthesis of structural components of the cytoplasmic and outer membranes.

Microbes growing in hydrothermal systems are adapted to very high temperatures

Some bacteria and archaea can grow at temperatures above 60°C. Such thermophilic organisms are found in the marine environment in areas of geothermal activity including shallow submarine hydrothermal systems, abyssal hot vents (black smokers), and active volcanic seamounts (*Figure 1.13*). In deep-sea vent systems, the temperature of seawater can exceed 350°C. As this superheated water mixes with cold seawater, a temperature gradient is established and diverse communities of thermophilic organisms with different temperature optima occur. Those organisms that can grow above temperatures of 80°C are termed hyperthermophiles. *Table 5.1* shows the temperature growth ranges of representative marine species. Most hyperthermophiles described to date belong to two major groups of the Archaea (*Chapter 5*), many of which have been isolated because of their biotechnological potential. Only a few major genera of bacteria, including *Thermodesulfobacterium*, *Aquifex*, and *Thermotoga* are

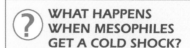

? WHAT HAPPENS WHEN MESOPHILES GET A COLD SHOCK?

While many mesophilic bacteria can protect themselves from sudden exposure to low temperatures by initiating a cold-shock response—rapid expression of a suite of cold-shock proteins (Csp) that mainly function at the post-translational levels as RNA chaperones. These are helper molecules that protect RNA from misfolding, or unwinding and degrading damaged RNA, whereas cold acclimation mechanisms in psychrophiles—evolved for living where cold temperatures are the norm—tend to be involved in the regulation of transcription and translation. Organisms may sense the temperature downshift due to changes in the supercoiling of DNA, which affects the recognition of the promoter regions of certain genes and leads to induction of *csp* genes (Phadtare, 2004).

hyperthermophilic. In both domains, hyperthermophiles occupy very deep branches of the phylogenetic tree, consistent with the idea that life evolved in high-temperature environments. Hyperthermophiles show a range of physiological types and can be aerobic or anaerobic, chemoorganotrophic, or chemolithotrophic. Their enzymes and structural proteins are adapted to show optimum activity and stability at high temperatures. The overall structure of proteins from hyperthermophiles often shows relatively little difference from that of homologous proteins in mesophiles. However, variation in a small number of amino acids at critical locations in the protein seems to affect the three-dimensional conformation, permitting greater stability and function of the active site of enzymes. Intracellular proteins from hyperthermophiles also contain a high proportion of hydrophobic regions and disulfide bonds, which improve thermostability. Enzymes from hyperthermophiles are useful in high-temperature industrial processes (*Chapter 14*). Adaptations of the cell membrane also occur to ensure stability and effective nutrient transport at high temperatures. The membranes of archaea contain ether-linked isoprene units and hyper-thermophiles usually possess monolayer membranes, which appear to be more stable at high temperatures.

Microbes that inhabit the deep ocean must withstand a very high hydrostatic pressure

Pressure increases by one atmosphere (atm = 0.101 megapascal, Mpa) for every 10 m water depth and over 75% of the ocean's volume is more than 1000 m deep. As we now know that bacteria and archaea are distributed in great numbers throughout the water column, as well as in marine sediments many hundreds of meters deep, growth under conditions of very high pressure is the normal state for the vast majority of marine microbes. Zobell and Morita pioneered the study of deep-sea bacteria in the 1940s and 1950s. Recent advances in sampling and cultivation methods together with the application of molecular techniques to the study of diversity and physiology are leading to some significant new insights. Bacteria and archaea that have been isolated from depths down to ~3000 m are usually found to be piezotolerant and can grow over a wide range of pressures up to 30–40 mPa when cultured in the laboratory, although increasing pressure above ~20 mPa results in reduced metabolic activity and growth rates. In contrast, many organisms isolated from the deep sea grow optimally at pressures >30 mPa; these are known as piezophiles. (The alternative terms barotolerant and barophilic may also be used, although the *piezo-* prefix is preferred, because it indicates pressure rather than *baro-* (weight)). Hyperpiezophiles display optimal growth rates at >60 mPa and many of these are unable to grow at lower pressures and may die if decompressed for more than a short period. By using special isolation techniques involving collection in pressurized chambers and cultivation in solid silica gel media, an increasing number of species of obligate piezophiles have been cultured in recent years, including some from the deepest habitats such as the Marianas Trench in the Pacific Ocean (10500 m). Genetic studies of extreme piezophiles indicate that many have a close resemblance to common psychrotolerant types (e.g. the gammaproteobacterial genera *Shewanella*, *Photobacterium*, *Colwellia*, and *Moritella*), although some unique taxa have also been discovered. At abyssal depths (>4000 m), psychrophiles appear to be ecologically dominant over bacteria from shallow waters carried there by sinking organic matter. The most abundant sources of piezophiles are nutrient-rich niches such as decaying animal carcasses or the gut of deep-sea animals. However, oligotrophic piezophiles adapted to very low-nutrient concentrations also occur in seawater. Some hyperpiezophilic and hyperthermophilic Archaea have been found near hydrothermal vents, including *Pyrococcus yayanossi*, which has optimal reproduction rates at 52 mPa and 98°C.

A range of adaptations seems to be present in deep-sea organisms. It is important to note that very low temperatures (apart from near hydrothermal vents) and very low-nutrient conditions (apart from the localized occurrence of concentrations of organic matter) characterize the deep sea, as well as high pressure. We must therefore consider the adaptations of piezophilic deep-sea microorganisms as a response to the combined effects of these factors. Protection of enzymes from the effects of pressure seems to be due mainly to changes in the conformation of proteins. Proteins in piezophiles seem less flexible and less subject to compression under pressure, owing to a decreased content of the amino acids proline and glycine. Cells grown under high pressure also contain high levels of osmotically active substances, which are thought to protect proteins from hydration effects of high pressure. The most well-studied adaptation

HIGH PRESSURE PRESERVED A DEEP-SEA SNACK

High pressure decreases the binding of substrates to enzymes, which explains why the metabolism of shallow-water or terrestrial psychrotolerant organisms is much slower when incubated in the laboratory under pressure. A remarkable demonstration of this effect occurred in 1968, when the submersible vessel *Alvin* sunk accidentally to a depth of 1500 m. While *Alvin* was being lowered into the sea for a dive, the cable snapped. As water poured through the hatch, the crew managed to escape but left their lunchboxes—containing meat sandwiches and apples—on board the vessel. When *Alvin* was recovered nearly a year later, these were found in almost perfect condition—looking fresh enough to tempt a couple of scientists to take a bite—because of the high pressure and temperature of ~2–4°C. At atmospheric pressure, growth of contaminating bacteria would have spoiled the food within a few days.

to high pressure and low temperature is a change in membrane composition. Membranes of piezophiles contain a higher proportion of polyunsaturated fatty acids and have a more tightly packed distribution of fatty-acyl chains. Pressure also affects DNA secondary structure.

Application of molecular genetics to the study of two piezophilic bacteria, *Photobacterium profundum* and *Shewanella* sp., has revealed some insight into the mechanisms of regulation of the pressure response. When *P. profundum* is shifted from atmospheric to high pressure, the relative abundance of two outer membrane proteins (OmpH and OmpL) is altered. These proteins act as porins for the transport of substances across the outer membrane. An increased production of OmpH probably provides a larger channel, suggesting that the pressure response enables the bacteria to take up scarce nutrients more easily (as would occur in the deep-sea environment). A pair of cytoplasmic membrane proteins regulate transcription of the *ompL* and *ompH* genes. Interestingly, these have a high-sequence homology to the genes encoding ToxR and ToxS proteins, which were first discovered in *V. cholerae,* where they detect changes in temperature, pH, and salinity during the transition from the aquatic environment to the host, leading to the expression of virulence factors (*Figure 12.1*). The homology indicates some common ancestry of this environmental sensing system that has evolved to perform different functions according to habitat. Some other genes have also been shown to be important in the response to pressure in a deep-sea *Shewanella* sp. and these may be grouped into pressure-regulated operons. Use of gene probes directed against these genes could aid in the identification of new species of piezophiles that we cannot currently culture.

Obligate aerobes always require the presence of O_2 and use it as the terminal electron acceptor in aerobic respiration. Facultative aerobes can carry out anaerobic respiration or fermentation in the absence of O_2 or aerobic respiration in its presence. Even though O_2 is not required, the growth of facultative organisms is better in its presence on account of the greater yield of ATP from aerobic respiration. Microaerophiles carry out aerobic respiration but require an O_2 level lower than that found in the atmosphere. Obligate anaerobes carry out fermentation or anaerobic respiration and many are killed by exposure to O_2, although some are aerotolerant and survive (but do not grow) in its presence. Examples of all categories occur in marine bacteria and archaea.

Oxygen can exist in various forms that are highly reactive and toxic to all cells unless they possess mechanisms to destroy them (reactive oxygen species, ROS). Singlet oxygen ($^1O_2$) is a high-energy state that causes spontaneous oxidation of cellular materials that forms during photochemical reactions; phototrophs usually contain carotenoid pigments, which convert singlet oxygen to harmless forms (quenching). For this reason, non-phototrophic marine organisms exposed to bright light (such as those inhabiting clear surface waters) are also often pigmented. Various other ROS form during respiration (*Figure 3.18*). Superoxide (O_2^-) and hydroxyl (OH^+) radicals are particularly destructive and react rapidly with cellular compounds. The evolution of mechanisms for the removal of ROS was a major step in the transition of the biosphere from anaerobic to aerobic following the development of oxygen-evolving photosynthesis. Organisms capable of aerobic growth usually contain the enzymes catalase ($2H_2O_2 \rightarrow H_2O + O_2$), superoxide dismutase ($2O_2^- + 2H^+ \rightarrow H_2O_2 + O_2$), and peroxidase ($H_2O_2 + NADH + H^+ \rightarrow 2H_2O + NAD^+$). Superoxide reductase is an enzyme originally found in *Pyrococcus furiosus* and thought to be unique to the Archaea, but genome sequence analysis has shown that it may be widely distributed in obligate anaerobes in place of superoxide dismutase. This enzyme reduces superoxide to H_2O_2 without the formation of O_2 ($O_2^- + 2H^+ + cyt$ $c_{red} \rightarrow H_2O_2 + cyt\ c_{ox}$).

Ultraviolet irradiation has lethal and mutagenic effects

Research into the effects of ultraviolet (UV) radiation on marine microbes is needed because of growing evidence that UV radiation is increasing at certain locations on Earth, as a result particularly of ozone depletion in the upper atmosphere and the formation of an ozone "hole" over Antarctica and the Southern Ocean. Restrictions in the use of chlorofluorocarbons (CFCs) is leading to partial closure of the hole over Antarctica, but the layer still appears to be thinning near the equator and middle latitudes. The lethal and mutagenic effects of UV radiation result from damage to DNA. UV-B causes direct damage to DNA through the

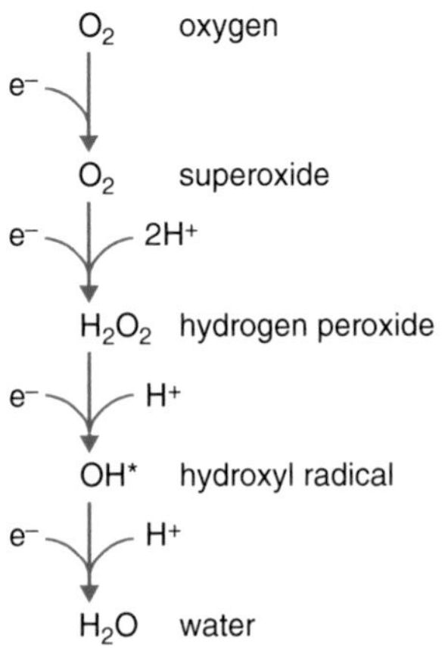

O$_2$ oxygen

e$^-$ ↓

O$_2$ superoxide

e$^-$ ↓ 2H$^+$

H$_2$O$_2$ hydrogen peroxide

e$^-$ ↓ H$^+$

OH*   hydroxyl radical

e$^-$ ↓ H$^+$

H$_2$O   water

Figure 3.18 Formation of toxic intermediates during reduction of oxygen.

formation of pyrimidine dimers, while the main effects of UV-A are due to formation of toxic oxygen and hydroxyl radicals. Various mechanisms for the repair of UV-induced damage exist, including nucleotide excision repair and light-activated enzyme repair (photoreactivation). Studies of DNA damage in bacteria in surface waters show that there is a pronounced effect over the course of the day, with maximal damage evident in the late afternoon and repair occurring during the night. We do not yet fully understand the ecological significance of these processes. Bacteria produce a range of UV-screening products such as mycosporine-like amino acids and scytonemin, a complex aromatic compound formed in the sheath of some cyanobacteria. Some bacteria isolated from corals in very clear surface waters show extreme resistance to the effects of UV radiation by enhancing the activity of NAD(P)H quinine oxidoreductase, a powerful antioxidative enzyme. These mechanisms could have significant biotechnological potential in human health, as products for skin-protection treatments and for overcoming the effects of oxidative stress during aging.

Bacterial bioluminescence may protect bacteria from ROS and UV damage

Some members of the gammaproteobacterial order *Vibrionales* exhibit the distinctive feature of bioluminescence. Bacterial bioluminescence occurs due to the action of the enzyme luciferase, which is a mixed-function oxidase that simultaneously catalyzes the oxidation of reduced flavin mononucleotide (FMNH$_2$) and a long-chain aliphatic aldehyde (RCHO) such as tetradecanal (*Figure 3.17C*). Blue-green light with a wavelength of about 490 nm is emitted because of the generation of an intermediate molecule in an electronically excited state. Luciferases from all bioluminescent bacteria are dimers of α (~40 kDa) and β (~35 kDa) subunits, encoded by the *luxA* and *luxB* genes that occur adjacently in the lux operon, along with other gene-encoding enzymes, leading to the synthesis of the aldehyde substrate via fatty-acid precursors. The overall process is dependent on ATP and NADPH. Bioluminescence is regulated by quorum sensing (*Figure 3.17A–B*).

Bioluminescent bacteria occur as free-living forms in seawater and on organic debris, as commensals in the gut of many marine animals, and as symbionts of the light organs of some squid and fish. The ecological benefits for animal hosts of harboring bioluminescent symbionts are obvious (*Box 10.3*), and selection pressure over many millennia must also have been a major evolutionary force in the development of such a complex process. However, it can consume up to 20% of the cell's energy and benefits for the bacteria are unclear. The discovery of complex mechanisms of regulation of bioluminescence by density-dependent quorum sensing adds a further dimension to this enigma. Numerous *Vibrio* species in coastal water contain *lux* genes but are not visibly luminescent in culture. Some experiments suggest that the cell's own light emission may function in the repair of DNA at night. UV damage causes formation of pyrimidine dimers that prevent DNA replication. In blue light, a process called photoreactivation occurs. The enzyme DNA photolyase binds to the dimers and excises them, and other enzymes then restore the damaged segment of DNA. Induction of the DNA repair process is initiated by a complex regulatory system called SOS. Mutants of *Vibrio harveyi* which do not emit light due to mutations in the *luxA* or *luxB* have lower survival when incubated in the dark after UV-irradiation. Thus, dark repair initiated by bioluminescence might promote survival of bacteria in surface waters. Other research indicates that bioluminescence may protect bacteria against the toxic effects of oxygen. As well as bacteria, bioluminescence is present in some fungi and diverse groups of animals and has multiple evolutionary origins—there is a very wide diversity of both enzymes and substrates for the bioluminescent reaction mechanism. Apart from light emission, the only common feature is the requirement for oxygen, and bioluminescent reactions may have evolved primarily as a mechanism for detoxification of the highly toxic derivatives of molecular oxygen (*Figure 3.18*), with the initial evolutionary driver being the nature of the substrates rather than the luciferase enzymes.

Microbes use various mechanisms to prevent osmotic damage

Several genera of the Archaea are extreme halophiles that grow at very high NaCl concentrations (15–35%) found in salterns, brine pockets within sea ice, and submarine brine pools. Extreme halophilicity is rare in the Bacteria, but *Salinibacter ruber* is an exception.

To protect themselves from dehydration due to loss of water from the cell to the external environment, marine microbes must maintain the concentration of intracellular solutes at a high level. One way of achieving this is by accumulating non-inhibitory substances known as compatible solutes or osmoprotectants. Usually, these types of sugars, alcohols, or amino acids are extremely soluble in water. For example, many Gram-negative bacteria synthesize compounds such as glycine-betaine, ectoine, glutamate, or a-glucosylglycerol. These substances are released when cells lyse, and some bacteria accumulate glycine-betaine from the environment rather than synthesizing it themselves. Most Gram-positive bacteria accumulate the amino acid proline as an osmoprotectant. In algae, DMSP is the main osmoprotectant and its production and release has major impacts on the ocean sulfur cycle (*Chapter 9*).

The extremely halophilic archaea use a different method to prevent water loss. They have an active mechanism for pumping K^+ ions into the cell until the internal concentration balances the high concentration of Na^+ outside. In some species, a large proportion of the proton-motive force for the ion pump derives from light-mediated generation of ATP via the pigment-containing protein bacteriorhodopsin. Extreme halophiles also have other adaptations for growth at high NaCl concentrations. Their enzymes and structural proteins have a high proportion of acidic amino acids, which protects the conformation from disruption by high salt concentrations. Internal cellular components, such as the ribosomes and DNA-replication enzymes, require high K^+ concentrations for their integrity and activity.

Conclusions

This chapter has explored the very wide range of metabolic activities in marine microbes, with a particular focus on bacterial and archaeal processes. We have seen a large diversity of mechanisms by which they obtain energy from light or chemical sources and various ways in which they use this to fuel their biosynthetic processes. Some types of metabolism, with major significance in global ecology, have been discovered only in the last few years and there are undoubtedly many more surprises to be revealed. Investigation of phenomena such as antagonism and intercellular communication is revealing information about the ways in which different microbial species interact with their environment and the interdependence of their metabolic activities. We are beginning to understand how the microscale effects of microbial processes can affect global processes in the oceans. In subsequent chapters, numerous examples of the activities of such communities in marine habitats will become apparent.

References and further reading

Phototrophy

Béjà, O., Aravind, L., Koonin, E. V. et al. (2000) Bacterial rhodopsin: evidence for a new type of phototrophy in the sea. *Science* **289**: 1902–1906.

Beja, O., Spudich, E.N., Spudich, J.L. et al. (2001) Proteorhodopsin phototrophy in the ocean. *Nature* **411**: 786–789.

Broda, E. (1977) Two kinds of lithotrophs missing in nature. *Z. Allg. Mikrobiol.* **17**: 491–493.

Bryant, D.A., and Frigaard, N.U. (2006) Prokaryotic photosynthesis and phototrophy illuminated. *Trends Microbiol.* **14**: 488–496.

Dubinsky, V., Haber, M., Burgsdorf, I. et al. (2017) Metagenomic analysis reveals unusually high incidence of proteorhodopsin genes in the ultraoligotrophic Eastern Mediterranean Sea. *Environ. Microbiol.* **19**: 1077–1090.

Ernst, O.P., Lodowski, D.T., Elstner, M. et al. (2014) Microbial and animal rhodopsins: structures, functions, and molecular mechanisms. *Chem. Rev.* **114**: 126–163.

Finkel, O.M., Béjà, O., and Belkin, S. (2013) Global abundance of microbial rhodopsins. *ISME J.* **7**: 448–51.

Frigaard, N.U., Martinez, A., Mincer, T.J. et al. (2006) Proteorhodopsin lateral gene transfer between marine planktonic *Bacteria* and *Archaea*. *Nature* **439**: 847–850.

Giovannoni, S.J., Bibbs, L., Cho, J.C. et al. (2005) Proteorhodopsin in the ubiquitous marine bacterium SAR11. *Nature* **438**: 82–85.

Gómez-Consarnau, L., Akram, N., Lindell, K., et al. (2010) Proteorhodopsin phototrophy promotes survival of marine bacteria during starvation. *PLoS Biol.* **8**: e1000358.

Gomez-Consarnau, L., Gonzalez, J.M., Coll-Llado, M. et al. (2007) Light stimulates growth of proteorhodopsin-containing marine Flavobacteria. *Nature* **445**: 210–213.

Graham, E.D., Heidelberg, J.F., and Tully, B.J. (2018) Potential for primary productivity in a globally-distributed bacterial phototroph. *ISME J.* **12**: 1861–1866.

Inoue, K., Kato, Y., and Kandori, H. (2015) Light-driven ion-translocating rhodopsins in marine bacteria. *Trends Microbiol.* **23**: 91–98.

Kumagai, Y., Yoshizawa, S., Nakajima, Y. et al. (2018) Solar-panel and parasol strategies shape the proteorhodopsin distribution pattern in marine Flavobacteriia. *ISME J.* **12**: 1329–1343.

McCarren, J., and DeLong, E.F. (2007) Proteorhodopsin photosystem gene clusters exhibit co-evolutionary trends and shared ancestry among diverse marine microbial phyla. *Environ. Microbiol.* **9**: 846–858.

Moran, M.A., and Miller, W.L. (2007) Resourceful heterotrophs make the most of light in the coastal ocean. *Nat. Rev. Microbiol.* **5**: 792–800.

Pinhassi, J., DeLong, E.F., Béjà, O. et al. (2016) Marine bacterial and archaeal ion-pumping rhodopsins: genetic diversity, physiology, and ecology. *Microbiol. Mol. Biol. Rev.* **80**: 929–954.

Sharma, A.K., Spudich, J.L., and Doolittle, W.F. (2006) Microbial rhodopsins: functional versatility and genetic mobility. *Trends Microbiol.* **14**: 463–469.

Sieradzki, E.T., Fuhrman, J.A., Rivero-Calle, S. et al. (2018) Proteorhodopsins dominate the expression of phototrophic mechanisms in seasonal and dynamic marine picoplankton communities. *PeerJ* **6**: e5798.

Steindler, L., Schwalbach, M.S., Smith, D.P. et al. (2011) Energy starved Candidatus Pelagibacter ubique substitutes light-mediated ATP production for endogenous carbon respiration. *PLoS One* **6**: e19725.

Wang, Z., O'Shaughnessy, T.J., Soto, C.M. et al. (2012) Function and regulation of *Vibrio campbellii* proteorhodopsin: acquired phototrophy in a classical organoheterotroph. *PLoS One* **7**: e38749.

Nitrogen metabolism

Casciotti, K.L., and Buchwald, C. (2012) Insights on the marine microbial nitrogen cycle from isotopic approaches to nitrification. *Front. Microbiol.* **3**: 00356

Dekas, A.E., Fike, D.A., Chadwick, G.L. et al. (2018) Widespread nitrogen fixation in sediments from diverse deep-sea sites of elevated carbon loading. *Environ. Microbiol.* **20**: 4281–4296.

Dekas, A.E., Poretsky, R.S., and Orphan, V.J. (2009) Deep-sea archaea fix and share nitrogen in methane-consuming microbial consortia. *Science* **326**: 422–426.

DeLong, E.F. (2002) All in the packaging. *Nature* **419**: 676–677.

Kuenen, J.G. (2008) Anammox bacteria: from discovery to application. *Nat. Rev. Microbiol.* **6**: 320–326.

Pereira, A.D., Cabezas, A., Etchebehere, C. et al. (2017) Microbial communities in anammox reactors: a review. *Environ. Technol. Rev.* **6**: 74–93.

Stahl, D.A., and de la Torre, J.R. (2012) Physiology and diversity of ammonia-oxidizing *Archaea*. *Annu. Rev. Microbiol.* **66**: 83–101.

Sulfur metabolism

Barton, L.L., and Fauque, G.D. (2009) Biochemistry, physiology and biotechnology of sulfate-reducing bacteria. *Adv. Appl. Microbiol.* **68**: 41–98.

Liu, Y., Beer, L.L., and Whitman, W.B. (2012b) Sulfur metabolism in archaea reveals novel processes. *Environ. Microbiol.* **14**: 2632–2644.

Lovley, D.R. (2017) Happy together: microbial communities that hook up to swap electrons. *ISME J.* **11**: 327–336.

Muyzer, G., and Stams, A.J.M. (2008) The ecology and biotechnology of sulphate-reducing bacteria. *Nat. Rev. Microbiol.* **6**: 441–454.

Muyzer, G., Yildirim, E., van Dongen, U. et al. (2005) Identification of "Candidatus Thioturbo danicus," a microaerophilic bacterium that builds conspicuous veils on sulfidic sediments. *Appl. Environ. Microbiol.* **71**: 8929–8933.

Rückert, C. (2016) Sulfate reduction in microorganisms—recent advances and biotechnological applications. *Curr. Opin. Microbiol.* **33**: 140–146.

Thar, R., and Kuhil, M. (2002) Conspicuous veils formed by vibrioid bacteria on sulfidic marine sediment. *Appl. Environ. Microbiol.* **68**: 6310–6320.

Wasmund, K., Mußmann, M., and Loy, A. (2017) The life sulfuric: microbial ecology of sulfur cycling in marine sediments. *Environ. Microbiol. Rep.* **9**: 323–344.

Methanogenesis and methylotrophy

Boetius, A., Ravenschlag, K., Schubert, C.J. et al. (2000) A marine microbial consortium apparently mediating anaerobic oxidation of methane. *Nature* **407**: 623–626.

Born, D.A., Ulrich, E.C., Ju, K.-S. et al. (2017) Structural basis for methylphosphonate biosynthesis. *Science* **358**: 1336–1339.

Carini, P., White, A.E., Campbell, E.O. et al. (2014) Methane production by phosphate-starved SAR11 chemoheterotrophic marine bacteria. *Nat. Commun.* **5**: 4346.

Chistoserdova, L., and Kalyuzhnaya, M.G. (2018) Current trends in methylotrophy. *Trends Microbiol.* **26**: 703–714.

Cui, M., Ma, A., Qi, H. et al. (2015) Anaerobic oxidation of methane: an "active" microbial process. *Microbiol. Open* **4**: 1–11.

DeLong, E.F. (2000) Resolving a methane mystery. *Nature* **407**: 577–579.

Karl, D.M., Beversdorf, L., Björkman, K.M. et al. (2008) Aerobic production of methane in the sea. *Nat. Geosci.* **1**: 473–478.

Lever, M.A. (2016) A new era of methanogenesis research. *Trends Microbiol.* **24**: 84–86.

Liu, Y., Beer, L.L., and Whitman, W.B. (2012a) Methanogens: a window into ancient sulfur metabolism. *Trends Microbiol.* **20**: 251–258.

Repeta, D.J., Ferrón, S., Sosa, O.A. et al. (2016) Marine methane paradox explained by bacterial degradation of dissolved organic matter. *Nat. Geosci.* **9**: 884–887.

Sun, J., Steindler, L., Thrash, J.C. et al. (2011) One carbon metabolism in SAR11 pelagic marine bacteria. *PLoS One* **6**: e23973.

Valentine, D.L., and Reeburgh, W.S. (2000) New perspectives on anaerobic methane oxidation. *Environ. Microbiol.* **2**: 477–484.

Yan, Z., Joshi, P., Gorski, C.A. et al. (2018) A biochemical framework for anaerobic oxidation of methane driven by Fe(III)-dependent respiration. *Nat. Commun.* **9**: 1642.

Sources and acquisition of nutrients

Adam, N., and Perner, M. (2018) Microbially mediated hydrogen cycling in deep-sea hydrothermal vents. *Front. Microbiol.* **9**: 02873.

Bergauer, K., Fernandez-Guerra, A., Garcia, J.A.L. et al. (2018) Organic matter processing by microbial communities throughout the Atlantic water column as revealed by metaproteomics. *Proc. Natl. Acad. Sci.* **115**: E400–E408.

Croft, M.T., Warren, M.J., and Smith, A.G. (2006) Algae need their vitamins. *Eukaryot. Cell* **5**: 1175–1183.

Decho, A.W., and Gutierrez, T. (2017) Microbial extracellular polymeric substances (EPSs) in ocean systems. *Front. Microbiol.* **8**: 922.

Johnstone, T.C., and Nolan, E.M. (2015) Beyond iron: non-classical biological functions of bacterial siderophores. *Dalt. Trans.* **44**: 6320–6339.

Kiørboe, T. (2003) Marine snow microbial communities: scaling of abundances with aggregate size. *Aquat. Microb. Ecol.* **33**: 67–75.

Kiørboe, T., and Jackson, G.A. (2001) Marine snow, organic solute plumes, and optimal chemosensory behavior of bacteria. *Limnol. Oceanogr.* **46**: 1309–1318.

Koch, A.L. (2001) Oligotrophs versus copiotrophs. *BioEssays* **23**: 657–661.

Lauro, F.M., McDougald, D., Thomas, T. et al. (2009) The genomic basis of trophic strategy in marine bacteria. *Proc. Natl. Acad. Sci.* **106**: 15527–15533.

Lipson, D.A. (2015) The complex relationship between microbial growth rate and yield and its implications for ecosystem processes. *Front. Microbiol.* **6**: 615.

Long, R.A., and Azam, F. (2001) Antagonistic interactions among marine pelagic bacteria. *Appl. Environ. Microbiol.* **67**: 4975–4983.

Mühlenbruch, M., Grossart, H.-P., Eigemann, F. et al. (2018) Mini-review: Phytoplankton-derived polysaccharides in the marine environment and their interactions with heterotrophic bacteria. *Environ. Microbiol.* **20**: 2671–2685.

Paerl, R.W., Bouget, F.-Y., Lozano, J.-C. et al. (2017) Use of plankton-derived vitamin B1 precursors, especially thiazole-related precursor, by key marine picoeukaryotic phytoplankton. *ISME J.* **11**: 753–765.

Reintjes, G., Arnosti, C., Fuchs, B.M. et al. (2017) An alternative polysaccharide uptake mechanism of marine bacteria. *ISME J.* **11**: 1640–1650.

Schönheit, P., Buckel, W., and Martin, W.F. (2016) On the origin of heterotrophy. *Trends Microbiol.* **24**: 12–25.

Sowell, S.M., Wilhelm, L.J., Norbeck, A.D. et al. (2009) Transport functions dominate the SAR11 metaproteome at low-nutrient extremes in the Sargasso Sea. *ISME J.* **3**: 93–105.

Stoecker, D.K., Hansen, P.J., Caron, D.A. et al. (2017) Mixotrophy in the marine plankton. *Ann. Rev. Mar. Sci.* **9**: 311–335.

Suffridge, C.P., Gómez-Consarnau, L., Monteverde, D.R. et al. (2018) B Vitamins and their congeners as potential drivers of microbial community composition in an oligotrophic marine ecosystem. *J. Geophys. Res. Biogeosciences* **123**: 2890–2907.

Tang, K., Jiao, N., Liu, K. et al. (2012) Distribution and functions of TonB-dependent transporters in marine bacteria and environments: implications for dissolved organic matter utilization. *PLoS One* **7**: e41204.

Traving, S.J., Thygesen, U.H., Riemann, L. et al. (2015) A model of extracellular enzymes in free-living microbes: which strategy pays off? *Appl. Environ. Microbiol.* **81**: 7385–7393.

Motility and taxis

Barbara, G.M., and Mitchell, J.G. (2003) Bacterial tracking of motile algae. *FEMS Microbiol. Ecol.* **44**: 79–87.

Barbara, G.M., and Mitchell, J.G. (2003) Marine bacterial organisation around point-like sources of amino acids. *FEMS Microbiol. Ecol.* **43**: 99–109.

Belas, R. (2014) Biofilms, flagella, and mechanosensing of surfaces by bacteria. *Trends Microbiol.* **22**: 517–527.

Fernandez, V.I., Yawata, Y., and Stocker, R. (2019) A foraging mandala for aquatic microorganisms. *ISME J.* **13**: 563–575.

Guttenplan, S.B., and Kearns, D.B. (2013) Regulation of flagellar motility during biofilm formation. *FEMS Microbiol. Rev.* **37**: 849–871.

Johansen, J., Pinhassi, J., Blackburn, N. et al. (2002) Variability in motility characteristics among marine bacteria. *Aquat. Microb. Ecol.* **28**: 229–237.

McCarter, L. (1999) The multiple identities of *Vibrio parahaemolyticus*. *J. Mol. Microbiol. Biotechnol.* **1**: 51–57.

Minamino, T., and Imada, K. (2015). The bacterial flagellar motor and its structural diversity. *TrendsMicrobiol* **23**: 267–274.

Mitchell, J.G., and Barbara, G.M. (1999) High speed marine bacteria use sodium-ion and proton driven motors. *Aquat. Microb. Ecol.* **18**: 227–233.

Mitchell, J.G., Pearson, L., and Dillon, S. (1996) Clustering of marine bacteria in seawater enrichments. *Appl. Environ. Microbiol.* **62**: 3716–3721.

Mitchell, J.G., Pearson, L., Dillon, S. et al. (1995) Natural assemblages of marine-bacteria exhibiting high-speed motility and large accelerations. *Appl. Environ. Microbiol.* **61**: 4436–4440.

Muramoto, K., Kawagishi, I., Kudo, S. et al. (1995) High-speed rotation and speed stability of the sodium-driven flagellar motor in *Vibrio alginolyticus*. *J. Mol. Biol.* **251**: 50–58.

Pallen, M.J., Penn, C.W., and Chaudhuri, R.R. (2005) Bacterial flagellar diversity in the post-genomic era. *Trends Microbiol.* **13**: 143–149.

Rusconi, R., Garren, M., and Stocker, R. (2014) Microfluidics expanding the frontiers of microbial ecology. *Annu. Rev. Biophys.* **43**: 65–91.

Schuergers, N., Lenn, T., Kampmann, R. et al. (2016) Cyanobacteria use micro-optics to sense light direction. *Elife* **5**: e12620.

Son, K., Guasto, J.S., and Stocker, R. (2013) Bacteria can exploit a flagellar buckling instability to change direction. *Nat. Phys.* **9**: 494–498.

Son, K., Menolascina, F., and Stocker, R. (2016) Speed-dependent chemotactic precision in marine bacteria. *Proc. Natl. Acad. Sci. U. S. A.* **113**: 8624–8629.

Stocker, R. (2011) Reverse and flick: Hybrid locomotion in bacteria. *Proc. Natl. Acad. Sci. USA* **108**: 2635–2636.

Stocker, R. (2012) Marine microbes see a sea of gradients. *Science* **338**: 628–633.

Stocker, R., Seymour, J.R., Samadani, A. et al. (2008) Rapid chemotactic response enables marine bacteria to exploit ephemeral microscale nutrient patches. *Proc. Natl. Acad. Sci. USA* **105**: 4209–4214.

Wilde, A., and Mullineaux, C.W. (2017) Light-controlled motility in prokaryotes and the problem of directional light perception. *FEMS Microbiol. Rev.* **41**: 900–922.

Xie, L., Altindal, T., Chattopadhyay, S. et al. (2011) Bacterial flagellum as a propeller and as a rudder for efficient chemotaxis. *Proc. Natl. Acad. Sci. USA* **108**: 2246–2251.

Quorum sensing, communication and multicellularity

Alberghini, S., Polone, E., Corich, V. et al. (2009) Consequences of relative cellular positioning on quorum sensing and bacterial cell-to-cell communication. *FEMS Microbiol. Lett.* **292**: 149–161.

Antunes, J., Leão, P., and Vasconcelos, V. (2018) Marine biofilms: diversity of communities and of chemical cues. *Environ. Microbiol. Rep.* doi: 10.1111/1758-2229.12694.

Dunny, G.M., Brickman, T.J., and Dworkin, M. (2008) Multicellular behavior in bacteria: communication, cooperation, competition and cheating. *Bioessays* **30**: 296–298.

Hense, B.A., Kuttler, C., Müller, J. et al. (2007) Does efficiency sensing unify diffusion and quorum sensing? *Nat. Rev. Microbiol.* **5**: 230–239.

Hmelo, L.R. (2017) Quorum sensing in marine microbial environments. *Ann. Rev. Mar. Sci.* **9**: 257–281.

Hmelo, L.R., Mincer, T.J., and Van Mooy, B.A.S. (2011) Possible influence of bacterial quorum sensing on the hydrolysis of sinking particulate organic carbon in marine environments. *Environ. Microbiol. Rep.* **3**: 682–688.

Krupke, A., Hmelo, L.R., Ossolinski, J.E. et al. (2016) Quorum sensing plays a complex role in regulating the enzyme hydrolysis activity of microbes associated with sinking particles in the ocean. *Front. Mar. Sci.* **1**: 55.

Matz, C., Webb, J.S., Schupp, P.J. et al. (2008) Marine biofilm bacteria evade eukaryotic predation by targeted chemical defense. *PLoS One* **3**: e2744.

Platt, T.G., and Fuqua, C. (2010) What's in a name? The semantics of quorum sensing. *Trends Microbiol.* **18**: 383–387.

Redfield, R.J. (2002) Is quorum sensing a side effect of diffusion sensing? *Trends Microbiol.* **10**: 365–370.

Tait, K., and Havenhand, J. (2013) Investigating a possible role for the bacterial signal molecules N-acylhomoserine lactones in *Balanus improvisus* cyprid settlement. *Mol. Ecol.* **22**: 2588–2602.

Tait, K., Joint, I., Daykin, M. et al. (2005) Disruption of quorum sensing in seawater abolishes attraction of zoospores of the green alga *Ulva* to bacterial biofilms. *Environ. Microbiol.* **7**: 229–240.

Ventura, R. (2019) Multicellular individuality: the case of bacteria. *Biol. Theory*: doi.org/10.1007/s13752-019-00317-7.

Wheeler, G.L., Tait, K., Taylor, A. et al. (2006) Acyl-homoserine lactones modulate the settlement rate of zoospores of the marine alga *Ulva intestinalis* via a novel chemokinetic mechanism. *Plant, Cell Environ.* **29**: 608–618.

VBNC bacteria

Bloomfield, S.F., Stewart, G.S.A.B., Dodd, C.E.R. et al. (1998) The viable but non-culturable phenomenon explained? *Microbiology* **144**: 1–3.

Oliver, J.D. (2005) The viable but nonculturable state in bacteria. *J. Microbiol.* **43**: 93–100.

Effects of pressure, temperature, ROS and UV radiation

Chénard, C., and Lauro, F.M. eds. (2017) *Microbial Ecology of Extreme Environments*. Springer International Publishing, Cham.

Counts, J.A., Zeldes, B.M., Lee, L.L. et al. (2017) Physiological, metabolic and biotechnological features of extremely thermophilic microorganisms. *Wiley Interdiscip. Rev. Syst. Biol. Med.* **9**: e1377.

Czyz, A., and Wegrzyn, G. (2001) On the function and evolution of bacterial luminescence. In: *Bioluminescence and Chemiluminescence* (ed. J. F. Case, P. J. Herring, B. H. Robinson, S. D. H. Haddock, L. J. Kricka and P. E. Stanley). World Science Publishing Company, Singapore, pp. 31–34.

Czyz, A., Wrobel, B., and Wegrzyn, G. (2000) *Vibrio harveyi* bioluminescence plays a role in stimulation of DNA repair. *Microbiology* 146: 283–288.

Durvasula, R., and Subba Rao, D. V. (2018) *Extremophiles: From Biology to Biotechnology*. CRC Press.

Fang, J., Zhang, L., and Bazylinski, D.A. (2010) Deep-sea piezosphere and piezophiles: geomicrobiology and biogeochemistry. *Trends Microbiol.* **18**: 413–422.

Haddock, S.d., Moline, M.A., and Case, J.F. (2010) Bioluminescence in the sea. *Ann. Rev. Mar. Sci.* **2**: 443–449.

Harrison, J.P., Gheeraert, N., Tsigelnitskiy, D. et al. (2013) The limits for life under multiple extremes. *Trends Microbiol.* **21**: 204–212.

Huber, H., and Stetter, K.O. (1998) Hyperthermophiles and their possible potential in biotechnology. *J. Biotechnol.* **64**: 39–52.

Kashefi, K., and Lovley, D.R. (2003) Extending the upper temperature limit for life. *Science* **301**: 934.

Kunzig, R. (2000) *Mapping the Deep: The Extraordinary Story of Ocean Science*. W.W. Norton & Co.

Martin, A., and McMinn, A. (2018) Sea ice, extremophiles and life on extra-terrestrial ocean worlds. *Int. J. Astrobiol.* **17**: 1–16.

Marx, J.-C., Collins, T., D'Amico, S. et al. (2007) Cold-adapted enzymes from marine Antarctic microorganisms. *Mar. Biotechnol.* **9**: 293–304.

Oger, P.M., and Jebbar, M. (2010) The many ways of coping with pressure. *Res. Microbiol.* **161**: 799–809.

Phadtare, S. (2004) Recent developments in bacterial cold-shock response. *Curr. Issues Mol. Biol.* **6**: 125–136.

Poli, A., Finore, I., Romano, I. et al. (2017) Microbial diversity in extreme marine habitats and their biomolecules. *Microorganisms* **5**: 25.

Rees, J. F., De Wergifosse, B., Noiset, O. et al. (1998) The origins of marine bioluminescence: turning oxygen defence mechanisms into deep-sea communication tools. *J. Exp. Biol.* 201: 1211–1221.

Siezen, R.J. (2011) Microbial sunscreens. *Microb. Biotechnol.* **4**: 1–7.

Tribelli, P., López, N., Tribelli, P.M. et al. (2018) Reporting key features in cold-adapted bacteria. *Life* **8**: 8.

Vezzi, A., Campanaro, S., D'Angelo, M. et al. (2005) Life at depth: *Photobacterium profundum* genome sequence and expression analysis. *Science* 307: 1459–1461.

Zeng, X., Birrien, J.-L., Fouquet, Y. et al. (2009) Pyrococcus CH1, an obligate piezophilic hyperthermophile: extending the upper pressure-temperature limits for life. *ISME J.* **3**: 873–876.

Zhang, Y., Li, X., Bartlett, D.H., and Xiao, X. (2015) Current developments in marine microbiology: high-pressure biotechnology and the genetic engineering of piezophiles. *Curr. Opin. Biotechnol.* **33**: 157–164.

Chapter 4

Diversity of Marine Bacteria

Although many organisms in the domain Bacteria can be grown and studied in the laboratory, the overwhelming majority are known only by genetic information obtained by direct analysis of DNA in environmental samples. Thus, there is a large discrepancy in the extent of our knowledge of the properties of the various members of the Bacteria. For those that cannot yet be cultured, we can infer many of their likely properties by considering their relationship to well-studied reference species, their habitats, and geochemical evidence relevant to their activities. Major advances in sequencing technology and bioinformatics mean that we can now predict the metabolic pathways and likely biogeochemical role through analysis of genomes reconstructed from environmental sequences. The first section of this chapter contains a synopsis of diversity and discusses the rapidly changing picture of the major phylogenetic groups of the Bacteria. This is followed by discussion of the properties of representative examples of major bacterial taxa, chosen with reference to activities of particular marine ecological or applied importance.

Key Concepts

- Phylogenetic and genomic methods have led to major changes in the systematics of the Bacteria and understanding of their diversity.

- Bacteria that are phylogenetically closely related can differ greatly in their metabolic and physiological characteristics.

- A relatively small number of major bacterial clades dominate most marine habitats, but extensive microdiversity enables colonization of specific niches.

- The problems of defining bacterial species are being resolved by application of the core genome and pangenome concept.

- There are probably millions of marine bacterial species, with many of them constituting the rare biosphere.

- Genomic analysis of cultured bacteria and metagenomic analysis of environmental samples leads to new insights into bacterial diversity, ecology, and biogeochemical roles.

OVERVIEW OF BACTERIAL DIVERSITY

Understanding of diversity has been revolutionized by phylogenetic and genomic techniques

Our understanding of the diversity of bacteria has undergone gradual evolution over the past 150+ years, since the first attempts to classify them by scientists such as Haeckel and Cohn. As classification schemes were developed in the 20th century, bacteria that could be isolated and grown in the laboratory were assigned into groups based on aspects of phenotype such as morphology (e.g. shape and grouping); structural features (e.g. Gram reaction); metabolism (e.g. nature of carbon and energy sources, enzymes, chemical products); behavior (e.g. motility, interactions); and ecology (e.g. habitats, pathogenicity). Biochemistry and genetics of bacteria gained increasing importance as knowledge of these aspects grew from the 1960s on, resulting in developments such as characterization of specific cell components (e.g. lipids, quinones, and DNA base composition). The methods underlying these traditional phenotypic approaches were described in *Chapter 2*; they are still widely used in culture-based microbiology—especially for bacterial identification—and they remain part of the recognized system by which bacteria are formally classified and named. However, these approaches are completely inadequate to provide a true picture of diversity.

As discussed in *Chapter 1*, the ideas developed by Carl Woese in the late 1970s heralded a revolution in our thinking about their relationships between different forms of life. Modern methods of classification attempt to group organisms by their presumed evolutionary relationships. Such phylogenetic systems of classification depend on comparisons of the genetic information contained in their macromolecules, especially nucleic acids and proteins. If two organisms are very closely related, we expect the sequence of the individual units (nucleotides or amino acids) in a macromolecule to be more similar than they would be in two unrelated organisms. Sequencing of the genes encoding the 16S rRNA molecules in the small ribosomal subunit (SSU) has been the most widely used tool in studies of microbial diversity, and the principles of this approach were discussed in *Chapter 2*. The development of direct sequencing of 16S rRNA genes in environmental samples in use since the late 1990s has produced huge databases of genetic information that are analyzed to identify groups of phylogenetically related sequences. With this technique, it soon became apparent that the marine environment contains organisms that are phylogenetically distinct from many previously known groups obtained by culture. By far the largest amount of bacterial diversity is represented by organisms that have not been cultivated but are known only by their genetic "signatures" observed in environmental samples. Microbiologists have long realized that there is a large discrepancy between the numbers of organisms counted using direct microscopic observation and those recovered on culture media (the "great plate count anomaly", p.34). The sequencing of 16S rRNA genes led to a complete reevaluation of the importance of marine members of the Bacteria (and Archaea, as discussed in the next chapter), which are both more abundant and more diverse than we could possibly have imagined before the advent of these techniques.

Our view of bacterial and archaeal diversity is now undergoing another major revolution as a result of analysis made possible by widespread use of high-throughput sequencing (HTS, p.51), leading to the ability to obtain many thousands of genetic sequences from individual fragments. We can now compare sets of multiple genes and obtain entire genome sequences from individual microbes. Researchers have developed vast databases of global genetic information (trillions of nucleotide bases) from different locations, depths, and habitats obtained via numerous individual and coordinated ocean sampling projects (*Figure 2.12*). These databases can be continually analyzed and reanalyzed—"mined" for information—leading to a new frontier in our understanding of diversity of marine microbes and their importance in ecological and biogeochemical processes. We are getting closer to answering some of the key questions: how many kinds of microbes are there and what are they doing? How do they interact with one another? How might they respond to future environmental changes? Some recent developments are discussed in *Box 4.1*.

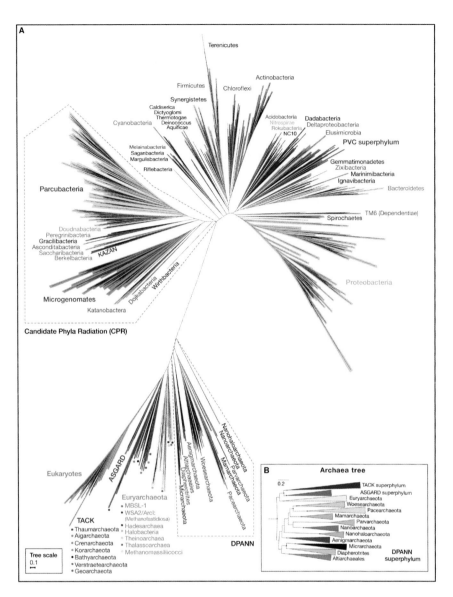

Figure 4.1 A view of the tree of life, encompassing the total diversity represented by sequenced genomes and illustrating the diversity of major bacterial lineages (upper section). The Candidate Phyla Radiation (CPR) is composed entirely of organisms without isolated representatives and are still in the process of definition at lower taxonomic levels. The lower section shows the eukaryotes as a single cluster and the Archaea as branches representing the phyla and newly designated super-phyla (see *Chapter 5*). This is an updated version of the tree devised by Hug et al. (2016) shown in *Figure 1.1C*. Credit: reprinted from Castelle and Banfield (2018) with permission from Elsevier.

Bacterial systematics is in transition due to application of genomic methods

Microbiologists have adopted the binomial system of Latin names (*Genus species*) first developed by Linnaeus for classification of plants and animals, but the concept of a bacterial species is very different. In plants and animals, the presence of distinct morphological differences, sexual reproduction, and geographic separation can all be used to underpin a theoretical explanation of the concept of species as a group of individuals that can produce fertile offspring and have evolved to be reproductively isolated from other species. This definition is biologically meaningless for bacteria, which are haploid and do not have sexual reproduction; therefore, the concept of a bacterial species is purely operational. Microbiologists use a polyphasic approach to classify bacteria into different taxa, based on observing a combination of numerous phenotypic, chemotaxonomic, and genotypic data. Members of a species should share a high degree of similarity in many independent characteristics and should be characterized by a distinctive phenotypic property (such as morphology, biochemical properties, or physiology) that distinguishes them from all other species. For a cluster of organisms to be considered members of the same species, they need to be genomically coherent, so phylogenetic information about the relatedness of properties of the strains must also be considered. This requirement led to the use of DNA-DNA hybridization (DDH) as a standard in the designation of a species: if two organisms show DDH of ≥70%

BOX 4.1   RESEARCH FOCUS

High-throughput DNA sequencing provides new insights into marine microbial diversity

The rare biosphere. The concept of the rare biosphere developed from the first use of HTS as part of the International Census of Marine Microbes Project (ICoMM, Amaral-Zettler et al., 2010). The method of tag 454-pyrosequencing (p.51) used enabled analysis of many thousands of sequences of a hypervariable region of 16S rRNA genes from a single sample. This overcame the bias inherent in previous studies of molecular diversity, in which dominant populations mask the detection of low-abundance types. Sogin et al. (2008) concluded that the number of different kinds of bacteria (operational taxonomic units, OTUs) in the oceans could exceed five to ten million and that the "rare biosphere is very ancient and may represent a nearly inexhaustible source of genomic innovation." It was later realized that the method used at this time can be prone to multiple errors in a few reads, which can lead to apparently unique sequences; Kunin et al. (2010) cautioned that careful analysis and interpretation of pyrosequencing data is needed to prevent overestimation of species richness. The Illumina technology that superseded 454-pyrosequencing produced shorter reads but had fewer errors. Researchers have continued to apply new DNA sequencing technology and develop new bioinformatic methods for data analysis to improve the reliability of estimates of richness and community structure (Pedros-Alio et al., 2018). Jousset et al. (2017) discuss the factors that drive the presence and persistence of rare microbes, which include competition and negative interactions between species. Also, specialization in the use of particular nutrients or preference for environmental conditions leads organisms to be rare in some environments and abundant in others. Grazing protists, bacterial predators, and phages are major factors in shaping community composition, because they tend to overconsume the most abundant prey, leading to trade-offs between organisms that can be explained by a theoretical framework known as the "kill the winner" hypothesis (Winter et al., 2010; also see p.202). Jousset et al. note that we may be overlooking a substantial part of the biosphere because (at that time) a standard procedure in processing the results of HTS was to eliminate sequences of singletons (sequences represented only once). They argue that low-abundance species should not be considered as "analytical annoyances" but treated as full members of microbial communities. They provide many examples of ways in which rare species can have a disproportionate role in biogeochemical cycles—"less is more"—with particular importance in response to nutrient cycling, pollutant degradation, and processes that affect levels of greenhouse gases (e.g. methane consumption and denitrification). Rare microbes can be more active than abundant types and may enhance community metabolic functions by providing a vast gene pool for unique or complementary metabolic pathways.

How many bacterial and archaeal species really exist? Surveys based on amplification and Sanger sequencing of 16S rRNA genes conducted since the 1990s soon revealed the diversity of ocean microbes. But can we put a numerical value on this diversity, rather than just concluding … "it's enormous"? Curtis et al. (2002) argued that it is not necessary to count every single species in a sample; instead, they showed how this can be achieved by measuring the area under the species abundance distribution (SAD) curve for that environment. This is a method frequently used in macroecology to measure the biodiversity

of insects, plants, fish, and other organisms. Ecological theory predicts that bacteria would show a log-normal species abundance distribution curve. Curtis et al. developed a model by which the number of individuals in a number of taxa can be derived by measuring the total number of individual bacteria in the community (N) and the number of individuals (N_{max}) of the most abundant species. Based on the limited data available at the time, and by speculating on larger scale diversity, they concluded that the entire bacterial diversity of the sea would likely be ~2×10^6 species. Subsequently, the use of HTS in the ICoMM survey mentioned above led Sogin et al. (2008) to raise that value to 5–10×10^6.

By 2016, the volume of microbial gene sequences in the databases had skyrocketed, prompting Locey and Lennon (2016) to analyze them again using these ecological scaling laws. For the first time, they combined data from biodiversity studies of microbes, plants, and animals. They concluded that they were successful in developing a scaling law that predicts the abundance of dominant species across 30 orders of magnitude (matching the estimated abundance of microorganisms on Earth). Using this scaling law, they predicted up to a trillion (10^{12}) species of microbes on Earth. Based on known distributions (Whitman et al., 1998) we might expect ~90% of these species to be marine. However, Willis (2016) criticized the methods used in the papers by Curtis et al. (2002) and Locey and Lennon (2016), arguing that extrapolating SAD curves has no predictive power for estimating true microbial diversity because it is statistically invalid. In response, Locey and Lennon argued that their method did *not* rely on extrapolation and was therefore valid, because they used values of N and N_{max} that matched the scale of predictions.

Yarza et al. (2014) analyzed the curated SILVA database for near full-length 16S rRNA sequences of Bacteria (1.4×10^6) and Archaea (5.3×10^4) and concluded that only near complete 16S rRNA sequences give accurate measures of taxonomic diversity. Schloss et al. (2016) showed that considerable biodiversity has been discovered since they conducted a similar census in 2004, but much of this had been biased towards particular phyla and environments. *Figure 4.2* shows the frequency of distribution of sequences from the various phyla. Schloss et al. also found that ~95% of new full-length bacterial and archaeal sequences belong to OTUs that have been observed more than once already. The rate of discovery of new species is slowing, as indicated by rarefaction curves (the number of species plotted against the number of samples) approaching saturation. Based on these results, the richness values of 10^{12} proposed by Locey and Lennon seem unrealistically high. Schloss et al. conclude that research to date has been effective at sampling the most abundant organisms, but it struggles to sample the rarer organisms. They emphasize the need to sample environments with high biodiversity using newer sequencing methods that will yield full-length sequences from these rare organisms. In discussion of this paper, Amann and Rosselló-Móra (2016) estimate that the census of bacterial and archaeal species might ultimately stop in the lower millions rather than the billions predicted earlier. They encourage microbiologists—rather than being dismayed at having "only" millions of species—to apply our creative minds to access the currently underexplored rare biosphere with strategies for sampling and analysis that

BOX 4.1   **RESEARCH FOCUS**

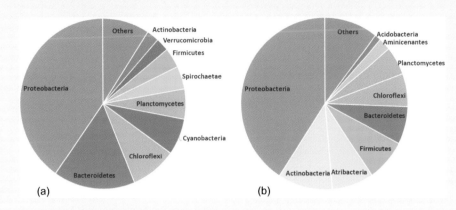

Figure 4.2 Frequency of sequences for bacterial phyla from (a) marine water samples and (b) marine sediments and hydrothermal vents. Data obtained from Schloss et al. (2016).

focuses on the unknown to seek new organisms. Without doubt, much new biology and biochemistry is still to be discovered.

New phylogenetic trees permit a new approach to bacterial taxonomy. The existing classification system for the Bacteria contains many inconsistencies with our goal of grouping organisms according to their evolutionary relationships. Many current taxa are polyphyletic; the application of 16S rRNA gene relationships has helped enormously but has not entirely resolved the problem. As discussed above, there is a huge diversity of unclassified rare taxa that are not currently affiliated with known clades. Some (many?) of these will undoubtedly possess hitherto unknown roles in marine processes. In order to focus on these unknown clades, Yilmaz et al. (2016) analyzed the SILVA database of bacterial and archaeal 16S rRNA gene sequences, using a new approach to construct phylogenetic trees that included aspects such as size, phylogenetic depth, and standard nomenclature. This involved the use of the candidate taxonomic unit (CTU) proposed by Yarza et al. (2014). Because the major known clades in SILVA have been annotated manually in a guide tree, Yilmaz et al. were able to reveal previously unknown clades. With this new phylogenetic and taxonomic framework, they performed a meta-analysis of the ICoMM marine water samples and other datasets. Because most of the new clades were interspersed by known taxa containing cultivated species whose genome has been sequenced, use of the bioinformatics software package PICRUSt could be used to predict the possible metabolic functions of the new clades. Thus, up to 92 unknown clades were characterized and classified. A different recent approach to incorporating uncultivated organisms into the tree of life was used by Hug et al. (2016). They argued that, despite its widespread use and enormous value to date, 16S rRNA gene analysis has some inherent limitations, so that many unknown groups escape our attention. To obtain higher resolution, they used an aligned and concatenated set of 16 ribosomal protein sequences deduced from available genomes for each organism (Bacteria, Archaea, and Eukarya were included) to construct a new view of the tree of life (*Figure 4.1*). The main conclusion of this study is that the domain Bacteria includes many more major lineages of organisms than either the Archaea *or* Eukarya, including the large branch termed the Candidate Phylum Radiation. Parks et al. (2018) took this approach further by developing a phylogeny inferred from concatenation of 120 ubiquitous, single-copy proteins covering ~95,000 genomes—of which 14% are assembled from metagenomes, or single-cell genomes, from uncultured organisms. This produced a new system and database (the Genome Taxonomy Database, GTDB), publicly available to guide reclassification and future classification of newly discovered bacteria. To validate their

system and to ensure that the GTDB taxa are monophyletic, Parks et al. tested their system with a variety of phylogenetic tools.

A new taxonomic system. This new approach has resulted in major reclassification of taxa above the rank of species—73% of the genomes currently assigned to a taxon within the NCBI database had one or more changes. The most changes occurred at the level of orders. Some of the most important changes at the higher ranks affect important and well-studied groups with culture-based classifications. These changes include: (a) reclassification of the class Betaproteobacteria as an order of the class Gammaproteobacteria; (b) removal of the Epsilonproteobacteria from the Proteobacteria and reclassification as a new phylum Epsilonbacteraeota; (c) reclassification of the order Desulfurellales (Deltaproteobacteria) to a new class within the Epsilonbacteraeota; (d) internal reclassification of the Firmicutes; and (e) transfer of the Chlorobi to the phylum Bacteroidetes. The GTDB system also allows the uncultivated microbes to be assigned to named candidate taxa. In particular, the large Candidate Phyla Radiation appears to be a monophyletic group for which Parks and colleagues propose the name Patescibacteria. It is hoped that maintenance of an up-to-date normalized genome-based classification will improve the analysis of microbial genome data and consistent reporting of diversity in scientific communication. But there will undoubtedly be continued debate about the merits of such a major overhaul of long-established taxonomic system.

Towards a new concept of bacterial species. It is now possible to envisage a genomic basis for defining species (and higher taxa) that takes account of the huge microdiversity observed and the difficulty of defining the boundaries of taxa. By comparing multiple genome sequences of related organisms, we can define the *pangenome*—the collection of all genes across the set of strains examined. We can then ask which genes are present in every strain—this is the *core* genome, which could be used as the proxy for defining a species. Similarity of genomes can then be quantified by "digital hybridization" of conserved regions of the query and reference sequences, or by comparing the average nucleotide identity (ANI) of the genome of the query strain with that of a refence strain (Richter and Rosselló-Móra, 2009). These methods could replace the current "gold standard" of laboratory determination of DNA-DNA hybridization, which is technically too demanding (Chun et al., 2019). Currently, the nomenclature rules require type cultures to be deposited in order for a species names to be validly published, but in future, genome type material may be permitted for uncultured organisms (Konstantinidis et al., 2017).

they are considered to be of the same species. DDH is a difficult technique and prone to errors, therefore comparison of 16S rRNA gene sequences became widely accepted as the mainstay of bacterial taxonomy. For many years, two strains were considered as belonging to distinct species if they shared 16S rRNA similarities <97% and different genera at <95%. The species threshold was raised to 98.6% in 2006. A variety of other indirect methods of genome comparison are shown in *Table 2.2*, and the currently preferred method for comparing genome sequences of different organisms is to calculate the overall genome relatedness, most commonly by calculating the ANI, where the cutoff value for the species boundary is 95–96%. The rapid development of relatively inexpensive and technically straightforward whole genome sequencing (WGS), plus a wealth of associated bioinformatics tools, means that taxonomy is shifting from its reliance on these indirect methods. WGS can provide a shortcut to direct evidence about the biochemical, physiological or structural features that are traditionally tested in polyphasic taxonomy. However, new classifications must be compatible with existing systems and there are many different opinions about how these novel approaches should be integrated.

While the designation of species is difficult and results in differences of opinion among experts, agreement about grouping species into higher taxa is even more difficult. Boundaries for defining taxa using 16S rRNA gene sequence identities are proposed at 94.5% for genus, 86.5% for family, 82.0% for order, 78.5% for class, and 75% for phylum. As discussed in *Box 4.1*, new genome-based criteria for the definition of species and higher taxa are now being implemented.

Although there is still no formal system of classifying bacteria, there are strict rules about the designation and naming of new taxa under a Code of Nomenclature regulated by the International Committee of Systematics of Prokaryotes. The main aims of the code are to ensure that names are stable and necessary and to avoid errors or confusion that may hamper scientific progress. Expert sub-committees keep the status of taxa in particular groups under review. A variety of characteristics must be considered before a proposed name for a new species, genus, or family can be defined as valid. Detailed rules about the published descriptions, etymology of the name, and diagnostic features are required. A key requirement is the designation of a "type strain" of a species and its deposit in internationally recognized culture collections, where it is archived and maintained in a viable form as a reference point for assigning other strains to the species. It is important to note that the Code only specifies rules about naming microbes and, like all biologists who specialize in taxonomy, microbiologists often have many differences of opinion about the assignment of organisms to taxa (taxonomists are often categorized as "lumpers" or "splitters"). There are currently over 15600 validly named species of Bacteria and Archaea.

Uncultured bacteria cannot be assigned to a valid species, because we do not have full information about their phenotype and there is no type strain, as required by the Code of Nomenclature. However, in some cases, it is possible to assign a provisional species name with *Candidatus* designation if genomic sequences point to significant differences from other recognized species in structure, metabolism, and other key properties. Names are written in the format "*Candidatus* Genus species" (note the use of quotation marks and italics); *Candidatus* may be abbreviated to *Ca*. This interim status has been extended to allow the designation of candidate higher taxa for uncultivated organisms.

OTUs and ASVs are used to represent diversity in community analyses

The difficulties of defining bacterial species means that we need a convenient way to identify groups of closely related organisms in gene-based surveys. Operational taxonomic units (OTUs) have been the traditional way of doing this, as they serve as a pragmatic proxy for species. Once 16S rRNA gene sequences have been obtained, computer algorithms can cluster them into OTUs on the basis of similarity—typically ≥98.6% to match the threshold for species. Sequences for other marker genes can be grouped in a similar way, using appropriate thresholds set by the researcher. This approach has been used in thousands of ecological investigations since the advent of environmental gene sequencing in the 1990s.

WHAT'S IN A NAME? BACTERIAL TAXA

A few guidelines may help to navigate the plethora of Latin bacterial and archaeal names. Orders and families take their name from the stem of the name of the type genus, and the suffix indicates the rank; families end in -aceae, orders in -ales, and classes in -ia or -ea. For example, *Vibrio cholerae* is the type species of the family Vibrionaceae, order Vibrionales, class Gammaproteobacteria. Genus and species names must always be written in italics (or underlined). The genus name can be abbreviated to single letter after first use, or a three-letter abbreviation if it would cause confusion. For example, *Aeromonas salmonicida* and *Aliivibrio salmonicida* would appear as *Aer. salmonicida* and *Ali. salmonicida* in a list. After a genus name, sp. and spp. (not italicized) indicate a single or multiple undesignated species, respectively. Names of all formal taxa start with a capital letter. According to the nomenclature codes, names of validly published higher bacterial, archaeal, and viral taxa should also be italicized but names of eukaryotic taxa are not. Publishers vary in application of this convention and you will see both forms in journal articles and books. In this book, names of all taxa above genus are not italicized.

A more recent approach is to compare all of the individual sequences present in a community, generating amplicon sequence variants (ASVs). (The term "exact sequence variants," ESVs, is sometimes used.) Until recently, this has been problematic because of small errors in amplification and sequencing of the marker gene, but recent improvements in control of these errors means that it is now possible to reliably distinguish sequences differing by as little as one nucleotide. The main advantage of ASVs is their improved resolution, because the species similarity threshold can mask considerable differences between microbes that have distinct phenotypic or ecological attributes. (As an example, see the discussion of genome differences in ecotypes of the species *Prochlorococcus marinus* on p.135.) A second advantage is that ASVs in different datasets can be directly compared against one another, which makes comparison between different studies more efficient. As with conventional taxonomy, there are differences of opinion—lumpers and splitters—and some consider that the lower resolution of OTUs can be advantageous for some purposes in analyzing diversity. There is a growing trend to encourage all ecological survey data to be reported using ASVs—if all datasets are reported in ASVs, the advantage of easy comparison is retained, but investigators can always recompute sequences as OTUs if they wish. Because most microbial community analyses to date have been reported in OTUs, this system is used in this chapter.

When considering microbial diversity, three general ecological concepts are used. Alpha diversity refers to the average diversity of OTUs found in a habitat or particular sample, i.e. at a local scale. It is usually expressed as richness, i.e. the number of different OTU types. Beta diversity is the ratio between diversity at a local scale (alpha) and a regional scale; in other words, the amount of change in the OTUs between the two habitats. Gamma diversity is a measure of the overall diversity of an ecosystem or large region. Richness is the simplest measure of diversity, but it does not tell us about evenness of distribution—the relative abundance of different OTUs. Ecologists have devised many ways to quantify these aspects of diversity into a single index, and discussion of the methods of calculation and merits of different approaches is beyond the scope of this book. Besides the simple measure of richness, two of the most important measures encountered in microbial ecology are the Shannon-Wiener index (a measure of evenness of populations within a community) and the Simpson index (a measure of dominance, weighted towards the abundance of the most common OTUs) and then used to measure proportional abundance.

Marine microbial communities show high alpha diversity

Using the new method of direct sequencing of nucleic acids from the marine environment, the first studies conducted in the early 1990s soon revealed that the most abundant clones of 16S rRNA bacterial gene sequences did not correspond to cultured species at all. When phylogenetic trees were constructed based on genetic similarity, more than three-quarters of the marine Bacteria revealed by this approach mapped onto about 20 phylogenetic groups with worldwide distribution. These formed clusters of related types, rather than single lineages, and genetic variability within these populations was observed to be very high. In some clusters, differences in 16S rRNA sequences were seen to be due to genetic variability associated with adaptation to different depths of water. Subsequent analyses showed that bacterial diversity in suspended particles and free picoplankton is very different and varies greatly with depth and environmental conditions.

Many of the initial studies were conducted in the Sargasso Sea, and new sequences that could not be linked to cultured bacterial groups were assigned to clades with the prefix SAR. Among the early discoveries was the abundance and wide distribution of the SAR11, SAR116, and *Roseobacter* clades, whose phylogeny indicated them to be members of the Alphaproteobacteria. The SAR 11 cluster of 16S rRNA sequences was found in almost every pelagic environment ranging from shallow coastal waters to depths of over 3000 m. It is probably the most abundant group of microorganisms in the sea and on average accounts for nearly a third of the cells present in surface waters and a fifth of the cells in the mesopelagic zone. SAR11 is a deeply branching member of the class Alphaproteobacteria and is phylogenetically distinct from all cultured members of the group. Some members of the SAR11 cluster have now been cultured and this has allowed study of genomic and physiological properties, as discussed in *Box 2.1.*

In these studies, members of the SAR86 clade (class Gammaproteobacteria) also emerged as ubiquitous inhabitants of the surface layers of the oceans and shallower coastal waters and were soon linked to a previously unknown type of photoheterotrophic metabolism in oligotrophic conditions (*Box 3.1*). These studies also showed that sequences matched to the phylum Cyanobacteria accounted for a large segment of the diversity in the photic zone. This was less surprising, since many can be cultured and some—notably the genera *Prochlorococcus*, *Synechococcus*, and *Trichodesmium*—had already been extensively studied in view of their importance in contributing to productivity through photosynthesis. By the turn of the century, the picture emerging of marine bacterial diversity was one in which over three-quarters of the sequences could be grouped into about 12 dominant clades.

Since these early investigations, many research groups have conducted studies in diverse marine habitats throughout the world and established databases containing many thousands of sequences from marine samples. Different investigators, using variations of the basic methods and sampling different geographic regions and depths have repeatedly found similar phylogenetic groups, which can be clustered into a number of major clades. The vertical structure of the water column emerged as being the most important factor in overall variation of diversity in the oceans. There are also high levels of beta diversity, with the composition of microbial communities greatly affected in time and space, with seasonal and geographic variations.

However, this "big picture" view hides a staggering amount of alpha diversity at the microscale. This finding was surprising, because ecological theory predicts that competition for the limited nutrient resources found in most of the ocean should lead to low levels of diversity. We can now explain this paradox through our realization of the importance of factors that are explored in other chapters. Firstly, patchy distribution and gradients of nutrients at the microscale, together with syntrophic and symbiotic interactions, results in heterogeneity of the environment, providing different niches that can support diverse microbes (*Chapter 3*). Secondly, top-down control of populations by protistan grazing and phage lysis have major influences on community structure and promote diversity by attacking the most abundant or active members of the community (*Chapters 6* and *7*).

As discussed in *Box 4.1*, the International Census of Marine Microbes (ICoMM) was established in 2004 to coordinate and catalog our knowledge of microbial diversity in the oceans, facilitated by high-throughput sequencing (HTS), emerging at this time through the development of the 454 Life Science sequencing technology (p.51). One of the most significant and surprising early discoveries of ICoMM was the realization that the abundance distributions of different OTUs are highly uneven, leading to the concept of the "rare biosphere." Despite their low abundance, rare microbes can have a disproportionally large effect on ecosystem functioning.

A TOUR OF THE BACTERIAL AQUARIUM

Even among the tiny fraction of all bacteria that we can culture and study in the laboratory, there are over 15000 recognized species with published names, together with several hundred organisms sufficiently well characterized to be given *Candidatus* species status. These are organized into hundreds of higher taxa, many of which contain organisms that cannot be grouped by phenotype. This makes it impossible for a book like this to provide a systematic description of all types of marine bacteria. Instead, I will imagine that we are visiting a virtual aquarium containing representatives of some of the most noteworthy and best-understood groups (a concept adapted from James W. Brown). Just as a real aquarium with a complete collection of all the world's fish would overwhelm the visitor, our bacterial aquarium contains selected examples that have particularly interesting characteristics or because they are important in marine ecology, biogeochemical cycles, or biotechnology. Mostly, the "tanks" are arranged by phyla, but in some places interesting properties from other phyla are mentioned in order to keep similar features together.

The discussion that follows includes information about the properties of some of these cultured, named species as well as information about related types that is gained from genomes constructed by metagenomic assembly or amplification from single cells extracted from the environment (see p.56 for a summary of methods). Note that the description of taxa given here mostly follows the current LPSN/NCBI organization, but some significant proposed changes to higher taxa (especially phyla) arising from the new genome-based GTDB classification (see *Box 4.1*) are included.

The Proteobacteria account for about half of all bacterial ocean diversity

Figure 4.2 shows the frequency of the most abundant sequences in marine water and sediment samples grouped by phyla. These show that the phylum Proteobacteria is the most abundant in 16S rRNA gene surveys and it is widely distributed in all major habitats. Almost all the main metabolic types (autotrophy, heterotrophy, anoxygenic phototrophy, methylotrophy, sulfate reduction, nitrogen fixation), requirements for oxygen and mechanisms of CO_2 fixation are represented, and many types of cell morphology are found. Cells have Gram-negative structure. The group has conventionally divided into six classes; namely the Alpha-Beta-, Gamma-, Delta-, Epsilon-, and Zetaproteobacteria (often abbreviated using the Greek letters α, β, γ, δ, ε, and ζ). Using the GTDB classification, only the Alpha-, Gamma-, and Zetaproteobacteria remain as classes within the phylum. The Betaproteobacteria is now an order within the Gammaproteobacteria class. Members of the Delta- and Epsilonproteobacteria are assigned to new phyla. Despite this reorganization, the biggest tank in our virtual aquarium is undoubtedly reserved for the phylum Proteobacteria and we begin our tour with this group (*Table 4.1*).

Members of the class Alphaproteobacteria are the most abundant marine bacteria

This class includes many types with important functions in marine habitats. Many have been isolated and cultured and *Table 4.2* contains a brief summary of representatives of the 434 cultivated genera in 15 families and 14 orders. This is a tiny fraction of the total diversity revealed by 16S rRNA gene surveys of euphotic and mesopelagic zones. Members display a wide range of metabolic types, evident in cultivated isolates as well as by evidence from metagenomes. The most abundant types, which include the dominant clades SAR11, SAR116, OCS116, and OM75, are slow-growing oligotrophs with small cells and streamlined genomes. SAR11 is often described as the most abundant type of bacteria on Earth, although this hides the fact that the clade can be divided into multiple subclades which form distinct ecotypes that show clear differences in dynamics due to environmental conditions. Many of the Alphaproteobacteria are metabolic generalists containing genes for utilizing light via proteorhodopsin and for metabolizing C1 compounds. Representatives of SAR11 and SAR116 have been cultured and given the *Candidatus* names "Pelagibacter ubique" and "Puniceispirillum marinum" respectively. These are closely related to the Rickettsiales while OM75 is related to the Rhodospirillaceae. The *Roseobacter* clade (family Rhodobacteriaceae) is also very abundant, sometimes representing up to 20% of bacterial cells in coastal ecosystems and 3–5% in the open ocean, blooming seasonally in association with algal growth. The genomes of many members of the *Roseobacter* clade have been sequenced, revealing enormous diversity of metabolic potential and regulatory systems that explain their ubiquity and success in many marine habitats. There is considerable variation in genome size—some are highly streamlined, while others are quite large, with over 4000 protein genes. "Ca. *Riegeria santtandreae*" is an intracellular symbiont with a genome of only 1.34 Mb, yet it retains a highly energy-efficient metabolism and serves as the primary energy source for its host, the flatworm *Paracatenula* (p.267).

The order Caulobacterales contains prosthecate bacteria

Members of the Caulobacterales are distinguished by the presence of extrusions or appendages of the cytoplasm called prosthecae. Unlike almost all other bacteria, members of this group show a life cycle with unequal cell division, in which the "mother cell" retains its shape and morphological features while budding off a smaller "daughter" cell. This occurs

Table 4.1 Representative members of the class Alphaproteobacteria. There are 14 named orders

Representative family[a,b]	Distinctive features	Representative marine genera or species[b]
ORDER Caulobacterales (2)		
Caulobacteriaceae (4)	Mostly stalked cells that produce a flagellated cell. Chemoheterotrophs attached to surfaces in soil, freshwater, and marine habitats.	*Caulobacter crescentus*
Hyphomonadaceae (15)	Some divide by binary fission with motile offspring, others have stalked cells. Oligotrophic chemoheterotrophs in a wide range of marine habitats.	*Algimonas, Hellea, Woodsholea, Oceanicaulis*
ORDER Rhizobiales (17)		
Aurantimonadaceae (4)	Small family of marine chemoheterotrophs, including the coral pathogen *Aurantimonas coralicida*.	*Aurantimonas, Fulvimarina, Martelella*
Bradyrhizobiaceae (10)	Mostly plant-associated nitrogen-fixers; *Rhodopseudomonas* is a marine phototroph. *Nitrobacter* is a widely distributed chemolithoautotrophic nitrifying bacterium.	*Nitrobacter, Rhodopseudomonas*
Methylobacteriaceae (3)	Pink-pigmented facultative methylotrophs growing on C1 compounds, as well as larger carbon compounds. Sometimes associated with human infections and contamination of fuel oils.	*Methylobacterium*
Methylocystaceae (10)	Type II methanotrophs utilizing methane and C1 compounds as source of carbon and energy; Distinguished by internal membranes containing methane monooxygenase around the periphery of the cell and assimilating C1 intermediates via the serine pathway. Widespread in sediments at the interface between CH_4 and O_2.	*Methylocystis, Methylosinus*
Rhizobiaceae (13)	*Rhizobium* is mostly known for symbiotic nitrogen fixation in plants. Some marine species have been isolated from marine sediments and coral mucus. *Methylobrevis* are facultative methylotrophs.	*Rhizobium, Methylobrevis*
ORDER "Pelagibacterales"		
"Pelagibacteriaceae"	Comprises at least five related groups of bacterioplankton found in marine and fresh waters that constitute up to half of all bacterial and archaeal cells in the open ocean. Includes the founder group SAR11. Oligotrophs with small cells and streamlined genomes, phylogenetically closely related to the *Rickettsiales* and the ancestor of the mitochondrion.	"Ca. **Pelagibacter** ubique"
ORDER Rhodobacterales (1)		
Rhodobacteriaceae (148)	Very large family with extensive metabolic and ecological diversity. Highly abundant in marine environments, with particular importance as aerobic photo- and chemoheterotrophs and in the carbon and sulfur cycles. Many form close associations or symbiotic interactions with other organisms (especially phytoplankton).	*Dinoroseobacter, Jannaschia, Pelagimonas, Phaeobacter, Rhodobacter, Roseobacter, Roseovarius, Ruegeria, Silicibacter*
ORDER Rhodospirillales (2)		
Rhodospirillaceae (37)	Diverse metabolism. including purple non-sulfur bacteria found on sediments or anaerobic niches of microbial mats; these can grow photoheterotrophically or anaerobically via fermentation or aerobically via respiration. Includes magnetotactic species found in salt marshes and fresh water.	*Rhodospirillum, Magnetospirillum,* "Ca. Riegeria santtandreae"
ORDER Sphingomonadales (2)		
Sphingomonadaceae (19)	Contain distinctive membrane sphingolipids. Widely distributed aerobic chemoorganotrophs, usually associated with rich sources of organic matter. Some species degrade polyaromatic hydrocarbons and toxic compounds, which may be important in bioremediation. Some species produce industrially useful polymers.	*Sphingomonas*

[a] Figures in parentheses indicate number of validly published taxa in the next lower rank (families in orders, genera in orders) from the List of Prokaryotic Names with Standing in Nomenclature (LPSN, www.bacterio.net). [b] Names in non-italic font and quotation marks are not validly published under the Bacteriological Code. Genera may have *Candidatus* status, but affiliation to higher taxa is not confirmed.

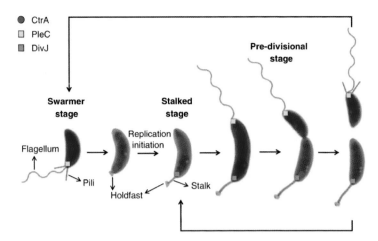

Figure 4.3 The dimorphic cell cycle of *Caulobacter crescentus*. Precise spatiotemporal patterning of the key regulatory proteins CtrA, PleC, and DivJ coordinate cell polarity, DNA replication, cell division, and flagellar biosynthesis. Cells are false colored red to indicate presence and distribution of active CtrA. Credit: reprinted from Hughes et al. (2012) with permission from Elsevier.

owing to polarization of the cell, whereby new cell wall material is grown from a single point rather than by intercalation as occurs in other bacteria. Some, such as *Hyphomicrobium* and *Rhodomicrobium*, bud off progeny cells from hypha-like extensions. Others, such as *Caulobacter*, have distinctive stalks by which they attach firmly to algae, stones, or other surfaces in aquatic environments. The progeny cells are motile and swim away to colonize fresh surfaces (*Figure 4.3*). *Caulobacter crescentus* has been intensively studied as a model of cellular differentiation. The cell cycle is controlled by a set of three master regulator proteins that determine the coupling of gene expression to cell-cycle events with extraordinary precision. The increased surface area: volume ratio of stalked bacteria probably enables them to thrive in nutrient-poor waters and the prosthecae also enable these aerobic bacteria to remain in well-oxygenated environments by avoiding sinking into sediments. These bacteria are often the first to colonize bare surfaces and are therefore important in the formation of biofilms, with significant consequences for larval settlement and biofouling (*Chapter 13*).

Several alphaproteobacterial genera show magnetotaxis

The genera *Magnetospira*, *Magnetospirillum*, and *Magnetovibrio* (Order Rhodospirillales) and *Magnetococcus* (Order Magnetococcales) show magnetotactic behavior, orienting themselves in the Earth's magnetic field. The rod-shaped or spiral cells contain chains of magnetic particles comprised of magnetite (Fe_3O_4) and greigite (Fe_3S_4) anchored within the cell (*Figure 4.4*). A magnetic field imparts torque on the chain of magnetosomes, so that the cell becomes aligned with magnetic field lines. The bacteria swim in this direction using polar flagella. The bacteria use this behavior in conjunction with aerotactic responses to locate a favorable zone in sediments of optimal O_2 and sulfide concentrations, which they require for growth. Although the best-studied examples are found in freshwater mud, magnetotactic bacteria are widespread in salt marshes and other marine sediments. Recent molecular studies indicate that magnetotaxis is not restricted to a small, specialized lineage as previously thought, and there are examples in the Deltaproteobacteria (see below). Magnetotactic bacteria can be isolated quite easily by applying a magnetic field to mixed environmental samples, although they are very difficult to cultivate. Analysis of the genome of these bacteria reveals that several genes are required for the acquisition of iron via siderophores and energy-dependent transport into the cell, followed by its conversion to magnetite or greigite within the magnetosomes. Microfossils of magnetite crystals found in deep-sea marine sediments are thought to originate from magnetotactic bacteria at least 50 MYA.

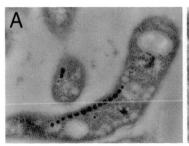
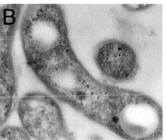

Figure 4.4 Transmission electron microscopy images of *Magnetospirillum* sp. AMB-1. A. Grown under iron-rich conditions, showing a chain of magnetite crystals. B. Grown in the absence of iron, showing empty magnetosome vesicles. Credit: reprinted from Komeili et al. (2004) with permission; copyright 2004 National Academy of Sciences, USA.

Figure 4.5 Electron microscopy images of "*Ca.* Magnetoglobus multicellularis". A. TEM showing the multicellular consortium consisting of about 40 bacterial cells containing lipid inclusions (asterisks) and magnetosomes surrounding an acellular space (Ac). The cells and the magnetosomes are specially arranged so that the total magnetic moment aligns the whole structure so it can swim along field lines. Bar = 1 μm. B. SEM showing the spherical morphology and the tightly bound cells. Bar = 2 μm. Credits: A. reprinted from Martins et al. (2007) with permission of John Wiley and Sons. B. reprinted from Silva et al. (2008). With permission of Elsevier.

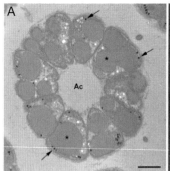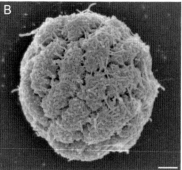

HOW DO BACTERIA MAKE NANO-SIZED MAGNETS?

The individual magnetosomes in magnetotactic bacteria are too small to behave as magnets on their own and must be aligned into a linear chain so that they function like a compass needle. The physical tendency of magnetic dipoles would be to agglomerate in order to lower the magnetostatic energy, so the magnetosomes are stabilized with an organic structure to anchor the chain in place. Scheffel et al. (2006) identified a key gene in *Magnetospirillum gryphiswaldense* that encodes the protein MamJ, which forms a novel filamentous structure. A MamJ deletion mutant showed reduced magnetic orientation; it produces normal magnetite crystals, but these assemble into clusters instead of chains. MamJ is laid down as a cytoskeleton-like structure from pole to pole of the cell, guiding the formation of the magnetosomes and magnetite crystals. Recently, Taoka et al. (2017) showed that daughter cells receive a functioning magnetic assembly at cell division, and the static chain-like arrangement of magnetosomes is needed to ensure this. The magnetosome gene clusters are highly conserved (Kolinko et al., 2016) and recent phylogenetic analysis based on metagenomic studies suggests that magnetotaxis evolved very soon after life began ~3 BYA (Lin et al., 2018).

Magnetotaxis is also found in other classes and phyla

There is growing evidence that the phenomenon of magnetotaxis is much more widespread than previously thought. In the Proteobacteria phylum, examples are now known from the Gamma-, Delta-, and Zetaproteobacteria classes. The phyla Nitrospirae and Planctomycetes, and the candidate phyla "Latescibacteria" and "Omnitrophica" also contain magnetotactic bacteria.

A unique bacterium named "*Ca.* Magnetoglobus multicellularis" forms a compact spherical aggregate of highly organized flagellated bacterial cells that swim in either helical or straight trajectories (*Figure 4.5*). The bacteria behave as a truly multicellular organism, as the aggregate grows by enlarging the size of its cells and doubling the volume of the whole structure. Cells divide synchronously and separate into two identical spherical aggregates. The classification of this bacterium is uncertain; it is currently in the Deltaproteobacteria and has not yet been assigned in the GTDB taxonomy.

The order Betaproteobacteriales includes many rare OTUs

This group has recently reclassified from a class to an order within the Gammproteobacteria in the GTDB system. It is metabolically diverse, containing 8 orders, 15 families, and ~192 cultivated genera. Many marine members are copiotrophic heterotrophs and they are frequently found in association with animal and plant surfaces. Others are chemolithoautotrophic, of which the most important types are sulfur-oxidizing bacteria (SOB) which use oxygen (or sometimes nitrate or nitrite) as terminal electron acceptors; some oxidize iron ($Fe^{2+} \rightarrow Fe^{3+}$). They occur at the interface between oxic and anoxic zones of sediments and mats (p.80). Among the best-known genera is *Thiobacillus*, although many of its species have recently been reclassified to other genera. OM43 is a significant uncultured clade of methylotrophs, which degrades C1 compounds (but not methane). Its genome has been sequenced and shown to be the smallest known for a free-living organism, but its taxonomic position in the class is not clear. Betaproteobacteria account for ~5–10% of retrieved 16S rRNA gene sequences in marine samples, many of which occur in very low abundance, as part of the rare biosphere.

The Gammaproteobacteria is a very large and diverse class

Cultivated members of this class comprise over 21 orders, 56 families, and 351 genera. It contains some of the most familiar bacteria encountered in laboratory microbiology as they are ubiquitous in aquatic and terrestrial habitats and found almost everywhere apart from extremely hot or alkaline environments; many are easily cultured with rapid growth; some are important as pathogens of plants, animals, and humans. All major types of metabolism are represented. In several large orders, notably the Alteromonadales, Oceanospirillales, and the Vibrionales, the species are almost entirely marine. Examples of cultivated representative marine genera are shown in *Table 4.2*. In marine environmental gene surveys, Gammaproteobacteria account for 25–30% of retrieved sequences and bacteria in many of the clades have not yet been cultured. A number of clades have been recognized in low to medium productivity ocean waters, of which the SAR86 clade is prevalent. This coexists with

Table 4.2 *Representative members of the class Gammaproteobacteria. There are 21 named orders*

Representative family[a, b]	Distinctive features	Representative marine genera[b]
ORDER Enterobacterales (1)		
Enterobacteriaceae (58)	Mainly occur as commensals or pathogens in the gut of warm-blooded animals; important as contaminants in sewage-polluted coastal waters which may infect marine mammals. *Serratia marcescens* from this source causes a coral disease. Facultatively anaerobic, usually motile by peritrichous flagella.	*Enterobacter, Escherichia, Salmonella, Serratia*
ORDER Vibrionales (1)		
Vibrionaceae (13)	Collectively known as "vibrios;" facultatively anaerobic, typically curved cells, usually motile by polar flagella. Copiotrophic heterotrophs, associated with surfaces, with a major role in biofilm formation. Includes bioluminescent symbionts of squid and fish. Many species are pathogenic (see text).	*Aliivibrio, Enterovibrio, Grimontia, Photobacterium, Salinivibrio, Vibrio*
ORDER Aeromonadales (2)		
Aeromonadaceae (6)	Aerobic or facultatively anaerobic heterotrophs. Some are denitrifiers with fermentative metabolism. Halotolerant, typically in freshwater and estuarine sediments. *A. salmonicida* is a pathogen in mariculture and *A. hydrophila* is an opportunistic human pathogen.	*Aeromonas, Oceanimonas, Zobellella*
ORDER Alteromonadales (10)		
Alteromonadaceae (16)	Strictly marine, obligately aerobic fast-growing heterotrophs readily isolated from marine particles or surfaces of algae and animals. Motile with polar flagellum. Large genomes encode multiple pathways for degradation of complex macromolecules (some degrade agar).	*Agarivorans, Alteromonas, Glaciecola, Marinobacter*
Colwelliaceae (3)	Strictly marine, aerobic or facultative. *Colwellia* includes obligate psychropiezophiles with biotechnological applications.	*Colwellia, Thalassomonas*
Pseudoalteromonadaceae (3)	Frequently isolated from the surface of algae and marine snow. Versatile metabolism with a wide range of hydrolytic enzymes and secondary metabolites with antimicrobial influences on biofilm and biofouling development.	*Pseudoalteromonas*
Shewanellaceae (2)	Facultative anaerobes. Mostly psychrophilic, some are extreme piezophiles from the deep sea. Psychrotolerant species are associated with fish surface microbiota and important in spoilage. Some species show dissimilatory reduction of metals with applications for microbial fuel cells.	*Psychrobium, Shewanella*
ORDER Cellvibrionales (5)		
Cellvibrionaceae (12)	Mostly found in sediments or in association with marine invertebrates. "*Ca.* Endobugula sertula" is an endosymbiont of the bryozoan *Bugula*. *Teredinibacter turneae* is a diazotrophic symbiont of wood-boring shipworms.	"*Ca.* Endobugula", *Teredinibacter*
Spongiibacteraceae (5)	Originally thought to be specific to marine sponges but appears to be a widely distributed, rare inhabitant of seawater. Motile, aerobic heterotrophs. Some contain proteorhodopsin.	*Dasania, Spongiibacter*
ORDER Chromatiales		
Chromatiaceae (31)	Mostly anaerobic photoautotrotphs known as the purple sulfur bacteria, using H_2S as electron donor for photosynthesis and depositing sulfur granules inside the cell. A major group found in many aquatic habitats, including the upper layers of sulfidic marine sediments and mats and as intracellular symbionts of nematodes and gutless worms. Some are photoheterotrophic. *Nitrosococcus oceani* is a widely distributed planktonic chemolithoautotroph oxidizing ammonia.	*Chromatium, Nitrosococcus, Phaeobacterium, Thiococcus, Thiospirillum,* "*Ca.* Thiosymbion"

(Continued)

Table 4.2 (Continued) *Representative members of the class Gammaproteobacteria. There are 21 named orders*

Representative family[a, b]	Distinctive features	Representative marine genera[b]
Ectothiorhodospiraceae (19)	Mostly marine purple sulfur bacteria with spiral cells; sulfur is deposited outside the cells. Some species are extremely halophilic and alkaliphilic, found in salterns, coastal lagoons, and salt lakes.	*Aquisalimonas, Ectothiorhodospira, Halorhodospira*
ORDER Methylococcales (3)		
Methylococcaceae (17)	Includes type I methanotrophs, able to use methane and other C1 compounds as sole carbon and energy sources. Contain methane monooxygenase in bundles of membrane vesicles in the cytoplasm and assimilate carbon via the ribulose monophosphate pathway. Membranes contain sterols. Play a critical role in oxidation of methane generated in sediments.	*Methyloprofundus*
ORDER Oceanospirillales (7)		
Alcanivoraceae (4)	Aerobic, hydrocarbon-degrading bacteria especially common in oil-contaminated environments (see text).	*Alcanivorax, Fundibacter, Kangiella*
Hahellaceae (6)	Aerobic, rod-shaped cells mainly associated with marine invertebrates including sponges, corals, molluscs, and other invertebrates as possible facultative symbionts. *Endozoicomonas* may be a member of the core microbiomes of corals but its functional role is unclear (p.283)	*Allohahella, Endozoicomonas, Hahella.*
Oceanospirillaceae (19)	Spiral, motile rods Halophilic and widely distributed in seawater, marine organisms, and sediment. Unique capabilities for hydrocarbon degradation (see text).	*Marinospirillum, Neptunomomas, Oceanospirillum*
ORDER *Thiotrichales* (4)		
Piscirickettsiaceae (8)	Highly diverse family of aerobic heterotrophs with varied characteristics. *Cycloclasticus* spp. degrade polycyclic aromatic hydrocarbons and are widely distributed in marine sediments, especially contaminated coastal sites and harbors. *Methylophaga* is a halophiic methylotroph; some are chemolithoheterotrophs. *Piscirickettsia salmonis* is a major pathogen in salmon aquaculture.	*Cycloclasticus, Methylophaga, Piscirickettsia*
Thiotrichaceae (9)	Cells are typically large and arranged in filaments arranged in large filaments rods; SOB or chemoheterotrophs prominent in sulfur-rich sediments (see text).	*Beggiatoa, Leucothrix, Thioploca, Thiomargarita, Thiothrix*

[a] Figure in parentheses indicate number of validly published taxa in the next lower rank (families in orders, genera in orders) from List of Prokaryotic Names with Standing in Nomenclature (LPSN, www.bacterio.net). [b] Names in non-italic font and quotation marks are not validly published under the Bacteriological Code. Genera may have *Candidatus* status, but affiliation to higher taxa is not confirmed.

SAR11 and the other alphaproteobacterial clades noted above, with similar metabolic potential and streamlined genome, although genome analysis—obtained by single-cell genome amplification—reveals it differs in the use of certain carbon substrates which lessens competition for nutrients. SAR86 can be divided into several ecotypes that show seasonal variations. The role of proteorhodopsin genes in ATP generation was first identified from SAR86 sequences (*Box 3.1*).

The Gammaproteobacteria includes many uncultivated species of sulfide-oxidizing bacteria (SOB)

The chemolithoautotrophic SOB are of major importance in sulfur cycling in sediments and mats, and as intracellular symbionts of invertebrates. The latter are mostly uncultivated and unclassified. They include the tubeworm intracellular symbiont "*Ca.* Endoriftia pachyptila," whose 3.4 Mbp genome possesses a wide range of genes permitting a free-living stage outside its host, while "*Ca.* Ruthia magnifica" has a highly reduced genome of 1.16 Mbp and is an obligate clam symbiont. Other unclassified SOB symbionts include "*Ca.* Thiodiazotropha spp." found in lucinid clams. These symbionts are discussed in *Chapter 10*.

The family Vibrionaceae includes many important pathogens and symbionts

Because of their importance, members of this family (commonly referred to as the vibrios) have been extensively studied and their classification has been frequently revised. The family is named after the type species *Vibrio cholerae*, which occurs in association with zooplankton in estuarine and coastal habitats but colonizes humans, causing epidemics of cholera. Other important human pathogens include *V. parahaemolyticus*, *V. vulnificus*, and *V. alginolyticus*, associated with seafood-borne or wound infections, as discussed in *Chapter 12*. A wide range of *Vibrio* spp. also cause economically important diseases of fish (e.g. *V. anguillarum*, *Aliivibrio salmonicida*, *Photobacterium piscicida*), molluscs (e.g. *V. splendidus*, *V. tubiashii*), and crustaceans (*V. harveyi*, *V owensii*, *V. nigropulchritudo*). The species *V. coralliilyticus*, *V. shilonii*, and *V. splendidus* are also opportunistic pathogens of corals. These diseases are explored in detail in *Chapter 11*.

The development of genomic methods such as MLST (p.45) led to a major overhaul of the phylogeny and classification of the Vibrionaceae and the advent of rapid genome sequencing led to further insights into the remarkable diversity within this family. A number of factors contribute to this diversity. Genome analysis of the vibrios shows that substantial gene acquisition and loss has occurred during their evolution, leading to enhancement of their physiological and ecological capabilities, facilitated by the presence of two chromosomes. The overall mean genome size is ~ 40% bigger than that of comparable marine bacteria. All of the genes recognized in complete genomes can be grouped into ~ 22000 gene families which constitute the pangenome of the Vibrionaceae, of which ~1630 probably constitute the core genome—genes that are present in all strains of all species examined. This represents huge genetic diversity; by comparison, the human genome contains about 25000 genes. Even within the same species as currently defined, the genomes of different strains can differ by as much as 20–30%. Thus, the pangenome represents the full genetic potential of the community of related organisms, but individual strains might have only a small fraction of these genes that equip it to occupy a particular niche. For example, isolates of *V. splendidus* obtained from a single coastal site can differ by as much as 1 Mbp in genome size and may contain at least a thousand distinct genotypes, each occurring at extremely low environmental concentrations.

Some members of the Vibrionales exhibit the distinctive feature of bioluminescence. These occur as free-living forms in seawater, on organic debris, as commensals in the gut of many marine animals, and as symbionts of the light organs of some squid and fish (*Chapter 10*). Most of the marine bioluminescent bacteria that can be isolated and cultured are members of the family Vibrionaceae—the commonest types being *Photobacterium phosphoreum*, *P. leiognathi*, *Aliivibrio fischeri*, *Vibrio campbellii*, and *V. harveyi*. An image of a massive ocean bioluminescent bloom—the "milky sea" phenomenon, probably caused by *V. harveyi* in association with *Phaeocystis* algae—is shown in *Figure 4.6*. The bioluminescent symbionts of flashlight and angler fish ("*Ca.* Enterovibrio luxaltus" and "*Ca.* E. escalota") have not been cultured; they have reduced genomes compared with other members of the genus and are probably obligate symbionts (*Box* 10.4). The reaction mechanisms (*Figure 3.17*) and possible functions of bacterial bioluminescence are discussed in *Chapter 3*.

Members of the order Oceanospirillales break down complex organic compounds

This group consists of uniquely marine types, as reflected in the names given to the various genera shown in *Table 4.2*. As the name suggests, bacteria in this order were originally characterized by their spiral cell shape, but they are very diverse in their physiology and ecology and the spirilla designation is not a reliable distinguishing feature, as several genera are rod-shaped rather than helical. New members of this group have been discovered in symbiotic association with *Osedax* worms and in sediments in the vicinity of whale falls (see

Figure 4.6 Modified rendition of a satellite image of a bioluminescent bloom "milky sea" observed in the northwest Indian Ocean in January 1995 and visualized by retrospective analysis of satellite data. The bloom was visible on three consecutive nights, covering an area of 15400 km². Credits: Main image, Gary Vora, US Naval Research Laboratory. Inset reprinted from Miller et al. (2005). with permission; copyright 2005, National Academy of Sciences, USA.

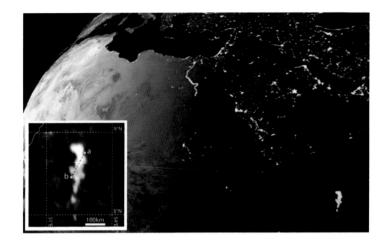

Chapter 10). Members of the type family Oceanospirillaceae, including Oceanospirillum spp. are aerobic and motile and play a major role in the heterotrophic cycling of nutrients in seawater. Some members in this group are important in the sulfur cycle, especially through degradation of dimethylpropionosulfonate (DMSP, p.90). Some, such as *Alcanivorax* spp., use alkanes exclusively as carbon and energy sources and are active in biodegradation of hydrocarbons at natural seeps and oil spills (*Box 13.2*). Genome analysis reveals multiple sequences coding for enzymes degrading different alkanes, as well as surfactants that emulsify the oil to aid its degradation. There is wide variation in physiological properties such as optimum growth temperature, halophilicity, and utilization of substrates. The taxonomy of these genera is in a state of considerable flux, and molecular methods indicate that, although related, they represent a number of deeper phylogenetic groups.

The family Thiotrichaceae includes some important SOB

Beggiatoa is very common in the top few millimeters of sulfur-rich marine sediments. They frequently show chemotaxis to seek out the desired gradient of oxygen and sulfur compounds and are also very prominent at hydrothermal vents and cold seeps. *Beggiatoa* and other filamentous forms commonly show gliding motility and become intertwined to form dense microbial mats, often with a complex community structure containing sulfate-reducers and phototrophs. Although *Beggiatoa* obtains energy from the oxidation of inorganic sulfur compounds, it does not possess the enzymes needed for autotrophic fixation of carbon dioxide and therefore uses a wide range of organic compounds as a carbon source. *Thioploca*, *Thiothrix*, and *Thiomargarita* are filamentous sulfur-oxidizing chemolithotrophs important in the oxidation of sulfide in anaerobic sediments.

Thioploca spp. are multicellular filamentous bacteria that occur in bundles surrounded by a common sheath (*Figure 4.7*) and contain granules of elemental sulfur. *Thioploca* is one of the largest bacteria known, with cell diameters from 15 to 40 µm and filaments many centimeters long, containing thousands of cells. Several species, including *T. chileae*, *T. araucae*, and *T. marina* have been described. Huge mats of *Thioploca* spp. were discovered along the Pacific coast of South America, where upwelling creates areas of nitrate-rich water, with bottom

Figure 4.7 (a) Filaments of *Thioploca* spp. extending from the sediment surface into the anoxic water column at 50–100 m depth at the shelf off the coast of Chile. The frame is 15 mm wide. (b) A bundle of *T. araucae* filaments extending out of their sheath. The appearance of a braid is seen where the filaments cross over, hence the name *Thioploca* ("sulfur braid"). Each filament is ~35 mm wide. Credit: reprinted from Jørgensen and Gallardo (1999) with permission from Oxford University Press.

(a)

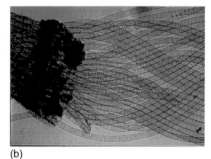

(b)

waters becoming anoxic. Blooms of *Thioploca* can be very dense, up to 1 kg m^{-2} (wet weight). Anoxic reduction of hydrogen sulfide is coupled to the reduction of nitrate. Each cell contains a very thin layer of cytoplasm around the periphery and a liquid vacuole that constitutes 80% of the cell volume. The vacuole stores very high concentrations of nitrate, which is used as an electron acceptor for sulfide oxidation. The bacteria can grow autotrophically or mixotrophically using organic molecules as a carbon source. In order to access the nitrate in the overlying water column, the filaments stretch up into the overlying seawater, from which they take up nitrate, and then glide down 5–15 cm deep into the sediment through their sheaths to oxidize sulfide formed by intensive sulfate reduction.

Thiomargarita namibiensis holds the record as the largest known bacterium. The spherical cells are normally 100–300 μm wide, but some reach diameters of 750 μm (*Figure 1.4C*). They occur in large numbers in coastal sediments off Namibia and occur in filaments with a common mucous sheath. Microscopic granules of sulfur reflect incident light, and the name derives from their resemblance to a string of pearls. The hydrographical conditions off this coast bring large quantities of nutrients to the surface and massive phytoplankton growth results in settlement of organic material to the seabed, where it is degraded by bacteria forming large amounts of hydrogen sulfide. *Thiomargarita* oxidizes sulfide using nitrate and, like *Thioploca*, the interior of the cell is filled with a large vacuole. The nitrate stored in the vacuole and the sulfur stored in the peripheral cytoplasm act as nutrient reserves that allow these bacteria to grow for several months in the absence of external nutrients.

The proposed phylum Desulfobacterota contains anaerobic sulfate- or sulfur-reducing bacteria (SRB)

The SRB play a major role in the sulfur cycle and are abundant in anoxic marine sediments where sulfate from seawater can diffuse several meters down. Eleven families and 63 genera (mostly sharing the prefix *Desulfo-*) have been validly described as members of the Deltaproteobacteria, but there has been substantial reclassification under the GTDB system. Together with methanogenic archaea, SRB are responsible for the mineralization of organic material in coastal and shelf sediments. As noted in *Chapter 3*, the activity of SRB and archaea changes with depth, due to competition for nutrients and penetration of sulfate, but the two groups can grow syntrophically when sulfate is depleted. SRB acquire energy for metabolism and growth by utilizing organic compounds or hydrogen as electron donors coupled to dissimilative sulfate reduction (*Figure 3.12*). SRB produce H$_2$S, resulting in the characteristic stench and blackening (due to deposits of FeS) of decomposition in anoxic mud, sediments, and decaying seaweed. H$_2$S is highly toxic and adversely affects many forms of marine life, but it can be used by a wide range of chemotrophic and phototrophic bacteria, as described in earlier sections, completing the cycling of sulfur. Numerous types of SRB have been isolated from marine sediments and these are classified on the basis of differing morphological and physiological properties, as well as 16S rRNA typing. Some genera, such as *Desulfobacter* and *Desulfonema* oxidize acetate and other fatty acids. Another group, which includes *Desulfovibrio* and *Desulfomonas*, can utilize lactate, pyruvate, and certain fatty acids, but not acetate. Sequences affiliated with this group can also be recovered from the water column, including clade SAR324 found in waters >500 m deep. These are thought to be particle-associated and genome analysis indicates metabolic flexibility, including oxidation of carbon monoxide and methylotrophy. It should be noted that some members of the Firmicutes and Archaea also carry out sulfate reduction.

The recently discovered cable bacteria, which couple oxygen reduction to sulfide oxidation via long-distance electron transport in sediments, have been identified as members of the family Desulfobulbaceae within this group (*Box 4.2*).

The proposed phylum Epsilonbactereota contains major contributors to productivity at hydrothermal vents

As noted in *Box 4.1*, new multi-gene analysis does not support the phylogenetic status of the Epsilonproteobacteria class as members of the Proteobacteria phylum, and they have been reassigned to the new proposed phylum Epsilonbactereota. There are a relatively

BOX 4.2   RESEARCH FOCUS

Cable bacteria unite to conduct electricity

A network of electrogenic bacterial filaments. Geoscientists at Aarhus University, Denmark noted unusual changes in the distribution of pH in marine sediments, which raised the notion that electric currents were running through the sea floor (Nielsen et al., 2010). The nature of this process was a mystery until the team, with collaborators from scientists at the University of Southern California, showed that gentle washing of the top 20 mm of sediment from sulfidic sediment revealed a tangled mass of filamentous multicellular bacteria responsible for long-distance electron transport (LDET, Pfeffer et al., 2012). Subsequently, Schauer et al. (2014) exposed sulfidic sediment to oxygen and witnessed the rapid growth and development of a network of filaments, reaching ~2 km per cm^2 after 21 days. Examination of the filaments showed that they can contain over 10^4 cells and be up to 70 mm long and 0.4–1.7 μm diameter, orienting themselves vertically in the upper sediment. Electron microscopic examination showed that the surface has 15–17 ridges 400–700 nm wide, with a filled 70–100 nm wide channel between the cytoplasmic and outer membrane (Pfeffer et al., 2012). There is a gap of 200 nm between adjacent cells in the filament, which is bridged by the ridge filling, creating a continuous channel along the filament. This is assumed to form the conductive "wires" transporting electrons. The whole structure is wrapped in a collective outer membrane that acts as insulation. Further detailed analysis of the architecture of the cell envelope using a range of advanced imaging techniques was recently described by Cornelissen et al. (2018). The cell envelope is built by a parallel concatenation of ridge compartments, each of which contains a ~50 nm diameter fiber in the periplasmic space (*Figure 4.8A*). These fibers continue across the junctions between cells in the filament, which have a distinctive cartwheel structure formed by invaginations of the outer membrane. As well as acting as conductive wires, the periplasmic fibers provide tensile strength to the filaments, which can withstand substantial stress before breaking.

Motility. Bjerg et al. (2016) used time-lapse microscopy to show that cable bacteria are motile, moving by gliding at a mean (highly variable) speed of 0.5 μm s^{-1}. The bacteria frequently moved in loops and sometimes twisted around themselves, indicating a helical rotation of the filament, but the mechanism by which this occurs is unknown. The bacteria show positive chemotaxis to increasing O$_2$ concentration in order to locate the oxic-anoxic interface.

Proof of electrical conduction in cable bacteria. Evidence that LDET occurs in the sediments came from high-resolution profiling of pore water chemistry and mapping of electron sources using different types of micro-electrodes (Nielsen et al., 2010; Risgaard-Petersen et al., 2014). Some ingenious micro-manipulation experiments by Pfeffer et al. (2012) provided further proof that the bacterial filaments act as solid conductors for LDET. Metabolic activity dropped rapidly when the filament network was cut by passing a thin tungsten wire through the sediment below the oxic-anoxic interface. When membrane filters were placed into sediments, electrical currents only occurred when pore sizes were large enough (>0.8 μm) for the bacteria to grow down vertically through the membranes to reach the underlying sediment, thus ruling out the possibility that electron transfer was due to dissolved or colloidal compounds. Also, Pfeffer et al. showed that when a layer of glass microspheres was inserted to replace a 5 mm thick layer of the sediment, LDET only occurred when filaments were able to bridge the non-conductive gap. In later experiments, Bjerg et al. (2018) placed filaments in microscope chambers with sulfide and O$_2$ at opposite ends. They used a laser microdissection beam to cut off the electron transport within individual filaments and measured the redox states of cytochromes using resonance Raman microscopy. A gradient of cytochrome redox potential was detected along the filaments, which broke down if O$_2$ was removed or the filaments were cut. The filaments showed a loss of electric potential (voltage) of 12–14 mV per mm of filament. Bjerg et al. calculated that the maximum distance for effective electron transport would be about 300 mm, beyond which filaments would have to use their gliding motility to remain in contact with the O$_2$- or S-rich zones.

Bacterial cells cooperate to perform different half reactions. The most remarkable feature of LDET is that different cells in the multicellular filament carry out the oxidative and reductive half-reactions separately, as shown in *Figure 4.8B* (Meysman, 2018). Separation of the half reactions over long distances occurs because of the electrical connection—like an electric cable—coupling

(a)

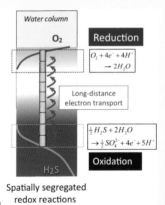

Water column

O$_2$

Reduction

$O_2 + 4e^- + 4H^+ \rightarrow 2H_2O$

Long-distance electron transport

$\frac{1}{2}H_2S + 2H_2O \rightarrow \frac{1}{2}SO_4^{2-} + 4e^- + 5H^+$

Oxidation

H$_2$S

Spatially segregated redox reactions

(b)

Figure 4.8 (a) Ultrastructure of cable bacterium. Upper panel shows a composite TEM image of the multicellular filament. Lower panel is a model of the filament, showing ridges containing periplasmic conductive fibers and internal structure at cell junctions. (b) The process of sulfur oxidation in cable bacteria, showing separation of the half reactions by LDET. (Credits: A. Rob Cornellisen, Hasselt University. B. Reprinted from Meysman (2018) with permission from Elsevier.

BOX 4.2 RESEARCH FOCUS

the supply of electrons from sulfide oxidation by the anodic cells to electron removal from oxygen reduction by the cathodic cells. Unlike typical SOB, which can only oxidize sulfide in the oxic-anoxic/O_2-H_2S interface or commute to collect oxidants from the surface sediment or water column, cable bacteria can "mine" sulfide from deeper layers. Cable bacteria can also use nitrate as an electron acceptor, meaning that oxidation of sulfide can occur by LDET to anoxic nitrate-rich water overlying the sediment (Marzocchi et al., 2014).

We have grown used to the dogma that all living cells—from unicellular microbes to those within complex multicellular organisms—are responsible for individually securing their own energy supply, which occurs by processing of electron donors and acceptors over very short distances (intracellularly, or within the immediate vicinity of the cell). To explain LDET, Meysman (2018) invokes a "thought experiment" imagining two twin brothers separated by 20 km—one of whom has access only to carbohydrates as food and the other who only has access to O_2 for respiration—who live by LDET through a connecting wire. The surprising discovery of physical separation of reactions between different, distant cells has created a paradigm shift in our understanding of biological electron transport.

Phylogeny of cable bacteria. To date, all attempts to isolate and cultivate cable bacteria have been unsuccessful. All specimens so far described have a similar morphology and 16S rRNA gene sequencing indicates that they belong to family Desulfobulaceae (Pfeffer et al., 2012). Using capillary glass hooks and a micromanipulator Trojan et al. (2016) isolated individual cable filaments from marine, saltmarsh, and freshwater sediments in Denmark, Japan, and the United States. Single-cell genomes were amplified and gene sequences encoding 16S rRNA and Dsr proteins (involved in oxidation of stored sulfur) were identified in the assemblies. By amplifying the genomes of individually picked filaments, Trojan et al. were able to directly correlate sequence information with phenotypic properties of the cable bacteria and their environmental origin, enabling them to identify the bacteria from all 16 samples as belonging to a monophyletic sister lineage of the genus *Desulfobulus*. Two major branches within this clade were identified and Trojan et al. proposed these as the novel genera "*Ca.* Electrothrix" (containing four novel candidate species from the marine and saltmarsh specimens) and "*Ca.* Electronema" (containing two novel candidate species from the freshwater specimens). Although genomic data is currently limited, it appears that the system of LDET evolved only once, and the responsible genes do not seem to have spread to other microbial groups.

Distribution and contributions to ecosystem function. Gene surveys show that cable bacteria occupy organic-rich coastal sediments with steep sulfide gradients, but do not feature strongly in sandy sediments with mechanical disruption by bioperturbating infauna. However, they are found in the sediment beneath intertidal beds of bivalve molluscs, which trap organic material and form a reef that lifts structures above the surrounding sediment level. Malkin et al. (2017) found high densities of cable bacteria in sediments accumulating within mussel and oyster reefs in the Wadden

Sea (Netherlands). They were also found, albeit at somewhat lower levels, in the surrounding bioperturbated sandy sediment. Despite frequent burial by the dynamic nature of these sediments, the bacteria continue to occupy the uppermost sediment layers, due to motility enabling them to keep the optimal connections between their separated electron donors and acceptors, as previously shown in the laboratory by Bjerg et al. (2016). Using microsensors, Malkin et al. showed that the cable bacteria are responsible for the generation of acid as a result of sulfate reduction. This has marked effects on the pore water chemistry by increasing the concentrations of dissolved calcium, manganese and iron. Burdorf et al. (2017) provide evidence that cable bacteria are globally distributed in different coastal habitats and are an important component of biogeochemical cycles. Recently, Kessler et al. (2018) showed that they have a marked effect on dissimilatory reduction of nitrate to ammonia (DNRA), which affects loss of bioavailable nitrogen from the system.

Protecting marine organisms from sulfide poisoning. The high levels of highly toxic H_2S produced in marine sediments by SRB are a potential threat to marine life, but as long as O_2 is present it is promptly oxidized by SOB and does not enter the water column. However, coastal waters often become hypoxic due to microbial respiration of decaying organic matter, eutrophication, stratification, and increased temperature. Nielsen (2016) notes that, during O_2 depletion, marine animals survive longer than predicted. Seitaj et al. (2015) conducted a four-year study of a seasonally hypoxic coastal marine lake with restricted water exchange with the North Sea. They observed that cable bacteria dominate the sediment in the winter, whereas mats of Beggiatoaceae colonize it during the summer. The electrogenic activity of the cable bacteria creates a store of iron oxides before summer hypoxia sets in, and these re-precipitate as a surface layer of reactive iron oxide that captures free H_2S in the surface sediment, preventing its escape to the water column (euxinia). Because of the wide distribution of cable bacteria, Seitaj et al. suggest that they may act as a "firewall" that explains why mass mortality in coastal waters due to euxinia is infrequently observed.

The activity of cable bacteria may partly explain why seagrasses are able to grow in anoxic sediments that have phytotoxic levels of sulfide. Martin et al. (2019) showed that young seagrass root tips leak O_2 into the surrounding environment. As well as oxidizing sulfide abiotically, this phenomenon may facilitate sulfide oxidation by cable bacteria, as these are found in higher abundance adjacent to the roots leaking O_2.

Future research directions. Considerable progress has been made in understanding this novel mechanism, but many intriguing questions remain. How are the energy and products of metabolism shared between cells? How are cell division and construction of the filaments controlled, and how do the filaments orient themselves in the sediment? What is the composition and electrical properties of the periplasmic fibers? What is the evolutionary history of cable bacteria and what is their role in the ecology of different sediments? How do they interact with other members of the microbial community? This newly discovered and fascinating lifestyle of bacteria is ripe for extensive investigation.

small number of cultivated representatives in this group, but environmental gene sequencing reveals much greater diversity. They carry out anaerobic respiration using organic or inorganic electron donors and acceptors to support chemoorganotrophic or chemolithoautotrophic growth. In marine habitats, Epsilonbacteraeota are particularly associated with hydrothermal vents and may be the dominant bacteria contribute the major fraction of the primary production via autotrophy in the waters between 25–60°C surrounding the geothermal emissions. They appear to play a key role in the initial colonization of newly formed vents. Here, the most common metabolism is oxidation of reduced sulfur compounds, hydrogen, or formate, using either oxygen or nitrate as a terminal electron acceptor and fixing CO_2 via the reverse TCA cycle. Examples of cultured genera include *Sulfurimonas*, *Nautilia*, and *Caminibacter*. As well as free-living forms, members of the Epsilonbacteraeota are involved in symbiotic associations with invertebrates at hydrothermal vents, notably *Rimicaris* shrimps, the Pompeii worm *Alvinella*, and the scaly-foot snail *Chrysomallon*, discussed in *Chapter 10*.

Myxobacteria have a complex life cycle

The group of related orders and families known as the myxobacteria are characterized by cells that have some of the largest bacterial genomes (typically ~9 Mb or more) which accounts for their complex life cycle with gliding motility and social behavior leading to the formation of reproductive fruiting bodies with production of myxospores. They are primarily associated with terrestrial soils, but they can be recovered from seawater. Many isolates are thought to have resulted from spores washed into the sea from coastal runoff, but some halotolerant and halophilic strains have been isolated and cultured from marine muds and sands and the surface of algae, seagrass, and invertebrates. These marine types form a distinct phylogenetic group with representative genera including *Plesiocystis*, *Nannocystis*, and *Haliangium*. The social behavior characteristic of soil myxobacteria probably does not occur in marine species. These were classified with the Deltaproteobacteria until recently but have been reassigned to the new GTDB phylum Myxococcota.

The Bdellovibrionales contains predatory bacteria

This order contains genera (e.g. *Bdellovibrio*, *Halobacteriovorax*, and *Vampirovibrio*) of small spiral cells with the unusual property of preying on other Gram-negative bacteria. The most studied type is *Bdellovibrio bacteriovorus*, whose life cycle is shown in *Figure 4.9*. The cells swim at high speed and use a chemosensory system to detect high concentrations

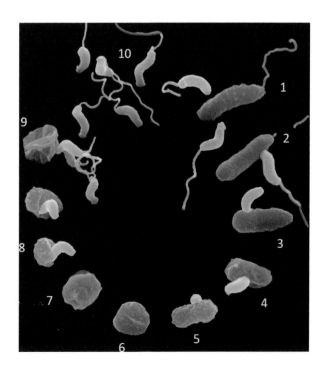

Figure 4.9 Composite of false-colored SEM images showing the life cycle of *Bdellovibrio bacterivorus* (yellow-colored cell) infecting a Gram-negative host bacterium (blue-colored cell). Infection begins by attachment of a flagellated *Bdellovibrio* cell (1, 2). During active penetration and entry to the host cell, the flagellum is shed (3–5). *Bdellovibrio* replicates in the periplasmic space, producing extracellular enzymes to liberate nutrients for growth and elongating into a large filamentous cell (6–7). When nutrients are exhausted, the *Bdellovibrio* differentiates into several motile cells (8–9), which then swim to seek out new hosts (10). Credit: Snejzan Rendulic, Juergen Berger, and Stephan Schuster; Max Planck Institute for Developmental Biology.

of prey. Different strains are highly specific in targeting particular hosts. It attaches to its prey and retractable fibers bring it into close contact with the bacterial cell. A cocktail of enzymes degrades proteins, lipids, and carbohydrates to create an opening in the cell wall. *Bdellovibrio* then enters the periplasmic space between the outer and inner membranes, where it replicates, producing extracellular enzymes to liberate nutrients for growth. The cytoplasm of the prey is consumed and the *Bdellovibrio* cell elongates into a large filamentous cell. When all the nutrients are exhausted, the filament differentiates into about 15 motile cells, which then seek out new hosts after digestion of the host cell. *Bdellovibrio*-and-like organisms (BALOs) are very widespread in aquatic environments and may exert top-down control of populations of other bacteria similar to phages, although their full ecological role is unknown.

Members of the Zetaproteobacteria are microaerophilic iron-oxidizers

The original isolate of bacteria in this class of Proteobacteria designated in 2010 was isolated from a deep-sea hydrothermal vent field off Hawaii and named *Mariprofundus ferrooxydans*. Subsequently, isolates assigned to this class have been found at a variety of Fe-rich brackish and marine sources, including in deep-sea microbial mats and brine seawater interfaces (DHABs, p.22). Experiments in certain coastal sediments have shown that they can colonize and corrode mild steel. All isolates so far described are microaerophilic Fe(II)-oxidizing chemolithoautotrophs found at opposing gradients of Fe(II) and O_2, requiring neutral pH and low O_2 concentrations. They are microaerophilic because they require just enough O_2 for use as an electron acceptor but not so much as to cause abiotic iron oxidation. Their growth on surfaces produces characteristic filamentous twisted stalks composed of Fe oxyhydroxide. Several other species have now been proposed, and related OTUs are abundant in 16S rRNA gene datasets from deep-sea samples, indicating that this group may have wider significance in biogeochemical cycling.

Members of the Cyanobacteria carry out oxygenic photosynthesis

The Cyanobacteria is a large and diverse group, and members are characterized by their ability to carry out photosynthesis in which oxygen is evolved, although some anoxygenic members have been described. Cyanobacteria contain chlorophyll *a*, together with accessory photosynthetic pigments called phycobilins. This group was formerly known as the blue-green algae because of the presence of the blue pigment phycocyanin together with the green chlorophyll. Designation as algae is inappropriate although the term persists in common usage and the organisms are still treated as the Cyanophyta division of the algae under the Botanical Code, even though the genetic and structural features of the Cyanobacteria mean that they clearly form one of the major phyla of the domain Bacteria. Many genera contain phycoerythrin pigments, which give the cells a red-orange color, rather than blue-green color.

Fossil evidence, in the form of morphological structures and distinctive biomarkers (hopanoids) typical of the group indicates that organisms resembling modern representatives of cyanobacteria may have evolved about 2.5 BYA. The evolution of oxygenic photosynthesis in ancestral cyanobacteria was responsible for transformation of the Earth's early atmosphere and the subsequent evolution of life on Earth, as discussed in *Chapter 1*.

Today, cyanobacteria play a major role in global biogeochemical cycles. They occupy very diverse habitats in terrestrial and aquatic environments, including extreme temperatures and hypersaline conditions. In the marine environment, cyanobacteria are major contributors to primary production as members of the phytoplankton, and occupy other important habitats including sea ice, shallow sediments, and microbial mats on the surface of inanimate objects, algae, or animal tissue. The Cyanobacteria show a range of metabolic properties besides oxygenic photoautotrophy—some can grow anaerobically using H_2, H_2S. or organic compounds as electron donors and some may be mixotrophic.

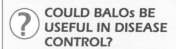

COULD BALOs BE USEFUL IN DISEASE CONTROL?

In the quest to overcome the threat posed by antibiotic resistance and the risks to human health, investigations are underway to investigate whether the predatory BALOs could be used as a live therapy. It has been shown to be effective as a potential alternative to antibiotics in local and systemic infections in experimental animals. It has also recently been suggested that introducing predators into the human gut could be a possible method of restoring a balanced, diverse microbiome in conditions such as Crohn's disease (Mosca et al., 2016), reminiscent of the successful "rewilding" of habitats to promote biodiversity in macroecology. Further developments depend on understanding how *Bdellovibrio* interacts with host phagocytes and the immune system (Raghunathan et al., 2019). BALOs are also under investigation for use in prevention of disease in aquaculture (also, see the discussion of phage therapy in *Box 14.1*).

A genome-based classification of the Cyanobacteria is under development

Until the use of 16S rRNA gene analysis established them as group within the Bacteria, "blue-green algae" were classified by botanists into about 150 genera and 1000 species based on morphological features. The Cyanobacteria are morphologically very diverse, ranging from small undifferentiated rods to large branching filaments showing cellular differentiation. Unicellular cyanobacteria divide by binary fission, while some filamentous forms multiply by fragmentation or release of chains of cells. Many types are surrounded by mucilaginous sheaths that bind cells together. The chlorophyll *a* is contained within lamellae called phycobilosomes, which are often complex and multilayered. Cyanobacteria show remarkable ability to adapt the arrangement of their photosynthetic membranes and the proportion of phycobilin proteins to maximize their ability to utilize light of different wavelengths.

Subsequent phylogenetic analysis shows that groupings based on morphology are very unreliable, and many current genera and higher taxa are polyphyletic. As with other groups of bacteria, the availability of large amounts of genomic data is propelling the development of a genome-based classification that will better reflect the evolutionary history, ecological, and physiological diversity of different cyanobacteria. Genomic analysis reveals a number of important factors that affect the diversity of cyanobacteria and their ability to occupy specific ecological niches. As with other bacteria, the types of transport systems for limiting nutrients such as P, N, and Fe are especially important. The components and mechanisms of the photosynthetic machinery are also highly variable. The structure and arrangement of photosynthetic membrane systems, the structure of carboxysomes, and the spectral tuning of pigments to different wavelengths of light have a major effect on the distribution and activity of different types of cyanobacteria. These factors are leading to major revision of existing schemes for classification and taxonomy of the group. Therefore, the remainder of this section will focus on a few selected genera without attempting to reflect their classification into higher taxa.

Prochlorococcus is the most abundant photosynthetic organism

Prochlorococcus is a very small cyanobacterium about 0.6 μm diameter (*Figure 4.10*), which is the dominant phototroph in oligotrophic ocean waters. It inhabits large parts of the oceans at depths of 25–200 m at a density between 10^5 and 10^6 mL^{-1}. The global population of 10^{27} cells means it is the most abundant photosynthetic organism on Earth. It fixes ~ 4Gt (Pg) of

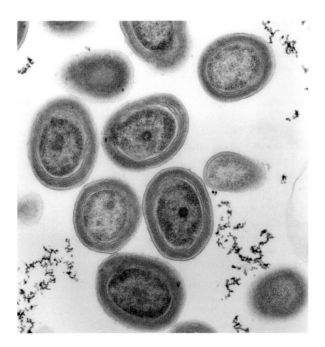

Figure 4.10 TEM of ultrathin section of *Prochlorococcus marinus* MED4 (artificially colored). Credit: Chisholm Lab, MIT.

carbon per year and contributes ~20% of O_2 to the atmosphere. Despite this, it was not discovered until 1985, using methods described on p.85. *Prochlorococcus* is most abundant in the subtropical gyres, occupying the region from 40°S to 40°N. It contains modified forms of chlorophyll (divinyl chlorophylls *a* and *b*), but lacks phycobilins.

Hundreds of strains have been isolated, cultured, and cataloged as the single species *Prochlorococcus marinus*, but this designation hides exceptional diversity. When samples of *Prochlorococcus* are isolated from different depths in the water column, it is clear that there are distinctly different ecotypes adapted for growth in different ecological niches in the photic zone. Beyond the upper mixed layer of the ocean, nutrient availability increases with depth, while light intensity and temperature decrease. *Prochlorococcus* ecotypes are adapted to life across all parts of the photic zone in the oligotrophic oceans with very low nutrient levels. Like many of the dominant heterotrophic bacteria discussed earlier, they have streamlined genomes and very small cells enabling efficient nutrient uptake and photosynthesis. Requirements for phosphorus, which is extremely limiting in the open ocean, are reduced by the lower DNA content and modified cell membranes that contain sulfolipids rather than phospholipids. Strains isolated from different depths have many differences, including the ratios of divinyl chlorophylls a_2 to b_2 and optimal irradiances for photosynthesis. Divinyl chlorophylls harvest longer wavelengths of blue light which penetrates deeper waters, allowing some ecotypes of *Prochlorococcus* to grow at depths where the amount of light is less than 1% of that at the surface.

Methods for culturing many strains of *Prochlorococcus* were developed allowing studies of their biochemical and physiological properties, showing that strains could be grouped into two broad categories—those adapted for high light (HL) and those adapted for low light (LL). Molecular phylogenetic analysis shows that this distinction appears to reflect an early evolutionary split into the HL group containing six clades and the LL group containing seven clades. These show differences in their acquisition of iron and other nutrients, and in their ability to adapt to different light and redox conditions.

Complete genome sequences have now been obtained for multiple strains of *Prochlorococcus* obtained from different depths varying in nutrient composition and light intensity/spectral quality. The genomes of HL-adapted strains are very much smaller (mean ~1.7 Mb) than those of other cyanobacteria. Genomes of LL-adapted strains are generally larger, but more variable in size, and one clade (LL IV) has genomes of ~2.7 Mb. For example, in the first study of *Prochlorococcus* genomes, the HL-adapted strain MED4 was found to have the smallest minimal genome (1716 genes) of any phototroph. Genes for systems such as chaperones, transport systems, and many other systems for signal transduction and environmental stress responses are absent. A small genome is less expensive for the cell to maintain but reduces the organism's ability to respond to changes in its environment. The LL-adapted strain SS120 found deep in the photic zone has 2275 genes with many genes conferring flexibility to utilize higher and more variable nutrient levels. These two representative strains share 1352 genes. As more genome sequences have been added, it has become apparent that the core genome containing the essential genes for cell structure, growth, and replication shared by all isolates is ~1000 genes. The remaining "flexible" genes are found in only a few genomes and account for the adaptation of each strain to its local environment. These flexible genes are clustered in highly variable genomic islands, indicating acquisition by horizontal gene transfer (HGT). The role of phages in this transfer is discussed in *Chapter 7*. The size of the pangenome—the total number of genes carried by all strains of *P. marinus*—has grown to over 85000. This represent a huge amount of diversity that clearly explains the ecological success of subpopulations of these organisms and their ability to vary in abundance according to seasonal and geographical light and nutrient gradients, although the function of three-quarters of the genes is still unknown.

Since they are responsible for so much carbon fixation in the oligotrophic ocean, *Prochlorococcus* has a major role in fueling the growth of heterotrophic bacteria with which they co-exist. Assessment of abundance of *Prochlorococcus* and SAR11 in the wild (based on metagenomic sampling) indicates that they co-exist at a median (highly variable) ratio of about three *Prochlorococcus* for every four SAR11. *In vitro* experiments show that SAR11 grows faster in co-culture with *Prochlorococcus*, but the growth of the latter is unaffected.

THE GLOBAL GENE FEDERATION OF PROCHLOROCOCCUS

The pangenome of *P. marinus* revealed by analysis of genomes from cultured cells or metagenomic assemblies contains about four times the number of human genes. That's impressive enough … but examining the cell by cell diversity of wild populations adds an even greater dimension to this genetic diversity. By amplifying the genomes of single cells (see p.56 for methods) in a single sample of seawater, Kashtan et al. (2014) showed that there are hundreds of subpopulations. Each of these has a distinctive "genetic backbone" of core gene alleles linked to a set of flexible genes. There seems to be a fine-scale co-variation between the core and flexible gene content, and Kashtan et al. suggest that these subpopulations diverged millions of years ago resulting in niche partitioning that is stable over long periods. Nucleotide variations in genes associated with cell surface appear to be particularly important. This study was done in a particular region (the BATS in the Sargasso Sea) and it will be interesting to see if different genotypes are spread globally via ocean circulation.

Some strains of *Prochlorococcus* meet SAR11's requirement for carbon in the form of glycine, but the two organisms probably compete for uptake of DMSP. In turn, it seems that close association with certain heterotrophs provides benefits in return. *Prochlorococcus* does not possess genes for catalase, which is needed to remove reactive oxygen species (ROS, p.108). It seems that this function is met by closely associated heterotrophs, although this has not yet been demonstrated experimentally. The significance of these cooperative versus competitive interactions is complicated by the discovery that *Prochlorococcus* may be mixotrophic. It has been known for some time that some strains of *Prochlorococcus* can grow heterotrophically in culture and that natural populations can assimilate amino acids and DMSP to meet their nitrogen and sulfur requirements. Recently, *Prochlorococcus* has also been shown to assimilate glucose and other sugars, which may be sufficient to provide energy and carbon for growth and cell maintenance. Also, *Prochlorococcus* has recently been found to release large numbers of outer membrane vesicles which probably have important functions in food webs and the biological carbon pump. This, together with their role as prey for grazing protists, is discussed in *Chapter 8*.

Synechococcus spp. dominate the upper photic zone

Synechococcus spp. have slightly larger cells (0.6–1.6 μm) than most *Prochlorococcus* strains and have a wider geographic distribution. Marine species tend to predominate in the well-lit top mixed layer (~25 m) of the water column and are more abundant (up to 10^6 cells mL^{-1}) in nutrient-rich coastal waters than in the ocean gyres (~10^3 cells mL^{-1}). Different species also occur in freshwater and are particularly abundant in the coastal plumes of major rivers. They contain chlorophyll *a* and ancillary pigments known as phycobiliproteins. The nature of these, together with habitat, morphology, and some genetic information have been used to designate ~50 species under the Botanical Code. Genome analysis reveals close similarities of the marine strains to *Prochlorococcus*, while freshwater strains are more distantly related. Such analysis also reveals that the genus as currently defined appears to be polyphyletic and a new genus *Parasynechococcus* has been proposed. Key differences between the genera include the nature of the carboxysome proteins and regulatory systems. Like *Prochlorococcus*, a core and flexible genome occurs and HGT again plays a large part in genome evolution.

About one-third of open ocean *Synechococcus* isolates exhibit a very unusual motility, moving through seawater at speeds of 5–25 μm s^{-1} (up to ten body lengths) while rotating at ~1 Hz although no flagella or changes in shape have been observed. Strains of *Synechococcus* only appear to show swimming motility if they produce a glycoprotein S-layer on their surface (p.69). One possibility is that proton motive force powers the movement of cargo-carrying motors along a continuous looped helical track anchored to the peptidoglycan layer of the cell wall. This is envisaged as driving rotation of the track which creates helical waves along the S-layer, causing the cell to rotate and move through the water. The role of swimming in ecology of these strains is unknown.

Another interesting feature of *Synechococcus* is its use as for investigation of circadian rhythms—the biological clock that controls the expression of genes that affect the physiological responses to daily environmental changes in light and temperature. Extensive studies of the circadian clock have been made using mutants of the model organism *S. elongatus*, revealing that a set of core oscillator proteins encoded by *kai* genes regulate global patterns of gene regulation, the timing of cell division, and compaction of the chromosome. As a simple approximation, about two-thirds of the genes are up-regulated around dawn and down-regulated at dusk, while the other third is switched off at dawn and on at dusk. The genes of the cyanobacterial clock appear to have little homology to those in other organisms and it seems that systems for circadian clocks have evolved independently in major groups of organisms, rather than having an ancient origin in ancestral bacteria, as once believed.

Some free-living and symbiotic cyanobacteria fix nitrogen

Nitrogen fixation is of fundamental significance in primary production in the oceans and many cyanobacteria carry out this process in addition to their important role in carbon fixation. As discussed in *Chapter 3*, the reduction of nitrogen to ammonia is an extremely energy-demanding process catalyzed by the key enzyme nitrogenase, which is strongly inhibited by oxygen. Nitrogen fixation is assumed to have evolved in the anoxic conditions of early Earth over 3 BYA. The cyanobacterial diazotrophs created a problem for themselves by generating oxygen by photosynthesis. As oxygen accumulated in the atmosphere, they evolved various mechanisms whereby the processes of nitrogen fixation and photosynthesis became segregated, either temporally or spatially.

Some filamentous cyanobacteria achieve the spatial separation of photosynthesis and nitrogen fixation by differentiation of specialized cells called heterocysts formed at intervals along the filament. These cells do not develop photosystem II, so they do not generate oxygen and are unable to fix CO_2; therefore, they cannot make the pyruvate needed as an electron donor for nitrogen fixation. As a result, they depend on neighboring cells in the filament to supply them with carbohydrate precursors. In return, they supply adjacent cells with fixed nitrogen in the form of the amino acid glutamine. The nitrogenase is also protected from oxygen because the heterocyst becomes surrounded by a gas-impermeable wall. Heterocyst-forming cyanobacteria are common in freshwater, estuaries, and the Baltic Sea, but are less commonly observed in the open ocean. However, two widely distributed cyanobacterial species, *Richelia intracellularis* and *Calothrix rhizosoleniae* form short chains with a terminal heterocyst and are widespread as symbionts inside the cells of diatom genera such as *Rhizosolenia* and *Hemiaulus*. The diatom *Chaetoceros* has nitrogen-fixing cyanobacterial partners that are closely associated with the cell, but outside the diatom frustule. The mutualistic nature of some of these associations has been clearly established—the diatom partners provide nutrients to increase the growth and metabolism of their cyanobacterial partners, which in turn provide the diatom host with fixed nitrogen (*Figure 4.11*). Symbiont-containing diatoms often form substantial blooms that sink rapidly, providing an important mechanism for rapid transport of fixed carbon and nitrogen to deep waters. It is not known if the intracellular cyanobacteria have a free-living stage, or how they are acquired or maintained during reproduction of the diatom hosts.

A common mechanism of separating the process of O_2 generation and N_2 fixation is to restrict the latter to the night when no photosynthesis occurs. This is the strategy employed by *Crocosphaera watsonii*, a large (2.5–6.0 μm) unicellular cyanobacterium present in tropical and subtropical waters, where it can be detected by the fluorescence of its pigment phycoerythrin or by demonstration of the *nif* nitrogenase genes in association with phytoplankton in the 3–10 μm size range. Surprisingly, metagenomic analysis indicates rather conserved genomes, unlike those of *Prochlorococcus* and *Synechococcus*. *C. watsonii* can be cultured and shown to possess a large genome, with a mean size of 5.9 Mb. Studies of gene expression

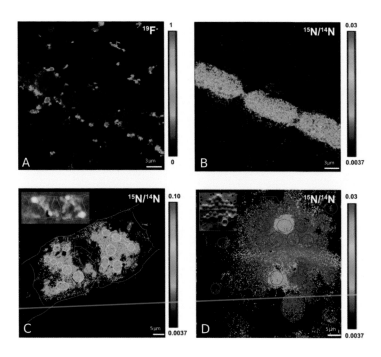

Figure 4.11 Nitrogen fixation and transfer in cyanobacteria revealed by NanoSIMS. A. *Cytophaga* bacteria attached to filamentous cyanobacteria, identified by HISH-SIMS. B. Fixed N is transferred directly to the bacteria. C. Symbiosis between the diatom *Hemiaulas* and the intracellular cyanobacterium *Richelia*. Inset is the epifluorescent image of cells prior to NanoSIMS analysis. N uptake is localized to the symbionts and also to host cells. D. A cluster of unicellular *Crocosphaera*, and a filament of *Trichodesmium* incubated in $^{15}N_2$. Note the differential uptake of N into only a few *Crocosphaera* cells. Credit: Rachel A. Foster, Birgit Adams, Tomas Vagner, Niculina Musat, and Marcel Kuypers; Max Planck Institute for Marine Microbiology, Bremen.

Figure 4.12 A. Image of a reddish-brown bloom of *Trichodesmium* between the Great Barrier Reef and the Queensland shore, viewed with the MODIS satellite. B. Aggregated filaments of *Trichodesmium*. Credits: A. NASA Earth Observatory. B. S. Kranz, Alfred Wegener Institute, Bremerhaven.

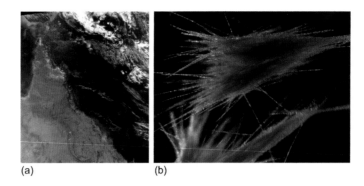

(a)   (b)

show that about one-third of the genome is subject to diel variation in expression. Many of the genes associated with nitrogen fixation are upregulated at the beginning of the dark period. During the transition from light to dark periods, another method ensures protection of the nitrogenase complex from the effects of oxygen. Photosystem II shuts down due to reduction in the pool of quinone intermediates, the rate of electron transfer, and deactivation of key proteins (see *Figure 3.3*).

The filamentous cyanobacterium *Trichodesmium* is probably the most abundant diazotroph in the open ocean and forms dense masses that are responsible for massive blooms that can cover many thousands of km^2 visible from space, especially in tropical seas (*Figure 4.12a*). Reddish-brown streamers of *Trichodesmium* are often described as scum or "sea sawdust" and were described in the journals of James Cook and Charles Darwin—"The whole surface of the water, as it appeared under a weak lens, seemed as if covered by chopped bits of hay, with their ends jagged." The colonies also provide a substrate for association of many other microbes and small invertebrates. Several species with different distributions can be distinguished morphologically. Genetically, these fall into two major clades: one containing *T. erythraeum* and another containing the species *T. tenue, T. hildebrandtii, and T. contortum*; also, the related genus *Katagnymene* should probably be reclassified into this group. *Trichodesmium* forms dense colonies in which the individual filaments appear braided together following self-assembly into aggregates (*Figure 4.12b*). The cells contain gas vacuoles that are used to control buoyancy, allowing the filamentous colonies to move up and down the water column to find optimum light conditions or nutrient levels, respectively. Analysis of the large (7.8 Mb) genome of *Trichodesmium* reveals that it contains several genes associated with heterocyst formation in other filamentous cyanobacteria, including the key regulatory gene *hetR*. However, these genes are not fully expressed and *Trichodesmium* does not produce heterocysts; instead, the filaments contain groups of specialized cells called diazocytes that contain the nitrogenase complex. These form at intervals along the filament and make up ~15% of the cells. Unlike heterocysts, these cells are not terminally differentiated and are able to divide, suggesting that their development may follow the same initial stages as heterocysts. In addition to this spatial separation, *Trichodesmium* also separates nitrogen fixation and oxygen production temporally, but unlike *Crocosphaera*, this is not by day-night separation—*Trichodesmium* fixes nitrogen during the daytime. Around midday, oxygen production from photosynthesis is lowered by enhancement of respiration (increased activity of cytochrome c oxidase) and the Mehler reaction (reduction of oxygen to hydrogen peroxide) in the diazocytes. Interestingly, the diel clock of *Trichodesmium* works to keep the energy-demanding processes of fixing carbon dioxide and nitrogen active during the day, while restricting cell division and diazocyte development to the night.

The final example of a nitrogen-fixing cyanobacterium is an obligate symbiont of picoeukaryotic algae. Gene surveys for the *nif* nitrogenase genes revealed widespread distribution of abundant small cyanobacteria termed UCYN-A that has defied all attempts at cultivation. Reconstruction of the genome showed that it is extremely reduced and lacks many gene systems normally found in cyanobacteria. The cyanobacterium cannot live independently and is dependent on its algal host to provide fixed carbon compounds and in return it provides fixed nitrogen. The uncultivated cyanobacterium has been given the name "*Ca.* Atelocyanobacterium thalassa" (the name means "incomplete marine cyanobacterium").

? UCYN-A—ARE WE WITNESSING THE EVOLUTION OF A "NITROPLAST?"

Studies conducted by Jonathan Zehr and colleagues using cell sorting and visualization by FISH (*Figure 4.13A*) showed that '*Ca.* Atelocyanobacterium thalassa' (UCYN-A) forms a close association on the exterior of its algal host, although the nature of the physical interactions and transfer of nutrients is unknown (review, Zehr et al., 2016). In separate studies, Hagino et al. (2013) discovered inclusion bodies in the calcified stage of *B. bigelow* (*Figures 4.13B, C*) and these appear to be intracellular UCYN-A, based on PCR amplification. These differences are currently enigmatic, but UCYN-A has some genomic similarities to intracellular symbionts of some freshwater diatoms. Could these associations be true endosymbiosis that might eventually lead to the evolution of a fully integrated nitrogen-fixing organelle? (*Figure 4.13D*). This occurred in the past with the evolution of the chloroplast from endosymbiosis of a cyanobacterial ancestor but, as far as we know, "nitroplasts" have not evolved. Further progress will depend on cultivating the bacterium and its host in order to examine membrane systems and study the physiology of exchange of carbohydrate and fixed nitrogen between the partners.

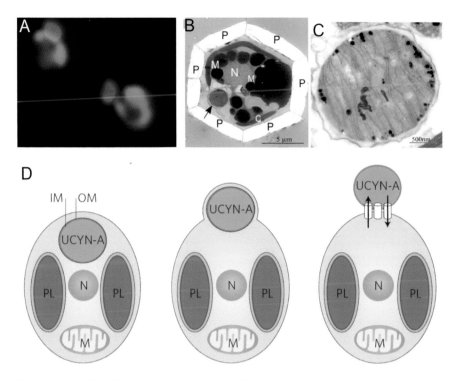

Figure 4.13 Association of nitrogen-fixing cyanobacterium UCYN-A with its prymnesiophyte host *Braarudosphaera bigelowii*. A. Epifluorescence micrograph visualized using CARD-FISH with probes directed against UCYN-A1 (red) and host cells (green) suggest attachment of the bacterium to the surface of the host cell. B. TEM of *B. bigelowii* specimen showing nucleus (N), chloroplasts (C), pentaliths (calcareous scales) (P), mitochondria (M), and a spheroid body (arrow), identified as UCYN-A by PCR amplification of 16S rRNA genes. C. TEM of intracellular SB showing internal lamellae (arrow) and envelope consisting of three layers possibly corresponding to outer membrane, peptidoglycan wall, and plasma membrane of gram-negative bacteria. D. Possible models of symbiotic interactions depending on whether there is full internalization of UCYN-A, leading to enclosure by a host membrane as a true endosymbiont (left), or surface attachment (right), in which case mechanisms for transfer of metabolites must exist. CB, cyanobacterium; N, host nucleus; PL, plastids; M, mitochondrion; OM, outer membrane; IM, inner membrane. (Credits: A. Ana M. Cabello, University of California, Santa Cruz. B, C. Reprinted from Hagino et al. (2013), CC-BY-4.0. D. Reprinted from Zehr et al. (2016) with permission from Springer Nature.)

Transcriptional studies show that there are distinct daily cycles for the expression of many genes, although it lacks two of the three circadian clock genes found in most cyanobacteria. It appears to use host-supplied carbohydrates and fix nitrogen in the daytime, with close coupling to the energy provided by host photosynthesis, but many aspects of the interaction remain enigmatic. The importance of the association is further discussed in *Chapter 9*.

Filamentous cyanobacteria are important in the formation of microbial mats

Complex stratified communities of microorganisms develop at interfaces between sediments and the overlying water. Filamentous cyanobacteria such as *Phormidium*, *Oscillatoria*, and *Lyngbya* are often dominant members of the biofilm in association with unicellular types such as *Synechococcus* and *Synechocystis*. Steep concentration gradients of light, oxygen, sulfide, and other chemicals develop across the biofilm. The mat becomes anoxic at night and H_2S concentrations rise. Anoxygenic phototrophs as well as aerobic and anaerobic chemoheterotrophs are also present. Cyanobacteria (and other motile bacteria in the biofilm) migrate through the mat to find optimal conditions. Gliding movement, up to 10 μm s^{-1}, occurs parallel to the cell's long axis and involves the production of mucilaginous polysaccharide slime. There are two possible mechanisms by which gliding occurs. One is the propagation of waves moving from one end of the filament to the other, created by the contraction of protein fibrils in the cell wall. The other mechanism is secretion of mucus by a row of pores around the septum of the cell. Some types, such as *Nostoc*, are only motile during certain stages of their life cycle, when they produce a gliding dispersal stage known as hormogonia.

Members of the Planctomycetes have atypical cell structure

The Planctomycetes is a deeply branching phylum of the Bacteria, with many unusual features of cell structure and morphology, dividing by polar budding and often attached to surfaces in rosette-like structures (*Figure 4.14*). The cytoplasm is divided into compartments, and some types contain organelle-like structures. When first discovered, they were originally classified as eukaryotic fungi (Planctomycetes means "floating fungi"), and it was not until the advent of 16S rRNA gene analysis that they were recognized as members of the Bacteria. They appear to form a monophyletic group with other phyla to form the Planctomycetes-Verrucomicrobia-Chlamydiae (PVC) superphylum. Besides those named, this also includes the phylum Lentisphaerae and the candidate phyla "Poribacteria" and "OP3." Although

Figure 4.14 Phase contrast photomicrograph showing the *Rhodopirellula* sp. strain P1 isolated from biofilm on the surface of kelp *Laminaria* hyperborean, displaying ovoid cells, budding and rosette formation. Reprinted from Bengtsson and Øvreås (2010), CC-BY-4.0.

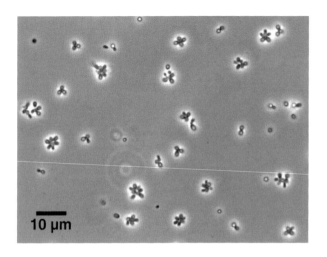

10 μm

monophyletic, the PVC superphylum contains bacteria with extremely diverse lifestyles and habitats, many of which are of biomedical or biotechnological importance.

The Planctomycetes (Planctomycetota in GTDB) phylum contains one class and three orders of cultivated types, which can be isolated from marine sediments, surfaces, and seawater. These are aerobic chemoheterotrophs and can be grown on simple organic media, although many grow very slowly. Representatives from fresh and brackish water are also known. The smallest order Phycisphaerales contains just three species isolated from marine algae. The major order Planctomycetales contains three families, of which the Planctomycetaceae includes the well-studied genera *Blastopirellula*, *Gemmata*, *Pirellula*, and *Rhodopirellula*, currently plus 13 other genera with many marine examples. In 16S rRNA gene surveys from coastal waters, OTUs affiliated with the Planctomycetes represent about 6% of the bacterioplankton and less than 1% in the open ocean. Although they are not one of the major groups of planktonic bacteria in pelagic waters, planctomycetes are especially associated with marine snow and often increase markedly in abundance in association with blooms of diatoms or other algae. Use of 16S rRNA gene-based methods suggests much greater diversity than the currently accepted classification based on culture; there may be at least 10 classes, 16 orders, and 43 families, which appear to show different distributions on medium and high productivity waters. Only 0.6% of the known diversity of the phylum at OTU level can be assigned to cultivated strains. Other molecular markers such as the *rpoB*, *carB*, and *acsA* genes give increased resolution for classification of new strains and biogeographical studies.

Genome analysis of cultivated isolates indicates that they possess rather large genomes in comparison with most marine heterotrophs. The average genome size is 7.6 Mb (range 3.60–12.4 Mb). They possess numerous genes encoding sulfatases, enzymes that could be responsible for the breakdown of complex sulfated heteropolysaccharides, which are produced in large quantities in marine environments (e.g. by fish and algae) and it is likely that they play an important role in global carbon cycling by turnover of these complex carbohydrates in marine snow and sediments, Remarkably, 57% of genes in the sequenced isolates are of unknown function.

A separate monophyletic clade within the Planctomycetes contains five uncultivated *Candidatus* genera identified as anammox bacteria. These bacteria possess a unique cell organelle, the anammoxosome (*Figure 3.9*), enabling anaerobic oxidation of ammonia. This process is of major importance in the marine nitrogen cycle, as discussed in *Chapter 9*. The "*Ca*. Scalindula" genus contains four species whose sequences have been identified in marine habitats.

The phylum Bacteroidetes has a major role in nutrient cycling via degradation of polymers

This collection of diverse, aerobic, facultative, or anaerobic chemoheterotrophs is now recognized as one of the major branches of the Bacteria. Members of the classes Sphingobacteria, Flavobacteriia, and Cytophaga contains many familiar cultivated species that are easily

---

? WHERE ARE THE PLANCTOMYCETES IN THE TREE OF LIFE?

Use of 16S rRNA gene phylogeny clearly links the Planctomycetes with major groups of Gram-negative bacteria. However, the unusual structure of planctomycetes cells has encouraged the idea that they blur the distinction between the Eukarya, Bacteria, and Archaea. They appeared to lack a peptidoglycan cell wall and possess membrane-bound cell compartments. Some types have an anammoxosome organelle or nuclear membrane and some studies suggested that planctomycetes might carry out endocytosis—a key feature of eukaryotes—supported by the presence of genes for eukaryote-like membrane coat proteins. There now seems to be enough evidence to dispel the idea of eukaryote-like properties. Multiple bioinformatic and biochemical analyses show that planctomycetes do possess a cell wall (Jeske et al., 2015). Boedeker et al. (2017) used advanced imaging techniques and new genetic methods to show that *Planctopirus limnophila* has a Gram-negative cell plan, in which the cytoplasmic membrane can show massive invaginations leading to an expandable periplasmic space. They also found that vesicle formation does not occur, although macromolecules are taken up by an unusual connection between the outer and cytoplasmic membranes.

isolated when marine samples are plated on laboratory media. As usual, this represents only a small fraction of the diversity shown in environmental gene surveys; members of the phylum can represent up to 20% of 16S rRNA sequences in coastal and open ocean waters, where they are mainly attached to phytoplankton particulate aggregates of organic material. They are also abundant in sediments, hydrothermal vents, and polar regions, and associated with marine animals.

Many of the key genera (*Cytophaga*, *Flavobacterium*, *Bacteroides*, *Flexibacter*, and *Cellulophaga*) are polyphyletic and their taxonomy is confused. Based on analysis of the *gyrB* gene, the new genus *Tenacibaculum* has been created to accommodate marine types formerly known as *Flexibacter*. Many marine isolates of the genera *Cytophaga*, *Cellulophaga*, and *Flavobacterium* have unusual flexirubin and carotenoid pigments and can be easily isolated as colored colonies on agar media inoculated from sediments, marine snow, and the surfaces of animals and plants and incubated aerobically at ambient temperatures (*Figure 2.6*). Agar is normally resistant to bacterial degradation, which is why it is an ideal gelling agent for culture plates but marine isolates belonging to this group often cause softening or the formation of craters on agar plates due to degradation. They produce a wide range of hydrolytic extracellular enzymes, which are responsible for degradation of carbohydrate polymers, chitin, and proteins. As noted on p.92, members of the Bacteroidetes use surface-associated enzymes to bind and partially degrade polysaccharides, ensuring efficient, "selfish" assimilation of nutrients. This is of major ecological significance in marine nutrient cycling, due to the degradation of complex organic materials, such as the cell walls of phytoplankton and exoskeletons of crustaceans. Some species are pathogenic for fish and invertebrates. Many are psychrophilic, being commonly isolated from cold-water marine habitats and sea ice. The normal habitat of the genus *Bacteroides* is the gut of mammals and they may be present in sewage-polluted waters and persist for quite long periods in the sea, prompting investigation of their use as indicators of water quality (p.366).

Members of the phylum Chloroflexi are widespread but poorly characterized

Bacteria in the phylum Chloroflexi are found in a wide range of habitats and members show diverse metabolism, including anoxygenic photosynthesis and aerobic or anaerobic heterotrophy. Some are thermophiles. They are also phylogenetically diverse, with nine classes, each containing only a few genera. Overall, only 26 genera have been cultivated from terrestrial hot springs, soils, and freshwater. However, in early marine 16S rRNA surveys the SAR202 clade, which is abundant and globally distributed in the aphotic zone, was shown to affiliate with the Chloroflexi. We now know that they account for about 9% of the total sequences and in the deep ocean they can comprise >40% of the total bacterial community. Because of the absence of cultured marine representatives, their metabolic activity and role in marine processes has been unclear. Recent analysis of MAGs from SAR202 combined with FISH measurement indicate that they appear to be free-living, large bacteria with genomes of ~3–4 Mbp, probably existing as at least two sub-clades. One group contains pathways for the metabolism of organosulfur compounds as a source of carbon, nitrogen, and sulfur, suggesting that they play a major role in deep ocean sulfur cycling. A group identified in other studies has genes encoding multiple families of oxidative enzymes that may be active in the degradation of cyclic hydrocarbons. This activity is postulated to be involved in the conversion of DOM to refractory DOM. If this proves correct, it means that the Chloroflexi play a critical role in the microbial carbon pump discussed in *Chapter 8*.

The phyla Aquificae and Thermotogae are deeply branching primitive thermophiles

In 16S rRNA phylogenetic trees, these phyla form branches closer to the root than other groups of bacteria. The Aquificae contains a small number of genera which are all thermophilic and mostly microaerophilic oxidizing hydrogen that use hydrogen, thiosulfate, or sulfur as the electron donor and oxygen as the electron acceptor, fixing carbon via the reductive TCA cycle.

Examples include *Aquifex aeolicus*, *A. pyrophilus*, *Hydrogenothermus marinus*, and *Persephonella marinus*, which are extremely thermophilic (maximum growth temperature can be as high as 95°C) and have a major role in primary production at marine hydrothermal vents. *A. aeolicus* was one of the first marine bacteria for which a full genome sequence was obtained in 1998; it was found to have a very small genome (1.6 Mbp). Other species have also been found to have small genomes, typically <2.0 Mbp.

The Thermotogae also forms a deeply branching, phylogenetically distant group of Bacteria. As well as evidence from gene sequences, the function of the ribosome is very different from other Bacteria and is not affected by rifampicin and other antibiotics that affect protein synthesis. The name of the genus derives from a unique outer membrane ("toga"), which balloons out from the rod-shaped cells. The cells are Gram-negative, but the amino acid composition of the peptidoglycan is unlike that of other Bacteria, and there are unusual long-chain fatty acids in the lipids. *Thermotoga* is widespread in geothermal areas and occurs in shallow and deep-sea hydrothermal vents. Different species vary in their temperature optima, with a range from 55°C up to 80–95°C for the hyperthermophilic species *T. maritima* and *T. neapolitana*. These are fermentative, anaerobic chemoorganotrophs and use a wide range of carbohydrates. They can also fix N_2 and reduce sulfur to H_2S. Like *Aquifex*, these organisms have considerable biotechnological potential, and genome sequences have been published for several species. A number of genomes have been sequenced (size 1.7–1.9 Mbp). Many of the genes are involved in the transport and utilization of nutrients, in keeping with their ability to use a wide range of substances for growth. The genomes of both *Aquifex* and *Thermotoga* contain a high proportion of genes with homology to archaeal genes, indicating that large-scale lateral gene transfer has occurred in the evolution of these bacteria.

The Firmicutes are a major branch of Gram-positive Bacteria

The large and diverse phylum Firmicutes contains unicellular or filamentous bacteria distinguished by a thick cell wall composed mainly of peptidoglycan, and genomes with a low G+C ratio, which distinguishes them from the other phylum of Gram-positive bacteria, the Actinobacteria (low G+C). The phylum contains seven classes, of which the most important are the Bacilli (two orders, 15 families, 212 genera) and Clostridia (four orders, 18 families, 146 genera). These are likely to be reclassified under the GTDB system.

Within the Bacilli, the family Bacillaceae contains over 62 genera and 460 species and are best known for their role in ecology of terrestrial soils and plant health, but they are also a major component of coastal, saltmarsh, and deep-sea sediments. Although relatively little information exists on their abundance and distribution in marine environments, some species have been named because of their initial isolation from marine sediments (e.g. *Bacillus marinus*, *Marinococcus*, *Oceanobacillus*, *Thalassobacillus*) and it is likely that great diversity in marine representatives remains to be discovered in these habitats. Members of the Firmicutes comprise ~4% of 16S rRNA sequences in the water column and ~8% in sediments, although it is possible that many of these sequences result from bacteria that enter the sea as dormant endospores with terrestrial soil runoff, especially near large river mouths. They are mostly aerobic or facultatively anaerobic chemo-organotrophs. They produce a range of extracellular hydrolytic enzymes that degrade polysaccharides, proteins, lipids, and nucleic acids which play an important role in organic decomposition in sediments. Their most distinctive feature is the production of extremely resistant endospores by all but a few taxa, which allow them to resist high temperatures, irradiation, and desiccation; endospores may persist for thousands of years. Thermophilic and halophilic species occur at hydrothermal vents and salterns, respectively. Many have biotechnological applications. For example, some *Bacillus* spp. bind and oxidize metals such as manganese and copper and the spores may have applications in remediation of polluted marine sediments. Some members of the related genus *Paenibacillus* produce powerful chitinolytic activity, which may be useful for the breakdown of shrimp and crab shell waste. Both genera are also used as probiotics in aquaculture (see p.403), as are *Lactobacillus* and *Pediococcus* in the family Lactobacillaceae. Other genera produce secondary metabolites including antibiotics with potential applications in biomedicine.

Other important genera of the order Bacillales include *Staphylococcus* and *Listeria*, which are occasionally isolated in marine samples, but are probably terrestrial contaminants or very minor members of the marine bacterial community. However, they can be important as agents of fish spoilage and food-borne intoxication following processing (p.360). *Streptococcus iniae* and some other species are pathogens in warm-water fish and dolphins.

The Clostridia are anaerobic and employ a range of pathways for fermentation of different types of carbohydrates, amino acids, and purines. As seen with Bacilli, they are best known for their activity in anaerobic soils, but many species also occur in anoxic marine sediments and are important in degradation of organic material. Some species reduce sulfate, sulfite, or thiosulfate. The toxin-producing *Clostridium botulinum* (type E) occurs in marine mud and fish guts and can cause the disease botulism in humans, occasionally associated with seafood (see p.341), or in waterfowl in estuaries or coastal lagoons.

One unusual group of clostridia are the uncultivated giant bacteria "*Ca.* Epulopiscium fishelsoni" and related species. These occur in large numbers in the intestinal tracts of certain species of herbivorous surgeonfish on the Great Barrier Reef and the Red Sea. They appear to be mutualistic symbionts, which aid in the breakdown of complex carbohydrate polysaccharides in the red and brown algae ingested by the fish. Different fish species appear to associate with particular subclades of Epulopiscium, leading to unique gut microbiota that permit specialization in feeding on certain algal types. Juvenile surgeonfish have been observed to be coprophagous (consuming the feces of adults) and presumably acquire symbionts through this behavior. Epulopiscium strains are some of the largest bacteria known; although morphology and cell size are variable, cells up to 400×80 μm are common and some rare cells can reach 600 μm. When originally discovered, they were thought to be a eukaryotic protist because of their large size and intracellular structure. They have a unique reproductive process; they are "viviparous," meaning that new cells are formed inside the parent cell, which undergoes a localized cell lysis to release the active progeny. One of the most remarkable features of the life cycle is that it follows a circadian rhythm. Samples of surgeonfish gut contents taken early in the morning show small internal offspring appearing near the tips of the mother cell. The offspring cells grow during the day and are released at night through partial lysis of the mother cell, which then dies. This process is speculated to have evolved from the process of endospore formation, which involves the partition of the spore from the parent cell (*Figure 4.15*).

Members of the Actinobacteria are a rich source of secondary metabolites, including antibiotics

The phylum Actinobacteria forms a second major group of Gram-positive bacteria, characterized by its high G+C content. It is highly diverse, containing 65 cultivated families and over 500 species. They are mostly free-living rod-shaped or filamentous organisms and many species have been cultivated from terrestrial soils, freshwater, and marine sediments. However, this represents only a small fraction of the diversity revealed by 16S rRNA sequence surveys,

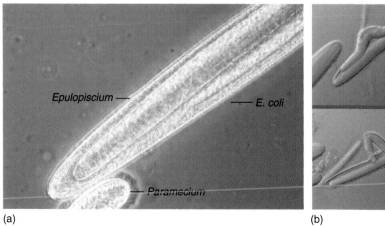
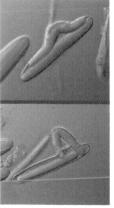

(a)   (b)

Figure 4.15 "*Ca.* Epulopiscium fishelsoni", one of the largest known bacteria. (a) Light micrograph showing comparative sizes of Epulopiscium, *Paramecium* (a protist), and *Escherichia coli* (a "typical" bacterium). (b) "Viviparous" release of daughter cell during reproduction. Credit: Esther R. Angert, Cornell University..

DO YOU LIKE SUSHI? *ZOBELLIA* GENES MAY HELP YOU TO DIGEST IT.

Humans do not produce the carbohydrate active enzymes (CAzymes) needed to digest polysaccharides—we depend on members of our gut microbiome to do this. During human evolution, we have acquired a diverse array of microbes, including many that degrade complex polysaccharides from terrestrial plants. Some bacteria, especially members of the Bacteroidetes can produce hundreds of specific glycoside hydrolases and polysaccharide lyases. Hehemann *et al.* (2010) identified a new CAzyme, porphyrinase, that specifically degrades porphyrin, a complex polysaccharide in the cell walls of red algae. They identified the substrate recognition site on the enzyme and showed that the gene sequence for this can be found in marine bacteria, including *Zobellia galactivorans* (Bacteroidetes), that grows on *Porphyra* red algae. It was also found in *Bacteroides plebeius* in the gut microbiome of a group of Japanese people investigated, but not in North Americans. Genomic analysis shows that acquisition of this gene likely occurred by HGT, because raw seaweed (nori) forms a major component of the Japanese diet.

especially those from coastal and estuarine waters, where they can represent about 10% of the bacterioplankton

The order Actinomycetales is a large and very diverse group of bacteria with various cell morphologies, ranging from coryneform (club-shaped, e.g. *Corynebacterium* and *Arthrobacter)* to branching filaments with reproductive conidiospores (e.g. *Streptomyces* and *Micromonospora*). Actinomycetes are widely distributed in marine sediments and particulate matter in the water column. Because of their high abundance in soil, it is possible that many actinomycetes in coastal sediments are derived from terrestrial runoff, but there are also halophilic species and they are also found in deep-sea samples. Their main ecological importance is in decomposition and heterotrophic nutrient cycling, owing to their production of diverse extracellular enzymes, which break down polysaccharides, proteins, and fats. Many of the actinomycetes have large genomes (up to 9 Mb), showing cellular differentiation that makes them an exceptionally rich source of secondary metabolites. Many of the most widely used antibiotics (e.g. aminoglycosides and tetracyclines) come from actinomycetes found in soil. Extensive surveys of the diversity of marine actinomycetes have been undertaken in the search for unique compounds with potential uses as antimicrobial or anticancer drugs (e.g. compounds from *Salinispora*). Marine *Micromonospora* are used in some aquaculture probiotics. These biotechnological aspects are considered in *Chapter 14*.

Mycobacteria are slow-growing, aerobic, rod-shaped organisms distinguished by unusual cell wall components, which render them acid-fast in a staining procedure. They are widely distributed as saprophytes on surfaces such as sediments, corals, fish, and algae.

The great majority of species in the Actinobacteria are harmless saprophytes, but a few species are pathogenic. Those with a marine connection include *Mycobacterium marinum* (a pathogen of fish and marine mammals that can be transmitted to humans) and *Renibacterium salmoninarum* (an obligate pathogen of salmonid fish).

Conclusions

This chapter has illustrated the great diversity of morphology, metabolism, physiology, and genetic makeup of members of the domain Bacteria. This diversity has enabled bacteria to colonize every conceivable habitat in the marine realm, and their activities affect all aspects of marine ecology and biogeochemistry. The evolutionary history and classification of the Bacteria remains problematic. Only a small fraction of the organisms known from molecular-based studies have been cultured and studied in detail, but genomic methods mean that completely new types of bacteria, biochemical pathways, ecological interactions, and biogeochemical effects are being discovered at ever-increasing rates. The oceans undoubtedly hold many more surprises yet to be revealed.

References and further reading

Principles of diversity, classification, and taxonomy

Achtman, M., and Wagner, M. (2008) Microbial diversity and the genetic nature of microbial species. *Nat. Rev. Microbiol.* 6: 431–440.

Amann, R., and Rosselló-Móra, R. (2016) After all, only millions? *MBio* 7: e00999–e00916.

Amaral-Zettler, L., Artigas, L.F., Baross, J., et al. (2010). A global census of marine microbes, In: *Life in the World's Oceans. Diversity, Distribution, and Abundance.* McIntyre, A.D. (ed.) Wiley-Blackwell, pp. 223–246.

Bergey's Manual Trust (N.D) *Bergey's Manual of Systematics of Archaea and Bacteria.* Wiley Online Library; available at https:// onlinelibrary.wiley.com/doi/book/10.1002/9781118960608. Accessed 23 April 2019.

Callahan, B.J., McMurdie, P.J., and Holmes, S.P. (2017) Exact sequence variants should replace operational taxonomic units in marker-gene data analysis. *ISME J.* **11**: 2639.

Castelle, C.J., and Banfield, J.F. (2018) Major new microbial groups expand diversity and alter our understanding of the tree of life. *Cell* **172**: 1181–1197.

Chun, J., Oren, A., Ventosa, A. et al. (2019) Proposed minimal standards for the use of genome data for the taxonomy of prokaryotes. *Int. J. Syst. Evol. Microbiol* **68**: 461–466.

Curtis, T.P., Sloan, W.T., and Scannell, J.W. (2002) Estimating prokaryotic diversity and its limits. *Proc. Natl. Acad. Sci. U. S. A.* **99**: 10494–10499.

Garrity, G.M. (2016) A new genomics-driven taxonomy of *Bacteria* and *Archaea*: are We There Yet? *J. Clin. Microbiol.* **54**: 1956–1963.

Hug, L.A., Baker, B.J., Anantharaman, K. et al. (2016) A new view of the tree of life. *Nat. Microbiol.* **1**: 16048.

Jousset, A., Bienhold, C., Chatzinotas, A. et al. (2017) Where less may be more: how the rare biosphere pulls ecosystems strings. *ISME J.* **11**: 853–862.

Konstantinidis KT, Rosselló-Móra R, and Amann R. (2017) Uncultivated microbes in need of their own taxonomy. *ISME J.* **11**: 2399–2406.

Kunin, V., Engelbrektson, A., Ochman, H. et al. (2010) Wrinkles in the rare biosphere: pyrosequencing errors can lead to artificial inflation of diversity estimates. *Environ. Microbiol.* **12**: 118–123.

Locey, K.J., and Lennon, J.T. (2016) Scaling laws predict global microbial diversity. *Proc. Natl. Acad. Sci. USA* **113**: 5970–5975.

Louca, S., Mazel, F., Doebeli, M. et al. (2019) A census-based estimate of Earth's bacterial and archaeal diversity. *PLOS Biol.* **17**: e3000106.

Parks, D.H., Chuvochina, M., Waite, D.W. et al. (2018) A standardized bacterial taxonomy based on genome phylogeny substantially revises the tree of life. *Nat. Biotechnol.* **36**: 996.

Pedros-Alio, C., Acinas, S.G., Logares, R. et al. (2018). Marine microbial diversity as seen by hight-throughput sequencing. In: Gasol, J.M. and Kirchman, D.L. (eds.) *Microbial Ecology of the Oceans.* John Wiley & Sons, pp. 47–98.

Richter, M., and Rosselló-Móra, R. (2009) Shifting the genomic gold standard for the prokaryotic species definition. *Proc. Natl. Acad. Sci. USA* **106**: 19126–19131.

Salazar, G., and Sunagawa, S. (2017a) Marine microbial diversity. *Curr. Biol.* **27**: R489–R494.

Schleifer, K.H. (2009) Classification of Bacteria and Archaea: Past, present and future. *Syst. Appl. Microbiol.* **32**: 533–542.

Schloss, P.D., Girard, R.A., Martin, T. et al. (2016) Status of the archaeal and bacterial census: an update. *MBio* **7**: e00201–e00216.

Schulz, F., Eloe-Fadrosh, E.A., Bowers, R.M. et al. (2017) Towards a balanced view of the bacterial tree of life. *Microbiome* **5**: 140.

Sogin, M.L., Morrison, H.G., Huber, J.A. et al. (2008) Microbial diversity in the deep sea and the underexplored "rare biosphere." *Proc. Natl. Acad. Sci.* **103**: 12115–12120.

Stackebrandt, E., and Ebers, J. (2006) Taxonomic parameters revisited: tarnished gold standards. Microbiol. Today **33**:152–155.

Waite, D.W., Vanwonterghem, I., Rinke, C. et al. (2017) Comparative genomic analysis of the class *Epsilonproteobacteria* and proposed reclassification to Epsilonbacteraeota (phyl. nov.). *Front. Microbiol.* **8**: 682.

Whitman, W.B., Coleman, D.C., Wiebe, W.J. et al. (1998) Prokaryotes: the unseen majority. *Proc. Natl. Acad. Sci. USA* **95**: 6578–6583.

Willis, A. (2016) Extrapolating abundance curves has no predictive power for estimating microbial biodiversity. *Proc. Natl. Acad. Sci. USA* **113**: E5096.

Winter, C., Bouvier, T., Weinbauer, M.G. et al. (2010) Trade-offs between competition and defense specialists among unicellular planktonic organisms: the "killing the winner" hypothesis revisited. *Microbiol. Mol. Biol. Rev.* **74**: 42–57.

Yarza, P., Yilmaz, P., Pruesse, E. et al. (2014) Uniting the classification of cultured and uncultured bacteria and archaea using 16S rRNA gene sequences. *Nat. Rev. Microbiol.* **12**: 635–645.

Yilmaz, P., Yarza, P., Rapp, J.Z. et al. (2016) Expanding the world of marine bacterial and archaeal clades. *Front. Microbiol.* **6**: 1524.

Alphaproteobacteria

Ferla, M.P., Thrash, J.C., Giovannoni, S.J. et al. (2013) New rRNA gene-based phylogenies of the *Alphaproteobacteria* provide perspective on major groups, mitochondrial ancestry and phylogenetic instability. *PLoS One* **8**: e83383.

Giovannoni, S.J., Hayakawa, D.H., Tripp, H.J. et al. (2008) The small genome of an abundant coastal ocean methylotroph. *Environ. Microbiol.* **10**: 1771–1782.

Greene, S.E., Brilli, M., Biondi, E.G. et al. (2012) Analysis of the CtrA pathway in *Magnetospirillum* reveals an ancestral role in motility in alphaproteobacteria. *J. Bacteriol.* **194**: 2973–2986.

Hughes, V., Jiang, C., and Brun, Y. (2012) *Caulobacter crescentus. Curr. Biol.* **22**: R507–R509.

Komeili, A., Vali, H., Beveridge, T.J. et al. (2004) Magnetosome vesicles are present before magnetite formation, and MamA is required for their activation. *Proc. Natl. Acad. Sci. USA* **101**: 3839–3844.

Kolinko, S., Richter, M., Glöckner, F.O. et al. (2016). Single-cell genomics of uncultivated deep-branching magnetotactic bacteria reveals a conserved set of magnetosome genes. *Environ. Microbiol.***18**: 21–37.

Lin, W., Zhang, W., Zhao, X. et al. (2018). Genomic expansion of magnetotactic bacteria reveals an early common origin of magnetotaxis with lineage-specific evolution. *ISME J.* **12**: 1508.

Luo, H., and Moran, M.A. (2014) Evolutionary ecology of the marine *Roseobacter* clade. *Microbiol. Mol. Biol. Rev.* **78**: 573–587.

Martins, J.L., Silveira, T.S., Abreu, F. et al. (2007). Grazing protozoa and magnetosome dissolution in magnetotactic bacteria. *Environ. Microbiol.* **9**: 2775–2781.

Rosenberg, E., DeLong, E.F., Lory, S. et al. (eds.) (2014) *The Prokaryotes. Alphaproteobacteria and Betaproteobacteria.* Springer Berlin, Heidelberg.

Scheffel, A., Gruska, M., Faivre, D. et al. (2006) An acidic protein aligns magnetosomes along a filamentous structure in magnetotactic bacteria. *Nature* **440**: 110–114.

Silva, K.T., Abreu, F., Keim, C.N. et al. (2008) Ultrastructure and cytochemistry of lipid granules in the many-celled magnetotactic prokaryote, 'Candidatus Magnetoglobus multicellularis.' *Micron* **39**: 1387–1392.

Taoka, A., Kiyokawa, A., Uesugi, C. et al. (2017) Tethered magnets are the key to magnetotaxis: direct observations of *Magnetospirillum magneticum* AMB-1 show that MamK distributes magnetosome organelles equally to daughter cells. *MBio* **8**: e00679–e00617.

Gammaproteobacteria

Frischkorn, K.R., Stojanovski, A., and Paranjpye, R. (2013) Vibrio parahaemolyticus type IV pili mediate interactions with diatom-derived chitin and point to an unexplored mechanism of environmental persistence. *Environ. Microbiol.* **15**: 1416–1427.

Jørgensen, B.B., and Gallardo, V.A. (1999) *Thioploca* spp.: filamentous sulfur bacteria with nitrate vacuoles. *FEMS Microbiol. Ecol.* **28**: 301–313.

Lin, H., Yu, M., Wang, X. et al. (2018) Comparative genomic analysis reveals the evolution and environmental adaptation strategies of vibrios. *BMC Genomics* **19**: 135.

Miller, S.D., Haddock, S.H.D., Elvidge, C.D. and Lee, T.F. (2005). Detection of a bioluminescent milky sea from space. *Proc. Nat. Acad.Sci USA* **102**: 14181–14184.

Thompson, C.C., Vicente, A., Souza, R.C. et al. (2009) Genomic taxonomy of vibrios. *BMC Evol. Biol.* **9**: 258.

Thompson, F.L., Gevers, D., Thompson, C.C. et al. (2005) Phylogeny and molecular identification of vibrios on the basis of multilocus sequence analysis. *Appl. Environ. Microbiol.* **71**: 5107–5115.

Cable bacteria, deltaproteobacteria and epsilonbacteriota

Bjerg, J.T., Boschker, H.T.S., Larsen, S. et al. (2018) Long-distance electron transport in individual, living cable bacteria. *Proc. Natl. Acad. Sci. U. S. A.* **115**: 5786–5791.

Bjerg, J.T., Damgaard, L.R., Holm, S.A. et al. (2016) Motility of electric cable bacteria. *Appl. Environ. Microbiol.* **82**: 3816–3821.

Burdorf, L.D.W., Tramper, A., Seitaj, D. et al. (2017) Long-distance electron transport occurs globally in marine sediments. *Biogeosci.* **14**: 683–701.

Cornelissen, R., and Bøggild, A., Thiruvallur Eachambadi, R. et al. (2018) The cell envelope structure of cable bacteria. *Front. Microbiol.* **9**: 3044.

Kessler, A.J., Wawryk, M., Marzocchi, U. et al. (2018) Cable bacteria promote DNRA through iron sulfide dissolution. *Limnol. Oceanogr.* **64**: 1228–1238.

Kuever, J. (2014a) The family *Desulfobulbaceae*. In: *The Prokaryotes. Deltaproteobacteria and Epsilonproteobacteria*. Rosenberg, E., DeLong, E.F., Lory, S. et al. (eds.) Springer Berlin, Heidelberg, pp. 75–86.

Kuever, J. (2014b) The family *Desulfovibrionaceae*. In: *The Prokaryotes. Deltaproteobacteria and Epsilonproteobacteria*. Rosenberg, E., DeLong, E.F., Lory, S. et al. (eds.) Springer Berlin, Heidelberg, pp. 107–133.

Malkin, S.Y., Seitaj, D., Burdorf, L.D.W. et al. (2017) Electrogenic sulfur oxidation by cable bacteria in bivalve reef sediments. *Front. Mar. Sci.* **4**: 28.

Martin, B.C., Bougoure, J., Ryan, M.H. et al. (2019) Oxygen loss from seagrass roots coincides with colonisation of sulphide-oxidising cable bacteria and reduces sulphide stress. *ISME J.* **13**: 707–719.

Marzocchi, U., Trojan, D., Larsen, S. et al. (2014) Electric coupling between distant nitrate reduction and sulfide oxidation in marine sediment. *ISME J.* **8**: 1682–1690.

Meysman, F.J.R. (2018) Cable bacteria take a new breath using long-distance electricity. *Trends Microbiol.* **26**: 411–422.

Nielsen, L.P. (2016) Ecology: Electrical cable bacteria save marine life. *Curr. Biol.* **26**: R32–R33.

Nielsen, L.P., Risgaard-Petersen, N., Fossing, H. et al. (2010) Electric currents couple spatially separated biogeochemical processes in marine sediment. *Nature* **463**: 1071–1074.

Pfeffer, C., Larsen, S., Song, J. et al. (2012) Filamentous bacteria transport electrons over centimetre distances. *Nature* **491**: 218–221.

Plugge, C.M., Zhang, W., Scholten, J.C.M.. et al. (2011) Metabolic flexibility of sulfate-reducing bacteria. *Front. Microbiol.* **2**: 81.

Risgaard-Petersen, N., Damgaard, L.R., Revil, A. et al. (2014) Mapping electron sources and sinks in a marine biogeobattery. *J. Geophys. Res. Biogeosci.* **119**: 1475–1486.

Schauer, R., Risgaard-Petersen, N., Kjeldsen, K.U. et al. (2014) Succession of cable bacteria and electric currents in marine sediment. *ISME J.* **8**: 1314–1322.

Seitaj, D., Schauer, R., Sulu-Gambari, F. et al. (2015) Cable bacteria generate a firewall against euxinia in seasonally hypoxic basins. *Proc. Natl. Acad. Sci.* **112**: 13278–13283.

Trojan, D., Schreiber, L., Bjerg, J.T. et al. (2016) A taxonomic framework for cable bacteria and proposal of the candidate genera Electrothrix and Electronema. *Syst. Appl. Microbiol.* **39**: 297–306.

Vetriani, C., Voordeckers, J.W., Crespo-Medina, M. et al. (2014) Deep-sea hydrothermal vent *Epsilonproteobacteria* encode a conserved and widespread nitrate reduction pathway (Nap). *ISME J.* **8**: 1510–1521.

Myxobacteria

Zhang, Y.-Q., Li, Y.-Z., Wang, B. et al. (2005) Characteristics and living patterns of marine myxobacterial isolates. *Appl. Environ. Microbiol.* **71**: 3331–3336.

Bdellovibrio

Mosca, A., Leclerc, M., and Hugot, J.P. (2016) Gut microbiota diversity and human diseases: Should we reintroduce key predators in our ecosystem? *Front. Microbiol.* **7**: 455.

Raghunathan, D., Radford, P.M., Gell, C. et al. (2019) Engulfment, persistence and fate of *Bdellovibrio bacteriovorus* predators inside human phagocytic cells informs their future therapeutic potential. *Sci. Rep.* **9**: 4293.

Rotem, O., Pasternak, Z., and Jurkevitch, E. (2014) Bdellovibrio and like organisms. In: *The Prokaryotes. Deltaproteobacteria and Epsilonproteobacteria*. Rosenberg, E., DeLong, E.F., Lory, S. et al. (eds.) Springer Berlin Heidelberg, Berlin, Heidelberg, pp. 3–17.

Zetaproteobacteria

Laufer, K., Nordhoff, M., Halama, M. et al. (2017) Microaerophilic Fe(II)-oxidizing zetaproteobacteria isolated from low-Fe marine coastal sediments: physiology and composition of their twisted stalks. *Appl. Environ. Microbiol.* **83**: e03118–e03116.

Moreira, A.P.B., Meirelles, P.M., and Thompson, F. (2014) The family *Mariprofundaceae*. In: *The Prokaryotes. Deltaproteobacteria and Epsilonproteobacteria*. Rosenberg, E., DeLong, E.F., Lory, S. et al. (eds.) Springer Berlin Heidelberg, Berlin, Heidelberg, pp. 403–413.

McAllister, S.M., Davis, R.E., McBeth, J.M. et al. (2011) Biodiversity and emerging biogeography of the neutrophilic iron-oxidizing Zetaproteobacteria. *Appl. Environ. Microbiol.* **77**: 5445–5457.

Cyanobacteria

Becker, J.W., Hogle, S.L., Rosendo, K. et al. (2019) Co-culture and biogeography of *Prochlorococcus* and SAR11. *ISME J.* **13**: 1506–1519.

Bergman, B., Sandh, G., Lin, S. et al. (2013a) *Trichodesmium* – a widespread marine cyanobacterium with unusual nitrogen fixation properties. *FEMS Microbiol. Rev.* **37**: 286–302.

Biller, S.J., Berube, P.M., Lindell, D. et al. (2015) Prochlorococcus: the structure and function of collective diversity. *Nat. Rev. Microbiol.* **13**: 13–27.

Biller, S.J., Schubotz, F., Roggensack, S.E. et al. (2014) Bacterial vesicles in marine ecosystems. *Science* **343**: 183–186.

Cohen, S.E., and Golden, S.S. (2015) Circadian rhythms in Cyanobacteria. *Microbiol. Mol. Biol. Rev.* **79**: 373–385.

Coutinho, F., Tschoeke, D.A., Thompson, F. et al. (2016) Comparative genomics of *Synechococcus* and proposal of the new genus *Parasynechococcus*. *PeerJ* **4**: e1522.

Dubinsky, Z., and Berman-Frank, I. (2001) Uncoupling primary production from population growth in photosynthesizing organisms in aquatic ecosystems. *Aquat. Sci.* **63**: 4–17.

Ehlers, K., and Oster, G. (2012) On the mysterious propulsion of *Synechococcus*. *PLoS One* **7**: e36081.

Farnelid, H., Turk-Kubo, K., Ploug, H. et al. (2019) Diverse diazotrophs are present on sinking particles in the North Pacific Subtropical Gyre. *ISME J.* **13**: 170–182.

Farnelid, H.M., Turk-Kubo, K.A., and Zehr, J.P. (2016) Identification of associations between bacterioplankton and photosynthetic picoeukaryotes in coastal waters. *Front. Microbiol.* **7**: 339.

Flombaum, P., Gallegos, J.L., Gordillo, R.A. et al. (2013) Present and future global distributions of the marine Cyanobacteria *Prochlorococcus* and *Synechococcus*. *Proc. Natl. Acad. Sci. U. S. A.* **110**: 9824–9829.

Foster, R.A., Kuypers, M.M.M., Vagner, T. et al. (2011) Nitrogen fixation and transfer in open ocean diatom-cyanobacterial symbioses. *ISME J.* **5**: 1484–1493.

Hagino, K., Onuma, R., Kawachi, M. et al. (2013) Discovery of an endosymbiotic nitrogen-fixing cyanobacterium UCYN-A in *Braarudosphaera bigelowii* (Prymnesiophyceae). *PLoS One* **8**: e81749.

Johnson, Z.I., Zinser, E.R., Coe, A. et al. (2006) Niche partitioning among *Prochlorococcus* ecotypes along ocean-scale environmental gradients. *Science (80-.).* **311**: 1737–1740.

Kashtan, N., Roggensack, S.E., Rodrigue, S. et al. (2014) Single-cell genomics reveals hundreds of coexisting subpopulations in wild *Prochlorococcus*. *Science* **344**: 416–420.

Kent, A.G., Dupont, C.L., Yooseph, S. et al. (2016) Global biogeography of *Prochlorococcus* genome diversity in the surface ocean. *ISME J.* **10**: 1856–1865.

Mishra, A.K., Tiwari, D.N., and Rai, A.N. (2019) *Cyanobacteria: From Basic Science to Applications.* Academic Press.

Moore, L.R. (2013). More mixotrophy in the marine microbial mix. *Proc. Nat. Acad. Sci. USA* **110**: 8323–8324.

Muñoz-Marín, M.D.C., Shilova, I.N., Shi, T. et al. (2019) The transcriptional cycle is suited to daytime N_2 fixation in the unicellular cyanobacterium "*Candidatus* Atelocyanobacterium thalassa" (UCYN-A). *MBio* **10**: e02495–e02418.

Paerl, H. (2017) The cyanobacterial nitrogen fixation paradox in natural waters. *F1000Research* **6**: 244.

Rocap, G., Larimer, F.W., Lamerdin, J. et al. (2003) Genome divergence in two *Prochlorococcus* ecotypes reflects oceanic niche differentiation. *Nature* **424**: 1042–1047.

Walter, J.M., Coutinho, F.H., Dutilh, B.E. et al. (2017) Ecogenomics and taxonomy of *Cyanobacteria* phylum. *Front. Microbiol.* **8**: 2132.

Zehr, J.P. (2011) Nitrogen fixation by marine cyanobacteria. *Trends Microbiol.* **19**: 162–173.

Zehr, J.P. (2015) How single cells work together. *Science* **349**: 1163–1164.

Zehr, J.P., Shilova, I.N., Farnelid, H.M. et al. (2016) Unusual marine unicellular symbiosis with the nitrogen-fixing cyanobacterium UCYN-A. *Nat. Microbiol.* **2**: 16214.

Planctomycetes and bacteroidetes

Boedeker, C., Schüler, M., Reintjes, G. et al. (2017) Determining the bacterial cell biology of Planctomycetes. *Nat. Commun.* **8**: 14853.

Fernández-Gomez, B., Richter, M., Schüler, M. et al. (2013). Ecology of marine Bacteroidetes: a comparative genomics approach. *ISME J.* **7**: 1026.

Hehemann, J.H., Correc, G., Barbeyron, T. et al. (2010). Transfer of carbohydrate-active enzymes from marine bacteria to Japanese gut microbiota. *Nature* **464**: 908.

Jeske, O., Schüler, M., Schumann, P. et al. (2015) Planctomycetes do possess a peptidoglycan cell wall. *Nat. Commun.* **6**: 7116.

Kartal, B., de Almeida, N.M., Maalcke, W.J. et al. (2013) How to make a living from anaerobic ammonium oxidation. *FEMS Microbiol. Rev.* **37**: 428–461.

Reintjes, G., Arnosti, C., Fuchs, B.M. et al. (2017). An alternative polysaccharide uptake mechanism of marine bacteria. *ISME J.* **11**:1640.

Wagner, M., and Horn, M. (2006) The *Planctomycetes, Verrucomicrobia, Chlamydiae* and sister phyla comprise a superphylum with biotechnological and medical relevance. *Curr. Opin. Biotechnol.* **17**: 241–249.

Wiegand, S., Jogler, M., and Jogler, C. (2018) On the maverick Planctomycetes. *FEMS Microbiol. Rev.* **42**: 739–760.

Žure, M., Fernandez-Guerra, A., Munn, C.B. et al. (2017). Geographic distribution at subspecies resolution level: closely related *Rhodopirellula* species in European coastal sediments. *ISME J.* **11**: 478.

Žure, M., Munn, C.B., and Harder, J. (2015) Diversity of *Rhodopirellula* and related planctomycetes in a North Sea coastal sediment employing carB as molecular marker. *FEMS Microbiol. Lett.* **362**: fnv127.

Bengtsson, M.M., and Øvreås, L. (2010). Planctomycetes dominate biofilms on surfaces of the kelp *Laminaria hyperborea*. *BMC Microbiol.***10**: 261.

Epulopiscium

Angert, E. R., and Clements, K. D. (2004) Initiation of intracellular offspring in *Epulopiscium. Molec. Microbiol.* 51: 827–835.

Kamanda Ngugi, D., Miyake, S., Cahill, M. et al. (2017) Genomic diversification of giant enteric symbionts reflects host dietary lifestyles. *Proc. Natl. Acad. USA* **1114**: E7592.

Chloroflexi

Hanada, S. (2014) The phylum Chloroflexi, the family *Chloroflexaceae*, and the related phototrophic families *Oscillochloridaceae* and *Roseiflexaceae*. In: *The Prokaryotes. Other Major Lineages of Bacteria and the Archaea*. Rosenberg, E., DeLong, E.F., Lory, S. et al. (eds.) Springer Berlin, Heidelberg, pp. 515–532.

Landry, Z., Swan, B.K., Herndl, G.J. et al. (2017) SAR202 Genomes from the dark ocean predict pathways for the oxidation of recalcitrant dissolved organic matter. *MBio* **8**: e00413–e00417.

Mehrshad, M., Rodriguez-Valera, F., Amoozegar, M.A. et al. (2018) The enigmatic SAR202 cluster up close: shedding light on a globally distributed dark ocean lineage involved in sulfur cycling. *ISME J.* **12**: 655–668.

Other taxa

Barka, E.A., Vatsa, P., Sanchez, L. et al. (2016) Taxonomy, physiology, and natural products of *Actinobacteria*. *Microbiol. Mol. Biol. Rev.* **80**: 1–43.

Garcia, R., and Müller, R. (2014) The family *Myxococcaceae*. In: *The Prokaryotes. Deltaproteobacteria and Epsilonproteobacteria*. Rosenberg, E., DeLong, E.F., Lory, S. et al. (eds.) Springer Berlin, Heidelberg, pp. 191–212.

McBride, M.J., Liu, W., Lu, X. et al. (2014) The family *Cytophagaceae*. In: *The Prokaryotes. Other Major Lineages of Bacteria and the Archaea*. Rosenberg, E., DeLong, E.F., Lory, S. et al. (eds.) Springer Berlin Heidelberg, Berlin, Heidelberg, pp. 577–593.

Gupta, R.S. (2014) The phylum *Aquificae*. In: *The Prokaryotes. Other Major Lineages of Bacteria and the Archaea*. Rosenberg, E., DeLong, E.F., Lory, S. et al. (eds.) Springer Berlin, Heidelberg, pp. 417–445.

Chapter 5

Marine Archaea

This chapter explores the diversity of the Archaea and their role in marine processes. As discussed in *Chapter 1*, the application of 16S rRNA gene sequencing methods first developed in the 1970s led to the establishment of the Archaea as a distinct domain of life, overturning the description of them at the time as an uncommon, specialized subset of bacteria ("archaebacteria") found only in extreme habitats. Other information, especially the nature of their ribosomes and mechanisms of transcription and translation has supported this view. Although many of the most thermophilic and halophilic microbes are indeed members of the Archaea, species belonging to this domain are very abundant and diverse, and can be found everywhere in the marine environment, especially the sub-photic ocean. Here, they play critical roles in the carbon and nitrogen cycles. Recent advances in DNA sequencing and genomic methods are revealing an unexpected diversity of novel archaeal lineages, leading to new ideas about the position of Archaea in the tree of life and the origins of eukaryotes.

Key Concepts

- The basic metabolism and organization of cells of Archaea resembles those of Bacteria, but there are important differences in membrane structure, replication of DNA, and protein synthesis.

- Numerous new phyla of the Archaea have been described following discovery in environmental gene surveys and this has led to reorganization of phylogenetic trees, providing support for the evolution of the Eukarya from within the Archaea.

- Production of methane by members of the Euryarchaeota is the final step in the anaerobic biodegradation of organic material and leads to formation of massive reserves of methane in deep-sea sediments.

- Anoxic methane oxidation is carried out by members of the Euryarchaeota by reverse methanogenesis, in syntrophic association with sulfate-reducing bacteria.

- Numerous types of chemolithotrophic and organotrophic Euryarchaeota and Crenarchaeota show adaptations for growth at very high temperatures and high salinity.

- Thaumarchaeota are abundant in the deep sea, where their main function is oxidation of ammonia coupled to chemolithoautotrophy, making a major contribution to dark carbon fixation.

ARCHAEA, ARCHAEA, EVERYWHERE

The discovery that Archaea are a major component of ocean microbiota ranks as one of the most significant surprises to emerge from the application of methods to directly sequence 16S rRNA isolated from plank-tonic biomass (see p.42). The first clues came in 1992, when Jed Fuhrman (University of Southern California) and Ed De Long (then at Woods Hole Oceanographic Institution) independently dis-covered archaeal sequences in seawater samples. Fuhrman et al. (1992) found sequences in pelagic water samples from the Pacific Ocean (100 and 500 m depth) that were only distantly related to those of any organisms previously characterized. The closest match for some of these sequences was to extreme thermophiles, which had been assigned to the new domain Archaea (at that time, often still referred to as archae-bacteria). DeLong (1992) also discovered widespread archaeal rRNA sequences in coastal surface waters off the east and west coasts of North America. Before these discoveries, it was assumed that Archaea are only found in extreme environments, inhospitable to other life forms. Within a few years, numerous studies had established these currently uncultivated Archaea are highly abundant in surface and deep surface waters in all the major ocean basins.

Several aspects of cell structure and function distinguish the Archaea and Bacteria

While the basic organization of the cells of bacteria and archaea is similar, there are some important structural and physiological differences. Indeed, some aspects of archaeal biology show closer similarities to those of the Eukarya than they do to the Bacteria, the significance of which will become clear when we consider their possible evolutionary relationships. As described in *Chapter* 3, there are major differences between bacterial and archaeal cell membranes, which contain fatty acid-based lipids and isoprenoid-based lipids, respectively. However, analysis of gene sequences has blurred this distinction; some of the recently discovered uncultivated bacterial taxa in the Candidate Phyla Radiation (*Figure 4.1*) appear to possess genes encoding the isoprenoid synthesis pathway that is typical of Archaea, while the newly discovered DPANN Archaea (see below) possess genes like those of Bacteria. Secondly, although archaeal cells possess small compact genomes like those in bacteria, organization of the nuclear material is more complex and variable in archaeal cells. In some species, packaging and stabilization of the DNA structure is achieved either by supercoiling (using the enzyme DNA gyrase), as in the Bacteria. However, other species possess positively charged proteins known as histones, which have amino-acid sequences homologous to those found in eukaryotes. Short sequences of DNA are wound around clusters of histones to form tetrameric clusters called nucleosomes. Some of the extremely thermophilic Archaea have an additional type of DNA gyrase that induces positive super-coiling in the DNA to protect it against denaturation. The process of DNA replication is also different from that in the Bacteria. Although most Archaea possess single circu-lar chromosomes, there are often multiple origins of replication, and the structure of the replication complex and polymerase enzymes again resembles that found in eukaryotes. Conversely, regulation of gene expression in the Archaea is very like that in Bacteria, involving the use of repressor or activator proteins to control the level of transcription. Translation of mRNA and protein synthesis in the Archaea shares many features with the Bacteria, but many of the component proteins of Archaea and Eukarya show closer homol-ogy than they do with those from Bacteria.

New phylogenomic methods have led to recognition of multiple phyla of the Archaea

As mentioned, the group we now know as the Archaea were first discovered in extreme environments, including hot springs, hydrothermal vents, high salinity, and low pH habi-tats. With the right conditions, it has proved relatively straightforward to bring these into culture, but almost none of the diverse archaeal taxa that occur in the oceans have been cultured. The application of 16S rRNA gene sequencing in the 1980s led to the early recogni-tion of two major branches (phyla) of the phylogenetic tree, namely the Euryarchaeota and Crenarchaeota. Later, the new phylum Thaumarchaeota was designated, and many former members of the Crenarchaeota have now been assigned to this. Despite the recognition of their abundance in the marine environment, we still know much less about the diversity of the Archaea than Bacteria. The number of researchers working with archaea is much smaller, there is a much shorter history of attempts to cultivate them, and environmental 16S rRNA gene surveys have been heavily biased towards bacterial sequences. Based on 16S rRNA sequences from marine habitats, the dominant phyla are the Thaumarchaeota and Euryarchaeota (*Figure 5.1*). As noted in previous chapters, despite the undoubted usefulness of 16S rRNA gene sequencing, reliance on a single gene to infer phylogenetic relationships can lead to problems. As we saw in the discussion of bacterial diversity and evolutionary relationships in *Chapter 4*, there have been major advances in phylogenetic and phyloge-nomic analysis, which has been extended by the study of other conserved key genes, such as those encoding proteins involved with transcription and translation, as well as compari-son of amplified genome sequences with genomes of cultivated representatives of different taxa. Over 2000 complete or draft archaeal genomes are now available—either from cultured species or metagenome assemblies—and the rate of discovery continues to increase. Some examples are shown in *Table 3.1*. This has led to sequential changes in the main branch of the tree of life that includes the Archaea and Eukarya, as shown in *Figures 4.1* and *5.2*. The current view (*Figure 5.2d*) shows four major clades of Archaea—designated as the phy-lum Euryarchaeota, plus the TACK, Asgard, and DPANN groupings ("superphyla"). Each of

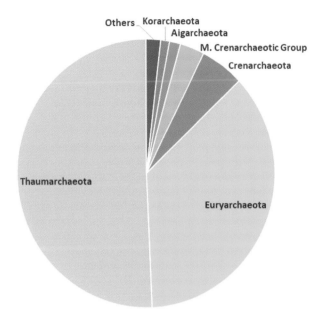

Figure 5.1 Frequency of sequences for archaeal phyla from samples of marine water, sediments, and hydrothermal vents. Data obtained from Schloss et al. (2016).

these contains several distinct phyla and the number of these—and their constituent classes and orders—is increasing rapidly with the recovery of more sequences from the marine environment and other habitats, such as deep sediments, hot and acidic springs, soil, and animal digestive tracts. These discoveries have important implications for theories on the origin of eukaryotes, discussed in *Box 5.1.*

It is suspected that many of these new species may exist in close associations with other bacteria or archaea and may form syntrophic partnerships. As our knowledge of archaeal diversity grows, it is likely that a wide range of novel metabolic functions will be uncovered, but the overwhelming importance of archaea in marine environments lies in their activities in the production and oxidation of methane and in autotrophy linked to nitrification, discussed below.

PHYLUM EURYARCHAEOTA

Many members of the Euryarchaeota produce methane

Production of methane (methanogenesis) is the final step in the anaerobic biodegradation of organic material. Methanogens show high physiological diversity in morphology, cell-wall constituents, and physiology, and are grouped into a number of orders and genera using these criteria (*Table 5.1*). Methanogens are mesophilic or thermophilic obligate anaerobes and can be isolated using strict anaerobic techniques from a wide range of habitats including anoxic sediments, decomposing material, and the gut of animals. They also occur as endosymbionts of anaerobic protozoa found especially in the hindgut of termites. It is likely that shipworms and other marine invertebrates that digest wood and cellulose also harbor ciliates with archaeal endosymbionts. Thermophilic methanogens are also important members of the microbial community at hydrothermal vents. The key features of methanogenesis as the final step in decomposition of organic material in anoxic environments were introduced in *Chapter 3*. Most genera utilize H_2 with CO_2 serving as both the oxidant for energy generation and for incorporation into cellular material using the reactions described on p.87.

Although they are strict anaerobes, methanogens can also be found in surface microbial mats and ocean waters, which can contain high levels of dissolved methane. Presumably, they exist in the anoxic interior of particles in which oxygen has been depleted by respiratory activity of other organisms. Methanogenesis is also significant in deeper waters with upwelling of nutrients, where intense heterotrophic oxidation of sinking organic matter leads to oxygen depletion.

The extremely thermophilic archaeon *Methanococcus jannaschii* has been recognized as one of the most important members of hydrothermal vent communities and has been studied

COULD GLOBAL WARMING TRIGGER RELEASE OF ANCIENT METHANE?

About 500–600 Mt of methane enters the atmosphere each year. Most comes from the "burps" of ruminant animals, natural wetlands, peat bogs, paddy fields, coal mining, and extraction of oil and gas. Emissions from ocean Archaea are only 1–2% of the total, because most methane is consumed by methanotrophs or trapped under pressure in sediments over millennia as solid methane hydrate (clathrate). There may be about ten times as much methane in clathrate as current natural gas reserves, making it an attractive energy source, but the logistics of extraction from deep marine sediments are formidable. Methane is 25 times more active than CO_2 in its greenhouse gas effects. There is strong geological evidence that the mass extinction events of 250 and 56 MYA were due to methane release and sudden global warming. There has been concern that rising sea temperatures in the Arctic could trigger catastrophic release of ancient methane from clathrate, but some recent research indicates that this might be less likely than feared (Ruppel and Kessler, 2017; Sparrow et al., 2018).

BOX 5.1   RESEARCH FOCUS

Evolution of an evolutionary pathway—changing views of Archaea in the tree of life

Woese and Fox (1977) recognized three distinct primary groupings of life on Earth. These ideas were consolidated in the highly influential paper by Woese et al. (1990)—it has >6500 citations to date—proposing the three domains: Bacteria, Archaea, and Eukarya (see *Chapter 1*). There was already considerable acceptance of the endosymbiosis theory, in which the eukaryotic cell type is believed to have been formed from a chimera of ancestral archaeal cells with ancestral bacterial cells, with strong genetic evidence that mitochondria are descended from alphaproteobacterial endosymbionts. However, the nature of the archaeal partner has proved much more difficult to establish and promoted much debate.

Eme et al. (2017) review the changing picture of diversity in the domain Archaea since it was first proposed, and how this has led to new ideas about the evolution of the Eukarya. *Figure 5.2a* shows the three-domain tree of life in which the Archaea and Eukarya each represent a monophyletic group sharing a unique common ancestor to the exclusion of Bacteria. An alternative view of a two-domain (Bacteria plus "Eocyta") tree of life (*Figure 5.2b*) was proposed early on (Lake et al., 1984; Rivera and Lake, 1992), based on evidence from comparison of the structural patterns of the ribosomes and the properties of genes encoding elongation factors—these are key cofactors during translation of mRNA to proteins. These authors favored a close relationship between the Eukarya and the Crenarchaeota. By 2010, the branch containing the Crenarchaeota was expanded to include three new phyla, following the discovery of new sequences from metagenomic data, forming the TACK superphylum. With this new phylogeny, Guy and Ettema (2011) proposed that the Eukarya branched within, or as a sister branch to, this clade (*Figure 5.2c*). Most recently, another branch of the Archaea, the phylum Lokiarchaota was discovered

from gene surveys of marine sediments near hydrothermal vents in the Arctic Ocean between Norway and Greenland. Based on the phylogeny of highly conserved protein genes, this soon emerged as the closest link to the Eukarya (Spang et al., 2015), supported by the fact that members of the Lokiarchaeota also contain genes encoding a group of eukaryotic signature proteins (ESPs). These ideas have not been universally accepted; for example, Nasir et al. (2016) and Da Cunha et al. (2017) used a different approach to interpret the phylogeny of Lokiarchaeota proteins and suggested that published genome data could be contaminated from distantly related organisms, concluding that they do not "bridge the gap" between Archaea and Eukarya. The phylum name was given after the sampling site, Loki's Castle black smoker vent, which in turn was named after the shape-shifting Norse god, Loki. In the study of Norse mythology, Loki has been the subject of many academic controversies, so Spang et al. (2015) chose a fitting name for an organism at the heart of ongoing debates about the origin of eukaryotes. Spang et al. (2018) gave several reasons why they considered the claims of Da Cunha et al. to be unfounded. The subsequent discovery and analysis of three related, but separate, phyla, the Odinarchaeota, Thorarchaeota, and Heimdallarchaeota—they are all named after Norse gods—provided further strong support for the origin of eukaryotes within the Asgard group or are a sister clade to them (*Figure 5.2d*). Genes for a wide range of ESPs involved in membrane trafficking and production of vesicles—defining features of eukaryotic cells, and previously thought to be unique to them—led Zaremba-Niedzwiedzka et al. (2017) to conclude that the ancestral host cells involved in initial endosymbiosis with the proteobacterial precursors of the mitochondrion already contained many of the key components that govern eukaryotic cellular complexity. Many unanswered questions remain about the sequence of evolutionary events before and

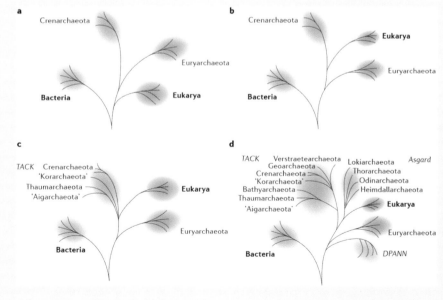

Figure 5.2 A schematic representation of changes in our understanding of the relationships between eukaryotes and archaea over the past 40 years. The DPANN group contains the Diapherotrites, Parvarchaeota, Aenigmarchaeota, Nanoarchaeota, and Nanohaloarchaeota phyla. It is shown as a monophyletic lineage, although this is a topic of debate. Reprinted from Eme et al. (2017) with permission from Springer Nature.

BOX 5.1   RESEARCH FOCUS

after acquisition of the bacterial symbionts that would become the mitochondria and several hypotheses are reviewed by Eme et al. (2017). They conclude that further study of Asgard archaea and deep-branching eukaryotic microbes might reveal even closer relatives and younger ancestors of eukaryotes, or perhaps reveal intermediate stages in the development of eukaryotic cell components. A recent major breakthrough has emerged from the successful cultivation of "*Ca.* Promethoarchaeum syntrophicum" by Imachi et al. (2019). This organism was recovered from deep marine sediments in 2006, long before the description of DNA sequences assigned to the Asgard archaea. After painstaking

work over 12 years, this extremely slow-growing organism—it has a lag phase lasting months and only doubles once every 20 days—was successfully cultivated in a bioreactor and recognized through genome sequencing as a member of the Lokiarchaeota. Physiological studies indicate it metabolizes organic acids through a syntrophic relationship with bacteria that consume the hydrogen generated. The cells have a remarkable structure, with long tendril-like protrusions. Based on these findings, Imachi et al. propose a new "Entangle–Engulf–Enslave" model to explain the archaeon-alphaproteobacterium symbiosis that led to evolution of the mitochondria in eukaryotic cells.

extensively. It is a primary consumer of H_2 and CO_2 produced by geochemical activity at the vents. Analysis of the complete genome sequence—the first from an archaeon, obtained in 1996—reveals a 1.7-Mb circular chromosome containing about 1700 genes. Sequence analysis of the genes encoding key metabolic pathways and cellular processes shows that these are similar to those found in Bacteria, whereas those encoding protein synthesis and DNA replication show more similarity to eukaryotic genes. However, a large number of genes in *M. jannaschii* have little homology to bacterial or eukaryotic genes, suggesting that many new cellular processes remain to be discovered. About 20 other genomes of methanogens have been sequenced; most are about the same size as *M. jannaschii*, but the genomes of Methanosarcina spp. are about four times as big, presumably reflecting more metabolic versatility. The function of many of the putative genes identified in these species is not known.

Methanopyrus is one of the most thermophilic organisms known, with rapid growth (generation time of 1 h) at a maximum temperature of 110°C. It is found in the walls of black smoker chimneys in hydrothermal vents. Although it is a methanogen, utilizing H_2 and CO_2 only, it is phylogenetically distant from the rest of this group and has very unusual membrane lipids and thermostabilizing compounds in the cytoplasm.

Anaerobic oxidation of methane (AOM) in sediments is carried out by syntrophic archaea

As described in *Chapter 3*, the enigmatic fate of methane produced in marine sediments was solved by the discovery of anaerobic methanotrophic Archaea (ANME) that carry out AOM—probably by reverse methanogenesis—in syntrophic consortia with sulfate-reducing bacteria (SRB), providing fixed nitrogen in return (*Figure 3.13*). Three clades of SRB-associated ANME archaea have been identified. The ANME-1 and ANME-2 are most widely distributed in anoxic sediments, while ANME-3 is mainly associated with mud volcanoes. They are thought to have very low growth rates, and none have been cultivated, but phylogenetic analysis shows they are related to the Euryarchaeota orders Methanomicrobiales (ANME-1) and Methanosarcinales (ANME-2 and ANME-3), and they show morphological similarities to the respective groups. In clade ANME-1, there is experimental evidence showing that syntrophic coupling occurs via direct electron transfer from the ANME archaea to the SRB, mediated by the production of extracellular cytochromes and/or conductive nanowires forming cell–cell connections. Genome analysis also shows that the SRB partners possess genes for multiheme cytochromes that are not present in free-living SRB, showing adaptations for syntrophy by both partners. Other similarities to methanogens include the presence of genes for the seven-step methanogenic pathway (in ANME-1 and ANME-2) and the same CO_2 fixation pathway as Methanomicrobiales (ANME-1). Furthermore, ANME-2 fixes N_2, as do members of Methanosarcinales. Close examination of the ANME now reveals that the associations with SRB are more diverse than originally envisaged. Different ANME archaea associate in a variety of ways with SRB—sometimes in loose associations, sometimes as structured consortia. ANME-1 archaea have been observed to exist as single cells,

Table 5.1 Properties of some representative methanogens found in marine sediments and hydrothermal vents

Order and genus	Morphology	Major substrates	Optimum temp. (°C)
Methanococcales			
Methanococcus	Irregular cocci	H_2, formate	35–40
Methanothermococcus	Cocci	H_2, formate	60–65
Methanocaldococcus	Cocci	H_2	80–85
Methanotorris	Cocci	H_2	88
Methanomicrobiales			
Methanoculleus	Irregular cocci	H_2, formate	20–55
Methanogenium	Irregular cocci	H_2, formate	15–57
Methanolacinia	Irregular rods	H_2	40
Methanospirillum	Spirilla	H_2, formate	30–37
Methanosarcinales			
Methanosarcina	Irregular cocci in groups	H_2, methylamines, acetate	35–60
Methanococcoides	Irregular cocci	methylamines	23–35
Methanopyrales			
Methanopyrus	Rods in chains	H_2	98

without an obvious SRB partner and ANME-2 and ANME-3 types occur with different kinds of bacteria, sometimes in mixed-species aggregates.

Since the initial discovery, other processes of AOM in marine methane-seep sediments have been described, in which Mn^{4+} or Fe^{3+} can serve as electron acceptors. The mechanisms of this reaction and the organisms responsible are unclear. AOM linked to reduction of NO_3^- or NO_3^-, has been observed in freshwater, but not marine, sediments.

The class Thermococci contains hyperthermophiles found at hydrothermal vents

This class contains three genera of hyperthermophiles: *Thermococcus*, *Pyrococcus*, and *Paleococcus*, each containing a small number of species. As shown in *Table 3.2*, these genera of Euryarchaeota have optimum growth temperatures in excess of 80°C. Their phylogenetic position as a deeply branching group resembles that of the hyperthermophilic genera Aquifex and Thermotoga in the Bacteria. It is interesting that extreme thermophiles in both domains branch very near the root of the tree, and this is taken as evidence that these organisms most closely resemble those that evolved first in the hotter conditions of the early Earth. Indeed, a common theory is that life may have evolved in submarine hydrothermal vents or subsurface rocks. The adaptation of hyperthermophiles to life at high temperatures is discussed on p.105.

Thermococcus celer forms highly motile spherical cells about 0.8 μm in diameter. It is an obligately anaerobic chemoorganotroph, using complex substrates such as proteins and carbohydrates, with sulfur as the electron acceptor with an optimum growth temperature of 80°C. *Pyrococcus furiosus* has similar properties to *Thermococcus*, but has an optimum growth temperature of 100°C and a maximum of 106°C. Both organisms have been investigated extensively because of their biotechnological potential, and their thermostable enzymes are used in the PCR (p.390). The genomes of several *Pyrococcus* species have been compared, revealing evidence of significant gene rearrangements in their recent evolution.

Archaeoglobus and Ferroglobus are hyperthermophilic sulfate-reducers and iron-oxidizers

The genus *Archaeoglobus* contains several species that are also extreme thermophiles (optimum temperature 83°C) found in sediments of shallow hydrothermal vents and around undersea volcanoes. They are strictly anaerobic organisms, which couple the reduction of sulfate to the oxidation of H_2 and certain organic compounds, resulting in the production of H_2S. Although *Archaeoglobus* forms a distinct phylogenetic group in the Euryarchaeota, it makes some of the key coenzymes used in methanogenesis, and analysis of its genome sequence shows that it shares some genes with the methanogens. However, it lacks the gene for one of the key enzymes, methyl-CoM reductase, so the origin of the small amount of methane produced is unknown. *Archaeoglobus* also occurs in oil reservoirs and has caused problems with sulfide "souring" of crude oil extracted from the North Sea and Arctic oilfields. Members of the related genus *Ferroglobus* also found in hydrothermal vents, do not reduce sulfate, but are iron-oxidizing/nitrate-reducing chemolithoautotrophs.

The Euryarchaeota contains extreme halophiles

Extreme halophiles (family Halobacteriaceae) grow in concentrations of NaCl greater than 9% and many can grow in saturated NaCl solutions (35%). They are found in salt lakes such as the Great Salt Lake (Utah, USA) and the Dead Sea (Jordan Rift Valley); as well as in hypersaline anoxic basins, such as those in the Mediterranean and Red Seas (see p.22). Extreme halophiles also occur in coastal regions in solar salterns, which are lagoons where seawater evaporates so that sea salt can be harvested. These operate as semi-continuous systems and maintain a fairly constant range of salinity throughout the year. Both conventional microbiological and 16S rRNA methods show that the overall diversity of microorganisms decreases as salinity increases. Up to about 11% NaCl, the range of bacteria is like that found in coastal seawater (most marine bacteria are moderately halophilic) while archaea are scarce. However, above 15% salinity, culturable members of the Archaea such as *Halorubrum*, *Halobacterium*, *Halococcus*, *Haloarcula*, and *Haloquadratum* become dominant, as well as archaea with novel gene sequences. The halophilic archaea occur as rods or cocci and in a variety of unusual shapes including flattened squares *Haloquadratum walsbyi*. They may contain very large plasmids constituting up to 30% of the genome. They are chemoorganotrophs that can use a wide range of compounds as sources of carbon and energy. The requirement for very high Na^+ concentrations is achieved by pumping large amounts of K^+ across the membrane, so that the internal osmotic pressure remains high and protects the cell from dehydration. The intracellular machinery of these organisms is adapted to high levels of salt, and so they are obligate extreme halophiles. Cells lyse in the absence of sufficient concentrations of Na^+, possibly because Na^+ stabilizes the high levels of negatively charged acidic amino acids in the cell wall.

Halobacterium salinarum and some other species use a membrane protein called bacteriorhodopsin to synthesize ATP using light energy. This compound is so called because it is structurally similar to the rhodopsin pigment in animal eyes. As noted in *Box 3.1*, rhodopsins consist of a protein covalently bound to a carotenoid pigment, retinal, which absorbs light and generates a proton motive force for the production of ATP. Bacteriorhodopsin absorbs light to provide energy for the proton pump required to pump Na^+ out and K^+ into the cell. Like proteorhodopsin, bacteriorhodopsin can provide enough energy for a low metabolic activity when organic nutrients are scarce. *H. salinarum* also contains other types of rhodopsin. Halorhodopsin captures light energy used for pumping Cl^- into the cell to counterbalance K^+ transport, while two other rhodopsin molecules act as light sensors that affect flagellar rotation, enabling chemotaxis towards light.

Uncultivated members of the Euryarchaeota are abundant in the plankton

In 16S rRNA gene surveys, four major groups of Archaea were identified in coastal, open ocean, and Arctic waters. The dominant Marine Group I (MG-I) comprises the phylum Thaumarchaeota (see below), while MG-II, MG-III, and MG-IV affiliate with

WHAT'S IN A NAME? RHODOPSINS

The extremely halophilic archaeon *Halobacterium* was given this name because at the time of its discovery it was considered to be a bacterium—because we were ignorant then of the true differences between the domains that we now know as the Archaea and Bacteria, they were known as the Archaebacteria and Eubacteria respectively. Therefore, when the rhodopsin-like pigment of *Halobacterium* was discovered, it was called bacteriorhodopsin, to distinguish it from the rhodopsin found in animal eyes. Another type of rhodopsin was discovered later, associated with sequences of Proteobacteria, so it was given the name proteorhodopsin (see *Box 3.1*). However, subsequent studies showed that this molecule is also present in many marine planktonic archaea as well as bacteria from other phyla! So, with hindsight, the molecules have been given inappropriate names. It's confusing, but unfortunately these names are now too well established in the literature to change.

the Euryarchaeota. Of the euryarchaeotal groups, all increase in abundance with depth (*Figure 5.3*). MG-II is most abundant in surface waters and may make up 4–20% of the total bacterial and archaeal community. Although no members of these groups have been cultivated, genome analysis of MG-II shows that some contain proteorhodopsin and other evidence of a motile, photoheterotrophic lifestyle, capable of degrading polymers in the surface ocean. A number of sub-clades have been identified, associated with different habitats. Some of the MG-II appear to be particle-associated and these show evidence of genomic adaptations typical of the copiotrophic lifestyle (p.93). In deeper waters, MG-II (and the rarer MG-III and MG-IV) archaea consume fresh organic matter and appear to bloom in response to pulses of organic matter.

PHYLUM CRENARCHAEOTA

Members of the Crenarchaeota are thermophiles occurring in hydrothermal vents

The phylum Crenarchaeota—part of the TACK superphylum—is phylogenetically very distinct from the Euryarchaeota described above, although many have similar physiological properties, including the ability to grow at extremely high temperatures. All cultured representatives are thermophilic, and most are hyperthermophilic, growing at >80°C (examples of minimum, optimum, and maximum temperature ranges are shown in *Table 5.2*). They were initially discovered from extensive studies of terrestrial hot springs, but several other species occur in submarine hydrothermal vents. They use a wide range of electron donors and acceptors in metabolism and can be either chemoorganotrophic or chemoheterotrophic. Most are obligate anaerobes. Members of the phylum all relatively closely related; there are just 31 cultivated, named genera belonging to six orders in the single class Thermoprotei.

Several species isolated from shallow and deep-sea hydrothermal areas belong to the order Desulfurococcales, of which the type genus, *Desulfurococcus*, is an obligate anaerobe with coccoid cells found in the vicinity of black smoker chimneys at hydrothermal vents. It oxidizes elemental sulfur using molecular hydrogen. *Pyrodictium* cells are disc-shaped and are connected into a mycelium-like layer attached to crystals of sulfur by very fine hollow tubules. Most *Pyrodictium* spp. are chemolithoautotrophs which gain energy by sulfide reduction, but *P. abyssi* grows chemoorganotrophically by fermenting peptides to CO_2, H_2, and fatty acids. *Pyrolobus fumarii* is found in the walls of black smoker chimneys in hydrothermal vents and has an optimum growth temperature of 106°C. In this very extreme environment, its production of organic compounds is a significant source of primary productivity. Cells are lobed cocci and the cell wall is composed of protein. It is a facultatively aerobic obligate chemolithotroph, using H_2 to reduce NO_3^- (to NH_4^+) or $S_2O_3^{2-}$ (to H_2S).

Staphylothermus marinus forms aggregates of cocci and is a chemoorganotroph, which, like *Pyrodictium abyssi*, ferments peptides to CO_2, H_2, and fatty acids. It is widely distributed in shallow and deep-sea hydrothermal systems and is a major decomposer of organic material. It is the largest known member of the Archaea—although normally about 1 μm in diameter, it can form very large cells up to 15 μm in high-nutrient concentrations. Unusually, cells also aggregate in clusters of 50–100 cells.

Igniococcus is an obligate hyperthermophile which has been detected in gene surveys of hydrothermal systems with global distribution. Three recognized species have been cultivated and characterized: *Igniococcus hospitalis*, *I. islandicus*, and *I. pacificus*. Growth of *Igniococcus* is anaerobic and occurs through a novel CO_2 fixation pathway, using sulfur reduction with molecular hydrogen as an electron donor. They can be cultured anaerobically using specialized high-temperature fermenters at 90°C. *Igniococcus* is one of only a few archaea known to have a double membrane and to lack an S-layer. The outer membrane is separated from the cytoplasmic membrane by a wide inter-membrane or periplasmic space (*Figure 5.5B*). This surrounds the cell as a loose sac enclosing a very large periplasmic space containing vesicles and tubes budding from the cytoplasmic membrane. These migrate and fuse to the outer membrane, forming a dynamic endomembrane system that appears to have secretory functions. As discussed below, these features are important aspects in the parasitism of *I. hospitalis* by *Nanoarchaeum equitans*.

? IS THERE AN UPPER TEMPERATURE LIMIT FOR LIFE?

Kashefi and Lovley (2003) discovered a new archaeon, related to *Pyrolobus* and *Pyrodictium*, from a hydrothermal vent at the Juan de Fuca Ridge in the Pacific, which they nicknamed "strain 121" because of its ability to grow up to 121°C. This is the standard temperature that is used in an autoclave for sterilization at atmospheric pressure. Strain 121 *grows* at this temperature and remains viable at 130°C for over 2 h. In reviewing the discovery, Cowan (2003) cautions that exact measurement of growth and survival at these temperatures is difficult. Microscopic examination of cross-sections of the black smoker chimney material from which the archaeon was isolated reveals diverse populations of microbes, indicating that microbes with even higher growth temperatures might exist. Such extreme thermophily depends on protein and membrane stability (see p.105), but Cowan suggests that the stability of small molecular-weight cofactors probably determines the upper limit for life. It is very likely that microbes that grow and survive even higher temperatures remain to be discovered, but some essential biological molecules such as ATP cannot exist above about 150°C.

In the order Thermoproteales, *Pyrobaculum* shows various modes of nutrition. Some species link sulfur reduction to anaerobic respiration of organic compounds, while others are anaerobic chemolithoautotrophs using H_2 to reduce NO_3^- or Fe^{3+}.

PHYLUM THAUMARCHAEOTA

A single clade of ammonia-oxidizing archaea comprises 20% of the picoplankton

One of the most unexpected discoveries arising from the application of ocean sampling of 16S rRNA genes in the 1990s was the discovery of widespread archaeal gene sequences in the marine water column, including polar oceans and very deep, cold waters. Initially,

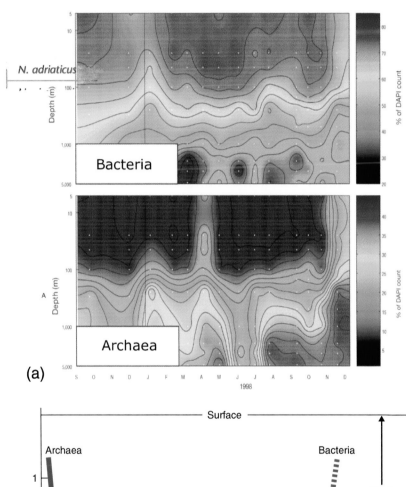

(a)

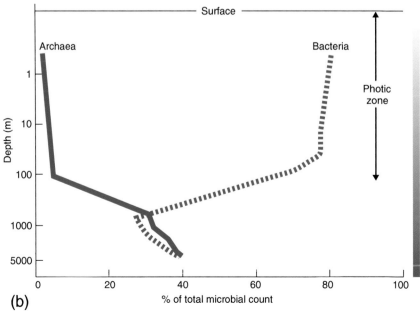

(b)

Figure 5.3 Distribution of Bacteria and Archaea in the ocean, showing mean annual depth profiles in the North Pacific subtropical gyre. Total cell abundance was measured using the DAPI nucleic acid stain and Bacteria and Archaea measured using whole-cell rRNA-targeted FISH with fluorescein-labeled polynucleotide probes. The counts of Archaea dominated the picoplankton below 1000 m and were almost entirely due to the clade now known as Thaumarchaeota; counts due to the Euryarchaeota were a few percent of the total throughout the water column. Redrawn from Karner et al. (2001) with permission from Springer Nature.

Table 5.2 Growth conditions of hyperthermophilic marine Archaea and Bacteria

Species	Growth temperature (°C)			Nutrition	Anaerobic/ Aerobic
	Min.	Opt.	Max.		
Archaea					
Archeoglobus fulgidus	60	83	95	CL	An
Ferroglobus placidus	65	85	95	CL	An
Igniococcus sp.	65	90	103	CL	An
Methanocaldococcus jannaschii	46	86	91	CL	An
Methanococcus igneus	45	88	91	CL	An
Methanopyrus kandleri	84	98	110	CL	An
Pyrobaculum aerophilum	75	100	104	CL	Ae/An
Pyrococcus furiosus	70	100	105	CO	An
Pyrodictium occultum	82	105	110	CL	An
Pyrolobus fumarii	90	106	113	CL	An
Staphylothermus marinus	65	92	98	CO	An
Thermococcus celer	75	87	93	CO	An
"Strain 121"	85	106	121	CL	An
Bacteria					
Aquifex pyrophilus	67	85	95	CL	Ae
Thermotoga maritima	55	80	90	CO	An

Abbreviations: Min., minimum; Opt., optimum; Max., maximum; CL, chemolithotrophic; CO, chemoorganotrophic; An, anaerobic; Ae, aerobic. Data from Huber and Stetter, 1998; Kashefi and Lovley, 2003.

TINY CELLS…TINY GENOME…ANCIENT PARASITE?

The minute cell size and minimal genome of *N. equitans* suggest that it is close to the limits for cellular life. Waters et al. (2003) showed that it contains only 536 genes, about one-third of which have no known homologs. Genes for components of many metabolic pathways and energy generation are missing. Loss of genes during evolution is common in obligate endosymbionts and parasites (see *Chapter 10*), but Waters et al. concluded that *Nanoarchaeum* is a genomically stable parasite that diverged anciently from an archaeal lineage. As an ectoparasite, *Nanoarchaeum* is not genetically isolated from the environment and it can acquire genes by HGT, so Nicks and Rahn-Lee (2017) argue that a different model of genome reduction may apply. They propose that the high-temperature lifestyle promoted genome streamlining—the *Igniococcus* genome is also small, only 1.30 Mb—and reduction in cell size, as observed in other thermophiles. Concurrently, the direct membrane association of the two thermophilic archaea led to sharing of metabolites, resulting in loss of biosynthetic genes in *Nanoarchaeum*. The evolution of smaller parasite cells would be favored because they place lower metabolic demands on the host that they inhabit.

analysis of these psychrophilic archaea (MG-1) identified them as members of the phylum Crenarchaeota, and they were classified as such until 2008. At this time, crenarchaeotes were known only as extreme thermophiles, and the discovery that these as yet uncultivated microbes are ubiquitous and so abundant in the water column was a great surprise.

One of the early changes in the phylogeny and classification of the archaea resulting from phylogenomic methods was the realization that this group of organisms formed a monophyletic group distinct from the Crenarchaeota. This followed the analysis of the genome sequence of an uncultivated archaeon "*Ca*. Cenarchaeum symbiosum" found in association with the cold-water sponge *Axinella mexicana* discovered off the coast of California. It accounts for 65% of the microorganisms associated with the tissues of the sponge and genomic analysis indicates that it oxidizes ammonia—it may have a symbiotic function by removing nitrogenous wastes from the sponge. When trees were calculated using both small and large subunit rRNA sequences and concatenated ribosomal protein sequences, it became clear that this organism was phylogenetically separate from the hyperthermophilic Crenarchaeota. In 2008, "Cenarchaeum" and the uncultivated mesophilic/psychrophilic ammonia-oxidizing archaea (AOA) identified in marine waters were therefore moved to the new sister phylum Thaumarchaeota (the name means "wonder" archaea), to be joined later by the Algarchaeota and Korarchaota to form the TACK superphylum (*Figure 5.1*). Recently, a further revision has been implemented under the genome-based taxonomy (GTDB, see p.117) and in this system, the AOA are now classified again as members of the phylum Crenarchaeota and assigned to a new class Nitrosophaeria. (At the time of writing, this change has not been formally accepted under the Code of Nomenclature).

Table 5.3 Properties of genomes from cultivated ammonia-oxidizing Thaumarchaeota

Species[1]	Source (enrichment culture)	GC (%)	Size (Mb)	Proteins	Growth temp. (° C)[2]	Ref.[3]
Nitrosopumilus						
N. maritimus	Tropical aquarium tank	34.2	1.65	1795	15–35 (32)	(1)
N. cobalaminigenes	Surface coastal water (50 m)	33.0	1.56	1817	10–30 (25)	(1)
N. oxyclinae	Surface coastal water (17 m oxycline)	33.1	1.59	1794	4–30 (33)	(1)
N. ureiphilus	Near-shore coastal sediment	33.4	2.17	2479	15–40 (25)	(1)
"N. koreensis"	Arctic marine sediment	34.2	1.64	1890	n.d.	(2)
"N. sediminis"	Arctic marine sediment	33.6	1.69	1949	n.d.	(3)
	Surface coastal water	33.4	1.80	2037	15–34 (30)	(4)
N. piranensis	Surface coastal water	33.8	1.70	2161	15–34 (32)	(4)
"N. salaria"	Estuarine sediment	36.8	1.27	1665	n.d.	(5)
"Nitrosopelagicus"						
"N. brevis"	Surface ocean (oligotrophic)	33.2	1.23	1419	15–34 (22)	(6)
"Nitrosocaldus"						
"N. islandicus"		41.5	1.61	1641	50–70 (65)	(7)

[1] Genus and species names in quote marks indicate *Candidatus* status. [2] Growth temperature range, optimum in parentheses. [3] References: (1) Qin et al. (2017); (2) Park et al. (2012a); (3) Park et al. (2012b); (4) Bayer et al. (2019); (5) Mosier et al. (2012); (6) Santoro et al. (2015); (7) Daebeler et al. (2018). Additional data courtesy of Wei Qin, University of Washington.

As shown in *Figure 5.3*, archaeal cells (mostly members of the phylum now known as Thaumarchaeota) comprise a large fraction of the picoplankton, especially in deeper waters. The total number of bacteria and archaea decreases with depth, from 10^5–10^6 ml^{-1} near the surface to 10^3–10^5 cells mL^{-1} below 1000 m. Bacteria are most prevalent in the upper 150 m of the ocean, but below this depth the fraction of archaea equals or exceeds that of bacteria. The pattern is consistent throughout the year. Combining the figures for cell density with oceanographic data for the volume of water at different depths indicates that the world's oceans contain approximately 1.3×0^{28} archaea and 3.1×10^{28} bacteria. About 1.0×10^{28} cells—20% of all the picoplankton—is dominated by a single clade of the pelagic Thaumarchaeota, suggesting that a common adaptive strategy has allowed them to radiate through the entire water column of the oceans. The physiology of these uncultured psychrophilic archaea was completely unknown until a combination of metabolic experiments and metagenomic evidence, together with study of the physiology of cultured isolates (*Table 5.3*) showed that they are mostly chemolithotrophic autotrophs that obtain energy from ammonia oxidation, therefore being the major contributor to nitrification and carbon fixation in the deep, cold, and dark ocean. These discoveries are discussed in *Box 5.2*.

BOX 5.2   RESEARCH FOCUS

Deep-sea FISHing—discovery of the role of archaea in carbon and nitrogen cycling

What are deep-water archaea living on? The MG-1 group of archaea, originally identified as members of the phylum Crenarchaeota and subsequently reassigned to the Thaumarchaeota, is the most abundant microbial group in the mesopelagic water column (*Figure 5.3*). The first clues that they might be autotrophs were indicated by stable isotope analysis (p.58) of sedimentary deposits laid down during the mid-Cretaceous period about 112 MYA (Kuypers et al., 2001). "Fossilized" cells of archaea could be identified by the presence of specific biphytanyl membrane lipids characteristic of the Crenarchaeota. The organic matter in these archaeal cells contained a higher ratio of ^{12}C to ^{13}C, indicating that they had fixed CO_2 by enzymic processes. Wuchter et al. (2003) detected these membrane lipids and archaeal 16S rRNA sequences in sea water. Using *in situ* radioactive tracer experiments in dark conditions, they showed that archaeal cells incorporate large amounts of ^{13}C from bicarbonate into the distinctive membrane lipids, thus confirming that they are chemolithoautotrophic. Teira et al. (2004) investigated this process further by developing a highly sensitive CARD-FISH method in which oligonucleotide probes directed against specific archaeal groups, and Herndl et al. (2005) coupled this with microautoradiography (MAR-FISH, p.58) to measure metabolic activity by incubating seawater samples with radioactive $[^3H]$-leucine or $[^{14}C]$-bicarbonate. After collecting the cells on a membrane filter, the incorporation of the radioactive tracer into individual cells (tagged by the FISH probe) was visualized microscopically after overlaying with photographic emulsion. Herndl et al. concluded that a large fraction of the Crenarchaeota (now reclassified in the Thaumarchaeota) at depths from 200–3000 m in the North Atlantic Ocean are metabolically active and that carbon fixation by their autotrophic growth makes a major contribution to productivity.

What is the energy source that fuels autotrophy? In the first metagenomic study conducted in the Sargasso Sea by Venter et al. (2004), a gene encoding ammonia mono-oxygenase (AMO), the key enzyme in ammonia oxidation, was associated with archaeal genome fragments. The enzyme AMO is composed of

three subunits, encoded by *amoA*, *amoB*, and *amoC*, and archaeal homologs of all three genes have subsequently been discovered in numerous metagenomic datasets, implicating the ocean archaea in nitrification by ammonia oxidation. But further proof was difficult, because no mesophilic/psychrophilic oxidizing archaea had been cultivated. The breakthrough came when Könneke et al. (2005) succeeded in cultivating an ammonia-oxidizing archaeon (AOA) from enrichment cultures prepared from seawater supplemented with ammonium chloride and gravel from a tropical aquarium tank. Laboratory experiments showed that cultures of the organism, which they named "*Ca.* Nitrosopumilus maritimus" strain SCM1, were exceptionally efficient at utilizing low levels of ammonia. Further evidence for the role of AOA-based autotrophy in marine systems was obtained in an enrichment experiment conducted by Wuchter et al. (2006), in which coastal sea water from the North Sea was kept in a mesocosm in darkness for six months without addition of any nutrients. During this time, the accumulation of characteristic archaeal membrane lipids was accompanied by the consumption of ammonia and production of nitrite. CARD-FISH showed that the water became enriched with archaea with 16S rRNA sequences that were very similar to the cultured "*Ca.* N. maritimus." In addition, when the *amoA* gene was isolated from the archaeal cells, one dominant sequence closely matched those of "*Ca.* N. maritimus" and the archaeal *amoA* gene from the Sargasso Sea metagenome. Following analysis of seawater samples throughout the year and use of Q-PCR (p.47) to detect levels of expression of archaeal *amoA* gene—this was 1–3 orders of magnitude higher than bacterial *amoA*—the authors concluded that nitrification by AOA in the upper 1000 m of the Atlantic system is more important than that carried out by the ammonia-oxidizing bacteria (AOB). Besides higher abundance of AOA than AOB, and higher expression of *amoA* of AOA, Martens-Habbena et al. (2009) showed that "*Ca.* N. maritimus" is also much more efficient at nitrification than AOB. They measured the removal of ammonia by cultures of this organism, which grew with a doubling time of 26 h, converting ammonia to nitrate until the ammonia levels were depleted below the detection limit of 10 nM. This is about 100

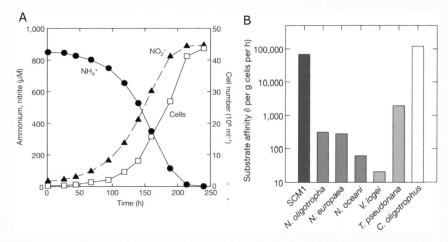

Figure 5.4 Efficiency of NH_4^+ oxidation by *Nitrosopumilus maritimus* SCM1. A. Growth in NH_4^+-limited seawater batch culture. B. Specific affinity of SCM1 (red), AOB (green), and the highest values for diatoms and heterotrophic bacteria (grey). For comparison, the open bar shows the highest recorded affinity of an organotrophic organism for its carbon substrate. Reprinted from Martens-Habbena et al. (2009) with permission from Springer Nature.

times lower than the concentration needed by ammonia-oxidizing bacteria. By measuring the kinetics of substrate binding, Martens-Habbena et al. showed that SCM1 sustains high specific oxidation rates at the exceptionally low ammonium concentrations found in the open oceans. SCM1 has a remarkably high specific affinity for NH_3/NH_4^+, suggesting that organisms like this would successfully compete with heterotrophic bacterioplankton and phytoplankton. This has been borne out by *in situ* open ocean estimates of community substrate affinity (Peng et al., 2016). Given the abundance of thaumarchaea in deep ocean waters, it is essential to know whether they are all capable of autotrophic nitrification, as this will affect our knowledge of carbon cycling as well as nitrogen budgets. The experiments of Herndl et al. (2005) showed that at least some of these archaea take up amino acids, suggesting that heterotrophic or mixotrophic nutrition is occurring, and this is supported by later genomic analysis (Qin et al., 2014). Various studies have shown that the AOA distribution varies with depth; they are in low and variable abundance at the surface (except in polar seas), reach maximal levels in the mesopelagic, and decrease below that (review, Santoro et al., 2018). The presence and abundance of different ecotypes of AOA is strongly affected by the flux of ammonia and dissolved organic compounds (*Figure 5.4*).

Isolation and cultivation of new species. Despite the advances in knowledge of metabolic capabilities obtained from metagenomic studies, culture of additional strains is needed to fully understand the physiological differences of diverse members of the AOA. Several strains isolated from soil and wastewater treatment plants have also been described, permitting comparison of genomes and metabolic characteristics. Within the genus *Nitrosopumilus*, there are considerable differences in the genes shared by different species and there are distinct ecotypes showing differences in growth rate, optimum temperatures, tolerance of oxygen, and ammonium concentrations, or other lifestyle characteristics. Several (including strain SCM1) have now been fully characterized with phenotypic, genotypic, and chemotaxonomic criteria and given formal species names (*Table 5.2*). Some appear to be fully autotrophic, and addition of a range of organic compounds has no effect on growth rates *in vitro*, while others show obligate mixotrophy, requiring organic compounds—including urea, a form of reduced nitrogen whose metabolism had previously been predicted from metagenomic datasets. Another candidate species "*Ca*. Nitrosopelagicus brevis" isolated from the open ocean has an extremely small genome and small cell size, suggesting that it has undergone similar streamlining to that seen in oligotrophic bacterioplankton like SAR11 (Santoro et al., 2015: see p.69). Despite this, it has the genetic machinery to make a range of vitamins and all amino acids. Like *N. pumilus*, all of these isolates show high rates of ammonia oxidation and CO_2 fixation in laboratory experiments, using a highly efficient variant of the hydroxypropionate/hydroxybutyrate cycle identified by Könneke et al. (2014). Like many other chemolithoautotrophs, the AOA grow slowly despite their comparatively efficient CO_2 fixation; most cultivated strains have laboratory doubling times in the range 19–34 h, although "*Ca*. N. brevis" doubles only every 4–4.6 days (Santoro and Casciotti, 2011).

What is the global biogeochemical impact of AOA? The autotrophic/mixotrophic metabolism of thaumarchaea clearly has great impact on the carbon and nitrogen cycles in the mesopelagic and deep ocean, particularly in areas with high export of organic matter, which provides the substrates for growth. These processes are especially important in supplying the nitrite or nitrate used by denitrifying or anammox bacteria in the oxygen-minimum zones (see *Chapter 9*). Denitrification leads to production of the greenhouse gas N_2O, and it is possible that this can also be formed as a by-product of ammonia oxidation. Elucidating full understanding of planktonic and particle-associated archaeal metabolism and *in situ* measurements of rates of growth, carbon fixation, and N_2O production is an exciting challenge for marine microbiologists and biogeochemists. This is especially important in connection with the microbial carbon pump and the capacity of the deep oceans for long-term storage of carbon, and the effects of ocean hypoxia and acidification due to rising CO_2 levels.

PHYLUM NANOARCHAEOTA

Nanoarchaeum is an obligate parasite of another archaeon

Nanoarchaeum equitans was discovered as a member of hyperthermophilic microbial communities near a submarine hot vent off Iceland. In cultures of *Igniococcus*, groups of tiny cocci about 400 nm in diameter were seen attached to the surface of its larger cells (*Figure 5.5A*). These were isolated, and 16S rRNA sequencing revealed unusual sequences that could not be matched to known members of the Archaea. *Nanoarchaeum* was therefore assigned to the novel phylum Nanoarchaeota. Its taxonomic position was uncertain for several years, and the status of this new phylum was questioned, but new phylogenomic evidence confirms this as a valid phylum within the DPANN supergroup. *Nanoarchaeum* appears to be an obligate parasite on the surface of *Igniococcus*, because all attempts to grow *N. equitans* in pure culture have been unsuccessful, even using extracts of *I. hospitalis*. However, the growth rate of *Igniococcus* does not seem to be strongly affected by the presence of *Nanoarchaeum* cells. Genome analysis shows that *Nanoarchaeum* has no genes to support the same chemolithoautotrophic physiology as its host. *Nanoarchaeum*

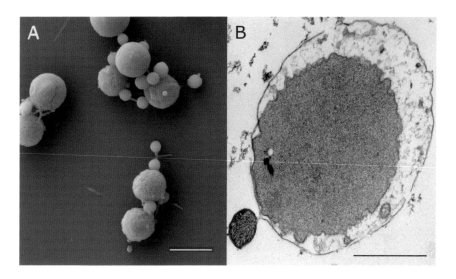

Figure 5.5 A. SEM showing larger cells of *Igniococcus* with attached cells of *Nanoarchaeum equitans*. Scale bar = 300 nm. B. TEM of ultrathin section showing close attachment of *Nanoarchaeum* to the outer membrane of *Igniococcus*. Credits: A. Mircea Podar, Oak Ridge National Laboratory. B. Reprinted from Cerdeño-Tárraga (2009) with permission from Springer Nature.

obtains energy and metabolites, including membrane lipids, amino acids, nucleotides, and cofactors from *Igniococcus*. As shown in *Figure 5.5B*, *Nanoarchaeum* attaches to this outer membrane via fibrillar structures, and at these contact points the periplasmic space of *Igniococcus* is greatly reduced, so that the inner and outer membranes are in contact with each other. Transcriptional studies of cultured *Igniococcus* show that the expression of membrane protein genes increases markedly following attachment of *Nanoarchaeum*, suggesting that the parasite exploits these structures. Advanced imaging techniques reveal direct contact between the cytoplasm parasite cells with the endomembrane system of the host, allowing *Nanoarchaeum* to acquire nutrients.

Conclusions

Members of the domain Archaea exhibit a wide variety of morphology, physiology, and metabolic types. In sediments, they play a major role in the carbon cycle in the production and oxidation of methane. At hydrothermal vents, chemolithotrophs and organotrophs are active at very high temperatures owing to the stability of their membranes, proteins, and intracellular constituents. Although exploration of marine archaea has lagged behind that of bacteria, there have been dramatic recent advances, revealing unexpected diversity, with many new phyla described. The discovery of the Thaumarchaeota and methods used to reveal their physiology, activities, and importance in nitrification and carbon fixation provides a important reminder of the value of efforts to cultivate new organisms, combined with a range of methodologies based on genomes, isotopes, and advanced imaging. Such integrated approaches will yield further advances in understanding the properties and functions of newly discovered archaea, revealing fresh insights into both the early evolution of life on our planet, and the impacts of human activities some three billion years later.

References and further reading

Archaeal phylogeny and evolution

Adam, P.S., Borrel, G., Brochier-Armanet, C., and Gribaldo, S. (2017) The growing tree of Archaea: New perspectives on their diversity, evolution and ecology. *ISME J.* **11**: 2407–2425.

Da Cunha, V., Gaia, M., Gadelle, D. et al. (2017) Lokiarchaea are close relatives of Euryarchaeota, not bridging the gap between prokaryotes and eukaryotes. *PLoS Genet.* **13**: e1006810.

Eme, L., Spang, A., Lombard, J. et al. (2017) Archaea and the origin of eukaryotes. *Nat. Rev. Microbiol.* **15**: 711–723.

Guy, L. and Ettema, T.J.G. (2011) The archaeal 'TACK' superphylum and the origin of eukaryotes. *Trends Microbiol.* **19**: 580–587.

Imachi, H., Nobu, M.K., Nakahara, N., et al. (2019) Isolation of an archaeon at the prokaryote-eukaryote interface. *bioRxiv*, 726976.

Kuypers, M.M.M., Blokker, P., Erbacher, J. et al. (2001) Massive expansion of marine Archaea during a Mid-Cretaceous oceanic anoxic event. *Science* **293**: 92–95.

Lake, J.A., Henderson, E., Oakes, M., and Clark, M.W. (1984) Eocytes: A new ribosome structure indicates a kingdom with a close relationship to eukaryotes. *Proc. Natl. Acad. Sci. USA* **81**: 3786–3790.

Nasir, A., Kim, K.M., Da Cunha, V., and Caetano-Anollés, G. (2016) Arguments reinforcing the three-domain view of diversified cellular life. *Archaea* **2016**: 1851865.

Ren, M., Feng, X., Huang, Y. et al. (2019) Phylogenomics suggests oxygen availability as a driving force in Thaumarchaeota evolution. *ISME J.* **13**: 2150–2161.

Rivera, M.C. and Lake, J.A. (1992) Evidence that eukaryotes and eocyte prokaryotes are immediate relatives. *Science* **257**: 74–76.

Schloss, P.D., Girard, R.A., Martin,T. et al. (2016) Status of the archaeal and bacterial census: An update. *MBio* **7**: e00201-16.

Spang, A., Saw, J.H., Jørgensen, S.L. et al. (2015) Complex archaea that bridge the gap between prokaryotes and eukaryotes. *Nature* **521**: 173–179.

Woese, C.R. and Fox, G.E. (1977) Phylogenetic structure of the prokaryotic domain: The primary kingdoms. *Proc. Natl. Acad. Sci.* **74**: 5088–5090.

Woese, C.R., Kandler, O., and Wheelis, M.L. (1990) Towards a natural system of organisms: Proposal for the domains Archaea, Bacteria, and Eucarya. *Proc. Natl. Acad. Sci.* **87**: 4576–4579.

Zaremba-Niedzwiedzka, K., Caceres, E.F., Saw, J.H. et al. (2017) Asgard archaea illuminate the origin of eukaryotic cellular complexity. *Nature* **541**: 353–358.

Euryarchaeota, methanogenesis and methanotrophy

Cui, M., Ma, A., Qi, H. et al. (2015) Anaerobic oxidation of methane: An 'active' microbial process. *Microbiol. Open* **4**: 1–11.

Evans, P.N., Boyd, J.A., Leu, A.O. et al. (2019) An evolving view of methane metabolism in the Archaea. *Nat. Rev. Microbiol.* **17**: 219–232.

Haro-Moreno, J.M., Rodriguez-Valera, F., López-García, P. et al (2017) New insights into marine group III Euryarchaeota, from dark to light. *ISME J.* **11**: 1102–1117.

Knittel, K., Losekann, T., Boetius, A. et al. Diversity and distribution of methanotrophic Archaea at cold seeps. *Appl. Environ. Microbiol.* **71**: 467–479.

Martin-Cuadrado, A.-B., Garcia-Heredia, I., Moltó, A.G. et al. (2015) A new class of marine Euryarchaeota group II from the mediterranean deep chlorophyll maximum. *ISME J.* **9**: 1619–1634.

McGlynn, S.E., Chadwick, G.L., Kempes, C.P., and Orphan, V.J. (2015) Single cell activity reveals direct electron transfer in methanotrophic consortia. *Nature* **526**: 531–535.

Orsi, W.D., Smith, J.M., Wilcox, H.M. et al. (2015) Ecophysiology of uncultivated marine euryarchaea is linked to particulate organic matter. *ISME J.* **9**: 1747–1763.

Ruppel, C.D. and Kessler, J.D. (2017) The interaction of climate change and methane hydrates. *Rev. Geophys.* **55**: 126–168.

Skennerton, C.T., Chourey, K., Iyer, R. et al. (2017) Methane-fueled syntrophy through extracellular electron transfer: Uncovering the genomic traits conserved within diverse bacterial partners of anaerobic methanotrophic archaea. *MBio* **8**: e00530-17.

Sparrow, K.J., Kessler, J.D., Southon, J.R. et al. (2018) Limited contribution of ancient methane to surface waters of the U.S. Beaufort Sea shelf. *Sci. Adv.* **4**: eaao4842.

Wegener, G., Krukenberg, V., Riedel, D. et al. (2015) Intercellular wiring enables electron transfer between methanotrophic archaea and bacteria. *Nature* **526**: 587–590.

Igniococcus and *Nanoarchaeum*

Cerdeño-Tárraga AM. (2009) Genome watch: What a scorcher! *Nat. Rev. Microbiol.* **7**: 408–409.

Heimerl, T., Flechsler, J., Pickl, C. et al. (2017) A complex endomembrane system in the archaeon *Ignicoccus hospitalis* tapped by *Nanoarchaeum equitans*. *Front. Microbiol.* **8**: 1072.

Nicks, T. and Rahn-Lee, L. (2017) Inside out: Archaeal ectosymbionts suggest a second model of reduced-genome evolution. *Front. Microbiol.* **8**: 384.

Waters, E., Hohn, M.J., Ahel, I. et al. (2003) The genome of *Nanoarchaeum equitans*: Insights into early archaeal evolution and derived parasitism. *Proc. Natl. Acad. Sci.* **100**: 12984–12988.

Thaumarchaeota

Bayer, B., Vojvoda, J., Reinthaler, T. et al. (2019) *Nitrosopumilus adriaticus* sp. nov. and *Nitrosopumilus piranensis* sp. nov., two ammonia-oxidizing archaea from the Adriatic Sea and members of the class *Nitrososphaeria*. *Int. J. Syst. Evol. Microbiol.* **69**: 1892–1902.

Brochier-Armanet, C., Boussau, B., Gribaldo, S., and Forterre, P. (2008) Mesophilic crenarchaeota: Proposal for a third archaeal phylum, the Thaumarchaeota. *Nat. Rev. Microbiol.* **6**: 245–252.

Brochier-Armanet, C., Gribaldo, S., and Forterre, P. (2011) Spotlight on the Thaumarchaeota. *ISME J.* **6**: 227–230.

Daebeler, A., Herbold, C.W., Vierheilig, J. et al. (2018) Cultivation and genomic analysis of "*Candidatus* Nitrosocaldus islandicus," an obligately thermophilic, ammonia-oxidizing thaumarchaeon from a hot spring biofilm in Graendalur Valley, Iceland. *Front. Microbiol.* **9**: 193.

Herndl, G.J., Reinthaler, T., Teira, E. et al. (2005) Contribution of Archaea to total prokaryotic production in the deep Atlantic Ocean. *Appl. Environ. Microbiol.* **71**: 2303–2309.

Karner, M.B., DeLong, E.F., and Karl, D.M. (2001) Archaeal dominance in the mesopelagic zone of the Pacific Ocean. *Nature* **409**: 507–510.

Könneke, M., Bernhard, A.E., de la Torre, J.R. et al. (2005) Isolation of an autotrophic ammonia-oxidizing marine archaeon. *Nature* **437**: 543–546.

Könneke, M., Schubert, D.M., Brown, P.C. et al. (2014) Ammonia-oxidizing archaea use the most energy-efficient aerobic pathway for CO_2 fixation. *Proc. Natl. Acad. Sci. USA* **111**: 8239–8244.

Martens-Habbena, W., Berube, P.M., Urakawa, H. et al. (2009) Ammonia oxidation kinetics determine niche separation of nitrifying Archaea and Bacteria. *Nature* **461**: 976–979.

Mosier, A.C., Allen, E.E., Kim, M. et al. (2012) Genome sequence of "*Candidatus* Nitrosopumilus salaria" BD31, an ammonia-oxidizing Archaeon from the San Francisco Bay estuary. *J. Bacteriol.* **194**: 2121–2122.

Park, S.-J., Kim, J.-G., Jung, M.-Y. et al. (2012a) Draft genome sequence of an ammonia-oxidizing archaeon, "*Candidatus* Nitrosopumilus sediminis" AR2, from Svalbard in the Arctic Circle. *J. Bacteriol.* **194**: 6948–6949.

Park, S.-J., Kim, J.-G., Jung, M.-Y. et al. (2012b) Draft genome sequence of an ammonia-oxidizing archaeon, "*Candidatus* Nitrosopumilus koreensis" AR1, from marine sediment. *J. Bacteriol.* **194**: 6940–6941.

Peng, X., Fuchsman, C.A., Jayakumar, A. et al. (2016) Revisiting nitrification in the Eastern Tropical South Pacific: A focus on controls. *J. Geophys. Res. Ocean.* **121**: 1667–1684.

Qin, W., Amin, S.A., Martens-Habbena, W. et al. (2014) Marine ammonia-oxidizing archaeal isolates display obligate mixotrophy and wide ecotypic variation. *Proc. Natl. Acad. Sci. USA* **111**: 12504–12509.

Qin, W., Heal, K.R., Ramdasi, R. et al. (2017) *Nitrosopumilus maritimus* gen. nov., sp. nov., *Nitrosopumilus cobalaminigenes* sp. nov., *Nitrosopumilus oxyclinae* sp. nov., and *Nitrosopumilus ureiphilus* sp. nov., four marine ammonia-oxidizing archaea of the phylum Thaumarchaeota. *Int. J. Syst. Evol. Microbiol.* **67**: 5067–5079.

Santoro, A.E. and Casciotti, K.L. (2011) Enrichment and characterization of ammonia-oxidizing archaea from the open ocean: Phylogeny, physiology and stable isotope fractionation. *ISME J.* **5**: 1796–1808.

Santoro, A.E., Dupont, C.L., Richter, R.A. et al. (2015) Genomic and proteomic characterization of "Candidatus Nitrosopelagicus brevis": An ammonia-oxidizing archaeon from the open ocean. *Proc. Natl. Acad. Sci. USA* **112**: 1173–1178.

Teira, E., Reinthaler, T., Pernthaler, A. et al. (2004) Combining catalyzed reporter deposition-fluorescence *in situ* hybridization and microautoradiography to detect substrate utilization by bacteria and archaea in the deep ocean. *Appl. Environ. Microbiol.* **70**: 4411–4414.

Venter, J.C., Remington, K., Hoffman, J. et al. (2004) Environmental genome shotgun sequencing of the Sargasso Sea. *Science* **304**: 66–74.

Wuchter, C., Abbas, B., Coolen, M.J.L. et al. (2006) Archaeal nitrification in the ocean. *Proc. Natl. Acad. Sci.* **103**: 12317–12322.

Wuchter, C., Schouten, S., Boschker, H.T.S., and Sinninghe Damste, J.S. (2003) Bicarbonate uptake by marine Crenarchaeota. *FEMS Microbiol. Lett.* **219**: 203–207.

Other articles

DeLong, E.F. (1992) Archaea in coastal marine environments. *Proc. Natl. Acad. Sci. USA* **89**: 5685–5689.

Fuhrman, J.A., McCallum, K., and Davis, A.A. (1992) novel major archaebacterial group from marine plankton. *Nature* **356**: 148–149.

Huber, H. and Stetter, K.O. (1998) Hyperthermophiles and their possible potential in biotechnology. *J. Biotechnol.* **64**: 39–52.

Kashefi, K. and Lovley, D.R. (2003) Extending the upper temperature limit for life. *Science* **301**: 934.

Santoro, A.E., Richter, R.A., and Dupont, C.L. (2019) Planktonic Marine Archaea. *Ann. Rev. Mar. Sci.* **11**: 131–158.

Chapter 6

Marine Eukaryotic Microbes

This chapter reviews the biology of protists and fungi, highlighting selected examples that are particularly important in marine ecology. The protists are highly diverse and play a major role in ocean food webs providing trophic links to zooplankton, both as primary producers in the photic zone and as consumers in all parts of the water column and the benthos. Due to their genomic complexity and associated technical difficulties, progress in application of molecular techniques for study of eukaryotic microbes has lagged behind that of bacteria and archaea, but new methods are revealing exciting new discoveries of hitherto unknown groups and their importance. The study of fungi (mycology) in the marine environment is considerably less developed than that of other microbes, but there is growing evidence of their importance in the carbon cycle as benthic and planktonic consumers of complex organic materials.

Key Concepts

- Microbial protists are highly diverse and range in size from <1 μm to >100 μm, with some types forming macroscopic colonies.

- Classification of protists is complex and has been subject to frequent change, but modern schemes combine morphological, functional, and molecular information in which multiple protist groups occur throughout the phylogenetic radiation of the eukaryotes.

- Many planktonic protists contain plastids and carry out photosynthesis, which contributes a large fraction of global carbon fixation; this primary production enters pelagic food webs at several levels because of the wide range of sizes of protists.

- Many protists are phagotrophic, feeding by predation (grazing) on bacteria, archaea, viruses, other protists, and even metazoa; they dominate the initial levels of food webs.

- Classical models of carbon and energy flow based on phototrophy or heterotrophy are confounded because mixotrophy is extremely common among protists.

- Seasonal blooms of some protists can have major impacts on ocean–atmosphere interactions and affect global climate processes; in other cases, blooms may have deleterious consequences for marine life and human health.

- Some protists (labyrinthulids and thraustochytrids) and marine fungi are saprotrophs, important as degraders of organic matter in coastal environments; their abundance and quantitative importance in carbon cycling has been underestimated.

- The protists and fungi include important parasites and parasites and pathogens of marine life.

MARINE PROTISTS

Protists are a highly diverse collection of unicellular eukaryotic microbes

Today, the term "protist" is widely accepted as a term of convenience to describe a wide assortment of unicellular eukaryotic microbes. There is no clear definition—in a way, it is easier to say that they are eukaryotic organisms that are *not* animals, plants, or fungi. The Fungi are not normally included in the protists because they form a separate monophyletic kingdom, although the distinctions between this group and some other forms are very blurred. In some cases, protists may form colonies or filamentous aggregates, but these do not show tissue differentiation, so are not regarded as multicellular organisms. The designation "microbial eukaryotes" may also be used as a collective term; this would include members of the Fungi. As members of the domain Eukarya, protists possess a defined nucleus bounded by a double nuclear membrane, which is continuous with a membranous system of channels and vesicles called the endoplasmic reticulum. This is the site of fatty acid synthesis and metabolism and is also lined with ribosomes responsible for protein synthesis. The Golgi apparatus (a series of flattened membrane vesicles) processes proteins for extracellular transport and is also responsible for the formation of lysosomes, membrane-bound vesicles containing digestive enzymes that fuse with vacuoles in the cell, either for the digestion of food in phagocytic vacuoles or for the recycling of damaged cell material.

Most protists demonstrate some form of sexual recombination as a regular component of their life cycle in which the fusion of the genomes of two cells and their nuclei is followed by meiosis. Unlike multicellular organisms, which are usually diploid (i.e. they contain two copies of each chromosome and reproduce by forming haploid gametes), protists vary greatly in whether the diploid or haploid state is the predominant phase in the life cycle.

The cytoskeleton is a system of hollow microtubules about 24 nm in diameter (composed of the protein tubulin) and microfilaments about 7 nm in diameter (composed of two actin monofilaments wound around each other). Protists often display a variety of other filaments. Microtubules often extend under the surface and provide the basic shape of the cell, or they may be grouped into bundles to support extensions of the cell. Microfilaments provide a strengthening framework for the cell. The cytoskeleton provides a mechanism for movement of intracellular structures within the cell, for example, in the separation of chromosomes during nuclear division. Another function of the cytoskeleton, which is of particular importance in the marine amoebozoa, radiolarians, and foraminifera, is amoeboid movement due to changes in the cross-linking state of the protein actin. Rearrangement of the membrane also enables the process of phagocytosis, which brings food particles or prey organisms into the cell, enclosed within a food vacuole, which subsequently fuses with lysosomes.

A defining feature of almost all eukaryotes is the presence of mitochondria—organelles with an outer membrane and an extensively folded inner membrane, which evolved from the primary endosymbiosis of an alphaproteobacterial ancestor with an ancestral eukaryotic precursor cell (see *Box 5.1*). The nature of the infoldings (cristae) is distinctive of different eukaryotic groups, and marine protists have either tubular cristae (as seen in alveolates and stramenopiles) or flattened cristae (as seen in euglenids and kinetoplastids). A few types of protists do not contain mitochondria but have alternative structures (mitosomes or hydrogenosomes). Most of these are animal parasites, but free-living protists lacking mitochondria may be important in marine anaerobic sediments, microbial mats, and particles.

Photosynthetic protists contain chloroplasts and are commonly referred to as algae. The term microalgae is often used to indicate unicellular algae of microscopic dimensions to distinguish them from larger multicellular types (macroalgae or seaweeds). Much evidence indicates that the chloroplasts evolved via primary endosymbiosis, in which an ancestral eukaryotic cell engulfed a cyanobacterium. Analysis of ribosomal RNA, the arrangement of the internal membranes, and the nature of photosynthetic pigments are used in classification of photosynthetic protists. Study of these features shows that chloroplasts have arisen on several occasions during evolution via secondary and tertiary symbiosis events.

Protists show enormous diversity and classification systems are regularly revised

As noted in *Table 1.1*, protistan microbes are found in the pico-, nano-, and microplankton size classes, ranging in size from ~1 to 200 μm. Early studies of marine protists using light microscopy revealed a wondrous diversity of forms, with many of the larger organisms in the microplankton size class showing complex and beautiful forms, such as the specimens from the 1872–1876 *HMS Challenger* expedition classified and drawn by Ernst Haeckel, which had important influences on early twentieth-century art and architecture and continues to inspire artists and designers (*Figure 6.1*).

The advent of scanning electron microscopy (SEM) allowed further exploration of these distinctive features, producing striking images of surface architecture that can be used for species identification (several examples occur later in this chapter). This enormous morphological diversity of the protists contrasts with what we have seen in discussion of the bacteria and archaea, where there are a small number of basic cell shapes, but an enormous metabolic diversity. That said, morphological features are of little use in distinguishing many species of the smaller picoeukaryotes. Classification systems, based largely on morphological features, were developed over many years by different groups of scientists who used two separate International Codes of Nomenclature for naming protozoa (Zoological Code) and algae, fungi, and plants (Botanical Code) leading to much confusion. Modern classification systems no longer include these terms as formal categories, although they persist because they are so embedded in general public understanding and are a convenient way of grouping certain organisms that share general features (even if they are not particularly closely related).

There have been many attempts to develop phylogenetic trees that depict the relationships between the many different eukaryotic lineages. Since the 1990s, the application of molecular methods has led to a fundamental shift in understanding of relationships between organisms. Sequencing of 18S rRNA provided a basis for developing new phylogenetic trees, but, as noted in *Chapter 1*, reliance on a single gene can produce many anomalous results, and evidence of horizontal gene transfer and secondary symbiosis events confounds the interpretation of relationships. By combining the sequences of various genes and amino acid sequences of key proteins, together with morphological and biochemical evidence, consensus phylogenetic trees grouping the eukaryotes into a number of supergroups were developed. The advent of high-throughput sequencing (HTS) has led to many new insights into phylogenetic relationships, especially in the higher taxa, resulting in further major revisions of classifications in 2005, 2012, and 2019. *Figure 6.2* is a graphical representation showing the phylogenetic relationships of the major groups of eukaryotes according to the 2019 formal

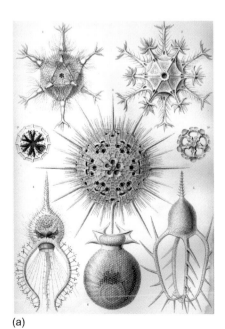

(a)

(b)

Figure 6.1 Artistic drawings of (a) radiolarians and (b) foraminifera published as prints by Ernst Haeckel between 1899 and 1904. From the collection *Kunstformen der Natur* (available online at caliban.mpiz-koeln.mpg.de).

Figure 6.2 Schematic representation of the phylogenetic tree of Eukarya. The branches (supergroups) are labeled with common names of major taxa; names in bold indicate protists and fungi discussed in this chapter. Adapted from Adl et al. (2019), CC-BY-4.0.

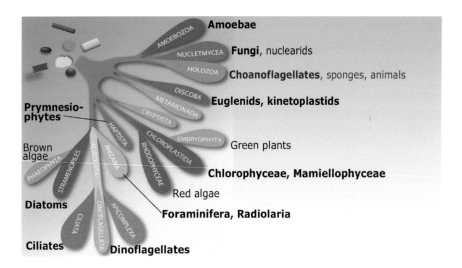

classification, coordinated by the International Society of Protistologists. Inevitably, there is still much disagreement among experts about the interpretation of phylogenetic relationships and classification schemes. In particular, designation of the taxonomic ranks phylum, class, and order is debatable, so the literature can be quite confusing. The tree indicates that evolutionary radiation occurred very early in the history of eukaryotes. Marine microbes appear in almost all the higher taxa envisaged in these classification systems. The huge diversity of protists makes a systematic coverage impossible in a book like this, so we will imagine a visit to the "Protist Aquarium" like the approach we used in the preceding chapters on Bacteria and Archaea, in this case grouping organisms under major functional or morphological themes, rather than attempting to present them in a formal classification scheme.

The -omics approaches have some limitations for understanding protist diversity

Like bacteria and archaea, most protists have not been cultivated and much of our knowledge about their diversity in the marine environment has come from direct sequencing of DNA. However, knowledge of the protists through the application of molecular methods has progressed much more slowly than for the Bacteria, for which the amount of sequence data is hundreds of times higher than it is for protists. There are several reasons for this. Most microbial eukaryotes tend to have smaller populations, and most have not been isolated and cultivated. Only a very small number of complete protist genomes are available (*Table 6.1*); in these, the genomes are many times bigger and more complex than bacteria and archaea (compare with *Table 3.1*). Nuclear genomes contain multiple chromosomes, repeat sequences, and large amounts of noncoding sequences which make interpretation difficult. Methods such as clone libraries and tag sequencing are often biased in favor of larger cells such as marine alveolates, including dinoflagellates and ciliates, which may have tens to hundreds of copies of 18S rRNA genes. Methods of sample preparation often result in contamination with DNA from fragmented cells of larger organisms. The niche specialization and ecological function of protists is much harder to explain by the presence of particular genes—unlike bacteria and archaea, where we can make reliable predictions based on the identification of genes conferring metabolic functions or adaptation to particular environmental conditions like temperature, salinity, or oxygen. Instead, eukaryotes have many complex structural adaptations, behaviors, and dynamic interactions with their environment and other organisms. Of course, we can use genomic information to predict that a protist is, say, photosynthetic, but we do not have any genetic indicators of what makes an organism a grazing predator or an infectious parasite, for example. Thus, the use of meta-omics approaches to predict ecological function and interactions *in situ* is inherently more difficult for protists, although advances are now being made (*Box 6.1*). To fully understand protist diversity, we will need a much larger body of information obtained by combining traditional biological observations with cultivation

Table 6.1 *Examples of marine protists for which whole genome sequences have been obtained*

Organism	Type[a]	Genome Size (Mb)[b]	Predicted genes[b]
Ostreococcus tauri	Prasinophyte (Mamiellophyceae) Photosynthetic	13.9	7932
Micromonas pusilla	Prasinophyte (Mamiellophyceae) Photosynthetic	21.9	10242
Bodo saltans	Euglenid (Kinetoplastea) Bacterivorous	39.9	18190
Monosiga brevicollis	Choanoflagellate (Salpingoecidae) Bacterivorous	41.7	9203
Schizochytrium sp.	Thraustochytrid (Labyrinthulomycetes) Osmotrophic	51.6	–
Fragilariopsis cylindrus	Diatom (Bacillariophyceae) Photosynthetic	76.2	18246
Emiliania huxleyi	Haptophyte coccolithophore (Isochrysidales) Photosynthetic	167.7	38544
Symbiodinium goreaui (type C1)	Dinoflagellate (Suessiales) Photosynthetic, coral symbiont.	1030	43403

[a] Common name of group, with class or order in parentheses. [b] Data from NCBI Genome Database.

and genome-sequencing of a wide range of species in order to establish a reference system that we can map our newly identified sequences against.

Picoeukaryotes play a major role in ocean food webs

Despite these reservations, environmental genomic approaches have led to some paradigm-shifting discoveries about the role of protists in marine processes. Most significant of these is the discovery in 2001 of a hitherto unknown extra level of diversity amongst picoeukaryotes (1 to 3 μm diameter). In these studies, DNA was extracted from filtered plankton cells and amplified via PCR using general primers for 18S rRNA genes, followed by construction of a clone library. Because eukaryotic plankton cells exist across such a broad range of sizes, from micrometers to millimeters, the method of specimen collection is very important, so the researchers used a combination of filters and prefilters to trap cells of different size fractions and verified their findings by flow cytometry. Novel organisms, discovered in different regions and at different depths, were affiliated to many diverse groups containing known, cultured organisms in higher size ranges, with a range of lifestyles including photoautotrophs (in the photic zone) and osmotrophic heterotrophs, grazing predators, and parasites throughout the water column. We now know that these are prevalent in surface waters at up to 10^4 cells mL^{-1}, decreasing to hundreds or tens of cells in the meso- and bathypelagic depths, respectively. This has led to a major shift in understanding the ecological role of protists in ocean food webs, both as primary producers through photosynthesis and as a trophic link through grazing on other members of the picoplankton. Recent breakthroughs in development of methods for analysis of protist communities by single-cell genomics are leading to new insights into their diversity and role, discussed in *Box 6.1*.

Heterotrophic flagellated protists play a major role in grazing of other microbes

Protists that possess one or more flagella for movement and feeding have traditionally been grouped under the general term "flagellates," although this characteristic has no phylogenetic significance and is found as a component at some stage of the life cycle in about half of the major protistan phyla. Eukaryotic flagella are composed of nine pairs of peripheral microtubules and two pairs of central microtubules. One pair of the tubules in each pair contains the protein dynein, which moves along an adjacent microtubule; rapid and repeated bending causes the flagellum to beat. The flagella are anchored in the cell via a basal body (kineto-some), which varies greatly in its ultrastructure and has been used in previous classification systems to distinguish different groups.

The introduction of epifluorescence microscopy and flow cytometry in the 1970s to 1980s led to realization of the abundance and importance of heterotrophic and mixotrophic flagellates in the nanoplankton (2–20 μm) size range as the major consumers of primary and secondary production in marine systems. These unicellular organisms swim through the water using flagella and ingest bacteria, archaea, and smaller protists via phagocytosis. Subsequently, as discussed above, flagellates were also shown to be abundant in the picoplankton size range. Grazing protists are also very abundant in benthic habitats such as mats and sediments. They are also important in sea ice and as members of the surface mucus and intestinal microbiota of invertebrates. Protistan grazing leads to the recycling of carbon, nitrogen, and phosphorus, since the smaller flagellates themselves are preyed upon by larger protists (e.g. ciliates and dinoflagellates), providing a trophic link to metazoan zooplankton in the food web (the microbial loop, discussed in *Chapter 8*). In addition, protists eliminate up to 30% of ingested cell material that has not been digested—this reenters the DOM/POM pool in the ocean. The flagellates are taxonomically very diverse, occurring in many distantly related different phyla. Their ecological and biogeochemical functions are probably also very diverse, and it is important to remember that grouping them as pico-, nano-, or macroplanktonic flagellates (often abbreviated to just picoflagellates, nanoflagellates, or microflagellates) is only a convenience based on the use of the logarithmic scale for the plankton size classes and the pico- and nano- prefixes do not have the same meaning as their usual scientific use (1 μm = 10^3 nm = 10^6 pm). Furthermore, the distinctions between these categories are blurred, and some authors consider the microflagellates as organisms larger than 15 μm, rather than 20 μm used in the Sieburth scale (*Table 1.2*). Examples of common grazing protists are shown in *Table 6.2* and *Figure 6.3*. Abundance and biomass vary greatly with seasonal and nutritional conditions. In temperate regions, abundance increases sharply due to higher numbers of prey associated with the spring bloom and decrease in early summer as the protists themselves become subject to intensive grazing by metazoan zooplankton.

Heterotrophic flagellated protists have different feeding mechanisms

Protists that feed on bacteria (bacterivores) by direct interception often show species-specific preference for selection and uptake of particular sizes or types of prey and vary in their mode of feeding. Different species may either be free-swimming or attached to particle such as algal debris, fecal pellets, and marine snow. As a rule, free-swimming forms tend to consume larger prey, while attached forms increase the probability of encounter with small prey by generation of water currents. Examples of genera of heterotrophic nanoflagellates (HNF) are shown in *Table 6.2*. Some of the most active HNF are the bicosoecids, which are members of the Stramenopiles. *Cafeteria roenbergensis* has been found in all ocean waters examined, occurring at densities up

Table 6.2 Examples of marine heterotrophic nano- and microflagellates

Taxonomic groups	Examples of common genera	Relative abundance	
		Pelagic	Benthic
Euglenids	*Petulamonas, Peranema*	+	++
Kinetoplastids	*Bodo, Caecitellus*	+	+++
Chrysomonads	*Spumella, Paraphysomonas*	+++	+
Bicosoecoids	*Cafeteria, Bicosoeca*	+++	+
Dinoflagellates	*Gymnodinium, Katodinium*	+	+
Choanoflagellates	*Monosiga, Diaphonoeca*	+++	+
Cercomonads	*Cercomonas, Bodomorpha*	+	+++

Data from Boenigk and Arndt (2002)

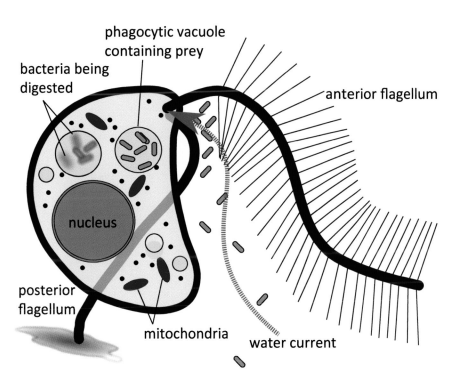

bacteria being digested

phagocytic vacuole containing prey

anterior flagellum

nucleus

posterior flagellum

mitochondria

water current

Figure 6.3 Schematic diagram of *Cafeteria roenbergensis*, illustrating ingestion of prey bacteria propelled towards the mouthparts at the base of the flagellum by a water current (red dotted line) generated by the anterior flagellum. The posterior flagellum trails during swimming or may be attached to a surface, as shown here. Adapted from image by Dennis Barthel (derivative zapyon) via Wikimedia Commons, CC-BY-3.0.

to 10^5 mL^{-1} in coastal waters where there are high concentrations of bacterial prey, and it exerts strong top-down control on bacterial populations. Its impact on populations of Archaea, which would be important in the deep sea, is less clear; however, experimental studies have shown that cultured *C. roenbergensis* can prey on both bacterial and thaumarchaeotal cells, although feeding and growth rates vary according to the prey type. The cells, typically 3–10 μm, have two flagella, one of which projects forwards to propel the cell in a spiral movement, while the other flagellum trails behind. It is a suspension feeder, filtering suspended bacteria and other particles from the water; the anterior flagellum beats about 40 times a second, creating a rapid water current that concentrates suspended particles and carries bacteria towards a "mouthparts" region at the base of the flagellum, where they are phagocytosed (*Figure 6.3*). This heterokont pattern of a "tinsel" anterior flagellum and a smooth posterior, "whiplash" flagellum is a distinguishing feature of the motile stage in most stramenopiles. In nonmotile species, the posterior flagellum attaches to the substrate. The cells have one nucleus and five mitochondria with tubular cristae, as found in all stramenopiles. Unlike most eukaryotes, *Cafeteria* has a highly compact mitochondrial genome, with a very small amount of noncoding DNA. A related species, *Pseudobodo tremulans*, is found mainly in sediments.

The group known as the euglenids provided one of the earliest challenges to traditional "zoo-" and "phyto-" based classification, since their nutrition may be phototrophic or heterotrophic, or combine features of both (mixotrophy). Thus, past taxonomic systems have separated them at phylum level on the basis that they are simple animals or plants respectively. They were classified as one of the main divisions of the supergroup known as the Excavata, although this is not now recognized as a monophyletic group. In modern systems, the phylum Euglenozoa and class Euglenida encompasses several orders of phagotrophic and mixotrophic types, as well as photosynthetic types (Euglenophycae). Also within the excavates, the kinetoplastids possess a unique structure termed the kinetoplast near the base of the flagellum, easily visible in the light microscope after applying DNA stains. This is a large, single mitochondrion containing a few interlocked large, circular chromosomes (maxicircles), which encode mitochondrial enzymes, plus thousands of smaller minicircles of DNA. The minicircles do not appear to encode any complete genes but are involved in post-transcriptional editing of mRNA. The bodonids (e.g. *Bodo* spp.) are kinetoplastids with oval cells about 4–10 μm long and are very common in coastal waters and sediments, often moving over surfaces using a trailing flagellum. A shorter flagellum propels water currents toward a cytopharynx lined with microtubules. The group of Euglenozoa protists known as the marine diplonemids has recently been recognized as being one of the most diverse and abundant lineages of heterotrophic protists in the oceans.

> ### WHAT'S IN A NAME? *CAFETERIA ROENBERGENSIS*
>
> Species names are often complex, and students struggle with pronouncing and retaining them, so it always helps if the name has some memorable feature. Humor is rarely used in the serious business of systematics (and scientific writing generally), but the marine ecologist Tom Fenchel and taxonomist David Patterson provide an interesting exception when they named a new species of planktonic protist (Fenchel and Patterson, 1988). They wrote: "We found a new species of ciliate during a marine field course in Rønberg [Denmark] and named it *Cafeteria roenbergensis* because of its voracious and indiscriminate appetite, after many dinner discussions in the local cafeteria."

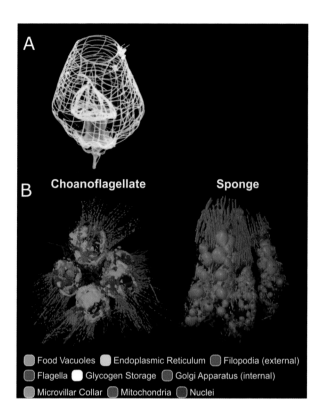

Figure 6.4 A. TEM of the choanoflagellate *Kakoeca antarctica*, showing the basket-like lorica. B. Schematic diagrams based on 3D reconstruction of serial ultrathin TEM, showing similar structure of choanoflagellate (left) and sponge (right) cells, but with differences in organization of membrane vesicles and other cellular components. Credits: A. Fiona Scott, Electron Microscopy Unit, Australian Antarctic Division, © Commonwealth of Australia. B. Davis Laundon, Marine Biological Association, Plymouth. (See Laundon et al., 2019).

Larger (>20 µm) predatory protists, described as heterotrophic microflagellates (HMF), include dinoflagellates, ciliates, and choanoflagellates, which are also very active grazers of bacteria and other protists; they are discussed below. These are especially important in the photic zone, where they are responsible for large-scale transfer of primary production from the phytoplankton in the microbial loop. In the deep sea also, protistan grazing is estimated to account for a daily turnover of up to 30% of the bacterial and archaeal cell stock.

Many protists are mixotrophic

The ability to combine the processes of feeding by phagotrophy with photosynthesis is extremely common in marine protists, found in diverse taxa throughout the photic zone and across the eukaryotic tree of life. Some organisms possess plastids permanently and supplement their photosynthetic metabolism by ingesting bacteria or other microbes. Endosymbiosis, involving extensive gene transfer and reorganization of cellular functions, occurred early in the evolution of eukaryotes, leading to the presence of organelles in several groups. So, it is not really surprising that many protists have retained the capacity for dual modes of nutrition. However, many mixotrophs may not permanently contain photosynthetic ability, but can "steal" it from phototrophs that they have ingested. This acquired phototrophy by transient retention of functional algae or their plastids for photosynthesis is known as kleptoplastidy. Usually, algal cells are broken up and digested in the food vacuole following phagocytosis. It is not known how the plastids remain intact and functional while the rest of the prey cell is digested. Many more persistent, facultative associations occur in other dinoflagellates, ciliates, foraminifera, and radiolarians. Kleptoplastidy does not seem to involve extensive adaptations to accommodate foreign cells or plastids—the hosts receive the benefit of additional carbon with few "upkeep" costs. This might explain the development of features such as rapid motility, complex surface structures, or food storage systems in mixotrophs. Acquired phototrophy is usually overlooked in ecological and biogeochemical models, and understanding the process is important to explain how aquatic ecosystems respond to changing environments.

The choanoflagellates have a unique morphology and feeding mechanism

About 100 species have been described from marine and estuarine waters throughout the world. The choanoflagellates resemble the feeding cells found in sponges and are of special interest to evolutionary biologists as they are phylogenetically the closest unicellular relatives to the obligately multicellular animals (members of the Holozoa, as shown in *Figure 6.2*). The choanoflagellates show strong similarities to the feeding cells (choanocytes) of sponges (*Figure 6.4*). Study of the choanoflagellate cell organization is providing important insights into the origins of multicellularity. The cells are ovoid or spherical, about 3–10 um diameter, with a single flagellum that propels free-swimming cells though the water and draws a current of water through a funnel-shaped collar of 30–40 microvilli around the top part of the cell, where they are trapped and ingested by the cell. Bacteria are trapped and taken into food vacuoles in the cell. Some choanoflagellates produce a delicate basket-like shell (lorica) composed of silica around the cell (*Figure 6.4*).

Dinoflagellates have several critical roles in marine systems

The most distinctive feature of the dinoflagellates is the presence of a transverse flagellum that encircles the body in a groove, and a longitudinal flagellum that extends to the posterior of the cell. This gives rise to a distinctive spinning motion during swimming, from which the name dinoflagellate is derived (from the Greek *dinein*, "to whirl"). Another distinctive feature of their cell structure is the presence of a layer of vesicles (alveoli) underlying the cell membrane. In many species of dinoflagellates, these alveoli contain cellulose, which forms a protective system of plates that fits together like a suit of armor. There are numerous forms of such armored dinoflagellates, mainly in the genera *Dinophysis* (200 marine species) and *Protoperidinium* (280 marine species), often with unusual morphology with spines, spikes, or horn-like projections. The geometry of the arrangement of plates is an important factor in identification. Many other naked or unarmored dinoflagellates also occur, including those that resemble species in the genera *Gymnodinium*, *Gyrodinium*, and *Katodinium*. Most dinoflagellates are in the 20–200 μm microplankton size range; however, several mixotrophic genera have been identified that appear to be below 20 μm, while *Noctiluca* can reach a size of 2 mm.

One of the most striking features of the dinoflagellates is their possession of huge genomes, which range is size from 1.2 to 272 Gb (for comparison, the human genome is 3.2 Gb). The huge diversity in genome size in different species cannot be explained in terms of differences in cell size or function. Genomes appear to contain large regions of repeated and interspersed noncoding DNA sequences and there is evidence of extensive horizontal gene transfer. Most of the plastid genes have been transferred from the organelles to the nuclear genome. Unlike other eukaryotes, the DNA of dinoflagellates occurs in a highly condensed form and most genes are not subject to transcriptional regulation; instead a process of mRNA editing occurs. The technical difficulties of working with such large genomes means that genetic information about the dinoflagellates is limited. Only a few genomes have been sequenced, from species with relatively small genomes. including the coral symbiont *Symbiodinium* (*Table 6.1*). This has revealed the presence of gene families related to establishment and maintenance of symbiosis (see *Chapter 10*).

At least 2000 dinoflagellate species are known, of which about 90% are marine, with different species adapted to life in many habitats per depth, temperature, and salinity. They occur as benthic stages associated with sediments, corals, seaweeds, and other surfaces and as free-floating members of the plankton. Representative images are shown in *Figure 6.6*. Photosynthetic dinoflagellates contain chlorophylls *a* and *c*, plus carotenoids and xanthophylls, and make a significant contribution to carbon dioxide fixation and primary productivity in the oceans. This is especially true in coastal waters in higher latitudes, where strong seasonality is affected by light, temperature, reduction in wind-driven turbulence, and input of fresh water from melting ice promoting spring blooms of dinoflagellates and diatoms.

The xanthophyll pigments confer the golden-brown color typical of many photosynthetic dinoflagellates and gives rise to the name zooxanthellae, used to describe the Symbiodiniaceae dinoflagellates that form symbioses in invertebrates such as corals, anemones and clams (see *Chapter 10*). Endosymbiotic interactions also occur within other protists, such as ciliates, foraminiferans, and colonial radiolarians.

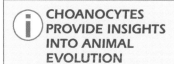

CHOANOCYTES PROVIDE INSIGHTS INTO ANIMAL EVOLUTION

Phylogenetic trees and comparison of genome sequences lead to the inference that the last common ancestor of multicellular animals would have been choanoflagellate-like. The actin-rich collar cells surrounding the whip-like flagellum are conserved across almost all animal phyla. In sponges, the closet relative of choanoflagellates, they are used for feeding, while in other animals they are adapted for sensory and osmoregulatory functions. To investigate the origins of multicellularity and cell differentiation, Davis Laundon and colleagues at the Marine Biological Association, Plymouth and University of California, Berkeley used TEM of ultrathin sections to generate 3D reconstructions of cells of *Salpingoeca rosetta* (Laundon et al., 2019). This choanoflagellate has a complex life cycle, transitioning between single and colonial collar cell types. This revealed morphologically and positionally distinct populations of intracellular vesicles and important differences between single-celled and multicellular choanoflagellates in structures connected with cellular energetics, membrane trafficking, and cell morphology (*Figure 6.4*). The discovery of cell-cell connections and cell differentiation in multicellular choanoflagellates reveals important insights into the early evolution of the animals.

BOX 6.1   RESEARCH FOCUS

Sailing towards a better understanding of heterotrophic ocean protists

As noted in the main text, our understanding of eukaryotic ocean microbes through the application of molecular methods in the oceans has been somewhat slower to emerge than knowledge of bacterial and archaea. Among the many exciting discoveries emerging from analysis of samples from the *Tara Oceans* expeditions (p.55) are significant recent breakthroughs in our understanding of protist diversity, functions, and importance, following the development of improved sampling methods and advanced molecular techniques.

Diversity studies using rRNA gene sequencing. For example, de Vargas et al. (2015) analyzed 18S rRNA gene sequences from 334 photic zone samples identifying ~150,000 OTUs in different size fractions, of which ~~~~~match in existing databases and about a third could not even be assigned to any of the known higher eukaryotic taxa. The greatest diversity occurred among heterotrophic protists, particularly those from groups known to include predators, parasites, or symbiotic hosts—suggesting the importance of complex ecosystems that are regulated mainly by interactions between organisms rather than competition for resources or space. In another study, Gimmler et al. (2016) analyzed >6×10^6 18S rRNA gene amplicons assigned to the phylum Ciliophora from surface and mesopelagic *Tara* samples. This study found a relatively low diversity of ciliates, with ~1300 OTUs, of which most were globally distributed. Ciliates are highly abundant in the oceans and more is known about them than other groups because they have been studied microscopically for many years, due to their larger size. Even so, more than half of the OTUs were distantly related to previously known groups.

Development of single-cell genomics. Unlike the advances made with bacteria and archaea, an understanding of ecological roles of eukaryotic microbes though analysis of genome sequencing has been impossible until the development of single-cell genomics (SCG). These methods are still in their infancy, but the problems of highly fragmented eukaryotic genomes are now being solved by co-assembly of single amplified genomes (SAGs) from several cells of the same population and integration of SCG with metagenomic and metatranscriptomic sequence data. Sieracki et al. (2019) analyzed 900 SAGs from *Tara* surface seawater samples collected across the Indian Ocean and Mediterranean Sea. They used improved methods of sample preservation and flow cytometry (FC) to sort cells based on chlorophyll fluorescence into those with plastids (likely phototrophs or mixotrophs) and those without plastids (likely heterotrophs). The taxonomic diversity of cells <5 μm in just hundreds of microliters of water matched that found in earlier surveys filtering many liters of samples. Sieracki et al. found 22 major taxa of aplastidic types, including Mamiellophyceae, Pelagophyceae, and Prymnesiophyceae, with varying relative abundance. Twelve major groups of aplastidic cells were found, dominated by marine stramenopiles (MAST), Chrysophyceae. Bicosoecida and marine alveolates (MALV). Gawryluk et al. (2016) linked SCG with microscopy to describe the diversity and abundance of marine diplonemids. SCGs generated from the *Tara* samples also led to new insights into the function of MALVs, confirming their importance as parasites of a wide range of hosts, including other protists, as well as diverse trophic strategies (Strassert et al., 2018). Network analysis of metagenomic data

Figure 6.5 Analysis of functional genes from marine heterotrophic SAG lineages shows that they form a functional group distinct from autotrophs and other heterotrophs. Non-metric multidimensional scaling (NMDS) projection of a Bray–Curtis distance matrix showing Pfam motif occurrences in various stramenopile genomes. Reprinted from Seeleuthner et al. (2018), CC-BY-4.0.

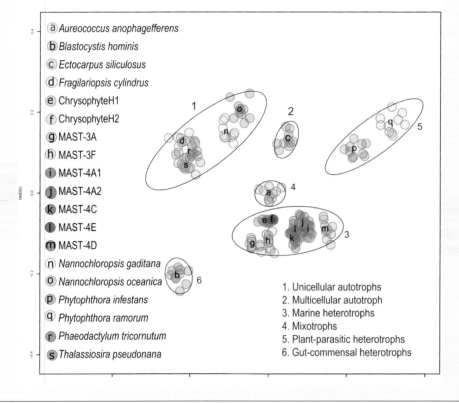

(a) *Aureococcus anophagefferens*
(b) *Blastocystis hominis*
(c) *Ectocarpus siliculosus*
(d) *Fragilariopsis cylindrus*
(e) ChrysophyteH1
(f) ChrysophyteH2
(g) MAST-3A
(h) MAST-3F
(i) MAST-4A1
(j) MAST-4A2
(k) MAST-4C
(l) MAST-4E
(m) MAST-4D
(n) *Nannochloropsis gaditana*
(o) *Nannochloropsis oceanica*
(p) *Phytophthora infestans*
(q) *Phytophthora ramorum*
(r) *Phaeodactylum tricornutum*
(s) *Thalassiosira pseudonana*

1. Unicellular autotrophs
2. Multicellular autotroph
3. Marine heterotrophs
4. Mixotrophs
5. Plant-parasitic heterotrophs
6. Gut-commensal heterotrophs

BOX 6.1   RESEARCH FOCUS

by Lima-Mendez et al. (2015) identified multiple interactions among grazers, primary producers, symbionts, and parasites.

Seeleuthner et al. (2018) assembled 900 SAGs from 40 single-cell representatives of FC-sorted heterotrophic protists in the 0.8–5.0 um size range. Before genome assembly, steps were taken to remove sequences from chloroplasts or mitochondria and more distantly related organisms to distinguish members of three stramenopile clades—chrysophytes, MAST-3, and MAST-4—known to be abundant in the pico- and nanoplankton. From the 40 SAGs sequenced and annotated, Seeleuthner et al. used bioinformatics software to predict the functional repertoires of these uncultured stramenopiles and compared them with genomes from previously cultured species. This showed that genomes clustered primarily per the trophic mode of the organisms, with the MAST lineages and chrysophytes clade-H clustering together, even though they are phylogenetically

distant (*Figure 6.5*). One interesting observation concerned genes encoding the flagellar protein components, a key gene that showed signs of relatively recent loss from some MAST-3 members. Perhaps these have adapted to a sessile or parasitic lifestyle. Some MAST-4 lineages possessed highly expressed proteorhodopsin genes indicating that they may have photoheterotrophic capacity (see *Box 3.1*). Seeleuthner et al. found that all SAGs showed a wide range of carbohydrate-active enzymes, but the gene content of chrysophytes and some MAST types suggested that they may not be bacterivorous. The predicted enzyme profile of many of these protists suggested that they also degrade chitin-containing organisms like fungi, diatoms, and crustaceans. Although the authors acknowledge some limitations to conclusions about geographical and seasonal distribution of these protists (because of sampling strategy of the *Tara* dataset), their finding clearly demonstrate that heterotrophic protists participate in various aspects of complex functional networks in ocean food webs.

Although traditionally grouped with the phytoplankton, heterotrophy or mixotrophy occurs in about half of known dinoflagellates and such species possess a variety of feeding mechanisms, including absorption of organic material, engulfment of other microbes by phagotrophy, or use of a peduncle to suck out the contents of larger prey. These can prey on bacteria and larger phytoplankton, as well as zooplankton or even fish eggs and larvae; they therefore have major effects on food web structure. Nonpigmented, heterotrophic dinoflagellates often comprise more than half the biomass of the microzooplankton size class (20–200 μm) and are frequently the most important grazers of blooms of diatoms and other microplankton. They also provide a major food source for copepods. They may also possess very different forms in their life cycle. The haploid stage, with cells reproducing by mitosis, appears to be the normal mode during active growth in optimal environmental conditions. During cell division, the thecal plates may also be divided, but in some species, they are lost completely before the naked cells divide and reform the plates. During sexual reproduction, haploid gametes fuse to form a motile diploid planozygote, which divides by meiosis to restore the vegetative stage. The planozygote, as well as haploid cells, can form resting cysts which may have thickened cell walls enabling them to withstand adverse conditions. Factors affecting the transition between sexual and asexual stages and the survival and germination characteristics of cysts are very variable and mechanisms are poorly understood. Endogenous clocks on an approximately annual cycle, nutrient composition and concentration, temperature, salinity, pollution, and hydrological conditions are all important factors in development of dinoflagellate blooms, and processes differ for benthic and planktonic forms. Understanding this process is especially significant for those dinoflagellates that are responsible for the formation of deleterious blooms, which may have damaging effects on marine life and humans; these are discussed in *Chapters 12* and *13*.

Dinoflagellates and other protists undertake diel vertical migration

In regions of the open ocean where there is little upwelling, most photosynthetic dinoflagellates and other motile protists such as raphidophytes are found in the epipelagic photic zone during the day and migrate to deeper nutrient-rich waters at night. This vertical diel (diurnal) migration in the water column can exceed 30 m in each direction; this is a remarkable two million times the cell diameter. The onsets of both ascent and descent are regulated by an endogenous biological clock, so that the cells "anticipate" sunrise and are in the optimum position to start photosynthesizing near the surface as soon as light is available. As light declines at dusk, they migrate to lower levels in the water column to take advantage of higher nutrient levels for the dark phase. The diel movement of phytoplankton such as dinoflagellates and raphidophytes is a tactic response to gravity, but during their daily migration, the cells encounter turbulence and gradients of temperature, nutrients, and the amount and spectral quality of light. All of these factors can act as input signals for regulating the circadian rhythm.

Figure 6.6 SEM images of dinoflagellates. A. *Protoperidinium defectum* B. *Protoperidinium incognitum*. C. *Protoperidinium antarcticum*. D. *Gonyaulax striata*. Credit: Rick van den Enden, Electron Microscopy Unit, Australian Antarctic Division, © Commonwealth of Australia.

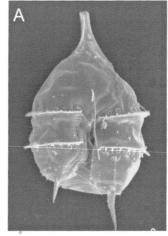
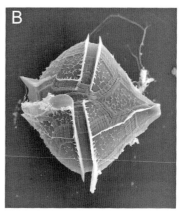

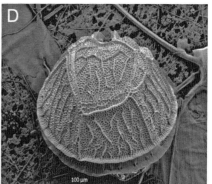

ⓘ **TOPSY-TURVY PHYTOPLANKTON CHANGE THEIR SHAPE**

Strong turbulence can cause damage to flagella or cell structure, and some organisms show adaptations to cope with this. For example, *Ceratocorys* alters its surface spines to enhance sinking to deeper, calmer layers, while *Alexandrium* forms chains to alter swimming behavior. But how do organisms respond to turbulence over short timescales? Sengupta et al. (2017) used an experimental millifluidic chamber that was repeatedly flipped vertically to assess the effects on upward swimming by *Heterosigma akashiwo*. Within minutes of overturning, the microscale turbulent eddies caused some cells to begin to swim downward, while others continued to swim upwards. This is due to changes in cell orientation relative to gravity. Reversal of swimming direction is caused by cells morphing from a pear-shaped cell to a more symmetric egg-shaped cell, which has greater stability and reorientates more quickly. This rapid behavioral response gives the phytoplankton precise control over their migratory behavior, increasing the likelihood that some members of the population will evade intermittent patches of turbulence in the water column.

Some dinoflagellates exhibit bioluminescence

About 2% of dinoflagellate species found in coastal waters are bioluminescent, with the best-known species being *Noctiluca scintillans* and *Lingulodinium polyedrum* (formerly *Gonyaulax polyedra*). They occur worldwide, but exceptionally high densities occur in certain tropical coastal waters and produce spectacular displays of sparkling light when the surface of water is broken at night. (This is often erroneously called phosphorescence, which is a different mechanism, caused by prolonged re-emission of radiation absorbed by fluorescent molecules.) Bioluminescence in dinoflagellates usually consists of brief flashes of blue–green light (wavelength about 475 nm), containing 10^8 photons and lasting about 0.01 s. *Lingulodinium* may also emit red light at 630–690 nm. The stimulus for light emission is deformation of the membrane due to shear forces, such as agitation of water by fish, breaking waves, or the wake of a boat. In the laboratory, bioluminescence can be stimulated by lowering the temperature or pH of the medium. The bioluminescent flash is preceded by an action potential, during which the inside of the membrane becomes negatively charged. This leads to acidification of vesicles in the vacuolar membrane containing the enzyme luciferase and the substrate luciferin. There are about 400 such vesicles (scintillons) in each cell and they occur as spherical evaginations of cytoplasm into the cell vacuole, each containing luciferin complexed with a special binding protein. A transient pH change results from the opening of membrane proton channels in the scintillons; this activates the reaction by release of the luciferin from its complexed state so that it can be oxidized in an ATP-mediated reaction similar to that in bacteria (*Figure 3.17B*). Experimental studies of *Lingulodinium* cultured under various light conditions have revealed that bioluminescence is regulated by a circadian rhythm and peaks in the middle of the dark period. The circadian expression of bioluminescence involves the daily synthesis and destruction of the scintillons and component proteins; this is regulated at the translational level and may involve a clock-controlled repressor molecule that binds to mRNA.

There are two main hypotheses for the ecological function of bioluminescence in dinoflagellates. One idea is that mechanical stimulation occurs when a predator such as a copepod approaches a dinoflagellate cell—stimulating a brief, bright flash. This is thought to startle the predator and lead to its disorientation. An alternative possibility, called "the burglar alarm hypothesis," is that light produced by luminescent prey attracts grazing predators, which in turn sends a signal to larger predators, so that the grazers themselves become prey. Grazing on bioluminescent dinoflagellates will decrease if the risk of consuming them results in reduction of the net benefit that consumers receive. Thus, bioluminescence could reduce grazing pressure by reduction of feeding efficiency and the species as a whole could benefit, even though some mortality of individuals occurs. Various experimental approaches have been used to determine which of these alternative hypotheses is correct, but there is no clear answer and progress in understanding requires methods to measure interactions *in situ* under natural conditions.

The ciliates are voracious grazers of other protists and bacteria

This group of protists, within the supergroup Alveolata is distinguished by the possession of cilia in at least one stage in the life cycle—these have the same basic structure as flagella, but the cilia often cover the cell surface or are arranged in groups; they beat synchronously to provide movement or to create water currents to channel particulate food into the cell. At least 8000 species of ciliates are known—many of which are marine— and, although highly diverse, the group appears monophyletic in modern classification systems. Marine ciliates are generally in the size range 15–80 μm, with some up to 200 μm. One group of ciliates, the tintinnids, produce a "house" called a lorica, which is constructed from protein, polysaccharides, and accumulated particulate debris collected from the water (*Figure 6.7*). Thus, tintinnid loricae are large enough to be collected in fine-mesh plankton nets and were, therefore, one of the first groups of marine ciliates to be studied. There has been intense interest in the ciliates in recent years because of growing recognition of the essential role that most types play in the microbial loop of ocean food webs by ingesting bacteria and other small protists and because they are large enough to be preyed upon in turn by larger protists and mesozooplankton. Selective grazing on particular prey types is an important factor in structuring the composition of microbial communities in food webs. Consequently, there have been numerous studies of the abundance and activity of marine ciliates, with a wide range of results. Typically, there are about 1–150 ciliates per milliliter in seawater, with the highest numbers in coastal waters; much higher concentrations occur in marine snow particles. Abundance varies greatly with water depth, temperature, and nutrient concentration. Water stratification during the different seasons is a major factor affecting ciliate numbers. Ciliates are also abundant in benthic sediments and microbial mats.

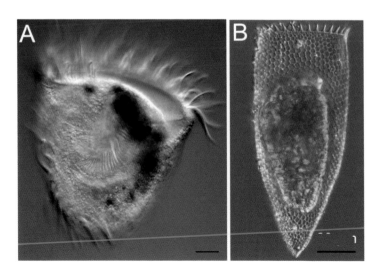

Figure 6.7 Light micrographs of ciliates. Scale bar = 20 μm. A. *Gastrocirrhus monilifer*. B. *Parafavella parumdentata* tintinnid. Credits: A. Bill Bourland, used by permission of David Patterson, http://micro-scope.mbl.edu/. B. Diane Stoecker, NOAA Fisheries Collection, via Flickr CC-BY-2.0.

The most common marine ciliates are spherical, oval, or conical cells with a ring of cilia surrounding the cytostome ("mouth"), which they use to filter bacteria and small flagellates from the surrounding seawater. Upon entering the cytostome, ingested food particles are engulfed by phagosomes, which then fuse with lysosomes in the cytoplasm. The phagosomes become acidified and enzymes contained in the lysosome lead to digestion. Nutrients pass into the cytoplasm and undigested waste material is egested. Although most genera are phagotrophic, some are strict photosynthetic autotrophs and have lost phagotrophic capability. The most important of these is *Mesodinium rubrum* (previously classified as *Myrionecta rubra* in the Zoological Code), which can occur in large blooms in coastal waters and may make a sizable contribution to primary production. The species is named for its association with red tides, which discolor coastal waters, although it is non-toxic. This ciliate contains functional organelles obtained from its prey, cryptophyte algae. Unlike the retention of plastids (kleptoplastidy) mentioned above and observed in a wide range of ciliates, *M. rubrum* also contains stolen mitochondria, cytoplasm, and a transcriptionally active nucleus, as well as chloroplasts. There have been suggestions that *M. rubrum* can retain intact cryptophyte cells as endosymbionts, although this is disputed. There are also many examples of mixotrophy in other ciliates, arising from transient retention of ingested plastids.

A distinctive feature of the ciliates is the possession of two types of nuclei. The larger macronucleus consists of multiple short pieces of DNA and is concerned largely with transcription of mRNA and growth processes. The smaller micronucleus is diploid and is responsible for an unusual type of sexual reproduction (conjugation), in which two cells fuse for a short period and exchange haploid nuclei derived from the micronuclei by meiosis. The macronuclei disappear during this process. The result of conjugation is that each partner ends up with one of its own haploid nuclei and one from its partner; these then fuse, and the cells separate. In each cell, the diploid nucleus thus formed divides and differentiates into micro- and macronuclei due to extensive rearrangement and amplification of DNA sequences.

The haptophytes (prymnesiophytes) are some of the most abundant components of ocean phytoplankton

The haptophytes, also known as prymnesiophytes, are very important members of the phytoplankton, some of which have profound influences on oceanic and atmospheric processes. There are about 500 living species in 50 genera, with many additional species identified as fossils in sedimentary rocks. Most species are marine and occur worldwide, with the greatest diversity and abundance in tropical waters.

Haptophytes are photosynthetic and contain one or two plastids, together with the yellow–brown accessory pigments diadinoxanthin and fucoxanthin. They often have a complex life cycle, alternating between motile and nonmotile haploid and diploid stages. The cells contain two slightly unequal (heterokont) flagella in one stage of their life cycle. They are also distinguished by the presence of a haptonema; this is a thin structure reminiscent of a flagellum, but with a different structure and unknown function. The haptonema is composed of six to seven microtubules in a ring or crescent, with a fold of endoplasmic reticulum extending out within the flagellum.

The best-known members of the haptophytes are the coccolithophores, which have been one of the most abundant forms of phytoplankton in the oceans for millions of years. Their distinguishing feature is the presence of external scales or plates, formed within the Golgi apparatus, that cover the cell surface in many species. The plates have distinctive shapes and surface architecture in different species.

The best-known and most abundant example of coccolithophores is *Emiliania huxleyi*, in which the typical cells are diploid, nonmotile, 4–5 μm in diameter, and covered with about 30 calcified plates (coccoliths) (*Figure 6.8B*). The haploid stage occurs as motile swarmer cells with noncalcified scales and another nonmotile naked cell type may also occur as a mutant diploid stage. Each stage can reproduce vegetatively. The haploid stage

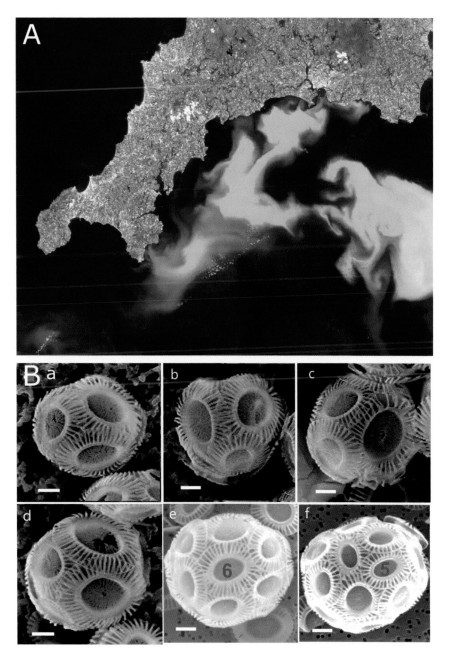

is very important in the resistance of *E. huxleyi* to viral infection, as discussed in *Box 7.1*. Calcification takes place intracellularly in a specialized vesicle of the Golgi apparatus and depends on a high flux of inorganic carbon (in the form of bicarbonate) and calcium. The calcium-binding protein is a key factor in the calcification process and the *gpa* gene that encodes this is often used as a genetic marker in studies of *E. huxleyi* community dynamics. Calcite (one of the crystalline forms of calcium carbonate) is precipitated onto the organic matrix of the vesicle and extruded to the surface to form a coccolith. The reasons why coccolithophores calcify their surface plates is unclear. It is an energy-demanding process and may have evolved to provide mechanical protection against grazing pressure, but it may provide additional benefits such as protection from photodamage. Calcification also enhances the efficiency of photosynthesis by lowering the internal pH of the cell, increasing the concentrations of dissolved CO_2 needed from maximum efficiency of the CO_2-fixing enzyme RuBisCO (p.83).

The coccoliths are arranged, usually in a single layer, to cover the entire surface area of the cell, which changes during growth. Observation of cultured *E. huxleyi* cells shows that the number of coccoliths formed keeps pace with the increase in cell size during growth, as shown in *Figure 6.8B*. To maintain full coverage of the surface, each coccolith interlocks with four to six others; small coccoliths interlock with fewer and larger coccoliths, and vice versa. The optimal configuration to enclose the volume of the cell is hexagonal tiling, following the same mathematical principles as the construction of bees' honeycombs or formation of soap film bubbles.

Each year, massive blooms of *E. huxleyi* occur in the upper photic layers of temperate and subtemperate waters. They are especially notable in the coastal waters of northern Europe and Scandinavia following spring diatom blooms, where they are highly visible as milky-white areas in satellite images (*Figure 6.8A*), because of the reflective properties of free coccoliths released from the cells as they disintegrate after death. Under favorable conditions, individual blooms may cover 10^5 km^2 and contain more than 10^{21} cells, accounting for almost 90% of all the phytoplankton in the upper water columns. The reflective "signature" of the coccoliths is unique and allows precise measurement of the fate of the bloom from satellite images. On a global scale, such dense blooms of *E. huxleyi* probably transiently cover up to 2×10^6 km^2 in a typical year. *E. huxleyi* is present at relatively constant, lower levels in many oligotrophic oceans, where they also contribute significantly to primary productivity.

E. huxleyi has global distribution and has been intensively studied in recent years because of its role in the global carbon cycle, being the largest global producer of calcium carbonate and, hence, a major sink for CO_2. However, there are several complications to the process and the exact contributions that *E. huxleyi* makes to carbon cycling and climate change are not clear. The light-scattering properties of the coccoliths may have a small albedo effect, reflecting solar radiation and helping to cool surface waters during a bloom. Calcification leads to rapid removal of inorganic carbon as the plates settle through the water column to form sediments. In fact, although *E. huxleyi* is the most dominant coccolithophore, other species with larger, heavier coccoliths may be more important in the flux of calcite. Various studies to estimate the global productivity of calcite in the modern ocean have been conducted, leading to a consensus value of about one gigatonne (petagrams) of calcite carbon a year. Only about half of this amount will be precipitated to the ocean floor because in deep waters (over about 3000 m), changes in the water chemistry due to high pressure cause the calcite to redissolve. Fossils of the coccolithophorid group first appeared in the Jurassic period about 290 MYA and reached their greatest abundance and diversity in the Late Cretaceous period, at the end of which a mass extinction of many genera occurred. Accumulation of coccolithophorid plates on the seabed contributed to the formation of ocean sediments and rocks such as the Mesozoic limestones and chalks. The famous White Cliffs of Dover in southern England are composed of a very fine-grained white chalk derived mainly from coccoliths. Most modern coccolithophores are smaller and less heavily calcified than ancient forms, but nevertheless remain highly diverse.

Dissolved carbon exists in three forms in seawater—dissolved CO_2 gas, bicarbonate, and carbonate, and the relative proportions of each depend on pH, temperature, and pressure. When coccolithophore cells remove bicarbonate to form coccoliths, the pH drops and shifts more of the carbon into the dissolved CO_2 form. Calcification can be represented by the formula:

$$Ca + 2HCO_3^- \leftrightharpoons CaCO_3 \leftrightharpoons H_2O + CO_2$$

Thus, coccolithophore blooms might increase global warming by causing release of CO_2 back to the atmosphere, but this is complicated by the fact that the association of heavy coccoliths with marine snow particles and the feces of grazing zooplankton could lead to more rapid settling of associated organic material, thus accelerating the sequestration of fixed carbon to the sediments. Understanding the fluxes involved is very important in view of rising atmospheric CO_2 levels leading to ocean acidification, but despite many laboratory and mesocosm experiments, there are no clear answers.

Two other prymnesiophytes, *Chrysochromulina* and *Phaeocystis*, have both been studied extensively because of their ability to produce large nuisance blooms under certain conditions, especially in northern Europe and Scandinavia. Blooms are linked to climatic conditions and the input of pollutants from land runoff. *Phaeocystis globosa* aggregates into large colonies surrounded by polysaccharide mucilage. This is responsible for the foam that commonly affects seashores during summer, and excessive production of DMS (see p.248) during the collapse of the blooms leads to sulfurous smells that beachgoers often wrongly confuse with sewage pollution. DMS is also suspected of affecting the migration behavior of fish, which seek to avoid it, while the mucilage has physical effects such as clogging of aquaculture nets and desalination plant filters and pipes.

Diatoms are extremely diverse and abundant primary producers in the oceans

The diatoms are one of most diverse groups of the protists, with in excess of 200 genera and about 10^5 current species, with probably about the same number of fossil species. In modern classification schemes, diatoms are recognized as unicellular microbes belonging to the stramenopiles in the class Bacillariophyceae. Diatoms range in size from 2–200 μm in length and possess either radial or bilateral symmetry; these forms are termed centric or pennate, respectively.

The defining feature of diatoms is enclosure of the cell within a hard but porous wall known as a frustule, composed of two overlapping plates (valves or thecae) of hydrated silica (silicon dioxide) that is polymerized intracellularly. The two valves overlap rather like the lid and base of a petri dish. These opal-like structures are highly patterned, forming a variety of shapes of great architectural beauty and are highly distinctive for identification (*Figure 6.9*). The mechanism by which diatoms create these wondrous structures is of great interest. A major goal is to link the genetic, biochemical, and physiological processes with mathematical models to explain the patterns of silica deposition, which has many potential applications in the nanofabrication of silicon-based materials for electronics. Like dinoflagellates, diatom genomes are complex and present many technical difficulties, but several examples have now been sequenced. They contain evidence of their chimeric origin from their algal and heterotrophic ancestors, as well as extensive horizontal gene transfer from bacteria.

The nature of the frustule enclosing the diatom cell results in a unique reproductive cycle. Reproduction is usually asexual, with each daughter cell constructing a new theca within the old one, resulting in a progressive reduction in cell size with each division cycle. At a critical point—this is usually when the diatom is about one-third of the original size—sexual reproduction occurs by formation of gametes via meiosis. The gametes fuse to form a zygote

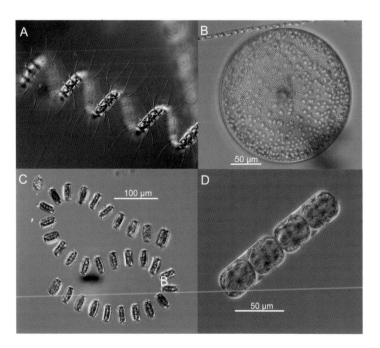

Figure 6.9 Light micrographs of marine diatoms. A. *Chaetoceros debilis*. B. *Melosira moniliformis*. C. *Thalassiosira nordenskiioeldi*. D. *Coscinodiscus centralis*. Credits: D. Cassis, T. Ivanochko, J. Shiller, B. Moore-Maley, J. Kim, S. Huang, A. Sheikh, G. Oka. Phytopedia: The Phytoplankton Encyclopedia Project. Published at www.eoas.ubc.ca/research/phytoplankton/

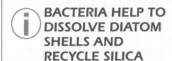

BACTERIA HELP TO DISSOLVE DIATOM SHELLS AND RECYCLE SILICA

When diatoms die, they aggregate and sink through the water column. Some of the silica dissolves to form silicic acid, which is then available through upwelling to support more diatom growth. The input of silicic acid from terrestrial runoff is balanced by deposition in sediments. Bacteria play a critical role in silica recycling by producing extracellular enzymes that break down cell wall polymers. Bidle and Azam (1999, 2001) showed that if bacteria are removed from seawater or inhibited by antibiotics, the dissolution of silica from diatom debris is very low. Diatom species vary in both the thickness of their silica shells and the extent of protection by glycoprotein coats. These factors determine the extent to which the shells of different species remain intact. Bidle (2002) showed that temperature strongly influences bacterial dissolution of silica. In polar regions, slow bacterial activity means that more carbon and silica is sequestered in the sediments. Understanding this process is important for modeling the role of the oceans in the removal of atmospheric CO_2 and the effects of global warming on geochemical cycles.

(auxospore), which increases in size and synthesizes new full-size thecae. Hence, sexual reproduction accomplishes genetic variability as well as allowing the cell line to regain maximum size before the next round of vegetative cell division.

Most diatoms are free-living in the plankton, but many attach to surfaces such as rocks, marine plants, mollusks, crustaceans, and larger animals. The skin of some whales has been shown to be covered with dense colonies of diatoms. Locomotion of some (mostly pennate) diatoms occurs in contact with surfaces, probably via the secretion of mucus through a slit (raphe) in their silica frustules but they are mostly nonmotile in the water column and their heavy cells sink rapidly, relying on turbulence in the surface layer to keep them in suspension. Some diatoms have pairs of thin spines projecting from the ends of the cells, which link with those of other cells to form long chains, thereby increasing buoyancy and forming large mats.

The major pigments in this group are usually chlorophylls *a* and *c* and the carotenoid fucoxanthin. It is thought that the chloroplast in stramenopiles, which contains four membranes, arose by a secondary endosymbiosis event about 700–1200 MYA, involving a red alga ancestor and a heterotrophic eukaryote. Genome sequencing reveals that genes from red algae have been transferred to the nucleus of diatoms, which provides an explanation for the unusual metabolic properties of diatoms. For example, unlike green plants, diatoms have a complete urea cycle and generate energy from the breakdown of lipids, and carbohydrates are stored as a β1,3-linked glucose polysaccharide.

For many years, diatoms have been recognized as the dominant member of the marine phytoplankton and therefore attributed with the major role in primary productivity and carbon export. They are estimated to contribute ~40% of the total ocean primary productivity. They are especially important in seasonal productivity in well-mixed, nutrient-rich coastal waters in higher latitudes, but we now realize that cyanobacteria and protists in the picoplankton size contribute a large fraction of photosynthetic productivity in open oceans. Environmental 18S rRNA gene datasets from the *Tara Oceans* surveys indicate that they are the most abundant group of obligately phototrophic marine eukaryotes. In polar regions, diatoms play a critical role in the food chain, especially as their mixotrophic metabolism allows them to survive long periods of darkness before emerging as blooms in the spring. In some Antarctic samples, diatoms represent 25% of the sequences. Like dinoflagellates, many open-ocean diatoms undertake a vertical migration from the nutrient-depleted photic zone to deeper waters in order to pick up nitrate and other nutrients. The complex ecological factors that determine community composition of the phytoplankton are discussed further in *Chapter 8*.

Diatoms are largely responsible for the spring bloom along the continental shelf in temperate waters and for seasonal blooms in regions of nutrient upwelling. As with other algae, a key factor determining the size of blooms is the availability of nutrients, especially nitrogen and phosphorus, but the concentration of silica is also a critical limiting factor for diatoms, and most will not grow at silica concentrations below about 2 µM. Some species make frustules with very high silica contents, and in some waters (e.g. the Antarctic circumpolar current), the shells are very resilient to dissolution as they settle through the water. Blooms are generally followed by the exhaustion of nutrients and aggregation and sinking of diatoms. The dynamics of diatom production and settlement are still poorly understood but are important in understanding ocean biogeochemistry.

Diatoms began to accumulate on the seabed about 100 MYA and reached a peak of abundance in the middle part of the Cenozoic period. Settlement of dead diatoms over millennia led to thick deposits and formation of siliceous sedimentary rocks, which are mined as diatomaceous earth. This has many industrial uses such as filtration compounds, abrasives, insulating agents, pharmaceutical products, and insecticides. The small fraction of organic matter that escaped remineralization accumulated in sediments, leading to the formation of petroleum hydrocarbons. Today, diatoms are being exploited in biotechnology, including applications in nanotechnology and the production of biofuels (see *Chapter 13*). Few diatoms are toxic, but an important exception is the genus *Pseudo-nitzschia*, which produces domoic acid. This is responsible for human illness associated with shellfish consumption and for mortalities in marine mammals and seabirds (p.320).

Other stramenopiles may cause harmful blooms

Raphidophytes are photoautotrophic stramenopiles related to the diatoms. They have large cells (50–100 μm) without cell walls and possess two flagella. Blooms of species such as *Heterosigma akashiwo*, *Chatonnella* spp., and *Fibrocapsa* spp. have caused "red tides" responsible for extensive fish mortalities in Japan, Scotland, and Canada. They are thought to kill fish as a result of the production of a range of toxins, including reactive oxygen species, hemolytic substances, and neurotoxins.

Other important stramenopile algae, with small cells about 2–3 μm in diameter, include *Aureococcus* spp., *Nannochloropsis* spp., and *Bolidomonas* spp. *Aureococcus anophageffferens* is notorious as a cause of regular "brown tides" in estuaries on the eastern US coast since the 1980s. It is always present at low concentrations and blooms to high concentrations under certain conditions, causing severe limitation of growth of seagrass (*Zostera marina*) beds, because of shading caused by the dense blooms. This affects larval recruitment of fish, scallops, and clams, with severe consequences for fisheries. *Aureococcus* is also suspected of producing a toxin that may affect ciliary function in bivalve mollusks, although this has not been isolated. *Nannochloropsis* has a very high content of polyunsaturated fatty acids, making it an important food source for fish larvae. This feature is being exploited in development of aquaculture feeds (see p.393).

Thraustochytrids and labyrinthulids are active degraders of organic matter

These two groups of closely related stramenopiles are characterized by the production of a slime net—a network of filaments produced as extensions of the cytoplasmic membrane that serve as tracks for the cells to glide along. They were originally grouped with slime molds in the Fungi, but their heterokont flagella, mitochondrial structure, and genetic features mark them out as stramenopiles. They have been known for many years, but their ecology and importance in marine systems are understudied.

Several studies have shown that thraustochytrids are widely distributed in the plankton in coastal and open ocean, although they seem to occur at very low densities, perhaps only 1–100 mL^{-1}. Along with bacteria, archaea, and fungi, these organisms are the only heterotrophic members of the plankton which feed by osmotic absorption (osmotrophy) of dissolved organic compounds rather than by phagocytosis of cells or particles. The slime net contains extracellular enzymes—including amylase, cellulase, lipase, protease, phosphatase, pectinase, and xylanase—that digest large polymeric organic compounds to smaller units for absorption. The slime net can penetrate particles, such as marine snow or clumps of decaying cells from algal blooms. These properties and the fact that their cells are many times bigger than bacterial or archaeal cells mean that, despite their low numbers, they may have a comparable biomass. This leads to a very significant and under-recognized role at the base of the food chain, especially since they can digest highly refractory organic matter and are readily eaten by ciliates and amoebae. This link to higher trophic levels is particularly important because the thraustochytrids produce high levels of omega-3 polyunsaturated fatty acids (PUFA), which are essential for the growth and reproduction of crustaceans and fish larvae. This feature is being exploited in aquaculture biotechnology. Thraustochytrids also degrade plant detritus such as algae and mangrove leaves, and a few species have been described as parasites of mollusks. Genome sequences of a number of strains have been obtained, providing information about the genetic basis of production of PUFA and hydrolytic enzymes. This reveals that some strains of thraustochytrids produce low levels of the enzymes responsible for degradation of refractory polymers and may depend on scavenging nutrients released by bacterial action.

Labyrinthulids are very similar in structure to the thraustochytrids but have not been found in plankton. Instead, they are usually associated with the surfaces of marine macroalgae, plants, or animals as parasites or commensals. One species, *Labyrinthula zosterae*, causes the "wasting disease" of the seagrass *Zostera marina* (p.324).

Photosynthetic prasinophytes are abundant members of the picoplankton

Many algae in the microplankton groups discussed above have been described since the studies of pioneer marine biologists and oceanographers, as their relatively large size and distinctive morphological features can be used for classification and identification. However, the full importance of photosynthetic protists as components of the picophytoplankton (<3 μm) has only been realized in the last few decades. Phototrophic types belonging to the green algae (or Chloroplastida), stramenopiles, and haptophytes (prymnesiophytes) occur in the photic zone of all the world's oceans. Together with the cyanobacteria *Prochlorococcus* and *Synechococcus* (see p.134), the picophytoplankton may constitute up to 80% of the chlorophyll-containing biomass in subtropical oceans. Studies in the Pacific Ocean have shown that, although phototrophic picoeukaryotes comprise only a small proportion of the total phytoplankton cells, they account for the majority of the net carbon production. They are subject to heavy grazing pressure by other protists, probably because they lack a cell wall. Thus, their role in marine food webs and the biological carbon pump (see *Figure 8.1B*) may be much more important than previously realized because of the efficient capture of carbon dioxide and its rapid transfer to food webs via grazing.

The most studied prasinophyte phototroph is *Ostreococcus tauri* in the family Mamelliaceae. Originally discovered and given its genome name because it was found in waters used for oyster (*Ostrea*) cultivation in France, *O. tauri* is now known to be ubiquitous in coastal waters and in the open ocean. This is the smallest eukaryotic organism known: the tiny cell (about 0.8 μm diameter) has no cell wall or flagella and contains the nucleus, only one mitochondrion, and one chloroplast, tightly packed within the cell membrane (*Figure 1.4B*). Genome analysis has revealed that it has a small and compact genome with 20 linear chromosomes (*Table 6.1*). It is thought that one chromosome may be a sex chromosome, as it contains genetic information for meiosis, although sexual reproduction has not been observed. Genes on some chromosomes show little homology to other genes and appear to have been laterally transferred; they may encode surface proteins to protect the cell against predation. One of the most important findings of genome analysis is the presence of genes for C_4 photosynthesis. This route of CO_2 fixation may be more efficient than the normal C_3 route when dense populations of cells are competing for available CO_2. As with the cyanobacterium *Prochlorococcus*, comparative genome analysis shows that different ecotypes of *O. tauri* are adapted to different light and nutrient niches.

Another prasinophyte, *Micromonas pusilla*, is a major component of the picoplanktonic community in several oceanic and coastal regions and shows summer blooms that may be terminated by viral infection. *Micromonas* has a pear-shaped cell (~ 2 μm) and is actively motile via a single flagellum, enabling it to swim toward light sources. Comparative genomic analysis of strains presumed to be of the same species, but isolated from different geographic regions, has shown remarkably large differences in genome sequences, with only about 90% of their genes shared and the taxon undoubtedly encompasses several species. The green algae were long thought to be a purely phototrophic group but *Micromonas*, like many other protists, is also phagotrophic. They are probably the dominant bacterivore in the Arctic Ocean, where studies have suggested that climate change is leading to higher temperatures and input of fresh water is lessening the mixing processes that deliver nutrients from deep to surface water. This seems to be favoring the picoplankton, as their smaller cells are more competitive in acquiring scarce nutrients.

Amoebozoa are important grazers of particle-associated bacteria

A wide range of phylogenetically diverse protists in various eukaryotic groups demonstrate a crawling or amoeboid movement using pseudopodia. Most of the free-living amoebae occur in the supergroup Amoebozoa in the eukaryotic phylogenetic tree (*Figure 6.2*). Hundreds of species have been described, including specifically marine lineages. Planktonic amoebae are sparsely distributed in the open-ocean water column, so their contribution to energy flow via preying on bacteria, microalgae, and other protists within pelagic food webs has largely been overlooked. However, they can reach high densities in more turbid estuarine and coastal waters and are particularly associated with surfaces such as flocs and marine snow particles, where the density can exceed that of the ciliated protists most commonly associated with bacterivory. Their plastic cell shape may facilitate removal of firmly attached bacteria by enabling them to penetrate crevices inaccessible to other grazers. Some bacteria have evolved the ability to survive the killing mechanisms in the phagocytic vacuole and can multiply within host amoebae, forming stable symbiotic or parasitic associations with their host.

Radiolarians and foraminifera have highly diverse morphologies with mineral shells

These two groups are classified as members of the supergroup Rhizaria (*Figure 6.2*). They are characterized by an amoeboid body form, using pseudopodia for locomotion and feeding. The cells are usually less than 1 mm in size, but some species are the largest protists known, with diameters up to several cm.

Radiolarians have existed since the beginning of the Paleozoic era about 600 MYA and produce a great diversity of beautiful, intricate shapes (*Figures 6.1, 6.10A*) used in identification of more than 4000 species. Cells are typically 0.1–0.2 mm. Two main groups known as the Polycystina and Acantharea are recognized. They are characterized by stiff, needle-like pseudopodia arranged in radial symmetry (from which they derive the name radiolarian) and internal skeletons made of silica. Larger species are often associated with surfaces and may contain algal symbionts that provide some nutrients to the cell, while the smaller types occur throughout the water column and in deep-sea sediments. Densities vary greatly, ranging from 10^4 cells cm^{-3} in some parts of the subtropical Pacific Ocean to less than 10 cells cm^{-3} in the Sargasso Sea. The silica skeletons of the polycystine radiolarians deposit as microfossils and are second only to diatoms as a source of silica in sediments. The cell body consists of a central mass of cytoplasm surrounded by a capsular wall. This contains pores through which the cytoplasm extrudes into an extracapsular

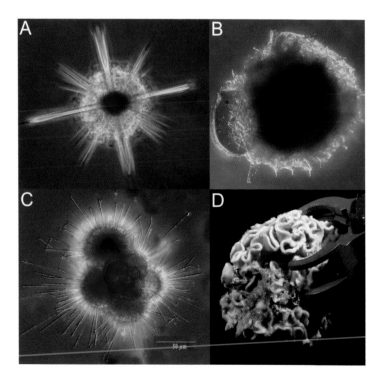

Figure 6.10 A. The radiolarian *Astrolithium cruciatum*. B. A radiolarian (unidentified) ingesting a tintinnid *Proplectella*. C. A globigerinid foraminiferan. D. A large (~20 cm) xenophyophore sampled from the seabed of the Atlantic Ocean, Mid-Atlantic Ridge. Credits: A–C. John R. Dolan, NOAA Fisheries Collection, via Flickr CC-BY-2.0. D. NOAA Photo Library; IFE, URI-IAO, UW, Lost City Science Expedition, 2005.

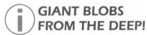

**GIANT BLOBS
FROM THE DEEP!**

A 2011 video exploration of the
Marianas Trench using free-falling
drop cameras led to identification of giant xenophyophores at a
depth of 10.6 km. The discovery
attracted much interest in the
popular press, who reported the
discovery with headlines implying
it was a new life form—reminiscent of 1950s horror movies. In
fact, xenophyophores were first
described in 1883 and are the
dominant megafaunal organism
on much of the deep ocean floor.
New research has been stimulated
by the need for baseline studies
during exploration of the eastern
equatorial Pacific, which includes
a large area being investigated by
deep-sea mining companies for
exploitation of metal nodules on
the seafloor. They are extremely
fragile and difficult to study, but
Gooday et al. (2018) examined the
internal structure of six species of
xenophyophores obtained from a
depth of ~4 km using non-destruc-
tive micro-CT 3D imaging. This
showed that the great majority of
the organisms' structure is composed of masses of waste material
and less than 5% of the "body" is
living cytoplasm, organized in a
network of organic tubes.

cytoplasm and forms the stiffened pseudopodia. The cytoplasm moves by streaming and
captures other protists and small zooplankton, which are then surrounded and digested in
food vacuoles. Hydrated, polymerized silicon dioxide (opal) is deposited within a framework of the cytoplasm. Reproduction occurs asexually by binary fission or sexually via
the production of haploid gametes. The Acantharea produce skeletons of strontium sulfate
(celestite) rather than silica.

The Foraminifera is a phylum or class of the Rhizaria whose members form a multi-cham-
bered shell within a cytoplasmic envelope produced by pseudopodia; the shapes of these are
highly distinctive for species identification (*Figure 6.10C*). The chambers are connected by
openings—from which forms derive their name from the Latin for "hole-bearers"—and have
sealed pores that face the external environment, through which cytoplasmic spines stretch
for long distances to form a net for the capture of prey, which can include bacteria, phytoplankton, and small metazoan animals. Cells are usually <1 mm in size, but larger species,
including multinuclear forms known as xenophyophores up to 20 cm across, occur in the
deep sea (*Figure 6.10D*). These have been found to be abundant on the sea floor of all ocean
basins, especially in submarine trenches on seamounts beneath highly productive waters,
where they can reach densities of 20 organisms m^{-2}. They appear to feed by scooping up
sediment particles using secreted mucus threads. These build up an agglutination of waste
material around the cells that traps particles, from which the group gets its name (Greek:
"bearer of foreign bodies"). The larger types are often found in association with animals such
as isopods, worms, and echinoderms and are thought to provide habitat structure promoting
"hotspots" of deep-sea diversity.

About 5×10^4 species of foraminifera have been documented, of which about four-fifths are
fossils. Many forams contain other protists as endosymbionts, including green algae, red
algae, diatoms, and dinoflagellates. Foraminifera make up only a very small component
of the plankton, but their discarded skeletons are the major source of calcareous deposits, accumulating as the "globigerian ooze" over vast areas of the ocean floor. Massive
deposits accumulated during the tertiary period, about 230 MYA, and these have become
uplifted over time and exposed as limestone beds in Europe, Asia, and Africa. The main
application of the study of the many diverse forms of foraminifera is in biostratigraphy,
a branch of geology which assigns relative ages of rock strata by recording fossil assemblages. This technique is widely used in oil exploration and drilling projects and for recon-
structing the history of oceanic conditions during geological eras (paleoceanography and
paleoclimatology).

MARINE FUNGI

The Fungi form a distinct monophyletic group on a branch within the Nucletmycea

The Fungi are a large and very diverse group of organisms ranging from microscopic forms
(chytrids, molds, and yeasts) to large hyphal networks. Many terrestrial fungi are familiar,
as mushrooms and toadstools formed as reproductive structures, but only the microscopic
forms occur in marine environments. In the phylogenetic tree of eukaryotes (*Figure 6.2*),
the fungi form a distinct monophyletic group on a branch within the Nucletmycea, part of
the Opisthokonta supergroup. They are a sister group to animal cells, which are believed to
have diverged ~1.5 BYA. The classification of fungi has traditionally been based on morphological features, especially the structure and properties of their systems for reproducing
asexually and sexually by the production of spores (sporulation). Most marine fungi belong
to three of the well-known major fungal phyla, namely Ascomycota, Basidiomycota, and
Chytridiomycota. The use of genetic analysis based on rRNA gene sequences in both the
small and large subunits, as well as the intergenic transcribed spacer (ITS) has led to a
revised classification of the Fungi into 18 phyla, although there is still considerable uncertainty about the placement of certain taxa within these divisions and novel marine taxa have
been identified.

Fungi are heterotrophic osmotrophs, feeding by the secretion of extracellular enzymes and absorbing the products of digestion of macromolecules. The majority are saprotophs, which decompose complex organic materials in the environment. Others are important parasites of algae and other protists, plants, and animals. A distinctive characteristic is the possession of a thick, tough cell wall composed of chitin, a polymer of N-acetyl glucosamine. Many fungi are filamentous, forming a multicellular network of hyphae that extends by growth at the tip. The hyphae may have cross-walls (septa) separating the individual cells, but in some species the filament is coenocytic, containing numerous nuclei formed by division without formation of septa. On aerial surfaces, the hyphae ramify over and into the substratum forming a network of branched filaments (mycelium) and produce branches bearing asexual spores (conidia). They may also form sexual spores in macroscopic fruiting bodies. Some of these distinctive reproductive structures are seen in marine fungi when they colonize surfaces such as submerged wood or vegetative litter, but very little is known about the organization of hyphae and reproductive structures in fungi in the water column. Alternatively, many species are unicellular (yeasts) and reproduce by budding cell division.

The Chytridiomycota (commonly referred to as chytrids) are characterized by the production of motile reproductive spores (zoospores) produced within a sac-like structure (their name is derived from the Greek for "little pot"). These zoospores swim rapidly using a posterior whiplash flagellum, and are rapidly dispersed in water and moist habitats, explaining their abundance in aquatic systems. Chytrids are mostly unicellular or may form colonies with short hyphae. The Basidiomycota are mostly filamentous forms, although some yeasts are common in marine habitats. They are named for their distinctive sexual reproductive structure—the basidium, meaning "little pedestal"—with external spores formed by meiosis. The largest phylum of the Fungi is Ascomycota, again named for their distinctive cup like sac, the ascus (meaning "sac" or "wineskin") that holds the sexual spores. In all groups, some species reproduce asexually.

Fungi are increasingly recognized to be major components of the marine microbiome

There are approximately 1.2×10^5 known species of fungi, most of which have been considered as being primarily terrestrial organisms. Until recently, their importance in marine systems had not been fully recognized and they have not received the depth of study afforded to other groups of marine microbes. Nevertheless, a small group of researchers dedicated to study of the fungi has cataloged the diversity of marine fungi over 150 years using painstaking culture methods and morphological examination, especially from coastal habitats such as salt marshes and tropical mangroves. Sea foam has also been a particularly rich source of inocula for marine mycologists, as fungal spores and hyphal fragments become aggregated and trapped in air bubbles in the polymeric matrix of foam formed by wind and wave action. *Figure 6.11* shows an example of some of the diverse fungal colonies obtained from a single sample. The advent of molecular identification techniques has improved our knowledge of marine fungi greatly. However, the methods employed so successfully in global surveys to assess diversity and distribution of bacteria and archaea have had some limitations when applied to fungi. Nevertheless, the importance of marine fungi is increasingly recognized for their role in nutrient cycling, as parasites of marine organisms, and as a source of natural products. Over 1100 marine species have now been described, which probably represents <1% of the true diversity. When concerted efforts using modern techniques are made to discover fungi in marine habitats, it becomes obvious that they occur almost everywhere we look. Examples of common genera are shown in *Table 6.3*. Many of these predominant genera were already familiar from soil, plant, and freshwater mycology studies. However, the use of HTS in marine samples has revealed many novel sequences and previously unknown taxa, especially those belonging to the Rozellomycota and Chytridiomycota, which are early divergent branches of the Fungi. Habitats range from the surface of coastal and ocean waters to the deep sea and subsurface sediments, from polar to tropical

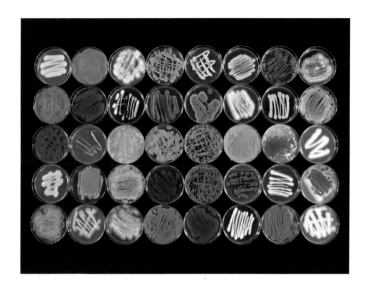

Figure 6.11 Morphological diversity of a collection of fungi isolated from a marine sponge, *Ircinia variabilis*, growing on agar plates. Reprinted from Amend et al. (2019), CC-BY-4.0.

Table 6.3 Predominant fungi in marine habitats determined by cultivation and cultivation-independent methods. (Data from Grossart et al., 2019)

Habitat	Dominant fungal group	Common genera (phylum, class in parentheses)
Deep sea	Filamentous fungi and yeasts	*Aspergillus, Cladosporium, Penicillium* (Ascomycota, Eurotiomycetes) *Mycosphaerella, Alternaria. Aureobasidium* (Ascomycota, Dothidiomycetes) *Cadophora, Candida, Pichia* (Ascomycota, Saccharomycetes) *Cryptococcus* (Basidiomycota, Tremellomycetes) *Rhodotorula, Rhodosporidium* (Basidiomycota, Ustilaginomycetes) *Malassezia* (Basidiomycota, Malasseziomycetes) *Ganoderma* (Basidiomycota, Agaricomycetes),
Sub-sea floor	Yeasts and filamentous fungi	*Exophiala* (Ascomycota, Eurotiomycetes) *Candida* (Ascomycota, Saccharomycetes) *Cryptococcus, Trichosporon* (Basidiomycota, Tremellomycetes) *Rhodotorula, Rhodosporidium* (Basidiomycota, Ustilaginomycetes) *Malassezia* (Basidiomycota, Malasseziomycetes)
Hydrothermal vents	Yeasts and chytridiomycetes	*Aureobasidium* (Ascomycota, Dothidiomycetes) *Malassezia* (Basidiomycota, Malasseziomycetes) *Rhodotorula* (Basidiomycota, Ustilaginmycetes) *Exophiala* (Ascomycota, Eurotiomycetes)
Coastal and ocean water	Ascomycota, Chytridiomycota, Basidiomycota and Rozellomycota	*Candida, Debaryomyces* (Ascomycota, Saccharomycetes) *Cryptococcus, Trichosporon* (Basidiomycota, Tremellomycetes) *Rhodotorula, Rhodosporidium* (Basidiomycota, Ustilaginomycetes) *Malassezia* (Basidiomycota, Malasseziomycetes) *Phaeosphaeria* (Ascomycota, Dothidiomycetes) *Aspergillus, Cladosporium* (Ascomycota, Eurotiomycetes) *Chytridium, Rhizophydium, Lobulomyces, Spizellomyces* (Chytridiomycota, Chytridiomycetes) Clade LKM11 (Rozellomycota)
Polar systems	Chytridiomycota, Rozellomycota, Basidiomycota and Ascomycota	*Rhizophydium, Lobulomyces, Podochytrium, Rhizoclosmatium, Cladochytrium, Cyclopsomyces, Mesochytrium, Polychytrium* (Chytridiomycota, Chytridiomycetes) Clade LKM11 (Rozellomycota) *Glaciozyma, Sporobolomyces* (Basidiomycota, Microbotryomycetes) *Mrakia* (Basidiomycota, Tremellomycetes) *Cadophora, Kluyveromyces, Candida* (Ascomycota, Saccharomycetes) *Cladosporium, Penicillium* (Ascomycota, Eurotiomycetes) *Rhodotorula* (Basidiomycota, Ustilaginomycetes) *Epicoccum, Aureobasidium* (Ascomycota, Dothidiomycetes) *Basidiobolus* (Zoopagomycota, Basidiobolomycetes)

regions, on surfaces and particles of organic material, and in association with a wide range of other organisms.

A major challenge has been defining fungal species that are truly marine. Many fungi associated with freshwater and terrestrial habitats like soil and plants can grow in seawater, for example, when decomposing detritus finds its way into the sea from coastal runoff or rivers. Terrestrial fungi also produce massive numbers of aerial spores, which can be blown into the sea. Indeed, many of the fungal species (especially ascomycetes and basidiomycetes) identified in marine samples are often well characterized from soil and plant sources. Such "accidental" introductions would not identify an isolate as "marine," although physiological and gene expression studies show that many of these fungi are equally at home on land and in the sea and might be considered as amphibious or aero-aquatic. In laboratory experiments, most fungi can grow at NaCl concentrations found in seawater since they have efficient systems for controlling cell turgor and function in hypertonic environment. Phylogenetic analysis shows that many marine fungi group with lineages from terrestrial habitats, suggesting that there have been repeated transitions from terrestrial to marine habitats. To be recognized as truly marine, a fungus must rely on the marine habitat for part or all of its life cycle. We need to demonstrate that it is regularly associated with marine habitats and is able to grow and reproduce (sporulate) in the sea. Further evidence of this comes from demonstration of *in situ* transcription and metabolic activity that can be directly associated with the fungus and from genomic evidence of adaptations to the marine environment (or to hosts, for symbionts and parasites). There is now sufficient evidence to conclude that fungi are a significant component of the marine microbiome, but despite growing evidence of the diversity of fungal species in marine samples, reliable estimates of their abundance, contribution to metabolic processes, and ecosystem functions remain limited. However, improved methodology and coordinated efforts by the research community are leading to rapid advances, some of which are discussed in *Box 6.2*.

Besides their activities in natural processes, fungi have important roles in anthropogenic pollution. In particular, their ability to degrade complex hydrocarbons means that they occur at natural oil seeps and show increased activity after accidental oil spills (*Box 13.2*). They are also involved in the colonization of plastic debris in the oceans; this can result in slow degradation or alteration of the physical properties of the plastic, which can have important influences on its long-term fate and effects on marine processes (*Box 13.3*).

 FROM DANDRUFF TO DEEP-SEA VENTS

Fungi in the genus *Malassezia* have been known for many years to be associated with human skin conditions such as dandruff and eczema. These fungi lack ability to synthesize fatty acids and use extracellular lipases to obtain these. It was thought that *Malassezia* spp. had evolved into a narrow niche associated with the skin of mammalian hosts but sequencing of environmental DNA samples from diverse environments has revealed that they are extremely widespread and hyper-diverse. As well as occurring in plants and soils, sequences related to *Malassezia* have been detected in deep-sea sediments, hydrothermal vents, corals, and other invertebrates (Amend, 2014). An obvious question facing researchers is whether these marine sequences might have arisen from contamination of samples by aerial skin flakes from the experimenters, but this was ruled out by the detection of distinctive sequences of *Malassezia* RNA (which degrades very rapidly) and by using stringent protocols to exclude contamination (Orsi et al., 2013). Amend (2014) concluded that members of the *Malassezia* lineage are among the most widespread fungi on the planet.

Conclusions

This chapter has shown that eukaryotic microbes are extremely diverse and have critical roles in ocean processes, as well as providing important insights to the evolutionary history of our oceans. Phototrophic protists are major contributors to primary productivity, while phagotrophic predators of other microbes and saprotrophic fungi provide conduits for the flow of nutrients into marine food webs. One of the most significant discoveries of recent years is the large extent to which many protists mix these nutritional strategies, with profound influences on ecology. Recent advances in methods for the application of molecular techniques to the study of protists and fungi is leading to greatly improved knowledge of their diversity and functional roles in ocean processes. As well as their role in food webs and biogeochemical cycles, the activities of protists and fungi have many significant implications for human welfare, both negative (e.g. diseases of marine life or harmful blooms) and positive (for biotechnological exploitation).

BOX 6.2   RESEARCH FOCUS

Exciting times for marine mycology

Chytrid parasites promote trophic links in aquatic food webs. Numerous studies have shown that marine fungi are associated with a variety of different organisms, including protists (Gutiérrez et al., 2016); macroalgae (Wainwright et al., 2017); and corals, sponges, and other invertebrates (Paz et al., 2010; Yarden, 2014; Ainsworth et al., 2017). Their role as parasites and pathogens is an area of active exploration. Although there are many examples in macroecology of the impact of infectious agents such as viruses on community structure and diversity due to mass mortalities, parasites are frequently missing or under-represented in food web models. Lafferty et al. (2008) provides examples of various ways in which parasites should be fitted into food webs and how they affect their topology and dynamics.

The association of chytrids with phytoplankton is an area of special interest, because chytrids are dominant obligate or facultative parasites in aquatic systems. They infect a variety of algae, other protists, and cyanobacteria (Frenken et al., 2017). Kagami et al. (2014) investigated the role of chytrids in aquatic food webs, particularly in the transfer of material from poorly grazed phytoplankton to zooplankton—a component of the food web that they termed the "mycoloop." When chytrid zoospores infect their host, they penetrate the cell and feed, before forming numerous motile zoospores. Agha et al. (2016) conducted experiments with freshwater cyanobacteria and the zooplankter *Daphnia*. In four out of five fitness parameters, *Daphnia* performed better when feeding on chytrid-parasitized filamentous cyanobacteria than in the absence of infection. The zoospores released from the infected phytoplankton provide a highly nutritional food source for the zooplankton. Also, chytrid infection breaks up the cyanobacterial filaments, which makes them more edible for *Daphnia* and the death and decay of the cyanobacteria increases the biomass of heterotrophic bacteria, which provide an additional food source. Such additional trophic links between primary and secondary production processes may be especially important when certain phytoplankton species bloom. In another recent study, Frenken et al. (2018) showed that rotifers declined in cultures of a large filamentous blooming cyanobacterium because they could not ingest them, but they grew rapidly if a chytrid parasite that infected the cyanobacteria was present. Again, this was due to the zooplankton feeding on the lipid-rich chytrid zoospores. To date, these experimental studies have been conducted

in freshwater systems. Model experimental systems to investigate interactions between marine fungal parasites, phytoplankton hosts, and zooplankton predators are now being developed.

Nevertheless, the results of several *in situ* surveys suggest that parasitism by marine chytrids does have similar food web effects. A major insight was obtained by Gutiérrez et al. (2016), who investigated the occurrence of infection of diatoms in the Humboldt current system off central Chile. This is one of the most productive marine ecosystems due to upwelling of nutrients, which sustains high primary production, mainly by blooms of large diatoms such as *Skeletonema* and *Thalassiosira*. Infection by chytrid parasites was observed using SEM and EFM, revealing attached spherical structures (sporangia) formed from recently settled zoospores, and branched rhizoid structures penetrating the diatom cell (*Figure 6.12A*). Measurements of abundance of these diatoms showed that they increased during the upwelling of cold, low-oxygen waters during the austral spring, peaking in November at ~5×10^4 cells per liter. Abundance of chytrids followed the same seasonal pattern, ranging from 2–391 per liter for attached, and from 111–395 per liter for detached sporangia. The ratio of attached to detached sporangia and the number of diatoms both showed a sharp decline at the beginning of summer. This coincided with an increase in the abundance of *Chaetoceros*, which became the dominant diatom at the start of summer. The decline of diatom blooms in upwelling systems is usually attributed to depletion of nutrients and grazing, but Gutiérrez et al. concluded that specificity of infection by chytrids partly accounts for the decline of particular diatoms. They calculated the biomass of detached zoospores at 3.7–10.1 µg C L^{-1} and suggest that these provide a significant conduit of energy within the pelagic trophic food web in this marine upwelling system, transferring organic carbon to zooplankton, which are inhibited from grazing on *Skeletonema* and *Thalassiosira* due to their production of toxic aldehydes.

Diatoms are also the most prominent and diverse component of algal blooms in polar sea ice systems (Gradinger et al., 2010). Hassett and Gradinger (2016) provided the first evidence linking diatom populations with parasitism by Chytridiomycota, which increased under higher light penetration due to reduced snow cover on the ice. The spatially constrained labyrinthine brine channels in sea ice

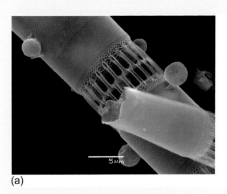
(a)

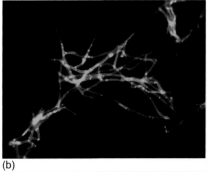
(b)

Figure 6.12 Fungi detected in the coastal upwelling ecosystem off Chile. A. SEM of chytrid sporangia attached to the diatom *Skeletonema* sp., showing penetration of filament by rhizoids. B. Epifluorescence microscopy of septate hyphae detected by Calcofluor staining. Credit: courtesy of Marcelo Gutiérrez, Universidad de Concepción.

BOX 6.2   RESEARCH FOCUS

(p.22) are thought to be a major reservoir of fungal diversity, and to investigate this, Hassett et al. (2017) enumerated chytrids in sea ice across the Western Arctic using microscopic counts. They also identified fungal diversity using HTS targeting the 18S rRNA gene, supplemented by deep sequencing of the 28S rRNA gene. Analysis of over 7 million unique sequences revealed a broad range of diverse Ascomycota, Basidiomycota, and Chytridiomycota. These Arctic chytrids were mainly assigned to the new order Lobulomycetales. In areas with light penetration due to low snow cover, chytrids were observed to be actively parasitizing large pennate diatoms, with the species *Pleurosigma elongatum* showing infection rates of ~25%, suggesting a major role in trophic transfer. Further studies by this group used novel methods of assessing the biomass of chytrid fungi in sea ice and seawater, by integrating information from CARD-FISH analysis using Chytridiomycota-specific probes with the concentrations of ergosterol, a major component of fungal membranes (Hassett et al., 2019). This confirmed that the biomass of chytrids in sea ice and seawater is similar to the biomass estimates of marine taxa considered necessary to support marine food webs and ecosystem functions.

The role of fungi in marine carbon cycling. Estimates of biomass composition in other marine habitats also confirms that the importance of fungi has been overlooked. Gutiérrez et al. (2010) used 0.22 µm membrane filtration to capture particles from seawater and stained them with a chitin stain (Calcofluor) before examining with epifluorescence microscopy (*Figure 6.12B*). This showed numerous hyphae and fungal aggregates, accounting for 0.03–0.12 ug C L^{-1} in the top 15 m of the water column in the Humboldt system, decreasing with depth. These authors also showed seasonal variability in fungal carbon associated with changes in phytoplankton biomass—fungal biomass was comparable to that of bacteria during upwelling (Gutiérrez et al., 2011). Mycelial growth increases the surface to volume ratio and gives fungi the possibility to mobilize intracellular content and substrates in heterogeneous environments. Hyphae also present a microhabitat for other microorganisms. Bochdansky et al. (2017) used gentle gravity filtration to collect slow-sinking marine snow from bathypelagic depths of ~1000–3000 m and used CARD-FISH with various eukaryotic primers to assess abundance of different groups. They found that approximately one fungal cell occurred for every 1000 bacterial or archaeal cells (revealed by DAPI epifluorescence). By extrapolation, this puts fungal biomass on a par with that of bacteria or archaea and suggests that fungi have a previously overlooked significant role in degradation of organic matter in the deep sea, especially because of their propensity to degrade complex organic compounds. Bochdansky et al. also suggest that penetration of the marine snow by hyphal filaments may help to stabilize these particles, facilitating their transport to deep waters. In a HTS study of the vertical distribution of eukaryotic microbes from the surface to the hadal zone of the Marianas Trench, Xu et al. (2018) found that fungal sequences were most abundant at 1759 m.

Further evidence that fungi form a major component of marine food webs comes from molecular dietary analysis of zooplankton using amplification of DNA with eukaryotic primers and blocking techniques that reveal the types of prey ingested. Maloy et al. (2013) showed that fungi constituted 16% of the diet of bivalve larvae and Hu et al. (2015) showed the presence of several genera of Ascomycota and Basidiomycota in coral reef copepods (although chytrids were not found in these studies).

Annual seasonal increases in abundance of chytrids linked to diatom blooms were also observed in a HTS- and qPCR-based study over six years of seasonal variation in coastal planktonic fungi at the Western English Channel Station L4, off the coast of Plymouth, conducted by Taylor and Cunliffe (2016). Assemblages of chytrids, and the more dominant Ascomycota and Basidiomycota, were dynamic and linked to fluctuations in particulate organic carbon and nitrogen, temperature, and salinity (linked to input from rivers). Although the method used to determine abundance (Q-PCR of the 18S rRNA gene) could not be used to produce carbon biomass estimates, the rapid changes in abundance again suggest that fungi are responsible for significant carbon turnover.

Most fungi are osmotrophic saprotrophs feeding by absorbing DOC generated by the action of the extracellular enzymes that they produce. Gutiérrez et al. (2011) had previously shown that abundance of mycoplankton in the Humboldt current system was positively correlated with processing of high molecular-weight polymers. In a follow-up to their study of the Western English Channel fungi, Cunliffe et al. (2017) sought to characterize mycoplankton that actively utilize algal-derived polysaccharides. They combined stable isotope probing (SIP) with analysis of 18S rRNA genes in L4 seawater incubated with ^{12}C and ^{13}C-labeled polysaccharide microgels produced from cultures of the diatom *Phaeodactylum tricornutum*. Comparison of the gene libraries from the ^{12}C (control) and ^{13}C incubations showed that the latter were enriched in fungal sequences relative to other plankton groups, indicating that they had assimilated carbon from the polysaccharide microgels. Dominant OTUs belonged to the genera *Malassezia* and *Cladosporium*. Cultures of *Cladosporium* F2 were isolated on agar plates containing a glucan identical to the algal polysaccharide laminarin and shown to possess strong β-glucosidase activity. Analysis of the 6-year dataset of fungal diversity from the L4 site showed that abundance of specific *Cladosporium*-related OTUs correlated with blooms of either *Leptocylindricus* or *Chaetoceros* diatoms, both of which are known TEP producers. Cunliffe et al. (2017) also analyzed the *Tara Oceans* dataset and found that *Cladosporium* OTUs accounted for 26% of the mycoplankton communities, almost always coinciding with high phytoplankton biomass.

Conclusions. These examples of research indicate that fungi are an abundant and highly active component of many marine habitats. In a review of a "state of the art" discussion of marine mycology, Amend et al. (2019) concluded: "We are in exciting times for marine fungal functional biology and ecology, and studies of these communities will very likely force us to rethink global biogeochemical cycles."

References and further reading

Protist—diversity, and systematics

Adl, S.M., Bass, D., Lane, C.E. et al. (2019) Revisions to the classification, nomenclature, and diversity of eukaryotes. *J. Eukaryot. Microbiol.* **66**: 4–119.

Ainsworth, T.D., Fordyce, A.J., & Camp, E.F. (2017) The other microeukaryotes of the coral reef microbiome. Trends Microbiol. 25: 980–991.

Caron, D.A., Countway, P.D., Jones, A.C. et al. (2012) Marine protistan diversity. *Annu. Rev. Mar. Sci.* **4**: 467–493.

de Vargas, C., Audic, S., Henry, N. et al. (2015) Ocean plankton. Eukaryotic plankton diversity in the sunlit ocean. *Science* **348**: 1261605.

Gast, R.J., Sanders, R.W. & Caron, D.A. (2009) Ecological strategies of protists and their symbiotic relationships with prokaryotic microbes. *Trends Microbiol.* **17**: 563–569.

Keeling, P.J. & Campo, J. Del (2017) Marine protists are not just big bacteria. *Curr. Biol.* **27**: R541–R549.

Lima-Mendez, G., Faust, K., Henry, N. et al. (2015) Determinants of community structure in the global plankton interactome. *Science* **348**: 1262073.

Millette, N.C., Grosse, J., Johnson, W.M. et al. (2018) Hidden in plain sight: The importance of cryptic interactions in marine plankton. *Limnol. Oceanogr. Lett.* **3**: 341–356.

Moreira, D. & López-García, P. (2002) The molecular ecology of microbial eukaryotes unveils a hidden world. *Trends Microbiol.* **10**: 31–38.

Moreira, D. & López-García, P. (2003) Are hydrothermal vents oases for parasitic protists? *Trends Parasitol.* **19**: 556–558.

Nissimov, J.I. & Bidle, K.D. (2017) Stress, death, and the biological glue of sinking matter. *J. Phycol.* **53**: 241–244.

Rocke, E., Pachiadaki, M.G., Cobban, A. et al. (2015) Protist community grazing on prokaryotic prey in deep ocean water masses. *PLoS One* **10**: e0124505.

Sherr, B.F., Sherr, E.B., Caron, D.A. et al. (2007) Oceanic protists. *Oceanography* **20**: 130–134.

Stoecker, D.K., Johnson, M.D., de Vargas, C., & Not, F. (2009) Acquired phototrophy in aquatic protists. *Aquat. Microb. Ecol.* **57**: 279–310.

Strassert, J.F., Karnkowska, A., Hehenberger, E. et al. (2018) Single cell genomics of uncultured marine alveolates shows paraphyly of basal dinoflagellates. *ISME J.* **12**: 304.

Worden, A.Z. & Not, F. (2008) Ecology and diversity of picoeukaryotes. In: *Microbial Ecology of the Oceans*, D.L. Kirchman (ed.). Wiley-Liss Inc., New York, pp. 159–205.

Heterotrophic flagellates and grazing

Boenigk, J. & Arndt, H. (2002) Bacterivory by heterotrophic flagellates: Community structure and feeding strategies. *Ant. van Leuevenhoek* **81**: 465–480.

De Corte, D., Paredes, G., Yokokawa, T. et al. (2018) Differential response of *Cafeteria roenbergensis* to different bacterial and archaeal prey characteristics. *Microb. Ecol.* **78**: 1–5.

del Campo, J., Not, F., Forn, I. et al. (2013) Taming the smallest predators of the oceans. *ISME J.* **7**: 351–358.

Fenchel, T. & Patterson, D.J. (1988) *Cafeteria roenbergensis* nov. gen., nov. sp., a heterotrophic microflagellate from marine plankton. *Mar. Microb. Food Webs* **3**: 9–19.

Gawryluk, R.M., del Campo, J., Okamoto, N. et al. (2016) Morphological identification and single-cell genomics of marine diplonemids. *Curr. Biol.* **26**: 3053–3059.

Gimmler, A., Korn, R., de Vargas, C. et al. (2016) The *Tara* Oceans voyage reveals global diversity and distribution patterns of marine planktonic ciliates. *Sci. Rep.* **6**: 33555.

González, J.M. & Suttle, C.A. (1993) Grazing by marine nanoflagellates on viruses and virus-sized particles: Ingestion and digestion. *Mar. Ecol. Prog. Ser.* **94**: 1–10.

Jezbera, J., Hornak, K., & Simek, K. (2005) Food selection by bacterivorous protists: Insight from the analysis of the food vacuole content by means of fluorescence in situ hybridization. *FEMS Microbiol. Ecol.* **52**: 351–363.

Johnson, M.D., Lasek-Nesselquist, E., Moeller, H.V. et al. (2017) *Mesodinium rubrum*: The symbiosis that wasn't. *Proc. Natl. Acad. Sci. USA* **114**: E1040–E1042.

Laundon, D., Larson, B.T., McDonald, K. et al. (2019) The architecture of cell differentiation in choanoflagellates and sponge choanocytes. *PLOS Biol.* **17**: e3000226.

Medina, L.E., Taylor, C.D., Pachiadaki, M.G. et al. (2017) A review of protist grazing below the photic zone emphasizing studies of oxygen-depleted water columns and recent applications of in situ approaches. *Front. Mar. Sci.* **4**: 105.

Pernice, M.C., Forn, I., Gomes, A. et al. (2015) Global abundance of planktonic heterotrophic protists in the deep ocean. *ISME J.* **9**: 782–792.

Seeleuthner, Y., Mondy, S., Lombard, V. et al. (2018) Single-cell genomics of multiple uncultured stramenopiles reveals underestimated functional diversity across oceans. *Nat. Commun.* **9**: 310.

Sengupta, A., Carrara, F. & Stocker, R. (2017) Phytoplankton can actively diversify their migration strategy in response to turbulent cues. *Nature* **543**: 555–558.

Sieracki, M.E., Poulton, N.J., Jaillon, O. et al. (2019) Single cell genomics yields a wide diversity of small planktonic protists across major ocean ecosystems. *Sci. Rep.* **9**: 6025.

Mixotrophy

Mcmanus, G.B., Schoener, D., & Haberlandt, K. (2012) Chloroplast symbiosis in a marine ciliate: Ecophysiology and the risks and rewards of hosting foreign organelles. *Front Microbiol* **3**: 321.

Qiu, D., Huang, L., & Lin, S. (2016) Cryptophyte farming by symbiotic ciliate host detected in situ. *Proc. Natl. Acad. Sci.* **113**: 12208–12213.

Stoecker, D.K., Hansen, P.J., Caron, D.A., & Mitra, A. (2017) Mixotrophy in the Marine Plankton. *Annu. Rev. Mar. Sci.* **9**: 311–335.

Stoecker, D.K., Johnson, M.D., de Vargas, C., & Not, F. (2009) Acquired phototrophy in aquatic protists. *Aquat. Microb. Ecol.* **57**: 279–310.

Ward, B.A. (2019) Mixotroph ecology: More than the sum of its parts. *Proc. Natl. Acad. Sci. USA* **116**: 5846–5848.

Coccolithophores

Balch, W.M. (2018) The ecology, biogeochemistry, and optical properties of coccolithophores. *Annu. Rev. Mar. Sci.* **10**: 71–98.

Monteiro, F.M., Bach, L.T., Brownlee, C. et al. (2016) Why marine phytoplankton calcify. *Sci. Adv.* **2**: e1501822.

Taylor, A.R., Brownlee, C., & Wheeler, G. (2017) Coccolithophore cell biology: Chalking up progress. *Annu. Rev. Mar. Sci.* **9**: 283–310.

Xu, K., Hutchins, D., & Gao, K. (2018) Coccolith arrangement follows Eulerian mathematics in the coccolithophore *Emiliania huxleyi*. *Peer J.* **6**: e4608.

Diatoms

Bidle, K.D. (2002) Regulation of oceanic silicon and carbon preservation by temperature control on bacteria. *Science* **298**: 1980–1984.

Bidle, K.D. & Azam, F. (1999) Accelerated dissolution of diatom silica by marine bacterial assemblages. *Nature* **397**: 508–512.

Bidle, K.D. & Azam, F. (2001) Bacterial control of silicon regeneration from diatom detritus: Significance of bacterial ectohydrolases and species identity. *Limnol. Oceanogr.* **46**: 1606–1623.

Gradinger, R., Bluhm, B.A., Hopcroft, R.R. et al. (2010) Marine life in the arctic. In: *Life in the World's Oceans*. Wiley-Blackwell, Oxford, UK, pp. 183–202.

Kröger, N. & Poulsen, N. (2008) Diatoms—From cell wall biogenesis to nanotechnology. *Annu. Rev. Genet.* **42**: 83–107.

Leblanc, K., Queguiner, B., Diaz, F. et al. (2018) Nanoplanktonic diatoms are globally overlooked but play a role in spring blooms and carbon export. *Nat. Commun.* **9**: 953.

Dinoflagellates

Akbar, M., Ahmad, A., Usup, G., & Bunawan, H. (2018) Current knowledge and recent advances in marine dinoflagellate transcriptomic research. *J. Mar. Sci. Eng.* **6**: 13.

Figueroa, R.I., Estrada, M., & Garcés, E. (2018) Life histories of microalgal species causing harmful blooms: Haploids, diploids and the relevance of benthic stages. *Harmful Algae* **73**: 44–57.

Gómez, F. (2012) A quantitative review of the lifestyle, habitat and trophic diversity of dinoflagellates (Dinoflagellata, Alveolata). *System. Biodiv.* **10**: 267–275.

Grossart, H.P., Levold, F., Allgaier, M. et al. (2005) Marine diatom species harbour distinct bacterial communities. *Environ. Microbiol.* **7**: 860–873.

Haddock, S.H.D., Moline, M.A. & Case, J.F. (2010) Bioluminescence in the sea. *Annu. Rev. Mar. Sci.* **2**: 443–493.

Hastings, J.W. (2007) The *Gonyaulax* clock at 50: Translational control of circadian expression. Cold Spring Harbor *Symp.* Quant. Biol. 72: 141–144.

Liu, H., Stephens, T.G., González-Pech, R.A. et al. (2018) Symbiodinium genomes reveal adaptive evolution of functions related to coral-dinoflagellate symbiosis. *Commun. Biol.* **1**: 95.

Valiadi, M. & Iglesias-Rodriguez, D. (2013) Understanding bioluminescence in dinoflagellates—How far have we come? *Microorganism* **1**: 3–25.

Wisecaver, J.H. & Hackett, J.D. (2011) Dinoflagellate genome evolution. *Annu. Rev. Microbiol.* **65**: 369–387.

Foraminifera and radiolarians

Anderson, O.R. (2001) Protozoa, radiolarians. In: *Encyclopedia of Ocean Sciences*, J. Steele, S. Thorpe & K. Turekian (eds.). Academic Press, pp. 2315–2319.

Gooday, A.J., Sykes, D., Góral, T. et al. (2018) Micro-CT 3D imaging reveals the internal structure of three abyssal xenophyophore species (Protista, Foraminifera) from the eastern equatorial Pacific Ocean. *Sci. Rep.* **8**: 12103.

Lampitt, R.S., Salter, I. & Johns, D. (2009) Radiolaria: Major exporters of organic carbon to the deep ocean. *Glob. Biogeochem. Cycles* **23**: GB1010. doi: 10.1029/2008GB003221

Murray, J. (2008) *Ecology and Applications of Benthic Foraminifera*. Cambridge University Press.

Fungi

Agha, R., Saebelfeld, M., Manthey, C. et al. (2016) Chytrid parasitism facilitates trophic transfer between bloom-forming cyanobacteria and zooplankton (Daphnia). *Sci. Rep.* **6**: 35039.

Amend, A. (2014) From dandruff to deep-sea vents: Malassezia-like Fungi are ecologically hyper-diverse. *PLoS Pathog.* **10**: e1004277.

Amend, A., Burgaud, G., Cunliffe, M. et al. (2019) Fungi in the marine environment: Open questions and unsolved problems. *MBio* **10**: e101189-18.

Bochdansky, A.B., Clouse, M.A. & Herndl, G.J. (2017) Eukaryotic microbes, principally fungi and labyrinthulomycetes, dominate biomass on bathypelagic marine snow. *ISME J.* **11**: 362–373.

Cunliffe, M., Hollingsworth, A., Bain, C. et al. (2017) Algal polysaccharide utilisation by saprotrophic planktonic marine fungi. *Fungal Ecol.* **30**: 135–138.

Frenken, T., Alacid, E., Berger, S.A. et al. (2017) Integrating chytrid fungal parasites into plankton ecology: Research gaps and needs. *Environ. Microbiol.* **19**: 3802–3822.

Frenken, T., Wierenga, J., van Donk, E. et al. (2018) Fungal parasites of a toxic inedible cyanobacterium provide food to zooplankton. *Limnol. Oceanogr*, **63**: 2384–2393.

Golubic, S., Radtke, G. & Campion-Alsumard, T.L. (2005) Endolithic fungi in marine ecosystems. *Trends Microbiol.* **13**: 229–235.

Grossart, H.-P., Van den Wyngaert, S., Kagami, M. et al. (2019) Fungi in aquatic ecosystems. *Nat. Rev. Microbiol.* **17**: 339–354.

Gutiérrez, M.H., Jara, A.M. & Pantoja, S. (2016) Fungal parasites infect marine diatoms in the upwelling ecosystem of the Humboldt current system off central Chile. *Environ. Microbiol.* **18**: 1646–1653.

Gutiérrez, M.H., Pantoja, S., Quiñones, R.A. & González, R.R. (2010) First record of filamentous fungi in the coastal upwelling ecosystem off central Chile. *Gayana* **74**: 66–73.

Gutiérrez, M.H., Pantoja, S., Tejos, E. & Quiñones, R.A. (2011) The role of fungi in processing marine organic matter in the upwelling ecosystem off Chile. *Mar. Biol.* **158**: 205–219.

Hassett, B.T., Borrego, E.J., Vonnahme, T.R. et al. (2019) Arctic marine fungi: Biomass, functional genes, and putative ecological roles. *ISME J.* **13**: 1484.

Hassett, B.T., Ducluzeau, A.-L.L., Collins, R.E. & Gradinger, R. (2017) Spatial distribution of aquatic marine fungi across the western Arctic and sub-arctic. *Environ. Microbiol.* **19**: 475–484.

Hassett, B.T. & Gradinger, R. (2016) Chytrids dominate arctic marine fungal communities. *Environ. Microbiol.* **18**: 2001–2009.

Hu, S., Guo, Z., Li, T. et al. (2015) Molecular analysis of in situ diets of coral reef copepods: Evidence of terrestrial plant detritus as a food source in Sanya Bay, China. *J. Plankton Res.* **37**: 363–371.

Kagami, M., Miki, T. & Takimoto, G. (2014) Mycoloop: Chytrids in aquatic food webs. *Front. Microbiol.* **5**: 166.

Lafferty, K.D., Allesina, S., Arim, M. et al. (2008) Parasites in food webs: The ultimate missing links. *Ecol. Lett.* **11**: 533–46.

Maloy, A.P., Culloty, S.C. & Slater, J.W. (2013) Dietary analysis of small planktonic consumers: A case study with marine bivalve larvae. *J. Plankton Res.* **35**: 866–876.

Nilsson, R.H., Anslan, S., Bahram, M. et al. (2019) Mycobiome diversity: High-throughput sequencing and identification of fungi. *Nat. Rev. Microbiol.* **17**: 95–109.

Orsi, W., Biddle, J.F. & Edgcomb, V. (2013) Deep sequencing of subseafloor eukaryotic rRNA reveals active fungi across marine subsurface provinces. *PLoS One* **8**: e56335.

Paz, Z., Komon-Zelazowska, M., Druzhinina, I.S. et al. (2010) Diversity and potential antifungal properties of fungi associated with a Mediterranean sponge. *Fungal Divers.* **42**: 17–26.

Taylor, J.D. & Cunliffe, M. (2016) Multi-year assessment of coastal planktonic fungi reveals environmental drivers of diversity and abundance. ISME J. **10**: 2118.

Sherwood, A.R., Smith, C.M. & Amend, A.S. (2017) Fungi associated with mesophotic macroalgae from the 'Au'au Channel, west Maui are differentiated by host and overlap terrestrial communities. *Peer J.* **5**: e3532.

Wainwright, B.J., Zahn, G.L., Spalding, H.L., et al. (2017) Fungi associated with mesophotic macroalgae from the 'Au 'au Channel, west Maui are differentiated by host and overlap terrestrial communities. *Peer J.* **5**: e3532.

Xu, Z., Wang, M., Wu, W. et al. (2018) Vertical distribution of microbial eukaryotes from surface to the hadal zone of the Mariana Trench. *Front. Microbiol.* **9**: 2023.

Yarden, O. (2014) Fungal association with sessile marine invertebrates. *Front. Microbiol.* **5**: 228.

Chapter 7

Marine Viruses

The true abundance and ecological importance of viruses in marine ecosystems and global processes has only been elucidated in the last few decades. Recent discoveries have led to the emergence of virus ecology as one of the most exciting and fastest developing branches of marine science. The world's oceans are estimated to contain more than 10^{30} viruses—they are the smallest and most abundant biological entities in marine ecosystems. All forms of cellular life—from bacteria to whales—are susceptible to viral infection. Through their interactions with all types of marine organisms, viruses play a critical role in the structuring of marine communities, in ocean processes, and in biogeochemical cycles. In this chapter, the focus is primarily on the properties and activities of marine viruses with respect to their interactions with other members of the plankton, especially bacteria, microalgae, and other protists. Viruses that infect other marine organisms, such as invertebrates, fish, marine vertebrates, or macroalgae are considered in *Chapter 11*. Human pathogenic viruses that may be introduced into the marine environment via pollution are discussed in *Chapter 12*.

Key Concepts

- Viruses contain either DNA or RNA and are obligate intracellular parasites that take over the biosynthetic machinery of host cells.

- There are more than 10^{30} viruses in the oceans; marine viruses are extremely diverse in host range, size, structure, and genomic composition.

- Viruses that infect bacteria (phages) and other planktonic microbes are responsible for the turnover of microbial communities, with consequent major effects on nutrient cycles and biogeochemical processes.

- Viruses are responsible for structuring the diversity of microbial communities and manipulating life histories through co-evolution with their hosts and widespread genetic exchange.

- Analysis of marine viral metagenomes (viromes) shows that groups of viral genotypes are globally distributed, although their relative abundance is affected by local selection factors.

- Studies of marine viromes are contributing to a reappraisal of the nature and evolution of viruses and cells.

Viruses are highly diverse non-cellular microbes

Viruses are non-cellular biological entities. They cannot be described as microorganisms, but they are included in the more encompassing term microbes. Virus particles, termed "virions," are composed of nucleic acid surrounded by a protein coat (capsid). A fundamental difference in the makeup of viruses compared with cells is that viruses contain only one type of nucleic acid, either DNA or RNA, whereas cells contain both. Also, they do not contain ribosomes and therefore cannot synthesize proteins. Therefore, viruses are obligate intracellular parasites, that rely on the biochemical machinery of the cell they infect to complete the flow of genetic information, synthesize proteins, and replicate their virions. Since all forms of cellular life are susceptible to virus infection, we can reasonably speculate that every type of marine organism—from bacteria to whales—is a host to at least one type of virus. Many organisms are known to be hosts to several different viruses. *Table 7.1* shows a list of representative marine virus families and their hosts; as can be seen, virions show great variation in size and exist in various morphological forms.

Various schemes have been developed to classify the different types of viruses. In the early days of virology in the mid-twentieth century, viruses were classified largely on the basis of their hosts (plants, animals, or humans). As knowledge of these accumulated in the 1960s and 1970s, the structure and replication of viruses became the main criteria for classification. The Baltimore classification scheme devised in the 1970s divided viruses into seven groups, based on the nature of their genome and their replication strategy. Since the 1990s, the International Committee on Taxonomy of Viruses (ICTV) has developed rules for classifying and naming viruses using taxonomic divisions and Latin species names, but these differ from the binomial format used for cellular organisms. The organization reflects the phylogenetic ancestry, structure, replication, hosts, and transmission vectors of different viruses. Thus, in the latest (2018) ICTV taxonomy, there are six classes (suffix -*viricetes*), 14 orders (-*virales*), 143 families (-*viridae*), 846 genera (-*virus*), and 4958 species of viruses. Some orders and families are divided into multiple sub-groups.

Viral genomes can be circular, linear, or segmented; with either a DNA or RNA genome, both of which can be either single- or double-stranded. A key factor in the replication strategy of single-stranded viruses—used in the Baltimore classification scheme—is whether the nucleic acid is positive or negative sense. Positive-sense RNA can serve as viral mRNA directly, whereas negative-sense RNA must be converted to a complementary mRNA by RNA-dependent RNA polymerase. As shown in *Table 7.1*, the size of virions in different viral groups varies greatly; the smallest are only ~20–30 nm in diameter. The largest known viruses belong to a clade known as the nucleocytoplasmic large DNA viruses (NCLDVs), and the family Mimiviridae contains the largest of them all—some virions of Tupanvirus are up to 2.3 μm in length. We will return to these so-called "giant viruses" later. The size of viral genomes is very variable, ranging from about 3.2 kb in the Hepapadnaviridae to >1200 kb in the Mimiviridae.

The virion contains identical protein subunits (capsomeres), which self-assemble to form the capsid, a shell-like structure surrounding the nucleic acid. The arrangement of the capsomeres and the overall morphology of the virion when visualized in the electron microscope is a key factor in the identification and classification of viruses. Helical viruses are rod-shaped or filamentous, with a single type of capsomere arranged around a central cavity. Negative charges on the nucleic acid bind it to positive charges on the protein helix. Many viruses have an icosahedral structure—a symmetry that is the optimum way of forming a closed shell-like structure for identical subunits. Enveloped viruses are surrounded by an outer membrane containing lipids and carbohydrates derived from the host and proteins encoded by the viral genome; the envelope is often involved in the infection of host cells by the virus. Many viruses have complex symmetry combining these different features. The best-known examples of this type are the tailed phages that have an icosahedral head containing the viral genome, which is bound to a helical tail with a base plate with protein fibers, which serves as a molecular syringe for delivery of the viral genome into the cell. An enormous variety of genomic structures can be seen among different viruses. Although we now know that there are millions of different virus types, only about 5000 have been described in any detail.

? WHAT'S IN A NAME?

As in other microbes with extensive genetic diversity, the concept of a "species" of virus is problematic. The ICTV provides a formal definition of a viral species as a monophyletic group whose properties can be distinguished from those of other species by multiple criteria. Examples of formal names for cultured viral species are *Cafeteria roenbergensis virus* (genus *Cafeteriavirus*, family Mimiviridae) and *Emiliania huxleyi virus 86* (genus *Coccolithovirus*, family Phycodnaviridae). Surprisingly, the nomenclature chosen by virologists does not follow the binomial *Genus species* system used in the rest of biology (see p.118), Virus strains may have informal names or acronyms reflecting their host. For example, the species of *Cafeteriavirus* and *Coccolithovirus* named above are usually referred to as CroV and EhV-86 respectively; RDJLΦ1 is an acronym for *Roseobacter virus* RDJL1 (genus *Xiamenvirus*, family Siphoviridae); and HaRNAV is short for *Heterosigma akashiwo RNA virus* (Genus *Marnavirus*, family Marnaviridae). The great majority of marine viruses cannot be cultured (because in most cases their hosts also cannot be cultured). This fact, and the vast amounts of sequence data emerging from metagenomic studies, presents a major challenge to the meaningful classification of viruses (Simmonds, 2015).

Table 7.1 Examples of viruses infecting marine organisms

Virus family	Morphology of virion	Size of virion (nm)[a]	Marine host(s)
Double-stranded DNA viruses			
Baculoviridae	Enveloped rods, some with tails	200–450 × 100–400	Crustaceans
Corticoviridae, Tectiviridae,	Icosahedral with spikes	60–75	Bacteria
Herpesviridae	Pleomorphic, icosahedral, enveloped	150–200	Mollusks, fish, corals mammals, turtles
Iridoviridae	Round, icosahedral	190–200	Mollusks, fish
Lipothrixviridae	Thick rod with lipid coat	40 × 400	Archaea
Mimiviridae	Icosahedral with microtubule-like projections	450–650[b]	Protists, corals (?), sponges (?)
Myoviridae	Polygonal head (icosahedral) with contractile tail (helical)	50–110 (head)	Bacteria
Nimaviridae	Enveloped, ovoid with tail-like appendage	120 × 275	Crustaceans
Papovaviridae	Round, icosahedral	40–50	Mollusks
Phycodnaviridae	Icosahedral	130–200	Algae
Podoviridae, Siphoviridae	Icosahedral with noncontractile tail	60 (head)	Bacteria
Single-stranded DNA viruses			
Microviridae	Icosahedral with spikes	25–27	Bacteria
Parvoviridae	Round, icosahedral	20	Crustaceans
Double-stranded RNA viruses			
Birnaviridae	Round, icosahedral	60	Molluscs, fish
Cystoviridae	Icosahedral with lipid coat	60–75	Bacteria
Reoviridae	Icosahedral, some with spikes	50–80	Crustaceans, molluscs, fish, protists
Totiviridae	Round, icosahedral	30–45	Protists
Single-stranded RNA viruses (positive sense)			
Caliciviridae	Round, icosahedral	35–40	Fish, mammals
Coronaviridae	Rod-shaped with projections	200 × 42	Crustaceans, fish, seabirds
Dicistroviridae	Round, icosahedral	30	Crustaceans
Leviviridae	Round, icosahedral	26	Bacteria
Marnaviridae	Round, icosahedral	25	Algae
Nodaviridae	Round, icosahedral	30	Fish
Picornaviridae	Round, icosahedral	27–30	Algae, crustaceans, thraustochytrids, protists (?), mammals
Togaviridae	Round, with outer fringe	66	Fish
Single-stranded RNA viruses (negative sense)			
Bunyaviridae	Round, enveloped	80–120	Crustaceans
Orthomyxoviridae	Round, with spikes	80–120	Fish, mammals, seabirds
Paramyxoviridae	Various, mainly enveloped, filamentous	60–300 × 1000	Mammals
Rhabdoviridae	Bullet-shaped with projections	45–100 × 100–430	Fish

[a] For rod-shaped virions, dimensions are shown as diameter × length. [b] The virion of Tupanvirus has a large cylindrical tail (450 × 550 nm) attached to the capsid (*Figure 7.8*). The average length of the complete virion is 1.2 μm, but some can reach 2.3 μm in length.

"PLANET VIRUS"— 75 MILLION BLUE WHALES, 10 MILLION LIGHT YEARS

The consensus estimates for viral density in different waters allow us to calculate the estimated total number of viral particles in the ocean. The total volume of the oceans is ~1.3×10^{21} liters, and assuming that the average abundance of viruses is ~3×10^{9} per liter, then the total number of viruses in the oceans is ~4×10^{30} (Suttle, 2005). This is about 10–15 times the total number of bacterial and archaeal cells, making viruses the most abundant biological entities in the ocean, comprising about 94% of all particles containing nucleic acids. Although their small size means that they only account for 5% of ocean biomass, the total carbon reservoir contained within marine viruses is 0.2 Gt (Pg)—the equivalent of ~75 million blue whales (Suttle, 2007). If the viruses were stacked end to end, they would span >10 million light years. Astonishingly, viruses attached to organic particles can also be swept into the atmosphere by aerosols—deposited at 8×10^{6} m² in the upper atmospheric boundary layer—and are transported long distances before falling back to the Earth's surface in rain and dust. Thus, genetically similar viruses are dispersed to all parts of the planet (Reche et al., 2018).

Early studies of free marine viruses in the plankton were made during the late 1970s and 1980s by direct observation of filtered seawater samples, using transmission electron microscopy (TEM). Samples can be concentrated onto grids by ultracentrifugation and negatively stained with uranyl acetate or another similar heavy-metal stain. Strictly, we should refer to structures observed in this way as "virus-like particles" (VLPs), because TEM cannot indicate whether the particles contain genetic information and thus whether they are capable of infecting a host cell. In addition, bacteria can produce large numbers of outer membrane vesicles which may have a similar size to virions. Although technically straightforward, observing and enumerating VLPs in seawater samples requires great care and control of variables such as the density of suspended particles. Despite many technical difficulties with sample preparation in early studies, this method led to the gradual realization in the late 1980s that viruses are highly abundant in the marine environment. The TEM method is also very valuable because it shows the morphology of viruses; the majority of free VLPs observed in seawater are phages (see below) having capsids with pentagonal or hexagonal icosahedral three-dimensional symmetry. Both tailed and non-tailed forms occur, and appendages such as capsid antennae or tail fibers can sometimes be seen (*Figure 7.1*). VLPs vary in size, with most phage particles being 30–100 nm in diameter, thus forming part of the fraction referred to as dissolved organic matter (DOM, see *Figure 1.6*). The giant viruses have particle sizes measuring hundreds of nanometers and are larger than many bacteria—the largest known (Tupanvirus) measures 1.25 µm. The method of sample preparation has a great influence on the observed morphology, and it is generally easier to obtain details of virus morphology from isolates in laboratory culture—in the few cases in which this is now possible—rather than from free VLPs obtained directly from seawater samples.

Another approach developed for enumerating VLPs in sea water is an adaptation of epifluorescence microscopy, in which the sample is treated with a fluorochrome that binds to nucleic acids (*Table 2.1*). Brightly fluorescent stains such as Yo-Pro-1 and SYBR Green 1® have been used successfully in many studies, with VLPs appearing as very small bright dots that can be distinguished from the larger stained cells of bacteria and other microscopic plankton (*Figure 7.2*). This is a relatively rapid and inexpensive method, although caution is needed to avoid overestimating viral abundance, because free nucleic acids bound to colloidal particles or within vesicles may also appear as fluorescent dots; conversely, large VLPs might be confused with bacteria, resulting in an underestimation. This method may also result in underestimations of viral density because RNA viruses and single-stranded DNA viruses do not stain well with current methods. Since viruses

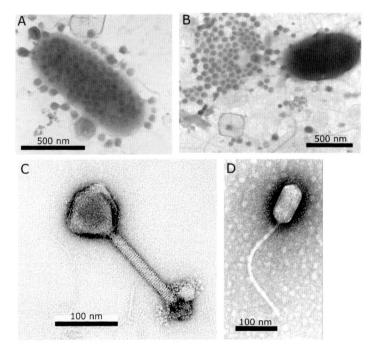

Figure 7.1 Transmission electron micrographs of phages. A. Mature phages inside a cultured bacterium before lysis. B. Free viruses after lysis. C. Cyanophage 9303-2 (Myoviridae). D. Cyanophage 9313-4 (Siphoviridae). Credits: A, B. Mikal Heldal and Gunnar Bratbak, University of Bergen. C, D. Bin Ni, and Matthew Sullivan, Chisholm Lab, MIT.

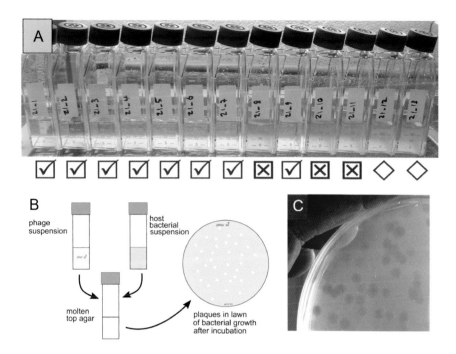

Figure 7.2 Assays for lytic phages. A. Cultures of *Ostreococcus tauri* host strain OTH95 inoculated with dilutions of seawater samples containing *O. tauri* virus. After incubation, lysis ☑ or no lysis ☒ is visible (◇ shows sham-inoculated controls). B. Schematic of plaque assay. The bacterial host is grown in broth culture and mixed with a sample containing the phage. This is resuspended in molten top agar, which contains a low concentration of agar so that phage can diffuse through the gel to infect adjacent bacteria when poured onto a plate of agar medium containing appropriate nutrients for growth of the host. After incubation, clear areas of lysis appear as plaques in the lawn of bacterial growth. By plating different dilutions of the phage suspension, the density of infective phage can be calculated as plaque-forming units (PFU) per mL, analogous to the viable colony count of bacteria. C. Example of plaque assay of *O. tauri* virus. Similar techniques can be used for the propagation of viruses infecting archaea, algae, and other protists, providing suitable media for growth are available. Credit: A, C: Derelle et al. (2008), CC-BY-2.0.

are below the limits of resolution of the light microscopy, fluorescence microscopy provides no information about the morphology of viruses—indeed, the only reason that they can be seen with this method is because of the halo of bright light emitted by the fluorochromes. Many experimental and field studies have used flow cytometry (*Figure 2.4*) as a high-throughput method to enumerate viruses alongside the community structure of their potential hosts in the plankton. It is possible to discriminate various host and viral populations based on their fluorescence and scatter signals after staining with fluorochromes, and this can be applied to analysis of seasonal and spatial variation in viral dynamics in mesocosm and open-water studies. Recent improvements include the use of virus-specific probes for detection of infected cells and the use of solid-phase single-molecule multiplex PCR (polony method).

Based on numerous studies of seawater from various geographic sites and different depths using TEM and epifluorescence microscopy since the 1990s, a consensus value for the density of viruses in seawater of about 10^7 per milliliter has become established. However, there is considerable seasonal and geographic distribution, and it is hard to generalize to all depths and locations. In general, the viral abundance increases with the productivity of the system. In the open ocean, virus density declines rapidly below a depth of 250 m to a relatively constant value of about 10^6 per milliliter. Counts in highly productive coastal waters are usually higher than in the open ocean and are not so dependent on depth, with typical densities of 10^8 per milliliter. Viruses are also found in high densities in marine sediments and sea ice. The distribution of viruses in the water column generally mirrors the productivity and density of host populations of bacterioplankton, as these are the most abundant hosts. As a rough approximation, it is generally assumed that there are about ten times as many viruses as bacteria in most seawater samples. In the photic zone of very oligotrophic waters, cyanophages infecting photosynthetically active cyanobacteria such as *Prochlorococcus* and *Synechococcus* form the dominant group. The distribution of viruses infecting eukaryotic algae also generally mirrors the abundance of the population of algal hosts, but this is a small proportion of the total. However, recent meta-analyses of previous virus abundance studies have questioned the validity of the assumed ~10:1 linear virus to host ratio and indicate that it is better represented as a power-law function. The estimates of VLPs to bacteria ratios are obtained from bulk counts of relatively large samples of water, which conceals large spatial variations in host and viral diversity. When host cell densities are high, the VLPs to bacteria ratios are often considerably less than ten. As noted previously, seawater is a highly heterogeneous environment, and microbes show microscale aggregation around nutrient sources.

Indeed, in studies of abundance at small spatial scales, changes in the density of VLPs are very dynamic, and large fluctuations can occur over short timescales as a result of synchronized lysis of host cells and different rates of degradation of released virus particles depending on environmental conditions.

Phages are viruses that infect bacterial and archaeal cells

When viruses that infect bacteria were discovered in 1915, they were called bacteriophages—a term meaning "bacteria-eaters." Subsequently, viruses that infect archaeal cells have also been discovered, but no specific term to denote them has been coined. Therefore, the abbreviated term "phage" is usually used to indicate viruses infecting members of either domain. Prefixes indicating a particular host range are also used: terms such as "vibriophages" (infecting *Vibrio* spp.), "cyanophages" (infecting cyanobacteria), or "roseophages" (infecting members of the *Roseobacter* clade) are used. The best-known isolated phages are dsDNA viruses comprising a head (nucleocapsid) connected to a tail structure used for adsorption to the cell and injection of the nucleic acid into the host cell. These phages may be classified into a number of major groups, namely the Myoviridae (with a contractile tail, *Figure 7.1C*); the Podoviridae (with a very short tail); and the Siphoviridae (with a long flexible tail, *Figure 7.1D*). However, a wide range of other morphologies occur in marine phages and global sampling of ocean waters and quantification by electron microscopy has concluded that non-tailed phages comprise 51–92% of VLPs and that the abundance of phages with RNA genomes has probably been significantly underestimated. Only a few hundred types of archaeal viruses have been isolated in culture, mostly associated with hyperthermophiles or halophiles. This small number of archaeal isolates contains a much higher level of morphological diversity and genome content than bacterial viruses, and it is speculated that viruses infecting ancestral cells diverged in early evolution due to differences in the cell envelope of bacteria and archaea.

Although the first marine phages were described in the 1950s, the significance of early findings was not appreciated, and it was not until the late 1980s that serious attention was paid to this field. The relatively slow development of this area of research may be linked to the long-held fallacy that microbial populations in the oceans are insignificant and that, by association, viruses must also be unimportant. As discussed in *Chapter 4*, early marine microbiologists seriously underestimated the abundance and diversity of bacteria because of reliance on inappropriate culture methods. The classical method of enumerating infective phages and viruses infecting protists is based on the formation of plaques of lysis in lawns of susceptible hosts grown on agar plates (*Figure 7.3*) or by dilution in liquid media and enumeration by the most probable number (MPN) method. In liquid culture, reduction in the turbidity of cultures can be measured by spectrophotometry as an indicator of cell lysis. These methods can be used with phages that infect easily cultivated heterotrophic and phototrophic bacteria but are not yet possible for the many bacteria and archaea which cannot yet be cultured. (Plaque assay and MPN methods can also be used to enumerate viruses of protists). It was not until reliable methods for direct observation and molecular analysis were developed in the study of both marine bacteria and their phages that it was realized that, far from being insignificant players in marine

Figure 7.3 Epifluorescence micrograph of a filtered (0.02-μm Anodisc) water sample stained with SYBR green for enumeration of virus-like particles (VLPs). The smallest bright dots are VLPs, the next largest are bacterial or archaeal cells. The two larger cells shown are protists. Credit: Jed Fuhrman, University of Southern California.

systems, phages play a critical and central role in ocean ecology and food web dynamics, although it must be remembered that these methods do not directly indicate the number of infective viruses.

The life cycle of phages shows a number of distinct stages

The first step in the life cycle of phages is adsorption to the host cell surface, which may be reversible, followed by irreversible binding to a specific receptor on the host cell surface. For those marine phages that have so far been propagated in cultivated bacteria, most show specificity for particular bacterial species and sometimes for particular strains. These findings are similar to those seen in other well-known viruses and are usually thought to be due to the molecular specificity of virus receptors on the host cell surface, the presence of restriction enzymes, or compatibility of the replication processes. However, there are indications that these highly specific interactions may be something of an artifact introduced by the assay system *in vitro*, and it is thought that some marine phages may have a relatively broad host range in the natural environment. For example, some cyanophages have been shown to infect both *Synechococcus* and *Prochlorococcus* and some vibriophages infect several species of *Vibrio*. If correct, this has significant implications for the possibility of genetic exchange between different organisms and for the role of phages in determining bacterial community structure. Very little is known about the nature of the receptors for phage adsorption to marine bacteria. It is possible that broad host range phages target conserved amino acid sequences of proteins on the cell surface of different bacterial types and that the tail fibers can recognize more than one type of receptor. Recent work suggests that some phages exploit membrane proteins of host cells as a mechanism for entry (see p.237).

In most cases, enzymes in the tail or capsid of the phage attack the bacterial cell wall, forming a small pore through which its nucleic acid enters. The phage genetic material then remains in the cytoplasm or is integrated into the host cell genome. In the lytic cycle (*Figure 7.5a*), this is followed by expression of phage proteins, phage genome replication, and formation of the capsids and other parts of the virion. When assembly is complete, most phages cause lysis of the host cell by producing enzymes that damage the cytoplasmic membrane and hydrolyze the peptidoglycan in the cell wall.

The "burst size" of phage infection is a term used to describe the average number of progeny virus particles released from the host cell at the time of lysis. Knowing this value is important for modeling the dynamics of viral infection of host populations. The larger the burst size, the smaller the number of host cells that need to be lysed to support viral production. Burst size can be measured directly by observing visibly infected cells using TEM (*Figure 7.1B*), or it can be calculated using models of virus to host ratios and theoretical rates of contact and infection required to maintain virus production. There is a wide variation in results, depending on the location of the study and the methods used, but values in the range of 10–50 progeny viruses per cell are typical for *in situ* studies in natural communities, with higher values occurring for copiotrophic bacterial hosts in nutrient-rich waters. For those phage-host systems that can be propagated in laboratory culture, the burst sizes are usually much larger (average about 180) because *in vitro* grown bacteria are bigger and support greater virus densities. In *Synechococcus*, a large burst size of about 100–300 occurs in culture.

Lysogeny occurs when the phage genome is integrated into the host genome

Lytic phages take over the host cell, replicate their nucleic acid, and cause lysis of the host cell after assembly of the virus particles. However, another outcome is seen when phages, known as temperate viruses, infect the cell (*Figure 7.5b*). The phage genome replicates along with the host DNA, but it is not expressed. Often, the silent viral genome is stably integrated into the bacterial genome; this latent state is known as a prophage. In other cases, the phage genome remains in the cytoplasm in a circular or linear form. Bacteria infected with these phages are known as lysogenic, because under certain conditions the bacteria enter the lytic cycle and release infective virus particles.

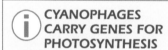

CYANOPHAGES CARRY GENES FOR PHOTOSYNTHESIS

Sequencing of phages infecting *Synechococcus* and *Prochlorococcus*, showed that some contain photosynthesis genes of cyanobacterial origin, including *psbA* (Mann et al., 2003; Lindell et al., 2004). In the host, the reaction center proteins are rapidly turned over due to light-induced damage and replacement by new proteins—this repair process is shut down during phage infection. It is "in the interest" of the phage for photosynthesis rates to remain high in order to provide energy for virion production. During infection, these phage genes are expressed along with essential capsid genes (Lindell et al., 2005). Presence of the genes may enhance the reproductive fitness and host range of the phage (Sullivan et al., 2006). Analysis shows that PSII genes have been transferred via HGT into phages on multiple occasions in evolution and that considerable genetic rearrangements have taken place between host and phage genes. Phages therefore serve as a reservoir of genetic diversity for their hosts. A wide range of such auxiliary metabolic genes (AMGs) involved in numerous functions have now been identified in marine phages (Breitbart et al., 2018).

BOX 7.1   RESEARCH FOCUS

Models for the co-evolution of bacteria and their phages

An "arms race" of co-evolution—the "Red Queen" hypothesis. According to ecological theory, competition between species in the same ecological niche should lead to extinction and reduced biodiversity. But the opposite occurs, leading Hutchinson (1961) to pose the "paradox of the plankton," questioning how so many plankton species can co-exist while competing for the same nutrients. We know now that this exceptional biodiversity can be partly explained by the microscale structuring of the marine environment (p.120), but the impact of phages as predators also has a major impact on promoting the exceptional diversity of microbial ecosystems. Since phages replicate more rapidly than their hosts, and lytic infection results in death of the host, why don't phages kill all the hosts and cause their own extinction in the process? A common explanation is that there is a constant co-evolution of the host and virus—an "arms race" via mutation and counter-mutation—resulting in continual selection pressure for the host to evolve resistance and for the virus to evolve to overcome that resistance. This is referred to as the "Red Queen Hypothesis," so named because in Lewis Carroll's *Alice in Wonderland* sequel, *Through the Looking Glass*, the Red Queen says, "It takes all the running you can do, to keep in the same place." Despite wide recognition of this model, co-evolutionary experiments indicate that this arms race does not continue indefinitely, due to genetic and metabolic constraints that lower the fitness of hosts. Resistance in the host and virulence in the phage both carry a fitness cost; for example, changes in surface phage receptors might affect host transport mechanisms associated with resistance and vice versa (Avrani et al., 2012). Therefore, this arms race may give way to fluctuating selection. In this model, fast-growing hosts will dominate the population, but the consequent increased contact rates with phages result in increased mortality and the host population will decline. Eventually, resistant mutant hosts will increase, despite their lower growth rate. Abundance of the dominant phage will also decrease, allowing the faster-growing susceptible hosts to increase in number again. Avrani et al. explain that, over time, this results in oscillation of resistant and susceptible hosts and virulent and less virulent phage in the community, due to alternating viral selection for resistant mutants and competitive selection for faster-growing hosts.

The "kill the winner" (KtW) hypothesis. Thingstad (2000) developed a highly influential concept based on a mathematical model of traditional host-parasite dynamics incorporating an idealized food web of bacterial and protist hosts, plus predatory grazing protists. The "winner" was conceived as the most *active* host population—not necessarily the most abundant. In this model, the outcome is that virulent phages reduce populations of their susceptible hosts to a low steady-state level, independent of the host's growth rate and allowing multiple species for each nutrient type (Maslov and Sneppen, 2017). Laboratory experiments involving enrichment or depletion of viruses have provided some support for the hypothesis, with strong effects on particular taxa of bacteria (e.g. Hewson and Fuhrman, 2006), but effects on overall bacterial community composition are less clear-cut, possibly because of the combined effects of phage-induced lysis and protistan grazing. Maslov and Sneppen propose a more dynamic interpretation of KtW, in which phage infections result in abrupt and severe collapses of bacterial populations when they become large. In their model, the total population of all species fluctuates around the carrying capacity of the environment and the overall diversity remains high. When viewed over a long period of time, this leads, counter-intuitively, to higher diversity of microbial communities that are exposed to frequent and severe collapses induced by phage infection. Maslov and Sneppen (2017) conclude that in natural ecosystems, this density-dependent selection will be supplemented by other drivers of diversity and modulated by the effects of spatial heterogeneity of nutrients.

Game of thrones—what determines the succession of hosts and phages? In the KtW model, new phages emerge to infect newly dominant hosts. Breitbart et al. (2018) propose an analogy

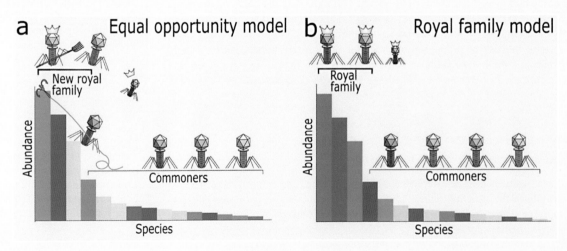

Figure 7.4 Proposed models for explaining the succession of marine phages. Reprinted from Breitbart et al. (2018) with permission from Springer Nature.

BOX 7.1 RESEARCH FOCUS

to conceptualize the hierarchy of bacterial and phage interactions. They imagine a "royal family" of dominant bacteria and phages—this will vary between different locations or "kingdoms"—and a population of all the other bacteria and phages that are the "commoners." On the "death" of the royal family—the decline of the bacterial monarch along with its associated phages—two possibilities exist for succession. In one scenario (*Figure 7.4a*), all bacterial hosts and their respective phage hosts have an equal opportunity to become royalty and to fill the empty niche. Each of the commoners is relatively rare, so it is unlikely that the same host and phage signature sequences would be abundant at different times. An alternative scenario (*Figure 7.4b*) assumes that members of the royal family are dominant because they are optimized to that specific niche ("blue blood?"). In this case, the bacterial successor to the monarch is much more likely to be a descendant of the royal family ("next in line to the throne?") than a commoner that has scaled the rank-abundance plot. Therefore, the next phages are likely to be variants of previous royal family phages that have overcome the host's newly developed resistance. Breitbart et al. note that this evolutionary arms race is supported by analysis of microdiversity in marine bacteria and their phages in culture-based experiments and in time-series datasets from natural marine systems. While further time-series studies are needed to test their hypothesis, Breitbart et al. conclude that initial evidence supports their royal family model, whereby hosts that are well-suited to the niche will stay in power, and any changes in the throne will favor bacteria belonging to royal lineages that have escaped their abundant phage attackers.

Switching to a temperate lifestyle—the "Piggyback-the-winner" hypothesis. The importance of the lytic and temperate cycles of phage biology must be factored into these attempts to understand phage and host selection and evolution. Knowles et al. (2016) conducted a meta-analysis of reports of virus to bacteria ratios across diverse habitats, showing a consistent trend for lower densities of VLPs at high host densities. Experimental analysis showed that viral densities were more consistent with temperate than lytic cycles at increasing host abundances. Knowles et al. included metagenomic analysis focusing on 24 coral reef viromes, showing a correlation between the abundance of host bacteria and abundance of genes that are hallmarks of temperate phages. Based on this finding of "more microbes, fewer viruses," they proposed a "Piggyback-the-Winner" (PtW) model, in which phages integrate into their hosts' genomes as prophages when microbial abundances and growth rates are high. Switching to the temperate life cycle reduces the control of phage predation on bacterial abundance and confers protection on the host cell from infection by a closely related phage (superinfection exclusion). This is in contrast with the conclusions of many previous studies on lysis and lysogeny.

The mechanisms that determine whether the phage enters the lytic or lysogenic cycle has been well studied in the lambda phage of *Escherichia coli* and a few other examples, where there appears to be a quorum-sensing like process (p.102) mediated by the phage and a peptide signal molecule. Similarly, there have been many studies of the process by which expression of the prophage genes controlling the lytic pathway switch are repressed until induced by the agents or treatments noted below. However, the molecular events of lysogeny in marine bacteria are largely unexplored and complete systems—temperate phage, lysogenic host, and uninfected host—have only been described for a few marine bacteria. Factors such as host growth rates, nutrient depletion, and phage density are thought to be major factors in the "decision" of some phages whether to enter the lytic or lysogenic cycle.

Little is known about the mechanism in marine phages by which the expression of the prophage genes controlling the lytic pathway switch is repressed. It is thought that about 1 in 10^5 lysogenic cells will revert to the lytic cycle without an obvious induction trigger, so that there is a constant low-level production of temperate phages. Experimental conditions such as exposure to ultraviolet light, temperature shifts, peroxides, antibiotics, or pollutants have all been shown to induce a large proportion of prophage-containing cells to enter the lytic cycle. These external stressors generally trigger the cell's SOS response, which is initiated by bacteria to prevent DNA damage. In laboratory experiments, the antibiotic mitomycin C is the most commonly used agent for inducing the lytic cycle. One method of assessing the proportion of the host population in the lytic and lysogenic states is to measure the abundance of genes that are characteristic of a temperate phage, such as the integrase and excisionase enzymes used at the start and end of the prophage integration process. These can be measured in metagenomic analyses which compare DNA libraries from ambient-free virus particles with libraries prepared after prophage induction.

Integration and excision of the phage genome has very important evolutionary consequences because it provides a natural mechanism, known as specialized transduction, by which specific host genes can be transferred from one cell to another when part of the host DNA becomes incorporated into the mature virus particle. The presence of a prophage also confers resistance to infection of the host bacterium by viruses of the same type (phage immunity) because repressor proteins, which prevent replication of the prophage genome, also prevent replication of incoming genomes of the same (or closely similar) virus. Prophage genes can also affect the phenotypic characteristics of the host cell in other ways, including effects on the general reproductive fitness of the host or alteration of the host phenotype through expression of different genes. An important example of phenotypic modification or lysogenic conversion is the role of phages in the acquisition of virulence factors enabling colonization and toxicity in *Vibrio cholerae*, a bacterium found in coastal and estuarine waters and responsible for the human disease cholera (p.333). Phage-mediated virulence of bacteria may be very widespread in other *Vibrio* spp. and other pathogenic bacteria. Lysogenic immunity and phage conversion both carry a strong selective value for the host cell and could provide phages with a strategy to survive periods of low host density or low metabolic activity when the probability of encountering a new host and initiating a lytic cycle factor is low. Since many marine environments contain low levels of slow-growing bacterioplankton and because free viruses are inactivated quite quickly (see below), virulent lytic phages could become rapidly depleted. For all these reasons, it would be reasonable to consider lysogeny to be a very common state of virus infections in nature. Indeed, some studies with cultivable bacteria have shown that lysogeny is more common in bacteria (up to 50%) in samples from offshore environments than nutrient-rich coastal environments, and in studies of deep-water communities the frequency of lysogenic cells appears to be negatively correlated with bacterial production. However, some other studies have produced conflicting results and we have no knowledge at all about the importance of lysogeny in bacteria that have not yet been cultured. Another form of hidden infection may occur, in which the viral nucleic acids remain in the host cell for an extended period, but the lysis of the host cell is delayed. Such pseudolysogeny (*Figure 7.5c*) may be due to nutrient deprivation and low metabolism of the host cell and so might also be a common trend in oligotrophic marine waters. It is hoped that the application of new molecular biology techniques will reveal the true importance of lysogeny and pseudolysogeny in marine virus-host interactions.

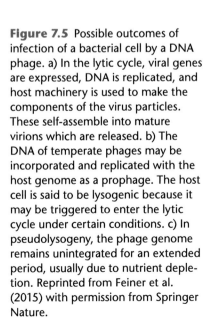

Figure 7.5 Possible outcomes of infection of a bacterial cell by a DNA phage. a) In the lytic cycle, viral genes are expressed, DNA is replicated, and host machinery is used to make the components of the virus particles. These self-assemble into mature virions which are released. b) The DNA of temperate phages may be incorporated and replicated with the host genome as a prophage. The host cell is said to be lysogenic because it may be triggered to enter the lytic cycle under certain conditions. c) In pseudolysogeny, the phage genome remains unintegrated for an extended period, usually due to nutrient depletion. Reprinted from Feiner et al. (2015) with permission from Springer Nature.

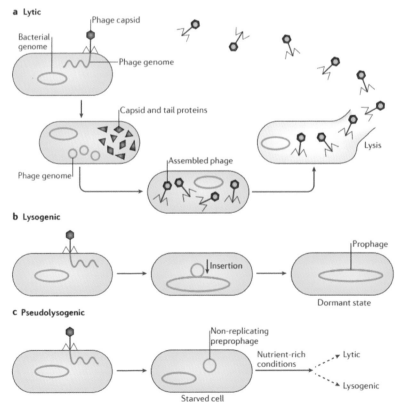

Loss of viral infectivity arises from damage to the nucleic acid or capsid

Usually, a virus will lose its infectivity before showing obvious signs of degradation. However, since most marine viruses are studied by microscopic or flow cytometric enumeration of VLPs, the term "virus decay" reflects the observation of a decline in numbers of VLPs over time in the absence of new viral production. Many of the studies of virus inactivation in water were originally carried out in connection with the health hazards associated with sewage-associated viruses (such as enteroviruses or coliphages; in waters for swimming or cultivation of shellfish (p.366). Subsequently, the results of these studies have been applied to the population dynamics of autochthonous marine viruses. A wide range of physical, chemical, and biological factors can influence virus infectivity and decay. Different studies have produced various estimates of decay rates, but a value of about 1% per hour is typical in natural seawater kept in the dark. Visible light and ultraviolet (UV) irradiation are by far the most important factors influencing virus survival, and in full-strength sunlight, the decay rate may increase to 3–10% per hour, and in some circumstances can be as high as 80%. UV light has its greatest effect in the upper part of the water column but is probably still effective down to about 200 m in clear ocean water. Even in very turbid coastal waters, viral inactivation by light can be observed down to several meters. Such high rates of inactivation would lead to the conclusion that there are no or very few infective viruses in the top layer of water. However, extensive repair of UV-induced damage can occur by mechanisms encoded either by hosts or their viruses. Another important factor in decay is the presence of particulate matter and enzymes such as proteases and nucleases produced by bacteria and other members of the plankton. These involve complex interactions, because adsorption of viruses to marine snow particles or TEPs can also afford some protection from decay. Virus inactivation does not proceed at a constant rate and it appears that there is variation in the resistance of viral particles to damaging effects, presumably because of minor imperfections in the capsid. Thus, over time, inactivation will lead to a slowly decaying low level of infective particles, but we do not know the rate of this for ecologically important viruses.

Measurement of virus production rates is important for quantifying virus-induced mortality

Many studies in viral ecology have attempted to measure the effect of viruses on microbial mortality and the rate of production of new virions. This enables estimates of the proportion of primary and secondary production that is "turned over" by viral lysis. It is possible to use filtration and high-speed centrifugation to obtain a pellet of planktonic microbes, which can be embedded in resin and sectioned for TEM. By examining such samples from various locations, it has been found that about 1–4% of microbial cells contain mature, fully assembled VLPs. Since VLPs can only be seen within infected host cells in the final stages of the lytic cycle of infection—this stage usually represents about a quarter of the life cycle—it is possible to estimate the total proportion of plankton that are infected at any one time. Another method used in early studies of virus production was to measure the incorporation of radioactively labeled precursors such as ^{3}H-thymidine, ^{14}C-leucine, or ^{32}P-phosphate into virus particles, which can be separated from cells and cell debris by filtration. The virus dilution (reduction) method has been widely used; in this method, free virus particles are removed from a water sample by filtration and the number of new particles occurring in the water over time is measured by epifluorescence microscopy or flow cytometry. After adjustments for the relative abundance of host cells in the sample, the production rate and the original percentage of infected cells can be calculated. Each of these methods has advantages and disadvantages and no single method gives precise estimates of virus-mediated mortality. (Obviously, these methods cannot detect prophage-infected cells). Nevertheless, despite variability depending on the method, location, and time of sampling, the unequivocal conclusion from various studies is that viral mortality has a highly significant impact on mortality of microbes that is at least as significant as grazing by protists and zooplankton. A consensus value is that about up to 40% of marine bacteria are killed each day by viral action. Between 20–30% of bacteria in the oceans are probably infected by phages at any one time, and an estimated 10^{23} viral infections occur every second. In the case of algae, laboratory experiments and mesocosm studies have indicated that viral infection can account for nearly 100%

of mortality of bloom-forming microalgae such as *E. huxleyi*. Using modified dilution methods to distinguish the effects of mortality due to viruses from mortality attributed to protistan grazing, viral mortality of *Micromonas pusilla* has been estimated at 9–25% standing stock per day and in this case, it is likely that there is a more stable co-existence of viruses with their microalgal hosts. As with bacteria, precise values for the effects of viral mortality on natural populations of microalgae are hard to obtain, but there is no doubt that it has a very significant impact on primary production, climate processes, and global biogeochemistry.

Viral mortality "lubricates" the biological pump

Viral infection of heterotrophic and autotrophic bacteria, archaea, fungi, microalgae, and other protists seems to be the most important factor influencing nutrient cycles in the oceans. This is important because it leads to the release of massive amounts of organic material and essential elements from these microbes into the dissolved organic pool from where it is metabolized by heterotrophs. Because the contents of cells lysed by viruses are rich in nitrogen and phosphorus, this "viral shunt" speeds up the recycling of nutrients, enhances the rate of microbial respiration, and reduces the amount of organic material available to higher trophic levels through protistan grazing in the microbial loop (see *Figures 8.3* and *8.4*). However, viral lysis is now recognized as having more complex effects, because it leads to the release of high molecular weight polymeric substances from cells. These contribute to the aggregation of algal flocs and marine snow (*Figure 1.9*), thus accelerating the export of carbon-containing compounds to the ocean floor via the biological pump. This has been termed the "viral shuttle" and seems to be particularly associated with viral lysis of certain microbial groups in oligotrophic ocean regions. The importance of these processes is considered further in *Chapter 8*.

Nucleocytoplasmic large DNA viruses (NCLDVs) are important pathogens of microalgae and other protists

One of the most studied family of viruses is the Phycodnaviridae, a family of large dsDNA viruses infecting a wide host range of freshwater and marine algae. Although genetically diverse, they share some morphological similarities, with a dsDNA-protein core usually surrounded by a lipid membrane and a capsid (120–220 nm diameter) with icosahedral symmetry. Genomes of several species have been obtained and range in size from 100–560 kb. There are currently six genera recognized in ICTV taxonomy. The best known of these is *Coccolithovirus*, with a single species that infects the important bloom-forming prymnesiophyte coccolithophorid *Emiliania huxleyi*. It causes cell lysis and contributes to the rapid collapse of blooms (*Figure 6.9*), with major ecological and biogeochemical consequences, as discussed in *Box 7.2*. *Raphidovirus* species infect a number of significant bloom-forming raphidophytes, including *Chrysochromulina*, *Aureococcus*, *Heterosigma*, *Heterocapsa*, and *Phaeocystis*. Apart from *Chrysochromulina*, all of these are the cause of harmful algal blooms (HABs). Virus-induced collapse of large blooms of *Phaeocystis* resembles that of the *E. huxleyi–Coccolithovirus* interaction and has similar important roles in carbon and sulfur cycling via DMSP release (*Box 9.2*). Collapse of blooms can also cause release of gelatinous polysaccharides leading to extensive formation of foam that has a nuisance (but not toxic) effect on beaches and in coastal waters. Of even greater significance is the large-scale shift in bacterial populations following bloom collapse, which can lead to anoxia. Members of the genus *Prasinovirus* infect picophytoplankton, including *Ostreococcus* (the smallest known eukaryote, (*Figure 1.3*) and *Micromonas*, which is a prominent member of the picophytoplankton in oceanic and coastal regions. Viruses specific to *Micromonas pusilla* cause lysis of up to 25% of the host's daily population but do not cause the "bloom and bust" dynamics seen in the *E. huxleyi–Coccolithovirus* interaction. There are nine species of *Phaeovirus*, which infect brown macroalgae (seaweeds), such as *Ectocarpus* and *Feldmannia*. In addition, there are more than 20 *Chlorovirus* species that infect the freshwater protists *Acanthocystis* and *Paramecium* and the cnidarian animal *Hydra viridis*.

Evolutionary analysis of the genomes of the phycodnaviruses shows that they belong to a monophyletic group of large DNA viruses infecting eukaryotes—the nucleocytoplasmic large DNA viruses (NCLDVs). Other viruses within this group include poxviruses, iridoviruses,

(i) THE DEEP IMPACT OF THE VIRAL SHUNT

Biogeochemical cycles in deep-sea ecosystems are determined by the activities of bacteria and archaea that largely depend on organic matter exported from the ocean surface. Microbial life is abundant in deep-sea sediments and constitutes a major fraction of the total microbial carbon on Earth, most of which appears to be unused by higher trophic levels. This paradox is explained by Danovaro et al. (2008) who studied viral activity at 232 sites at different locations and depths, showing that over 80% of the heterotrophic production is shunted by viral lysis into labile dissolved organic material (DOM), producing between 0.37 and 0.63 Gt (Pg) of carbon a year. Viruses therefore play a major role in deep-sea processes by injecting huge amounts of organic material into the ecosystem—stimulating microbial metabolism and accelerating biogeochemical processes. In a recent study of sediments from the Black Sea, Cai et al. (2019) showed high viral production 37 m below the seafloor in sediments ~6000 years old, with densities up to 1.8×10^{10} viruses cm^{-3}.

and mimiviruses, including a range of plant and animal pathogens. These viruses replicate inside the nucleus and/or cytoplasm and their large genomes encode hundreds of genes of unknown function. Unlike many viruses, NCLDVs can carry genetic information for entire biosynthetic pathways. Genome analysis of the NCLDVs shows that they have a very ancient origin and co-evolution with their hosts—divergence of the major families of NCLDVs appears to have occurred prior to the radiation of the major eukaryotic lineages 2–3 billion years ago. *Figure 7.7* shows examples of NCDLV infections of plankton.

The mechanism of infection of host cells differs amongst the NCLDVs, with the mechanisms of entry and exit showing large variations. Among the phycodnaviruses, *Chlorovirus* possesses a lipid-containing membrane (underlying the capsid), which fuses with the host cell membrane after enzymic digestion of the algal cell wall, followed by injection of the viral DNA and associated proteins. *Phaeovirus* also fuses with the host cell, in this case infecting only the spores or gametes of its seaweed host, which lack a cell wall. *Coccolithovirus* infects *E. huxleyi* cells using a mechanism that resembles that seen in enveloped animal viruses, involving entry of the entire nucleocapsid into the cytoplasm following either fusion of the envelope with the cell membrane or an endocytotic process.

Remarkable new insights into the origin and evolution of these NCDLV viruses and their importance in marine ecosystems have followed the description of the giant virus *Mimivirus* and the discovery that related viruses are very abundant in marine ecosystems as pathogens of heterotrophic eukaryotes, as well as the photosynthetic plankton currently described.

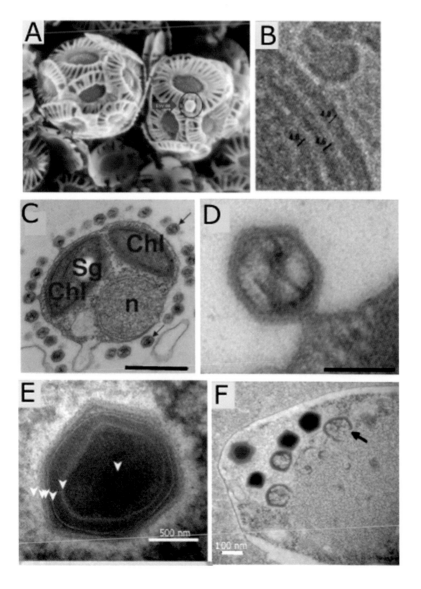

Figure 7.7 Electron microscopy of NCLDV infections of plankton cells. A. SEM of *E. huxleyi* cells (~5 mm diameter) and shed coccoliths, showing a virion of *Coccolithovirus* EhV-86 (~180 nm diameter). B. Cryotomography image of an isolated virion; three lipid bilayer-like membranes surrounding the capsid (average width = 4.5 nm. C. Multiple virions of *Prasinovirus* OtV attaching to a cell of *Ostreococcus tauri*. D. Intact OtV virion (containing DNA) attached to cell membrane of *O. tauri*, 30 min after infection. E. Mature virion of the *Klosneuvirus* BsV. The particle contains at least six layers, with a DNA-containing core surrounded by a core wall and inner membrane, with a putative membrane under a double-capsid layer. The top vertex of the virion contains a possible stargate structure like that in Mimivirus ApMV. F. BsV virion assembly and maturation in cell of *Bodo saltans*: lipid vesicles migrate through the virion factory where capsid proteins attach for the proteinaceous shell. Vesicles burst and accumulate at the virus factory periphery where the capsid assembly completes (black arrow). Once the capsid is assembled, the virion is filled with the genome and detaches from the virus factory. Internal structures develop inside the virion in the cell's periphery where mature virions accumulate until the host cell bursts. Credits: A. reprinted from Michaelson et al. (2010) with permission of Elsevier; B. reprinted from Schatz et al. (2014), CC-BY-3.0; C., D. reprinted from Derelle et al. 2008, CC-BY-2.0; E., F. reprinted from Deeg et al. (2018), CC-BY-4.0.

BOX 7.2   RESEARCH FOCUS

"Bloom and bust"—the life cycle of *Emiliania huxleyi* and *Coccolithovirus* (EhV)

An arms race between host and virus. Massive natural blooms of the coccolithophorid prymnesiophyte alga *Emiliania huxleyi* develop annually in many temperate and sub-temperate coastal and oceanic waters, before collapsing suddenly with the release of millions of coccoliths their calcium carbonate shells (cocoliths) due to lysis following infection by coccolithoviruses (*Figure 6.9*). Because of the importance of this process in global ocean biogeochemical cycles, this virus-host system has been explored extensively in the natural setting, in laboratory experiments, and in large-scale mesocosms (*Figure 2.14*). Several different strains of *E. huxleyi* viruses (EhVs) have been characterized from different locations (Schroeder et al., 2002; Wilson et al., 2002, Nissimov et al., 2017) and genome sequences are available for 13 strains. Extensive studies of community dynamics and genetic diversity structure in the open ocean and mesocosm studies were carried out in different years using PCR-DGGE analysis of genes encoding the calcium-binding protein (GPA) of *E. huxleyi*, and the major capsid protein (MCP) and DNA polymerase genes of EhVs (Martinez et al., 2007, 2012; Schroeder et al., 2003). These studies revealed a periodic annual succession of identical *E. huxleyi* genotypes, which in turn determined the succession of viral genotypes. By monitoring the MCP and GPA genes every 4 h during a bloom, Sorensen et al. (2009) showed that the viral community was highly dynamic, with fluctuations in detectable levels of different viral genotypes occurring at very short intervals. Thus, there is a constant "Red Queen"-style arms race in which there is continual selection pressure for the host to evolve resistance and for the virus to evolve to overcome that resistance, like that for bacteria–phage interactions discussed in *Box 7.1*.

Escape from infection. Until the study by Frada et al. (2008), most studies on *E. huxleyi* had focused on the diploid cells, which are coated in calcified coccoliths. Noncalcified motile haploid cells also occur but these were overlooked until flow cytometric analysis showed the appearance of smaller cells in populations of *E. huxleyi* undergoing viral lysis (Jacquet et al., 2002). Frada et al. (2008) showed that three strains of cultured diploid *E. huxleyi* were all sensitive to five EhV strains, whereas haploid stages were not infected. Electron microscopy and PCR amplification of the gene encoding MCP revealed that viruses did not adsorb to haploid cells, perhaps because a different cell surface structure prevents the enveloped virus from fusing with the cytoplasmic membrane. Additional experiments showed that the motile, haploid cells only appeared in the late stages of infected cultures, suggesting that viral infection triggers meiosis (Frada et al., 2017). The authors called this strategy of escape from infection the "Cheshire Cat" phenomenon—another allegorical reference to *Alice in Wonderland*, in which the fictional cat disappears and appears without its head. The authors discuss the implications of releasing host evolution from pathogen pressure, acting to counter the co-evolutionary arms race. Perhaps this is why *E. huxleyi* populations can survive bloom and bust dynamics—the escape strategy creates a reservoir of resistant haploid cells that may persist in the environment for some time, eventually mating to produce new diploid *E. huxleyi* genotypes. Thus, selection for sexual reproduction as an antiviral mechanism helps to maintain high host diversity that helps to buffer the effects of virus infection and environmental change. A further twist to the tale of this cat is supplied by Mordecai et al. (2017), who invoked the analogy of Schrödinger's Cat (from the famous "simultaneously dead and alive" thought experiment in quantum theory). Although the haploid *E. huxleyi* cells are thought to escape infection, they contain viral lipids and viral RNA, but not viral DNA—hence the virus cannot progress to capsid formation and lysis in this stage, which they designate a haplococcolithovirocell. To complicate things further, a recent study by Johns et al. (2019) also indicates that the calcification state of diploid *E. huxleyi* cells indeed plays a protective role from infection by serving as a physical barrier from virus particles. Nevertheless, where infectious viruses go between blooms—when host densities are extremely low and distance between cells and EhV particles are extremely vast—remains an enigma!

EhVs manipulate host lipid metabolism. One of the most remarkable features of the EhV genome is the presence of a cluster of seven genes encoding an almost complete sphingolipid biosynthesis pathway (SBP) (Wilson et al., 2005; Monier et al., 2009). Virally encoded glycosphingolipids (vGSLs) are essential for EhV infection (Vardi et al., 2009; Rosenwasser et al., 2014; Ziv et al., 2016) and transcriptional studies show expression of viral SPB genes during infection in laboratory (Allen et al., 2006) and natural mesocosm studies (Pagarete et al., 2009). Electron microscopy studies have revealed that EhV infects cells using a mechanism that resembles that of enveloped animal viruses, involving entry of the entire nucleocapsid into the cytoplasm, following either fusion of the viral envelope with the cell membrane or an endocytotic process (Mackinder et al., 2009). During viral replication, capsids are transported through and assembled in the host cytoplasm, and leave the cell via a budding process. These processes require extensive manipulation of lipid metabolism during host–virus interactions (Pagarete et al., 2009, 2011; Malitsky et al., 2016; Ziv et al., 2016) and indeed, vGSLs comprise >80% of the lipid content of the EhV envelope (Fulton et al., 2014)). In addition, purified vGSL from an EhV virion suppresses host growth in a dose-dependent manner (Vardi et al., 2009) and EhVs use vGSL enriched membrane lipid domains called "lipid rafts" for entry and exit from the host cell (Rose et al., 2014). In view of the importance of vGSLs in the infection process, and previous observations (Nissimov et al., 2016) that closely-related EhVs can outcompete one another during infection, Nissimov et al. (2019) characterized the dynamics, diversity and catalytic production of vGSLs in a range of EhV strains. Interestingly, the laboratory infections suggest that fast-infecting virulent EhV strains with higher rates of vGSL production and utilization of intracellular substrates for sphingolipid production have a competitive advantage at high host densities in laboratory experiments. However, slow-infecting EhVs appear to co-exist and sometimes dominate their faster counterparts in natural North Atlantic populations. Nissimov et al. (2019) conclude that although the biochemical diversity of glycosphingolipid biosynthesis is a major driver of the competitive ecology of EhV strains, additional factors in the natural environment likely influence the relative abundance

BOX 7.2   RESEARCH FOCUS

of fast- and slow-infecting EhV genotypes, which can influence affect their co-existence. Coupled with mathematical ecosystem modeling, it was concluded that some of these factors likely include differences in removal rates of fast-infecting viruses compared to slow-infecting ones into the deep ocean, and access to a larger array of susceptible hosts available to slower-infecting EhV genotypes.

Assisted suicide? The sudden crash—from densities up to $\sim10^5$ cells mL^{-1}—that is so characteristic of *E. huxleyi* blooms is usually attributed to viral infection. Mature EhV virions are released by budding, so why do infected cells undergo such rapid lysis? In fact, Mimivirus unicellular phytoplankton undergo senescence and cell death as a natural process of their life cycle due to cell aging and adverse conditions such as nutrient depletion, high light, oxidative stress, or excessive salinity. A "cellular suicide" phenomenon known as programmed cell death (PCD) occurs—this is a fundamental developmental process with ancient origins in all forms of life (Bidle, 2016). Inhibitors of these enzymes were shown to prevent production of EhV virions (Bidle et al., 2007). The major processes are apoptosis (morphological changes including cell shrinkage, DNA fragmentation, and cell membrane blebbing) and autophagy (in which phagosomes interact with lysosomes to engulf and digest cellular constituents). A complex system of regulation

occurs by surveillance of the stress molecules nitric oxide and reactive oxygen species and cell signaling pathways. The importance of PCD in infection of *E. huxleyi* by *Coccolithovirus* is illustrated in *Figure 7.6*. The virus hijacks the process of PCD in the host cell by triggering the upregulation of host metacaspase enzymes, which are highly specific proteases that play a central role as "executioners" in PCD (Bidle, 2016). In particular, the process of autophagy is controlled by the vGSLs and is essential for virus assembly and exit from the cells. Schatz et al. (2014) used a variety of microscopic and biochemical methods to show that EhV initiate upregulation of autophagy-related genes, which are essential for propagation of the virus and generation of a high burst size. Specific inhibitors of autophagy did not affect EhV DNA replication but led to a large drop in the assembly and production of extracellular virions. Nissimov and Bidle (2017) highlight the great significance of PCD in ocean biogeochemical cycles as it relates to stress, because it can determine whether the products of photosynthesis by phytoplankton are transferred predominantly to the microbial loop for regeneration in surface waters, or exported to deep waters via the biological pump (see *Chapter 8*). Further complexity is added by bacteria associated with the *E. huxleyi* phycosphere, whose pathogenic effects are mediated by DMSP, as discussed in *Box 9.2*.

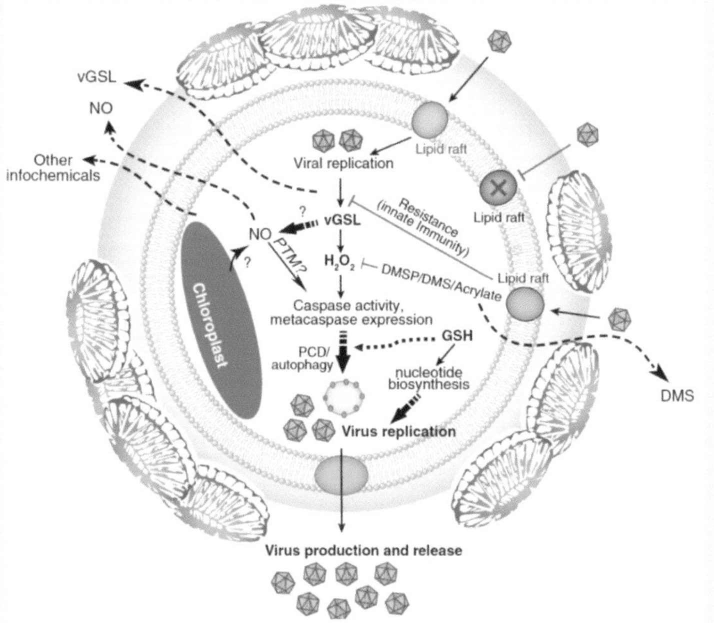

"Cross-kingdom thievery and metabolic thuggery". It has been known for some time that SBP genes were transferred from the host to EhV by horizontal gene transfer (HGT) (Wilson et al., 2005; Monier et al., 2009), but the comparative analysis of the 13 EhV genomes by Nissimov et al. (2017) revealed evidence of even more extensive HGT. They showed that 84% of the EhV coding sequences (CDSs) have close similarities to those in the Eukarya. In addition, results of Nissimov et al. (2017) indicate that many EhV genes may have had their origin in a wide range of other protists, some phylogenetically very distant from the Isochrysidiales (the order to which *E. huxleyi* belongs). Perhaps EhVs (or their ancestors) have (or had, in their evolutionary past) alternative hosts. Furthermore, 16% of EhV genes were found to be very similar to those in the Bacteria. Indeed, many bacteria exist in close association with *E. huxleyi* cells and their phycosphere, and colonization of the exopolymers exuded by the algal cell during virus infection may facilitate evolutionary relationships such as HGT between the algal cell, its virus, and the co-habiting, co-occurring bacterial community. Nissimov et al. (2017) conclude by highlighting the central role of viruses in genetic transfer across the three domains of life, but acknowledge the challenge of unwinding these relationships to provide further insight into the ecology and evolution of viruses and their hosts.

NO nitric oxide; vGSL viral glycosphingolipid; GSH reduced glutathione
DMSP dimethylsulfoniopropionate; DMS dimethyl sulfide
PCD programmed cell death; PTM post-translational modification

Figure 7.6 Importance of PCD in interactions between an *E. huxleyi* calcified diploid cell and *Coccolithovirus*. Credit: reprinted from Bidle (2016) with permission from Elsevier.

Other giant viruses are abundant pathogens of heterotrophic protists

The designation "giant virus" was first applied following the discovery of a giant, dsDNA virus with a 650 nm icosahedral capsid and a genome of 1.2 Mb, which was isolated from *Acanthamoeba polyphaga* (La Scola et al., 2003). It was originally thought to be a bacterium due to its size and was therefore named *Mimivirus* (derived from mimicking microbe). Several related giant viruses were isolated soon afterward, including *Mamavirus* from another *Acanthameoba* species and *Megavirus* from marine sources. These viruses have sufficient information to allow them to carry out most, but not all, parts of their life cycle. They possess some genes for energy production but are dependent on the host ribosomes for translation of mRNA into proteins (Suzan-Monti et al., 2006). The life cycle thus resembles that of *Rickettsia* but is clearly viral in nature as it does not replicate through cell division, but by assembly of preformed units. Most DNA viruses of eukaryotes insert their DNA into the nucleus of host cells, where it is replicated. However, after infection of *Acanthamoeba*, *Mimivirus* enters an eclipse period, during which host cells are instructed to build a large organelle-like "virion factory" in which metabolic processes leading to the synthesis of new virus particles are coordinated. The function of these intracytoplasmic virion factories has been compared with that of a cell nucleus, because they both contain the apparatus for DNA replication and transcription but lack the ability for energy production and mRNA translation. This discovery has led to revival of the hypothesis that evolution of the eukaryotic nucleus may have occurred by infection of an ancestral bacterial cell by a virus. Other evidence supports the concept that DNA viruses are the origin of replication proteins in eukaryotic cells.

After the discovery of *Mimivirus*, reexamination of the Sargasso Sea and GOS metagenomes (see p.55) revealed large numbers of sequences homologous to NCLDV genes and found that members of the family Mimiviridae are the second most abundant group after phages, suggesting that they commonly infect plankton. Further analysis showed multiple sequences of specific genes that clustered with four of the six NCLDV families. Because these metagenomes were derived from filters designed to capture particles between 0.1 and 0.8 μm, it was presumed that the sequences must have come either from free giant viruses of similar size to *Mimivirus* or from infected picoeukaryotes.

A very large (300 nm capsid diameter) lytic virus infects the abundant protist *roenbergensis*. This is a motile phagotrophic flagellate with a major role as a grazer of bacteria, viruses, and other picoplankton (p.171). The genome of the *C. roenbergensis virus* (CroV) was obtained by pyrosequencing and *de novo* assembly in 2010. The 0.73 Mb genome contains 544 putative protein-coding sequences (CDS), most of which are of unknown function. The diverse array of genes includes a large number associated with DNA replication, translation (including tRNAs), and transcriptional control mechanisms, indicating that replication and propagation of CroV is relatively autonomous in comparison with many other viruses. Like other mimiviruses, elements of the genome are thought to have evolved by HGT from eukaryotic hosts and bacteria.

BsV is a giant virus infecting the heterotrophic kinetoplastid flagellate *Bodo saltans* virus. This has a similar role to *C. roenbergensis*, and the two genera were initially confused. Based on the recruitment of sequences from metagenomes, BsV appears to be one of the most abundant members—and the first to be isolated and cultured—of a subfamily of the Mimiviridae (proposed name Klosneuvirinae), which are the most abundant group of giant viruses in ocean metagenomes. It has an icosahedral particle size ~300 nm and has a genome size of 1.39 Mb with 1227 CDS and complex replication machinery (*Figure 7.7*). Almost all components of translation have been lost, including tRNAs, but it does carry tRNA repair genes making it likely that it depends on the host's tRNAs during infection. The genome reveals an array of toxins and DNA cutting enzymes, which likely prevent replication of other competitor viruses. It has also acquired a membrane fusion system from its host via HGT, which is thought to enable release of the viral genome into the cytoplasm. Recently, strains of a new genus called Tupanvirus (a sister clade to the amoebal mimiviruses) have been isolated from soda lake and deep ocean sediments. Besides the exceptional size of their virions (*Figure 7.8*), they are remarkable because the genome encodes an almost complete apparatus for the translation of all 20 standard amino acids in proteins. Astonishingly, there

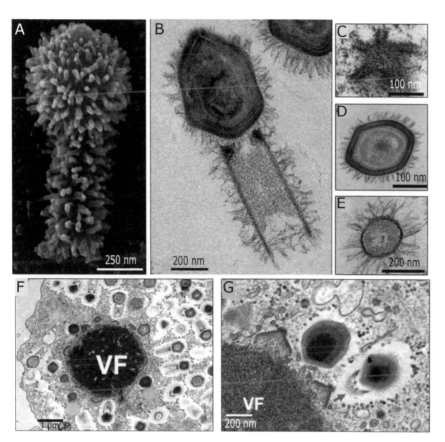

are ORFs for up to 70 tRNA, 20 ORFs related to aminoacylation and transport, 11 factors for all translation steps, and factors related to tRNA/mRNA maturation and ribosome protein modification. Only the genes encoding the ribosome itself are missing. Tupanvirus appears to have a broad host range and infects amoebas and other protists in experimental laboratory systems, although the natural hosts of Tupanvirus and its abundance and distribution in the ocean environment are currently unknown.

RNA viruses also infect protists

It is generally assumed that DNA viruses are the most dominant viruses in the ocean, but this assumption may be influenced by methodological constraints because RNA viruses with small genomes are more difficult to observe and enumerate using electron microscopy or flow cytometry, and metagenomic methods for their detection have not been fully developed. Nevertheless, RNA viruses appear to be very diverse, and numerous ecologically important examples occur in the major virus groups, as shown in *Table 7.1*. PCR-based screening of seawater samples using conserved sequences of the RNA-dependent RNA polymerase gene has demonstrated that distinct groups of picorna-like viruses are abundant, widespread, and persistent. Metagenomic analysis based on reverse-transcribed shotgun sequencing has shown that most sequences are unrelated to known RNA viruses. Some RNA viruses are well-known because of their economic importance as pathogens in marine fish and shellfish aquaculture (*Chapter 11*). However, until recently, there has been little information about the role of RNA viruses in marine ecology as pathogens of planktonic microbes. Only a small number of RNA phages infecting bacteria are known, but several RNA viruses infecting protists have now been described. A ssRNA virus called HaRNAV in the family Reoviridae infects the raphidophyte *Heterosigma akashiwo*. HaRNAV has been investigated extensively in view of the importance of this virus in controlling blooms of this organism, which is responsible for large fish kills. Electron microscopy of infected cells shows that they may contain as many as 2×10^4 viral particles. Genome sequencing led to the definition of a new virus family called the Marnaviridae. These viruses seem to be widespread and may play an important role in population dynamics of other marine protists. A dsRNA virus has also been isolated from the prasinophyte *Micromonas pusilla*.

Diatoms are of major importance in ocean food chains, being one of the main components of the phytoplankton responsible for primary productivity, especially in coastal waters and at high latitudes. One might imagine that the silica frustule of diatoms would protect them from viral infection, but some viruses have been identified as lytic agents of bloom-forming diatoms. Small viruses presumably enter via pores in the frustule. Novel ssRNA viruses have been isolated from *Rhizosolenia setigera* in temperate coastal waters. Another ssRNA virus has also been described in *Chaetoceros* spp., where infection leads to a burst size of more than 10^4 virions per cell. As infection of diatoms is likely to have a major impact on primary production and ocean processes, further study of the role of viruses in population dynamics is necessary. For example, we do not know the extent to which viral infection is involved in the collapse of diatom blooms, which is a major contributor to the formation of marine snow and silicon cycling. Nevertheless, there is evidence to suggest that, as in *E. huxleyi*–EhV infections, diatom viruses also contribute to the aggregation of material following infection and potentially influence the vertical flux of carbon into deep waters.

Dinoflagellates form another major group of the phytoplankton of special importance in coastal waters, with many forming harmful blooms. Two distinct types of ssRNA virus (HcRNAVs) (as well as the dsDNA Phycodnaviridae virus HcV mentioned above) have now been described from *Heteroplasma circularisquama*. There appears to be a clear correlation between density of HcRNAV and host population dynamics. Again, the burst size of infected cells seems to be very large. VLPs have been observed at high densities in heterotrophic and mixotrophic protists in Arctic sea-ice brines. Densely packed small icosahedral VLPs have been observed in myxosporean parasites of fish, and viruses have been described as putative pathogens of sarcodines, rhizopods, and radiolarians.

A novel icosahedral ssRNA virus (possibly a picornavirus) that affects the thraustochytrid protist *Schizochytrium* sp. has also been described. Since thraustochytrids have special importance in marine ecology because of their high content of omega-3 polyunsaturated fatty acids, viral infection of this group could be very significant in ocean food webs and biotechnological applications.

Viral mortality plays a major role in structuring diversity of microbial communities

As well as influencing microbial diversity by HGT and genetic rearrangements between host and virus, viruses influence host diversity at ecological scales. The discovery in the 1990s that phages are so numerous and responsible for high levels of bacterial mortality led to the development of models to explain how this might be influencing community structure. These approaches subsequently extended also to interactions between viruses and algae. The probability of an encounter between a host cell and a virus is more likely at high host densities because increased contact rates between a virus and its host are determined by their relative abundances. Also, some viruses may be very specific, and small changes in host cell surface receptors or replication processes could make them resistant to infection by a particular virus genotype (but note the earlier discussion suggesting that some phages probably have a broad host range). The density dependence and host range factors mean that viruses should preferentially infect the most common hosts: abundant hosts are more susceptible and rare hosts are less susceptible. Therefore, viruses may control excessive proliferation of hosts that have an advantage in nutrient acquisition and growth, encouraging a high diversity in the host population so that less competitive, but virus-resistant, microbes survive. This has been developed into a mathematical model termed the "kill the winner" hypothesis, discussed in *Box 7.1*.

BOX 7.3

New ideas about the nature and evolution of viruses

Perhaps the most wondrous discovery in virology in the past decade has been that of the giant viruses. With each new giant virus discovered—many of them in the marine environment—new surprises about their structure, replication, and interaction with their hosts has emerged. During the same period, the widespread use of high-throughput sequencing and metagenomics has confirmed the enormous diversity of viruses, and deciphering their genomes has given rise to new ideas about the nature and evolution of viruses and how this has influenced the evolution of all other forms of life.

Are viruses alive? This question provokes great differences of opinion whenever it is discussed—ask a group of virologists and they will probably split into three roughly equal groups answering "alive," "not alive," or "something in between." There is no clear-cut answer, simply because there is no completely satisfactory definition of what we mean by "life." In the early days of virology at the end of the nineteenth and early twentieth century, viruses were considered to be infectious agents—originally thought to be liquid, but subsequently shown to be small particles—that behaved like bacteria, and they were perceived as simple living entities. However, in the 1930s, the view of viruses changed to one in which they were regarded as purely chemical structures composed of proteins and nucleic acids. By focusing our attention on the virus particles (virions), the dogma grew that they cannot be "alive" in the usual sense of the word. However, when a virus infects a cell, it is clear that it is part of a living system, even though it is not self-sustained and depends on the host cell to enable it to replicate.

We can regard life as an emergent, complex state—a collection of non-living molecules which need to reach a critical level of complexity and interactions before it achieves the status of "living." Villareal (2004) gives the analogy of human consciousness—for which a whole fully functional brain is needed, not just the nerve cells—as another example of an emergent complex system. Villareal argues that this is more than just a philosophical question, because the tendency to categorize viruses as non-living has had the unintended, serious consequence of leading a generation of biologists to ignore their key importance in evolution—a major oversight from which we are only just escaping.

Viruses, virions, and virocells. An important concept to grasp is that a virus can only reveal its biological (living?) nature when it is replicating inside a host cell. A viral particle (virion) might be inactive and incapable of infecting a cell—this is one of the problems of attempting to enumerate "viruses" by fluorescence microscopy or flow cytometry. Forterre (2012) introduced the concept of the "virocell" to describe a cell (a bacterium, archaeon, or eukaryote) that has been transformed by viral infection so that its major function becomes the propagation of viral genes, rather than cell division. (In some cases, normal cellular processes can continue alongside viral replication). With this concept, the virocell is a cellular organisms, so it can clearly be regarded as the "living form" of the virus, whereas the virions are akin to the seeds or spores of multicellular organisms—packages of genetic information that will only be expressed under the right environmental conditions. Forterre

(2013) emphasizes the importance of the virocell as "cradles of new genetic information" and the idea that cells are giant "pickpockets of viral genes" during replication or recombination—a phenomenon of immense importance that is overlooked by many evolutionary biologists. Furthermore, as Forterre emphasizes, we can only truly understand the ecology and evolution of viruses by studying virocells to elucidate the entry, intracellular, and exit stages of the viral life cycle—complementing the vast bodies of metagenomic information. This adds further weight to the importance of overcoming the "unculturability" of most marine microbes discussed in *Box 2.1*. Rosenwasser et al. (2016) also highlight the realization that "virocell metabolism is a unique metabolic state, determined by modulation of host-derived metabolic genes and the introduction of virus-encoded auxiliary metabolic genes (vAMGs)" (see *Box 7.2*) and that study of virocells will "expand the current virion-centric approaches to quantify the impact of marine viruses on microbial food webs."

The origin of viruses revisited. One view for the origin of viruses is that they developed from degeneration of cells that were intracellular parasites of other cells; another is that they have evolved within cells via the escape of nucleic acids that have become partially independent of their cellular origin as "selfish" genetic entities. Koonin et al. (2006) reviewed the evidence for an alternative scenario—that viruses developed before the evolution of cellular organisms. Interestingly, this idea had been first put forward in the 1920s by Felix d'Herelle following his discovery of phages, and by J.B.S. Haldane in his 1928 treatise *The Origin of Life*. Koonin et al. (2006) showed how analysis of genes with key roles in viral replication and capsid formation can be used to identify viral hallmark genes that are shared by apparently unrelated viruses, but never found in cellular organisms (except as proviruses integrated into the genome). They use this to construct a concept of an ancient "virus world" with extensive gene mixing that predated the emergence of cellular organisms. Claverie (2006) proposed a model in which the initial step was infection of an RNA-based cell to form a primitive nucleus, which becomes transformed into a DNA-based cell because of the selective advantages of DNA biochemistry. He proposed that new pre-eukaryotic viruses were created by rapid reassortment of genes from the viral and cellular pools before the evolution of a stable eukaryotic cell with a fully developed DNA nuclear genome. Forterre (2006) suggested that the three domains of life originated from different RNA-based cell precursors containing DNA viral genomes. The discovery that mimiviruses contain the DNA encoding the molecules needed to translate mRNA into proteins prompted the idea that giant viruses descended from an ancient free-living cell that no longer exists—a "fourth domain of life" (Colson et al., 2012). These ideas have provoked much debate and alternative explanations have been proposed (e.g. Moreira and López-García, 2009). Recently, Schulz et al. (2017) discovered a new group of giant viruses (Klosneuviruses) whose genomes indicate an expanded translation system, including aminoacyl tRNA synthetases for all 20 amino acids. Schulz et al. considered that the evolutionary history of these translation proteins is incompatible with an origin in an ancient cellular ancestor or fourth domain of life and suggest that they have been acquired relatively recently in

BOX 7.3   RESEARCH FOCUS

a piecemeal fashion, but disagreement about interpretation of the results has not settled the debate (Reardon, 2017). The first member of the klosneuviruses to be cultivated was the *Bodo saltans* virus (BsV, Deeg et al., 2018) and genome analysis provides further evidence that the translational machinery of the klosneuviruses has not been derived from an ancient common ancestor. The size of the BsV genome appears to have grown rapidly by the duplication of genes at the end of the genome—a feature noted in other mimiviruses and dubbed the "genomic accordion" (Boyer et al., 2011). However, BsV appears to have lost most of the genes for its translational machinery, including all tRNAs. This loss might be explained by the fact that BsV infects the kinetoplastids, which have unusual RNA modifications. Deeg et al. suggest that strong evolutionary pressure is placed on the giant viruses by evolutionary virus-host arms races, with competition between related viruses for shared hosts. By contrast, Abrahão et al. (2018) isolated the Tupanviruses, which have an even more complete translation machinery than the klosneuvirus genomes studied by Schulz et al. (2017). The giant viruses continue to provide surprises with each new discovery.

Did the eukaryotic cell nucleus evolve from a virus? Most DNA viruses of eukaryotes insert their DNA into the nucleus of host cells, where it is replicated. However, it was shown that after infection of

their *Acanthamoeba* hosts, mimiviruses enter an eclipse period, during which host cells are instructed to build a large organelle-like "virion factory" in which metabolic processes leading to the synthesis of new virus particles are coordinated (La Scola et al., 2003). The function of these intracytoplasmic virion factories has been compared with that of a cell nucleus because they both contain the apparatus for DNA replication and transcription but lack the ability for energy production and mRNA translation. Bell (2001) had previously proposed that the eukaryotic nucleus may have evolved from an infection of an ancestral bacterial cell by a virus, and his controversial viral eukaryogenesis hypothesis was revamped in the light of the mimivirus discovery. Bell (2009) suggests that "the first eukaryotic cell was a multimember consortium consisting of a viral ancestor of the nucleus, an archaeal ancestor of the eukaryotic cytoplasm, and a bacterial ancestor of the mitochondria." Mimiviruses and other NCDLV possess several other features that are common to the eukaryotic nucleus (Bell, 2013; Villarreal and DeFilippis, 2000) and the recent discovery of the formation of a nucleus-like compartment for viral replication during infection of bacterial cells (*Pseudomonas*) infected with the very large phage 201φ2-1 (Chaikeeratisak et al., 2017) lends further support to the viral eukaryogenesis hypothesis.

Marine viruses show enormous genetic diversity

As with other microbial groups, the application of culture-independent methods has led to major advances in our understanding of the level of diversity of marine viruses. One of the problems in the study of viral diversity is that there is no universal marker like the ribosomal RNA genes used for other microbes. However, some success has been made by using representative signature genes that are sufficiently conserved to be used as markers of particular viral groups. For example, extensive virus diversity studies focused on the variation in the sequences of structural proteins, such as the g20 capsid protein gene in cyanophages or the major capsid protein (MCP) in phycodnaviruses. Primers for parts of the viral DNA or RNA polymerase genes are also used in PCR amplification reactions for phycodnaviruses or picornaviruses, respectively. Such approaches can be used for identifying and "fingerprinting" uncultured viruses in environmental samples, for example by PCR-DGGE or TGGE analysis (p.??). Construction of clone libraries and RFLP analysis and quantitative PCR (p.??) has also given some useful results for community analysis. These studies show that there is a very high level of diversity of single genes, even within these restricted viral groups. Attempts to overcome limitations of assessing diversity based on single genes (as previously discussed for other microbes in *Chapters 4–6*) have been made using microarrays (59) or pulsed field gel electrophoresis (48).

Major advances in the study of viral diversity were also made possible using high-throughput metagenomic sequencing. The genome sequences of a large number of cultured marine phages have been obtained, providing insight into metabolic functions and replication strategies. In addition, direct sequencing of environmental samples is helping us to understand the extent of viral diversity and biogeography in their natural setting. New developments in single-cell genomics and metagenome assembled genomes (p.56) of viruses is leading to vast increases in the amount of sequence data, which in turn catalyzes further exploration of viral diversity. In most cases however, the hosts of newly identified viruses remain unknown. As noted earlier, there is increasing evidence that many marine viruses have a relatively wide host range and co-infection by different viruses may be common.

Viromes are creators of genetic diversity and exchange

As noted previously, besides affecting microbial population dynamics, viruses influence diversity in the marine environment because of genetic exchange. Analysis of viral metagenomes (viromes) has unexpectedly revealed numerous genes involved in metabolic pathways, indicating that they are a reservoir of genetic information, which is important in the evolution and adaptation of their hosts to different ecological niches. Lysogeny of bacteria by temperate phages may lead to the introduction and expression of new genes into a host and excision of a prophage can lead to transduction of genes to new hosts. In addition, a process known as generalized transduction can occur, in which the enzymes responsible for packaging viral DNA into the capsid may mistakenly incorporate host DNA. These virions are defective and cannot induce a lytic infection, but DNA can be passed from one host to another and may recombine with the DNA of the recipient host. We have known for some time that viruses are vectors for HGT, but we now know that this has powerful effects on the evolution of both the microbial hosts and the viruses. Genetic information is moved by viruses from organism to organism and throughout the biosphere. As discussed in *Box 7.2*, genes encoding entire metabolic pathways may move between viruses and their hosts.

Unlike defective phages, which are "genuine" virions that are empty or have mistakenly packaged host DNA, another class of virus-like entities called gene transfer agents (GTAs) has recently been discovered in *Rhodobacter* and some other members of the Alphaproteobacteria. Some authors also refer to these as generalized transducing agents. GTAs are different to the defective phages described above because they seem to function only to transfer random fragments of DNA between cells. They are smaller than phages (about 30–40 nm head diameter), and they carry less DNA (about 4 kb) than would be required to encode the protein components of the virion. There are no negative effects associated with gene transfer to the recipient. It appears that some species of bacteria may produce these GTA particles as a dedicated mechanism of gene transfer and it has been suggested that they are cellular structures akin to flagella or pili. The gene cluster encoding the GTA is widespread in the Alphaproteobacteria and may have evolved from a defective prophage—through loss of replication, regulatory, and lysis genes—long before the divergence of the major phylogenetic groups of bacteria.

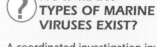

? HOW MANY TYPES OF MARINE VIRUSES EXIST?

A coordinated investigation involving metagenomic assemblies of 3.95 terabytes of sequencing data resulted in the Global Ocean Viromes 2.0 dataset, with sampling from >80 sites, from the surface to 4000 m, including new sequences collected as part of the *Malaspina* and *Tara Oceans* studies. Nearly 2×10^5 populations of DNA viruses were found, with the most diverse communities in tropical and temperate surface waters, as well as the Arctic Ocean. Because of the difficulties of defining viral species, Gregory et al. (2019) used a method of genotypically clustering sequenced viruses sharing >95% of their DNA. More than 90% of the viral "species" designated in this way couldn't be identified as belonging to any currently named virus family. Communities appeared to be mapped into five ecological zones based on temperature and depth, although viral macro- and microdiversity did not follow the latitudinal diversity gradient. Like most studies, this investigation did not evaluate the diversity of RNA viruses. Culley (2018) reviews the new sampling, methodological, and bioinformatics advances that will be needed to reveal this hidden diversity.

Conclusions

This chapter has illustrated the rapid pace at which the field of marine virology is moving. Viruses have emerged to take "center stage" in marine ecology and biogeochemical processes, through their effects on plankton composition and production. On a more fundamental level, recent metagenomic studies and the genomic analysis of isolated viruses have revealed that they provide an unprecedented reservoir of genetic diversity and play a major role in the evolution of life. The discovery of extensive gene transfer and the metabolic effects of viral infection provides important insights into evolution and adaptation to environmental change. Viruses feature again in the next three chapters, which include further discussion of their involvement in nutrient cycles, in specific biogeochemical cycles, and as disease agents of marine organisms other than microbes.

References and further reading

Diversity, ecology, and importance

Avrani, S., Schwartz, D.A., & Lindell, D. (2012) Virus-host swinging party in the oceans: Incorporating biological complexity into paradigms of antagonistic coexistence. *Mob. Genet. Elem.* **2**: 88–95.

Baran, N., Goldin, S., Maidanik, I., & Lindell, D. (2018) Quantification of diverse virus populations in the environment using the polony method. *Nat. Microbiol.* **3**: 62.

Brum, J.R., Schenck, R.O., & Sullivan, M.B. (2013) Global morphological analysis of marine viruses shows minimal regional variation and dominance of non-tailed viruses. *ISME J.* **7**: 1738–1751.

Cai, L., Jørgensen, B.B., Suttle, C.A. et al. (2019) Active and diverse viruses persist in the deep sub-seafloor sediments over thousands of years. *ISME J.* **13**: 1857–1864.

Chaikeeratisak, V., Nguyen, K., Egan, M.E., et al. (2017) The phage nucleus and tubulin spindle are conserved among large Pseudomonas phages. *Cell Rep.* **20**: 1563–1571.

Culley, A. (2018) New insight into the RNA aquatic virosphere via viromics. *Virus Res.* **244**: 84–89.

Danovaro, R., Dell'Anno, A., Corinaldesi, C. et al. (2008) Major viral impact on the functioning of benthic deep-sea ecosystems. *Nature* **454**: 1084–1087.

Derelle, E., Ferraz, C., Escande, M.L., et al. (2008) Life-cycle and genome of OtV5, a large DNA virus of the pelagic marine unicellular green alga Ostreococcus tauri. *PLoS One* **3**: e2250.

Feiner, R., Argov, T., Rabinovich, L., et al. (2015) A new perspective on lysogeny: Prophages as active regulatory switches of bacteria. *Nat. Rev. Microbiol.* **13**: 641.

Gregory, A.C., Zayed, A.A., Conceição-Neto, N. et al. (2019) Marine DNA viral macro- and microdiversity from pole to pole. *Cell* **177**: 1109–1123.e14.

Hewson, I. & Fuhrman, J.A. (2006) Viral impacts upon marine bacterioplankton assemblage structure. *J. Mar. Biol. Assoc. UK* **86**: 577–589.

Johns, C.T., Grubb, A.R., Nissimov, J.I., et al. (2019) The mutual interplay between calcification and coccolithovirus infection. *Environ. Microbiol.* **21**: 1896–1915.

Knowles, B., Silveira, C.B., Bailey, B.A., et al. (2016) Lytic to temperate switching of viral communities. *Nature* **531**: 466.

Martínez, J.M., Schroeder, D.C., & Wilson, W.H. (2012) Dynamics and genotypic composition of Emiliania huxleyi and their co-occurring viruses during a coccolithophore bloom in the North Sea. *FEMS Microbiol. Ecol.* **81**: 315–323.

Maslov, S. & Sneppen, K. (2017) Population cycles and species diversity in dynamic Kill-the-Winner model of microbial ecosystems. *Sci. Rep.* **7**: 39642.

Michaelson, L.V., Dunn, T.M., & Napier, J.A. (2010) Viral trans-dominant manipulation of algal sphingolipids. *Trends Plant Sci.* **15**: 651–655.

Munn, C.B. (2006) Viruses as pathogens of marine organisms—From bacteria to whales. *J. Mar. Biol. Assoc. UK* **86**: 1–15.

Nissimov, J.I. & Bidle, K.D. (2017) Stress, death, and the biological glue of sinking matter. *J. Phycol.* **53**: 241–244.

Reche, I., D'Orta, G., Mladenov, N. et al. (2018) Deposition rates of viruses and bacteria above the atmospheric boundary layer. *ISME J.* **12**: 1154.

Rohwer, F. & Thurber, R. V (2009) Viruses manipulate the marine environment. *Nature* **459**: 207–212.

Simmonds, P. (2015) Methods for virus classification and the challenge of incorporating metagenomic sequence data. *J. Gen. Virol.* **96**: 1193–1206.

Soler, N., Krupovic, M., Marguet, E. & (2015) Membrane vesicles in natural environments: A major challenge in viral ecology. *ISME J.* **9**: 793–796.

Suttle, C.A. (2005) Viruses in the sea. *Nature* **437**: 359–361.

Suttle, C.A. (2007) Marine viruses—Major players in the global ecosystem. *Nat. Rev. Microbiol.* **5**: 801–812.

Thingstad, T.F. (2000) Elements of a theory for the mechanisms controlling abundance, diversity, and biogeochemical role of lytic bacterial viruses in aquatic systems. *Limnol. Oceanog.* **45**: 1320–1328.

Winter, C., Bouvier, T., Weinbauer, M.G. & Thingstad, T.F. (2010) Trade-offs between competition and defense specialists among unicellular planktonic organisms: The "killing the winner" hypothesis revisited. *Microbiol. Mol. Biol. Rev.* **74**: 42–57.

Phages

Biller, S.J., Schubotz, F., Roggensack, S.E. et al. (2014) Bacterial vesicles in marine ecosystems. *Science* **343**: 183–186.

Breitbart, M., Bonnain, C., Malki, K., & Sawaya, N.A. (2018) Phage puppet masters of the marine microbial realm. *Nat. Microbiol.* **3**: 754–766.

Breitbart, M. & Rohwer, F. (2005) Here a virus, there a virus, everywhere the same virus? *Trends Microbiol.* **13**: 278–284.

Krishnamurthy, S.R., Janowski, A.B., Zhao, G. et al. (2016) Hyperexpansion of RNA bacteriophage diversity. PLoS Biol. **14**: e1002409.

Lindell, D., Jaffe, J.D., Johnson, Z.I. et al. (2005) Photosynthesis genes in marine viruses yield proteins during host infection. *Nature* **438**: 86–89.

Lindell, D., Sullivan, M.B., Johnson, Z.I. et al. (2004) Transfer of photosynthesis genes to and from Prochlorococcus viruses. *Proc. Natl. Acad. Sci. USA* **101**: 11013–11018.

Mann, N.H., Cook, A., Millard, A. et al. (2003) Marine ecosystems: Bacterial photosynthesis genes in a virus. *Nature* **424**: 741.

Ofir, G. & Sorek, R. (2018) Contemporary phage biology: From classic models to new insights. *Cell* **172**: 1260–1270.

Paul, J.H. (2008) Prophages in marine bacteria: Dangerous molecular time bombs or the key to survival in the seas? *ISME J.* **2**: 579–589.

Sullivan, M.B., Lindell, D., Lee, J.A. et al. (2006) Prevalence and evolution of core photosystem II genes in marine cyanobacterial viruses and their hosts. *PLOS Biol.* **4**: 1344–1357.

Coccolithovirus infection of *E. huxleyi*

Allen, M.J., Forster, T., Schroeder, D.C. et al. (2006) Locus-specific gene expression pattern suggests a unique propagation strategy for a giant algal virus. *J. Virol.* **80**: 7699–7705.

Bidle, K.D. (2016) Programmed cell death in unicellular phytoplankton. *Curr. Biol.* **26**: R594–R607.

Bidle, K.D., Haramaty, L., Barcelos e Ramos, J. & Falkowski, P. (2007) Viral activation and recruitment of metacaspases in the unicellular coccolithophore, *Emiliania huxleyi. Proc. Natl. Acad. Sci. USA* **104**: 6049–6054.

Frada, M., Probert, I., Allen, M.J. et al. (2008) The "Cheshire Cat" escape strategy of the coccolithophore *Emiliania huxleyi* in response to viral infection. *Proc. Natl Acad. Sci. USA* **105**: 15944–15949.

Frada, M.J., Rosenwasser, S., Ben-Dor, S. et al. (2017) Morphological switch to a resistant subpopulation in response to viral infection in the bloom-forming coccolithophore *Emiliania huxleyi*. PLoS Pathogens **13**: e1006775.

Fulton, J.M., Fredricks, H.F., Bidle, K.D. et al. (2014) Novel molecular determinants of viral susceptibility and resistance in the lipidome of *Emiliania huxleyi*. *Environ. Microbiol.* **16**: 1137–1149.

Jacquet, S., Heldal, M., Iglesias-Rodriguez, D. et al. (2002) Flow cytometric analysis of an *Emiliania huxleyi* bloom terminated by viral infection. *Aquat. Microb. Ecol.* **27**: 111–124.

Laber, C.P., Hunter, J.E., Carvalho, F. et al. (2018) Coccolithovirus facilitation of carbon export in the North Atlantic. Nat. Microbiol. **3**: 537.

Mackinder, L.C.M., Worthy, C.A., Biggi, G. et al. (2009) A unicellular algal virus, *Emiliania huxleyi* virus 86, exploits an animal-like infection strategy. *J. Gen. Virol.* **90**: 2306–2316.

Malitsky, S., Ziv, C., Rosenwasser, S. et al. (2016) Viral infection of the marine alga Emiliania huxleyi triggers lipidome remodeling and induces the production of highly saturated triacylglycerol. *New Phytol.* **210**: 88–96.

Martinez, J.M., Schroeder, D.C., Larsen, A. et al. (2007) Molecular dynamics of *Emiliania huxleyi* and cooccurring viruses during two separate mesocosm studies. *Appl. Environ. Microbiol.* **73**: 554–562.

Monier, A., Pagarete, A., de Vargas, C. et al. (2009) Horizontal gene transfer of an entire metabolic pathway between a eukaryotic alga and its DNA virus. *Genome Res.* **19**: 1441–1449.

Mordecai, G.J., Verret, F., Highfield, A. et al. (2017) Schrödinger's Cheshire Cat: Are haploid *Emiliania huxleyi* cells resistant to viral infection or not? *Viruses* **9**: 51.

Nissimov, J.I., Napier, J.A., Allen, M.J., & Kimmance, S.A. (2016) Intragenus competition between coccolithoviruses: An insight on how a select few can come to dominate many. *Environ. Microbiol.* **18**: 133–145.

Nissimov, J.I., Pagarete, A., Ma, F. et al. (2017) Coccolithoviruses: A review of cross-kingdom genomic thievery and metabolic thuggery. *Viruses* **9**: 52.

Nissimov, J.I., Talmy, D., Haramaty, L. et al. (2019) Biochemical diversity of glycosphingolipid biosynthesis as a driver of *Coccolithovirus* competitive ecology. *Environ. Microbiol.* **21**: 2182–2197.

Nissimov, J.I., Vandzura, R., Johns, C.T. et al. (2018) Dynamics of transparent exopolymer particle production and aggregation during viral infection of the coccolithophore, Emiliania huxleyi. *Environ. Microbiol.* **20**(8): 2880–2897.

Pagarete, A., Allen, M.J., Wilson, W.H. et al. (2009) Host-virus shift of the sphingolipid pathway along an *Emiliania huxleyi* bloom: Survival of the fattest. *Environ. Microbiol.* **11**: 2840–2848.

Pagarete, A., Le Corguillé, G., Tiwari, B. et al. (2011) Unveiling the transcriptional features associated with coccolithovirus infection of natural *Emiliania huxleyi* blooms. *FEMS Microbiol. Ecol.* **78**: 555–564.

Rose, S.L., Fulton, J.M., Brown, C.M. et al. (2014) Isolation and characterization of lipid rafts in *E miliania huxleyi* : A role for membrane microdomains in host-virus interactions. *Environ. Microbiol.* **16**: 1150–1166.

Rosenwasser, S., Mausz, M.A., Schatz, D. et al. (2014) Rewiring host lipid metabolism by large viruses determines the fate of *Emiliania huxleyi*, a bloom-forming alga in the ocean. *Plant Cell* **26**: 2689–2707.

Schatz, D., Shemi, A., Rosenwasser, S. et al. (2014) Hijacking of an autophagy-like process is critical for the life cycle of a DNA virus infecting oceanic algal blooms. New Phytol. **204**: 854–863.

Schroeder, D.C., Oke, J., Hall, M. et al. (2003) Virus succession observed during an *Emiliania huxleyi* bloom. *Appl. Environ. Microbiol.* **69**: 2484–2490.

Schroeder, D.C., Oke, J., Malin, G. & Wilson, W.H. (2002) *Coccolithovirus* (*Phycodnaviridae*): Characterisation of a new large dsDNA algal virus that infects *Emiliania huxleyi*. *Arch. Virol.* **147**: 1685–1698.

Sorensen, G., Baker, A.C., Hall, M.J. et al. (2009) Novel virus dynamics in an *Emiliania huxleyi* bloom. *J. Plankton Res.* **31**: 787–791.

Vardi, A., Van Mooy, B.A.S., Fredricks, H.F. et al. (2009) Viral glycosphingolipids induce lytic infection and cell death in marine phytoplankton. *Science* **326**: 861–865.

Wilson, W.H., Allen, M.J., Schroeder, D.C. et al. (2005) Complete genome sequence and lytic phase transcription profile of a *Coccolithovirus*. *Science* **309**: 1090–1092.

Wilson, W.H., Tarran, G.A., Schroeder, D. et al. (2002) Isolation of viruses responsible for the demise of an *Emiliania huxleyi* bloom in the English Channel. *J. Mar. Biol. Assoc. UK* **82**: 369–377.

Yamada, Y., Tomaru, Y., Fukuda, H., & Nagata, T. (2018) Aggregate formation during the viral lysis of a marine diatom. *Front. Mar. Sci.* **5**: 167.

Ziv, C., Malitsky, S., Othman, A. et al. (2016) Viral serine palmitoyltransferase induces metabolic switch in sphingolipid biosynthesis and is required for infection of a marine alga. *Proc. Natl. Acad. Sci. USA* **113**: E1907–E1916.

Giant viruses and virophages

Abrahão, J., Silva, L., Silva, L.S. et al. (2018) Tailed giant Tupanvirus possesses the most complete translational apparatus of the known virosphere. *Nat. Commun.* **9**: 749.

Boyer, M., Azza, S., Barrassi, L. et al. (2011) Mimivirus shows dramatic genome reduction after intraamoebal culture. *Proc. Natl. Acad. Sci. USA* **108**: 10296–10301.

Deeg, C.M., Chow, C.-E.T., & Suttle, C.A. (2018) The kinetoplastid-infecting *Bodo saltans* virus (BsV), a window into the most abundant giant viruses in the sea. *Elife* **7**: e33014.

Fischer, M.G., Allen, M.J., Wilson, W.H., & Suttle, C.A. (2010) Giant virus with a remarkable complement of genes infects marine zooplankton. *Proc. Natl. Acad. Sci. U S A* **107**: 19508–19513.

Fischer, M.G. & Suttle, C.A. (2011) A virophage at the origin of large DNA transposons. *Science* **332**: 231–4.

Forterre, P. (2010) Giant viruses: Conflicts in revisiting the virus concept. *Intervirology* **53**: 362–378.

Koonin, E.V. & Yutin, N. (2010) Origin and evolution of eukaryotic large nucleo-cytoplasmic DNA viruses. *Intervirology* **53**: 284–292.

La Scola, B., Desnues, C., Pagnier, I. et al. (2008) The virophage as a unique parasite of the giant mimivirus. *Nature* **455**: 100–4.

La Scola, B., Robert, C., Jungang, L. et al. (2003) A giant virus in amoebae. *Science* **299**: 2033.

Santini, S., Jeudy, S., Bartoli, J. et al. (2013) Genome of *Phaeocystis globosa* virus PgV–16T highlights the common ancestry of the largest known DNA viruses infecting eukaryotes. *Proc. Natl. Acad. Sci. USA* **110**: 10800–10805.

Sobhy, H. (2018) Virophages and their interactions with giant viruses and host cells. *Proteomes* **6**: 23.

Suzan-Monti, M., La Scola, B. & Raoult, D. (2006) Genomic and evolutionary aspects of Mimivirus. *Virus Res.* **117**: 145–155.

Viruses and evolution

Bell, P.J.L. (2001) Viral eukaryogenesis: Was the ancestor of the nucleus a complex DNA virus? *J. Molec. Evol.* **53**: 251–256.

Bell, P. (2013) Meiosis: Its origin according to the viral eukaryogenesis theory. *In Meiosis*. IntechOpen. doi: 10.5772/56876.

Broecker, F. & Moelling, K. (2019) What viruses tell us about evolution and immunity: Beyond Darwin? *Ann. N. Y. Acad. Sci.* **1447**: 53–68.

Chaikeeratisak, V., Nguyen, K., Khanna, K. et al. (2017) Assembly of a nucleus-like structure during viral replication in bacteria. *Science* **355**: 194–197.

Claverie, J.M. (2006) Viruses take center stage in cellular evolution. *Genome Biol.* **7**, 110.

Colson, P., De Lamballerie, X., Fournous, G., & Raoult, D. (2012) Reclassification of giant viruses composing a fourth domain of life in the new order Megavirales. *Intervirology* **55**: 321–332.

Forterre, P. (2006) Three RNA cells for ribosomal lineages and three DNA viruses to replicate their genomes: A hypothesis for the origin of cellular domain. *Proc. Nat. Acad. Sci. USA* **103**: 3669–3674.

Forterre, P. (2012) Virocell concept. In: *The Encyclopedia of Life Science*. John Wiley & Sons, Ltd.

Forterre, P. (2013) The virocell concept and environmental microbiology. *ISME J.* **7**: 233–236.

Hutchinson, G.E. (1961) The paradox of the plankton. *Am. Nat.* **95**: 137–145.

Koonin, E.V., Senkevich, T.G., & Dolja, V.V. (2006) The ancient Virus World and evolution of cells. *Biol. Dir.* **1**: 29.

Moelling, K. & Broecker, F. (2019) Viruses and evolution—Viruses first? A personal perspective. Front. Microbiol. **10**: 523.

Moreira, D. & López-García, P. (2009) Ten reasons to exclude viruses from the tree of life. *Nat. Rev. Microbiol.* **7**: 306–311.

Reardon, S. (2017) Giant virus discovery sparks debate over tree of life. *Nature News* (online) doi: 10.1038/nature.2017.21798.

Rosenwasser, S., Ziv, C., Van Creveld, S.G., & Vardi, A. (2016) Virocell metabolism: Metabolic innovations during host–virus interactions in the ocean. *Trends Microbiol.* **24**: 821–832.

Schulz, F., Yutin, N., Ivanova, N.N. et al. (2017) Giant viruses with an expanded complement of translation system components. *Science* **356**: 82–85.

Villarreal, L.P. (2004) Are viruses alive? *Sci. Am.* **291**: 100–105.

Villarreal, L.P. & DeFilippis, V.R. (2000) A hypothesis for DNA viruses as the origin of eukaryotic replication proteins. *J. Virol.* **74**: 7079–7084.

Villarreal, L.P. & Witzany, G. (2010) Viruses are essential agents within the roots and stem of the tree of life. *J. Theor. Biol.* **262**: 698–710.

Wessner, D.R. (2010) Discovery of the giant mimivirus. Nat. Ed. **3**: 61.

Chapter 8

Microbes in Ocean Processes—Carbon Cycling

The aim of this chapter is to bring together areas explored in the preceding chapters on eco-physiology and diversity of marine microbes, emphasizing their importance in carbon cycling in the oceans and their global biogeochemical significance. *Chapters 3–6* have discussed the activities of various individual types of marine microbes in major processes including pho-totrophy, chemolithotrophy, nitrogen fixation, heterotrophic breakdown of organic material, and oxidation–reduction transformations of major elements. These processes occur in seawa-ter and sediments, as well as in specialized habitats such as hydrothermal vents, cold seeps, and epi- or endobiotic associations. In this chapter, and in *Chapter 9*, the focus is primarily on the role of planktonic microbes. This exciting area of work, in which the activities of micro-biologists and chemical and physical oceanographers come together, has led to spectacular paradigm shifts in our view of the importance of ocean microbes.

Key Concepts

- Microbes play a central role in carbon cycling in the oceans, both as primary producers and as consumers.

- Fixed carbon from primary production is released as dissolved organic material (DOM) which is metabolized by heterotrophic bacteria, archaea, and fungi.

- Aggregates of particulate organic matter (POM) are exported to the deep ocean and sediments via the biological pump.

- Degradation of labile components of DOM results in a large residue of refractory DOM that has accumulated over millennia in the deep oceans via the microbial carbon pump.

- Grazing of heterotrophs by protists and filter feeding by tunicates leads to the transfer of fixed carbon to higher trophic levels via the microbial loop.

- Viral lysis of bacteria, archaea, and eukaryotic organisms catalyzes nutrient regenera-tion in the upper ocean by diverting the flow of carbon from the food chain into a semi-closed cycle of bacterial uptake and release of nutrients, as well as promoting carbon flux due to release of aggregative polymers.

- Marine phytoplankton are responsible for about half of the global CO_2 fixation, with cyanobacteria and picoeukaryotes having a dominant role in oligotrophic gyres and with blooms of eukaryotic phytoplankton being most important in mixed, turbulent waters.

- There is increasing recognition of the importance of chemolithoautotrophic bacteria and archaea in "dark" CO_2 fixation.

BOX 8.1   RESEARCH FOCUS

The role of microbes in the ocean carbon cycle

Milestones in discovery. This *Box* highlights a small selection of milestone research papers that have led to major paradigm shifts in microbial ecology and oceanography. Space does not permit a full historical treatment of the development of modern ideas about carbon cycling and ocean food webs, but I will outline a few of the most important highlights in the last few decades. These built on the pioneering work by early marine microbiologists including the German scientists Ernst Haeckel, Bernard Fischer, and Wilhelm Benecke, who first described marine microbes in the late nineteenth and early twentieth century. A succession of scientists in the mid-twentieth century—including Haldane Gee, Claude Zobell, Selman Waksman, Holger Jannasch, E. J. Ferguson Wood, Cornelius B. van Niel, and John Sieburth—established research programs in the USA at Scripps Institute for Oceanography, Hopkins Marine Station, and Woods Hole Oceanographic Institution (WHOI), which laid the foundations for the modern subject. In this abbreviated account of developments, I name a few individuals whose work is especially associated with these conceptual advances, but it is important to recognize that progress in this field has only been possible by the collaborative, interdisciplinary research of numerous microbiologists, oceanographers, biogeochemists, mathematical modelers, and other scientists that underpins our knowledge.

New methods reveal the importance of microbes in carbon cycling. Until the mid-1970s, marine bacteria were regarded as of little importance other than as decomposers of detritus. The classic view of trophic interactions in the oceans was of a simple food chain, in which primary production by photosynthesis is due mainly to algae like diatoms and dinoflagellates that are large enough to be trapped in traditional plankton nets. A simple food chain was envisaged, in which these algae are consumed by copepods (a diverse group of small crustaceans), which in turn are consumed by larger zooplankton, eventually reaching fish at the end of the food chain. In this scenario, phytoplankton was assumed to be consumed as rapidly as it is produced, with all the primary production going through herbivorous zooplankton. Bacteria did not feature at all in this food chain. Estimates of bacterial abundance were orders of magnitude lower than we now know to be the case, fitting with the prevalent view at that time that most of the oceans are biological "deserts" with low-nutrient fluxes, low biomass of phytoplankton, and low productivity. In 1977, John Hobbie of WHOI published one of the most important methods papers in marine microbiology (Hobbie et al., 1977; >5085 citations) describing the use of controlled pore-size filters and epifluorescence microscopy for the visualization of marine bacteria. This also led to the discovery of *Synechococcus* (Waterbury et al., 1979; >933 citations), a previously overlooked cyanobacterium now recognized as one of the major contributors to primary productivity in a large part of the oceans. We quickly realized that the oceans are, in fact, teeming with microbes—typically 10^6–10^7 mL^{-1}—in the picoplankton size class (<2–3 μm). Until the early 1980s, the perceived small populations of bacteria were thought to be largely dormant and this perspective only changed as a result of the application of new techniques, as discussed in *Chapter 2*. A paper in 1974 by Lawrence Pomeroy of the University of Georgia, entitled "The ocean's food web: a changing paradigm" (Pomeroy, 1974; >1423 citations) is widely credited as being one of the most influential advances in our thinking about the role of microbes in the movement of energy and nutrients in marine systems. The main arguments in this paper were: (1) that the main primary producers in the oceans are small phototrophs <60 μm in size rather than the larger phytoplankton previously recognized from studies with traditional plankton net sampling; (2) that microbes are responsible for most of the metabolic activity in seawater; and (3) that dissolved and particulate organic matter (DOM, POM) form an important source of nutrients consumed by heterotrophic microbes in the marine food web. In the 1980s, new analytical techniques to measure *in situ* bacterial activity were developed. Jed Fuhrman and Farooq Azam (Scripps) pioneered the measurement of the incorporation of radiolabeled amino acids in field experiments, showing that a large proportion of marine bacterial communities are metabolically active, consuming up to half of the photosynthetically fixed carbon (Fuhrman and Azam, 1982; >1575 citations). These new methods were also used by Barry and Evelyn Sherr of the University of Georgia to reveal the importance of the consumption of a large fraction of bacterial production by small phagotrophic protists (Sherr et al., 1989; >229 citations), providing evidence of a direct link between bacterial production and higher components of the food web.

The "microbial loop" concept. Evidence about the role of the heterotrophic bacterioplankton and protistan grazing steadily accumulated until a series of seminal papers were published in the early 1980s by Farooq Azam (Scripps), together with other microbial oceanographers and ecologists from Denmark, South Africa, Norway, and Germany. The concept of the "microbial loop" was developed to explain the flow and cycling of dissolved organic matter (DOM) in the oceans (Azam et al., 1983). The paper has had one of the greatest impacts of all papers in the aquatic sciences (>5432 citations) and led to a true paradigm shift—a sudden and dramatic change in our thinking about the role of ocean microbes. Twenty-five years after publication, Tom Fenchel (University of Aarhus, Denmark) one of the original authors on the paper, reflected on why the paper was so influential (Fenchel, 2008: 222 citations). He points out that the general idea had been expressed more or less explicitly in several earlier papers and attributes the impact of the paper to the fact that the simple descriptive name "microbial loop" was used for the first time—suddenly clarifying and connecting the various studies and catalyzing further research.

The original microbial loop concept has been continuously developed as a result of new discoveries made possible by technical advances, involving numerous teams of microbiologists and oceanographers. For example, the development of flow cytometers that could be used on research vessels led to the discovery of photosynthetic *Prochlorococcus* by Penny Chisholm of MIT (Chisholm et al., 1988; <808 citations). We now know that this is one of the major primary producers in the oceans, and possibly the most abundant photosynthetic organism on Earth, yet it escaped detection

BOX 8.1   RESEARCH FOCUS

until 1988. Since the late 1980s, the large biomass and diversity of marine microbes has been revealed through the application of molecular techniques described in earlier chapters, and the pace of discovery has been dramatic. The previously unimagined importance of archaea, picoeukaryotes, and mixotrophs as components of the plankton has emerged in the last few years, whilst metagenomic techniques have led to the discovery of new organisms, new energy-generating mechanisms, and the unraveling of metabolic processes in organisms that have not yet been cultured.

The "viral shunt" concept. A highly significant addition to the microbial loop concept occurred in the 1990s, with recognition of the role of viruses in controlling community structure in the plankton and nutrient cycling. As discussed in the preceding chapter, the importance of marine viruses was unknown until the late 1970s, when pioneering studies by Jed Fuhrman's group (University of Southern California) first recognized their role in controlling bacterial populations in the sea (Proctor and Fuhrman, 1990; >986 citations). Growing knowledge of the role of viruses in nutrient cycles led Steven Wilhelm and Curtis Suttle (University of British Columbia) to assign the moniker "viral shunt" to the function of viral lysis in transferring organic carbon from microbes back into the DOM pool (Wilhelm and Suttle, 1999; >894 citations).

The "microbial carbon pump" concept. The most recent major paradigm shift in the ocean carbon cycle grew from extensive research carried out in the 1990s—largely driven by growing

concerns about the effects on ocean processes of increasing atmospheric CO_2 levels. Studies sought to answer questions about the ability of the ocean to absorb increasing CO_2 and the effects of associated ocean acidification and global warming. This led to the realization that the integrated activity of the microbial loop and viral shunt has the effect of altering the nutrient composition of DOM as it cycles through these pathways. Heterotrophic bacteria, archaea, and fungi break down labile (i.e. readily transformed) organic compounds, leading to a residue of increasing amounts of "inedible" material which is unavailable as a source of energy for further microbial respiration. This material is known as recalcitrant DOM (RDOM) and includes complex polymers such as those from cell walls, or compounds that have been modified by photochemical action. It accumulates in the deep ocean where it is sequestered for hundreds or thousands of years. Various multi-disciplinary teams published results from analysis of the origins, distribution, and analysis of RDOM collected during research cruises and this led to the realization that there is a huge, ancient reservoir of organic carbon compounds in the deep ocean. Again, many scientists were involved in this breakthrough, but Nianzhi Jiao of Xiamen University, China is credited with making the conceptual advance explaining how microbes "pump" bioavailable carbon into this reservoir of relatively inert compounds. A seminal paper (Jiao et al., 2010); >675 citations) published with other leading microbial oceanographers and ecologists first described the concept of the microbial carbon pump.

Physical factors and biotic processes determine the fate of carbon in the oceans

The unique chemical properties of carbon mean that it is present in many different inorganic and organic forms, playing a fundamental role in abiotic and biotic ocean processes. The major reservoir of carbon occurs as carbonate minerals in sedimentary rocks in the Earth's crust and in organic compounds such as coal, oil, and natural gas. The turnover of this carbon by natural geological processes is very slow, occurring over timescales of millennia. However, in the last few hundred years, human activities have accelerated the flux of carbon to the biosphere because of the use of fossil fuels and changes in land use. The atmosphere currently contains ~858 Gt (Pg) of carbon and the concentration has risen rapidly since pre-industrial times to ~405 ppm (parts per million), its highest level in the past 800,000 years. The oceans are estimated to contain ~662 Gt (Pg) of dissolved carbon, about three-quarters of that in the atmosphere. Understanding the role of marine microbes in the fate of carbon in the oceans is a critical component of assessing the impact and possible mitigation of climate change caused by anthropogenic CO_2 emissions.

Gaseous CO_2 is highly soluble in water and forms carbonic acid ($H_2CO_3^-$). At the current natural pH of seawater (7.8–8.2), this mostly dissociates rapidly to form bicarbonate (HCO_3^-, p.12). Absorption of CO_2 at the interface between ocean and atmosphere and its transfer to deeper layers is driven to a large extent by physical processes—circulation of water occurs as a result of tilted density gradients caused by changing temperature and salinity at the surface, wind forcing and mixing, and the Coriolis effect due to the Earth's rotation. In the North Atlantic and Southern Oceans, cold, saline, and dense water with a high CO_2 concentration sinks vertically to depths of 2000–4000 m and then distributes through the ocean basins. The mass of water sinking to the deep ocean is replaced by upwelling to the surface elsewhere. As the water warms, CO_2 escapes to the atmosphere again. This process for CO_2 circulation in the oceans is called the "solubility pump."

Figure 8.1 Biological processes in the sequestration of atmospheric CO_2 and transfer of organic material to the deep ocean. In the biological pump, CO_2 is fixed by phytoplankton and sinks to the deep sea as particulate organic material (POM). In the microbial carbon pump (MCP), dissolved organic matter (DOM) is taken up by bacteria and archaea, which enter food webs via protist grazing (microbial loop). Viral lysis of microbes returns POM to the DOM pool (viral shunt). Transformation of DOM by microbial activity results in the formation of recalcitrant (refractory) RDOM, which persists for 100s or 1000s of years in the deep ocean. Reprinted from Zhang et al. (2018) by permission of Oxford University Press; image courtesy of Glynn Gorick.

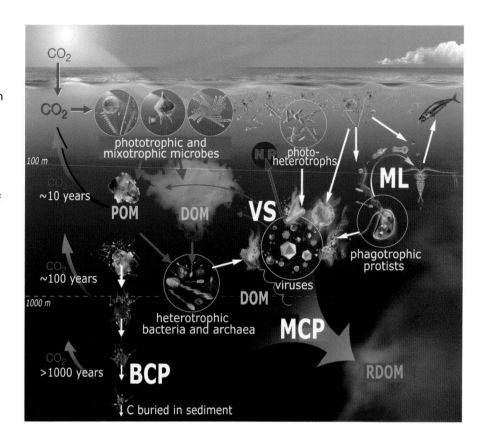

In addition to purely physical factors, the "biological carbon pump" is responsible for massive redistribution of organic carbon in the oceans (*Figure 8.1*). Fixation of CO_2 by phytoplankton in the upper ocean leads to primary production by incorporation of carbon into cellular material. Phytoplankton cells may be consumed by zooplankton or fish, so some of that carbon enters the food chain directly. Significant amounts of fixed carbon are released as DOC from phytoplankton cells by exudation and when they die and break up, either by natural senescence or due to viral lysis. Polymeric substances (TEPs and EPS, see p.12) and cell debris formed in the photic zone form POM, which aggregate (together with calcium carbonate and silica shells of protists) to form larger particles of marine snow (also see *Figures 1.7–1.8*). These settle into deeper waters by gravity, vertical migrations of zooplankton, and subduction of water masses. On its passage through the water column, a large fraction of the fixed organic carbon is respired by microbes and will be returned to the atmosphere over a 10–1000+ year timescale, varying with depth, as shown in *Figure 8.1*. About 25% of the primary production reaches the seafloor. There, most of the carbon compounds are broken down by microbial fermentation and respiration, fueling sulfate-reducing bacteria and methanogenic archaea. About 0.3% of primary production will be buried in deep sediments and over many millennia, this residue is transformed by increasing heat and pressure to form vast reservoirs of hydrocarbons.

The microbial loop, viral shunt, and microbial carbon pump components of the carbon cycling system illustrated in *Figure 8.1* provide both a *sink* for fixed carbon, as well as a *link* to higher trophic levels. About half of the DOM resulting from phytoplankton production passes through the microbial loop, acting as a source of energy, carbon, and other nutrients for heterotrophic organisms. This results in greater retention of dissolved nutrients in the upper ocean but, as mentioned above, microbial use of the carbon compounds alters the elemental composition and results in the RDOM, which is transported to the deep ocean. These processes are discussed in more detail below.

Marine phytoplankton are responsible for about half of the global primary production

Primary production can be defined in terms of the total amount of CO_2 fixed into cellular material during daylight (gross primary production) or as the fraction that remains after losses due to respiration by the phytoplankton, which occurs both day and night (net primary production). Another concept, net community production, has been introduced to encompass the effects of respiration by heterotrophic organisms. Oceanographers also use various other ways of measuring productivity (the rate of production) in terms of new, export, and regenerated (recycled) production. Both the biomass of phytoplankton and the rate of gross and net production are affected by a range of factors, which vary greatly over both long-term and short-term timescales, owing to the interplay of hydrographic conditions (currents, upwelling, and turbulence) with light conditions (latitude, season, cloud cover) and nutrient availability (especially nitrates, iron, and phosphorus). Temperature increases both photosynthetic and respiratory rates; but this effect is complicated in the oceans because, although most upwelling waters have a low temperature, the stimulation of photosynthetic rates by their increased nutrient levels overcomes the inhibitory effects of cooling.

Quantifying photosynthetic productivity has been a central question in biological oceanography for more than a century. Various methods are available, and these may produce different results depending on the parameter measured and the timescale employed. The most common laboratory experiments measure incorporation of radioactively labeled bicarbonate ($H^{14}CO_3^-$) into organic material, comparing the effects of incubation in light and dark conditions. Oxygen concentration can be measured by high-precision titration, and changes indicate O_2 evolution by oxygenic photosynthesis and O_2 consumption by respiration. Alternatively, specialized equipment can be used to follow $^{18}O_2$ evolution from $H_2^{18}O$, indicating photosynthesis without interference from respiratory processes. Clearly, the discovery of widespread anoxygenic phototrophic bacteria (p.77) means that earlier assumptions based on the use of O_2 evolution need to be reconsidered, as these bacteria are thought to be responsible for about 5% of primary production. All methods have certain drawbacks, and it is necessary to make allowances for isotope effects and differences in the photosynthetic quotient (ratio of HCO_3^- assimilated to O_2 generated), which varies according to species and nutrient availability. It is important to recognize that growth of phytoplankton is not a simple process concerned only with these processes; growth also depends on assimilation of other nutrients and many metabolic transformations to produce macromolecules in the cell. Measuring the physiological rates of photosynthesis in a natural setting is difficult. Most experiments are conducted in laboratory cultures with model representative organisms, or by enclosing seawater containing natural communities in bottles, which are then held at various depths in the field. Photosynthetic rates are measured over time under different environmental conditions, including light. Batch culture and bottle experiments obviously have inherent limitations because nutrients are quickly depleted in the absence of diffusion and artifacts can be introduced during deployment and retrieval of samples, by contamination, or by alteration of the spectral qualities of light entering the containers. Mesocosm experiments in large natural enclosures containing several cubic meters of seawater, or semi-continuous and chemostat culturing in the laboratory can overcome some of these problems (see p.61).

Since the distribution of primary production is highly dynamic and influenced by many factors, reliable estimates of productivity throughout the oceans have only been possible since the advent in the 1970s of remote sensing, using the spectral scanners on satellites, which measure ocean color. The NASA SeaWiFS satellite-borne sensor, which operated from 1977–2010, and its successors operated by international space agencies, have produced a vast database of global ocean color and other measurements that can be used to track regional and temporal variability in the phytoplankton communities, allowing estimates of primary production, as shown in the example data in *Table 8.1* and *Figure 8.2*. In clear waters, with little suspended material, the main factor influencing the ocean color is light absorption by the phytoplankton pigments, especially chlorophyll. Therefore, water appears bluer when there are low densities of phytoplankton, and greener when there are high densities. By measuring the different absorbance properties of chlorophyll *a* and other photosynthetic pigments, it is possible to map surface phytoplankton biomass daily (cloud cover permitting). The relationships between chlorophyll concentration, biomass of phytoplankton, and gross and net productivity are complex. Furthermore, the depth of the mixed layer, which can vary fivefold, has to be accounted for when estimating pools and fluxes. Mathematical models and different algorithms based on *in situ* and experimental measurements of the various parameters are then used to compute these values.

The total net annual primary production of the world's oceans is ~45–50 Gt (Pg) of carbon per year, which is similar to the total productivity of all terrestrial ecosystems. However, on an area basis, the annual global marine productivity is about 50 g C m^{-2}, which is only one-third that of terrestrial productivity. This discrepancy is due to the lower utilization of solar radiation by ocean phytoplankton than terrestrial plants, largely because of differences in the availability of nutrients and the effect of suspended particles—including the phytoplankton cells themselves—in absorbing light.

There are wide geographical and seasonal variations in primary production

As shown in *Table 8.1* and *Figure 8.2*, the most productive regions are upwelling regions associated with currents flowing along the west coasts of continents; namely, the Canary Current (off Northwest Africa), the Benguela Current (off southern Africa), the California Current (off California and Oregon), the Humboldt Current (off Peru and Chile), and the Somali Current (off Western India). Because of this, all these currents support major fisheries. Around the equator, winds and the Coriolis effect generate a divergence current resulting in upwelling of denser, nutrient-rich water. There is also large-scale upwelling in the Southern Ocean resulting from replacement of the northward flow of water driven by strong winds and the Coriolis effect. The least productive areas are the subtropical gyres of the ocean basins.

Around coastlines, photosynthetic activity is generally high owing to the input of nutrients from rivers, sediments, and wind-blown dust. Seaweeds, seagrasses, and salt-marsh plants in the coastal zone contribute about 5% of total global primary production. Tropical coral reefs are locally very productive because of the dense concentrations of photosynthetic dinoflagellates in many corals, but because they occupy such a small area, they probably contribute less than 1% to global primary production.

Although productivity on a global basis is fairly constant across the year, it is highly seasonal in some oceanic regions. High latitude temperate regions, especially the North Atlantic, characteristically show a spring bloom. Deep convective mixing occurs as the surface layer cools

Table 8.1 Net primary production of the oceans, estimated from SeaWiFS data using a vertically generalized production model. There is a slight discrepancy in the total production estimates between the geographical and seasonal estimates. (ICMS, 2000).

	Gt (Pg) C fixed	% of total
Annual production in different oceans		
Pacific	19.7	42.8
Atlantic	14.5	31.5
Indian	8.0	17.3
Antarctic	2.9	6.3
Arctic	0.4	0.9
Mediterranean	0.6	1.2
Total global annual production	46.1	100.0
Seasonal estimates of global ocean production		
March–May	10.9	23.0
June–August	13.0	28.2
September–November	12.3	26.7
December–February	11.3	22.1
Total global annual production	47.5	100.0

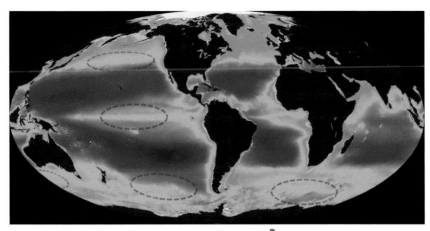

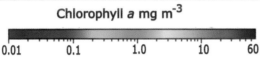

Chlorophyll *a* mg m^{-3}

Figure 8.2 Average global concentration of surface chlorophyll *a* measured by the SeaWiFS spectral scanner from 1997–2003. Chlorophyll concentrations (logarithmic scale) are higher near coastlines due to terrestrial runoff and nutrient upwelling, and higher in the northern than the southern hemisphere. Very low levels occur in the ocean gyres. Dotted ellipses indicate approximate locations of high-nutrient low-chlorophyll (HNLC) regions (see p.236). Credit: SeaWiFS Project, NASA Earth Observatory.

during the cold, dark, windy winter—bringing nutrients to the surface layers and promoting a rapid increase in photosynthesis as light levels increase during spring. Increased stratification and nutrient depletion lead to reduction in productivity during summer, even though light levels are at their highest. During autumn, vertical mixing occurs again and a small secondary peak of production results. In tropical seas, seasonal effects are much less pronounced, except where there is a seasonal upwelling of nutrients, such as that occurring in the Arabian Sea.

Remote sensing can now also differentiate size classes among the phytoplankton, based on the models that allow estimation of photosynthetic rates relevant to cell sizes combined with *in situ* measurements of abundance using flow cytometry. This reveals differences in geographic distribution and the relative contribution of different groups to the total production. Microphytoplankton (larger diatoms and dinoflagellates) are abundant mainly at mid to high latitudes and are patchily distributed. Nanophytoplankton (smaller diatoms and flagellates) are moderately abundant globally, but concentrations are higher at equatorial and mid to high latitudes and relatively low in subtropical gyres. Picophytoplankton (cyanobacteria and picoeukaryotes) mainly dominate oligotrophic gyres. Detailed analysis of monthly and annual variations in distributions of different size classes and correlation with other oceanographic parameters reveals how phytoplankton biomass and production vary over different periods—such information is a critical component of models to predict the impact of climate change on ocean processes.

Carbon is rarely a limiting factor in productivity, since HCO_3^- is abundant in seawater and most phytoplankton have mechanisms to concentrate it to improve photosynthetic efficiency (p.83), but nitrogen, phosphorus, silicon, trace metals, and some trace organic compounds such as B vitamins can all be limiting nutrients, thus affecting production. The availability of iron has special significance, as discussed in *Chapter 9*.

The penetration of water by light of different wavelengths limits photosynthesis to the upper 100–200 m of clear ocean waters, and considerably less in the presence of suspended material. Different phototrophic organisms are adapted to use light of different wavelengths due to different types of chlorophyll and ancillary pigments, and this affects their distribution in the water column. Indeed, the traditional definition of the limit of the euphotic zone as the depth at which light intensity is reduced to 1% of its surface level is no longer valid, since some ecotypes of *Prochlorococcus* are photosynthetic at light levels less than 0.1% of surface irradiance. Very high irradiances can inhibit photosynthesis, especially due to damage to the photosystem by ultraviolet light, and organisms possess photoprotective mechanisms to overcome these effects (p.107). The water depth at which the light intensity is just enough to balance the incorporation of CO_2 by photosynthesis and its loss due to respiration is known as the compensation depth. This varies according to species composition of the phytoplankton, geographic region, season, light penetration during the day, and nutrient availability.

Clouds and atmospheric dust have a marked effect on light penetration, and the density of phytoplankton itself is a major factor due to self-shading. Usually, maximum primary production does not occur at the very surface of the ocean, but several meters deeper where light and nutrient levels are optimal.

Dark ocean carbon fixation makes a major contribution to primary production

Although photosynthesis by the phytoplankton in the surface layer of the oceans is the most obvious source of marine primary production, experiments measuring incorporation of $^{14}CHO_3^-$ and ^3H leucine showed that there is substantial fixation of dissolved inorganic carbon (DIC) in the mesopelagic ocean, perhaps reaching a level half that of phytoplankton export production. As discussed above, much of the DOC transferred to the deep ocean is refractory and is therefore unable to support heterotrophic microbial growth. Furthermore, flux measurements of the transfer of particulate organic carbon (POC) show that it is not enough to meet the energy and biosynthesis demands of the standing stock of deep ocean microbes. Some of this shortfall is made up by the ability of heterotrophs to incorporate DIC by anaplerotic reactions—these are "top-up" mechanisms used to maintain the supply of intermediates in the TCA cycle. But this can only supply a small fraction of the carbon requirements of heterotrophs. Therefore, the recent discovery of chemoautotrophy as a source of new fixed carbon provides an explanation for the enigma that most of these microbes are metabolically active and clearly finding plenty of nutrients to live on.

The discovery that about a fifth of all the picoplankton in the deep ocean are chemolithoautotrophic archaea (phylum Thaumarchaeota, Class Nitrosphaera), which fix CO_2 using energy from ammonia oxidation (AOA, see *Box 5.2*) provided a possible explanation. However, calculations of the amount of ammonia reaching the mesopelagic zone—the only source is the breakdown of sinking POM—show that it is much lower than that needed to fuel the levels of autotrophy observed. The amount of carbon fixed by AOA is only about 10% of the estimate for dark ocean autotrophy.

Recently, it was discovered that the organisms responsible for this missing 90% of new organic carbon production are nitrite-oxidizing bacteria (NOB). Screening of single amplified genomes (SAGs) from the North Atlantic found that the major group of NOB in the deep ocean belongs to the phylum Nitrospinae. These bacteria contain genes encoding nitrite oxido-reductase and metagenomic and metatranscriptomic evidence supports nitrite oxidation as the only energy source for these bacteria. Experiments measuring $H^{14}CO_3^-$ incorporation indicate that they could fix ~1 Gt (Pg) C per year globally—about a tenth as much as the export production from the surface waters.

Classic food chain and microbial loop processes occur in the epipelagic

As noted earlier, the classic view of marine trophic interactions was a simple pyramidal food chain, roughly summarized as microalgae (mainly diatoms) → copepods → fish and whales. Development of the microbial loop concept has placed much greater emphasis on the importance of DOM and the activity of bacteria, archaea, grazing protists, fungi, and viruses, culminating in our modern view of the ocean food web represented schematically in *Figure 8.3*. This shows that there are trophic interactions at multiple levels, with microbial processes occupying center stage. However, elements of the classic model are still valid, and their relative importance depends on hydrographic and environmental conditions. In weakly stratified water masses, with a high degree of mixing and turbulence—such as polar and coastal temperate seas during the spring bloom—the food web is often based strongly on the dominance of organisms in the microplankton size class or greater (i.e. diatoms, dinoflagellates, colonial prymnesiophytes, and crustacean grazers). By contrast, the large areas of the ocean in the tropics and subtropics that have highly stratified waters with constant low-nutrient levels have food webs that are strongly dependent on microbial loop processes dominated by the activities of nanoplankton (small protists), picoplankton (bacteria and very small protists), and viruses. As a generalization, the relative

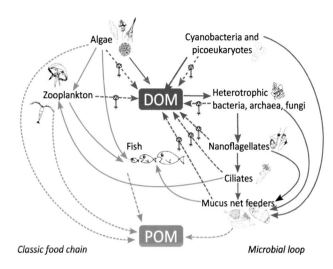

Figure 8.3 Schematic representation of ocean food webs, showing the central importance of dissolved organic material (DOM) and the microbial loop. The left side of the figure represents the "classic" food chain, whilst the right side represents microbial loop processes. Solid arrows show the main routes for trophic transfer of fixed carbon. Dashed blue lines indicate the release of cellular material as DOM by viral lysis of plankton, release of exudates, or cell breakage during feeding by zooplankton. Dashed orange lines indicate the formation of particulate organic material (POM) from aggregation of fecal pellets, polymers and cellular debris. Mucus net feeders such as salps and larvaceans can feed directly on very small microbes, including virus particles. For clarity, not all trophic connections and groups of organisms are shown. Images of organisms and viruses are illustrative only, and not to scale.

influence of the pico- or nanoplankton size classes of microbes is greatest in oligotrophic waters, because conditions of low-nutrient flux exert positive selection for small size (see p.71). The relative importance of activity of larger particle-feeding protists, zooplankton, and fish becomes progressively greater as nutrient levels in the water column increase. A useful analogy may be made between the behavior of food webs in nature and the behavior of laboratory cultures. Food webs dominated by the microbial loop tend to behave like chemostat (steady state) cultures. Despite rapid multiplication rates of some of the component organisms, species abundance and the concomitant geochemical processes in the oligotrophic ocean gyres seem remarkably stable if viewed over long timescales. By contrast, food chains dominated by microalgae and copepods typical of spring bloom events have dynamics that resemble batch laboratory cultures. Biomass, photosynthetic activity, and respiration increase rapidly and then decline owing to nutrient depletion, cell death, and the pressure of copepod grazing, leading to a greater export of photosynthetically fixed carbon to deeper waters through the sedimentation of phytoplankton aggregates, zooplankton fecal pellets, and cell debris. Certain hydrographic, nutritional, and climatic conditions can lead to the very sudden development of massive, dense blooms, such as dinoflagellates or diatoms and prymnesiophytes, which then disappear equally suddenly, often as a result of viral lysis (see *Chapter 7*).

The microbial loop results in retention of dissolved nutrients

About 50% of the daily net production from photosynthesis enters the ocean system as DOM, which supports the growth of heterotrophic microbes via the microbial loop, resulting in greater retention of dissolved nutrients in the upper layers of the ocean. Some DOM is formed from direct extracellular release of carbohydrates, amino acids, lipids, and organic acids from phytoplankton cells as they grow. The amounts released by this route are very variable but are generally highest in the most photosynthetically active regions at high light intensities. Extracellular release might result from overproduction of photosynthate when CO_2 fixation provides more organic material than can be incorporated into growing cells because of nutrient limitation. There is probably also a constant leakage of low molecular weight (MW) compounds across cell membranes owing to the steep concentration gradient caused by very low solute concentrations in seawater. Under some circumstances, phytoplankton cells can lose 5–10% of cell mass a day. The greatest contribution to the DOM pool comes from the release of dimethylsulfoniopropionate (DMSP) from microalgae—it has been suggested that no other single compound contributes as much to the DOM pool in surface waters—the formation and fate of this compound is discussed in the next chapter. A large amount of DOM and POM is released by protistan grazing on plankton. Partially digested particles of prey, unabsorbed small molecules, and digestive enzymes are released as colloidal material during the egestion process, when food vacuoles fuse with the external membrane. DOM and POM are also released during grazing as a result of "sloppy feeding," when phytoplankton cells are broken apart by the mouthparts of zooplankton such as copepods. However, by far the most important source of DOM is the lysis of phytoplankton and heterotrophic microbes by viruses. For example, up to 80% of the total photosynthetic production can be released as DOM during the collapse of phytoplankton blooms.

The biological pump transports fixed carbon to the deep ocean and sediments

The fecal pellets of zooplankton and fish, together with the fragments and polymeric substances from organisms that have died due to senescence or viral lysis contribute to the formation of aggregates, and these settle to deeper waters as marine snow. Thus, some of the carbon fixed in the upper layers is removed to deeper waters where the turnover time increases greatly. As it sinks, some of the carbon in POM is remineralized to CO_2 by the respiratory activities of microbes, zooplankton, or fish following ingestion of particles. As shown in *Figure 1.7*, viral lysis and the activity of extracellular enzymes leads to a substantial release of DOM from particles as they sink. The vertical movement of organic matter is accentuated by the daily migration of plankton over hundreds of meters in the water column.

Salps, larvaceans, and other mucus net feeding animals also provide a mechanism for export of POC produced by small phytoplankton. These are small, pelagic barrel-shaped tunicates that pump large volumes of seawater through their body for movement by jet propulsion and trap food particles in a sticky mucus net that is rolled up and passed to the digestive tract. Salps can trap and consume submicron particles, including cyanobacteria and heterotrophic bacteria, at very high rates. The role of salps in the food chain has been overlooked until recently, but we now know that they are a major food source for many species of fish, marine mammals, turtles, and seabirds—stable isotope analysis shows that they are the dominant source of nutrition for some species. Thus, they provide a direct trophic link between microbial processes and higher organisms. In addition, salps package undigested material into large, dense fecal pellets rich in carbon that provide a route for rapid sequestration of POC to deep waters, amplified by the vertical migration of salps to a depth of 600–800 m at night. The remains of dead salps also provide a large fraction of the carbon reaching the sea floor.

A large amount of fixed carbon—possibly as much as a quarter of net primary production—is also removed to the ocean floor as $CaCO_3$ from the walls of calcifying plankton such as coccolithophores and foraminifera. In waters deeper than 3000 m, some of this $CaCO_3$ dissolves to HCO_3^-, but in shallower waters the remains of these calcifying organisms lead to the formation of calcareous sediments. Together, all the processes described constitute the biological pump depicted in *Figure 8.1*, which results in the transport of fixed carbon from the atmosphere to deep ocean waters.

Carbon export of primary production may change due to ocean warming and acidification

Many studies conducted in the past few years have documented changes in the distribution of plankton, which appear to be occurring due to the warming of the surface ocean due to increasing levels of atmospheric CO_2. This leads to changes in stratification, salinity, and pH of the water column, which is expected to have marked effects on the production and export of fixed carbon.

Alterations in seawater chemistry as a result of ocean acidification (OA, p.12) will have a large effect on the process of calcification by organisms such as coccolithophores (p.180). Some studies based on projections of future OA scenarios indicate that such calcifying organisms could begin to experience difficulties in maintaining their $CaCO_3$ skeletons as early as 2050, especially in polar oceans where the state of $CaCO_3$ saturation is lower. However, research has also indicated that there is great variation in the responses of different species (and strains within species) to changes in ocean pH and CO_2 levels. Because coccolithophores fix CO_2 during photosynthesis, but also release CO_2 when they make the plates of calcite (coccoliths) that surround the cell, the balance between these processes is very important in the global carbon cycle. Increased CO_2 levels may enhance photosynthetic carbon fixation of some types of phytoplankton and it is possible that calcifying plankton might benefit at the expense of some other groups. Some mesocosm studies of the response of *Emiliania huxleyi* blooms to high CO_2 levels simulating end of the century conditions have shown a reduction in calcification rates when they were applied, but other evidence based on analysis of geological sediment cores suggests that coccolithophores have gradually increased their overall calcification rates in response to past increases in CO_2 levels. Some experimental studies have shown that some strains of *E. huxleyi* increased both photosynthesis and calcification rates at high CO_2 levels. Enhancement of photosynthesis in this and other phytoplankton might result in stimulation of the microbial loop through release of excess carbon as DOC and the microbial carbon pump might be effective in sequestering this to deep waters as RDOM. Alternatively, greater formation of sticky, polysaccharide-containing, transparent exopolymer particles (TEPs) might contribute to the formation of marine snow that settles more quickly, transporting POC and $CaCO_3$ to the seafloor via the biological pump. Thus, there are two mechanisms by which increased productivity might provide a partial negative feedback mechanism whereby some of the increased CO_2 levels dissolved in seawater is removed. Further evidence for a possible negative feedback mechanism comes from observations of the cyanobacterium *Trichodesmium*, which increases its rate of nitrogen fixation very markedly at high CO_2 levels. This could enhance the productivity of oligotrophic oceans that are currently limited by nitrogen and increase the flux of carbon in the biological pump.

An alternative scenario is that warming, freshening (due to melting of polar ice), and changes in stratification of surface waters will lead to a shift towards the dominance of small phototrophs, whose consumption by grazing heterotrophic protists results in retention of fixed carbon in the surface layers rather than transfer to higher trophic levels. This would result in a positive feedback mechanism in which microbial respiration remineralizes much of the DOC to CO_2. Numerous studies have shown an increasing abundance of picophytoplankton such as cyanobacteria and prasinophytes replacing microplankton such as diatoms and coccolithophores. This is generally predicted to lead to a reduction in carbon export via the biological pump due to lower sinking rates of picophytoplankton, which are smaller and lighter (possibly with positive buoyancy) than the negatively buoyant silica-containing diatoms or calcite-containing coccolithophores. However, we know now that aggregation of small phytoplankton due to production of TEPs means that particles can be large enough to be consumed by zooplankton and removed to deeper waters by vertical migration and the formation of fecal pellets containing undigested carbon compounds.

Ingestion of bacteria by protists plays a key role in the microbial loop

The consumption of bacteria by heterotrophic and mixotrophic protists (*Chapter 6*) helps to explain the regulation of bacterial biomass at near-constant levels. Bacterivory constitutes a mechanism by which nutrients contained in very small bacterial cells are made available to larger planktonic organisms. In addition to its importance as a "top-down" control on bacterial production involved in heterotrophic recycling, grazing of cyanobacteria by protists has a direct influence on primary production. It should be noted that, for many species of grazing protists, bacteria are not the sole diet. Feeding experiments and analysis of the food vacuole contents of protists show that many can feed on larger photosynthetic and heterotrophic protists as well as inanimate organic particles (by phagotrophy) or dissolved compounds (by osmotrophy). As discussed in *Chapter 6*, the most active and abundant bacterivorous protists are flagellates in the pico- and nanoplankton size classes, with most being less than 5 μm. Flagellated protists, including dinoflagellates, cryptomonads, euglenids, and ciliates in the microplankton (20–200 μm) class are also active bacterivores. These organisms feed by the generation of water currents with cilia and flagella and they can process hundreds of thousands of body volumes per hour. As noted in *Chapter 6*, many of these protists are mixotrophic. Larger protists such as radiolarians and foraminifera are also bacterivorous, but their importance is less well quantified. (*Chapter 6* contains micrographs of representative examples of these organisms.)

Some metazoan zooplankton can also consume bacteria directly. The larvae and juvenile stages of copepods have been shown to consume labeled bacteria (1–5 μm) in experimental studies, but they are possibly less efficient grazers of the very small bacteria that dominate

ARE SALPS A FAST TRACK FOR CARBON EXPORT?

The discovery that salps can feed on very small picoplankton, which includes primary producers, as well as heterotrophic bacteria and picoeukaryotes (Sutherland et al., 2010), means that large populations of salps might divert a substantial fraction of the carbon and energy from primary production away from heterotrophic remineralization by transferring it rapidly to deep waters, providing a mechanism for sequestering CO_2. Warming of the Southern Ocean (SO) is predicted to cause an increase of salps over krill. Salps can form large swarms and may therefore be important in the vertical export of POM. Iversen et al. (2017) concluded that a large proportion of salp fecal pellets would probably break up before reaching the deep ocean—"floaters" rather than "sinkers"—but their increased activity here might make production from picophytoplankton more accessible to other zooplankton, which can feed on undigested material in the pellets. In a linked study, Cabanes et al. (2017) concluded that increased dominance of salps in the SO could lead to greater export of POC, but this would be associated with greater removal of iron that would reduce its availability as a required nutrient for photosynthesis.

pelagic systems. As noted above, tunicates such as larvaceans trap bacteria in fine mesh structures constructed of gelatinous mucus, and the discarded structures and fecal pellets make a major contribution to the formation of marine snow. Indeed, since many pelagic bacteria and picoeukaryotes associate with particles of marine snow rather than exist as freely suspended forms, they may be consumed by a wide range of larger metazoan zooplankton and small fish when they ingest these particles, although they be unable to feed directly on individual cells.

An additional key component of the microbial loop is the release of DOM through protistan grazing. Large amounts of DOM—up to 25% of ingested prey carbon—are egested by protists, and some of this is readily metabolized and enters the DOM pool for further recycling by heterotrophs, whilst the remainder enters the sink of long-term refractory material. The relatively high respiration and excretion rates of protistan grazers mean that the cycling of "lost" photosynthetic products through the microbial loop is rather inefficient.

Detailed measurements of grazing rates are technically difficult. In one type of experimental study, the uptake of fluorescently labeled prey or ^{3}H– or ^{14}C–labeled prey bacteria and their accumulation in food vacuoles is followed using microscopy. Another approach is the dilution method, in which a dilution series of seawater is prepared, so that the natural growth rate of the prey bacteria is unaltered, whilst the predator to prey ratio is reduced. The rate of change in bacterial density is then monitored during incubation at different dilution levels. Separation of bacteria and their grazers in different size classes may also be achieved by filtration. Using such techniques, many studies have attempted to evaluate the impact of protistan grazing on the density, dynamics, and structure of bacterial populations in the oceans. In low productivity waters, grazing seems to be the dominant factor in balancing bacterial production. However, in higher productivity waters, viral lysis will be favored because of the higher bacterial population densities, as discussed below. Other factors such as removal by benthic filter-feeding animals may also be important in controlling bacterial numbers in coastal regions and estuaries.

Up to a certain limit, which depends on their own size, protistan grazers show a preference for larger prey as food. Therefore, one consequence of grazing pressure is to encourage the domination of microbial assemblages by small cells. As discussed on p.71, most bacteria in oligotrophic pelagic environments are less than 0.6 μm, with many less than 0.3 μm, and many of the smallest bacteria may be in a state of metabolic maintenance rather than active cell division. Thus, the interesting concept arises that protistan grazers are preferentially removing the larger, actively growing and dividing bacteria, leaving a large stock of smaller bacteria that are growing slowly, if at all. However, selective bacterivory of larger, active bacteria seems to stimulate the growth of other bacteria by the release of regenerated nutrients such as ammonium and phosphate. Use of high-resolution video techniques shows that, as well as chance encounters, flagellates may actively select their prey. Species-specific differences in the processing of food particles explain the coexistence of various bacterivorous nanoflagellates in the 3–5 μm size range and indicate the existence of specific predation pressure on different bacteria. Motility, surface characteristics, and toxicity can all affect the outcome of bacterium-protist interactions at the various stages of capture, ingestion, and digestion.

Bacteria are also susceptible to predation by other bacteria. The only well-studied example is *Bdellovibrio* (p.132), but it is likely that many other types remain to be discovered; these could have considerable ecological importance in nutrient dynamics in marine systems.

The viral shunt catalyzes nutrient regeneration in the upper ocean

When the microbial loop was first conceived, we had little idea of the abundance and activity of marine viruses. Now, we know that viral lysis of primary producers and heterotrophic organisms act as a short circuit that disrupts the flow of nutrients into higher trophic levels—the so-called viral shunt. As illustrated in *Figure 8.1* and *8.3*, viral lysis leads to the release of cell contents, much of which enters the DOM pool and is readily recycled by heterotrophs. However, cell fragments and high MW polymers are more recalcitrant to breakdown. For example, phage lysis of bacteria leads to release of cell envelope fragments containing embedded bacterial outer membrane proteins, which are relatively resistant to proteolytic degradation. Some components, such as cell wall polymers of algal cells lysed by viruses are also very refractory. The quantity and dynamics of these less labile components resulting from viral lysis are not yet fully known. Mathematical models can be constructed to compare the effects of different levels of viral mortality on nutrient budgets. At a level of 50% bacterial mortality from viruses, the overall level of bacterial production and respiration rate is increased by about a third, compared with that occurring with zero mortality due to viruses. In these models, a high level of protistan grazing leads to carbon input to the higher trophic levels of the food chain (animals), whereas a high level of viral lysis diverts the flow of carbon from the food chain into a semi-closed cycle of bacterial uptake and release of organic matter. Because cell fragments, viruses, and dissolved substances do not sink—unless they aggregate into larger particles—one effect of viral lysis is to maintain carbon and inorganic nutrients such as nitrogen, phosphorus, and iron in the upper levels of the ocean. Viral lysis also contributes to the microscale heterogeneity of seawater through the release of polymeric substances and the dissolution of material from marine snow. As noted in *Box 7.2*, viral infection of *E. huxleyi* blooms also directly enhances the efficiency of the biological pump by stimulating TEP production, TEP aggregation, TEP sinking, and zooplankton grazing. There is growing recognition that this production of sticky viral lysates and promotion of aggregation plays a greater role in carbon flux than previously realized—a "shuttle" rather than a "shunt" effect. As discussed in *Chapter 7*, both lytic and lysogenic cycles of infection occur when phages infect marine bacteria. Although the lysogenic state is common and has important consequences for genetic transfer, induction in natural marine systems seems to be relatively low, so the majority of viral infection seems to occur via encounter of bacteria with active virions that initiate the lytic cycle. This process is highly dynamic and is affected by the virus to host ratio, the contact rates between virus and host cells, the rate of virus adsorption, the rate of viral replication, the degree of host resistance, the physiological state of infected host cells, the availability of nutrients for the production of viable virus particles, the burst size of virus progeny, and the rates of viral decay. As discussed in *Chapter 7*, estimation of the impact of viral lysis on bacterial and microalgal populations produces very variable results, depending on the techniques used, but there is no doubt that viruses are responsible for a significant turnover of ocean microbes, perhaps 20–40% per day overall. As noted above, viral lysis is relatively more important in eutrophic coastal waters, whereas protistan grazing is more important in oligotrophic ocean waters.

Microbial processes alter the composition of DOM

Labile DOM from primary production by autotrophs is rapidly metabolized by heterotrophic microbes using extracellular or cell-associated ectoenzymes (p.91). Different organisms vary greatly in their preference for particular substrates depending on their possession of membrane transport systems; for example, among the dominant Alphaproteobacteria, SAR11 favors uptake of amino acids, whilst the *Roseobacter* clade favors carbohydrates. Degradation of semi-labile material such as complex polymers occurs more slowly and often depends on activity of specific organisms such as Bacteroidetes (p.92), fungi, or thraustochytrids (p.183) or the syntrophic activity of consortia of microbes. Different amounts of specific types of organic compounds can have a major influence on community structure of bacterioplankton.

Organic material is thus incorporated into microbial cells until it is returned to the DOM pool by viral infection and lysis, as discussed above. This secondary material will be less easily metabolized—an effect which will be amplified with each cycle of absorption, incorporation, and release by viral lysis, ultimately resulting in RDOM which is recalcitrant to further degradation—RDOM is usually defined as having a lifetime of >100 years. Different researchers have used various methods to calculate the annual rate of formation of this refractory carbon, resulting in a consensus value around 0.2 Gt (Pg) per year, with a total pool of ~662 Gt (Pg). Studies of the composition of RDOM shows that at least 25% is of bacterial origin, with much of it occurring as refractory forms of neutral sugars, amino sugars, and amino acids. The D-isomers of amino acids are particularly prevalent since they are major components of bacterial cell walls and require conversion by racemase enzymes before they can be metabolized. Certain types of microbial lipids

Figure 8.4 Schematic representation of the viral shunt. The viral shunt moves material from phototrophs (green arrows) and heterotrophs (red arrows) and into particulate organic matter (POM) and dissolved organic matter (DOM). During this process there is a stoichiometric effect, such that the C:N:P ratio (numbers of atoms in parenthesis) of the POM and DOM pools are altered. Material that is exported to deeper waters by the viral shunt is more carbon rich than the material from primary production, increasing the efficiency of the biological pump. Redrawn from Suttle (2007) by permission of Springer Nature.

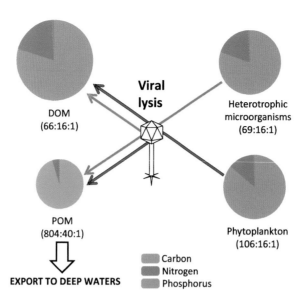

and derived alkane compounds are probably the most resistant and oldest forms of DOM and are very resistant to degradation. They persist in sediments for millions of years—for example, the use of crenarchaeol as a biomarker was used to establish the importance of archaeal autotrophy by Thaumarchaeota in the deep ocean (*Box 5.2*). Degradation products of carotenoid pigments (carboxyl-rich alicyclic molecules) are also a significant component of ancient DOM, with a probable lifespan of >1500 years. Abiotic processes, especially photochemical reactions at the ocean surface, also participate in conversion of DOM to a refractory form.

The combined activity of heterotrophic microbes and the recycling of nutrients via cycles of viral lysis leads to transformation in the elemental composition of DOM, because the residue after repeated cycles is enriched in carbon content relative to the ratios of nitrogen and phosphorus (*Figure 8.4*).

Eutrophication of coastal waters stimulates microbial activity

Nutrient enrichment of estuaries and coastal waters is a growing problem as a result of anthropogenic sources of pollutants such as runoff from heavily fertilized land, sewage, and animal wastes from agriculture and aquaculture. The impact of eutrophication depends on the source, nature, and level of nutrient inputs, as well as hydrographic factors (especially tidal flushing and mixing) and other physical factors (especially light and temperature). Increased nutrient loading—particularly from nitrate and phosphate fertilizers—stimulates phytoplankton growth beyond the point at which it is controlled by zooplankton grazing and/ or viral infection. For example, massive cyanobacterial blooms (e.g. *Nodularia*, *Microcystis*, and *Oscillatoria*) occur regularly in the Baltic Sea, and eutrophication is probably a major factor in the increased occurrence of harmful dinoflagellate blooms (p.319) as well as excessive growth of macroalgae. Active microbial loop processes convert excess primary production, but these too may be overwhelmed, and large amounts of decaying detritus and particles of organic material will sink toward the seafloor. Bacterial decomposition leads to a heavy demand for oxygen, resulting in mass mortality of benthic animals and fish. The number of such "dead zones" in shallow coastal waters has approximately doubled since the 1960s. Ironically, schemes to reduce the use of fossil fuels by production of bioethanol from crops such as corn may increase the spread of coastal dead zones unless runoff from increased fertilizer use can be controlled. Bacterial decomposition results in oxygen depletion, and the overlying water column may become hypoxic, with oxygen levels too low to support fish and many invertebrate animals. Even slightly reduced oxygen levels may disrupt food chains or cause stress-induced disease in economically important fish and shellfish. As well as these local effects, the expansion of dead zones may have serious consequences for climate change. Heterotrophic denitrification leads to the production of the greenhouse gas nitrous oxide

(p. 244); elevated atmospheric concentrations of this gas could further exacerbate the effects of global warming and contribute to ozone loss, causing an increase in exposure to harmful ultraviolet radiation. This is discussed further in *Chapter 9*.

Conclusions

This chapter has shown how research has revealed the central role and inter-connectedness of microorganisms and viruses through their activities as components of carbon cycling systems—the biological pump, microbial loop, viral shunt/shuttle, and microbial carbon pump. Whilst 'omics studies have led to major advances in understanding of processes in the surface ocean, knowledge of deep ocean microbiology (especially the functional biology of planktonic archaea) is still sparse and extensive, long-term microbiome time-series studies at different locations and depths are needed. Increased use of autonomous sampling and monitoring devices and development of large-scale mesocosms for experimental studies will be very valuable. This is an area of intensive research activity and coordinated methodological approaches are leading to further insights into possible future trends. Better understanding will come from improved knowledge of the rates of primary production in different size classes and taxonomic groups of phytoplankton, and from better quantification of the activities of heterotrophic microbes in the remineralization of DOM or its transformation to RDOM. We also need better knowledge of the decay rates of marine viruses, and how they will be affected by climate change and other altered conditions. Increased efforts are now needed to gain further understanding of the integration of these systems, and how they will respond to changes in the physical, chemical, and hydrographic properties of the oceans as a result of pollution by CO_2, microplastics, other pollutants, and terrestrial nutrient input.

References and further reading

Azam, F., Fenchel, T., & Field, J.G. (1983) The ecological role of water-column microbes in the sea. *Mar. Ecol.–Prog. Ser.* **10**: 257–262.

Cabanes, D.J.E., Norman, L., Santos-Echeandía, J. et al. (2017) First evaluation of the role of salp fecal pellets on iron biogeochemistry. *Front. Mar. Sci.* **3**: 289.

Carlson, C.A., del Giorgio, P.A., & Herndl, G.J. (2007) Microbes and the dissipation of energy and respiration: From cells to ecosystems. *Oceanography* **20**: 89–100.

Cermeño, P., Chouciño, P., Fernández-Castro, B. et al. (2016) Marine primary productivity is driven by a selection effect. *Front. Mar. Sci.* **3**: 173.

Chisholm, S.W., Olson, R.J., Zettler, E.R. et al. (1988) A novel free-living prochlorophyte abundant in the oceanic euphotic zone. *Nature* **334**: 340–343.

Diaz, R.J. & Rosenberg, R. (2008) Spreading dead zones and consequences for marine ecosystems. *Science* **321**: 926–929.

Dittmar, T. & Arnosti, C. An inseparable liaison: Marine microbes and nonliving organic matter. In: *Microbial Ecology of the Oceans*, 3rd edition (ed. Gasol, J.M. & Kirchman, D.L.), Wiley-Blackwell, pp. 189–288.

Dutkiewicz, S., Hickman, A.E., Jahn, O. et al. (2019) Ocean colour signature of climate change. *Nat. Commun.* **10**: 578.

Fenchel, T. The microbial loop—25 years later. *J. Exp. Mar Bio. Ecol.* **366**: 99–103.

Fuhrman, J.A. & Azam, F. (1982) Thymidine incorporation as a measure of heterotrophic bacterioplankton production in marine surface waters: Evaluation and field results. *Mar. Biol.* **66**: 109–120.

Hansell, D.A. & Carlson, C.A. (eds.) (2014) *Biogeochemistry of Marine Dissolved Organic Matter*. Academic Press.

Henschke, N., Everett, J.D., Richardson, A.J., & Suthers, I.M. (2016) Rethinking the role of salps in the ocean. *Trends Ecol. Evol.* **31**: 720–733.

Herndl, G.J., Reinthaler, T., Teira, E. et al. (2005) Contribution of Archaea to total prokaryotic production in the ocean. *Appl. Environ. Microbiol.* **71**: 2303–2309.

Hirata, T., Aiken, J., Hardman-Mountford, N. et al. (2008) An absorption model to determine phytoplankton size classes from satellite ocean colour. Remote Sensing Environ. **112**: 3153–3159.

Hobbie, J.E., Daley, R.J., & Jasper, S. (1977) Use of nuclepore filters for counting bacteria by fluorescence microscopy. *Appl. Environ. Microbiol.* **33**: 1225.

ICMS (2000) ICMS Ocean Productivity Study. Rutgers, The State University of New Jersey. Institute of Marine and Coastal sciences. http://marine.rutgers.edu/opp/

Iversen, M.H., Pakhomov, E.A., Hunt, B.P. et al. (2017) Sinkers or floaters? Contribution from salp pellets to the export flux during a large bloom event in the Southern Ocean. *Deep Sea Res. Part II Top Stud. Oceanogr.* **138**: 116–125.

Jiao, N., Herndl, G.J., Hansell, D.A. et al. (2010) Microbial production of recalcitrant dissolved organic matter: Long-term carbon storage in the global ocean. *Nat. Rev. Microbiol.* **8**: 593–599.

Jürgens, K. & Massana, R. (2008) Protistan grazing on marine bacterioplankton. In: *Microbial Ecology of the Oceans*, 2nd edition (ed. Kirchman, D.L.). Wiley, pp. 383–342.

Kolber, Z. (2007) Energy cycle in the ocean: Powering the microbial world. *Oceanography* **20**: 79–88.

Laber, C.P., Hunter, J.E., Carvalho, F. et al. (2018) Coccolithovirus facilitation of carbon export in the North Atlantic. Nat. Microbiol. 537–547.

Legendre, L., Rivkin, R.B., Weinbauer, M.G. et al. (2015) The microbial carbon pump concept: Potential biogeochemical significance in the globally changing ocean. *Prog. Oceanogr.* **134**: 432–450.

Madin, L.P., Kremer, P., Wiebe, P.H. et al. (2006) Periodic swarms of the salp *Salpa aspera* in the slope water off the NE United States: Biovolume, vertical migration, grazing, and vertical flux. Deep-Sea Res. I: Oceanogr. Res. Pap. **53**: 804–819.

Nagata, T. (2008) Organic matter-bacteria interactions in seawater. In: *Microbial Ecology of the Oceans*, 2nd edition (ed. Kirchman, D.L.). Wiley, pp. 207–242.

Pachiadaki, M.G., Sintes, E., Bergauer, K. et al. (2017) Major role of nitrite-oxidizing bacteria in dark ocean carbon fixation. *Science* **358**: 1046–1051.

Pernthaler, J. (2005) Predation on prokaryotes in the water column and its ecological implications. Nat. Rev. Microbiol. **3**: 537–546.

Pernthaler, J. & Amann, R. (2005) Role and fate of heterotrophic microbes in pelagic habitats: Focus on populations. *Microbiol. Mol. Biol. Rev.* **69**: 440–461.

Pomeroy, L.R. (1974) The ocean's food web: A changing paradigm. *Bioscience* **24**: 409–504.

Pomeroy, L.R., Williams, P.J.L., Azam, F., & Hobbie, J.E. (2007) The microbial loop. *Oceanography* **20**: 28–33.

Proctor, L.M. & Fuhrman, J.A. (1990) Viral mortality of marine-bacteria and cyanobacteria. *Nature* **343**: 60–62.

Rahlff, J. (2019) The virioneuston: A review on viral–bacterial associations at air–water interfaces. *Viruses* **11**: 191.

Reinthaler, T., van Aken, H.M., & Herndl, G.J. (2010) Major contribution of autotrophy to microbial carbon cycling in the deep North Atlantic's interior. *Deep Sea Res. Part II Top. Stud. Oceanogr.* **57**: 1572–1580.

Richardson, T.L. (2019) Mechanisms and pathways of small-phytoplankton export from the surface ocean. Ann. Rev. Mar. Sci.**11**: 57–74.

Robinson, C. (2008) Heterotrophic bacterial respiration. In: *Microbial Ecology of the Oceans*, 2nd edition (ed. Kirchman, D.L.). Wiley, pp. 299–327.

Schiffer, J.M., Mael, L.E., Prather, K.A. et al. (2018) Sea spray aerosol: Where marine biology meets atmospheric chemistry. *ACS Central Sci.* **4**: 1617–1623.

Sherr, B.F., Sherr, E.B., & Pedrós-Alió, C. (1989) Simultaneous measurement of bacterioplankton production and protozoan bacterivory in estuarine water. *Mar. Ecol. Prog. Ser.***54**: 209–219.

Sherr, E. & Sherr, B. (2008) Understanding roles of microbes in marine pelagic food webs: A brief history. In: *Microbial Ecology of the Oceans*, 2nd edition (ed. Kirchman, D.L.). Wiley, pp. 27–44.

Sutherland, K.R., Madin, L.P., & Stocker, R. (2010) Filtration of submicrometer particles by pelagic tunicates. *Proc. Natl Acad. Sci. USA* **107**: 15129–15134.

Suttle, C.A. (2007) Marine viruses—Major players in the global ecosystem. *Nat. Rev. Microbiol.* **5**: 801–812.

Talmy, D., Beckett, S.J., Taniguchi, D.A. et al. (2019) An empirical model of carbon flow through marine viruses and microzooplankton grazers. *Environ. Microbiol.* **21**: 2171–2181.

Talmy, D., Beckett, S.J., Zhang, A.B. et al. (2019) Contrasting controls on microzooplankton grazing and viral infection of microbial prey. *Front. Mar. Sci.* **6**: 182.

Waterbury, J.B., Watson, S.W., Guillard, R.R.L., & Brand, L.E. (1979) Widespread occurrence of a unicellular, marine, planktonic, cyanobacterium. *Nature* **277**: 293–294.

Wilhelm, S.W. & Suttle, C.A. (1999) Viruses and nutrient cycles in the sea. *Bioscience* **49**: 781–788.

Yamada, Y., Tomaru, Y., Fukuda, H., & Nagata, T. (2018) Aggregate formation during the viral lysis of a marine diatom. *Front.Mar. Sci.* **5**: 167.

Zhang, C., Dang, H., Azam, F. et al. (2018) Evolving paradigms in biological carbon cycling in the ocean. *Natl. Sci. Rev.* **5**: 481–499.

Zubkov, M. & Hartmann, M. Ecological significance of microbial trophic mixing in the oligotrophic ocean: The Atlantic Ocean case studies. In: *Microbial Ecology of the Oceans*, 3rd edition (ed. Gasol, J.M. & Kirchman, D.L.), Wiley-Blackwell, pp. 99–122.

Chapter 9

Microbes in Ocean Processes—Nitrogen, Sulfur, Iron, Phosphorus, and Silicon Cycling

This chapter integrates closely with the discussion of carbon and energy cycling in the preceding chapter and builds on the knowledge of ecophysiological processes and transformations by the different microbial groups introduced in *Chapters 3–6*. In the first section, the role of nutrients as limiting factors in ocean productivity is considered, with particular emphasis on the importance of iron. Then, the significance of nitrogen cycling in ocean processes is discussed, with emphasis on recent discoveries of new nitrogen-fixing organisms, anaerobic ammonia oxidation, and the role of archaea in nitrification, which have led to major revision of conventional ideas. Sulfur cycling is considered in the third section, mainly from the perspective of the transformations of organic sulfur compounds, the implications for climate control, and the structuring of microbial communities. This is followed by a brief overview of recent developments in our understanding of the uptake of inorganic and organic phosphorus. The final section considers the role of silica in the life of diatoms. In all cases, major advances in our understanding have occurred in the past few years due to the integration of biogeochemical studies with genomic and metagenomic approaches.

Key Concepts

- Phytoplankton growth may be limited by the availability of nitrogen, phosphorus, and silicon; co-limitation and effects on different community members often occurs.

- The availability and distribution of iron is a critical factor in ocean productivity.

- Oceanic nitrogen fixation is a greater contributor to the nitrogen cycle than previously thought and knowledge of the diversity, distribution, and importance of diazotrophic cyanobacteria is increasing.

- Fixed nitrogen is returned to the inorganic pool by ammonification and nitrification, which fuel growth of heterotrophs and phytoplankton.

- Nitrate reduction, denitrification, and anammox reactions return nitrogen to its elemental form and other gases; the balance of these processes carried out by networks of nitrogen-transforming microbes has important effects in oxygen minimum zones.

- A key component of the marine sulfur cycle is the production of dimethylsulfonio-propionate (DMSP) by algae which is consumed by numerous heterotrophic bacteria; products from its breakdown have many effects on community structure and ocean ecology and contribute to climatic effects.
- Marine microbes have diverse adaptations to phosphorus limitation; these are important factors in microbial evolution and community dynamics.
- Silicon is an essential limiting nutrient for the growth of diatoms and its availability affects competition between diatoms and other phytoplankton, with major influences on operation of the biological pump for carbon export from the photic zone.

NUTRIENT LIMITATION

Key elements act as limiting nutrients for phytoplankton

Carbon, nitrogen, and phosphorus are required by all phytoplankton and silicon is required by diatoms and some other organisms as major constituents of cellular material. Therefore, their availability affects the production of organic material that enters the food webs discussed in *Chapter 8*. Carbon availability is not normally a limiting nutrient, because of the high concentration of HCO_3^- in seawater, but most surface waters have low levels of inorganic N, P, and Si. All organisms also require iron, as it is a key component of many enzyme systems and electron transfer proteins. Although cells incorporate only miniscule amounts of Fe, its low concentration in seawater is often the most important factor determining productivity. Liebig's Law of the Minimum is a principle derived from crop science, holding that growth is not dependent on the total amount of resources available, but by the single resource in shortest supply. Study of nutrient cycling in the oceans reveals more complex interactions. Limitation of phytoplankton may occur by multiple nutrients, as well as light. Such co-limitation can operate at the level of individual cells, at the population level, or at the community level. Diverse species of plankton may be limited by different nutrients and may be affected differently at various stages of their life cycle. Furthermore, the use of these and other nutrients is closely linked in mixed natural communities, so it is not easy to untangle these complex interactions.

Productivity of surface waters shows marked geographical variations

Early ideas that available nitrogen is the most important limiting nutrient in the oceans arose largely because most of the initial studies of biological oceanography were carried out in temperate North Atlantic waters. During the spring bloom, dominated by diatoms, investigators observed that the increase in phytoplankton biomass and concentration of nitrate were inversely proportional. They showed that phytoplankton growth in seawater increased with experimental addition of ammonium, but not of phosphate. In addition, some phosphate remained even when the nitrate levels reached zero. At the onset of the spring bloom in temperate waters, phytoplankton biomass—as indicated by chlorophyll concentrations—is uniform throughout the mixed layer, but as stratification develops, the maximum chlorophyll levels occur in deeper waters where nitrate levels are higher. New production is that portion of primary production dependent on new sources of nitrogen to the photic zone, including nitrate from the deep layers, nitrogen fixed by diazotrophs, and terrestrial runoff. Regenerated production is that portion of primary production supported by nitrogen recycled by heterotrophic decomposition in the photic zone.

This view of limiting nutrients has been challenged by detailed study of open-ocean ecosystems. In particular, remote sensing has confirmed the existence of ocean expanses which have much lower phytoplankton biomass (indicated by chlorophyll concentrations <0.5 µg L^{-1} chlorophyll) than would be expected considering the relatively high (>2 µM) concentrations of nitrate and phosphate. These high-nutrient, low-chlorophyll (HNLC) expanses occupy about 20% of the surface area of the oceans in the eastern equatorial and subarctic regions of the Pacific and the whole of the Southern Ocean *Figure 8.2*. Nitrate and phosphate are never significantly depleted, and the low-chlorophyll anomaly is largely explained by the

---

? WHAT IS THE REDFIELD RATIO?

During a series of cruises from WHOI, Alfred Redfield analyzed thousands of marine samples from different ocean regions. In a keynote paper published in 1934, he reported that the average elemental ratios of C:N:P in POM and dissolved nutrients are nearly constant at 106:16:1. These findings have been hugely influential in our understanding of biogeochemical cycles. Many subsequent studies have revealed that, although the average C:N:P ratio is generally stable, this can conceal significant differences in temporal and geographical deviations and differences between phytoplankton species. Some organisms are adapted to sustained growth under low-resource conditions, while others respond to high-nutrient concentrations with rapid growth (blooming).

low levels of available iron. The trace metals zinc and cobalt may also be co-limiting. These regions tend to be dominated by pico- and nanophytoplankton size classes; larger phytoplankton are less prevalent, especially diatoms and the cyanobacterium *Trichodesmium*. In some areas, such as the eastern Mediterranean, phosphorus appears to be the key limiting nutrient, while silicon availability can be limiting for diatom growth in coastal waters and may lead to harmful algal blooms (p.254).

Ocean microbes require iron

Bioavailable iron occurs in extremely low concentrations in seawater due to its low solubility at ~pH 8. Nitrogen fixation and photosynthesis are especially dependent on iron because the key components in these processes—the nitrogenase enzyme complex and the photosystem-cytochrome complex respectively—rely on iron-containing proteins. Over 99.9% of "dissolved" iron is tightly bound to organic compounds; it occurs mainly as colloidal $Fe(OH)_3$, which is very insoluble, precipitates rapidly and adsorbs strongly to organic particles (*Figure 9.1*). Microbes need special mechanisms to acquire this iron and many bacteria achieve this by production of siderophores (p.92), which bring iron into the cell via binding of the organic complexes to surface receptors and specialized transport mechanisms to bring the iron into the cell. Some bacteria are also able to acquire the iron bound to siderophores produced by other species. Photochemical reactions with α-hydroxy acid-containing siderophores (such as aquachelins) may lead to an increased bioavailability of the siderophore-complexed iron. In the nitrogen-fixing cyanobacterium *Trichodesmium*, iron deprivation leads to formation of the puff-like trichome colonies from individual filaments (*Figure 4.12*) and these serve to collect and concentrate iron-containing dust particles. Although some novel mechanisms for iron acquisition have been identified in some members of the Gammaproteobacteria, their general significance for marine bacteria is not clear, and almost nothing is known about iron uptake in archaea. Eukaryotic phytoplankton do not appear to synthesize siderophores but can take up iron via a cell-surface ferrireductase enzyme that liberates iron bound to organic compounds such as porphyrins, which are released from cells through zooplankton grazing and viral lysis. Phagotrophic eukaryotes such as flagellates can acquire iron from ingested bacteria. Depending on the chemical nature of available iron complexes, microbes could therefore be competing for iron and the outcome will affect the separation of ecological niches and community composition. This will have consequences for the fate of carbon in iron-limited waters. Siderophore-bound iron may be the major source of the element in regions dominated by cyanobacterial photosynthesis and the microbial loop (such as the oligotrophic tropical and subtropical oceans), while porphyrin-complexed iron may be the major source in coastal regions dominated by diatoms and zooplankton grazing. Another important factor may be the production of storage proteins to sequester scarce elements like iron within cells. We know that some bacteria produce ferritin-like compounds to store iron when it is more abundant than is needed for immediate use, but these compounds have not yet been detected in aquatic bacteria. Microalgae are known to synthesize phytochelatins, which can store a range of trace metals, but these do not seem to sequester iron. An interesting observation is that domoic acid—a toxin produced by the diatom *Pseudo-nitzschia* (p.320)—binds iron and is produced in greater amounts when cultures are grown with high iron concentrations.

Terrestrial runoff, dust, and volcanic ash are major sources of iron input

Low concentrations of iron have been proposed to explain the paradox of the low productivity of HNLC regions. As noted above, these oceanic regions show a much lower rate of photosynthesis than would be expected from the availability of nitrate. The major source of iron is terrestrial, and it follows that most coastal regions receive regular input from rivers, runoff from weathering of rocks, and wind-borne transport of dust. The extent of this input depends on the geology of the land, and some upwelling coastal regions can be iron-deficient. By contrast, the HNLC regions occur at great distances from the continents and will only receive iron inputs via wind-blown dust or from remobilization of iron in sediments and upwelling of deep waters. Episodic input of iron also occurs following volcanic eruptions. Iron concentrations in surface water (<200 m) decrease with distance from land. Some of the experimental

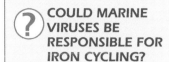

? **COULD MARINE VIRUSES BE RESPONSIBLE FOR IRON CYCLING?**

Many phages recognize and infect their bacterial hosts via the cell-surface TonB receptors for siderophores. As iron is of such importance as an essential micronutrient, receptors are highly conserved, favoring their use by phages as an entry route. The tail fibers of phages of enteric bacteria are known to contain small numbers of iron ions, which are responsible for recognition of the TonB binding sites. It seems probable that marine phage tails also contain iron, leading Bonnain et al. (2016) to propose a mechanism dubbed the "Ferrojan Horse Hypothesis"—derived from the "Trojan Horse" metaphor, whereby the target bacterium "invites" the phage attacker with a "gift" of iron. Bonnain et al. calculate that the ~10^7 phage per mL could contain as much as 70% of the colloidal iron (0.02–0.2 µm) in the surface ocean and that cellular iron stocks are recycled into the tail fibers of new phage, which therefore act as organically complexed "dissolved" iron. If supported by further research, this concept will have major impacts on ocean biogeochemistry.

Figure 9.1 Iron cycling in the oceans. Arrows show the transfer of iron and the acquisition mechanisms employed by marine microbes. POM = particulate organic matter.

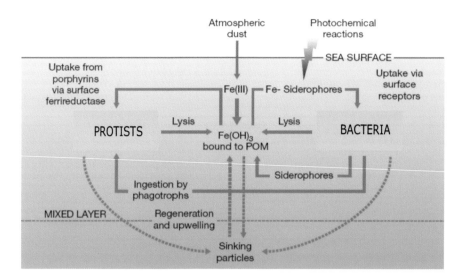

evidence in support of the iron-limitation theory is discussed in *Box 9.1*, together with controversial proposals to use iron fertilization of the oceans in geoengineering, as a way of removing excess CO_2 from the atmosphere as a countermeasure for global warming.

Hydrothermal vents and glacial melting also supply iron to the oceans

Recently, iron input sources other than wind-borne dust have been discovered. There is an annual flux of ~10^9 mol Fe of dissolved iron from hydrothermal vent plumes. This becomes associated with particulate organic complexes that can be transported to provide a major source of iron in the deep ocean. Bacteria in the region of these vents show high expression of genes associated with siderophore synthesis and uptake, leading to the hypothesis of a "microbial iron pump" that transfers inorganic iron to the organic phase, which may be dispersed throughout the oceans (*Figure 9.2*).

Iron also enters the ocean via meltwater from sediment released from the base of glaciers and icebergs. Ice sheets at the poles contain large amounts of iron accumulated from dust in snow over many millennia. The current melting of the Greenland ice sheet is estimated to release ~0.4–2.5 Mt (Tg) of bioavailable particulate iron into the North Atlantic each year, while the Antarctic ice sheet is releasing ~0.1 Mt (Tg). It is conceivable that this additional iron could boost phytoplankton growth, leading to sequestration of atmospheric CO_2—thus acting as a potential feedback mechanism to limit further warming. However, its impact would be affected by the input of large amounts of fresh melt water, which is likely to alter plankton communities and ocean circulation with unpredictable effects.

Figure 9.2 The proposed microbial iron pump in deep-sea hydrothermal plumes. Uptake of Fe (primarily Fe(III)) is conducted by dominant chemosynthetic, methanotrophic and heterotrophic populations. Subsequent dispersal of Fe may occur as Fe-siderophore complexes or via whole cells or POC or DOC produced through cell lysis. Bacterial siderophores may also be dispersed away from the plume and contribute to the pool of strong iron-binding ligands in the deep ocean. Reprinted from Li et al. (2014), with permission of Springer Nature.

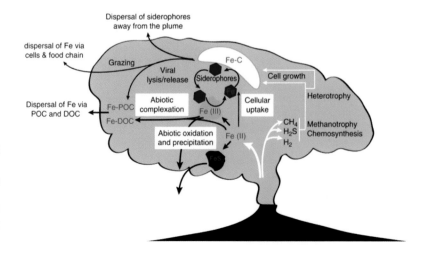

BOX 9.1   RESEARCH FOCUS

Can artificial fertilization of the oceans with iron help to mitigate global warming?

The iron hypothesis. In 1990, John Martin of the Moss Landing Marine Laboratory, California proposed that climatic changes in glacial and interglacial periods of Earth's history were caused by changing patterns of deposition of iron-rich dust, resulting in modification of CO_2 exchange between the atmosphere and oceans. He developed the ultraclean analytical techniques necessary to study trace metals and used bottle experiments to show that addition of small amounts of iron led to rapid growth of phytoplankton in water from HNLC regions of the ocean. Martin is credited with a famous quip made in a lecture: "Give me half a tanker of iron and I will give you an ice age" (Weier, 2010), which provoked intense scientific controversy and public interest. Martin planned experiments to enrich areas of the ocean with iron, but he died shortly before his hypothesis could be tested. His work was carried on, and the first experiment (IronEx I) in artificial ocean iron fertilization (aOIF) was conducted in 1993 in the equatorial Pacific off the Galapagos Islands, by addition of ferrous sulfate dispersed in the wake of a ship's propeller as it sailed over the area. In this patch, the surface of the sea turned from blue to green—the idea that addition of iron would stimulate phytoplankton growth in the ocean had been vindicated. Oceanographers developed plans for more experiments.

Iron enrichment experiments—what have they shown? Thirteen mesoscale aOIF experiments have been conducted between 1990–2012—four in the northwest Pacific, two in the equatorial Pacific, and seven in the Southern Ocean (SO)—and a detailed analysis of these is provided in the review by Yoon et al. (2018). Any significant gains in CO_2 sequestration would have to come from the SO because it has very low iron levels, and the largest reservoirs of unused macronutrients in the upper and deep mixed layers. Most experiments employed similar methodology, in which hundreds or thousands of kg of iron sulfate plus an inert tracer (sulfur hexafluoride) were added over patches of ocean a few kilometers in diameter. Over periods of a few days to weeks, scientists on board research vessels measured biomass and diversity of phytoplankton, together with a wide range of biochemical, geochemical, hydrographic, and physical parameters. In most experiments, phytoplankton blooms dominated by diatoms have been stimulated with up to 15-fold increases in the chlorophyll content of surface waters, providing strong evidence that iron is a growth-limiting factor. However, estimates of the amount of carbon drawn down into deeper water as blooms sink from the surface have been variable and interpreted as much lower than predicted; although the SOIREE, SEEDS I, and EisenEx experiments were terminated for logistical reasons after 2–3 weeks duration—too soon to detect sinking. EIFEX was the only experiment in which vertical flux could be followed in a vertically coherent water column from surface to bottom. Recycling of fixed carbon by microbial loop processes may return CO_2 to the atmosphere after a short period, defeating the goal of long-term CO_2 sequestration. Watson et al. (2008) reviewed the limitations of these first-generation experiments and concluded that the timescale and spatial extent of the experiments conducted were severely hampered by logistical difficulties and expense associated with carrying out this type of work —large scientific teams are needed for shipboard experiments, many miles from land. Smetacek and Naqvi (2008) noted that the total ship time dedicated to the first 10 years of aOIF experiments was only 11 months, compared with more than 2500 months to other oceanographic research cruises in this period. These authors argued

that the iron enrichment experiments provided much useful information about ocean processes, but the design and timescale of the experiments were simply not appropriate to determine if aOIF could be a successful strategy for carbon sequestration. They proposed that the next generation of experiments should be conducted on a much larger scale, over thousands of square kilometers, and with observations using a wide range of techniques made over periods of months rather than days or weeks. In addition, they proposed that experiments should take place in semi-closed areas of water formed by eddies—stable circulating water masses, vertically coherent down to the bottom with limited exchange with the surrounding ocean—and new approaches to the introduction of iron compounds should be developed in order to provide better evidence for the effect of iron fertilization on carbon sequestration. The most convincing evidence that aOIF might be effective comes from the EIFEX trial, conducted in 2004 in a 150 km^2 patch in the core of a mesoscale eddy of the Antarctic Circumpolar Current (ACC). By using an extended period of observation and detailed analysis of the depletion of dissolved and particulate carbon in the trophic zone, Smetacek et al. (2012) provided powerful evidence that over half the biomass stimulated by iron addition in the mixed layer sank to a depth >1000 m, due mainly to sinking aggregates formed by the diatom *Chaetoceros dichaeta*.

Possible risks of OIF. Since its inception, there has been strong opposition to OIF. Environmental pressure groups argue that the deliberate alteration of marine ecology—stimulating blooms of relatively large phytoplankton that are usually not abundant—could shift the food web in unpredictable ways and distort biogeochemical and ecological relationships throughout the system (Strong et al., 2009). Some scientists argue that aOIF could stimulate the growth of toxic algae such as *Pseudo-nitzschia*, producing the toxin domoic acid (DA, p.320) that could potentially lead to levels of DA sufficient to cause mortalities in seabirds and mammals or necessitate the closure of fisheries to protect public health (Trick et al., 2010). There have been concerns that production of other climate-relevant gases (N_2O, DMS, CH_4, and volatile organic compounds) could be enhanced by aOIF, or rapid bacterial decomposition of stimulated blooms could lead to anoxic "dead zones" (reviewed by Yoon et al., 2018). Iron addition might also disturb the nitrogen-fixing activity of *Trichodesmium*, with potential major effects on ocean ecology (Walworth et al., 2018). Many scientists argue strongly that these risks are not worth taking, because there is not enough evidence that OIF would be effective in making a significant impact on atmospheric CO_2 levels. The Royal Society (2009) considered the feasibility of aOIF as part of a wider review of planetary-scale geoengineering approaches to control increased CO_2 levels and global warming. They concluded that to have even a modest impact, we would need to remove several Gt (Pg) of CO_2 per year, maintained over decades and more probably centuries. A report by the National Research Council, USA (2015) also argued strongly against OIF, while supporting investigation of some other approaches to geoengineering.

Commercial interests and the moratorium on aOIF experiments. Early models based on the potential of fertilization of low productivity regions were promoted enthusiastically by some scientists and entrepreneurs. They suggested that the technique might accelerate the removal of billions of tons of atmospheric CO_2—helping to control

BOX 9.1   RESEARCH FOCUS

the effects of global warming, as well as having potential benefits for enhancement of fisheries—while providing direct economic benefits. Following United Nations initiatives for reduction of CO_2 emissions, the concept of a global market in the trading of "carbon offset credits" emerged. Subsequent treaties on climate change imposed by governments have placed strict caps and taxes on emissions, encouraging entrepreneurs to establish companies with the aim of profiting from ocean fertilization while mitigating the consistent rise in atmospheric CO_2 levels. Such iron fertilization takes place in international waters and is subject to treaties and laws administered by the London Convention through the International Maritime Organization. In 2007, they ruled that commercial iron "dumping" and carbon trading based on OIF ocean iron fertilization should be prohibited unless research provides the scientific foundation to evaluate risks and benefits. This moratorium was subsequently supported by nearly 200 countries at the UN Convention on Biological Diversity in 2008.

Scientific opinion about aOIF remains divided. Leading oceanographers and marine ecologists have presented conflicting views of aOIF. Strong et al. (2009) argued that it is "time to move on" because further iron enrichment experimentation "will not resolve any remaining debate about the risks of iron fertilization for geoengineering … and is both unnecessary and potentially counterproductive, because it diverts scientific resources and encourages … inappropriate commercial interest." By contrast, Smetacek and Naqvi (2008) and Buesseler et al. (2008) argue that continued multiagency research of large-scale experiments is necessary to study the relationship between phytoplankton growth and the effects of grazing and breakdown of biomass, before we can determine whether significant amounts of CO_2 are removed from the atmosphere and whether large-scale iron fertilization could be useful. They argue that future experiments must be designed to produce proper evidence of effective long-term CO_2 sequestrations so that society can weigh up the risks aOIF against the likely benefits. Smetacek (2018) deplores the "vilification of OIF … due to the negative public perception of

geoengineering and the potential threat of runaway commercialization." He makes an impassioned case for further experiments to explore the ecology of sinking diatoms in the ACC.

The future of aOIF and other ocean solutions to address climate change. Interest in aOIF continues, and new proposals for enhanced experiments emerge from time to time, but opposition remains strong. Yoon et al. (2018) of the Korea Polar Research Institute conducted the most comprehensive review to date of past aOIF experiments and published detailed design guidelines for a planned long-term experiment in the SO (*Figure 9.3*), modeled on the successful EIFEX experiment. The open peer review comments published with this paper provide valuable insight into current thinking. However, the funding for this experiment was withdrawn. The Oceaneos Marine Research Foundation (Canada) is planning a large-scale release of iron off the coast of Chile, emphasizing its potential value in ocean restoration ("seeding") on a local scale to stimulate depleted fisheries, rather than global geoengineering designed to remove CO_2; however, some scientists suspect that the foundation is seeking to profit from an unproven and potentially harmful activity (Tollefson, 2017). Emerson (2019) has recently proposed research to investigate the feasibility of using aircraft to distribute a fine dust containing biogenic hydrous iron oxides (produced by cultivating iron-oxidizing bacteria in shallow ponds), timed to coincide with seasonal phytoplankton blooms in meso-scale eddies in HNLC regions. Gattuso et al. (2018) assess 13 global- and local-scale ocean-based measures, including aOIF, that could significantly reduce the magnitude and rate of ocean warming, ocean acidification, and sea-level rise, as well as their impacts on marine ecosystems and ecosystem services. A high-level review by a UN expert advisory body on environmental protection proposes urgent development of a streamlined, coordinated framework for assessing and regulating future aOIF and other marine geoengineering initiatives, emphasizing the importance of a distinction between research to assess feasibility and the possible commercial benefits from their deployment (Boyd and Vivian, 2019a,b).

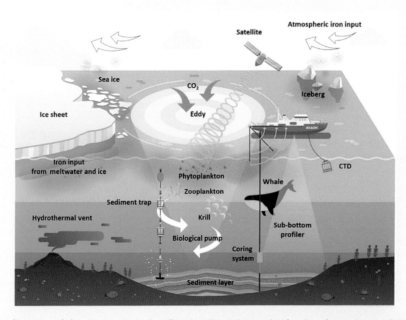

Figure 9.3 Schematic diagram of the Korean Iron Fertilization Experiment in the Southern Ocean (KIFES) representing the experiment target site (eddy structure) and survey methods (underwater sampling systems, multiple sediment traps, sub-bottom profilers, sediment coring systems, and satellite observations). Reprinted from Yoon et al. (2018), CC-BY-3.0.

Whales and seabirds play a major role in supply of iron to phytoplankton

In the last decade, there has been growing recognition of the important role that whales play in nutrient cycling in the oceans. By their diving and surfacing activities, the large bodies of whales induce physical vertical mixing of the water column, which is important in areas of highly stratified ocean, while bottom-feeding results in suspension of sediments and recycling of nutrients. Sperm whales feed mostly on cephalopods at great depths and only defecate or urinate when they surface to breathe. The excreta are rich in nitrogen, phosphate, and iron from their prey and this "whale pump"—operating in the opposite direction to the normal biological pump—serves as a natural source of fertilization by transporting nutrients from deep to surface waters. Because much of the body mass of these whales is blubber, adult animals only incorporate a very small proportion of the iron from their prey and the feces therefore contain high concentrations of iron. This has particular importance in HNLC areas of the Southern Ocean, where it will stimulate new phytoplankton production. A sperm whale defecates ~50 Mt tonnes of iron as dispersible liquid plumes into the photic zone each year, so the current population of ~12,000 sperm whales could facilitate the sequestration of 0.5 Mt (Tg) of C yr[1]. Although this is only ~0.05% of atmospheric CO_2, deaths of whales ("whale fall" – also see p.269) along migration routes also provides significant transfer of carbon to the ocean floor. The current whale population is probably <10% of the historical level and as low as 1% for the giant blue whales. This is due to the impact of whaling—especially the massive commercial operations in the nineteenth and twentieth centuries which resulted in the killing of millions of whales of many species. Therefore, depletion of whale stocks has undoubtedly had a major impact on ocean productivity and carbon sequestration, providing a strong argument for conservation through prohibition of whaling, leading to restoration of this process and other ecosystem functions. Other marine mammals such as seals and seabirds such as penguins, petrels, and albatrosses also contribute to iron and phosphate cycling and this can be especially important in sustaining productivity in regions where their population density is high (*Box 9.2*).

THE NITROGEN CYCLE

There have been major shifts in our understanding of the marine nitrogen cycle

Nitrogen is a key element that constitutes about 12% by dry weight of all living cells because of its role in the structure of proteins, nucleic acids, and several other cell materials. In the environment, nitrogen exists in various oxidation states from -3 in ammonium (NH_4^+) to +5 in nitrate (NO_3^-); the role of microbes in these oxidation-reduction transformations was introduced in *Chapter 3*. While the principal aspects of the nitrogen cycle have been understood for more than 120 years, major changes in our understanding of how it functions in the marine environment have resulted from recent discoveries of previously unknown microbial processes. As shown in *Figure 9.4*, six nitrogen-transforming processes can be recognized, namely: nitrogen fixation, assimilation, ammonification, nitrification, denitrification, and anerobic ammonia oxidation (anammox). The key metabolic reactions involved in these processes were discussed in *Chapter 3*. One of the biggest challenges facing oceanographers is determining a budget for the input and output of nitrogen via nitrogen fixation and nitrate reduction processes, respectively. As can be seen from the estimated fluxes shown in *Figure 9.4*, marine processes have a dominant role in the global nitrogen cycle.

Diazotrophs incorporate atmospheric nitrogen into organic material

Nitrogen gas (N_2, N≡N) forms 78% of the modern atmosphere and must be fixed in a reduced state into organic material before it can used in cellular processes. Only a small number of microbes (diazotrophs) are capable of carrying out this transformation (see *Figure 3.9*). For many years, the rate of N_2 fixation in the surface waters of the oceans was thought to be negligible compared with other sources of N_2 such as upwelling of nitrate from deep waters; it was also thought to be only a small fraction of the N_2 fixed in soils on land. It has always been a mystery that there seem to be so few ocean organisms that can exploit the abundant supplies of gaseous N_2 under the evolutionary pressure of severe shortages of dissolved inorganic nitrogen compounds, although it must be recognized that diazotrophy is energetically expensive and depends on a large use of ATP. However, biogeochemical evidence indicates that diazotrophy in the large portion of the biosphere occupied by the tropical and subtropical oceans is much more important than previously realized and is currently estimated at ~180-200 Mt (Tg) y⁻¹, exceeding the flux due to terrestrial microbes and industrial production of ammonia for fertilizers (*Figure 9.4*). In regions where primary production is N-limited, diazotrophy supports up to half of new production. In the 1960s, classical microbiological and physiological techniques using isotope tracer techniques led to the suggestion that the large filamentous cyanobacterium *Trichodesmium*—highly abundant as colonial blooms in tropical seas (see *Figure 4.12*)—could carry out N_2 fixation. For some time, this was discounted because it was not clear how *Trichodesmium* could fix nitrogen without forming heterocysts, structures known to protect the nitrogenase enzyme from O_2 (p.136). However, use of a sensitive acetylene reduction assay technique that could be used in the field proved that *Trichodesmium* makes a significant contribution to the global N_2 fixation by temporal and spatial separation of N_2 fixation and photosynthetic O_2 production. However, *Trichodesmium* diazotrophy did not seem to be high enough to match estimates of global N_2 fixation. A major breakthrough came from the application of molecular biological approaches to detect the presence of nitrogenase (*nif*) genes in water samples and the discovery of new groups of very small, abundant unicellular diazotrophic cyanobacteria occurring as free-living forms, or in association with heterocystous diatoms (*Figure 4.11*), tintinnids, and radiolarians. A discovery of special interest is the symbiosis between members of the uncultivated cyanobacterial clade UCYN-A ("*Ca.* Atelocyanobacterium thalassa"). This forms a symbiotic association with a single-celled prymnesiophyte algal host *Braarudosphaera bigelowii* and related species (*Figure 4.13*). The cyanobacterium, which has an extremely reduced genome (*Table 3.1*) is dependent on its algal host to provide fixed carbon compounds. In return, the host receives fixed nitrogen. UCYN-A appears to use host-supplied carbohydrates and fix nitrogen in the daytime, with close coupling to the energy provided by host photosynthesis. Diverse lineages of the organism have now been identified, including a separate type UCYN-A2, whose genome has the same gene deletions as UCYN-A, but has some marked differences. It appears that there are distinct pairs of hosts and UCYN-A strains, with differences in the size and number of UCYN-A cells associated with the host. UCYN-A, together with *Trichodesmium* and the diatom-diazotroph associations, makes a major contribution to ocean N_2 fixation. Diazotrophic Alpha- and Gammaproteobacteria, as well as Planctomycetes, have also been identified in ocean metagenomes, but little is currently known about their contribution to nitrogen budgets. Nitrogen fixation is also thought to be carried out by particle-associated anaerobic bacteria and archaea, including those inhabiting the gut of copepods; as these are highly abundant members of the zooplankton, this could be a significant source of fixed nitrogen into ocean food webs. As discussed in *Chapter 10*, diazotrophic bacteria are also very important as symbionts of bivalves in sediments and seeps.

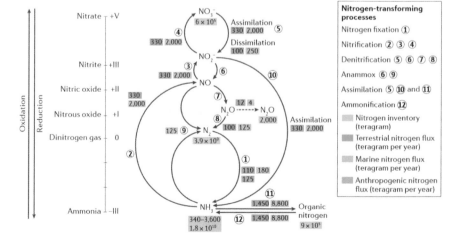

Figure 9.4 Overview of nitrogen inventory and transformation, showing marine, terrestrial, and anthropogenic fluxes. These processes do not form one balanced nitrogen cycle, as often depicted. Reprinted from Kuypers et al. (2018), with permission of Springer Nature.

The input of new organic nitrogen is balanced by losses due to denitrification (see below) and understanding of how these processes are coupled has proved elusive. Many complex factors determine the geographical and seasonal distribution of the various types of diazotrophic bacteria, including temperature and the availability of either phosphorus or iron as limiting nutrients. Recent research has shown that most marine N_2 fixation takes place in the subtropical ocean gyres, which have low levels of nutrients; whereas water column denitrification dominates in the eastern tropical Pacific and Arabian Sea, as shown in *Figure 9.5*. However, UCYN-A is not limited to subtropical waters and has recently been shown to be present and active in N_2-fixation in the Western Arctic and Bering Seas. Because of the rapid changes in Arctic ecosystems due to global warming, it is possible that selection may be occurring for UCYN-A and other unicellular diazotrophs normally associated with warmer waters.

Fixed nitrogen is returned to the inorganic pool by ammonification and nitrification

When organisms die and decompose, the amino groups of proteins, nucleotides, and other organic N-containing compounds are returned to the inorganic pool via remineralization. The excretory products of animals also contain urea or uric acid. The first stage in this process is ammonification, carried out by many types of heterotrophic bacteria and fungi, resulting in the production of NH_3 or NH_4^+. Because it is highly reduced, ammonium can be easily assimilated by many microbes and used to synthesize amino acids, which are then transformed to other compounds via transamination reactions. This is a less energy-demanding process than the uptake of nitrate, which requires large amounts of NADPH and ATP because of its high oxidation state.

Nitrification is usually a two-stage process, consisting of (1) oxidation of ammonia to nitrite and (2) oxidation of nitrite to nitrate, carried out by different organisms and enzymes (p.81). Recently, members of the bacterial genus *Nitrospira* have been shown to be capable of complete ammonia oxidation ("comammox") to nitrate. Ammonia-oxidizing archaea (AOA) and bacteria (AOB) are highly diverse, abundant, and active in areas critical for global nitrogen cycling, including the base of the photic zone, mesopelagic and sub-oxic waters, and coastal sediments. As a group, the marine AOA (Thaumarchaeota, such as *Nitrosopumilus*) have a high degree of metabolic versatility, and in addition to their role in the nitrogen cycle, their contribution to carbon fixation via autotrophy means that measurement of their activities in different geographical regions and at different depths is a high priority for research. AOA dominate environments with low levels of ammonia, because of their high affinity for this substrate (see *Box 5.1*), whereas NOB appear more prevalent in environments with high ammonia concentrations. Nitrite-oxidizing bacteria are diverse and widely distributed.

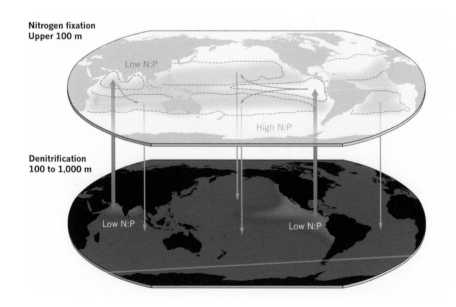

Nitrogen fixation Upper 100 m

Low N:P

High N:P

Denitrification 100 to 1,000 m

Low N:P

Low N:P

Figure 9.5 Links between marine nitrogen fixation and denitrification. Nitrogen fixation occurs mainly in the subtropical gyres (yellow areas) downstream of the tropical upwelling regions in the upper 100 m while denitrification occurs primarily at depths of 100–1000 m in the Indian Ocean and the eastern tropical Pacific Ocean (red areas). The high levels of nitrogen fixation in the gyres contribute to the formation of biomass that has high N:P ratios, which sinks to the ocean's interior (blue arrows) and compensates for the loss of nitrogen caused by denitrification. Reprinted from Gruber (2019) with permission of Springer Nature.

Assimilation of ammonium and nitrate fuels growth of phytoplankton and other microbes

Nitrate is a major nitrogen source for protists, fungi, bacteria, and archaea, many of which possess assimilatory nitrate reductases in their cytoplasm and ATP-dependent transporters in their membranes. In the photic zone, phytoplankton assimilate ammonia as their major source of nitrogen for cell growth. They are thus competing for available ammonia with heterotrophic microbes and with AOA and AOA, with the consequence that ammonia is consumed almost as rapidly as it is formed. However, most phytoplankton, including diatoms and *Synechococcus*, respond to the high-nitrate concentrations found at high latitudes and in coastal and open-ocean upwelling, with rapid growth leading to the formation of seasonal or transient blooms. They can also utilize nitrate, if it is available, using assimilatory nitrate reductase enzymes. When nitrate is plentiful due to upwelling, community shifts favor the preferential use of nitrate. This network of nitrogen-transforming microbes is illustrated in *Figure 9.6a*.

Nitrate reduction, denitrification, and anammox reactions return nitrogen to its elemental form and other gases

Denitrification is an anaerobic respiratory pathway resulting in the removal of combined nitrogen from the ocean, returning N_2 to the atmosphere. Nitrous oxide and nitrogen dioxide can form under certain conditions. The first stage in this process is dissimilatory reduction of nitrate to nitrite. Although the denitrification process only occurs under low-oxygen conditions, potential denitrifiers belonging to a very wide diversity of heterotrophic microbes can be identified in almost all environments in both oxic and anoxic waters and sediments, and many of these organisms can also grow aerobically using oxygen as an alternative electron

Figure 9.6 Potential nitrogen-transforming microbial networks in marine ecosystems. (a) The open-ocean gyres are vast nutrient-limited regions in which nitrogen flux is nearly balanced due to extensive recycling. In the sunlit surface waters, cyanobacteria mainly assimilate ammonium and/or DON for growth. Viral lysis and zooplankton grazing release DON, which is subsequently mineralized back to ammonium by heterotrophic bacteria, protists, and animals. Nitrogen-fixing bacteria provide additional ammonium. In deeper waters, ammonium is oxidized to nitrate. Some of this nitrate diffuses up into the surface waters and is assimilated by phytoplankton. (b) In OMZs, upwelling stimulates primary productivity in the surface waters. Aerobic mineralization of sinking POM depletes oxygen in the underlying waters. Nitrifying communities adapted to low-oxygen conditions oxidize ammonia to nitrite and nitrate. The OMZs are major regions of nitrogen loss owing to the activity of anammox and denitrification, which involves complex communities of bacteria and archaea. Reprinted from Kuypers et al. (2018), with permission of Springer Nature.

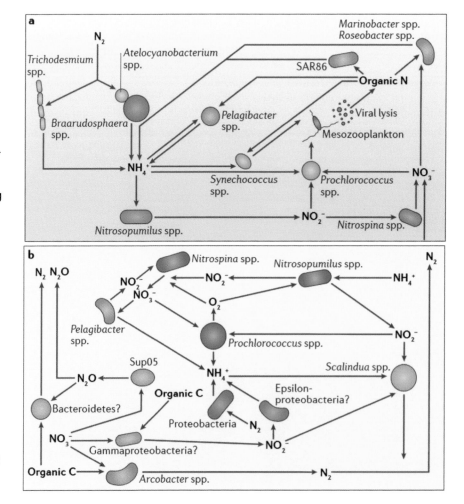

acceptor. Some abundant organisms such as *Beggiatoa* reduce nitrate fully to ammonium, while many (including SAR11) only reduce it to nitrite.

Since the 1960s, measurements of the levels of ammonium in anoxic basins have indicated that there are much lower levels of ammonium than might be expected to accumulate during the intense remineralization (either as aerobic or N-dependent respiration) supposed to occur there, suggesting that there is some unknown oxidation mechanism for its removal by anaerobic microbial activity. The discovery of anammox (anaerobic ammonia oxidation) explains this deficit and has necessitated a major reevaluation of the ocean nitrogen cycle. Only a very narrow phylogenetic group of bacteria in the class Planctomycetes are known to carry out this process, which depends on a specialized cellular structure with a membrane that protects the cytoplasm from hydrazine, a highly reactive intermediate formed during the oxidation of ammonia to nitrogen using nitrite as the electron acceptor (*Figure 3.9*). Anammox bacteria grow extremely slowly—perhaps doubling only every 14 days or so—and many oceanographers believed that they would be unlikely to thrive in natural environments. However, several studies have indicated that, in sediments and in some anoxic basins, anammox is responsible for at least as much nitrogen loss as occurs by denitrification. It is likely that nitrate reduction and anammox are coupled and some of the ammonium needed for the anammox reaction can be provided via the process known as dissimilatory nitrate reduction (DNRA). Another source is remineralization of organic matter linked to sulfate reduction.

Diverse microbial metabolic processes occur in oxygen minimum zones (OMZs)

Up to half of the global loss of nitrogen from the oceans due to the processes of denitrification and anammox occurs in the nitrite-rich oxygen minimum zones (OMZs). In most parts of the ocean, a suboxic layer (usually defined as <20 μM dissolved oxygen) typically occurs at intermediate depths (usually 1000–1500 m). This usually corresponds to the pycnocline, where there is change in density and accumulation of high levels of particulate organic matter, resulting in a zone of intense heterotrophic bacterial activity which depletes oxygen. In some areas of the ocean, most notably the eastern part of the Pacific Ocean and in the Arabian Sea, high rates of phytoplankton productivity results in intense microbial respiration creates an extensive oxygen-deficient layer of water at a much shallower depth (typically 50–200 m). Here the oxygen concentrations at their core are often about 50 times lower than those at the "classical" oxygen minimum. Defining the extent of the OMZs depends on the definition adopted; if defined by an oxygen concentration below 5 μM, they occupy about 0.05% of the volume of the world's oceans, whereas if defined at the suboxic concentration of below 20 μM, they occupy about 1% of the volume. OMZs are almost devoid of multicellular organisms, except for a few with special adaptations, and are dominated by anaerobic microbes. The major permanent open-ocean OMZs, in which extremely low-oxygen levels and extensive loss of nitrogen occur, are the eastern tropical North Pacific (ETNP), the eastern tropical South Pacific (ETSP) near the equator off Chile and Peru, and the northern Indian Ocean, as shown in *Figure 9.7*. Recently, another permanent OMZ has been identified

> ### ⓘ DIFFERENCES IN NITRATE USAGE BY *PROCHLOROCOCCUS*
>
> Surprisingly, most cultivated strains of *Prochlorococcus*—the dominant member of the phytoplankton in subtropical and tropical oligotrophic oceans—were found to be unable to grow using nitrate because they lack genes for assimilatory nitrate reductase, leading to the assumption that they depend entirely on regenerated nitrogen through growth using NH_4^+. However, analysis of ocean metagenomes revealed micro-diverse lineages that possess nitrite and nitrate assimilation genes (Martiny et al., 2009). The genes are more prevalent in some regions than others, and Martiny et al. speculate that this is related to nitrogen availability. There may be selection of lineages with the ability to acquire nitrate in regions with low N concentrations, whereas in high N regions, this may be outweighed by the advantage of possessing a smaller genome. Berube et al. (2015) later derived axenic culture of *Prochlorococcus* strains that can grow using nitrate as sole source of N (~17% slower than with NH_4^+). Analysis of 41 *Prochlorococcus* genomes showed that genes for nitrate assimilation have been obtained multiple times and exist in distinct lineages.

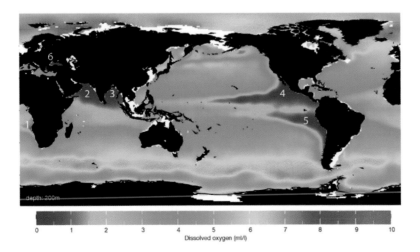

Figure 9.7 Location of the major oxygen minimum zones (OMZs). The scale bar shows the oxygen concentration at 200 m depth. 1, Southwest African continental margin; 2, Arabian Sea; 3, Bay of Bengal; 4, eastern tropical North Pacific; 5, eastern tropical South Pacific; 6, Black Sea. Credit: NOAA, National Oceanographic Data Center, World Ocean Atlas.

Analyses of NH_4^+ production by DNRA and remineralization of organic matter in oxygen minimum zones (OMZs) have revealed an enigma—these processes do not seem to generate enough ammonium to account for the level of anammox activity observed. Bianchi et al. (2014) show that this can be explained by the diurnal vertical migration (DVM) of zooplankton and micronekton. These authors used acoustic measurements to show that the animals spend the daytime in the oxygen-deficient waters of the OMZ, where they excrete products of NH_4^+ and urea (which is quickly converted to NH_4^+ by bacterial metabolism). Although the total amount of primary production transported from the photic zone is low, it is concentrated within a limited depth range of anoxic water. Bianchi et al. devised models to show that this input accounts for the "missing" NH_4^+ that supports the levels of anammox observed. Subsequent modeling of the global effects of DVM suggest that it also drives a significant and efficient flux of carbon to the mesopelagic zone, of similar magnitude to the transport of DOC (Aumont et al., 2018).

in deeper waters of the eastern subtropical North Pacific (25°N–52°N) off the west coast of North America. In addition, OMZs form during the winter in higher latitudes, such as the western Bering Sea and the Gulf of Alaska. Oxygen concentration is not uniform throughout the OMZ. A core region occupies about 10% of the volume of the whole OMZ where oxygen levels can be so low that they are undetectable using the most sensitive instruments (<3 nM). Some authors refer to this as an anoxic marine zone (AMZ), others designate it an oxygen depleted zone (ODZ). There is usually a narrow oxycline (about 10–20 m thick) at the top and bottom of the OMZ.

Detailed studies of OMZs reveal that they may not be uniformly anoxic, because physical process such as eddies can introduce low concentrations of oxygen, creating niches that allow aerobic microbial processes to continue in apparently anoxic conditions. There can be sufficient light below the oxycline to support growth of oxygenic *Prochlorococcus*, leading to a secondary chlorophyll maximum, as shown in *Figure 9.8*. The oxygen produced is quickly scavenged by aerobic microbes. Methanotrophs, feeding on methane produced in continental shelf sediments that diffuses into the OMZ, are also active. These are usually regarded as strict aerobes, although some clades have been recognized that use nitrate or nitrite as terminal electron acceptors. Others can obtain oxygen from the dismutation of nitric oxide.

The anoxic or almost anoxic conditions result in reduction of nitrate, leading to high concentrations of nitrite. This is further reduced to the gases nitrous oxide (N_2O) via the process of denitrification or to N_2 via DNRA/anammox. Analysis of metagenomes and SAGs has revealed that OMZs may contain high proportions (up to 40%) of bacteria belonging to subclades of SAR11. These are normally considered to be aerobic heterotrophs (see p.41), but lineages able to utilize nitrate as a terminal oxidant instead of oxygen have evolved in the OMZs.

Many studies have investigated the relative importance of the processes of denitrification and anammox in OMZs, because N_2O is a greenhouse gas with a global warming potential nearly 300 times that of CO_2 over 100 years. Since the two processes are regulated by different

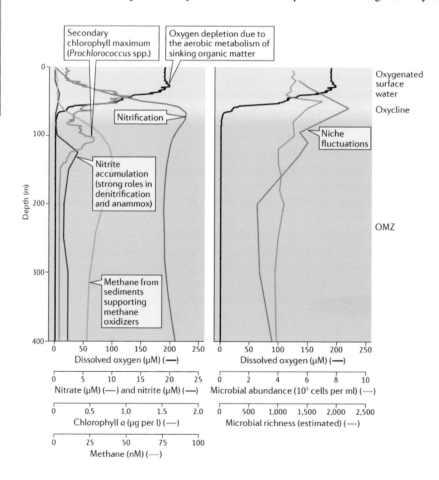

Figure 9.8 Chemical and biological gradients in the core of the ETNP OMZ, showing representative profiles of oxygen, nitrate, nitrite, methane, and chlorophyll a concentrations (left) and microbial richness and abundance (right). Reprinted from Bertagnolli and Stewart (2018), with permission of Springer Nature.

factors and are likely to be affected differently by climate change, this could lead to major shifts in ocean biogeochemistry and ecology. Results of different studies have produced variable results, with some concluding that anammox is the dominant process in the ETSP and Arabian Sea OMZs, accounting for up to 80% of removal of fixed nitrogen. Other studies suggest denitrification can be dominant. As well as differences due to variations in methods, the process of denitrification is likely to have a much patchier temporal and spatial distribution, because it is fueled by high input of organic matter. Surprisingly, nitrite oxidation can also occur under nearly anoxic conditions at the core of the OMZ and novel species of the phylum Nitrospinae have recently been implicated in this process.

All of the known nitrogen transformation processes work alongside each other in the OMZs, forming a complex network, as shown in *Figure 9.6b*. Studies of the diversity of microbial adaptations to the niche conditions found in OMZs and their influences on biogeochemical cycles are increasingly important to predict the possible effects of ocean warming and ocean acidification.

Microbial processes in sediments are a major contributor to nitrogen cycling

Nitrogen-containing POM that reaches the seafloor may be permanently buried, while a range of anaerobic heterotrophs are responsible for production of NH_4^+, which is available for regeneration by other organisms, or may be coupled to nitrate production in the oxic layers of sediments. POM exported to sediments in deep waters usually has a much higher C:N ratio than plankton in the upper ocean because of the extensive recycling of labile nitrogen-rich compounds in the microbial loop and viral processes (see *Figure 8.4*). Nitrogenous compounds reaching the seafloor are predominantly proteins, chitin, and other polymers with a relatively high molecular weight that have escaped degradation by bacterial enzymes during their descent to the seafloor. Organisms in the sediment may also absorb nitrate, and this process is especially important in coastal waters receiving high levels of nitrate from agricultural runoff. As noted on p.128, in some regions nitrate can serve as an electron acceptor for bacteria such as *Thioploca* and *Thiomargarita*. These form dense colonies underneath anoxic waters with high levels of nitrate from natural upwelling, and form filaments that stretch up to collect nitrate which is then reduced to NH_4^+ using sulfide as electron donor. Sediments on the seafloor are usually anoxic below the top few millimeters of the surface and are therefore a major site for the anammox and denitrification processes previously described for the loss of fixed nitrogen. Bioturbation by worms, crustaceans, echinoderms, and molluscs plays a major role in altering the structure and chemical composition of the sediment through burrowing and irrigation. Such organisms are often described as "ecosystem engineers." In particular, burrows introduce fresh oxygenated water into the sediments, which greatly stimulates heterotrophic decay of organic matter, with concomitant effects on nitrogen cycling. Microbial community composition varies greatly, depending on the sediment type and the species of bioturbating animals. Nitrogen cycling also occurs on the surface of sediments in coastal and shelf seas in relatively shallow waters, where the penetration of sufficient light allows establishment of communities of microalgae and cyanobacteria. Although nitrate is the main source of nitrogen for these communities, some nitrogen fixation may take place within microbial mats. Benthic nitrogen fixation by symbiotic or closely coupled diatoms, cyanobacteria, and other diazotrophs helps to explain the high productivity of some benthic systems, such as seagrass beds and coral reefs, as discussed in *Chapter 10*.

THE SULFUR CYCLE

The oceans and sediments contain large quantities of sulfur compounds

Sulfur is an essential element for all organisms, constituting about 1% of cellular mass, mainly occurring in the amino acids methionine and cysteine and in various coenzymes and metalloproteins. The oceans and sediments contain a high concentration of inorganic sulfur compounds in the various oxidation states. The most important oxidation states are −2 (sulfhydryl R-SH and sulfide HS⁻), 0 (sulfur S^0), and +6 (sulfate SO_4^{2-}). The oceans are

EXPANSION OF COASTAL OMZS

Rising sea temperatures mean that seawaters hold less oxygen. In coastal areas, this is compounded by massive inputs of nitrates, phosphates, and organic nutrients from sewage and fertilizers in terrestrial runoff. This drives the microbial processes leading to oxygen depletion and denitrification, with consequent reductions in biodiversity. The volume of anoxic water bodies in OMZs is estimated to have quadrupled since the 1950s, and oceanographic models predict further declines in the twenty-first century, even with ambitious declines in fossil fuel use (Breitburg et al., 2018). The coastlines of many industrialized and developing countries now have numerous hypoxic zones (O_2 <2 mg L⁻¹) and some areas, including the Gulf of Mexico, estuaries on the US East Coast, and the Baltic Sea, have large areas where O_2 concentrations are so low that they are designated "dead zones." Upwelling of anoxic waters due to changes in atmospheric forcing can result in mass mortalities of sea life close to shore, such as the emergence of a dead zone off the Oregon coast in 2006, where there had been no previous records of O_2 deficits (Chan et al., 2008). Oxygen depletion is leading to the inability of an increasing fraction of the oceans to support diverse animal assemblages.

DOES THE BREAKDOWN OF DMSP HAVE A DEFENSIVE FUNCTION?

Wolfe et al. (1997) concluded that grazing of *Emiliania huxleyi* by protists results in mixing of the enzyme DMSP lyase and its substrate DMSP, producing DMS and acrylate. When different protistan grazers were presented with a mixture of *E. huxleyi* strains possessing high and low levels of DMSP lyase, they preferentially selected strains with low enzyme activity. The highly concentrated acrylate that is produced in this reaction was thought to act as a deterrent in view of its toxic effects. However, Strom et al. (2003) concluded that dissolved DMSP itself inhibits grazing and suggested that this may have exerted strong selective pressure in the evolution of high levels of DMSP production in bloom-forming algae such as *E. huxleyi*. Strains of *E. huxleyi* with high DMSP lyase activity also appear resistant to *Coccolithovirus* infection (Schroeder et al., 2002). Evans et al. (2006) showed that both DMS and acrylic acid affected viral infectivity and concluded that the DMSP system functions as an antiviral defense, protecting remaining members of the population from infection. The mechanism is unknown.

the largest reservoir of sulfur (in the form of sulfate) in the biosphere. As noted on p.87, it is necessary to distinguish between dissimilative and assimilative sulfate reduction. The dissimilative process, for energy generation, leads to the production of hydrogen sulfide and is restricted to the sulfate-reducing bacteria (including many members of the Desulfobacterota (Deltaproteobacteria, see p.129), which are a key part of the community in anaerobic sediments. By contrast, many protists, bacteria, and archaea can assimilate sulfide produced during ATP-driven reduction of sulfate and incorporate it into organic sulfur compounds. The metabolic processes involved in microbial transformations of these states, and examples of key taxa involved in sulfate reduction and sulfide oxidation in sediments, mats, vents, and seeps were discussed in detail in *Chapters 3* and *4*, respectively. Sulfur oxidation–reduction reactions are also of major importance in symbiotic associations of bacteria with marine invertebrates (*Chapter 10*).

Metabolism of organic sulfur compounds is especially important in surface waters

Organisms need sulfur for the synthesis of sulfur-containing amino acids and other essential compounds. Despite the abundance of sulfate in seawater, most marine microbes obtain sulfur from recycling of organic sulfur compounds produced by phytoplankton, especially dimethylsulfoniopropionate (DMSP, $(CH_3)_2S^+CH_2CH_2COO^-$), which is produced from the amino acid methionine by many algae, especially dinoflagellates and prymnesiophytes. Some of the bloom-forming prymnesiophytes, such as *Phaeocystis globosa* and *Emiliania huxleyi* (see p.181) and some dinoflagellates, are particularly potent producers of the compound, with intracellular concentrations of up to 300 mM. DMSP appears to have many functions. It acts as a compatible solute that provides protection against osmotic stress and has cryoprotectant, antioxidant, and defensive functions. It may have evolved as a metabolic relief mechanism when algae are undergoing unbalanced growth, in order to eliminate excess reducing power and reduced sulfur. Synthesis of DMSP and another compatible solute, glycine betaine, may be regulated by nitrogen levels. Some DMSP is exuded naturally from healthy algae, but this probably represents only 1–10% of the cellular contents—most DMSP enters the water column as result of disruption of the cells by zooplankton grazing or by viral lysis. Although widespread, not all phytoplankton produce DMSP, and its intracellular concentrations vary considerably in different groups, even within genera. The enzymes for DMSP synthesis are found in the mitochondria and chloroplasts. Until recently, DMSP production was thought to be a unique property of phytoplankton, but the key enzyme DsyB is now thought to have originated in bacteria, as discussed in *Box 9.2*.

The annual global input of sulfur by DMSP production and release from phytoplankton has been estimated at over 1000 Gt (Pg). DMSP is the main vehicle for the cycling of organic sulfur compounds in the oceans. The high carbon content of DMSP means that it also plays a very important role in the flow of carbon in the marine food web—it is probably the most important single organic compound transferring fixed carbon to the microbial loop processes (*Figure 8.3*).

Many algae also incorporate large amounts of sulfur into sulfated polysaccharides such as mucus and cell-wall components. These compounds have important functions in defense against grazing and the sequestration of metals. Because they are quite recalcitrant to microbial attack, these compounds form a significant component of marine snow aggregates.

DMSP production leads to release of the climate-active gas dimethyl sulfide (DMS)

The fate of DMSP is a very important factor in marine ecology and ocean–atmosphere interactions. When DMS was first discovered, it was assumed that phytoplankton produced the gas directly, but we now know that it is generated by the action of a group of DMSP lyase enzymes, which are widespread in many marine microbes and break down DMSP into DMS and acrylate (*Figure 9.9*). DMSP-producing algae often contain DMSP lyase themselves, but this is in a separate cellular compartment, so it is thought that the enzyme only encounters the substrate when cells are broken apart by grazing or viral lysis.

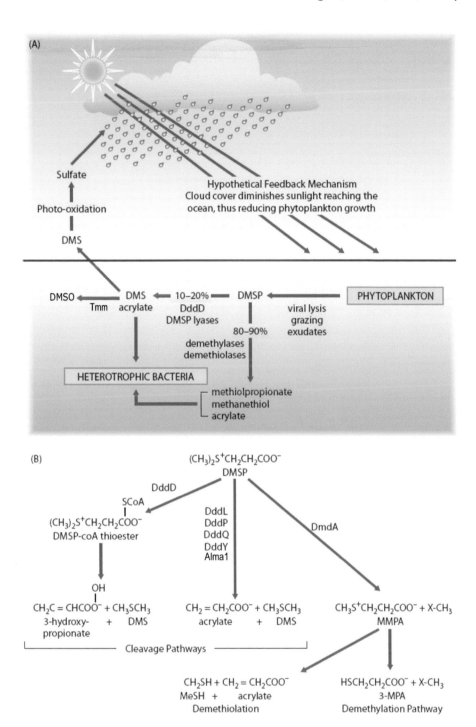

Figure 9.9 Microbial transformations and the fates of dimethylsulfoniopropionate (DMSP) in the ocean, resulting in production of dimethyl sulfide (DMS) and products that are assimilated by different bacterial groups for cell growth. A. Representation of the hypothetical mechanism by which release of DMS gas to the atmosphere results in the formation of cloud condensation nuclei, producing cloud albedo that may constitute a feedback process that regulates phytoplankton growth. Figures show the estimated percentage of DMSP transformed at each stage. B. Principal transformations of DMS by microbial activity. Arrows are labeled with the principal enzymes known for the cleavage and demethylation pathways. DddD, a class III acyl CoA transferase enzyme, which is thought to perform the initial step in DMS release, as an alternative to DMSP lyase; MeSH, methanethiol; MMPA, 3-methiolpropionate; MPA, 3-mercaptopropionate; X-CH3, unidentified intermediate with terminal methyl group. Most of the DMSP is broken down by microbial activity; 80–90% is demethylated and enters the microbial loop, while 10–20% is metabolized by the DddD cotransferase enzyme or degraded by DMSP lyases to DMS and acrylate, which are also metabolized by bacteria.

The flux to the atmosphere is ~21 Gt (Pg) y^{-1} (similar to S emissions from burning of fossil fuels) where it is oxidized to sulfates, which act as nuclei for condensation of water vapor causing cloud formation. Therefore, DMS has an albedo function affecting temperature and light availability to surface waters. DMS production was found to be related to phytoplankton growth and is therefore controlled by light, temperature, and nutrient availability. This led to the "CLAW hypothesis," in which algal DMS production is envisaged as providing a feedback mechanism that regulates climate. This idea became one of the "planks" for the influential Gaia Hypothesis developed by James Lovelock and Lynn Margulis, which proposes that biological and physical components of Earth interact to maintain the climatic and biogeochemical conditions of the planet as relatively stable (in comparison with the history of other planets in our solar system). As shown in *Figure 9.9A*, one version of the model envisages a decline in phytoplankton growth as cloud increases owing to a rise in the level of DMS, causing a reduction in phytoplankton production until cloud formation diminishes again.

BOX 9.2   RESEARCH FOCUS

Transformations of DMSP are major drivers of diverse ocean processes

The smell of sea. Remember those summer holidays? Getting close to the beach, we can detect a very distinctive tangy odor—the evocative "smell of the sea." This is largely caused by the gas DMS, produced from DMSP from marine algae. As well as evoking seaside memories, DMSP and DMS have far-reaching effects in biogeochemical processes and marine ecology.

DMSP lyases. The nature of the enzymes that degrade DMSP varies greatly among different microbes (*Figure 9.9B*). Seven gene families encoding DMSP lyases have been identified; these all produce DMS and acrylate, but they seem to be completely unrelated as they have almost no sequence similarity and differ in their requirement for metal cofactors. The DddP and DddQ enzymes are the most widely distributed and occur mainly in members of the marine *Roseobacter* clade (alphaproteobacterial order Rhodobacterales), but they also occur in other groups due to HGT. The respective genes are highly represented in marine metagenomes. In cultures of *Ruegeria pomeroyi* and *Roseovarius nubinhibens*, expression of *dddP* and *dddQ* was induced by the presence of the substrate DMSP (Bullock et al., 2017). Algae also produce DMSP lyase; the aspartate racemase protein Alma1 was first isolated from *E. huxleyi* and subsequently found in a wide range of phytoplankton, and some bacteria (Alcolombri et al., 2015). The DMSP lyase enzymes appear to have different catalytic mechanisms for carrying out the same reaction, indicating separate evolutionary paths to this activity, suggesting that there were multiple ancestral enzymes and that there are strong selective pressures to maintain this function in different groups of organisms (Bullock et al., 2017).

Alternative routes for DMSP catabolism. A significant fraction of DMSP catabolism is diverted away from DMS production and towards degradation and incorporation by chemoheterotrophic bacteria. DMSP is estimated to supply up to 15% of the total bacterial carbon demand and nearly all of the bacterial demand in the surface waters of the ocean—probably no other single compound contributes as much carbon to the DOM pool (Yoch, 2002). The gene *dmdA* encoding the enzyme responsible for DMSP demethylation was first recognized in *Roseobacter* and SAR11 clades in the

Alphaproteobacteria (Howard et al., 2006; Reisch et al., 2008) and subsequently found in other clades, with indications of transmission by HGT. Demethylation is now recognized as the major DMSP catabolism pathway, converting >80% of dissolved DMSP. Again, DmdA is most prevalent in members of the Alphaproteobacteria, and analysis of metagenomes shows that more than half of the bacteria in marine surface waters contain genes for DMSP methylation; many of these also have photoheterotrophic potential via proteorhodopsins (Moran et al., 2012; see *Box 3.1*). In a transcriptomic study, Vila-Costa Vila-Costa et al. (2010) showed that enrichment of surface water with DMSP led to an increased expression of genes supporting heterotrophic activity. Lidbury et al. (2016) showed that *R. pomeroyi* and other marine roseobacters associated with algal blooms can also use the enzyme trimethylamine monooxygenase (Tmm), in the presence of methylated amines, to oxidize DMS (produced by DMSP lyases) to dimethyl sulfoxide (DMSO).

Interactions between DMSP-producing algae and phycosphere-associated bacteria. Barak-Gavish et al. (2018) observed that inoculation of *E. huxleyi* cultures with a consortium of bacteria isolated from grazing copepods collected during a bloom in the North Atlantic led to rapid death of the algae. Application of antibiotics prevented cell death, implicating an algicidal effect by bacteria. A strain of *Sulfitobacter* was isolated and shown to cause death and lysis of 90% of the algal cells in co-culture, whereas uninfected *E. huxleyi* only showed ~40% mortality after 15 days. Electron microscopy revealed membrane blebbing in the early stages of cell death. Coculturing with a strain of *Marinobacter* resulted in similar proliferation of bacteria, but no increase in algal cell death. Barak-Gavish et al. showed that the *Sulfitobacter* was closely associated with the dynamics of naturally occurring bloom. The experimenters identified the production of volatile methanethiol (MeSH), DMS, and dimethyl disulfide in infected cultures, whereas noninfected control cultures produced only DMS (from the activity of DMSP lyase). Further analysis of the interplay between DMSP production and the growth and pathogenicity of *Sulfitobacter* suggests that the assimilation of reduced sulfur, via MeSH, into methionine and cysteine leads to synthesis of bacterial algicidal compounds;

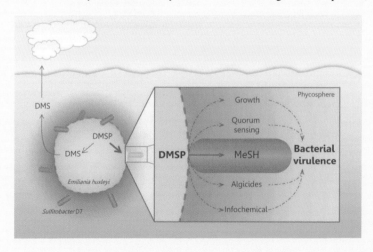

Figure 9.10 Model of the possible routes by which algal DMSP promotes bacterial virulence in the *E. huxleyi* phycosphere. During interaction, *Sulfitobacter* D7 consumes *E. huxleyi*–derived DMSP and transforms it into MeSH, which facilitates bacterial growth. DMSP and its metabolic products can promote production of QS molecules which may enable chemoattraction to algal cells and production of bacterial algicides. Reprinted from Barak-Gavish et al. (2018), CC-BY-4.0.

BOX 9.2   RESEARCH FOCUS

this has also been shown in the phycosphere-associated roseobacter *Phaeobacter*, which also kills *E. huxleyi* during senescence (Wang et al., 2016). Thus, the balance between competing DMSP catabolic pathways, driven by microbial interactions, may regulate oceanic sulfur cycling and DMS flux to the atmosphere (*Figure 9.10*).

Production of DMSP in bacteria and algae. Curson et al. (2017) demonstrated the production of DMSP in a range of cultivated marine bacteria and identified the *dysB* gene encoding the key methyltransferase enzyme responsible for its biosynthesis. Curson et al. (2018) identified numerous functional *dysB* homologs (DSYB) in the genomes or transcriptomes of all marine prymnesiophytes, most dinoflagellates, and some diatoms. Phylogenetic analysis of the DsyB/DSYB proteins led Curson et al. to suggest that DMSP production originated in bacteria and was transferred to eukaryotes, either through endosymbiosis at the time of mitochondrial origin from the alphaproteobacterial ancestor, or more recently by HGT. Using NanoSIMS tracking of $^{34}SO_2$, they showed that DMSP is produced in the chloroplasts and mitochondria. This discovery sheds important light on the possible evolution of DMSP production in the dinoflagellates, prymnesiophytes, and diatoms, which first evolved about 250 MYA (Falkowski et al., 2004). Bullock et al. (2017) hypothesize that the proposed roles for DMSP as an antioxidant, osmolyte, cryoprotectant, defensive molecule, and cellular energy-balancing mechanism may explain the evolution and dominance of these phytoplankton groups, by enabling their rapid adaptation to extreme environmental changes prevalent at the time of their radiation. Bullock et al. also provide evidence for the co-evolution of the DMSP-producing phytoplankton and the marine *Roseobacter* clade.

The role of DMSP and DMS in coral biology. As discussed in *Chapter 10*, corals are holobionts consisting of complex communities of the coral animal and various microbes, often including Symbiodiniaceae dinoflagellates (zooxanthellae). These produce high intracellular levels of DMSP, and the very high densities of zooxanthellae within the tissue of many corals means that the concentration of DMSP produced by corals may be a highly significant component of the global sulfur cycle, despite the restricted distribution of reefs. Release of large quantities of DMSP in coral mucus provides a major link to microbial loop processes in the nutrient-deficient waters inhabited by most tropical corals. This has been investigated on the Great Barrier Reef by scientists from the Australian Institute of Marine Science. Raina et al. (2009) isolated a variety of alpha- and gammaproteobacterial species capable of degrading DMSP from *Acropora* and *Monipora* corals. The dominance of *Spongiibacter* correlated with the observations of Bourne et al. (2007), who showed the decline of this bacterium in corals during bleaching—when loss of *Symbiodinium* results in lower DMSP synthesis—and its reappearance as the corals recover and resume DMSP production. Raina et al. (2010) undertook bioinformatic analysis of an extensive collection of marine metagenomes and showed that the *dmdA* gene was consistently found in reef water samples, supporting the idea that DMSP demethylation is an important source of sulfur and carbon for microbial communities in coral reefs. Many of these genes were associated with phage sequences, and Raina et al. suggest that they may have originated from host bacteria and are carried by phages as additional information to ensure effective propagation, as occurs in viruses infecting cyanobacteria and *E. huxleyi*. Raina et al. (2010) suggest that sulfur cycling plays a major role in structuring the microbial communities associated with the coral holobiont and that DMS production provides a local feedback mechanism by increasing cloud cover and albedo effect over coral reefs (as in *Figure 9.9A*). In 2013, this group made the surprising discovery that coral juveniles produce DMSP in the absence of algal symbionts, a finding which "overturn[s] the paradigm that photosynthetic organisms are the sole biological source of DMSP, and highlight[s] the double jeopardy represented by worldwide declining coral cover, as the potential to alleviate thermal stress through coral-produced DMSP declines correspondingly" (Raina et al., 2013).

We smell the seaside … seabirds smell lunch. Experimental studies by Gabrielle Nevitt and coworkers at the University of California–Davis have shown that foraging procellariform seabirds such as albatrosses, shearwaters, and petrels fly in response to concentration gradients of DMS in the atmosphere (Nevitt and Bonadonna, 2005; Nevitt et al., 2008). "Hotspots" of DMS production occur at regions where there are high levels of phytoplankton, with accompanying grazing by krill. Higher-order predators like fish and squid will also be attracted to the area, so high DMS levels indicate a good place for birds to forage for food. Models can be used to link climatological data linking the distribution of seabirds with DMS or types of DMS-producing phytoplankton. Besides birds, many organisms including zooplankton, reef fishes, whale sharks, turtles, and baleen whales have been shown to respond to dissolved DMS as a feeding strategy. A fascinating example of this infochemical effect of DMS on marine ecology and large-scale processes comes from the research by Savoca and Nevitt (2014) who analyzed a 50-year database of Southern Ocean (SO) seabirds to test the hypothesis that DMS emanating from phytoplankton mediates a mutualistic trophic interaction, If this is the case, then carnivorous seabirds that are attracted to DMS should specialize in feeding on primary consumers of phytoplankton that produce the gas. Indeed, krill comprised 80% of the diet of DMS-tracking seabirds, whereas non-DMS-tracking species fed mainly on secondary consumers like fish and squid. Savoca and Nevitt made a connection between phytoplankton, krill, and seabirds, postulating a three-way trophic mutualism based on the recycling of iron in surface waters. Krill contains about a quarter of the bioavailable iron in surface waters of the SO. By feeding on krill, seabirds make large local deposits of iron via their feces, which stimulates further phytoplankton growth. Many other bird species are visually attracted to krill swarms by the activity of the DMS-sensitive birds. Since seabirds only return to land briefly to breed, most of their feces will be deposited at sea. Savoca and Nevitt calculated that there are ~2.5 x 10^8 seabirds in the SO, with a biomass similar to that of blue whales. In a further twist to the DMS story, Savoca et al. (2016) showed that after less than a month of exposure to seawater, common plastics acquire a DMS signature at concentrations that attract procellariform seabirds. This may explain why these birds and other animals that use olfactory feeding cues consume so much plastic, with devastating consequences (*Box 13.1*), although Dell'Ariccia et al. (2017) question some of the data that Savoca et al. used to support their hypothesis.

THE PHOSPHORUS CYCLE

Phosphorus is often a limiting or co-limiting nutrient

Phosphorus is an essential element for all life, being a key constituent of nucleic acids and lipids and a crucial component of ATP in energy transfer reactions. Inorganic phosphorus also has a key role in photosynthesis. In the oceans, there are various forms of dissolved inorganic phosphorus (DIP), of which orthophosphate PO_4^{2-} is the most abundant. A large fraction of the phosphorus in surface oligotrophic waters is present as dissolved organic phosphorus (DOP) compounds. The sole input and output of phosphorus to the oceans is via terrestrial runoff and burial in sediments, respectively. Thus, biogeochemical cycling of phosphorus is closely linked with carbon flux and the availability of phosphorus has a major limiting or co-limiting effect on oceanic primary production rates and microbial community composition. This phosphorus limitation has been studied extensively in oligotrophic regions of the North Pacific, western North Atlantic (Sargasso Sea), and the Mediterranean Sea, where analyses show that DIP is turned over very rapidly. There is some evidence that climate change over the last few decades is leading to stratification of the subtropical Pacific gyre, resulting in an increase in nitrogen fixation and changing the system toward limitation by phosphorus. The organic and inorganic forms of phosphorus are transformed continuously by microbial activity. In the surface ocean, phytoplankton assimilate DIP and incorporate it into cellular material, which is ingested by zooplankton or enters the microbial loop via exudation or cell lysis, where it is taken up and recycled by heterotrophic bacteria. Heterotrophic and autotrophic microbes compete for available phosphorus, which affects productivity. Recycling of DOP occurs throughout the water column and regeneration of inorganic phosphorus occurs in deeper waters. Although phosphate seems to be the preferred source of phosphorus for phytoplankton, they can also hydrolyze organic phosphorus compounds when phosphate does not meet their demands.

Marine microbes are adapted to low and variable levels of phosphorus

As discussed in *Chapter 3*, planktonic bacteria demonstrate one of two lifestyles. Many are oligotrophic and are adapted to subsist on a relatively fixed diet of constant low levels of nutrients. This is often reflected in very small cell sizes and streamlined genomes, as exemplified by photoautotrophs like *Prochlorococcus* and chemoheterotrophs like SAR11 bacteria. Others are copiotrophic or opportunistic and respond to inputs of increased levels of nutrients because they have larger genomes and more metabolic versatility. Such organisms may respond to nutrient starvation by initiating a program of changes in cellular composition and activity. Since DNA and phospholipids incorporate large amounts of phosphorus, small cells with reduced genomes and less membrane content require less and are at an advantage in conditions of phosphorus limitation. In fact, some phytoplankton (both cyanobacteria and eukaryotes) growing in low-phosphorus environments reduce their demand further by substituting phospholipids in their membranes with lipids containing sulfur and sugars.

The mechanisms of uptake of phosphorus compounds have been extensively studied in laboratory cultures of bacteria like *Escherichia coli*, but less so in marine bacteria. Homologs of PstS, the high-affinity phosphate-binding protein in *E. coli*, are produced by *Synechococcus*, *Prochlorococcus*, and *Trichodesmium*, and multiple copies of the gene are present in the genome sequences of cultured strains. These genes are also abundant in marine metagenomes. High-affinity phosphate-uptake genes are also present in the genomes of viruses infecting cyanobacteria and *E. huxleyi*, and their expression is upregulated under conditions of phosphate deficiency—presumably increased phosphate uptake ensures efficient viral replication. Other phosphate transporter systems have also been identified in marine microbes. Cell lysis leads to the release of enzymes such as nucleases, lipases, and esterases that degrade organic compounds to regenerate phosphate.

To obtain phosphorus from organic phosphorus compounds, bacterial cells must hydrolyze them to orthophosphate. The Pho regulon studied in *E. coli* contains a number of genes associated with regulation, uptake, and hydrolysis of phosphorus compounds. Pho gene homologs are abundant in genomes and metagenomes of marine bacteria, which appear to possess several different mechanisms for transporting and hydrolyzing phosphate esters. The most common type of enzyme is alkaline phosphatase, whose expression is repressed by high phosphorus levels and is often used as a marker for phosphate stress. These enzymes are grouped into several families, including PhoA, PhoD, and PhoX, which differ in their substrate specificity, cellular location, and requirement for metals as cofactors. Based on bioinformatic analysis of marine metagenomes, it seems that a large proportion of coastal and oceanic bacteria contain phosphatases within the cytoplasm, suggesting that small DOP compounds are carried across the cytoplasmic membrane, providing carbon (and possibly nitrogen) to the cell, as well as phosphorus.

The phosphonates are another group of organic phosphorus compound whose contributions to marine nutrient cycles have been relatively little studied until recently. Phosphonates are characterized by a stable C–P bond, rather than the C–O–P bond found in esters, and they are generally very resistant to chemical and thermal decomposition. Naturally occurring phosphonates are produced by a range of organisms as membrane components and therefore are found in the DOP pool, but interest in these compounds has increased following the use of synthetic phosphonates as herbicides, insecticides, detergent additives, and other applications. Tens of thousands of tonnes of these long-lived compounds are released each year into the environment. Analysis of metagenomic databases for phosphonate utilization genes (*phn*) has revealed that they are widespread and abundant among diverse bacterial phyla that are common in marine bacterioplankton, including members of the Proteobacteria, Planctomycetes, and Cyanobacteria. *In situ* studies have shown that expression of these genes and phosphonate degradation occurs in phosphate-limited regions. Since up to 25% of DOP may occur as phosphonates, these molecules appear to be a significant phosphorus source for marine microbes.

THE SILICON CYCLE

Silicon in one of the most abundant elements on Earth. Calcium and magnesium silicate minerals form the major component of rocks in the Earth's crust. Weathering of these minerals by dissolved CO_2 results in the formation of dissolved silicate (DSi, i.e. ortho-silicic acid, H_4SiO_4) which is transported through soil and enters the ocean via rivers. Other inputs include weathering of the seafloor and hydrothermal vents. DSi plays a major role in the growth of phytoplankton and the export of biogenic Si (BSi) carbon in coastal zones and deep-sea sediments via the biological pump. A range of marine microbes incorporate DSi across their cell membrane and control its precipitation in specific patterns as particulate hydrated silica (SiO_2) glass or opal-A, forming a structural component of their cell walls. The most important siliceous organisms that require silicon as an essential nutrient are the diatoms and radiolarians, but some flagellates, choanoflagellates, and cyanobacteria (as well as non-microbial organisms such as sponges and gorgonian corals) also incorporate significant quantities of silica. The discussion here focuses on the role of diatoms in silicon cycling.

Silicification of diatoms is an economic process for construction of a cell wall

Diatom species differ greatly in their level of silicification of the frustules (p.181). Some species, such as *Fragilaria*, have very thick, strong shells that offer resistance to cracking and crushing by the mouthparts of mesozooplankton during grazing, meaning that viable cells may pass through the gut undigested. Other species, such as *Thalassiosira*, have thin shells that are easily broken and are subject to heavy grazing pressure, but this can be compensated by rapid proliferation under favorable light and nutrient conditions. However, there is limited direct evidence that silicification evolved as an anti-predatory function. Silicification is a relatively energy-efficient process, and diatoms need to expend a much smaller proportion of their energy intake making a silica wall than one composed of polysaccharides. Diatoms that grow slowly under non-limiting silicon conditions have plenty of time to take up silicon during the reproductive cycle, resulting in thicker cell walls; by contrast, rapidly growing diatoms are likely to have thinner walls.

Diatom blooms depend on the availability of silica in the environment

Large blooms of diatoms typically occur in spring and summer at high latitudes and in eutrophic regions. The average Si:N ratio of diatom cells is ~1:4, so silica is often the limiting nutrient. Actively growing diatoms remain in the photic zone due to buoyancy through gas vacuoles, but when conditions become unfavourable, they sink rapidly from the surface waters due to aggregation into large masses through the production of mucilage or the formation of heavy resting spores. Therefore, there is little recycling of silica in the photic zone and it becomes exhausted before nitrogen or phosphorus, thus ending the diatom bloom and providing other phytoplankton groups with a competitive advantage to proliferate using the remaining nutrients. Areas characterized by intense spring blooms typically have low Si:N ratios (*Figure 9.11*). Most sinking cells are transported to the seafloor and will be buried in sediments via the biological pump, but refuge populations remain in the water column or the surface of sediments and will seed the next seasonal bloom via vertical mixing. As noted on p.182, bacterial-mediated degradation of the organic matrix covering diatom shells affects the rate of dissolution of BSi and its return to the DSi pool. In some regions, a second short bloom can occur in the autumn, caused by the breakdown of summer stratification, providing vertical mixing of nutrients while light levels are still sufficiently high to support growth. These bloom cycles mean that diatoms play a major role in the export of POC to the seafloor via the biological pump—about 40% of oceanic carbon sequestration (~1.5–2.8 Gt (Tg) y⁻¹) can be attributed to diatoms. Blooms of diatoms have a significant difference from those of the other major bloom-forming group, the coccolithophores. Because CO_2 is released during the formation of calcite coccoliths (p.180), coccolithophore dominance leads to reduction in the sequestration of CO_2 from the atmosphere to the ocean floor, whereas diatom blooms increase carbon flux.

Eutrophication alters the silicon balance in coastal zones

As with other biogeochemical cycles, anthropogenic activities are having a major impact on the silica cycle. As noted in earlier sections of this chapter, input of nitrogen and phosphorus into rivers and coastal zones from terrestrial runoff of fertilizers has increased greatly, leading to stimulation of algal growth. The construction of dams also leads to longer residence time in rivers, which results in trapping of DSi and reduced export to the sea. The increased N:Si or P:Si ratios mean that diatom growth is limited, whereas non-siliceous phytoplankton such as flagellates can flourish. This alters estuarine and coastal food webs, because diatoms are replaced by organisms that have less nutritional value or are less acceptable as prey for zooplankton. This has a major impact on the productivity of fisheries. Furthermore, blooms of non-diatom species increasingly lead to harmful algal blooms and anoxic dead zones discussed above.

Figure 9.11 Monthly mean maps of global distributions of silica and nitrogen and distribution of main phytoplankton groups. Upper panels show ambient Si: N ratios in June and December: values <2 (yellow) represent areas where dissolved silica becomes depleted before inorganic nitrogen; values >2 (green) represent areas where inorganic nitrogen is a growth-limiting nutrient. Lower panels show the prevailing phytoplankton groups in June and December. White regions indicate areas with no data. Reprinted from Pančić and Kiørboe (2018) with permission by John Wiley & Sons.

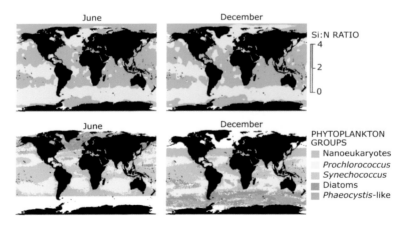

Conclusions

This chapter has illustrated how recent discoveries have led to dramatic shifts in our thinking about the role of microbes in the major nutrient cycles and their consequences for global ocean processes. It is important that we have a clear understanding of underlying mechanisms because of the increased effects of anthropogenic inputs on the composition of our atmosphere and oceans. Increasing concentrations of CO_2 and sulfur from the use of fossil fuel and industrial activity, inputs of nitrogen and phosphorus nutrients from agriculture and aquaculture, as well as shifts in silica levels due to deforestation and alteration of river flows and groundwater are all causing shifts in microbial community composition and altering the balance of marine microbial nutrient cycles. Indeed, major shifts in the nitrogen and sulfur nutrient cycles might catalyze further climatic changes in a runaway effect, for example, through increased output of greenhouse gases like nitrous oxide or depletion of the albedo effect due to changes in DMS emissions. However, predicting the outcomes of these changes is very difficult unless we have a clear understanding of underlying processes. In the last couple of decades, modern techniques have led to the discovery of previously unknown processes and major new surprises are currently appearing every year. As seen in other chapters, exploring these cycles has also revealed the variability and patchiness of the chemical and physical properties of the oceans, and the consequent huge diversity and adaptability of microbes. It is becoming increasingly difficult to categorize microbial taxa according to possession of a particular metabolic function, such as nitrifiers, denitrifiers, aerobes, anaerobes etc., because there are now so many examples of microbial clades crossing such functional boundaries. Organisms that appear closely related through their core genome form a diverse federation due to the presence of variants that possess genes for pathways that allow them to succeed in particular niches with very different physical and chemical conditions. Networks of competition and cooperation in the utilization of key nutrients add further layers of complexity to the microbial ecology of the oceans.

References and further reading

Iron, ocean fertilization, and other geoengineering

Bhatia, M.P., Kujawinski, E.B., Das, S.B. et al. (2013) Greenland meltwater as a significant and potentially bioavailable source of iron to the ocean. *Nat. Geosci.* **6**: 274–278.

Bonnain, C., Breitbart, M., & Buck, K.N. (2016) The Ferrojan Horse Hypothesis: Iron-virus interactions in the ocean. *Front. Mar. Sci.* **3**: 82.

Boyd, P.W. & Vivian, C.M.G. (eds.) (2019a) High level review of a wide range of proposed marine geoengineering techniques. *Rep. Stud. GESAMP 98*: 144.

Boyd, P.W. & Vivian, C.M.G. (2019b) Should we fertilize oceans or seed clouds? No one knows. *Nature* **570**: 155.

National Research Council (2015) *Climate Intervention: Carbon Dioxide Removal and Reliable Sequestration.* National Academies Press, Washington, DC.

Ratnarajah, L., Nicol, S., & Bowie, A.R. (2018) Pelagic iron recycling in the Southern Ocean: Exploring the contribution of marine animals. *Front. Mar. Sci.* **5**: 109.

Roman, J., Estes, J.A., Morissette, L. et al. (2014) Whales as marine ecosystem engineers. *Front. Ecol. Environ.* **12**: 377–385.

Roman, J. & McCarthy, J.J. (2010) The whale pump: Marine mammals enhance primary productivity in a coastal basin. *PLoS One* **5**: e13255.

Royal Society (2009) Geoengineering the climate: Science, governance and uncertainty. Online at https://royalsociety.org/topics-policy/publications/2009/geoengineering-climate/ (accessed 5 July 2019).

Rubin, M., Berman-Frank, I., & Shaked, Y. (2011) Dust-and mineral-iron utilization by the marine dinitrogen-fixer *Trichodesmium. Nat. Geosci.* **4**: 529.

Shoenfelt, E.M., Sun, J., Winckler, G. et al. (2017) High particulate iron(II) content in glacially sourced dusts enhances productivity of a model diatom. *Sci. Adv.* **3**: e1700314.

Smetacek, V. (2018) Seeing is believing: Diatoms and the ocean carbon cycle revisited. *Protist* **169**: 791–802.

Smetacek, V., Klaas, C., Strass, V.H. et al. (2012) Deep carbon export from a Southern Ocean iron-fertilized diatom bloom. *Nature* **487**: 313–319.

Smetacek, V. & Naqvi, S.W. (2008) The next generation of iron fertilization experiments in the Southern Ocean. *Philos. Trans. Roy. Soc. A* **366**: 3957–3967.

Strong, A., Chisholm, S., Miller, C., & Cullen, J. (2009) Ocean fertilization: Time to move on. *Nature* **461**: 347–348.

Tollefson, J. (2017) Iron-dumping ocean experiment sparks controversy. *Nature* **545**: 393–394.

Trick, C.G., Bill, B.D., Cochlan, W.P. et al. (2010) Iron enrichment stimulates toxic diatom production in high-nitrate, low-chlorophyll areas. Proc. Natl. Acad. Sci. **107**: 5887–5882.

Tzubari, Y., Magnezi, L., Be'er, A., & Berman-Frank, I. (2018) Iron and phosphorus deprivation induce sociality in the marine bloom-forming cyanobacterium *Trichodesmium. ISME J.* **12**: 1682–1693.

Watson, A.J., Boyd, P.W., Turner, S.M. et al. (2008) Designing the next generation of ocean iron fertilization experiments. *Mar. Ecol. Prog. Ser.* **364**: 303–309.

Weier, J. (2010) *On the Shoulders of Giants* (Obituary, John Martin). NASA Earth Observatory, online at https://earthobservatory.nasa.gov/features/Martin/martin_4.php (accessed 13 Sep 2019).

Yoon, J.-E., Yoo, K.-C., Macdonald, A.M. et al. (2018) Reviews and syntheses: Ocean iron fertilization experiments—Past, present, and future looking to a future Korean Iron Fertilization Experiment in the Southern Ocean (KIFES) project. *Biogeoscience* **15**: 5847–5889.

Nitrogen cycle, oxygen minimum zones

Aumont, O., Maury, O., Lefort, S., & Bopp, L. (2018) Evaluating the potential impacts of the diurnal vertical migration by marine organisms on marine biogeochemistry. *Glob. Biogeochem. Cycles* **32**: 1622–1643.

Benavides, M., Bonnet, S., Berman-Frank, I., & Riemann, L. (2018) Deep into oceanic N$_2$ fixation. *Front. Mar. Sci.* **5**: 108.

Bertagnolli, A.D. & Stewart, F.J. (2018) Microbial niches in marine oxygen minimum zones. *Nat. Rev. Microbiol.* **16**: 723–729.

Berube, P.M., Biller, S.J., Kent, A.G. et al. (2015) Physiology and evolution of nitrate acquisition in *Prochlorococcus. ISME J.* **9**: 1195–1207.

Bianchi, D., Babbin, A.R., & Galbraith, E.D. (2014) Enhancement of anammox by the excretion of diel vertical migrators. *Proc. Natl. Acad. Sci. USA* **111**: 15653–15658.

Breitburg, D., Levin, L.A., Oschlies, A. et al. (2018) Declining oxygen in the global ocean and coastal waters. *Science* **359**: eaam7240.

Caputo, A., Stenegren, M., Pernice, M.C., & Foster, R.A. (2018) A short comparison of two marine planktonic diazotrophic symbioses highlights an un-quantified disparity. *Front. Mar. Sci.* **5**: 2.

Chan, F., Barth, J.A., Lubchenco, J. et al. (2008) Emergence of anoxia in the California current large marine ecosystem. *Science* **319**: 920–920.

Dalsgaard, T., Thamdrup, B., Farías, L., & Revsbech, N.P. (2012) Anammox and denitrification in the oxygen minimum zone of the eastern South Pacific. *Limnol. Oceanogr.* **57**: 1331–1346.

Delmont, T.O., Quince, C., Shaiber, A. et al. (2018) Nitrogen-fixing populations of Planctomycetes and Proteobacteria are abundant in surface ocean metagenomes. *Nat. Microbiol.* **3**: 804.

Farnelid, H., Turk-Kubo, K., Muñoz-Marín, M., & Zehr, J. (2016) New insights into the ecology of the globally significant uncultured nitrogen-fixing symbiont UCYN-A. *Aquat. Microb. Ecol.* **77**: 125–138.

Foshtomi, M.Y., Braeckman, U., Derycke, S. et al. (2015) The link between microbial diversity and nitrogen cycling in marine sediments is modulated by macrofaunal bioturbation. *PLoS One* **10**: e0130116.

Gruber, N. (2016) Elusive marine nitrogen fixation. *Proc. Natl. Acad. Sci. USA* **113**: 4246–4248.

Gruber, N. (2019) Consistent patterns of nitrogen fixation identified in the ocean. *Nature* **566**: 191–193.

Hagino, K., Onuma, R., Kawachi, M., & Horiguchi, T. (2013) Discovery of an endosymbiotic nitrogen-fixing cyanobacterium UCYN-A in *Braarudosphaera bigelowii* (Prymnesiophyceae). *PLoS One* **8**: e81749.

Harding, K., Turk-Kubo, K.A., Sipler, R.E. et al. (2018) Symbiotic unicellular cyanobacteria fix nitrogen in the Arctic Ocean. *Proc. Natl. Acad. Sci. USA* **115**: 13371–13375.

Henke, B.A., Turk-Kubo, K.A., Bonnet, S., & Zehr, J.P. (2018) Distributions and abundances of sublineages of the N$_2$-fixing cyanobacterium "*Candidatus* Atelocyanobacterium thalassa" (UCYN-A) in the New Caledonian coral lagoon. *Front. Microbiol.* **9**: 554.

Kuypers, M.M.M., Marchant, H.K., & Kartal, B. (2018) The microbial nitrogen-cycling network. *Nat. Rev. Microbiol.* **16**: 263–276.

Laverock, B., Gilbert, J.A., Tait, K. et al. (2011) Bioturbation: Impact on the marine nitrogen cycle. *Biochem. Soc. Trans.* **39**: 315–320.

Laverock, B., Smith, C.J., Tait, K. et al. (2010) Bioturbating shrimp alter the structure and diversity of bacterial communities in coastal marine sediments. *ISME J.* **4**: 1531–1544.

Martiny, A.C., Kathuria, S., & Berube, P.M. (2009) Widespread metabolic potential for nitrite and nitrate assimilation among Prochlorococcus ecotypes. *Proc. Natl. Acad. Sci. USA* **106**: 10787–10792.

Messer, L.F., Mahaffey, C., Robinson, C. et al. (2016) High levels of heterogeneity in diazotroph diversity and activity within a putative hotspot for marine nitrogen fixation. *ISME J.* **10**: 1499–1513.

Muñoz-Marín, M.D.C., Shilova, I.N., Shi, T. et al. (2019) The transcriptional cycle is suited to daytime N$_2$ fixation in the unicellular cyanobacterium "*Candidatus* Atelocyanobacterium thalassa" (UCYN-A). *MBio* **10**: e02495-18.

Otero, X.L., De La Peña-Lastra, S., Pérez-Alberti, A. et al. (2018) Seabird colonies as important global drivers in the nitrogen and phosphorus cycles. *Nat. Commun.* **9**: 246.

Pedersen, J.N., Bombar, D., Paerl, R.W., & Riemann, L. (2018) Diazotrophs and N$_2$-fixation associated with particles in coastal estuarine waters. *Front. Microbiol.* **9**: 2759.

Shiozaki, T., Ijichi, M., Isobe, K. et al. (2016) Nitrification and its influence on biogeochemical cycles from the equatorial Pacific to the Arctic Ocean. *ISME J.* **10**: 2184–2197.

Sun, X., Kop, L.F.M., Lau, M.C.Y. et al. (2019) Uncultured *Nitrospina*-like species are major nitrite oxidizing bacteria in oxygen minimum zones. *ISME J.* **13**: 2184–2197.

Ulloa, O., Canfield, D.E., DeLong, E.F. et al. (2012) Microbial oceanography of anoxic oxygen minimum zones. *Proc. Natl. Acad. Sci.* **109**: 15996–16003.

Walworth, N.G., Fu, F.-X., Lee, M.D. et al. (2018) Nutrient-colimited *Trichodesmium* as a nitrogen source or sink in a future ocean. *Appl. Environ. Microbiol.* **84**: e02137-17.

Wannicke, N., Frey, C., Law, C.S., & Voss, M. (2018) The response of the marine nitrogen cycle to ocean acidification. *Glob. Change Biol.* **24**: 5031–5043.

Zehr, J.P. (2015) How single cells work together. *Science* **349**: 1163–1164.

Zehr, J.P. & Kudela, R.M. (2011) Nitrogen cycle of the open ocean: From genes to ecosystems. *Ann. Rev. Mar. Sci.* **3**: 197–225.

Zehr, J.P., Shilova, I.N., Farnelid, H.M. et al. (2017) Unusual marine unicellular symbiosis with the nitrogen-fixing cyanobacterium UCYN-A. *Nat. Microbiol.* **2**: 16214.

Zerkle, A.L. & Mikhail, S. (2017) The geobiological nitrogen cycle: From microbes to the mantle. *Geobiology* **15**: 343–352.

Sulfur, DMSP, climate regulation

Alcolombri, U., Ben-Dor, S., Feldmesser, E. et al. (2015) Identification of the algal dimethyl sulfide-releasing enzyme: A missing link in the marine sulfur cycle. *Science* **348**: 1466–1469.

Amato, P., Joly, M., Besaury, L. et al. (2017) Active microorganisms thrive among extremely diverse communities in cloud water. *PLoS One* **12**: e0182869.

Barak-Gavish, N., Frada, M.J., Ku, C. et al. (2018) Bacterial virulence against an oceanic bloom-forming phytoplankter is mediated by algal DMSP. *Sci. Adv.* **4**: eaau5716.

Bourne, D., Iida, Y., Uthicke, S., & Smith-Keune, C. (2007) Changes in coral-associated microbial communities during a bleaching event. *ISME J.* **2**: 350–363.

Bullock, H.A., Luo, H., & Whitman, W.B. (2017) Evolution of dimethylsulfoniopropionate metabolism in marine phytoplankton and bacteria. *Front. Microbiol.* **8**: 637.

Charlson, R.J., Lovelock, J.E., Andreae, M.O., & Warren, S.G. (1987) Oceanic phytoplankton, atmospheric sulphur, cloud albedo and climate. *Nature* **326**: 655–661.

Curson, A.R.J., Liu, J., Bermejo Martínez, A. et al. (2017) Dimethylsulfoniopropionate biosynthesis in marine bacteria and identification of the key gene in this process. *Nat. Microbiol.* **2**: 17009.

Curson, A.R.J., Williams, B.T., Pinchbeck, B.J. et al. (2018) DSYB catalyses the key step of dimethylsulfoniopropionate biosynthesis in many phytoplankton. *Nat. Microbiol.* **3**: 430–439.

Dell'Ariccia, G., Phillips, R.A., Van Franeker, J.A. et al. (2017) Comment on "Marine plastic debris emits a keystone infochemical for olfactory foraging seabirds" by Savoca et al. *Sci. Adv.* **3**: e1700526.

Evans, C., Malin, G., Wilson, W.H., & Liss, P.S. (2006) Infectious titers of *Emiliania huxleyi* virus 86 are reduced by exposure to millimolar dimethyl sulfide and acrylic acid. *Limnol. Oceanogr.* **51**: 2468–2471.

Falkowski, P.G., Katz, M.E., Knoll, A.H. et al. (2004) The evolution of modern eukaryotic phytoplankton. *Science* **305**: 354–360.

Howard, E.C., Henriksen, J.R., Buchan, A. et al. (2006) Bacterial taxa that limit sulfur flux from the ocean. *Science* **314**: 649–652.

Jones, G., Curran, M., Deschaseaux, E. et al. (2018) The flux and emission of dimethylsulfide from the Great Barrier Reef region and potential influence on the climate of NE Australia. *J. Geophys. Res. Atmos.* **123**: 13835–13856.

Lenton, T.M., Daines, S.J., Dyke, J.G. et al. (2018) Selection for Gaia across multiple scales. *Trends Ecol. Evol.* **33**: 633–645.

Lenton, T.M. & Latour, B. (2018) Gaia 2.0. *Science* **361**: 1066–1068.

Lidbury, I., Kröber, E., Zhang, Z. et al. (2016) A mechanism for bacterial transformation of dimethylsulfide to dimethylsulfoxide: A missing link in the marine organic sulfur cycle. *Environ. Microbiol.* **18**: 2754–2766.

Moran, M.A., Reisch, C.R., Kiene, R.P., & Whitman, W.B. (2012) Genomic insights into bacterial DMSP transformations. *Ann. Rev. Mar. Sci.* **4**: 523–542.

Pančić, M. & Kiørboe, T. (2018) Phytoplankton defence mechanisms: Traits and trade-offs. *Biol. Rev.* **93**: 1269–1303.

Quinn, P.K. & Bates, T.S. (2011) The case against climate regulation via oceanic phytoplankton sulphur emissions. *Nature* **480**: 51–56.

Raina, J.B., Dinsdale, E.A., Willis, B.L., & Bourne, D.G. (2010) Do the organic sulfur compounds DMSP and DMS drive coral microbial associations? *Trends Microbiol.* **18**:101–108.

Raina, J.-B., Tapiolas, D.M., Forêt, S. et al. (2013) DMSP biosynthesis by an animal and its role in coral thermal stress response. *Nature* **502**: 677–680.

Raina, J.-B.B., Tapiolas, D., Willis, B.L., & Bourne, D.G. (2009) Coral-associated bacteria and their role in the biogeochemical cycling of sulfur. *Appl. Environ. Microbiol.* **75**: 3482–3501.

Reisch, C.R., Moran, M.A., & Whitman, W.B. (2008) Dimethylsulfoniopropionate-dependent demethylase (DmdA) from *Pelagibacter ubique* and *Silicibacter pomeroyi*. *J. Bacteriol.* **190**: 8018–8024.

Savoca, M.S. & Nevitt, G.A. (2014) Evidence that dimethyl sulfide facilitates a tritrophic mutualism between marine primary producers and top predators. *Proc. Natl. Acad. Sci. USA* **111**: 4157–4161.

Savoca, M.S., Wohlfeil, M.E., Ebeler, S.E., & Nevitt, G.A. (2016) Marine plastic debris emits a keystone infochemical for olfactory foraging seabirds. *Sci. Adv.* **2**: e1600395.

Schiffer, J.M., Mael, L.E., Prather, K.A. et al. (2018) Sea spray aerosol: Where marine biology meets atmospheric chemistry. *ACS Central Sci.* **4**: 1617–1623.

Seymour, J.R., Simó, R., Ahmed, T., & Stocker, R. (2010) Chemoattraction to dimethylsulfoniopropionate throughout the marine microbial food web. *Science* **329**: 342–345.

Vila-Costa, M., Pinhassi, J., Alonso, C. et al. (2007) An annual cycle of dimethylsulfoniopropionate-sulfur and leucine assimilating bacterioplankton in the coastal NW Mediterranean. *Environ. Microbiol.* **9**: 2451–2463.

Wang, R., Gallant, É., & Seyedsayamdost, M.R. (2016) Investigation of the genetics and biochemistry of roseobacticide production in the *Roseobacter* clade bacterium *Phaeobacter inhibens*. *MBio* **7**: e02118.

Yoch, D.C. (2002) Dimethylsulfoniopropionate: Its sources, role in the marine food web, and biological degradation to dimethylsulfide. *Appl. Environ. Microbiol.* **68**: 5804–5815.

Phosphorus

Dyhrman, S., Ammerman, J., & Van Mooy, B. (2007) Microbes and the marine phosphorus cycle. *Oceanography* **20**: 110–116.

Dyhrman, S.T. (2018) Microbial physiological ecology of the marine phosphorus cycle. In *Microbial Ecology of the Oceans*, 3rd edition (Gasol, J.M & Kirchman, D.L., eds.), Wiley-Blackwell, pp. 377–434.

White, A. & Dyhrman, S. (2013) The marine phosphorus cycle. *Front. Microbiol.* **4**: 105.

Silicon

Baines, S.B., Twining, B.S., Brzezinski, M.A. et al. (2012) Significant silicon accumulation by marine picocyanobacteria. *Nat. Geosci.* **5**: 886–891.

Conley, D.J., Frings, P.J., Fontorbe, G. et al. (2017) Biosilicification drives a decline of dissolved Si in the oceans through geologic time. *Front. Mar. Sci.* **4**: 397.

Struyf, E., Smis, A., Van Damme, S. et al. (2009) The global biogeochemical silicon cycle. *Silicon* **1**: 207–213.

Tang, T., Kisslinger, K., & Lee, C. (2014) Silicate deposition during decomposition of cyanobacteria may promote export of picophytoplankton to the deep ocean. *Nat. Commun.* **5**: 4143.

Tréguer, P.J. & De La Rocha, C.L. (2013) The world ocean silica cycle. *Annu. Rev. Mar. Sci.* **5**: 477–501.

Other articles

Arrigo, K.R. (2005) Marine microorganisms and global nutrient cycles. *Nature* **437**: 349–355.

Bristow, L.A., Mohr, W., Ahmerkamp, S., & Kuypers, M.M.M. (2017) Nutrients that limit growth in the ocean. *Curr. Biol.* **27**: R474–R478.

Hutchins, D.A. & Fu, F. (2017) Microorganisms and ocean global change. *Nat. Microbiol.* **2**: 17058.

Richardson, T.L. (2019) Mechanisms and pathways of small-phytoplankton export from the surface ocean. *Annu. Rev. Mar. Sci.* **11**: 57–74.

Schneider, S.H. (2004) *Scientists Debate Gaia: The Next Century*, MIT Press.

Chapter 10

Microbial Symbioses of Marine Animals

In the broadest definition, the term symbiosis can be used to describe any close, long-term relationship between two different organisms, which can range from commensalism (a loose association in which one partner gains benefit but does no harm to the host) to parasitism (in which the parasite benefits at the expense of the host). However, in common usage, the term symbiosis often refers to a mutualistic relationship, in which both partners benefit through increased fitness in evolution. We now know that many marine animals obtain their food directly from the products of photosynthetic or chemosynthetic bacteria living in or on their bodies. Other microbial symbionts may provide their host with behavioral or reproductive benefits. These interactions are very widespread and have had a major influence on the evolution of host animals and their associated microbes. This chapter provides examples of mutualistic interactions with multicellular animals that illustrate their importance in marine ecology. (Symbioses also occur between different types of microbes—examples of these are considered in *Chapters 4–6*.) In addition, we now know that all animals contain diverse microbes in or on their bodies (the microbiome) and this may have less obvious influences that we are just beginning to understand. In particular, disruption of the balanced microbiome can lead to disease, and these effects are discussed in *Chapter 11*.

Key Concepts

- A wide range of marine invertebrates obtain some or all their nutrition from chemosynthetic or photosynthetic microbes, which may occur as intracellular endosymbionts within the tissues or as epibionts on internal or external surfaces.

- Microbial symbioses are fundamental to the function of specific animals as ecosystem engineers.

- The behavior of some fish and invertebrates depends on bioluminescence produced by bacterial symbionts that colonize specialized light organs.

- Symbiotic hosts have evolved complex adaptations of recognition, tissue development, and internal structure in order to maximize the benefits of the relationship.

- Most symbioses in marine animals depend on horizontal transmission, in which free-living microbes are acquired from the environment by the larvae. In a few cases, symbionts are transmitted vertically from the parent host; this usually leads to an obligate relationship in which the symbiont undergoes genome reduction.

- Environmental stress often leads to breakdown of the symbiotic relationship.

- Study of marine symbioses has resulted in the recognition that animals and their resident microbiomes should be regarded as holobionts or meta-organisms. This has led to models for understanding animal development and immunity, with important applications in biotechnology and human health.

DIFFERENT TERMS FOR THE LOCATION OF SYMBIONTS

We can consider symbioses in terms of the degree of "intimacy" of the association. Some microbes are intracellular organisms with special adaptations enabling them to live inside the host cells—these are endosymbionts (from *endo-*, meaning inside). However, this term is often used to describe extracellular microbes living inside the animal body on the surface of the internal ducts or body cavities, such as the intestinal tract. These are better described as episymbionts (from *epi-*, meaning upon) to indicate that they are living on the surface of the host cells. Episymbionts are also commonly found on the external surface of many animals or algae, in which case they can be described as ectosymbionts (from *ecto-*, meaning outside).

Symbioses occur in many forms

The concept of symbiosis between different organisms developed at the end of the nineteenth century and had a profound influence on the development of biology and evolutionary theory. The term symbiosis is derived from the Greek for "living together," and encompasses a broad range of interactions. Our usual definition of symbiosis depends on the concept of *mutual* benefit that the partners receive—but benefit can be very hard to assess and is often rather one-sided. This creates a spectrum of possible interactions. In several of the examples discussed in this chapter, we will see that the microbial symbionts have been shown to have a free-living stage. Therefore, this can be described as a *facultative* relationship, because the symbiosis is not *obligate* for the microbes as they can clearly live independently. However, some hosts are completely dependent on their microbes' activities for their growth and development beyond the initial stages of their life cycle, so for them the relationship *is* obligate. In other cases, which are less common, the co-evolution of the host and symbiont has led to the symbiont losing the potential for independent existence and the partners are obligately interdependent throughout their life cycles. Advances in DNA sequencing and high-resolution imaging technologies are leading to great insights into the complexities of microbe-animal symbioses. We now recognize that the development, function, health, and evolutionary fitness of animals is hugely influenced by microbes, which have been present throughout the history of life on Earth. This has led to the realization that we should no longer think about hosts and their microbial partners separately as independent units, but rather as an integrated whole—a "holobiont" or "meta-organism." The study of symbioses in marine invertebrates provides invaluable models that underpin the holobiont concept and are influencing our understanding of animal development and evolution and the effects of environmental pressures. Furthermore, some of these models have direct applications for studies relevant to the advancement of human health.

Chemosynthetic bacterial endosymbionts of animals were discovered at hydrothermal vents

Chapter 1 introduced the dense undersea communities of giant tubeworms, bivalve molluscs, and other creatures growing around deep-sea hydrothermal vents (*Figure 1.15B*), whose discovery at the Galapagos Rift in the Pacific Ocean in 1977 was such a great surprise. How can such a highly productive ecosystem be sustained? Vent habitats are completely dark, under enormous pressure, and at such great depth that organic material from the upper layers of water settles too slowly to support this level of growth. At first, investigators thought that the vent animals might feed by filtration of planktonic chemosynthetic bacteria that exist in the water around the vents, concentrated by local warm-water currents created by the vent activity. This idea became untenable when equipment was developed for bringing intact animals to the surface. It was then discovered that the filter-feeding mechanism and digestive tract of the giant white clams (named as *Calyptogena magnifica*, family Vesicomyidae) found at the vents was greatly reduced. Investigation of the giant tubeworm *Riftia pachyptila*, family Siboglinidae) showed that it lacked a mouth and gut, so it was assumed at first that it must feed by simply absorbing nutrients from the surrounding water. However, the rate of uptake of organic material is insufficient to explain the extremely high metabolic rates of these animals. Thus, it was realized that the nutrition of these animals and the flow of energy through the vent food web depends directly on chemosynthetic bacteria within the tissues of the animals.

DIFFERENT TERMS FOR CHEMOSYNTHESIS

Chemosynthesis is a widely used general term to describe the formation of organic compounds which includes chemolithoautotrophy and methanotrophy. Chemolithoautotrophs use CO_2 as the sole carbon source, using energy derived from the oxidation of reduced inorganic substrates such as sulfide, hydrogen, or ammonium. Bacteria which oxidize hydrogen sulfide, thiosulphate, and other reduced-sulfur compounds are also commonly described as thiotrophic (although many of these bacteria can also use hydrogen). Methanotrophs oxidize methane, which acts as a source of reducing power and as the source of carbon. The reactions involved are described in *Chapter 3*.

Various lines of evidence supported this conclusion. First, stable isotope analysis (p.58) of the tissue of the clam *C. magnifica* showed their cellular material does not have its ultimate origin in photosynthesis. The ratios of ^{13}C to ^{12}C reflect the efficiency with which different enzymes deal with the different isotopes. The ratio found in these animals is well outside the value of photosynthetically fixed carbon, proving that they are not feeding on material derived from CO_2 fixed in the upper photic zone.

When specimens of the tubeworm *R. pachyptila* were dissected, it was found that the trunk is filled with an organ supplied with many blood vessels (*Figure 10.1a*). Electron microscopy of this structure (the trophosome) showed that it contained granules of elemental sulfur and large numbers of structures that strongly resembled bacterial cells. The trophosome tissue was also found to contain large amount of lipopolysaccharide (LPS), indicating that it is packed with endosymbiotic Gram-negative bacteria. These are contained within host cells called bacteriocytes. Further biochemical analysis showed the presence of enzymes associated with sulfur metabolism, including ATP-reductase and ATP-sulfurylase and high levels of RuBisCO (p.83), providing further proof of endogenous autotrophic metabolism. Subsequent investigations showed that the products of chemosynthesis are transferred across the host membrane surrounding the bacteria to the animal tissue for growth. This is a truly remarkable animal-bacterium symbiosis; there are about 10^9 bacteria per gram of tissue, occupying up to a quarter of the volume of the trophosome. *Riftia* has evolved into a truly autotrophic animal, with sophisticated physiological mechanisms to ensure the efficient transport of oxygen and sulfide to the symbionts. The endosymbionts were soon identified by 16S rRNA gene sequences as members of the Gammaproteobacteria and free-living bacteria corresponding to the symbiont phylotype can be detected in seawater using FISH with 16S rRNA gene-targeted probes, as far as 100 m from tube worm colonies. Although all attempts to culture the *Riftia* symbiont have failed, the high densities of an apparently pure population of bacteria within the trophosome has enabled separation of host and symbiont cells and construction of a composite metagenome. The findings explain why the bacterium is successful in the two very different environments, living autotrophically within host cells and heterotrophically when free-living, as shown in *Figure 10.1b*. This dual nature is recognized in the species name given to the symbiont—"*Ca*. Endoriftia persephone" (from Persephone in Greek mythology, who was the goddess of fertility as well as queen of the underworld). Further insights into the acquisition and dissemination of the symbionts are described in *Box 10.1*.

In the clam *C. magnifica*, microscopy revealed that the symbiotic bacteria are contained in large cells of the gills that are exposed to the seawater on one side and the blood supply on the other, although the anatomical adaptations to ensure efficient transport of nutrients to the symbionts are not as extensive as those of *Riftia*. Once again, enzyme assays show the thio-autotrophic nature of the bacteria. The symbionts of *Calyptogena* spp. are transmitted vertically. Intuitively, one might expect this to be a common mechanism to ensure that the bacteria are passed on directly to their offspring, but this is one of only a few examples currently known in marine chemosynthetic invertebrates. The genomes of *Calyptogena* symbionts are very small—just 1.0–1.2 Mb, encoding only about 1000 genes. Genome analysis shows that the major chemoautotrophic pathways are present, as well as routes for the production of vitamins, co-factors, and amino acids required by the clam. However, many of the genes

Figure 10.1 Schematic diagrams illustrating symbiosis between the tube worm *Riftia pachyptila* and its endosymbiotic bacterium "*Ca*. Endoriftia persephone." (a) An extensive capillary system transports O_2, H_2S, and CO_2 bound to hemoglobin (Hb) to the trophosome where host cells (bacteriocytes) are packed with endosymbiotic bacteria. (b) Within the bacteriocytes, endosymbionts oxidize sulfide via the APS pathway (see *Figure 3.3*), yielding ATP and NADPH via an electron transport system (ETS). CO_2 fixation occurs via the CBB and reductive tricarboxylic acid (r-TCA) cycles (see *Figure 3.4*). Organic compounds made by symbionts pass to the host via translocation of simple organic compounds and digestion of symbiont cells (open arrows). Nitrogen for the symbiont's biosynthesis is derived from urea excreted by host cells, or from reduction of nitrate absorbed from seawater. (c) In their free-living state, the bacteria are thought to live heterotrophically by absorbing dissolved organic compounds in seawater. These may be respired via glycolysis and the oxidative (ox-)TCA cycle. Free-living bacteria may also use the rTCA cycle for autotrophic metabolism. Based on data from Stewart et al. (2005); Harmer et al. (2008); Robidart et al. (2008).

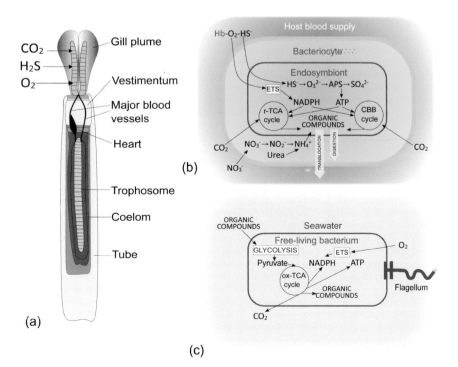

(a) (b) (c)

BOX 10.1   RESEARCH FOCUS

Genomic insights into host–symbiont interactions: infection, colonization, and dissemination

As we learn more about symbioses of marine invertebrates through the application of modern techniques, the interactions show many parallels with the interactions between pathogenic bacteria and their hosts. Two areas of active research in this field are discussed here.

How do symbiotic bacteria infect their tubeworm hosts? The bacterial symbionts found in *Riftia pachyptila* and related siboglinid tubeworms have genomes of similar size (~3.4–4.6 Mb) and composition to their free-living counterparts (Robidart et al., 2008; Li et al., 2018) and there is no evidence of co-speciation with their hosts (McMullin et al., 2003). They are acquired from the environment at an early stage of larval development. Hydrothermal vents are very ephemeral, and the success of these tubeworms depends on larval dispersal over great distances. Since the adult worms depend absolutely on the symbionts for their nutrition, it seems a risky strategy for the larvae not to contain symbionts. However, horizontal transmission may offer more opportunities for the host to acquire specific types of bacteria that might be best adapted to local conditions and this is thought to have played a major role during evolution (Fisher et al., 2017). To investigate how the bacteria colonize the worms, Nussbaumer et al. (2006) constructed specially designed chambers with grooved plates that were placed at a hydrothermal vent using the submersible *Alvin*. A year later, the tiny (0.2–2 mm) and very fragile bodies of newly settled larvae and juvenile animals of three different species (*Riftia pachyptila*, *Tevnia jerichonana*, and *Oasisia alvinae*) were collected and examined microscopically.

Bacteria that had colonized the host were detected using symbiont-specific FISH probes (p.36). When they first settle, the larvae have an obvious digestive tract, including a mouth and anus. The bacteria can be detected when the larvae reach a length of about 250 μm. However, even though the larvae feed actively on microbes, the FISH micrographs showed that the symbionts enter via the skin (where they become enclosed in vacuoles), not via the mouth. Fewer than 20 bacteria are needed for infection, but these rapidly migrate through the layers of host tissue, multiply, and initiate differentiation of the mesodermal tissue. Their presence then provokes massive apoptosis (programmed cell death) of the epidermis, muscles, and mesoderm, followed by trophosome development (Nussbaumer et al., 2006). Further insight into the infection and colonization process has been provided by Li et al. (2018), who compared genomes of the symbionts from several vent and seep-dwelling tubeworms. In all symbionts, they identified a range of genes that have previously been associated with infection processes in pathogenic bacteria; these include adhesion, secretion systems, virulence proteins, and overcoming oxidative stress. In addition, the presence of genes for flagellar motility, chemotaxis, and adhesive pili indicate the role of these processes in initial contact with the host larvae, which are thought to secrete a mucus that attracts potential symbionts (Nussbaumer et al., 2006). Li et al. (2018) also identified genes encoding cytolytic proteins, chitinase, and collagenase, which probably enable the bacteria to migrate through the trophosome tissue during its early stages of development.

How do the symbionts seed the environment? To ensure colonization of host larvae by horizontal transmission, there must be a mechanism by which the symbionts seed the environment. However, they are contained inside the bacteriocytes deep inside the tissue and there are no openings to the exterior, through which they might escape. The trophosome is a highly structured organ with an extremely rapid growth rate (comparable to cancer or wound healing) in which symbiont cells within the bacteriocytes go through a coordinated cell cycle. The density of symbionts is controlled through apoptosis and digestion at the periphery and replacement by dividing symbionts at the center (Pflugfelder et al., 2009). To investigate the possibility that the symbiont Endoriftia is released from *R. pachyptila* when it dies, Klose et al., 2015 developed high pressure incubation chambers in which trophosome tissue, dissected from the worms, can be maintained in the laboratory under conditions simulating the deep-sea environment at a

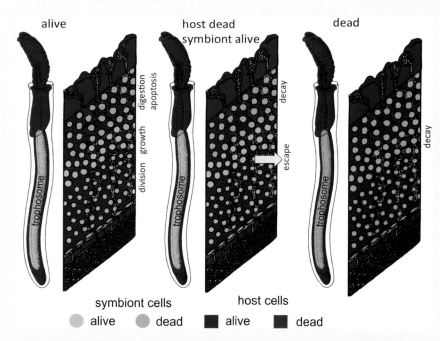

Figure 10.2 Symbiont viability in the course of host degradation *of Riftia* tissue in experimental conditions. Upon host death, symbiont digestion ceases allowing escape of viable symbionts before complete autolysis. Adapted from Klose et al. (2016) CC-BY-4.0.

BOX 10.1   RESEARCH FOCUS

hydrothermal vent, with controlled supplies of sulfide and oxygen and temperatures simulating the vent fluid (~22°C) or surrounding seawater (−4°C). Escaping bacteria were collected into specially designed porous chambers and identified using FISH. Only living symbiont cells were detected following host death due to starvation from lack of sulfide, indicating an active escape process through the skin. Based on the average size of *Riftia*, Klose and colleagues estimate that a typical clump of tubeworms, dying as result of cessation of vent flow, could release over 10^9 symbionts within a very short period. Scavenging by crabs on dead tubeworms—frequently observed at vents—could increase the release of symbionts further. In follow-up experiments, (Klose et al., 2016) monitored the death and lysis of host cells, using analytical techniques similar to those used in forensic investigation of putrefaction. They concluded that, following host cell death from starvation, the continued degradation of symbionts ceases rapidly (*Figure 10.2*). Video monitoring of *in situ* tubeworm communities shows that tubeworms have a short life, typically less than two years, which suggests regular seeding of the environment by released Endoriftia to facilitate colonization of nearby vents. However, the mechanism by which *Rifta* larvae and their Endorifta symbionts migrate to newly developed vent fields that may be hundreds of kilometers away is something of a mystery. Klose et al. (2015) conclude that the temporary association between the symbiont and its *Riftia* host must provide a fitness benefit, since only 20 or so bacteria are needed to infect the host at its larval stage, yet a dying worm will release almost a million living bacteria.

How do symbiotic bacteria infect *Bathymodiolus* mussels? Despite relying on chemosynthetic symbionts for a large part of their nutrition, *Bathymodiolus* spp. have a fully functional filtering system in their gills and are thus exposed to many different bacteria in their environment. However, each species appears to harbor a specific phylotype of either a sulfur-oxidizing or methane-oxidizing symbiont or have a dual symbiosis of both types (Dubilier et al., 2008; Duperron et al., 2013). Again, the symbionts are contained in bacteriocytes and transmission is horizontal, but the mechanism of colonization is unknown. The gills of bivalve molluscs grow throughout their life cycle and Wentrup et al. (2014) investigated whether the symbionts are present in undifferentiated growth zones before new gill filaments develop, or whether the newly differentiated gill filaments are freshly infected. Using FISH, immunohistochemistry, fluorescence, and electron microscopy, they showed that symbionts were present in the newly-formed filaments but not in the growth zones, indicating that gill colonization occurs throughout the host's lifetime. Interestingly, this pattern was also observed in the vesicomyid clam *Calyptogena ponderosa*, in which transmission of the symbiont is vertical, via the eggs. Wentrup and colleagues concluded that this likely occurs via self-infection from older gill tissues, rather than re-infection with bacteria from the environment and/or co-occurring mussels. Infection of the newly formed gill tissue results in morphological changes, including the loss of microvilli. It appears that the bacteria escape the colonized bacteriocytes without causing lysis of their membranes and infect neighboring cells.

Insights into the evolution of symbionts and pathogens. This ability to enter the cells, avoid immediate digestion, and have an escape mechanism is reminiscent of that shown by many intracellular pathogens, which have been extensively studied in diseases of humans and other vertebrates. In this case, the mussel symbionts are related to free-living marine bacteria with no known pathogenic relatives. By comparing the genomes of symbionts and their free-living relatives, Sayavedra et al. (2015) showed that *Bathymodiolus* sulfur-oxidizing symbionts have undergone extensive genome rearrangements and horizontal gene transfer and that many of the genes found only in the symbionts are homologous to classes of toxin-related genes (TRGs) that are involved in the virulence of pathogens. Transcriptomic and proteomic analyses showed that these genes are expressed and translated in the gill tissue. Sayavedra and colleagues conclude that these genes have probably been acquired via horizontal gene transfer after the divergence of the symbionts from their free-living relatives. Some of the TRGs may have beneficial interactions, such as attachment and suppression of host rejection or the inhibition of parasites and suggest that the *Bathymodiolus* symbionts have "tamed" the genes for beneficial use in the host. Alternatively, they raise the possibility that the genes actually evolved as "symbiosis factors … commandeered for use in harmful interactions" by pathogens (Sayavedra et al., 2015). (Interestingly, the Endoriftia genomes analyzed in the study mentioned above, contained only a small number of one class of TRGs.) In a follow-up study by this group, Sayaveedra et al. (2019) examined the distribution of TRGs of symbionts in a wider range of *Bathymodiolus* species and in two deep-sea sponges. Again, they found a wide array of toxin-related and secretion system genes, but only one toxin class, MARTX, was common to all symbionts and might therefore be essential for recognition, attachment, and symbiont uptake. The authors conclude that the variations observed in the other TRGs may indicate convergent evolution, in which free-living bacteria adopted different paths in their evolution as intracellular symbionts (Sayaveedra et al., 2019).

encoding processes required for survival outside the host including motility, DNA repair mechanisms, and stress responses are missing. Such extreme genome reduction is also seen in intracellular obligate parasites (see p.72) and occurs because of genetic bottlenecks due to reductions in effective population sizes, with reduced opportunities for recombination and horizontal gene transfer. In addition, positive selection pressure results in loss of genes for metabolic functions that are provided by the host. In *Calyptogena* spp., the symbiotic bacteria are predominantly passed from one generation to the next via vertical transmission in the eggs. DNA sequence evolution in the symbiont is closely coupled with the DNA in host mitochondria (which are only inherited in the cytoplasm of the eggs), confirming that transmission is vertical, through the maternal line. However, nothing is known about the development of the symbiosis in the larvae and how the gill filaments become colonized. The association between these clams and their microbial symbionts is relatively young in phylogenetic terms—less than 50 million years—so it is tempting to think that we are witnessing an early stage in the evolution of this symbiosis toward full enslavement of the bacteria to become organelles through further genome reduction. This can be seen in many endosymbionts of insects and it is how mitochondria and chloroplasts developed in the original evolution of eukaryotes. However, it is important to note that most marine chemoautotrophic symbionts provide all, or almost all, of the host's nutritional requirements, whereas in insects the symbiont often supplies only a few key nutrients or enzymes. Therefore, genome reduction of insect symbionts can evolve to be much more extensive.

A wide range of other chemosynthetic endosymbioses occurs in the deep sea

Many other species of siboglinid tubeworms also form dense colonies at hydrothermal vents and seeps. These include *Tevenia, Ridgea, Oasisa,* and *Lamellibrachia.* These may occupy different sites, according to the local physicochemical properties of the vent or seep fluids, although the same type of symbiont may occur in different animal species. Some examples are given below.

Large mytilid mussels (up to 20 cm long) of the genus *Bathymodiolus* are found in dense colonies attached to rocks adjacent to hydrothermal vents and also at cold seeps (*Fig. 1.16*), which emit fluids containing combinations of methane, other hydrocarbons, and hydrogen sulfide, in differing proportions depending on geology. Some mussels appear to contain only a single type of symbiont, either thiotrophic as in *B. thermophilus* or methanotrophic as in *B. childressii*, while several other species including *B. azoricus* and *B. puteoserpentis* have a dual symbiosis, containing both one (or more) thiotroph(s) and one (or more) methanotroph(s), all members of the Gammaproteobacteria. The ratio of the two types of endosymbiont within the host reflects the relative abundance of sulfide and methane in the site at which the animal is growing. In some cases, there is an even more remarkable diversity of endosymbionts with four or more phylogenetically distinct symbionts, providing additional metabolic versatility. For example, *B. heckerae* growing in petroleum-rich seeps contain hydrocarbon-degrading bacteria related to *Cycloclasticus*. There is a well-developed filter-feeding apparatus in *Bathymodiolus* spp., enabling them to gather food particles from the water column, so they probably obtain their nutrition from a mixture of autotrophic and heterotrophic routes. This, combined with their greater mobility, explains the fact that these mussels can tolerate a wider range of habitats at the vent sites than the tubeworms and vesicomyid clams.

Snails of the genus *Alvinoconcha*, found at vents in the Indian and south-western Pacific Oceans, harbor dense populations of intracellular bacteria in their enlarged gill tissue and are thought to provide the bulk of the animal's nutrition via chemosynthesis. Most of the five known species of *Alviniconcha* contain dominant populations of bacteria from either the Gammaproteobacteria or Epsilonbactereota, but one species holds approximately equal populations of two distinct Gammaproteobacteria. Some specimens have also been found to contain smaller numbers of heterotrophic bacteria provisionally identified as *Endozoicomonas*, belonging to the Oceanospirillales, an order whose members are known for their ability to degrade complex organic material (p.127). These bacteria are contained in vacuoles; however, it is not known whether they have a mutualistic function as a secondary symbiont, perhaps involved in providing carbon compounds or in the cycling of sulfur from organic compounds. A different species, *Ifremeria nautilei*, is one of the most abundant snails found at hydrothermal vents in the Western Pacific. It contains gammaproteobacterial symbionts (one methanotroph and one thiotroph), as well as two alphaproteobacterial phylotypes of unknown function.

Other deep-sea symbiotic associations have been described in which episymbiotic bacteria occur in addition to endosymbionts. The scaly foot snail (*Chrysomallon squamiferum*), found at hydrothermal vents in the Indian Ocean at ~2500 m deep, is encased in an unusual shell armored with iron sulfides (greigite Fe_3S_4 and pyrite FeS_2) and the foot is covered in overlapping scales (sclerites) of proteinaceous material containing sulfide (*Figure 10.3*).

? ARE MUSSEL EPISYMBIONTS IN EVOLUTIONARY TRANSITION?

In addition to the gammaproteobacterial chemosynthetic endosymbionts, *Bathymodiolus* spp. have been shown to harbor bacteria belonging to the Epsilonbactereota phylum (recently reclassified, see p.129). This was revealed by the presence of characteristic gene sequences in metagenomic libraries from two distantly related and geographically separated species (Assié et al., 2016). Subsequent analysis of DNA from seven species around the world and transcriptomic analysis of mussel showed that they all contained members of a novel group of Epsilonbactereota. FISH, in conjunction with electron microscopy, identified these as filamentous epibionts on the surface of the gill. They are similar to epibiotic bacteria found in hydrothermal vent crustaceans (see p.265), so Assié and colleagues suggest that may also be sulfur-oxidizers. Interestingly, the epibionts were not found in mussels which lacked sulfide-oxidizing endosymbionts, leading to speculation on the possibility that the epibiotic symbionts are in a phase of evolutionary transition towards endosymbiotic integration (Assié et al., 2016).

Figure 10.3 The scaly-foot snail *Chrysomallon squamiferum.* Credit: Ching Chen, JAMSTEC.

This is thought to provide further protection against predators attempting to attack its foot, enabling the snail to stay firmly attached to vertical rocks with good feeding locations, rather than retracting into its shell. The sclerites are covered by various epibiotic bacteria dominated by sulfate-reducing bacteria in the Deltaproteobacteria and Epsilonbactereota classes, which are thought to be responsible for their mineralization with iron sulfide. The snail may secrete mucus secretions which stimulate colonization by these bacteria. The snail has a highly reduced digestive tract and appears to obtain all its nutrition from another set of chemosynthetic bacteria, this time endosymbionts belonging to the Gammaproteobacteria, contained within bacteriocytes in an esophageal gland, reminiscent of the *Riftia* trophosome.

Another vent animal that has been the subject of intensive research is *Alvinella pompejana*, an oligochaete worm about 9 cm long and 2 cm wide, which lives in dense masses in the walls of black smoker chimneys on the East Pacific vents. This animal is known as the "Pompeii worm" in recognition of the association with volcanic activity and its remarkable temperature gradient across its body; the worm's posterior end is in galleries within the chimney wall (~85°C), while the anterior end projects into the surrounding cooler water (~20°C). Hair-like projections secreted from mucous glands on the dorsal surface are covered in episymbiotic bacteria affiliated with members of the Epsilonbactereota, forming a dense fleece-like structure. They are chemoautotrophic, possessing genes for sulfide oxidation and denitrification, and fix CO_2 via the reductive TCA cycle. It is not clear whether the host obtains any nutritional benefit. The main benefit to the worm appears to be protection against extremes of temperature and detoxification of metals in the vent fluids, rather than nutrition. The bacterial proteins predicted from the metagenomic sequences appear to have special adaptations to variable temperature and chemical conditions.

Crustaceans have also been shown to host chemosynthetic episymbionts. When hydrothermal vents on the mid-Atlantic Ridge were first investigated in 1985, large populations of a type of shrimp (*Rimicaris exoculata*) were discovered. They occur in large swarms, with rapid and continuous movement, appearing to graze the walls of the vent (*Figure 10.4B*). These animals have specially adapted mouthparts and an enlarged gill chamber, containing a large mixed population of up to 10^7 episymbiotic bacteria dominated by members of the Gammaproteobacteria and Epsilonbactereota, with lesser numbers of Delta- and Zetaproteobacteria. The varied metabolic capabilities of the mixed symbiont community allow hydrogen, sulfide, methane, or iron to be used as reductants for chemosynthesis. The symbionts have a free-living stage and the same phylotypes of bacteria are also present in large numbers on and around the vent chimneys. It seems likely that the shrimps obtain much of their nutrition by feeding on their episymbionts. The shrimps provide the bacteria with the optimum concentrations of sulfide and oxygen needed for chemosynthesis by scraping metal

COULD THE SCALY-FOOT SNAIL INSPIRE NEW FORMS OF ARMOR?

The protective shell of the scaly-foot snail discovered by Warén et al. (2003) and named *Chrysomallon squamiferum*s by Chen et al. (2015) has a unique three-layered shell. Unlike most snail shells that are composed solely of calcium carbonate, the shell of *C. squamiferum* is fortified with iron compounds and has very unusual mechanical properties that dissipate energy, as shown experimentally by Yao et al. (2010). They used diamond-tipped indenters to measure the enormous mechanical strength of the shell. Materials scientists and engineers are being inspired by such examples from the natural world that may lead to new forms of body armor, sporting equipment, and protective surfaces.

sulfides from the chimney wall and by fanning water currents over the bacteria with their rapid movements.

Kiwa hirsuta is a species of decapod crustacean discovered in 2005 at a 220 m deep hydrothermal vent on the Pacific-Atlantic Ridge. It was nicknamed the "yeti crab" because it is covered in silky blond filaments (setae), resembling fur (*Figure 10.4A*). Molecular phylogenetic analysis revealed the setae to be covered by dense populations of bacteria in the phyla Gammaproteobacteria, Epsilonbactereota, and Bacteroidetes. Key enzymes for sulfide oxidation and carbon dioxide fixation indicate a thioautotrophic lifestyle.

Chemosynthetic symbioses are also widespread in shallow-water sediments

The discovery of chemosynthetic symbionts in animals from hydrothermal vents was quickly followed by their detection in methane cold seeps. This then led biologists to search for examples in other environments, such as coastal sediments with sources of methane or sulfide provided by the activities of sediment sulfate-reducing bacteria and methanogenic archaea. (see p.21) It was quickly realized that chemosynthetic symbioses are not restricted to a few "exotic" animals in deep-sea habitats, but occur in almost all sediments where reduced compounds able to act as an electron donor (especially sulfides, methane, or hydrogen) are available. Such habitats include seagrass beds, coral reefs, marshes, and mangrove swamps where there is a redox gradient between the top oxygenated layer of sediment and the anaerobic bottom layers.

Chemosynthetic symbioses have now been described in hundreds of species in at least seven phyla of invertebrates. A few examples of these are discussed below.

Investigation of different hosts soon revealed that many—probably most—involve interactions with more than one partner. This enhances the advantage to the host through the acquisition of different metabolic capabilities. Perhaps the most remarkable example of multiple symbionts discovered so far occurs in the small oligochaete worm *Olavius algarvensis* and related species in the family Siboglinidae. This worm contains both sulfate-reducing and sulfur-oxidizing bacteria, members of the Deltaproteobacteria (see note on reclassification, p.129) and Gammaproteobacteria respectively. These tiny worms, only 0.1–0.2 mm in diameter and up to 4 cm long (*Figure 10.5A*), are completely devoid of a gut system and rely almost entirely on the production of nutrients by the thioautotrophic bacteria. Unlike the previous examples, the bacteria are not contained within specialized cells, but occur extracellularly just below the cuticle of the worm (*Figure 10.5B*). The bacteria show cooperative metabolism (syntrophy): the deltaproteobacterial symbionts reduce sulfate—in the seawater permeating the sediment and absorbed through the worm's cuticle—to sulfide, which is used by the gammaproteobacterial symbionts as an electron donor for the autotrophic fixation of CO_2. Because the worm's habitat has very low external concentrations of sulfide, the internal generation of sulfide by the deltaproteobacterial symbionts is essential for sustaining the autotrophy by the gammaproteobacterial symbionts (*Figure 10.5C*). When first described, it was thought that the worm contained one population of each type, but metagenomic analysis has revealed that there are two genetically different types of symbionts affiliated with Deltaproteobacteria and two types affiliated with Gammaproteobacteria. Some individuals also contain up to five other symbionts from different bacterial groups, with unknown metabolism. Analysis of the metagenome shows that one of the symbionts possesses an alternative mechanism for production of electrons for CO_2

DANCING CRABS FARM THEIR BACTERIA

A new species similar to the yeti crab was discovered at a methane seep off the coast of Costa Rica by Thurber et al. (2011). *Kiwa puravida* was shown to have similar thiotrophic epibiont communities to those found by Goffredi et al. (2008) in *K. hirsuta*. Thurber and colleagues provide powerful evidence that *K. puravida* obtains its nutrition by "farming" the chemosynthetic bacteria on its surface. Analysis of stable isotopes and fatty acid biomarkers in the crab tissue indicated that the epibiotic bacteria are the main source. Thurber et al. (2011) also provided video footage of the crabs' behavior, filmed during submersible dives to the methane seep. They can be seen to wave their chelipeds (front legs) back and forth in a rhythmic fashion. This is assumed to increase the supply of oxygen to the symbionts needed for oxidation of sulfide. The crab also has specialized structures on its mouthparts that it uses to scrape the bacteria from the setae covering its body and transfers them to its mouth. Thus, *K. puravida* has both behavioral and anatomical adaptations that enable it to farm and harvest its epibionts for food.

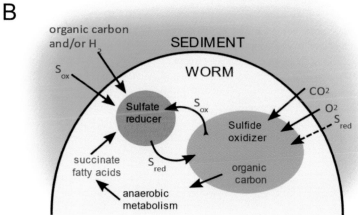

Figure 10.5 A. An adult gutless worm *Olavius algarvensis* isolated from sediment, Isle of Elba, Italy. B. Model showing syntrophic cycling of oxidized (S_{ox}) and reduced (S_{red}) sulfur compounds between SOB (green) and SRB (red) endosymbionts within the tissue of the worm. Credits: A. Alexander Gruhl, MPI Marine Microbiology, Bremen. B. Redrawn from Dubilier et al. (2001) with permission of Springer Nature.

> ⓘ **BUCKETS OF MUD AND "STONE AGE" SEQUENCING**
>
> The herculean task in unraveling the interactions in the multi-partner consortium of the worm *O. algarvensis* is only partly evident in the paper by Woyke et al. (2006). However, in a follow-up paper revisiting this investigation, Kleiner et al. (2011) reveal some of the complexities of their work. This involved collaboration between teams led by Nicole Dubilier of the MPI for Marine Microbiology and Eddy Rubin of the Joint Genome Institute, at a time when metagenomic studies were just beginning. For lead author Tanja Woyke, fieldwork to collect fresh worms by scuba diving off the Mediterranean Isle of Elba must have seemed appealing, but the reality involved collection and sifting of bucketloads of sediment searching for the tiny microlitre-volume worms. It took many weeks to collect enough material—about 1000 worms—for DNA sequencing. The team had to construct large-insert clone libraries and take special steps to enrich the proportion of microbial DNA before developing novel algorithms to assign sequences to the different symbionts, depending on the relative frequency of nucleotides. These binned sequences were then assembled and annotated to construct the genome of four symbionts, allowing their metabolic potential to be predicted. Kleiner et al. (2011) comment that, at the time (2006), this was a very large-scale project, while modern metagenomic analyses easily provide gigabases of data—"our study ... already belongs to the stone age of Sanger sequencing."

fixation, using nitrate or fumarate in the absence of O_2. The deltaproteobacterial symbionts possess genes for CO_2 fixation via the reductive tricarboxylic acid (rTCA) cycle, as well as numerous genes for heterotrophic metabolism of organic compounds in aerobic conditions. Thus, as the worm moves through the oxic and anoxic regions of the coastal sediments that it inhabits, electron donors will be available at all levels and the symbiotic consortium provides the worm with an optimal energy supply. As in *Riftia*, it is likely that *Olavius* digests the bacteriocytes to obtain fixed carbon compounds. *Olavius* is also thought to have an adapted hemoglobin that transports oxygen and sulfide for metabolism by the thiotrophic symbionts; there are also large amounts of a storage protein for oxygen that ensures a supply of oxygen in anoxic environments. Furthermore, the symbionts also recycle the host's waste products of ammonia and urea, which explains why the worms can live with a completely reduced excretory system. Further complexities have been revealed by metaproteomic analyses, showing that the SRB can use hydrogen as an energy source and that three of the symbionts use carbon monoxide (CO) as an energy source—this is very surprising given its toxicity to animal life. Examination of related species from other parts of the world reveals similar symbiotic consortia. The relationship is clearly obligate for the host's nutrition, but how the worms acquire the symbionts is still unclear. There is some evidence for vertical transmission, in which the symbionts may be transferred to the eggs via "smearing" during egg-laying, as occurs in some insects.

Flatworms in the genus *Paracatenula* (phylum Platyhelminthes) are also completely devoid of a digestive system and rely on mutualistic symbiosis with the bacteria identified as "*Ca.* Riegeria santandreae" (family Rhodospirillaceae). To date, this is the only known chemosynthetic symbiont in the Alphaproteobacteria, containing genes encoding sulfur oxidation and CO_2 fixation pathways for chemolithoautotrophy. Despite possessing one of the most highly reduced genomes known in marine bacteria (1.34 Mb, see *Table 3.1*), it has retained genes for a versatile carbon metabolism, including a complete TCA cycle, which allow it to use sugars and fatty acids an energy source. However, these are not used for heterotrophic metabolism, but for internal recycling and storage, as shown by proteomic and transcriptomic analysis. Remarkably for a chemoautotroph with such a reduced genome, the symbiont has pathways for the synthesis of the storage carbohydrates polyhydroxyalkonoates, trehalose, and glycogen, which accumulate as inclusion bodies and make up a large part of the volume of bacterial cells (*Figure 10.6*). For the bacterium, these reserves provide an energy sink and carbon store when the worm moves into anoxic regions of the sediment (necessary to acquire sulfide, as discussed above for *Olavius*), but they also serve as the main source of nutrition for the host. The symbiont does not possess sufficient transport mechanisms for translocation of nutrients to its host, and *Paracatenula* does not digest the symbionts by lysosomal fusion as occurs in many other nutritional symbioses (including *Bathymodiolus* mussels and insects). Indeed, *Paracatenula* does not produces lysozyme enzymes—except under conditions of prolonged starvation—which would be necessary for degradation of the bacterial peptido-glycan cell wall. Instead, the symbiont builds up a stockpile of substantial stores of nutrient reserves which provide the primary energy storage for the bacterium themselves and for the host, by transfer in outer membrane vesicles, as shown by gene expression and ultrastructural studies (*Figure 10.6*). This fascinating symbiosis also provides valuable insights into the evolution of mutualistic symbioses. Phylogenetic analysis indicates that *Paracatenula* and its Riegeria symbionts have been co-evolving for ~500 million years. Although the bacterium has undergone extensive genome reduction, it has retained many essential functions and shows no signs of the continued gene loss that would result in it becoming a functional organelle of the host, as has occurred with many insect endosymbionts, over much shorter periods. This can be explained by the fact that *Paracatenula* relies entirely on the symbiont for all its nutrition, whereas symbionts of insects often only supplement the host diet with one or a few nutrients, such as vitamins or essential amino acids. This is often complemented by secondary endosymbiosis, so that restrictions on continued gene loss in the bacterium are removed and it eventually becomes "enslaved" as an organelle.

Figure 10.6 Symbiosis of "*Ca.* Riegeria santandreae" with the flatworm *Paracatenula*. (A) Habitus of *Paracatenula* from Sant'Andrea Bay, Elba. The white trophosome contains endosymbionts, while the anterior and transparent part of the worm (rostrum) is bacteria-free. B. Differential interference contrast image of Riegeria symbionts indicating multiple intracellular inclusions. C. Confocal laser-scanning image of CARD-FISH combined with lipophilic staining (Nile Red) on a transverse section of a *Paracatenula* specimen in the symbiont-bearing region. Overlay of DAPI signal (blue), Nile Red (red), and probe targeting the symbionts (green, EUB I–III) is shown. D. Trace illustration of TEM of the trophosome region with the status of the intracellular bacterial symbiont indicated in dark blue (OMVs detected), light blue (no OMVs detected), and orange (digested—lysosomal bodies). E-G. Representative image outcrops of the OMV-secreting symbiont population. CR, "*Ca.* R. santandreae"; hc, host cytosol; om, OMVs; ph, storage inclusion (PHA). H. A symbiont cell that has undergone lysosomal digestion (ly) in a starved worm. The enclosed structure in the lysosome resembles a PHA inclusion (ph). Reprinted from Jäckle et al. (2019), CC-BY-NC-ND 4.0.

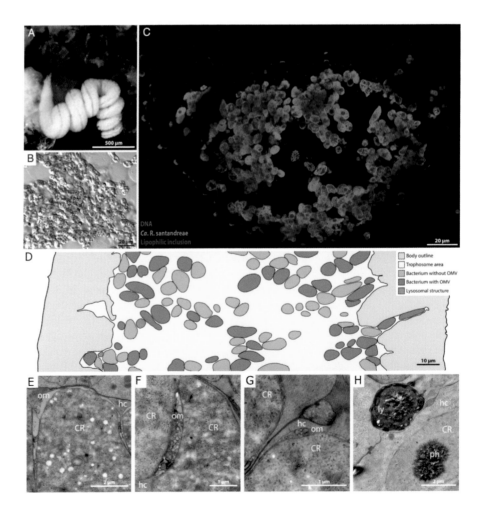

Figure 10.7 A. Part of the skeleton of a gray whale carcass discovered in 2002 in Monterey Canyon at >3000 m deep, densely covered with *Osedax* worms. B. Close-up of *Osedax* trunk and palps emerging from a whalebone. Credits: A. Copyright MBARI, reprinted with permission. B. Nick Higgs, University of Plymouth.

Animals colonizing whale falls depend on autotrophic and heterotrophic symbionts

Large whales weigh 30–160 tons, so the amount of organic carbon from a single carcass reaching the seafloor can be the equivalent of thousands of years of marine snow falling from the surface. The occasional, but huge, pulse of organic matter to the ocean floor supports diverse animal communities. After initial removal of soft tissue by sharks, hagfish, crabs, and other scavengers, followed by colonization of the surrounding sediments, the whale skeleton can support a rich population of animals sustained by thioautotrophic microbes oxidizing sulfur obtained from the bones. Many of these associations resemble those found at hydrothermal vents and cold seeps. Some scientists have speculated that whale carcasses provide "evolutionary stepping stones" for dispersal of animals harboring autotrophic symbionts as they have radiated from shallow- to deep-water habitats, but other researchers do not support this idea.

In 2002, scientists from the Monterey Bay Aquarium Research Institute who were following the decay of a gray whale carcass that had sunk to a depth of nearly 3000 m off the coast of California, discovered *Osedax*, a new genus of gutless sibliognid worms (the name is derived from the Latin for "bone devourer"). The animals covered the skeleton, penetrating the bones with a root system developed from an enlarged egg sac and richly supplied with blood vessels (*Figure 10.7*). In electron micrographs, large bacterial cells were observed inside bacteriocytes in the host cells; 16S rRNA gene sequencing showed these to be very different to the chemoautotrophic bacteria found in other gutless worms like *Riftia*. The dominant symbionts belong to the Oceanspirillales group (p.127) which are heterotrophs that break down the collagen, cholesterol, and lipids, present in large amounts in the whale bone. The host derives all its nutrition from the bacteria, probably by digesting the bacteriocytes. Analysis of lipids in the worm tissue confirms that they are derived from bacterial metabolism. The animals are biologically very unusual since the males are contained within the female body; little is known about the nutrition of the males. Free-living Oceanospirillales bacteria occur in the bone tissue and are acquired by the larvae during settlement. It is likely that signaling between the bacteria and host triggers the development of the root system of the worm, but nothing is currently known about this. Presumably, the symbionts benefit by

 HOW LONG HAVE *OSEDAX* **WORMS BEEN DEVOURING BONES?**

The exceptional diversity of *Osedax* worms and its wide geographic distribution at depths from 21 to 4000 m suggests a very long evolutionary history and poses something of a mystery. Using micro-computed tomography (CT), paleontologists can detect the characteristic traces of penetration by *Osedax* roots in fossilized bones of early whale species from ~30 MYA, suggesting that they have been living on whale skeletons since the first whales evolved. Danise and Higgs (2015) also found these traces in fossil bones from 100 MY-old plesiosaurs. Therefore, before the evolution of whales, marine reptile carcasses played a key role in the evolution and dispersal of *Osedax* and its symbiotic bacteria. The fossil traces found in this study also showed that the whole family of gutless worms is much older than previous estimates. The Cretaceous period was thus a key period for the evolution of symbiotic associations.

access into the bone via the worm's roots, receiving plentiful supplies of oxygen for aerobic breakdown. The wide geographic distribution of *Osedax* is something of a mystery, given the rarity of carcasses of whales since their massive depletion by whaling in the nineteenth century. However, a large whale carcass should sustain a community for many years and, based on calculation of natural mortality rates, significant deposition of carcasses could still occur, especially along migration routes. Thus, whale falls still play a major role in the ecology of deep-sea benthic processes, sustained by long-term microbial activity. We now know that there are numerous species of *Osedax* and many of these are found in much shallower habitats, where they feed on bones of other vertebrates, as well as whales.

Large inputs of organic material to the floor of the deep sea are also provided in the form of sunken logs, masses of kelp or large accumulations of dead fish or invertebrates. Wood-boring bivalves (Xylophaginae) containing symbiotic bacteria digest the wood. Because of the difficulties of study, little is known about the nature and diversity of the symbionts, It might be expected that they are like the heterotrophs found in the shallow-water bivalves known as shipworms (see p.358), which produce enzymes for the degradation of cellulose and other carbohydrates, as well as providing fixed nitrogen to their host. Thiotrophic symbionts may also be involved, using sulfides generated by sulfate-reducing bacteria in the anoxic sediments surrounding the wood.

Sea squirts harbor photosynthetic bacteria

In their adult stage, ascidians (commonly known as sea squirts) are colonial sessile filter-feeding invertebrates belonging to the subphylum Tunicata, which are characterized by a tough polysaccharide coat or "tunic." Ascidians have assumed special importance because they are transmitted over great distances by attachment to ship's hulls and floating marine debris and non-native species have appeared as invasive species in harbors and marinas all over the world, where they quickly become established as invasive species that can affect the ecology of sub-tidal waters or cause damage to docks or other floating structures. This occurs despite their sedentary nature and limited mechanisms of dispersal. The adult stage of some species harbors the photosynthetic cyanobacterium *Prochloron*, and cells of the symbionts are transmitted vertically during late stages of embryonic development of the larvae. Some species of ascidians also harbor a diverse community of bacteria, including some types that carry out aerobic anoxygenic phototrophy (AAnP, p.77). This unexpected finding indicates a new type of light-based symbiosis very different from the oxygen-generating activities of zooxanthellae and symbiotic cyanobacteria (discussed below). It will be interesting to find out whether the host derives nutritional benefit from the association and, if so, the mechanisms involved. As occurs in many other marine invertebrates, *Endozoicomonas* may form a facultative association with tunicates, with a key role in degradation of mucus. It appears that different ascidians contain stable bacterial communities with a high degree of host specificity, and this may be a factor in their ability to adapt to new environments, explaining their success as invasive species.

Endosymbionts of bryozoans produce compounds that protect the host from predation

Bryozoans are tiny colonial animals that resemble moss coating the surface of rocks. One species, *Bugula neritina*, has been the subject of much interest because it produces complex polyketides (bryostatins), which is effective in the treatment of some cancers (p.394). We know now that an endosymbiotic bacterium synthesizes bryostatin and the genes responsible have been identified. The bryostatin molecules coat the surface of the larvae at high concentrations and seem to function in protecting the larvae from predation, as they are unpalatable to fish. They may also play a role in attracting conspecific larvae to settle to form colonies. The bacterium has never been cultivated, but genetic analysis shows that it is a gammaproteobacterium, and it has been designated "*Ca.* Endobugula sertula." Several other bryozoan symbionts, including an alphaproteobacterial example in *Watersipora* spp., have now been described. Since the Bryozoa show extensive diversity in structure and ecology, it is likely that symbiosis has evolved independently and may have different physiological functions in different groups.

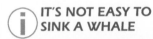

IT'S NOT EASY TO SINK A WHALE

Although some knowledge of the biology of whale falls came from chance discovery of carcasses, most of our knowledge of the succession of events in the decay of whale skeletons comes from manipulative experiments. Craig Smith (University of Hawaii) developed methods for towing carcasses of beached whales to drop sites (mapped precisely with GPS coordinates) and sinking them to the seafloor (Smith and Baco, 2003). Apart from the logistic problems of arranging transport and support at short notice, sinking a whale carcass bloated with gases after many days of towing behind a ship is a difficult and unpleasant task. Many such "whale-drops" have now been conducted, especially by scientists from MBARI. Manned submersibles and ROVs are used for long-term video recording and sampling of the carcass over many years. Deliberate creation of sunken wood falls and long-term monitoring of their decay and colonization may be slightly less challenging, but still requires careful logistical planning, as described by Ristova et al. (2017).

BOX 10.2

Chemosynthetic bacteria fuel ecosystem-engineering bivalves

We know now that chemosynthetic symbioses occur in many shallow-water habitats, as well as the deep sea. However, the ecological role of these interactions has been poorly understood until recently. Genomic and proteomic approaches are revealing the importance of chemotrophic symbionts in fixation of carbon and nitrogen, underpinning the ecology and productivity of coastal ecosystems.

Lucinid clams detoxify seagrass sediments and fix nitrogen.
Molluscs belonging to the family Lucinidae inhabit various habitats, including mangroves, coral reef sediments, and the dense beds (meadows) of seagrasses—flowering plants attached to sand or mud bottoms in shallow coastal waters. There are numerous seagrass species found throughout the world, and they have great ecological importance as "ecosystem engineers" due to their high production from photosynthesis and the trapping of sediment and reduction of water movement and coastal erosion. They harbor highly diverse and abundant animal communities and they provide nursery areas for fish and a direct food source for turtles, dugongs, manatees, seabirds, and other animals. The lucinid clam *Codakia orbicularis*, which lives in the sediment of *Thalassia testudinium* seagrass beds in the Caribbean, contains a single species of thioautrophic symbiont in its gills (*Figure 10.8*). The bacteria are housed in bacteriocytes and appear to be digested, although this may only be significant during periods of starvation when normal filter feeding is insufficient (König et al., 2015). However, in this case a more important ecological role for the symbionts is detoxification of sediments by removal of sulfide. The high productivity of seagrass beds is highly dependent on nitrogen fixation in the sediment, which is largely carried out by anaerobic sulfate-reducing bacteria during the breakdown of

organic material. This results in production of large amounts of sulfide, which is highly toxic to the seagrass roots and to animals. The activity of the thiotrophic symbionts in the lucinids reduces this toxic effect, resulting in enhanced seagrass production, while the clams and their symbionts benefit from increased organic matter and oxygen from the seagrass roots—a three-stage symbiosis (Heide et al., 2012). The symbiont cannot be cultured, so König et al. (2016) separated bacterial DNA from *C. orbicularis* tissue and carried out genomic and proteomic analysis. As expected, the appropriate genes encoding enzymes in the sulfide oxidation and Calvin–Benson–Bassham (CBB) carbon fixation pathways were all present. Unexpectedly, they also found that the genome contained a complete *nif* gene cluster for nitrogen fixation, including the nitrogenase complex, which they showed to be functionally active using an acetylene reduction assay (see p.242). By using an antibody detection method, König and colleagues confirmed the presence of the protein NifH in the bacterial endosymbionts inside the bacteriocytes of clam's gill tissue. Seagrass bed sediments in which the *C. orbicularis* grow have low levels of NH_4^+, so nitrogen fixation by the symbionts would be advantageous to the clams, by supplementing the supply of nitrogen compounds. The symbiont belongs to the class Gammaproteobacteria and has been named "*Ca.* Thiodiazotropha endolucinda" and appears similar to several other uncharacterized symbionts from lucinid clams. In a separate study, Petersen et al. (2017) showed the presence of NifH genes in genome sequences obtained from the thiotrophic symbiont of another lucinid species, *Loripes lucinales*. Again, they identified a member of the Gammaproteobacteria, which they named "*Ca.* Thiodiazotropha endoloripes". PCR analysis of *L. lucinales* specimens from around the world provided evidence

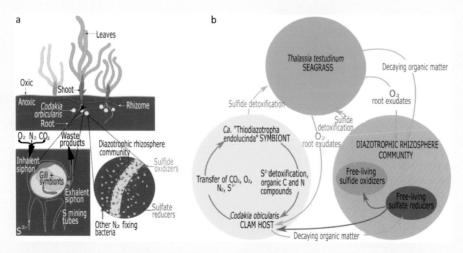

Figure 10.8 Model of interactions in a seagrass bed between *Thalassia testudinum*, *Codakia orbicularis*, and its symbionts. (a) Seagrass roots are inhabited by a complex microbial community, including diazotrophic SRB. (b) SRB produce sulfide from the decay of dead seagrass and other organic detritus. Bacterial endosymbionts and free-living sulfide-oxidizing bacteria detoxify sulfide to the benefit of plants and animals. Reprinted from König et al. (2017) CC-BY-4.0.

BOX 10.2   RESEARCH FOCUS

of genetically similar *nifH* sequences. Petersen and colleagues used metatranscriptomic analysis of clam gill tissue to show the expression of these genes. Further to the suggestion that seagrasses grow better when they contain large populations of lucinid clams (van der Heide et al., 2012), Petersen et al. (2017) suggest that excess fixed nitrogen released into the sediments could be taken up by the seagrass roots.

Chemosynthetic products feed lobsters. Higgs et al. (2016) showed that, besides its obvious ecological importance, this symbiosis provides a rare example of direct input of chemosynthetic production into marine food webs. The clams are a major food source for the Caribbean spiny lobster, supporting commercial fisheries of high economic worth in the Caribbean. By analyzing the C, N, and S stable isotope ratios of spiny lobsters, Higgs et al. (2016) concluded that they obtain 20–35% of their nutrition from organic material produced by chemosynthesis.

Thyasirid clams oxygenate marine sediments with help from magnetotactic, thiotrophic symbionts. Members of the diverse bivalve family Thyasiridae are widely distributed in seeps and coastal sediments. Although small (usually <1 cm), they create extensive ramifying burrows, oxygenating the sediment (Dufour and Felbeck 2003). This bioturbation has large impacts on the ecology of the sediments that they inhabit. Many contain thiotrophs in their gills, but in this case the bacteria are extracellular. The molluscs promote sulfide oxidation by their deep burrowing, "mining" the sediment for sulfide, and provide active oxygenation by villi in the gill filaments. The host obtains some nutrition by engulfing and digesting the extracellular symbionts, but

it is not completely dependent on its chemosynthetic symbionts. Batstone et al. (2014) found wide differences in the structure of the gills and the presence of symbionts, even within the same species, which probably reflect local differences in sulfide concentration. In early microscopic studies, dark intracellular inclusions had been seen inside the gill cells and these were assumed to be viruses. However, using specialized electron microscopy techniques, Dufour et al. (2014) showed that these inclusions are magnetosomes (*Figure 10.9A*). They form octahedral bodies containing iron, characteristic of magnetotactic bacteria (p.123), and rRNA gene sequencing confirmed their relationship to this group. It seems strange that such bacteria are symbiotic, as magnetotactic behavior can have no advantage inside the host. Dufour and colleagues suggest that in their free-living state these bacteria possess flagellar motility and use magnetotaxis in conjunction with aerotaxis to track the oxic-anoxic interface. They would thus be attracted to the *Thyasira* burrows, which provide optimal conditions for chemolithotrophic sulfur oxidation; they could then become trapped in mucus and accumulate in the gills. Once inside the host, they lose their flagella. Do these symbionts provide advantage to their host—are the hosts "gardening" their symbionts? We know that the bacteria are endocytosed by the host, but the iron inclusions are not digested and accumulate in the gill tissue (*Figure 10.9B*). Using a model to estimate the dynamic energy budget of *Thyasira*, Mariño et al. (2018) provide evidence that the presence of the symbionts constitutes an adaptation to buffer fluctuations in the availability of food obtained by feeding on organic particles. Therefore, the host does not need to build up large energy reserves, which improves its maturation and reproductive success.

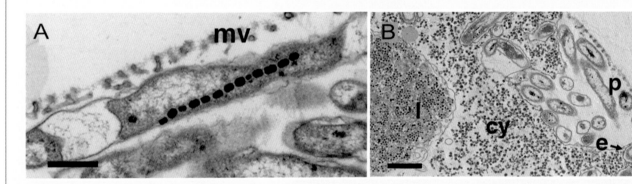

Figure 10.9 TEMs of a gill-associated symbiont of *Thyasira* cf. *gouldi*. A. Symbiont showing a chain of magnetosomes, located close to the microvillar boundary (mv) of the bacteriocyte. B. View of symbionts in an extracellular 'pocket' (p), with some undergoing endocytosis (e). A large aggregate of lysed products of symbiont digestion (l), including magnetosomes, is visible in the host cytoplasm (cy). Dark inclusions may be resistant particles of iron. Reprinted from Dufour et al. (2014), with permission of Springer Nature.

Sponges contain dense communities of specific microbes

Sponges (Phylum Porifera) are the oldest group of multicellular animals, comprising more than 8000 species in a wide variety of tropical, temperate, and cold-water marine habitats, with a smaller number of freshwater examples. They have a simple three-layered body structure with an internal system of pores and channels lined by flagellated cells (choanocytes, *Figure 6.4*) that pump water through the body. They feed by filtering bacteria and microalgae from the surrounding water and digesting them with specialized phagocyte cells. Sponges filter huge volumes of water to concentrate food particles from the surrounding water—a modest-sized sponge filters tens of thousands of liters of water a day—and they can remove over 90% of the bacteria in the inhalant water. The sponge body (mesohyl) is a gelatinous matrix strengthened by fibers and spicules of calcium carbonate or silica. In most sponges, the mesohyl is packed with a diverse community of many different microbes, which can constitute up to half of the volume of the sponge. Most microbes in the mesohyl are extracellular, but intracellular bacteria and microalgae are sometimes found.

Original interest in the study of the microbial content of sponge tissue was largely driven by the interest in natural products such as antibiotics and antitumor compounds, which they often contain. Only a tiny fraction of sponge microbes has been cultured, but suspicion that the microbes (rather than the host's own metabolism) produce a wide range of bioactive compounds has been confirmed by recognition of microbial genes for their biosynthesis. This offers new possibilities for biotechnological exploitation, which is discussed in *Chapter 14*.

In addition to this biotechnological focus, there is inherent interest in the interaction between microbes and sponges from a biological and evolutionary perspective. Sponges have a major role in marine ecosystem functioning, including the creation and stabilization of reef communities and the coupling of pelagic and benthic processes. Recent efforts to understand the microbiota of sponges have been stimulated by the need to understand the effects of pollution, climate change, ocean acidification, and other pressures on sponge health and disease. In the past few years, there has been intensive application of molecular methods to investigate sponge microbes, and many thousands of rRNA gene sequences have been described.

These studies reveal that sponge microbes are very diverse, with at least 40 different phyla or candidate bacterial or archaeal phyla identified. All sponge species examined contain members of numerous different phyla, with Gamma- and Alphaproteobacteria dominant in most sponge species (*Figure 10.10*). A wide diversity of fungi has also been described; most of these are ascomycetes related to terrestrial fungi, but some fungal species may be specialist associates of particular sponge species and can be identified at different geographical locations. The variability in the richness of their microbiota between hosts of the same species is generally low, indicating that there is species selectivity for particular microbial partners. Some microbes can be detected in even distantly related sponges isolated from a wide geographic range while others appear to be specialists that are present in only a few sponge species. There is evidence of a stable core microbiome, where particular microbial types are abundant in all members of the sponge species, while others are much more variable. One explanation for this is that specific microbes became associated with sponges very early in their evolution, more than 600 million years ago, and remained associated with the sponges as they underwent evolutionary radiation. This hypothesis requires that the microbes are passed vertically from one generation to the next, and there is now considerable evidence (largely from microscopic studies) that this occurs during sexual reproduction, with bacteria and yeasts observed in the eggs, embryos, or larvae of many species. During asexual reproduction, microbes could be passed via tissue buds.

The almost ubiquitous occurrence of symbiotic microbes and their probable presence throughout evolution of sponges indicates strongly that these are mutualistic associations. Most sponges are heterotrophic, consuming microbes harvested from the seawater. Some sponge species rely on photosynthetic cyanobacteria, dinoflagellates, or diatoms for up to half of the sponge's energy requirements and some species of sponges growing at methane

EXCEPTIONAL DIVERSITY OF SPONGE MICROBES

Various investigations have been consolidated into the Global Sponge Microbiome Project, as part of the Earth Microbiome initiative (www.earthmicrobiome.org/). Building on previous large-scale analyses of sponge microbiota (Schmitt et al., 2012; Taylor et al., 2013), this collaborative venture between different laboratories used standardized methods of DNA extraction and amplification of the V4 region of the 16S rRNA genes from different sponge species, as well as seawater and sediment controls. In addition, standardized bioinformatics protocols for identifying and clustering retrieved sequences were employed (Thomas et al., 2016; Moitinho-Silva et al., 2017). The technical standardization has resulted in large publicly available datasets that can be used reliably and consistently for research into sponge host-microbe specificity and the effects of environmental factors on microbiome structure. Thomas et al. (2016) concluded that sponge-associated communities are a major contribution to the total microbial diversity of the oceans.

Figure 10.10 Microbial diversity (OTU richness) in sponge-associated microbial communities at phylum level detected by the Global Sponge Microbiome project. Redrawn from Pita et al. (2018), CC-BY-4.0.

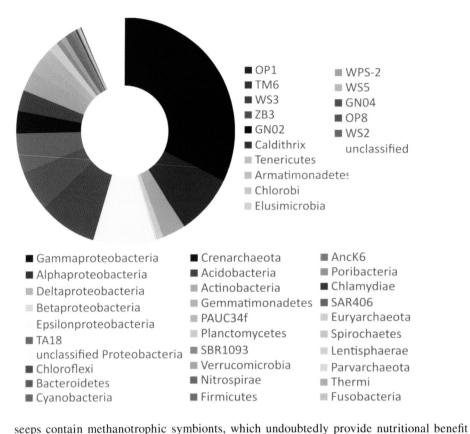

OP1, TM6, WS3, ZB3, GN02, Caldithrix, Tenericutes, Armatimonadetes, Chlorobi, Elusimicrobia, WPS-2, WS5, GN04, OP8, WS2, unclassified

Gammaproteobacteria, Alphaproteobacteria, Deltaproteobacteria, Betaproteobacteria, Epsilonproteobacteria, TA18 unclassified Proteobacteria, Chloroflexi, Bacteroidetes, Cyanobacteria, Crenarchaeota, Acidobacteria, Actinobacteria, Gemmatimonadetes, PAUC34f, Planctomycetes, SBR1093, Verrucomicrobia, Nitrospirae, Firmicutes, AncK6, Poribacteria, Chlamydiae, SAR406, Euryarchaeota, Spirochaetes, Lentisphaerae, Parvarchaeota, Thermi, Fusobacteria

COULD SPONGES FILTER RARE BACTERIA FROM SEAWATER?

Although the idea of a very ancient symbiosis with specific microbes is favored by most sponge microbiologists, alternative explanations involving acquisition of some organisms from the environment are possible. One of the main reasons given in favor of specific associations is that many types of sponge microbes can never be detected in the environment and these have been assigned to a candidate phylum "Poribacteria" (Fieseler et al., 2004). However, because a sponge can filter tens of thousands of liters each day, even if a very rare bacterium was present at only one cell per liter—and therefore unlikely to be detected by commonly used methods such as PCR amplification of genes in water samples—a sponge could still acquire tens of thousands of this type every day. Indeed, later studies using deep sequencing showed that many of the previously described "sponge-specific bacteria" gene clusters are widespread in seawater, albeit at very low abundances (Taylor et al., 2013).

seeps contain methanotrophic symbionts, which undoubtedly provide nutritional benefit to their host. Many sponge-associated microbes are active degraders of complex carbohydrates such as algal polysaccharides. In some sponges, ingestion and breakdown of the bacteria in the mesohyl has been observed, leading some scientists to suggest that sponges cultivate bacteria as a food source. Bacteria and archaea certainly seem to play a key role in nitrogen cycling within sponges, particularly by conversion of toxic ammonia—produced in large amounts by the sponge tissue—by oxidation to nitrate by nitrification. Anaerobic processes of denitrification or anammox can occur when sponge tissue becomes anoxic when pumping activity stops temporarily. Nitrogen fixation by cyanobacteria may also be important in some sponges. The production of large amounts of polysaccharides by the mesohyl bacteria is also thought to contribute to the structural integrity of the tissue. Sponge symbionts have numerous genes involved in synthesis of vitamins and vital amino acids, which complement the host's metabolism. One the most important beneficial functions of the microbiome is the production of bioactive compounds that may protect the sponge against the harmful effects of ultraviolet radiation or oxidative stress, or act as defense compounds to prevent predation.

Recognition of the association between the sponge organisms and their specific microbial inhabitants allows us to consider how the activities of these holobionts becomes integrated into larger communities and the ecosystem. Thus, activities of particular microbes can have ecosystem-wide effects. The sponge microbiome carries out functions that are amplified to affect the productivity, nutrient cycles, and food webs of the sponge habitat, as illustrated in *Figure 10.11*.

Given that sponges filter-feed on microbes, how is it that most of the bacteria can live within the tissues, adjacent to phagocytic cells? It is likely that the symbionts possess surface properties such as production of grazing deterrents, altered cell walls, or defensive proteins that protect them from recognition and ingestion by the phagocytes. As yet, firm conclusions about the origin of the association between sponges and their resident microbes are not possible. Further application of metagenomics, metatranscriptomics, metaproteomics, and other advanced techniques will aid our understanding of the specificity and interdependence of the host–symbiont interactions, which is such a fascinating aspect of sponge microbiology.

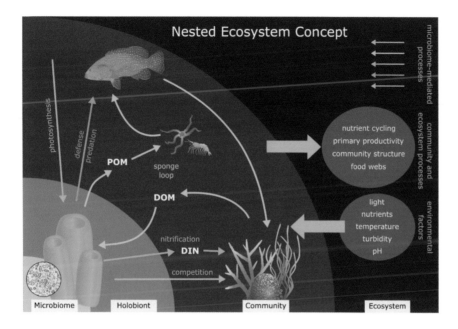

Many marine invertebrates depend on photosynthetic endosymbionts

Biologists first recognized associations between marine invertebrates and photosynthetic microbes more than a century ago and introduced several descriptive terms at this time, based largely on the coloration of the tissue due to pigments from the symbionts. The best known and most important group is the zooxanthellae—the term refers to the golden-brown color (*Figure 10.12*)—which are dinoflagellates occurring in a wide range of Cnidaria (corals, anemones, jellyfish, and zoanthids), as well as some molluscs, sponges, and flatworms. Green zoochlorellae are members of the Chlorophyceae, occurring mainly in sponges, coelenterates, and flatworms, while blue–green zoocyanellae belong to the Cyanobacteria (including *Prochloron*) found in some sea squirts and molluscs. Many tropical sponges also rely on cyanobacteria for more than half of their energy requirements. Zooxanthellae also occur in the protist group foraminifera.

Zooxanthellae (Symbiodiniaceae) show extensive genetic diversity and host specificity

Zooxanthellae are members of the family Symbiodiniaceae (class Dinophyceae). Although differences in structure, physiology, biochemistry, and behavior were recognized, until the 1970s all zooxanthellae were considered members of the single species *Symbiodinium*

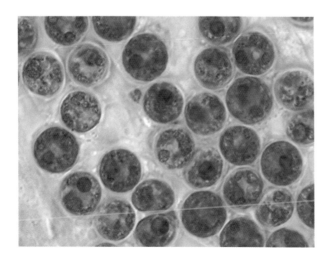

Figure 10.12 Zooxanthellae isolated from tentacles of upside-down jellyfish, *Cassiopea* sp. showing the distinctive golden-brown color. Individual cells are ~6-10 μm diameter. Credit: Todd LaJeunesse, Pennsylvania State University.

microadriaticum. However, molecular analysis has revealed a very high level of phylogenetic diversity. *Symbiodinium* has been divided into nine major groupings (clades A–I) based on sequence analysis of 18S rRNA genes. However, more discriminating methods based on gene sequencing of the variable internal transcribed (ITS) region show that *Symbiodinium* has significant intraclade diversity, and there are several hundred genetically distinct types, of which some have been formally described as named *Symbiodinium* species. However, the genetic differences between clades of the genus is much greater than that between higher dinoflagellate taxa, suggesting that they should belong to different genera. Whole genome analysis has proved difficult due to the very large size of dinoflagellate genomes and only a few genomes have been assembled, but multigene analysis of key markers has yielded valuable insights. Using a variety of morphological, physiological, and genetic criteria (including molecular clock analysis to determine evolutionary relationships), seven of the clades have now been formally named as genera within the family Symbiodiniaceae. Only members of clade A have been retained in the genus *Symbiodinium (Table 10.1)*. These taxonomic changes should become quickly accepted, although readers should be aware that the designation *Symbiodinium* as a general descriptor may continue to be used by some authors. Correlative studies of distribution and depth zonation of different clade types, together with numerous experimental studies, have shown that different clade types differ in their physiological characteristics, especially sensitivity of photosynthesis to light and temperature. Some Symbiodinaceae species associate with a broad range of hosts, while others are found in only a few host species.

Table 10.1 shows that Symbiodinaceae have been described in five different animal phyla. In their hosts, zooxanthellae usually form coccoid cells surrounded by a series of algal membranes, within a host vacuole called a symbiosome; however, in clams the zooxanthellae lie extracellularly in specialized channels in the gill mantle. Many symbiotic dinoflagellates can be isolated and maintained in culture. The morphology and life cycle of the free-living and symbiotic forms are very different; in particular, the dinoflagellate loses its flagella inside the host. The symbionts carry out photosynthesis, harvesting light energy via a complex of chlorophyll *a*, chlorophyll c_2, and the protein peridinin. Light energy is used to fix CO_2 which occurs mainly via the C3 pathway (CBB cycle) using the enzyme RuBisCO. This can be shown by enzyme assays and by incubating with radiolabeled bicarbonate and measuring the incorporation of ^{14}C into the tissues. Some species may also use a C4 route employing phospho-enol pyruvate carboxylase as the key enzyme. Animals that contain zooxanthellae clearly derive nutritional benefit from the relationship in the form of photosynthetically fixed carbon compounds, although views differ on the exact contribution that each partner makes to the acquisition of other nutrients—especially nitrogen and phosphorus. Radiolabeling experiments show that the zooxanthellae release a high proportion of photosynthetically fixed carbon as small molecules, including glycerol, glucose, and organic acids. It is possible that some essential amino acids are also provided by the zooxanthellae. The density of zooxanthellae varies widely in different hosts; high numbers are often associated with a reduced dependence on feeding by capture of plankton, manifested by smaller tentacles or reduced digestive systems.

We know little about how these intimate relationships came about, but fossil evidence shows that coral reefs first evolved in the mid-Triassic period (~250 MYA). In all the animal phyla that harbor dinoflagellates, the final step in digestion is intracellular. Cells of the animal's digestive tract may have retained algal cells that were resistant to digestion, and subsequent evolution has refined this association. A key step in establishment of the symbiosis appears to be due to the zooxanthellae arresting the maturation of phagosomes. The importance of the association to the host is emphasized by anatomical and behavioral adaptations observed in the various animal groups, which have evolved in order to expose the zooxanthellae to light.

Many corals are dependent on zooxanthellae for nutrition

Corals and anemones belong to the Anthozoa class of the phylum Cnidaria. Corals typically form colonies containing numerous identical (clonal) polyps. They feed on prey—varying from small plankton to small fish—using tentacles and stinging cells for capture, but in many

Table 10.1 Lineages of Symbiodiniaceae, showing redesignation of *Symbiodinium* clades to new genera

Clade	Host taxa[a]						New genus designation and key features[b]
	Foraminifera	Ciliophora	Porifera	Cnidaria	Platyhelminthes	Mollusca	
A				+	+	+	*Symbiodinium.* Adapted to variable or high light conditions.
B			+	+		+	*Breviolum.* Small cells. Primarily associated with cnidarians. Prevalent in Caribbean.
C	+	+		+	+	+	*Cladocopium.* Large number of species. Broad host range. Ecologically abundant and broad geographic distribution.
D	+		+	+		+	*Durusdinium.* Adaptations to survive fluctuations in temperature or turbidity; more resistant to environmental disturbance (e.g. coral bleaching).
E							*Effrenium.* Exclusively free-living. Grazes on bacteria and unicellular eukaryotes.
F	+			+			*Fugacium* (Subclade F5). Found in foraminifera. Several non-symbiotic species. Occur as transient, low-abundance densities in cnidarians. Other subclades of clade F not yet fully characterized.
G	+		+	+			*Gerakladium.* Basal lineage of family. Ecologically rare. Certain species form association with excavating sponges and black corals.
H	+						Not yet fully characterized.
I	+						Not yet fully characterized.

[a] Data from Pochon et al. (2014). Representatives of most clades have also been detected in the free-living environment; some sub-clades/species may be exclusively free-living. [b] Data from LaJeunesse et al. (2018). Only clade A is retained in the genus *Symbiodinium*.

species, much of their nutrition comes from high densities (~10^6 cm^{-2}) of the endosymbiotic zooxanthellae (Symbiodiniaceae). This is especially true of the best known and most-studied scleractinian corals in shallow tropical and sub-tropical waters, which are responsible for building coral reefs through formation of a calcium carbonate skeleton. The zooxanthellae occur in the gastroderm (innermost cell layer) of the tissue (*Figure 10.15*). Reef-building (hermatypic) corals lay down their skeleton over a very fine organic matrix and field observations show that a high content and activity of zooxanthellae is essential for the rapid building of the reef structure. Under optimum growth conditions, zooxanthellae may incorporate up to 100 times more photosynthate than they need for their own growth and reproduction—most of this excess is transferred to the coral, where it is mainly respired. Excess carbon is also secreted in the mucus, which provides a major source of nutrients sustaining the microbial loop and benthic processes, explaining why coral reefs form highly productive "oases" surrounded by nutrient-poor waters. The coral produces signal molecules, which alter the control of photosynthesis and stimulate the release of organic compounds from the zooxanthellae by altering membrane permeability. When photosynthesis is inhibited, for example, by restriction of light, the uptake of calcium and subsequent secretion of calcium carbonate is reduced. Excretion of lipids by the zooxanthellae is also important in construction of the skeleton.

As in other symbioses, a key question is how the host acquire their symbiotic partners. The larvae of most corals do not contain zooxanthellae (they are described as aposymbiotic), and they must obtain zooxanthellae from the environment. In almost all cases, this occurs by horizontal transmission in the early larval stages, but it can also occur in adults, especially during adaptation to environmental change (see below). Brooding corals show vertical transmission, transferring zooxanthellae in the eggs or brooded planula larvae. Mechanisms by which the initial selection and uptake of specific symbionts occurs is still unclear, although this undoubtedly involves recognition of microbe-associated patterns on the zooxanthellae by host cell receptors. Free-living motile Symbiodiniaceae are thought to be attracted to the coral coelenteral mouth and are then ingested by the gastrodermal cells. Juvenile colonies appear to be relatively non-specific, with uptake of a mixture of symbiont types, which may reflect the status of the host immune system. Later, a "winnowing" process ensures that one symbiont type increases in abundance, so that particular clades can be characteristic of certain coral species. At this stage, the host somehow marks the less desirable Symbiodiniaceae, so that they are recognized by lysosomes and digested. However, some corals maintain stable relationships with multiple phylotypes (evident at clade or sub-clade level) present within the colony. This is often considered to be an adaptive feature, as it means that the host "cultivates" the symbiont type with the maximal physiological performance under the prevailing environmental conditions.

Sunlight is obviously essential for maximal photosynthesis, but the high UV irradiation in clear, shallow waters is highly damaging. To minimize this damage, corals accumulate mycosporine-like amino acids, which act as sunscreens. Within the symbiosome, the intracellular zooxanthellae depend on a supply of CO_2 to carry out photosynthesis. The coral host has efficient ATP-dependent mechanisms for capturing CO_2 from the surrounding seawater. Thus, the symbionts must "pay" the host with photosynthetic products in order to receive a continuing supply of CO_2 for carbon fixation. The host has several other controls for tight regulation of photosynthesis by the zooxanthellae, including fluorescent pigments that limit photoinhibition of photosystem II by high light levels.

Coral bleaching occurs when the host–symbiont interactions are uncoupled

The importance of zooxanthellae to tropical corals is shown dramatically by the phenomenon of coral bleaching, due to a breakdown in the tight coupling of host–symbiont processes. The combined effects of elevated sea temperature and high solar irradiation are the major triggers for mass coral bleaching. Degradation of the photosynthetic pigments and expulsion of the zooxanthellae makes the coral tissue transparent, revealing the white skeleton (*Figure 10.13*). Loss of zooxanthellae does not necessarily lead to immediate death of the corals, because they can still obtain some nutrition by capture of plankton. During bleaching of some corals, the activity of endolithic algae harbored in the skeleton increases, and they may provide the coral with significant nutrition from photoassimilates via this source. Bleached corals are able to regain a fresh population of zooxanthellae when conditions improve. However, the health of the colony is severely impaired, and the corals quickly become susceptible to disease or overgrowth by algae, especially when prolonged high temperatures prevail in successive years.

Because of its importance, numerous investigations of the mechanisms of coral bleaching have been undertaken, revealing complex interactions and diverse physiological factors that affect the process. The most favored explanation for expulsion of the zooxanthellae is that it results from photo-oxidative stress under elevated temperature, which damages the photosynthetic system and leads to leakage of reactive oxygen species into the cytoplasm of the host.

? WHAT'S IN IT FOR THE ZOOXANTHELLAE?

Is the association between corals and their Symbiodiniaceae partners really of mutual benefit? The nutritional benefits for the host are obvious, but the benefits that the zooxanthellae receive in return are less clear. It is often stated that the zooxanthellae within the host have stable access to a favorable light regime and receive a reliable source of nitrogen compounds produced by the host and associated microbes. However, free-living Symbiodiniaceae "make big sacrifices"—loss of cell wall and flagella, reduction in growth rate, and tight controls on their metabolism and reproduction—when they are acquired by the host to become zooxanthellae. And when things go wrong (in bleaching), the zooxanthellae are ejected. Wooldridge (2010) argues that regarding this as mutualism is illogical as it is like saying that "dairy cattle and humans have a comparable relationship, because we provide them with food and shelter and regulate their population before we harvest (exploit) their milk production." He proposes that the relationship would be better regarded as a "controlled parasitism in which the functioning of symbionts is constrained and manipulated."

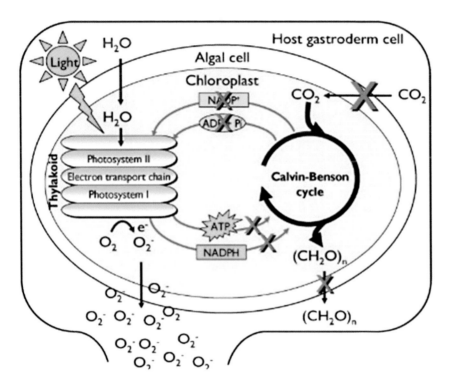

Figure 10.13 Coral bleaching. Upper panel shows early (left) and late (right) stages of bleaching on the Great Barrier Reef, resulting in the loss of pigmented zooxanthellae. Images courtesy of Mary Wakeford, Australian Institute of Marine Science. Lower panel shows a representation of internal carbon cycling in the coral-zooxanthellae symbiosis. Photosynthesis takes place within the algal chloroplast, with the 'light reactions' occurring in the thylakoid membranes, and the 'dark reactions' (CBB cycle) in the stroma. Typically, about 95% of the photosynthates [$(CH_2O)_n$] are transferred to the coral host. Breakdown of the symbiosis (zooxanthellae expulsion) is triggered by a limitation of CO_2(aq) substrate for the dark reactions of zooxanthellae photosynthesis, resulting in inability to turn over ATP and NADPH (indicated by red crosses). As a result, the photosynthetic electron transport chain becomes blocked, which damages the light-sensitive photosystems and generates damaging reactive oxygen species (O_2^-). Reprinted from Wooldridge (2010), with permission of John Wiley & Sons.

However, recent research indicates that nutritional mechanisms also regulate bleaching. Nutrient availability, especially the forms and ratios of N and P can shift the relationship between Symbiodiniaceae and the host from mutualistic to parasitic. Stable metabolic compatibility between the coral host and algal symbiont can ameliorate bleaching and increase resilience to environmental stress.

It is still unclear how the damaged zooxanthellae are removed. Possible mechanisms include expulsion of zooxanthellae by exocytosis; induction of apoptosis, in which programmed cell death is initiated; and symbiophagy, in which the vacuole containing the zooxanthellae is transformed into a digestive organelle. Infection by bacteria and viruses has also been implicated as the cause of some types of coral bleaching, as discussed in *Box 11.1*.

It is possible that the coral holobiont can adapt to changing environmental conditions by "shuffling" or reassortment of the mixture of symbionts already present, or by acquisition of new types of symbiont better suited to the new situation. This "adaptive bleaching hypothesis," was first proposed by Buddemeier and Fautin in 1993, and there is now considerable experimental evidence to support many aspects of the concept. For example, aquarium experiments have shown that *Symbiodinium* clade D (now reclassified as the genus *Durusdinium*) has a better quantum yield of photosynthesis than other clades, but only at elevated temperatures. Thus, the physiological cost of harboring a thermotolerant clade may override the apparent benefits. Field experiments, such as transplantation of corals from cooler to warmer locations on the Great Barrier Reef support the idea that adult corals may acquire more thermotolerant zooxanthellae that enable them to be more resistant to bleaching.

Although zooxanthellae are vitally important in the reef-building corals, it is important to bear in mind that most corals obtain much of their nutrition by feeding on zooplankton via their tentacles, and some do not contain any symbionts (these are described as azooxanthellate). For example, an azooxanthellate variant of the Mediterranean coral *Oculina patagonica* occurs in undersea caves. In particular, thousands of species of azooxanthellate cold-water corals, such as *Lophelia pertusa*, form massive colonies in the deep ocean, hundreds of meters below the photic zone.

The coral holobiont contains multiple microbial partners

In addition to the zooxanthellae, the tissues and secreted mucus layer of healthy corals support a diverse and dynamic community of other microbes, including other protists, bacteria, archaea, fungi, and viruses. These contribute to coral nutrition and the health of the coral. Bacteria have been isolated from corals since the 1970s, but rapid advances in this field did not take off until 2002 with the application of culture-independent DNA-based methods. Numerous studies of multiple coral species in different geographic regions have now revealed a high diversity (hundreds or thousands of phylotypes) and abundance of coral-associated bacteria.

The surface mucus layer supports particularly high populations of diverse bacteria (10^6–10^8 cells mL^{-1}) with numerous extracellular enzymes for digestion of the rich proteins and carbohydrates secreted by the coral. Their activities probably benefit the host through provision of vitamins, co-factors and other essential nutrients. This breakdown of coral mucus also provides a major contribution to the productivity of the reef, by making released DOM available to planktonic microbes (p.25) Many of these bacteria are also important in protecting the coral from colonization by opportunist pathogens, although shifts in community composition can lead to dysbiosis and disease, as discussed in *Chapter 11*. In some coral species, microscopy of thin sections of coral tissue reveals colonization of the epithelial and gastrodermal tissues by bacteria-like structures which are sometimes aggregated into closely packed groups (illustrated in *Figure 10.14*), although it is not clear if they are contained within a host membrane like the bacteriocytes found in other endosymbioses. Application of FISH with bacteria-specific probes is the ideal technique to confirm the identity of these structures; however, this has been difficult due to background autofluorescence and non-specific binding by various structures within coral tissue. Refinements of techniques to overcome these limitations are now yielding new insights into these aggregates. The bacteria most commonly found belong to the Alpha- and Gammaproteobacteria, Cyanobacteria, Bacteroidetes, and Actinobacteria. Besides bacteria, various other microbes are found. Archaea have been less well studied, but members of the Crenarchaeota and some other groups have been found to have important roles in some corals. Endolithic fungi are regularly identified in the skeleton, and ciliates and other alveolate protists often occur in the mucus and surface tissue. Distinguishing the normal inhabitants of healthy corals and those associated with the onset of disease can be very difficult, as discussed in the next chapter. Many early investigations showed a high specificity of bacterial associations with particular coral species, across broad geographic distances. However, as with the perceived host specificity of particular Symbiodiniaceae clades, more focused examination of coral colonies shows that this is not so straightforward. The diversity and relative abundance of different bacterial phylotypes can be affected by the physiological status of the coral host, location within the individual polyps or colony, and environmental conditions. Therefore, recent analyses have attempted to identify the "core microbiome," which can be defined as the phylotypes that appear to be stable and consistent in all cases, irrespective of whether they are dominant or rare, discussed in *Box 10.3*. Identifying the core microbiome

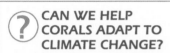

CAN WE HELP CORALS ADAPT TO CLIMATE CHANGE?

Some coral biologists have predicted that major reef systems will be irreversibly damaged during this century (e.g. Hoegh-Guldberg et al., 2009). Rohwer and Youle (2010) remind us that corals have survived previous major environmental upheavals over millennia and inject a note of optimism with the view that "The possibility of adaptation by the holobiont, thanks to its algal, viral and microbial partners, is a source of hope." Madeleine van Oppen of the Australian Institute of Marine Science and the late Ruth Gates of the Hawaii Marine Laboratory championed the proposal that we might "assist evolution" by breeding corals that are more resilient to future climate change. Although this requires a careful risk–benefit analysis, van Oppen et al. (2015) compare this approach to the successful use of genetic and epigenetic modification in commercial agriculture and aquaculture. Torda et al. (2017) and Donelson et al. (2018) set out the experimental approaches needed to determine whether such "transgenerational plasticity" can protect corals over several generations. Many groups are now working on different approaches, and guidelines to evaluate the risks and benefits of novel ecological, genetic, and environmental interventions to help corals survive have been agreed (NAS, 2019).

is particularly important to understanding the metabolic roles and core functions provided by the microbes in the holobiont.

Recent metagenomic analyses have confirmed a very high diversity of metabolic functions associated with coral microbes, including nitrogen cycling, sulfur cycling, photosynthesis, breakdown of complex proteins, and polysaccharides. Genes for stress responses, virulence, DNA repair, and antibiotic resistance also seem to be important. The chemical properties of coral mucus, as well as intercellular signaling (such as quorum sensing, p.102) and antagonistic interactions between the microbes themselves, are important factors in the establishment of specific microbiota, although the protective function is easily disturbed, leading to disease (see *Box 11.1*). Of particular significance is the discovery that a diverse group of resident nitrogen-fixing bacteria (including Cyanobacteria and vibrio-like Gammaproteobacteria) are present in many corals. Provision of nitrogen compounds to the zooxanthellae may help to explain the high productivity of corals in the apparently nitrogen-limited environments that they inhabit. The role of coral bacteria in the global cycling of sulfur has also recently been shown to be significant (see *Box 9.2*). Although present in low abundance, archaea seem to be especially important in recycling waste products in the holobiont, by nitrification and denitrification of ammonia in the mucus layer. Study of cold-water, deep-sea corals such as *Lophelia pertusa* is much more difficult, but it appears that they also have specific microbiota, including archaea.

Fungi are also an important component of the coral holobiont, being particularly associated with the surface layer and growing within the calcium carbonate skeleton of hard corals. These endolithic fungi are mostly related to terrestrial fungi and metagenomic studies have revealed a high diversity of chytridiomycetes, ascomycetes, and basidiomycetes. Some studies have identified obligate marine ascomycetes. Heterokont microalgae, labyrunthulids, and thraustochytrids are also common and grow profusely in coral mucus.

Viruses are also abundant in corals, with over 60 different virus families detected in coral viromes. Many of the viruses are phages that must play a key role in healthy corals by regulating the bacterial population, while others have been found associated with the zooxanthellae and coral animal tissue. There is also evidence for widespread integration of phages into the genomes of their hosts (prophages, see p.203). Environmental changes induce major shifts in the abundance and activity of viruses, and this is often associated with the onset of disease as discussed in *Chapter 11*.

Zooxanthellae boost the growth of giant clams

Anyone who has dived or snorkeled on the Great Barrier Reef or Pacific Islands will have marveled at the size and beautiful colors of the shell mantles of the giant clams such as *Tridacna gigas* and related species, which can reach enormous sizes up to 300 kg. How can these animals grow so rapidly when the waters they inhabit are very poor in nutrients? The siphon tissue around the mantle is packed with endosymbiotic zooxanthellae. In this case, there are additional pigments of many different colors that provide protection to the zooxanthellae from the high ultraviolet irradiation in the clear waters that the clams inhabit. *Tridacna* has evolved remarkable anatomical and behavioral modifications to optimize the benefits from their photosynthetic symbionts. This type of clam is unusual among bivalve molluscs because it lives above the sediment. During evolution, the orientation of the internal organs has twisted through 180°, allowing the siphon to be at the top of the body so that the maximum surface area colonized by the zooxanthellae is exposed to the light. Photosynthesis is further enhanced by specialized hyaline structures, which focus light onto the zooxanthellae. Experimental studies have shown that, under the right conditions, the zooxanthellae can provide all the clam's carbon requirements through the release of small molecules such as glycerol and organic acids. However, as in many corals, acquisition of nutrients by heterotrophic feeding is also important. Tridacnid clams have particularly efficient filter-feeding mechanisms and can extract significant quantities of plankton from the water, even though the plankton is in low concentrations because of the oligotrophic nature of the environment. Thus, it seems that under natural conditions, the clams rely on both autotrophic and heterotrophic sources for their nutrition, with between 35% and 70% of the carbon requirements coming from photosynthetic activity of the zooxanthellae.

THE HOLOGENOME CONCEPT OF EVOLUTION IS CONTROVERSIAL

Margulis (1991) first used the term holobiont to describe an assemblage of different species of organisms that function collectively as an ecological unit. The term was little used until Rohwer et al. (2002) used it for the association of coral hosts with diverse microbes and it quickly became widely accepted. The hologenome concept also emerged from coral microbiology research, as an extension of the coral probiotic hypothesis (see p.294). Rosenberg et al. (2007) proposed that all the genetic components of a host and its microbial partners must be considered as a unit of natural selection in evolution. This has provoked considerable differences of opinion. For example, Moran and Sloan (2015) and Douglas and Werren (2016) point out the confusion caused by inconsistent interpretations and the importance of mutualistic and antagonistic interactions (fitness conflicts) that undermine the concept. However, Rosenberg and Zilber-Rosenberg (2018) cite considerable theoretical and experimental evidence in support of their theory from microbiome research in numerous animals. Morris (2018) reviews the ongoing controversy between proponents and critics of the concept, considering that it is stimulating the development of testable hypotheses "to produce a way of looking at life which is simultaneously exciting, confusing, and challenging."

BOX 10.3   RESEARCH FOCUS

Do corals have a core microbiome?

Diversity of the coral holobiont is revealed by high throughput sequencing (HTS). We now know that corals associate with hundreds or thousands of different kinds of highly diverse bacteria (and to a lesser extent, archaea, fungi, and protists). There is strong evidence that some of these microbes play critical functional roles in the coral holobiont, through functions including photosynthesis, protection against pathogens, and nitrogen and sulfur cycling. In recognition of the fact that the microbiome plays a key role in the health of corals and in their resistance and resilience to environmental pressures, numerous research groups have used HTS to catalogue the microbes associated with diverse coral species. Only some of these have been cultured, so diversity is usually assessed by grouping related sequences into OTUs—often interpreted as the equivalent of a species (see p.129). Recently, refinement of sequencing methods allows finer resolution and there are proposals to abandon OTUs in gene surveys in favor of exact or amplicon sequence variants (ASVs, Callahan et al., 2017). A key aim is to identify a core set of microbes that is always present and likely to have a particularly critical functional role. However, Blackall et al. (2015) and Hernandez-Agreda et al. (2017) emphasize the difficulties in interpreting results due to a huge disparity in methods of sampling, replication, and DNA amplification and sequence analysis.

Corals contain niche habitats for microbial colonization. Many profiling studies have not fully recognized the fact that a coral colony provides a variety of very different niche habitats for microbial colonization, as shown in *Figure 10.14*. Due to exposure to very different environmental conditions of light, temperature, or water flow, variability occurs between individual colonies and different parts of a colony, as well as between the different compartments of individual polyps. Ainsworth et al. (2015) extracted DNA from replicate samples of three species of Pacific corals, using PCR amplification and 454 tag sequencing (p.51) of 16S rRNA genes. Laser microdissection was used to separate the niche microhabitats of host tissues, and significant differences were evident between the whole coral community (holobiont), the skeleton-associated tissue, and the community in cells containing endosymbiotic Symbiodiniaceae. In all corals sampled, the relative abundance of phylotypes within the different niche habitats differed significantly. Results for the coral *Acropora granulosa* showed the communities to be very diverse, with 1508 bacterial phylotypes (from 16 bacterial families) found in the whole community. Of these, 159 were identified as the core microbiome (defined as present in >30% of all samples), with 76 and 71 phylotypes identified in the symbiotic tissue and endosymbiont microbiomes, respectively. Only 15 phylotypes were found in all three microbiomes, but 41 were found exclusively in the endosymbiont microbiome. Further analysis showed that most of the 159 phylotypes in the *A. granulosa* microbiome had very low relative abundance. With higher cut-off values used to define the microbiome, only 0.09% of all phylotypes identified were present in 90% of the coral hosts, 0.5% in 75% were in 50% of the hosts, and 2.3% were in 50% of the hosts. Thus, the vast majority

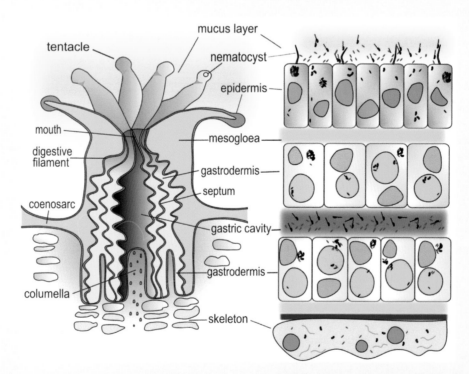

Figure 10.14 Diagrammatic representation of a coral polyp (left) with cross-section showing the habitats of microbes within the mucus and tissues (not to scale). Zooxanthellae are represented as large gold cells within the gastroderm. Aggregates of bacteria observed in some corals are shown as black clusters.

BOX 10.3   RESEARCH FOCUS

of phylotypes amplified from individual corals were not associated with all colonies. Further analysis of the data showed that different phylogenetic groups are restricted to particular habitats within the holobiont. In all three corals examined, members of the Rhizobiales, Caulobacterales, and Burkholderiales—groups that contain known symbionts in other metazoans and plants—were associated with the symbiotic microbiome, but not the whole coral community. In contrast, bacteria belonging to the Rickettsiales and Rhodobacterales were amplified from the whole coral microbiome, but not from symbiotic tissue. The authors identified just seven bacterial phylotypes that were universally present in the core microbiome of the three corals analyzed, even though the samples spanned a wide range of depth and geographic area. Another study by Hernandez-Agreda et al. (2016) found a core of eight persistent phylotypes in the widely distributed Pacific coral *Pachyseris speciosa* at different scales and depth gradients, using an 80% cut-off. This core contained members of the classes Actinobacteria, Alpha-, Delta-, Epsilon-, and Gammaproteobacteria, and the phylum Bacteroidetes. The study by van de Water et al. (2016) found a core of 12 bacterial OTUs accounting for 95% of the overall community that were consistently associated with the gorgonian *Corallium rubrum* from five reef sites persisting over a three-month period. In *Corallium* the core was dominated by members of the Spirochaetales and a relatively unknown family in the Oceanospirillales. Ainsworth et al. (2015) and Hernandez-Agreda et al. (2017) discuss the precautions needed when choosing the parameters that define the core microbiome. Using high or low relative abundance, without considering persistence, can be greatly affected by methodological and biological factors.

From their study of *P. speciosa* Hernandez-Agreda et al. (2016) concluded that this generalist coral, which occurs in many different reef habitats, can be associated with a vast number of distinct bacterial phylotypes. Interestingly, they found that samples from mesophotic reefs (60–80 m depth) consistently associated with a higher number of bacterial phylotypes. They proposed that there is a ubiquitous core microbiome of a very small number of symbiotic host-selected bacteria, a microbiome of <100 phylotypes that fills specific functional niches, and a highly variable community of thousands of phylotypes.

Is *Endozoicomonas* a core associate? A bacterium of particular interest in microbiome studies is *Endozoicomonas*, because numerous previous studies have identified this as a potential core symbiont of different corals, as well as many other marine invertebrates (Neave et al., 2017). Neave and colleagues showed the presence of *Endozoicomonas* by CARD-FISH deep within the tissues of two different species of coral (*Pocillopora verrucosa* and *Stylophora*

pistillata) sampled across seven major geographical regions. This intimate association in 85% and 79%, respectively, of these corals led Neave and colleagues to consider it a member of the core microbiome. They identified fine-scale specificity of association with different *Endozoicomonas* types and speculated that this bacterial-coral association might resemble the Symbiodiniaceae-coral partnership, in which shuffling—or acquisition of different symbiont strains—is associated with response to different environmental conditions. Surprisingly, in their study Ainsworth et al. (2015) amplified sequences affiliated with *Endozoicomonas* in the holobiont genome, but not in the internal tissues, implying that the bacterium resides in the surface mucus or skeleton. In their study of *Corallium*, van de Water et al. (2016) found that *Endozoicomonas* only accounted for 3.4% of the core microbiome. In a meta-analysis of coral-associated bacterial sequences from numerous studies, Blackall et al. (2015) found that the relative abundance of *Endozoicomonas* varied substantially, from below detection to 43% of the community, confirming that its presumed role as universal member of the core microbiome of corals is dubious. It appears to be rare or undetectable in deep-sea corals. The functional role of *Endozoicomonas* is unknown, although various possibilities include nutrition, sulfur cycling, and production of protective compounds.

Co-evolution of the host genome and microbiome. The huge diversity of corals and their associated microbiomes presents a major challenge to understanding how these relationships evolved. How are the microbial communities in the different coral anatomical compartments shaped by phylogeny and functional traits of the hosts and by environmental variables? Is there evidence of phylosymbiosis, i.e. a correlation between host phylogenetic relationships and composition of the microbial community—does the microbiome recapitulate the host's evolution? To answer these questions, Pollock et al. (2018) used standardized sampling, sequencing, and complex computational methods to analyze 16S rRNA gene libraries from the mucus, tissue, and skeletons of diverse Australian corals across a broad geographic range and to compare them to host phylogenetic trees based on mitochondrial genes. They found strong evidence for phylosymbiosis between corals and their microbiomes reflecting the speciation of modern corals that began 25–65 MYA. Surprisingly, the endolithic skeleton communities were richer in microbial diversity than the tissue and showed the strongest signals of long-term phylosymbiosis. Host specificity in the mucus microbiome seemed to be due to relatively recent divergences. Further analysis of a wider, global range of corals in different habitats, including deep-sea azooxanthellate species, and refinement of the analytical methods will reveal further insights into the evolution of coral holobionts.

Some fish and invertebrates employ symbiotic bacteria to make light

Another type of cooperative association exists between bacteria and animals, in which the activities of the bacteria confer a behavioral and ecological benefit to the host, rather than a direct nutritional one. Bioluminescence is the emission of blue or green light from oxygen-utilizing reactions via the luciferase enzymes (*Figure 3.17*). Bioluminescence occurs very widely in the oceans, particularly in animals that inhabit deep waters where no sunlight penetrates or in animals that are active in shallow waters at night. Bioluminescent animals use light emission for avoidance or escape from predators, attracting prey, or as a means of communication (such as mate recognition). Bioluminescence seems to have multiple evolutionary origins, as there is a very wide diversity of enzymes and substrates for the bioluminescent reaction mechanism. Apart from light emission, the only common feature is the requirement for oxygen, and it is possible that bioluminescence evolved originally as an antioxidative mechanism. Bioluminescent animals often possess complex structures such as lenses, filters, and shutters to control or modify the light emitted. In most bioluminescent animals, the light is generated by the animal itself within specialized cells of the animal. However, in some invertebrates and fish, the light is produced by symbiotic bacteria living within specialized organs.

In some cases, these bioluminescent bacteria can be isolated and will emit light in culture, but some have not yet been cultured and have been identified only by their genetic sequences. Another clue to the bacterial origin of bioluminescence is that bacteria usually emit light continuously over a period, whereas light produced by eukaryotic enzymes tends to occur as brief flashes. Final proof of the role of bacteria in light emission by animals is the use of an assay for luciferase in the presence of reduced flavin mononucleotide, as this reaction is unique to bacterial bioluminescence. Bacterial bioluminescence occurs in a wide range of phylogenetically distinct animals, and the associations range from relatively unspecialized facultative colonization of the intestinal tract or organs derived from it, through to obligate interactions with specialized external light organs.

The bacteria associated with symbiotic bioluminescence have probably evolved via adaptive radiation with at least 17 origins in fish and two origins in squid over the last few hundred million years. Over 460 host species are known to harbor bioluminescent bacterial symbionts. In all cases, dense communities of extracellular bacteria occur in specialized light organs contained in tubules that communicate with the intestine or the external environment. Release of bacteria is an important component in the control of the populations of these symbionts. All known types of bacteria in bioluminescent symbioses are members of the Vibrionaceae. The best known are *Photobacterium phosphoreum*, *Photobacterium leiognathi*, and *Aliivibrio fischeri* (formerly *Vibrio fischeri*; see p.103), which are facultative symbionts and are readily culturable. They occur as free-living bacteria in numerous habitats and are transmitted horizontally, being acquired by host larvae from the environment by each generation. In the animal host, these bacteria grow more slowly than in culture. It is in the host's interest to maximize bioluminescence but to minimize diversion of nutrients to the bacteria; restriction of oxygen supply or iron limitation may be important factors in this process. In other cases (see below), the bacterial symbionts have never been cultivated and are probably obligate symbionts.

Bacterial light organs in flashlight fishes (members of the family Anomalopidae) are the largest in any fish relative to body size. The light organs in these strictly nocturnal tropical reef fish are located below the eyes and emit a constant blue light (*Figure 10.16*). In very dark conditions, the forward illumination is bright enough to allow the fish to seek out their zooplankton prey. Rapid blinking, which is affected by the ambient light, may serve a function in communication with other members of the species (possibly for group interactions or mating).

Anglerfishes are a diverse group of species belonging to nine families in the suborder Ceratioidei. Most are solitary animals found in the deep sea. The females have a light organ (esca), encased in a crystalline reflector, at the end of a "fishing rod" projecting from a modified dorsal fin ray on the head (*Figure 10.15*). This acts as a lure to attract prey near to the

---

ⓘ DEEP-SEA CORALS ALSO CONTAIN MICROBIAL ASSEMBLAGES

Many coral species do not rely on zooxanthellae. Some occur in cold, dark conditions up to 2000 m deep. They occur worldwide, with the best studied locations being large mounds in the North Atlantic, where they form a critical, threatened habitat for many organisms. Because of the difficulty of sampling, much less is known about their associated microbes than other corals, but genomic analyses have revealed their diversity and function (e.g. Penn et al., 2006; Kellogg et al., 2009). Using ^{13}C and ^{15}N tracer techniques, Middelburg et al. (2015) demonstrated chemoautotrophy, N_2 fixation, and recycling by microbial symbionts in one of the most-studied species, *Lophelia pertusa*. Gene analysis provided further evidence that these functions are carried out by a core microbiome (Kellogg et al., 2017). Some deep-water corals found in the Gulf of Mexico and Red Sea contain hydrocarbon-degrading bacteria which may contribute to coral nutrition by provision of organic compounds (Kellogg et al., 2017; Röthig et al., 2017).

BOX 10.4   **RESEARCH FOCUS**

Genomic insights into the evolution of bioluminescent symbionts of fish

Flashlight fish symbionts have a free-living phase. It has long been known from 16S rRNA gene sequences, that the bioluminescent symbionts of anglerfishes and flashlight fishes are members of the Vibrionaceae. However, repeated attempts to culture these symbionts have failed. Does this mean that they are obligate symbionts? Hendry and Dunlap (2011) sought to answer this question by extracting bacterial DNA from the light organs of *Anomalops katopraton* (*Figure 10.15*). The bacterium was identified as a clade nested within the Vibrionaceae, with high divergence from other members and named "*Ca*. Photodesmus katoptron." Next, the authors assembled its draft genome and found that it was only 1 Mb, about a fifth of the size of other Vibrionaceae genomes and lacking almost all genes for amino acid synthesis and energy-yielding pathways (Hendry and Dunlap, 2014). This supports the idea that that it is an obligate symbiont that has become dependent on its host for nutrition. Surprisingly, however, the genome retains almost all the genes for synthesis of a robust cell wall, flagellar motility, and chemotaxis. These are almost always lost in previously described obligate symbionts and pathogens. It seems unlikely that these gene pathways would be retained, while so many others have been lost, unless they are required for part of the bacterium's lifestyle. The idea that they are used outside the host fits with previous observations that the symbionts are shed from pores in the light organ and can be found for a short time in seawater. Although the absence of so many metabolic genes would prevent the released bacteria from growing and multiplying, the robust cell wall would permit survival in the seawater and motility and chemotaxis could be used to seek and infect larvae to establish a new symbiosis. Hendry and Dunlap (2014) review the knowledge of the life history and behavior of *Anomalops* gained from earlier studies by Haygood (1993) and others, which suggest a pseudovertical transmission due to proximity of eggs, larvae, and adults when the fish group together in caves during daylight. This provides a possible explanation for evolution of symbiont gene loss due to population bottlenecks resulting from symbiont co-divergence at the level of the population rather than the individual.

Angler fish symbionts. Similarly, numerous attempts to culture the symbiotic bacteria of ceratioid deep-sea anglerfish have been unsuccessful. Hendry et al. (2018) isolated bacterial DNA from the light organs of specimens of two distantly related species of anglerfish, *Melanocetus johnsonii* (*Figure 10.16*) and *Cryptopsoras coesii*. Comparison of the symbionts' genome sequences placed them within the *Enterovibrio* clade, and they were given the *Candidatus* names "Enterovibrio luxaltus" and "Enterovibrio escacola," respectively. Hendry and colleagues noted that these species are separated from other taxa by long branches, indicating that they may be evolving at a rate several times faster than their relatives. This was supported by molecular clock analysis of genes and predicted amino acid sequences. The total genome sizes (2 and 2.6 Mb) of both anglerfish symbionts are about 50% smaller than those of their free-living relatives, but if only predicted functional protein-coding genes are considered, the difference is even more striking. As found in the flashlight fish symbiont, Hendry et al. (2018) conclude that this genome reduction is due to relaxed selection and high genetic drift due to the close association of the bacteria within the light organ of their hosts. Also, as seen with the "Photodesmus" flashlight fish symbiont, the anglerfish symbionts have a highly reduced complement of genes encoding amino acid synthesis and energy metabolism, membrane transport, and cell signaling, but have retained a large complement of genes for cell wall synthesis, motility, and chemotaxis. Both symbiont genomes contain a very large number of pseudogenes and show large proliferation of transposable elements that appear to have occurred at different times. Based on this, Hendry et al. (2018) concluded that genome reduction is still going on in these symbiont genomes. It is surprising that this rapid evolution towards tighter relationship is still occurring, even though the fish themselves evolved about 100 MYA and the bacteria are not constrained by a purely host-associated, intracellular lifestyle. How the symbionts are transmitted between generations remains a mystery—the pseudovertical transmission suggested for the flashlight fish symbiont seems untenable, since anglerfishes live a very solitary existence in the deep sea.

Figure 10.15 (a) Side view of *Anomalops katoptron* showing position of the light organ. This appears as a white patch because of the guanine crystal reflector on the backside of the light organ and flash photography. (b) Enlarged view of light organ. (c) Front view of both sub-ocular light organs of showing bioluminescence during the night. Scale bars=5 mm. Reprinted from Hellinger et al. (2017), CC-BY-4.0.

Figure 10.16 The anglerfish *Melanocetus johnsonii*. Image courtesy of Edie Widder, ORCA.

jaws, and light emission is controlled by the supply of oxygenated blood to the esca. Some species have several escae and filaments. In these groups of fishes, the symbionts have never been cultured and appear to be obligate symbionts, although recent work suggests the bacteria are released into the seawater and in some cases are transmitted to larvae via a pseudovertical mode, as discussed in *Box 10.4*).

The bobtail squid uses bacterial bioluminescence for camouflage

The Hawaiian bobtail squid (*Euprymna scolopes*) inhabits shallow reefs, hiding in the sediment during the day and emerging to feed at night (*Figure 10.17A*). The squid emit light from their ventral surface, adjusted to match the intensity of moonlight. This counter-illumination camouflage helps them to be less visible to predators from below. A highly specific association with a certain strain of bioluminescent *Aliivibrio fischeri* is responsible. (Note that many papers on this topic still use the original classification *Vibrio fischeri*.) Newly hatched squid are aposymbiotic and acquire this particular bacterium exclusively from the surrounding seawater within a few hours, even though it is present there at very low densities—just a few hundred cells per mL—comprising just ~0.01% of the total bacterial population. During respiration and movement, ventilation by the tiny squid flushes a miniscule amount of seawater—just 1 μL or so—in and out of their mantle cavity about twice per second, and this brings bacteria into contact with epithelial fields covered in cilia. The host responds to the presence of bacteria by the production of mucus which leads to an aggregation of cells over three tiny pores leading into each side of the nascent light organ (*Figure 10.17B*). Motile *A. fischeri* cells out-compete other members of this mixed population and swim through a duct against the outward flow of water and mucus produced by ciliated cells before reaching the crypts of the light organ. Elucidation of the full genome sequence of *A. fischeri* and subsequent mutation studies have revealed the mechanisms involved in this critical stage of colonization. Of special importance is the regulation of expression of *syp* genes associated with polysaccharide synthesis, biofilm formation, and aggregation outside the squid light organ. A complex network of interconnected regulatory systems controls colonization, with multiple interactions between genes for structural (flagella, pili, polysaccharides) and regulatory (two-component regulators, quorum sensing, and other signaling) processes in the formation

Figure 10.17 The *Euprymna scolopes-Aliivibrio fischeri* symbiosis. A. The adult host. B. Outline of the host's body, superimposed over a laser-scanning confocal micrograph (LSM) of the nascent light organ, indicating its relative size and position within the mantle cavity. The organ is circumscribed by the posterior portion of the excurrent funnel (dotted white lines). Ventilatory movements of the host draw ambient seawater (blue arrows and lines) containing *A. fischeri* cells into the mantle cavity. The water travels into the funnel where, before being vented back into the environment, it encounters complex ciliated fields (bright green) on the lateral surfaces of the organ. The fields entrain water into the vicinity of pores on the light organ surface. C. Higher-magnification LSM of one side of a hatchling light organ, showing the location of the three pores (arrows) that lie at the base of the appendages of each ciliated field. Credits: A. Reprinted from McFall-Ngai (2014), CC-BY-4.0. B, C. Reprinted with permission from Nyholm et al. (2000). Copyright (2000) National Academy of Sciences, USA.

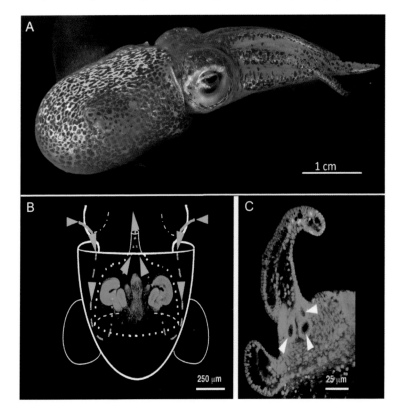

of biofilms. Genome analysis has revealed the importance of a single gene encoding a master regulator of these pathways in determining the host specificity of the particular strains of *A. fischeri* that colonize *Euprymna*. Production of nitric oxide, reactive oxygen species, and other antimicrobial responses of the host also play a key role in this selection process. In the crypt spaces, peptidoglycan and lipopolysaccharide from the bacterial cell wall initiate a series of changes to the development of the growing light organ, starting with apoptosis of the ciliated epithelium. The monoculture of *A. fischeri* within the larval host is established within as little as 2 h from hatching. After adhesion to the crypt cells, the bacteria grow rapidly (about three doublings per hour) for the first 10–12 h until they reach a density in the crypt fluid of about 10^{11} bacteria mL^{-1}. Once this critical density is reached, an autoinducer of bioluminescence accumulates, leading to initiation of light emission via quorum-sensing regulation of the *lux* operon (see *Figure 3.17*). Once established, the bacteria grow more slowly (about 0.2 doublings per hour). The host is able to detect the production of light by its symbionts and experimental infection with non-light-emitting mutants results in their rejection by the host. Even more surprising, however, is the finding that the bioluminescent strains of *A. fischeri* isolated from *E. scolopes* do not emit light outside their host, raising the possibility that the host has some mechanism for actively selecting for these "dim" strains.

The established symbiosis in the adult squid has a distinct circadian rhythm. Each morning, in response to daylight, the squid squeezes out the contents of the light organ—this is a thick paste of mucus, bacteria, and macrophage-like cells—so that over 90% of the bacteria are expelled. This behavior of the squid ensures that it maintains a "fresh" active culture of *A. fischeri* that will build up to the high density required for bioluminescence, coinciding with the onset of darkness when the squid emerges from its hiding place. It also provides regular seeding of the environment with the specific strain of *A. fischeri*, which ensures horizontal transmission of the symbiont to juvenile squid. This has a clear competitive advantage for the bacterium, which can maintain larger population sizes than if it were entirely free-living.

We could say that the adult squid host "tolerates" the presence of *A. fischeri* for a while, but the daily expulsion of the bacteria from the light organ indicates that a process akin to immunological responses to the presence of a pathogen takes over. Numerous *A. fischeri* genes closely resemble those encoding toxins and surface-associated virulence factors associated with pathogenicity in other vibrios, including *Vibrio cholerae*, *V. parahaemolyticus*, and *V. vulnificus* (see *Chapter 12*). Studies of gene expression show that a key set of virulence-like bacterial genes are switched on just before dawn, and this correlates with the appearance in the host tissue of blebbing and effacement of the light organ crypt epithelial surface, very similar to what is seen during infection of the human gut by some pathogens. The bacteria vary the expression of genes associated with utilization of different substrates throughout the diel cycle. After the dawn expulsion, the remaining bacteria regrow, upregulating genes for the anaerobic metabolism of glycerol derived from the host membranes. After 12 h, genes associated with fermentation of chitin are upregulated.

LITTLE SQUID— BIG IMPACT

The pioneering study of the squid–vibrio symbiosis has been carried out since the 1990s through partnership between Ned Ruby, who works on the bacterium, and Margaret McFall-Ngai, who studies the developmental biology of the host. Their many graduate students have gone on to form their own research teams investigating this topic—described as the "F1 and F2 generations" by McFall-Ngai (2014). This exceptional network of collaboration has yielded many remarkable results, and hundreds of research papers have been published on diverse aspects of the host–symbiont interactions. It has provided an ideal model for the study of symbiosis because the host is easily maintained in culture and the symbiont can be readily analyzed and manipulated by genetic techniques. This little squid and its bacterial partner have provided a model that has transcended marine biology and led to major changes in the way that we think about the broader context of the role of microbes in health, disease, development, and evolution of animals, including humans.

Conclusions

Since the discovery of hydrothermal vent communities in 1977 and the realization that the animals found there depend for their nutrition on chemosynthetic endosymbiotic bacteria, knowledge of the range and diversity of mutualistic associations between microbes and animals has expanded enormously. The examples provided here should convince the reader that such symbioses, far from being an exceptional occurrence in specialized habitats, occur in diverse habitats and in many animal phyla. Frequently, multiple microbial partners are present, providing the host with fixed carbon and nitrogen or performing other essential functions. As the use of genomic techniques to explore these relationships is expanded, we are gaining deep insight into their molecular basis and the evolutionary adaptations of both the hosts and their microbial partners, as well as the importance of these animal–microbe associations in the function of ecosystems.

References and further reading

Books and general reviews

Apprill, A. (2017) Marine animal microbiomes: Toward understanding host–microbiome interactions in a changing ocean. *Front. Mar. Sci.* **4**: 222.

Bright, M. & Bulgheresi, S. (2010) A complex journey: Transmission of microbial symbionts. *Nat. Rev. Microbiol.* **8**: 218–230.

Douglas, A.E. & Werren, J.H. (2016) Holes in the hologenome: Why host-microbe symbioses are not holobionts. *mBio* **7**: e02099-15.

Duperron, S. (2017) *Microbial Symbioses.* ISTE Press/Elsevier, Oxford.

Fisher, R.M., Henry, L.M., Cornwallis, C.K. et al. (2017) The evolution of host-symbiont dependence. *Nat. Commun.* **8**: 15973.

Hurst, C.J. (ed.) (2016) *The Mechanistic Benefits of Microbial Symbionts.* Springer International Publishing, Switzerland.

Margulis, L. (1991) Symbiosis as a source of evolutionary innovation: Speciation and morphogenesis. In: Margulis, L., Fester, R., eds. *Symbiogenesis and Symbionticism.* MIT Press, Cambridge, pp. 1–14.

Moran, N.A. & Sloan, D.B. (2015) The hologenome concept: Helpful or hollow? *PLoS Biol.* **13**: e1002311.

Morris J.J. (2018) What is the hologenome concept of evolution? *F1000Research* **7**: 1664.

O'Brien, P.A., Webster, N.S., Miller, D.J., & Bourne, D.G. (2019) Host-microbe coevolution: Applying evidence from model systems to complex marine invertebrate holobionts. *MBio* **10**: e02241-18.

Raina, J.B., Eme, L., Pollock, F.J. et al. (2018) Symbiosis in the microbial world: From ecology to genome evolution. *Biol. Open* **7**: bio032524.

Rosenberg, E., Koren, O., Reshef, L. et al. (2007) The role of microorganisms in coral health, disease and evolution. *Nat. Rev. Microbiol.* **5**: 355–362.

Rosenberg, E. & Zilber-Rosenberg, I. (2018) The hologenome concept of evolution after 10 years. *Microbiome* **6**: 78.

Schwartzman, J.A. & Ruby, E.G. (2016) Stress as a normal cue in the symbiotic environment. *Trends Microbiol.* **24**: 414–424.

Chemosynthetic symbioses

Assié, A., Borowski, C., van der Heijden, K. et al. (2016) A specific and widespread association between deep-sea bathymodiolus mussels and a novel family of epsilonproteobacteria. *Environ. Microbiol. Rep.* **8**: 805–813.

Batstone, R.T., Laurich, J.R., Salvo, F., & Dufour, S.C. (2014) Divergent chemosymbiosis-related characters in *Thyasira* cf. *gouldi* (Bivalvia: Thyasiridae). *PLoS One* **9**: e92856.

Chen, C., Rogers, A.D., Linse, K. et al. (2015) The "scaly-foot gastropod": A new genus and species of hydrothermal vent-endemic gastropod (Neomphalina: Peltospiridae) from the Indian Ocean. *J. Mollusc. Stud.* **81**: 322–334.

Dubilier, N., Bergin, C., & Lott, C. (2008) Symbiotic diversity in marine animals: The art of harnessing chemosynthesis. *Nat. Rev. Microbiol.* **6**: 725–740.

Dubilier, N., Mulders, C., Ferdelman, T. et al. (2001) Endosymbiotic sulphate-reducing and sulphide-oxidizing bacteria in an oligochaete worm. *Nature* **411**: 298–302.

Dufour, S.C. & Felbeck, H. (2003) Sulphide mining by the superextensile foot of symbiotic thyasirid bivalves. *Nature* **426**: 65–67.

Dufour, S.C., Laurich, J.R., Batstone, R.T. et al. (2014) Magnetosome-containing bacteria living as symbionts of bivalves. *ISME J.* **8**: 2453–2462.

Duperron, S., Gaudron, S.M., Rodrigues, C.F. et al. (2013) An overview of chemosynthetic symbioses in bivalves from the North Atlantic and Mediterranean Sea. *Biogeoscience* **10**: 3241–3267.

Goffredi, S.K., Jones, W.J., Erhlich, H. et al. (2008) Epibiotic bacteria associated with the recently discovered Yeti crab, *Kiwa hirsuta. Environ. Microbiol.* **10**: 2623–2634.

Grzymski, J.J., Murray, A.E., Campbell, B.J. et al. (2008) Metagenome analysis of an extreme microbial symbiosis reveals eurythermal adaptation and metabolic flexibility. *Proc. Natl. Acad. Sci. USA* **105**: 17516–17521.

Harmer, T.L., Rotjan, R.D., Nussbaumer, A.D. et al. (2008) Free-living tube worm endosymbionts found at deep-sea vents. *Appl. Environ. Microbiol.* **6**: 3895–3898.

Heide, T. van der, Govers, L.L., de Fouw, J. et al. (2012) A three-stage symbiosis forms the foundation of seagrass ecosystems. *Science* **336**: 1432–1434.

Higgs, N.D., Newton, J., & Attrill, M.J. (2016) Caribbean spiny lobster fishery is underpinned by trophic subsidies from chemosynthetic primary production. *Curr. Biol.* **26**: 3393–3398.

Jäckle, O., Seah, B.K., Tietjen, M. et al. (2019) Chemosynthetic symbiont with a drastically reduced genome serves as primary energy storage in the marine flatworm *Paracatenula. Proc. Natl. Acad. Sci. USA* **116**: 8505–8514.

Kleiner, M., Woyke, T., Ruehland, C., & Dubilier, N. (2011) The *Olavius algarvensis* metagenome revisited: Lessons learned from the analysis of the low-diversity microbial consortium of a gutless marine worm. In *Handbook of Molecular Microbial Ecology II.* John Wiley & Sons, Inc., Hoboken, NJ, pp. 319–333.

Klose, J., Aistleitner, K., Horn, M. et al. (2016) Trophosome of the Deep-sea tubeworm *Riftia pachyptila* inhibits bacterial growth. *PLoS One* **11**: e0146446.

Klose, J., Polz, M.F., Wagner, M. et al. (2015) Endosymbionts escape dead hydrothermal vent tubeworms to enrich the free-living population. *Proc. Natl. Acad. Sci. USA.* **112**: 11300–11305.

König, S., Gros, O., Heiden, S.E. et al. (2016) Nitrogen fixation in a chemo-autotrophic lucinid symbiosis. *Nat. Microbiol.* **2**: 16193.

König, S., Le Guyader, H., & Gros, O. (2015) Thioautotrophic bacterial endosymbionts are degraded by enzymatic digestion during starvation: Case study of two lucinids Codakia orbicularis and C. orbiculata. *Microsc. Res. Tech.* **78**: 173–179.

Li, Y., Liles, M.R., & Halanych, K.M. (2018) Endosymbiont genomes yield clues of tubeworm success. *ISME J.* **12**: 2785–2795.

Mariño, J., Augustine, S., Dufour, S.C., & Hurford, A. (2018) Dynamic Energy Budget theory predicts smaller energy reserves in thyasirid bivalves that harbour symbionts. *J. Sea Res.* **143**: 119–127.

Mcmullin, E.R., Hourdez, S., Schaeffer, S.W., & Fisher, C.R. (2003) Phylogeny and biogeography of deep sea vestimentiferan tubeworms and their bacterial symbionts. *Symbiosis* **34**: 1–41.

Nussbaumer, A.D., Fisher, C.R., & Bright, M. (2006) Horizontal endosymbiont transmission in hydrothermal vent tubeworms. *Nature* **441**: 345–348.

Petersen, J.M., Kemper, A., Gruber-Vodicka, H. et al. (2017) Chemosynthetic symbionts of marine invertebrate animals are capable of nitrogen fixation. *Nat. Microbiol.* **2**: 16195.

Pflugfelder, B., Cary, S.C., & Bright, M. (2009) Dynamics of cell proliferation and apoptosis reflect different life strategies in hydrothermal vent and cold seep vestimentiferan tubeworms. *Cell Tissue Res.* **337**: 149–165.

Robidart, J.C., Bench, S.R., Feldman, R.A. et al. (2008) Metabolic versatility of the *Riftia pachyptila* endosymbiont revealed through metagenomics. *Environ. Microbiol.* **10**: 727–737.

Sayavedra, L., Ansorge, R., Rubin-Blum, M. et al. (2019) Horizontal acquisition followed by expansion and diversification of toxin-related genes in deep-sea bivalve symbionts. *BioRxiv* 505386.

Sayavedra, L., Kleiner, M., Ponnudurai, R. et al. (2015) Abundant toxin-related genes in the genomes of beneficial symbionts from deep-sea hydrothermal vent mussels. *eLife* **4**: e07966.

Stewart, F.J., Newton, I.L.G., & Cavanaugh, C.M. (2005) Chemosynthetic endosymbioses: Adaptations to oxic-anoxic interfaces. *Trends Microbiol.* **13**: 439–448.

Thurber, A.R., Jones, W.J., & Schnabel, K. (2011) Dancing for food in the deep sea: Bacterial farming by a new species of yeti crab. *PLoS ONE* **6**: e26243.

Warén, A., Bengtson, S., Goffredi, S.K., & Van Dover, C.L. (2003) A hot-vent gastropod with iron sulfide dermal sclerites. *Science* **302**: 1007.

Wentrup, C., Wendeberg, A., Schimak, M. et al. (2014) Forever competent: Deep-sea bivalves are colonized by their chemosynthetic symbionts throughout their lifetime. *Environ. Microbiol.* **16**: 3699–3713.

Woyke, T., Teeling, H., Ivanova, N.N. et al. (2006) Symbiosis insights through metagenomic analysis of a microbial consortium. *Nature* **443**: 950–955.

Yao, H., Dao, M., Imholt, T. et al. (2010) Protection mechanisms of the iron-plated armor of a deep-sea hydrothermal vent gastropod. *Proc. Natl. Acad. Sci. USA* **107**: 987–992.

Deep-sea wood and whale falls

Danise, S. & Higgs, N.D. (2015) Bone-eating *Osedax* worms lived on Mesozoic marine reptile deadfalls. *Biol. Lett.* **11**: 20150072.

Little, C.T. (2017) The prolific afterlife of whales. *Sci. Am.* **26**: 14–19.

Ristova, P.P., Bienhold, C., Wenzhöfer, F. et al. (2017) Temporal and spatial variations of bacterial and faunal communities associated with deep-sea wood falls. *PLoS ONE* **12**: e.0169906.

Smith, C.R. & Baco, A.R. (2003) Ecology of whale falls at the deep-sea floor. *Oceanogr. Mar. Biol.* **41**: 311–354.

Ascidians, bryozoans and sponges

Anderson, C.M. & Haygood, M.G. (2007) α-proteobacterial symbionts of marine bryozoans in the genus *Watersipora*. *Appl. Environ. Microbiol.* **73**: 303–311.

Evans, J.S., Erwin, P.M., Shenkar, N., & López-Legentil, S. (2018) A comparison of prokaryotic symbiont communities in nonnative and native ascidians from reef and harbor habitats. *FEMS Microbiol. Ecol.* **94**: fiy139.

Fieseler, L., Horn, M., Wagner, M., & Hentschel, U. (2004) Discovery of the novel candidate phylum "Poribacteria" in marine sponges. *Appl. Environ. Microbiol.* 70: 3724–3732.

Karagodina, N.P., Vishnyakov, A.E., Kotenko, A.L. et al. (2018) Ultrastructural evidence for nutritional relationships between a marine colonial invertebrate (Bryozoa) and its bacterial symbionts. *Symbiosis* 75: 155–164.

López-Legentil, S., Turon, X., Espluga, R., & Erwin, P.M. (2015) Temporal stability of bacterial symbionts in a temperate ascidian. *Front. Microbiol.* **6**: 1022.

Moitinho-Silva, L., Nielsen, S., Amir, A. et al. (2017) The sponge microbiome project. *GigaScience* **6**: 1–7.

Pita, L., Rix, L., Slaby, B.M. et al. (2018) The sponge holobiont in a changing ocean: From microbes to ecosystems. *Microbiome* **6**: 46.

Schmitt, S., Tsai, P., Bell, J. et al. (2012) Assessing the complex sponge microbiota: Core, variable and species-specific bacterial communities in marine sponges. *ISME J.* **6**: 564–576.

Schreiber, L., Kjeldsen, K.U., Funch, P. et al. (2016) Endozoicomonas are specific, facultative symbionts of sea squirts. *Front. Microbiol.* **7**: 1042.

Taylor, M.W., Tsai, P., Simister, R.L. et al. (2013) "Sponge-specific" bacteria are widespread (but rare) in diverse marine environments. *ISME J.* **7**: 438–443.

Thomas, T., Moitinho-Silva, L., Lurgi, M. et al. (2016) Diversity, structure and convergent evolution of the global sponge microbiome. *Nat. Commun.* **7**: 11870.

Zooxanthellae and the coral holobiont

Ainsworth, T.D., Krause, L., Bridge, T. et al. (2015) The coral core microbiome identifies rare bacterial taxa as ubiquitous endosymbionts. *ISME J.* **9**: 2261–2274.

Blackall, L.L., Wilson, B., & Oppen, M.J.H. van (2015) Coral—The world's most diverse symbiotic ecosystem. *Mol. Ecol.* **24**: 5330–5347.

Bourne, D.G., Morrow, K.M., & Webster, N.S. (2016) Insights into the coral microbiome: Underpinning the health and resilience of reef ecosystems. *Annu. Rev. Microbiol.* **70**: 317–340.

Callahan, B.J., McMurdie, P.J., & Holmes, S.P. (2017) Exact sequence variants should replace operational taxonomic units in marker-gene data analysis. *ISME J.* **11**: 2639.

Davy, S.K., Allemand, D., & Weis, V.M. (2012) Cell biology of cnidarian-dinoflagellate symbiosis. *Microbiol. Mol. Biol. Rev.* **76**: 229–61.

Donelson, J.M., Salinas, S., Munday, P.L., & Shama, L.N.S. (2018) Transgenerational plasticity and climate change experiments: Where do we go from here? *Glob. Change Biol.* **24**: 13–34.

Fine, M. & Loya, Y. (2002) Endolithic algae: An alternative source of photoassimilates during coral bleaching. *Proc. Roy. Soc. B* **269**: 1205–1210.

Hernandez-Agreda, A., Gates, R.D., & Ainsworth, T.D. (2017) Defining the core microbiome in corals' microbial soup. *Trends Microbiol.* **25**: 125–140.

Hernandez-Agreda, A., Leggat, W., Bongaerts, P., & Ainsworth, T.D. (2016) The microbial signature provides insight into the mechanistic basis of coral success across reef habitats. *MBio* **7**: e00560-16.

Hoegh-Guldberg, O., Hughes, T., Antoiony, K. et al. (2009) Coral reefs and rapid climate change: Impacts, risks and implications for tropical societies. *IOP Conf. Ser.: Earth Environ. Sci.* **6**: 302004.

Hughes, T.P., Barnes, M.L., Bellwood, D.R. et al. (2017) Coral reefs in the Anthropocene. *Nature* **546**: 82–90.

Hughes, T.P., Kerry, J.T., Baird, A.H. et al. (2018) Global warming transforms coral reef assemblages. *Nature* **556**: 492–496.

LaJeunesse, T.C., Parkinson, J.E., Gabrielson, P.W. et al. (2018) Systematic revision of Symbiodiniaceae highlights the antiquity and diversity of coral endosymbionts. *Curr. Biol.* **28**: 2570–2580.

Morris, L.A., Voolstra, C.R. Quigley, K.M. et al. (2019) Nutrient availability and metabolism affect the stability of coral-Symbiodiniaceae symbioses. *Trends Microbiol.* **27**: 678–689.

NAS (2019) National Academies of Sciences Engineering and Medicine. *A Research Review of Interventions to Increase the Persistence and Resilience of Coral Reefs.* The National Academies Press, Washington, DC.

Neave, M.J., Rachmawati, R., Xun, L. et al. (2017) Differential specificity between closely related corals and abundant *Endozoicomonas* endosymbionts across global scales. *ISME J.* **11**: 186–200.

Pochon, X., Putnam, H.M., & Gates, R.D. (2014) Multi-gene analysis of *Symbiodinium* dinoflagellates: A perspective on rarity, symbiosis, evolution. *PeerJ* **2**: e394.

Pollock, F.J., McMinds, R., Smith, S. et al. (2018) Coral-associated bacteria demonstrate phylosymbiosis and cophylogeny. *Nat. Commun.* **9**: 4921.

Raghukumar, S. (2017) *Fungi in Coastal and Oceanic Marine Ecosystems: Marine Fungi.* Springer International.

Rohwer, F., Seguritan, V., Azam, F., & Knowlton, N. (2002) Diversity and distribution of coral-associated bacteria. *Marine Ecol. Prog. Ser.* **243**: 1–10.

Rohwer, F. & Youle, M. (2010) *Coral Reefs in Microbial Seas.* Plaid Press, San Diego.

Torda, G., Donelson, J.M., Aranda, M. et al. (2017) Rapid adaptive responses to climate change in corals. *Nat. Clim. Change* **7**: 627.

van de Water, J.A.J.M., Allemand, D., & Ferrier-Pagès, C. (2018) Host-microbe interactions in octocoral holobionts - recent advances and perspectives. *Microbiome* **6**: 64.

van de Water, J.A.J.M., Melkonian, R., Junca, H. et al. (2016) Spirochaetes dominate the microbial community associated with the red coral *Corallium rubrum* on a broad geographic scale. *Sci. Rep.* **6**: 27277.

van Oppen, M.J.H., Oliver, J.K., Putnam, H.M., & Gates, R.D. (2015) Building coral reef resilience through assisted evolution. *Proc. Natl. Acad. Sci.* **112**: 2307–2313.

Wada, N., Ishimochi, M., Matsui, T. et al. (2019) Characterization of coral-associated microbial aggregates (CAMAs) within tissues of the coral *Acropora hyacinthus*. *Biorxiv, BioRxiv*: 576488.

Wada, N., Pollock, F.J., Willis, B.L. et al. (2016) *In situ* visualization of bacterial populations in coral tissues: Pitfalls and solutions. *PeerJ* **4**: e2424.

Webster, N.S. & Reusch, T.B.H. (2017) Microbial contributions to the persistence of coral reefs. *ISME J.* **11**: 2167–2174.

Wegley, L., Edwards, R., Rodriguez-Brito, B. et al. (2007) Metagenomic analysis of the microbial community associated with the coral *Porites astreoides*. Environ. Microbiol. **9**: 2707–2719.

Wooldridge, S.A. (2010) Is the coral-algae symbiosis really "mutually beneficial" for the partners? *BioEssays* **32**: 615–625.

Deep-sea corals

Kellogg, C.A., Goldsmith, D.B., & Gray, M.A. (2017) Biogeographic comparison of *Lophelia*-associated bacterial communities in the western Atlantic reveals conserved core microbiome. *Front. Microbiol.* **8**: 796.

Kellogg, C.A., Lisle, J.T., & Galkiewicz, J.P. (2009) Culture-independent characterization of bacterial communities associated with the cold-water coral *Lophelia pertusa* in the northeastern Gulf of Mexico. *Appl. Environ. Microbiol.* **75**: 2294–2303.

Middelburg, J.J., Mueller, C.E., Veuger, B. et al., (2015) Discovery of symbiotic nitrogen fixation and chemoautotrophy in cold-water corals. *Sci. Rep.* **5**: 17962.

Penn, K., Wu, D., Eisen, J.A., & Ward, N. (2006) Characterization of bacterial communities associated with deep-sea corals on Gulf of Alaska seamounts. Appl. Environ. Microbiol. **72**: 1680–1683.

Röthig, T., Yum, L.K., Kremb, S.G., et al. (2017) Microbial community composition of deep-sea corals from the Red Sea provides insight into functional adaption to a unique environment. *Sci. Rep.* **7**: 44714.

Bioluminescent symbioses

Haygood, M.G. (1993) Light organ symbioses in fishes. *Crit. Rev. Microbiol.* **19**: 191–216.

Hellinger, J., Jägers, P., Donner, M. et al. (2017) The flashlight fish *Anomalops katoptron* uses bioluminescent light to detect prey in the dark. *PLOS ONE* **12**: e0170489.

Hendry, T.A. & Dunlap, P.V. (2011) The uncultured luminous symbiont of *Anomalops katoptron* (Beryciformes: Anomalopidae) represents a new bacterial genus. *Mol. Phylogenet. Evol.* **61**: 841–843.

Hendry, T.A. & Dunlap, P. V (2014) Phylogenetic divergence between the obligate luminous symbionts of flashlight fishes demonstrates specificity of bacteria to host genera. *Environ. Microbiol. Rep.* **6**: 331–8.

Hendry, T.A., Freed, L.L., Fader, D. et al. (2018) Ongoing transposon-mediated genome reduction in the luminous bacterial symbionts of deep-sea ceratioid anglerfishes. *MBio* **9**: e01033-18.

McAnulty, S.J. & Nyholm, S.V. (2017) The role of hemocytes in the Hawaiian bobtail squid, *Euprymna scolopes*: A model organism for studying beneficial host-microbe interactions. *Front. Microbiol.* **7**: 2013.

McFall-Ngai, M. (2014) Divining the essence of symbiosis: Insights from the squid-vibrio model. *PLoS Biol.* **12**: e1001783.

Nawroth, J.C., Guo, H., Koch, E. et al. (2017) Motile cilia create fluid-mechanical microhabitats for the active recruitment of the host microbiome. *Proc. Natl. Acad. Sci.* **114**: 9510–9516.

Norsworthy, A.N. & Visick, K.L. (2013) Gimme shelter: How *Vibrio fischeri* successfully navigates an animal's multiple environments. *Front. Microbiol.* **4**: 356.

Nyholm, S.V. & McFall-Ngai, M.J. (2004) The winnowing: Establishing the squid-vibrio symbiosis. *Nat. Rev. Micriobiol.* **2**: 632–642.

Nyholm, S.V., Stabb, E.V., Ruby, E.G., & McFall-Ngai, M.J. (2000) Establishment of an animal–bacterial association: Recruiting symbiotic vibrios from the environment. *Proc. Natl. Acad. Sci. USA* **97**: 10231–10235.

Chapter 11

Microbial Diseases of Marine Organisms

Marine biologists have become increasingly aware of the effects of infectious microbial diseases on populations and communities of marine organisms, which can sometimes result in major changes at the ecosystem level. Infectious diseases caused by viruses, bacteria, fungi, and protists have been characterized in marine mammals, fish, and turtles, as well as in a wide range of invertebrates, including molluscs, crustaceans, sponges, echinoderms, and corals. Infections of multicellular algae and seagrasses also have important ecological and economic effects. Toxins produced by diatoms and dinoflagellates also cause disease in marine organisms. This chapter explores the various types of microbial diseases and their importance in natural ecosystems and aquaculture, with special consideration of the impact of climate change, pollution, global transport, and other anthropogenic factors. Aspects of coral disease link closely with the discussion of the microbiome in *Chapter 13*. Biotechnological aspects of disease control are further considered in *Chapter 14*.

Key Concepts

- The development of disease depends on complex interactions between the host (and its associated microbiota), the pathogen, and environmental factors.

- It is not possible to define single causative agents for many marine diseases, and disturbance of the microbiome is a major factor allowing development of opportunistic pathogens.

- Incidence of disease in many groups of organisms has increased in natural marine ecosystems in the past few decades, with global climate change, pollution, and other anthropogenic factors being contributory factors.

- Diseases of foundation species, keystone predators, and ecosystem engineers can lead to serious alterations to marine ecosystems.

- Development of intensive marine aquaculture has led to increased levels of bacterial and viral diseases in molluscs, crustaceans, and fish; this has major economic effects in the industry as well as impacts on natural ecosystems.

- Bacteria (especially vibrios) are strongly associated with diseases of corals, molluscs, crustaceans, and fish.

- Viruses are responsible for major epizootics in molluscs, crustaceans, fish, reptiles, and mammals, some of which seriously affect natural populations.

Diseases of marine organisms have major ecological and economic impact

There is an enormous diversity and abundance of marine parasites and pathogens, which are a natural part of the life cycle of all organisms, regulating food webs and shaping the structure and function of ecosystems. In the last few decades, there has been increasing concern about the spread of infectious diseases and outbreaks of mass mortality in wild marine organisms. In some cases, disease outbreaks in foundation species, keystone species, or ecosystem engineers lead to substantial long-term changes in ecosystem structure and function. With "emerging diseases," it is always difficult to decide whether there is a genuine increase in incidence, or whether it simply reflects more intensive observation. There is no doubt that interest in monitoring the health of marine habitats has increased greatly and there is now a large community of sub-aqua diving scientists, which partly explains the increased reporting of disease. However, it is generally accepted that diseases are increasing as a result of recent disturbance of balanced marine ecosystems caused by human activities. Pollution, destruction of habitats, climate change, global sea traffic, and intensive aquaculture all affect the complex interactions between hosts, pathogens, and the environment in the disease process (*Figure 11.1*).

While epidemiology of diseases in humans and terrestrial plants and animals is a well-developed science, the dynamics of the origins, spread, and cycles of disease in the oceans are less well understood. In part, this is due to the greater host and pathogen diversity, but it must also be noted that many marine organisms such as corals, sponges, and ascidians are colonies of genetically identical individuals, which facilitates the spread of disease. Furthermore, marine systems are more open, allowing rapid dispersal of pathogens over large distances and opportunities to survive in reservoirs allowing reinfection. It is vital that we develop a better understanding of the role of microbes in the ecology of marine infectious diseases. Thus, research efforts are directed towards developing new epidemiological models to explain, predict, and manage marine infectious diseases. The direct costs to animal health in aquaculture are considerable and quantifiable, but disease also has considerable economic impacts in natural ecosystems due to adverse direct and indirect effects on fisheries, food security, coastal management, and tourism. Of course, this is in addition to obvious concerns about the loss of biodiversity and the short- and long-term ecological impact of these events.

DISEASES OF CORALS, SPONGES, AND ECHINODERMS

Infectious diseases threaten the survival of corals

As discussed in *Chapter 10*, coral-associated microbes play a vital role in the health of corals and the reef ecosystem, through provision of fixed nitrogen and carbon, recycling of waste products, antibiotic defense, and contribution to food webs. However, the balanced

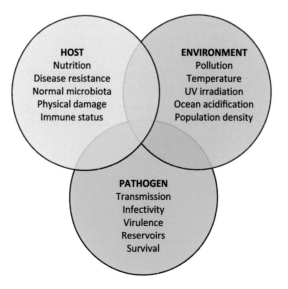

Figure 11.1 The disease process in marine organisms depends on complex interactions of the host, pathogen, and environmental factors.

microbiome in healthy corals is easily disturbed by shifts in environmental conditions. Many of the world's tropical and subtropical reefs are at severe risk from a multitude of threats, including overfishing, coastal development, physical damage from boating and diving activities, pollution, ocean acidification, and temperature-induced bleaching. There is now increasing recognition of the importance of coral diseases associated with microbial infection as a factor in their decline. Coral diseases were first studied intensively in the Caribbean, where successive outbreaks of disease have contributed to the decline of the major reef-building species, especially *Acropora*, resulting in major ecosystem changes. Diseases have now been described in all the major reef systems of the world, involving several hundred different species of coral.

One of the problems in evaluating the historical importance of coral diseases is that the early literature contains many poorly documented reports of disorders for which details of pathology and evidence of specific causation are incomplete. Diagnosis of disease in corals is primarily based on characteristics such as the loss of tissue, change in color, and exposure of coral skeleton. These limited macroscopic descriptions lead to imprecise names for the conditions—for example, in the Caribbean, the terms "white plague," "white disease," "white pox," and "white band" have all been used to represent different syndromes, without clear definitions—and this results in inaccurate subsequent diagnosis. The term "white syndrome" is a collective term that encompasses the range of coral pathologies observed in the Indo-Pacific in which a spreading zone of tissue loss exposes the white coral tissue. It is likely that there are multiple diseases involving different pathogens, but methods are currently not robust enough to distinguish them.

Although several studies have reported fulfilment of Koch's Postulates, which are the usual criteria needed to prove the etiology of disease (see information box on p.305), there is usually no evidence specifically linking cellular responses of the host to the presence of the pathogen and development of the disease. Thus, there is only a small number of diseases that have been sufficiently well characterized for us to be confident of a clear and *possibly* causative association with specific bacteria, fungi, or protistan microbes (*Table 11.1*). Even so, the situation is further complicated by the fact that the same pathogen may cause different signs in different hosts; alternatively, different pathogens may cause the same signs in a particular host. Furthermore, organisms established as the causative pathogen of a disease are often not found in all cases and may change with time. For most diseases, knowledge of the mechanisms of pathogenesis and the host responses remains very limited and some coral biologists remain skeptical about the role of microbes as primary causative agents of disease in uncompromised hosts (see *Box 11.1*). As we saw in *Chapter 10*, it is increasingly important to consider hosts and their associated microbes together as *holobionts*, and disease must be considered within this context.

Vibrios are associated with many coral diseases

It can be seen from *Table 11.1* that *Vibrio* spp. are particularly prominent as possible causative agents of well-recognized coral diseases. Furthermore, other members of the family Vibrionaceae, including *Enterovibrio* and *Photobacterium*, form a significant proportion of the microbiota of "healthy" corals (i.e. lacking any obvious disease signs) and some contribute to nitrogen fixation, defense against pathogens, and other functions. The *Vibrio* spp. in *Table 11.1* possess an array of enzymes, toxins, and other virulence factors, which supports them being primary pathogens, but in many cases bacteria must be regarded as "opportunistic" pathogens that are only able to overwhelm the natural defenses of the host under certain conditions. Environmental factors can alter the immune status of the host or the probiotic effects of resident microbiota. All vibrios are versatile heterotrophs capable of rapid growth when nutrients are, so they can quickly take advantage of these changes and digest damaged tissue, leading to signs of disease.

As discussed in *Chapter 10*, bleaching is a highly damaging process affecting coral reefs. It is generally perceived to be due to physiological disturbance resulting in disruption of the symbiosis between the host and the photosynthetic zooxanthellae, most often as the result of increased seawater temperatures. The concept that some types of bleaching might be due to bacterial infection stems from the comprehensive studies of the research group led by Eugene

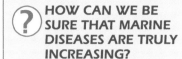

HOW CAN WE BE SURE THAT MARINE DISEASES ARE TRULY INCREASING?

In the late 1990s, concerns were raised about dramatic increases of diseases in the marine environment. Is there truly an increase in disease rates, or is it simply that more scientists are conducting studies in this field? To answer this question, Ward and Lafferty (2004) analyzed 5900 papers (1970–2001) reporting disease in nine taxonomic groups. They used a sophisticated normalization method to determine the proportion of disease reports about a given taxonomic group relative to the total number of reports about any subject in that group. The data were also adjusted to remove the effect of multiple reports of the same outbreak, or the effect of particularly prolific authors investigating disease. Reports of disease were increased for all groups over this period, but after removing the skewing effects, Ward and Lafferty concluded that there was a clear increase in disease among turtles, corals, mammals, urchins, and molluscs. Reports of diseases in fishes actually decreased, which they suggested was caused by drastic reductions in population density due to overfishing, presenting fewer opportunities for transmitting infection.

Table 11.1 *Diseases of corals for which there is a strong association with microbial infection. (Data from Sweet et al., 2011; Mera and Bourne, 2018). Changes in our understanding of the causative agents of coral diseases may occur because more than one pathogen causes the same signs on different hosts, and etiologies may change through time (the "moving target hypothesis" of Sutherland et al., 2016).*

Disease	Associated pathogen(s)	Commonly affected species	Location[a]
Aspergillosis	*Aspergillus sydowii*	*Gorgonia ventalina* and other octocorals	CS
Black band disease	Mixed consortium; *Phormidium corallyticum, Roseofilum reptotaenium, Trichodesmium erythraeum, Desulfovibrio* spp. and *Beggiatoa* spp. are key components	Wide range of corals including *Acropora* spp. *Montipora* spp., *Favia* spp., *Porites* spp., *Siderastrea* spp.	CS, IP, MS
Bleaching	*Vibrio shilonii* (originally classified as *V. shiloi*)	*Oculina patagonica*	RS.MS
Bleaching and tissue lysis	*Vibrio coralliilyticus* and related *Vibrio* spp.	*Pocillopora damicornis,*	MS, IP
White band	*Vibrio carchariae* (= *V. harveyi*) and other *Vibrio* spp.	*Acropora* spp.	CS
White plague-like disease	*Thalassomonas loyana*	*Favia favus*	RS
White plague type II	*Aurantimonas coralicida*	*Montastrea* spp. and other scleractinian corals	CS
White pox	*Serratia marcescens*	*Acropora palmata*	CS
White syndrome	*Vibrio* spp. including *V. coralliilyticus, V. owensii, V. harveyi*: Rhodobacteriaceae-associate; *Philaster lucinda*	Acropridae, Pocilloporidae, *Montipora capitata*	IP
Yellow blotch/band	Consortium of *Vibrio* spp.: *V. rotiferianus, V. harveyi, V. alginolyticus, V. proteolyticus*	*Acropora* spp., *Orbicella*[b] spp., *Diploastrea* spp., *Fungia* spp.	CS, IP

[a] CS, Caribbean Sea; IP, Indo-Pacific Ocean; MS, Mediterranean Sea; RS, Red Sea. [b] Formerly *Montastrea*.

Rosenberg of Tel Aviv University. In the 1990s, they showed that *Vibrio shiloi* (subsequently renamed as *V. shilonii*) is the causative agent of summer bleaching in the invasive coral *Oculina patagónica* on the Israel coast of the Mediterranean Sea. In a series of experiments, researchers showed that when *O. patagonica* was infected with *V. shilonii* at 29°C, bleaching occurred rapidly, whereas at 20 and 25°C, the rate of bleaching was slower and less complete and none was observed at 16°C or in uninoculated controls at any temperature. Also, addition of antibiotics prevented bleaching. Virulence was attributed to temperature-regulated production of an adhesin to a carbohydrate receptor on the coral surface and the production of a peptide toxin that inhibits photosynthesis in the zooxanthellae, leading to bleaching. Subsequently, the marine fireworm *Hermodice carunculata* was implicated as a winter reservoir and vector of the pathogen. Extrapolating from these findings to a general hypothesis that bacteria cause bleaching provoked considerable controversy at the time, because the temperature shifts involved are extreme (16–29°C), whereas mass bleaching events in the Caribbean, Pacific, and Indian Oceans typically involve rises of just 1–2°C above normal sea-surface temperatures. The *O. patagonica–V. shilonii* interactions became an elegant model system in Rosenberg's laboratory and, until 2002, all strains of pathogenic *V. shiloi* held in the laboratory caused bleaching in aquarium experiments. However, the same strains are now incapable of infecting the coral. Because coral lack adaptive immunity, the "Coral Probiotic Hypothesis" was proposed as an explanation for this phenomenon, which was later developed as the "Hologenome Hypothesis" (see p.281). The deliberate manipulation of the coral microbiome by introduction or enhancement of beneficial microbes as "probiotics" has recently been proposed as a method of treating diseased corals or preventing the spread of disease. This idea is considered in *Chapter 14*.

Further support for the role of bacteria in coral bleaching was provided by the discovery of a new species, *Vibrio coralliilyticus*, first isolated from the coral *Pocillopora damicornis* in the Indian Ocean. Subsequently, strains of *V. coralliilyticus* have been associated with a variety of disease syndromes in corals and other invertebrates in many different regions. Recent research on the virulence of this versatile marine pathogen is discussed in *Research Focus Box 11.1*.

The fungus *Aspergillus sydowii* caused mass mortality of sea fans in the Caribbean Sea

Beginning in 1995, one of the most devastating epizootics yet seen in the marine environment occurred in sea fans (*Gorgonia ventalina*) in the Caribbean Sea. A new fungal species, *Aspergillus sydowii*, was identified as the causative agent responsible for massive tissue destruction. The disease was shown to be transmissible through contact with healthy corals. Since 1998, the severity of the disease has declined and it now occurs only sporadically, although many of the largest sea fans have been killed, especially in the Florida Keys. Investigations reveal that there is a complex relationship between temperature, growth of the pathogen, and host resistance due to production of antifungal compounds. The host attempts to limit spread of the pathogen by encapsulation of necrotic tissue, resulting in the characteristic purple lesions (*Figure 11.3*). Other microbes associated with surface mucus also differ between healthy and diseased sea fans and between healthy and diseased parts of the same individual. *A. sydowii* has also been implicated as the causative agent of the virtual eradication of sea urchins (*Diadema antillarum*) from parts of the Caribbean Sea in the 1980s, although there is no direct evidence for this. There is recent evidence that the fungus can be isolated from apparently healthy *G. ventalina* and other species of octocorals, suggesting that it is an opportunist pathogen that causes disease only under certain conditions.

Black band disease of corals is a disease of corals worldwide

Black band disease (BBD) was the first coral disease to be described. It takes its name from the characteristic black band (from about 10 mm to several cm wide), which is a thick microbial mat that migrates over the coral colony by as much as 2 cm a day (*Figure 11.4*). Healthy coral tissue is killed, and the band moves on leaving the exposed skeleton, which usually becomes overgrown by turf algae. BBD usually occurs in the summer when water temperatures rise. It can arise on apparently pristine reefs but is more often associated with reefs receiving sewage and runoff from the coast. BBD can have severe effects because it often attacks large corals such as *Orbicella* (formerly *Montastrea*), which are important in building the framework of reefs. Infection can be transmitted via contact, and damaged colonies are more susceptible. Since its first description in the Caribbean in the 1970s, BBD has been observed in a wide range of coral species throughout the world.

BBD has been intensively studied but establishing the etiology of the disease has been extremely difficult. Microscopic examination of diseased tissue consistently shows the presence of large, gliding, filamentous cyanobacteria, including *Phormidium corallyticum*, *Leptolyngbya* spp., *Geitlerinema* spp., and *Pseudoscillatoria coralii*, together with numerous other heterotrophic bacteria, sulfide-oxidizing bacteria (SOB), and sulfate-reducing bacteria (SRB, especially *Desulfovibrio* spp.); the SRB are responsible for the characteristic black pigment. Protists and members of the Archaea have also been identified. It is generally recognized that BBD is a polymicrobial disease caused by a succession of microbial virulence factors in microenvironments produced by activities of the mixed consortium. Anoxic and sulfide-rich conditions at the base of the band result in death of the coral tissue as it spreads across the coral. Molecular analysis of the lesions as they develop in the field has yielded new information about the causes and progression of the disease.

Increases in seawater temperature, water quality, presence of nutrients from sewage, damage from divers or boat anchors, and other factors have all been linked to the incidence of BBD. Pathogens are transmitted to other corals in the reef via water currents and vectors such as snails, worms, and reef fishes. A similar condition called red band occasionally affects hard star and brain corals in the Caribbean and GBR. It also seems to involve similar cyanobacteria and other members of the microbial consortium; the reasons for difference in pigmentation are not clear.

DEVELOPMENT OF THE CORAL PROBIOTIC HYPOTHESIS (CPH)

After 2002, *V. shilonii* no longer infected *O. patagonica*. Reshef et al. (2006) proposed the CPH to explain this: "the coral animal lives in a symbiotic relationship with a diverse metabolically active population of microorganisms ... When environmental conditions change ... the relative abundance of microbial species changes [to] allow the coral holobiont to adapt to the new condition." We cannot go back in time to investigate what changes may have occurred in *O. patagonica*, but one clue comes from the fact that several bacteria isolated from the coral in 2008 show strong activity against *V. shilonii* (Nissimov et al., 2009). Alternatively, the pathogen may have changed; analysis indicates that modern strains lack genomic islands that might encode virulence factors (Reshef et al., 2008). Seasonal bleaching *of O. patagonica* still occurs and Ainsworth et al. (2008) questioned whether *V. shilonii* was ever the primary cause of bleaching. However, Rosenberg et al. (2009) provide strong counter-arguments to this and Mills et al. (2013) show that other bacteria may currently play important roles in both causing and preventing bleaching.

BOX 11.1

The pathogenicity of *Vibrio coralliilyticus*

Discovery of a versatile marine pathogen. The discovery by Ben-Haim and Rosenberg (2002) of *V. coralliilyticus* as the cause of temperature-dependent bleaching and necrosis in *Pocillopora* gave support to the bacterial bleaching hypothesis, because this coral is widely distributed and susceptible to bleaching throughout the world. Laboratory experiments by Ben-Haim et al. (2003) showed that infection and tissue lysis proceed rapidly at 29°C, but no tissue damage occurs at 25°C or lower. We now know that *V. coralliilyticus* shows global distribution, and numerous strains have been isolated in association with a wide range of diseased hard and soft corals. Besides corals, *V. coralliilyticus* also infects a range of molluscs, crustaceans, and fish, although caution is needed in interpretation of some experimental "model systems" used to investigate pathogenicity by Austin et al. (2005). Natural infections of bivalve larvae by *V. coralliilyticus* and the closely related *V. tubiashii* are responsible for high mortalities in oyster hatcheries and extensive cultivation (Genard et al., 2013). Whole-genome sequencing has revealed that several strains associated with disease in bivalve larvae and originally identified as *V. tubiashii* are in fact *V. coralliilyticus*, confirmed by PCR assays based on detection of the gene *vcpA* encoding a zinc metalloprotease implicated in virulence (Wilson et al., 2013).

Role in coral white syndromes. Sussman et al. (2008) used a combination of molecular and culture-based techniques to isolate putative pathogens from various species of corals during epizootics of white syndrome (WS) on the Great Barrier Reef (GBR) and other areas of the Indo-Pacific. They showed that two novel strains of *V. coralliilyticus* and four closely related *Vibrio* spp. are causative agents of WS. Virulence is highly dependent on temperature, with the most important factor being a powerful zinc metalloprotease (MP). Sussman et al. (2009) tested the effects of the metalloprotease on photosynthetic activity in zooxanthellae isolated from different corals which had varying degrees of sensitivity to WS. The MP inactivated photosystem II of zooxanthellae isolated from hosts susceptible to WS, followed by spreading lesions leading to death of the coral tissue. Despite the strong link between MP and the development of disease signs, Sussman et al. (2009) were careful to point out that WS is a multifactorial disease. In culture, protease production is highly dependent on bacterial cell density, and accumulation of sufficient levels of protease to cause cell damage in the field would require similar high densities; these might occur at the band that is the progressing interface between dead and living tissue, characteristic of WS. The genomes of two *V. coralliilyticus* (strains P1 and Vc450, from the GBR and near Zanzibar, respectively) show many similarities, with gene distribution on two chromosomes typical of that observed in other vibrios, the presence of large plasmids, and a large proportion of shared genes (Santos et al., 2011; Kimes et al., 2012). However, there are variations in genes, including important virulence factors, pathogenicity islands, and phage genes that could explain different physiological characteristics, suggesting that the two strains could be regarded as distinct ecotypes or subspecies. Santos et al. (2011) found the *vcpA* metalloprotease gene to be very similar to

the hemagglutinin/protease genes described as major virulence factors in other vibrios, including *V. cholerae*, *V. splendidus*, and *V. tubiashii*. Surprisingly, there was no difference in pathogenicity to zooxanthellae and two model animal systems of a wild type and a mutant strain (in which *vcpA* was deleted). This led Santos et al. (2011) to conclude that "the pathogenicity of vibrios in marine animals is a complex interplay of multiple genetic factors and unlikely the result of one determinant." Some caution is perhaps necessary in the interpretation of such model systems for experimental determination of pathogenicity; for example, *V. coralliilyticus* and its extracellular products were previously shown to induce high mortalities of artemia and fish, even when cultured at 18°C (Austin et al., 2005). In their study of the Vc450 genome, Kimes et al. (2012) showed that a number of virulence factors involved in motility, host degradation, secretion, antimicrobial resistance, and transcriptional regulation factors are upregulated at 27°C, correlating with the induction of bleaching and tissue damage previously observed with this strain. Interestingly, temperature (23°C or 27°C) did not affect virulence of a different strain, ON008, shown to be responsible for acute WS in *Montipora capitata* in Hawaii despite this strain sharing many temperature-regulated genes with strain BAA-450 (Ushijima et al., 2014). However, the true cause (or causes) of WS remains elusive. Pollock et al. (2017) concluded that WS in acroporids on the GBR is not due to infection by vibrios, based on an 18-month field study. They found no significant differences in the abundance of *Vibrio* sequences between healthy and diseased samples and found no sign of these bacteria in the winter, despite year-round disease progression. Pollock et al. found elevated levels of Rhodobacteriaceae sequences in WS lesions and suggested that these bacteria are opportunists, which are normally regulated by antibiotic activities of other members of the microbiome.

Motility and chemotaxis. Meron et al. (2009) first demonstrated the importance of flagellar motility and chemotaxis of *V. coralliilyticus* in virulence, by showing that a mutant strain defective in the flagellar gene *fhlA* was unable to swim towards mucus and adhere to the tissue of *P. damicornis*. Further advances in studying the behavior of *V. coralliilyticus* have been made by the application of microfluidics, in which very small volumes of fluid (microliters to picoliters) can be controlled within micro channels, often mounted on a microscope slide. Garren et al. (2014) used a microfluidics device to create a 400 μm thick layer of mucus adjacent to a 1 mm thick suspension of *V. coralliilyticus* in seawater. The behavior of the bacteria in the diffusive gradients was measured by high-speed video microscopy. *V. coralliilyticus* showed a very rapid and intense response to the presence of mucus, and further analysis showed that the key component that initiates this response is DMSP, produced by zooxanthellae and coral cells (p.253). By tracking the movement of individual bacteria, it was found that *V. coralliilyticus* demonstrates both chemotaxis (biased swimming direction in response to a chemical gradient) and chemokinesis (increased swimming speed when moving up an attractant gradient) in response to DMSP. Garren et al. surmised that the

BOX 11.1   RESEARCH FOCUS

rare presence of these two processes is an adaptation that maximizes the likelihood of *V. coralliilyticus* reaching the coral surface and remaining close to it. Surprisingly, *V. coralliilyticus* does not appear to utilize DMSP as a nutrient source, but the pathogen could be using it as an infochemical to target stressed hosts, because DMSP levels are higher in heat-stressed corals. In follow-up experiments, Garren et al. (2016) recorded the individual swimming behavior of thousands of individual *V. coralliilyticus* cells at different temperatures. At temperatures >23°C, chemotaxis towards coral mucus increased by >60% and at 30°C chemokinesis increased by >57%. Q-PCR analysis of shifts in bacteria associated with *P. damicornis* during heat shift experiments showed dramatic increases of *Vibrio* spp., especially *V. coralliilyticus* (Tout et al., 2015). Thus, rising temperatures affect both the behavior of *V. coralliilyticus* and the chemical ecology of the coral, and the interaction between these processes favors colonization of heat-stressed hosts by the pathogen.

Studying early stages of bacterial bleaching in real time.
Despite the presence of powerful tissue-degrading enzymes and rapid disease progression, it is rather surprising that experimental infection of corals in aquaria requires a high inoculum. However, it is not known what concentration is required for natural infection and whether physical injury or a specific environmental stressor is necessary (Ushijima et al., 2014). Part of the problem in elucidating the infection mechanism is the complexity of the multi-partner coral holobiont and the difference in scale of the partners. Also, traditional cellular pathology methods are only useful at advanced stages of the infection process. Further insights into the pathogenicity of *V. coralliilyticus* in bleaching of *P. damicornis* are emerging from use of a "coral on a chip" system, combining microfluidics, live-imagining microscopy, and NanoSIMS (p.59) to connect macro-scale behavioral responses of the coral with the micro-scale metabolic interactions between the host and its symbiotic partners (*Figure 11.2*). Micro-propagated coral fragments are placed in a microchip with 250 μL chambers with individual inlet and outlet tubes. Continuous collection of exudates from the

coral fragments allowed biochemical and microbial analyses to be performed in real time, alongside video microscopy to follow disease progression. This experimental infection sytem has been used to challenge coral fragments with *V. coralliilyticus* (labeled with red-fluorescent protein for microscopic tracking). Gavish et al. (2018) showed that soon after infection, the bacteria accumulated in the coral pharynx. Over the next few hours, the polyps released large amounts of mucus containing the pathogen. About two-thirds of the infected fragments showed distinct pathology, with tearing of the coenosarc and subsequent loss of colony integrity and necrosis, accompanied by pathogen multiplication and increased MP. If the coral tissue had previous damage, accumulation of *V. coralliilyticus* occurred at the lesion edges resulting in tissue necrosis. The control coral colonies (inoculated with only seawater or non-pathogenic *V. fischeri*) were unaffected. In follow-up experiments, Gibbin et al. (2018) inoculated *P. damicornis* with ^{15}N-labeled *V. coralliilyticus* and used NanoSIMS and TEM to visualize penetration and dispersal of the bacteria and their products. They identified that the mesentery filaments had hotspots of ^{15}N enrichment, suggesting that phagosomal cells in these tissues collect and digest pathogenic bacteria. Gibbin et al. (2019) then used ^{13}C-labeled seawater to determine if the presence of the pathogen affects carbon assimilation in the zooxanthellae and its transfer to the host. Infection led to reduced ^{13}C-assimilation in zooxanthellae and increase in the host. ^{13}C-turnover was reduced in zooxanthellae by *Vibrio* infection at night, but host ^{13}C turnover was not affected. Gibbin and colleagues suggest that the coral "under attack" needs extra energy from this translocated carbon for the mucus spewing defensive response, and that the zooxanthellae might increase their rate of carbon fixation to provide this. The authors emphasize the need for measurement of other parameters such as photosynthesis, respiration, and symbiont density, but these micro-scale techniques offer clear benefits for following the coral's immune, behavioral, and nutritional responses to infection. and determining the "tipping point" between health and disease.

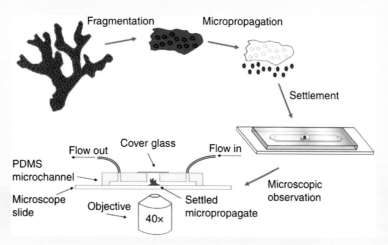

Figure 11.2 Schematic illustration of the "coral on a chip" microfluidics system. Reprinted from Shapiro et al. (2016).

? THE WIND IS IN FROM AFRICA ... DID IT BRING THE SEA FAN PATHOGEN?

One long-held explanation for the sudden emergence of the *Aspergillus sydowii* epizootic in sea fans is that the fungus is of terrestrial origin and is associated with airborne dust transported from North Africa and deposited in the Caribbean. It seems unlikely that this is the only source, as Rypien (2008) failed to find *A. sydowii* in samples of dust and sediments from Africa, the Caribbean, or Cape Verde Islands. Rypien et al. (2008) also used molecular markers to study strains of *A. sydowii* collected from many different regions, hosts, and environments and concluded that a single source introduction of the fungus is unlikely. Rypien and Baker (2009) found evidence from isotopic labeling studies that a reef snail *Cyphoma gibbosum* may act as a vector. The fungus is salt tolerant and can grow in the sea. Interestingly, a massive dust storm in Australia in 2009 deposited huge amounts of spores and hyphae in coastal waters, but no infections were observed (Hallegraeff et al., 2014). *A. sydowii* has also been detected in gorgonians in Ecuadorian Pacific coral, although no disease signs were found (Soler-Hurtado et al., 2016).

As is also observed in microbial mats the distribution in the band of sulfide (produced by SRB) and oxygen (produced by cyanobacterial photosynthesis) varies according to light levels. The SOB migrate vertically with this diurnal variation to locate the optimal $S:O_2$ interface for their metabolism. Field studies of cyanobacterial patches on corals show there is a succession of events; formation of anoxic conditions, proliferation of SRB, and deep anoxia and sulfide-rich conditions produced by degradation of the host tissue. Metagenomic and transcriptomic analysis of the changes in microbial types and the activity of metabolic genes has confirmed this sequence of events, as illustrated in *Figure 11.5*.

White plague and white pox are major diseases affecting Caribbean reefs

White plague disease (WPD) of corals was first described in the 1970s as a disease of large encrusting and branching corals in the Florida Keys. This form (subsequently designated WPD Type I) did not appear to cause major problems, but a new virulent outbreak in 1995 caused high mortalities in many species of coral, including the dominant reef-building *Orbicella* species, and by 2001 outbreaks had spread throughout the Caribbean. The more virulent form was designated WPD Type II. It starts at the base or edge of the coral and progresses up to 2 cm per day, with the destruction of tissue (*Figure 11.6*). Type III of the disease spreads at >2 cm per day. Unlike BBD, there seems to be a sharply defined line between healthy and diseased tissue, suggesting that toxins or enzymes are excreted by the advancing microbial population. The bacterium *Aurantimonas coralicida* was named as the causative agent of WPD-II in the coral *Dichocoenia stokesii*, but the pathogen has not been detected in other WP infected corals and, conversely, it has been detected in healthy but not WP-diseased

Figure 11.4 A colony of symmetrical brain coral, *Diploria strigosa*, affected by black-band disease (BBD), Florida Keys. Credit: Christina Kellogg, USGS.

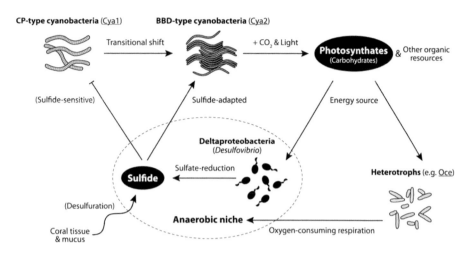

Orbicella annularis corals. In 1999, another form of the disease, designated WPD Type 3, which spreads at >2 cm per day, appeared in the northern Florida Keys. WPD-2 often causes mass mortality after temperature-induced bleaching, with which it can be confused. Electron microscopy and metagenomics studies have suggested a possible role for viruses in WPD, with a group known as small circular ssDNA viruses being consistently associated with diseased *O. annularis*. However, no evidence for a causal relationship has been found. The role of viruses in coral health and disease is further considered below. Further uncertainty arises because a different bacterial pathogen, *Thalassomonas loyana*, was shown to be the causative agent of WP-like disease in the Red Sea, and bacterial etiology was supported by the successful use of phage therapy to control the disease (see *Box 14.1*).

Since the mid-1990s, the disease termed white pox (WPX) has had a devastating effect on the ecology of the shallow-water elkhorn coral, *Acropora palmata*, in the Florida Keys and throughout the Caribbean. It is now considered an endangered species in the USA, with losses approaching 90% in living cover in the Florida Keys; its function as a foundation reef-builder has been lost. Rapid tissue loss resulting in patches of coral skeleton occurs in the summer. Disease signs (*Figure 11.6*) are very different from the tissue loss occurring due to WBD, bleaching, or scars from snail predation. The disease spreads quickly to neighboring colonies, implicating an infectious agent. Using a combination of cultural and molecular techniques, the causative agent of WPX was identified as *Serratia marcescens*. Koch's postulates were fulfilled in experimental infection in 2002 (although a very high inoculum was required, raising some doubts about the ecological significance of the conclusions). Since this bacterium is commonly associated with the human gut, it appeared that the infection originated from sewage pollution and this was proved by demonstrating that a specific human strain from wastewater acted as a coral pathogen. Non-host corals and reef snails can serve as reservoir of the pathogen. Other human enteric bacteria can be isolated from the mucus of corals in the Keys; it is likely that the emergence of WPX is linked to expansion of marinas and coastal dwellings in the region and the widespread use of septic tanks, which release sewage runoff to the sea. However, ongoing monitoring of diseased corals suggests that other pathogens besides *S. marcescens* also contribute to WPX signs.

Protistan parasites may cause tissue necrosis and skeletal erosion

A condition known as Brown Band Disease has been observed in necrotic tissue of *Acropora* spp. and other corals on the GBR. Colonies show a characteristic brown band which is followed by bare skeleton as the band progress across the coral branch. Under the microscope, a mass of ciliates can be seen actively moving over the surface of the coral at more than 1 mm per day, devouring the tissue. Large numbers of ingested zooxanthellae can be seen inside the ciliate cells, giving rise to the brown color. These are thought to remain photosynthetically active and may provide nutrients to the ciliate, or perhaps more importantly provide oxygen that may offset hypoxia that arises from the intense metabolism of the ciliate mass.

ⓘ USING BETTER METHODS TO DIAGNOSE CORAL DISEASES

Some leading coral biologists criticize the experimental basis for claims of specific bacterial etiology, concluding that "most common coral 'diseases' are a result of opportunistic, nonspecific bacteria that exploit the compromised health of the coral after exposure to environmental stressors" (Lesser et al., 2007). However, Work et al. (2008) argue that current knowledge is insufficient either to support or refute this claim. They set out a model for improved investigation, based on standard biomedical and veterinary principles. A decade later, this requirement is still unfulfilled. Analysis of 492 papers investigating coral disease showed that only one used microscopic pathology (Work and Meteyer, 2014). Sweet and Bythell (2017) describe seven steps combining "traditional C19th approaches with C21st technological advancements" that researchers should follow to prove causation. Burge et al. (2016) provide several case studies showing successful combination of traditional and modern diagnostic methods for diseases of corals and other marine organisms.

Figure 11.6 Elkhorn coral (*Acropora palmata*) infected with white pox disease. Credit: NOAA Fisheries, SE Regional Office.

The ciliate is a previously undescribed member of the class Oligohymenophorea, subclass Scutcociliata.

Another ciliate, *Halofolliculina corallasia*, has been associated with a condition called Skeletal Eroding Band—a moving dark band observed on a wide range of branching and massive corals on Indo-Pacific reefs. In the skeleton behind the moving front, the ciliates are packed into an indistinguishable dark mass due to the black housing (loricae) of the ciliate. It kills the tissue and damages the skeleton. Protistan parasites are also frequently found in aquarium corals. A characteristic "brown jelly" is often observed to move across the surface of the coral. A ciliated protist, possibly *Helicostoma nonatum*, has been found in large numbers actively feeding on coral tissue.

Viruses have a pivotal role in coral health

Given the ubiquitous role of viruses in diseases of other animals and plants, it is surprising that their role in diseases of corals and other benthic invertebrates was overlooked for many years, with only a small number of investigators studying the topic until recently. Early studies mostly relied on the use of transmission electron microscopy to demonstrate virus-like particles (VLPs) in corals, some of which are believed to infect the coral animal tissue and some the zooxanthellae. This led to the hypothesis that the zooxanthellae may contain latent viruses that are triggered into a lytic infection by environmental changes associated with bleaching, such as elevated temperature, UV irradiation, or pollution. Other microscopy studies indicate that a diverse range of viruses infect coral tissues and they can also be observed in the surface mucus layer, but identifying viruses based on capsid size and shape is unreliable. More useful evidence comes from recent application of metagenomic and metatranscriptomic techniques, which have revealed about 60 virus families associated with corals, of which 9–12 families form a core coral virome. Tailed phages with a head ~ 60 nm (dsDNA, Caudovirales) in the families Siphoviridae, Podoviridae, and Myoviridae are highly abundant—these infect the bacterial members of the coral holobiont. Nucleocytoplasmic large DNA viruses (especially Phycodnaviridae, Mimiviridae, Poxviridae, Iridoviridae, and Ascoviridae) and herpes-like viruses occur in all or most coral samples investigated – these infect eukaryotic cells, either the coral host or the zooxanthellae. Other common viral groups encountered include ssDNA Circoviridae and ssRNA Retroviridae. Several metagenomic studies comparing the virome of healthy, bleached, and diseased corals have now revealed shifts in the relative abundance of different virus groups. Temperature and nutrient stress can increase the abundance of particular types, such as herpes-like viruses. For example, Circoviridae are most abundant in corals showing signs of WPD and Phycodnaviruses increase in bleached corals, supporting the latent virus hypothesis for bleaching. Small ssRNA viruses related to HcRNAV (p.213) may also infect the zooxanthellae. Further evidence comes from examination of whole transcriptomes of cultures of Symbiodiniaceae species, which reveals upregulation of virus-like gene expression following stress experiments. At present, there is no clear evidence of viral etiology for any specific coral disease, but there is growing evidence for a connection between the activity of viruses and the health of reefs; not just through overt bleaching, tissue loss and disease, but also through modifying the flux of nutrients and biogeochemical processes in the reef water column, through a "viral vortex" of reef decline driven by anthropogenic drivers such as eutrophication and climate change, as shown in *Figure 11.7*.

Sponge disease is a poorly investigated global phenomenon

Sponges are highly important members of tropical, temperate, polar and deep-water marine ecosystems. As discussed in *Chapter 10*, sponges are host to a highly diverse microbial population comprising heterotrophic bacteria, cyanobacteria, archaea, protists, and fungi, the cells of which can comprise almost half of the tissue volume in some species. This makes it especially difficult to determine the involvement of pathogenic microbes in disease. Although sponge diseases have been reported worldwide for more than 100 years, outbreaks have increased dramatically in recent years. Most diseases appear as white or colored spots and patches on the sponge surface, followed by necrosis and damage to the internal structures due to bacterial digestion of internal fibers—the sponge body crumbles as a result of water pressure. One of the best-documented major epizootics occurred in the Caribbean in 1938, resulting in loss of 70–95% of sponges in some parts. The commercial collection of sponges in the Mediterranean and Ligurian Seas suffered massive losses in the 1980s and in 2008–2009 and some aquaculture systems for sponge cultivation have also experienced disease outbreaks. The microbiological investigation of sponge disease remains limited and the etiological agents and environmental factors are poorly identified. A novel strain of *Pseudoalteromonas agarivorans* has been identified as a pathogen of the GBR sponge *Rhopaloeides odorabile*. Disease in deep-sea sponges has also been recorded and there are concerns that this could impact the functioning of benthic systems in Norwegian fjords. As in corals, it seems likely that sponge disease is initiated by imbalance in the holobiont microbiome, followed by opportunistic or polymicrobial infections. Systematic surveys of the health status of sponges in reef ecosystems, accompanied by detailed microbiological investigation, are being conducted to complement our growing knowledge of coral diseases.

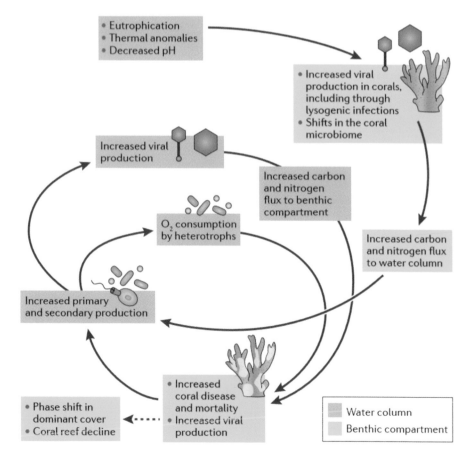

Figure 11.7 Overview of the virus-mediated vortex of coral reef decline. Coral reef viruses are envisaged as having a substantial effect on ecosystem function because nutrients that are released through viral lysis are readily transferred between the reef benthic (BC) and water column (WC) compartments, as if through a "revolving door" Stressful conditions can trigger excessive microbial growth and viral production in coral colonies resulting in the release of coral cellular contents (e.g. mucus exudates) into the WC. Some of these materials move from the BC to the WC in which they can stimulate primary and ultimately secondary production. Increased abundances of heterotrophic bacteria can lead to O_2 drawdown in the WC adjacent to corals, which contributes to hypoxic conditions and colony mortality. Concurrently, viral lysis of autotrophic and heterotrophic microbial blooms in the WC results in additional C and N flux, which can be transferred to the BC and cause further shifts in microbial communities and viral production, and increased coral disease and mortality. The nutrients that are subsequently released then feedback on the ecosystem (for example, to increase primary and secondary production in the WC) and further degrade reefs. Stony coral die-off on reefs can ultimately result in a shift in dominant cover on reefs from scleractinian corals to macroalgae, sponges or soft corals (dashed arrow). Reprinted from Thurber et al. (2017) with permission of Springer Nature.

BOX 11.2   RESEARCH FOCUS

Caribbean corals under threat from a new disease syndrome

Stony Coral Tissue Loss Disease (SCTLD). A major new coral disease is currently spreading throughout the Caribbean, resulting in extremely rapid tissue loss with up to 100% mortality in the most susceptible species affected. The syndrome is characterized by rapid tissue loss that progresses as a band from the base of the colony to the margins or as multiple coalescing blotches (*Figure 11.8A*). Small colonies often die within a few weeks to months, while infections can persist on larger colonies for several seasons. The rapidity of disease progression and the range of coral species affected distinguishes it from other previously characterized syndromes like WPD and WPX. SCTLD was first identified in 2014 at the northern end of the Florida Keys near a site with increased sedimentation due to dredging (Miller et al., 2016; Precht et al., 2016) and by 2019 has spread throughout the entire Florida reef tract, the Mexican Caribbean, Turks and Caicos, Dominican Republic, and Belize. A survey by Walton et al. (2018) found losses of over 30% of coral colony density and 60% of live tissue area in the Southeast Florida reef tract and concluded that ecosystem function in some areas has been altered to a point where recovery is greatly challenged.

Infection has been described in 22 coral species. Typically, the maze coral *Meandrina meandrites*, the star coral *Dichocoenia stokesii*, and the pillar coral *Dendrogyra cylindrus* are the first to show signs of infection at a new site. This is followed quickly by infection of other highly susceptible species, including the brain corals *Colpophyllia natans*, *Diploria labyrinthiformis*, *Pseudodiploria strigose*, and *P. clivosa*, and the smooth flower coral *Eusmilia*

fastigiata. Intensive efforts are underway to collect tissue samples from different coral species and geographical areas for histopathological, microscopic, and molecular analysis. Sentinel sites have also been established to monitor rates of transmission and spatial progression of the disease (Sharp and Maxwell, 2018).

What is causing the disease? Many scientists investigating the disease suspect that waterborne transmission and bacterial infection is responsible but identifying the causative agent(s) has so far been unsuccessful. As with other marine diseases, it is very difficult to determine the role of primary and opportunist pathogens and the effects of disturbance of the normal microbiome (dysbiosis) The situation is made even more complex by the large number of different coral species affected—indeed, we cannot be sure that all species are suffering from the same disease, due to the lack of diagnostic tools. The first detailed analysis of the microbiomes of affected corals was undertaken by Meyer et al. (2019) who sequenced 16S rRNA PCR amplicons from tissue and mucus samples taken from diseased and apparently healthy tissue from multiple samples of the coral species *Montastraea cavernosa*, *Orbicella faveolata*, *Diploria labyrinthiformis*, and *Dichocoenia stokesii*. As in other microbiome studies of coral diseases (Huggett and Apprill, 2019), there was evidence of a stochastic shift in microbial diversity present in diseased and apparently healthy tissue, resulting in unique altered microbiomes. This complexity was compounded by differences observed in samples taken at different times of the year. Meyer et al. were unable to identify any single potential pathogen, but five unique ASVs (see p.282) were enriched in the SCTLD

Figure 11.8 A. Rapid progression of SCTLD across a colony of brain coral, *Pseudodiploria strigosa*. B. Removal of an unaffected star coral, *Dichocoenia stokesii*, for the Reef Rescue project. C. Long-term maintenance of corals at Florida Aquarium's Center for Conservation. D. Experimental "firebreak" procedure to prevent disease progression in *Orbicella faveolata* (see text). Credits: A–C. Florida Fish and Wildlife Conservation Commission; D. Brian Walker, Nova Southeastern University.

lesions of three of the corals. These were classified as belonging to the orders Flavobacteriales, Clostridiales, Rhodobacterales, Alteromonadales, and Vibrionales. Some of these ASVs were exact matches for sequences isolated from other disease syndromes like WPD and BBD. Other groups of opportunistic colonizers were also enriched in diseased corals—these are presumed to be saprophytic, feeding on the dead tissue Although this study is unable to reveal specific pathogens responsible for the disease, it lays the foundation for further studies, including isolation and culture of pathogens and better investigation of control methods such as antimicrobial treatments, probiotics, or phage therapy (see p.404). Some preliminary experimental treatments in mesocosms and in the field have involved application of antibiotic pastes, with some success in halting the progress of disease lesions (FKNMS, 2018). Another approach has been the creation of a trench chiseled into the coral

skeleton, which is filled with chlorinated epoxy resin to halt disease progression (analogous to a firewall) as illustrated in *Figure 11.8D*. The possibility that viruses or protistan parasites are also involved cannot yet be excluded.

The Florida Reef Rescue Project. The growing severity of the epizootic has prompted a multi-agency initiative with the aim of collecting thousands of specimens of unaffected corals of different species ahead of the disease front (*Figure 11.8B, 11.8C*). These are shipped to land-based holding facilities at the University of Miami and Nova Southeastern University, where samples are taken for genotyping to aid individual tracking of each coral. Samples are being distributed to over a dozen aquaria around the USA to serve as long-term breeding stock for possible future reef restoration (FFWCC, 2019).

Mass mortalities of echinoderms have caused major shifts in reef and coastal ecology

The long-spined black sea urchin *Diadema antillarum* occurs throughout the Caribbean and in tropical regions of the Atlantic Ocean. It is omnivorous, grazing on live coral, seagrasses, and its preferred food source, benthic turf algae. In 1983, a mass mortality event began near Panama, and quickly spread east and west at up to 3000 km per year, eventually affecting 3.5 million km² of the entire Caribbean. Mortality was over 90% in all locations and the population was completely wiped out in many areas. The deposition of dust from North Africa, following severe drought, has been suspected as a possible trigger for the outbreak (see p.298). Recovery has been extremely slow, and only small populations at greatly reduced densities survive. Although a host-specific waterborne infectious agent was suspected, the causative agent was never discovered. The urchin played a key role in maintaining the balance of the reefs by grazing on turf algae and massive increases in algal cover occurred within weeks of the epizootic. Freed from this top-down control, the Caribbean reefs underwent a major ecological phase shift from a coral-dominated to an algal-dominated system from which they have never recovered—with the situation compounded by subsequent coral diseases discussed above, pollution, and overfishing.

In the summer of 2013, diseased asteroids (sea stars) began washing up onto the shores of the Olympic Peninsula and British Columbia on the Pacific north-west coast. The animals showed loss of body turgor, missing limbs, and tissue decay, quickly leading to a disintegrated mush littering the seabed (*Figure 11.9*). The condition became known as "sea star wasting disease" (SSWD) and intensive efforts were mobilized to elucidate the cause and investigate the ecological effects of the epizootic. Over 20 species of asteroids have been affected, with the ochre star (*Pisaster ochraeus*) and sunflower star *(Pycnopodia heliamthoides)* being particularly hard hit. Within a year, surveys showed that populations of *P. ochraeus* were drastically reduced along the entire Pacific coast from Alaska to southern California, with declines of over 75% in all areas and up to 99% in southern California. Sustained high seawater temperatures in 2014 and 2015 are thought to have exacerbated the disease's impact. The outlook for recovery of sea star populations is uncertain. *P. ochraeus* is a keystone predator and it is likely that its reduction or disappearance in some regions will result in major ecological changes to subtidal and intertidal systems.

The causative agent(s) of SSWD are still unclear, despite many attempts to identify viruses, bacteria, or fungi that might be responsible. The best lead has come from the finding that 0.22 μm-filtered extracts of infected tissue reproduced disease signs in experimental infection. A ssDNA virus in the Parvoviridae family (sea star associated densovirus, SSaDV) was identified and associated with the disease in some species. However, the virus has been found

Figure 11.9 The purple ochre sea star *Pisaster ochraeus* disintegrating as it dies from sea star wasting syndrome. Credit: Elizabeth Cerny-Chipman, Oregon State University via Wikimedia Commons, CC BY 2.0.

to be present in some asymptomatic asteroids and in specimens preserved long before SSWD was observed. Definitive proof of etiology—as we have seen with so many other marine diseases—is inconclusive. Indeed, detailed analysis shows densoviruses are only consistently associated at a population level with one species, *P. helanthioides*, and disease signs differ subtly between different species. This, the term SSWD may wrongly imply common etiology of a single condition, and there may be a complex repertoire of interactions between environmental conditions and different pathogenic agents.

DISEASES OF MOLLUSCS

Bacteria are a major cause of disease in molluscs

Numerous *Vibrio* spp. have been described as part of the normal microbiota and as disease agents in bivalve and gastropod molluscs. Vibrios are the major cause of disease in intensive hatcheries, where mortalities of larvae and juveniles can reach 100%. Until the introduction of genomic methods, classification and identification of this group has been very confused, but several species have been particularly implicated, namely *V. alginolyticus*, *V. anguillarum*, *V. splendidus*, *V. aesturianus*, *V. neptunius*, *V. tubiashii*, and *V. coralliilyticus*. These organisms are all found as members of the normal microbiota of seawater and in biofilms on marine surfaces. Growth is encouraged by accumulation of organic matter, so careful monitoring of water quality and temperature is essential to prevent disease outbreaks in hatcheries. Larvae of different bivalve species seem quite variable in their sensitivity to vibrios and there are also marked differences in the virulence of bacterial isolates. Extracellular toxins (hemolysins, proteases, and ciliostatic factors) have been identified.

Disease resulting from vibrio infections is less of a problem in adult molluscs, but important exceptions in mariculture include regular epizootics in cultured Portuguese oyster (*Crassostrea gigas*) caused by *V. splendidus* and related species, and high mortalities of the European abalone (*Haliotis tuberculata*) caused by *V. harveyi*, which also infects cultured pearl oysters (*Pinctada maxima*) in north-western Australia. *V. tapetis* causes brown ring disease in cultured Manila clams (*Ruditapes philippinarum*); the pathogen attaches to the clam tissue, causing abnormal thickening and a characteristic brown ring along the edge of the shell. Mass mortalities due to *V. tapetis* have caused severe economic losses on the entire Atlantic coast of Europe since the 1980s. The persistence of the bacteria in molluscan tissues depends partly on their resistance to the bactericidal activity of the hemolymph, a natural defense mechanism. The effect of environmental factors (especially temperature) and stress on the neuroendocrine systems of the host are very important in determining whether these interactions are benign or lead to disease.

Juvenile oyster disease (JOD) results in seasonal mortalities of hatchery-produced juvenile oysters (*Crassostrea virginica*) raised in the north-eastern United States. The disease first appeared in the late 1980s, and losses exceeding 90% of total production have since occurred in the states of Maine, New York, and Massachusetts. There are several similarities to brown ring disease, indicating a bacterial etiology. Growth rate is reduced, and the shell becomes fragile and uneven, with proteinaceous deposits (conchiolin) on the inner shell surfaces, followed by sudden high mortality (*Figure 11.10a*). Infected animals are heavily colonized by a previously undescribed species of the Roseobacter group of the Alphaproteobacteria, named as *Roseovarius crassostreae*. Attachment of the bacteria to the oyster tissue via polar fimbriae is thought to be a key factor in pathogenesis (*Figure 11.10b*). The disease is now known as *Roseovarius* oyster disease (ROD) to emphasize its specific etiology.

Gliding bacteria of the *Cytophaga–Flavobacterium–Bacteroides* (CFB) group can infect the hinge-ligament of the bivalve shell, leading to liquefaction via the production of extracellular enzymes, and interference with respiration and feeding. CFB appear to be weak opportunist pathogens, and infection is precipitated by poor nutrition of the animals, rising water temperatures, or other environmental stresses. *Nocardia crassostreae* causes high mortalities of cultivated *C. gigas* and *Ostrea edulis* oysters in Japan and on the west coast of Canada and the United States.

Intracellular rickettsia-like and chlamydia-like pathogens are widespread and have been reported in at least 25 species of bivalves. These infections frequently produce little evidence of tissue damage, and mortality in adult animals is usually low, except in conditions of environmental stress such as sudden temperature change. The larvae are usually very susceptible, leading to problems in aquaculture hatcheries of scallops (*Argopecten irradians*) and other bivalves. The morphology of the bacteria within the cells of the digestive gland and gills is very similar to that of known rickettsias and chlamydia, but detailed identification and taxonomic studies are limited because these are obligate intracellular pathogens that can only be grown in suitable cell cultures. *Francisella halioticida* has been recognized as the cause of mass mortalities in some cases. An uncultured rickettsia-like organism, "*Ca.* Xenohaliotis californiensis," has been identified as the cause of a lethal disease (withering syndrome) of the black abalone *Haliotis cracherodii* at high seawater temperatures. This large edible gastropod was once highly abundant on the west coast of North America and was an important food source harvested for ~10,000 years. Withering syndrome was first reported on the Californian coast in the 1990s and spread rapidly. Stocks were already heavily depleted by overfishing, and the mollusc is now critically endangered.

Several protistan diseases affect culture of oysters and mussels

The most widespread and destructive protistan disease, described since the 1970s, is bonamiasis caused by *Bonamia ostrae*, *B. exitiosa*, *B. perspora*, and *B. roughleyi*, affecting a wide range of oyster species along the entire Atlantic coast of Europe, Canada, USA, New Zealand, and SE Australia. The disease is characterized by yellowing of tissue, lesions on gill and mantle, and breakdown of connective tissue, with mortalities up to 90%. Aber disease in *Ostrea edulis* and *Mytilus galloprovincialis* culture on the European Atlantic and Mediterranean coasts is caused by *Marteilia refringens*, leading to a pale digestive gland, emaciation, and tissue necrosis. A closely related species, *M. sydneyii* has recently emerged as a seasonal infection of rock oysters farmed on the east coast of Australia; QX disease leads to necrosis of the digestive glands, loss of condition, and reabsorption of the gonad. Starvation

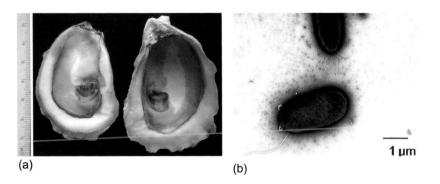

(a)   (b)

FULFILLING KOCH'S POSTULATES FOR AN OYSTER DISEASE

The "gold standard" evidence for proving disease etiology is fulfilment of Koch's Postulates (KP). The following conditions must be met: (1) the microbe suspected of causing the specific disease must be associated with all cases; (2) it must be grown in pure culture; (3) the disease must be reproduced by inoculating healthy individuals with the pure culture; and (4) the specific microbe must be re-isolated from experimentally infected hosts. The investigation of juvenile oyster disease (JOD) by Maloy et al. (2007) is a rare example of a rigorous, model approach to prove KP for a marine disease. Maloy et al. showed that *Roseovarius crassostreae* is consistently associated with diseased animals (>90% of the total bacterial community) but is never found in healthy oysters. They developed a laboratory model to reproduce the disease; exposure to *R. crassostreae* caused JOD-like mortality including the characteristic pathology. The timing of bacterial colonization paralleled the development of disease signs in a natural setting. The bacteria were detected simultaneously to the first microscopic disease signs in actively growing oysters, providing strong evidence that they are the primary pathogen.

Figure 11.10 *Roseovarius* oyster disease (ROD). (a) Shell of infected oyster showing deposition of conchiolin on the inner surface of both valves. (b) TEM of negatively stained cell of *Roseovarius crassostreae* showing one bacterium with a flagellum and a partical cell with a tuft of polar fimbriae. Credit: Katherine Boettcher, University of Maine.

1 µm

THE NUCLEAR OPTION—BACTERIAL PARASITES OF DEEP-SEA MUSSELS

Although symbiotic interactions between bacteria and molluscs found at vents and seeps have been studied intensively (see *Chapter 10*), little is known about parasitic interactions. Virus-like and rickettsia-like inclusions have been observed in TEM of *Bathymodiolus* mussels and may be linked to mortality (Ward et al., 2004; Mills et al., 2005). Using 16S rRNA analysis, Zielinski et al. (2009) described a parasitic bacterium, which they named "*Ca.* Endonucleobacter bathymodioli." FISH probes showed that the bacterium has a very atypical lifestyle, invading the nuclei of intercalary gill cells between the bacteriocytes that contain the chemosynthetic symbionts. Infection of a nucleus begins with a single rod-shaped bacterium, which grows to an aseptate filament of up to 20 μm in length, and then divides repeatedly until the nucleus becomes greatly swollen with up to 8×10^4 bacteria. The nuclear membrane then bursts, destroying the host cell. The bacteriocytes never become infected, suggesting that the symbiotic bacteria in these cells protect their host nuclei against the parasite.

and death of the oyster follow when energy reserves are depleted. A polychaete worm has been identified as an intermediate host. Dermo disease in the eastern oyster *Crassostrea virginica* is caused by *Perkinsus marinus*. This infects the hemocytes of oysters, causing severe emaciation, loss of condition, and high mortality rates, depending on temperature and salinity. It is naturally prevalent in the Gulf of Mexico, south-eastern United States, and Brazil. Culture of *C. virginica* on the Pacific coast of Mexico and California has led to introduction of the disease there. Waterborne spores of the parasite are transmitted over large distances.

Virus infections are a major problem in oyster culture

France is one of the leading producers of cultivated oysters and has been particularly susceptible to disease caused by viruses, as well as the vibrio infections previously described. A major epizootic of gill necrosis erupted in France in the 1960s in Portuguese oysters (*Crassostrea angulata*), with tissues displaying the presence of large inclusion bodies containing icosahedral iridovirus particles in the tissues. This disease spread rapidly along the Atlantic coast of Europe, virtually destroying the European oyster fishery. Pacific oysters (*Crassostrea gigas*) were introduced to replace the lost stock, but these also suffered high mortalities. A serious outbreak of another viral disease killed over eight billion oysters in France in 2008. This was shown to be due to the ostreid herpesvirus (OsHV1), which had previously been associated with epizootics in *C. gigas* in Japan, the USA, and Australia since the development of high-density culture methods in the 1940s. It is thought that the 2008 outbreak followed a mild, wet winter with high nutrient levels, which encouraged the young oysters to invest more energy in the development of sex organs, weakening their natural defense mechanisms against viral attack. Highly virulent variants of OsHV-1 have now spread globally and are causing major losses on the US west coast and in Australia, especially Tasmania. Susceptibility to the disease has a genetic basis and it may be possible to breed more resistant varieties of *C. gigas*.

Another group of viruses known as the marine birnaviruses (MABVs) are widely distributed as disease agents in marine invertebrates (as well as in fish; see below). In Japan, infection by MABVs causes considerable losses in the culture of pearl oysters (*Pinctada fucata*). PCR screening shows that 60% of bivalves and 35% of gastropods from a range of wild species in Japanese waters are positive for MABVs and that zooplankton may act as a vector.

Papova-like viruses have been described in the gonads of several bivalve species, especially oysters, leading to hypertrophy of gametocytes. Viruses commonly cause secondary infections and may be associated with infection by parasites. Many other VLPs have been observed in marine bivalves and are often associated with disease signs, although proof of etiology and study of pathogenesis is usually lacking, except in cultured species. In most cases, identification of VLPs in electron micrographs has been on the basis of size, morphology, and intracellular location; molecular characterization and propagation are needed.

Changes in temperature or salinity, human handling, and increasing stocking density are particularly important factors contributing to susceptibility to bacterial, protistan, and viral diseases of molluscs. There is currently considerable research on the immune defense mechanisms of molluscs in the hope that genetic improvement of disease resistance will lead to better disease control.

DISEASES OF CRUSTACEANS

Bacteria cause epizootics with high mortalities in crustaceans

Various *Vibrio* species are again the major causes of devastating losses in hatcheries and grow-out stages of tropical shrimp and prawn culture, the most important of these being *V. harveyi*. Because of its bioluminescence, a large *V. harveyi* outbreak in prawns can result in a spectacular greenish light in infected ponds, leading to the name "luminous vibriosis." In hatcheries, vibrios attach to the feeding appendages and oral cavity of the larvae, which become weakened and swim erratically. The virulence of different strains of *V. harveyi* varies greatly. Some have a minimum lethal dose (100% mortality) of only 100 CFU mL^{-1}, whereas other strains are nonvirulent even at 10^6 CFU mL^{-1}. Virulent strains produce two lethal protein exotoxins, which probably act in the larval intestinal tract on the gut epithelial cells by facilitating passage across the gut and colonization of other tissues. The closely related *V. nigripulchritudo* and *V. penaeicida* are major disease agents of prawn and shrimp culture in New Caledonia, Japan, and southwest India and *V. owensii* has been identified associated with diseased cultured crustaceans in Australia. There is some evidence to suggest that acquisition of a plasmid may be responsible for the emergence of highly virulent *V nigripulchritudo*, which produce an exceptionally toxic protein, nigritoxin. This is lethal for crustaceans and also insects and may find application as a novel insecticide.

In lobsters and crabs, the exoskeleton is covered by a thin protective lipoprotein layer, and damage to this layer enables various bacteria (especially vibrios) to attach to the chitin exoskeleton. It is possible that biofilm growth of lipolytic bacteria may facilitate penetration resulting in "bacterial shell disease," in which pits in the exoskeleton are produced by bacterial chitinase activity. These can erode to form deep lesions and the bacteria may penetrate the tissues. The disease has a major impact on some populations of the American lobster (*Homarus americanus*), with consequent serious effects on the lobster fishery of the north-eastern coast of the United States and Canada. Another very damaging pathogen of lobsters is gaffkemia, caused by the Gram-positive *Aerococcus viridans* var. *homari* (previously known as *Gaffkya homari*). The disease is usually found in holding tanks or ponds and is of considerable concern to lobster fishers because of the high value of their catch. It is highly contagious and is probably acquired from a reservoir in wild animals, which show an infection rate of 5–10%. Schemes to restock depleted fisheries have been badly affected by this disease, and there is considerable evidence that disease spread is linked to the movement of animals caught from infected wild stocks and transported over large distances. The bacterium gains entry to the lobster hemolymph via abrasions in the shell and multiplies very rapidly at higher water temperatures (>10°C). This explains differences in severity of the disease observed during summer and winter impoundment. Virulence seems to be strongly connected with a capsule that prevents agglutination in the hemolymph. Antibiotics such as oxytetracycline are sometimes used as a preventive measure in holding ponds, but there is concern about antibiotic residues in treated lobsters; a withdrawal period of at least 30 days should be observed.

The intracellular rickettsias and mycoplasmas can cause epizootics with high mortalities in crustaceans such as crabs, lobsters, and penaeid prawns, owing usually to infection of the hepatopancreas. Such epizootics can have a marked effect on marine ecology, for example, the outbreaks caused by currently unidentified rickettsias affecting Dungeness crabs (*Cancer* or *Metacarcinus* spp.) in Europe and the USA. Diagnosis of disease is difficult, being based largely on histopathology, but the recent introduction of molecular diagnostic techniques is leading to improved understanding of their ecology.

Expansion of crustacean aquaculture is threatened by viral diseases

A number of crustacean viral diseases are important in both wild and cultured populations of decapod crustaceans such as lobsters, crabs, prawns, and shrimp. Intensive and semi-intensive aquaculture of prawn and shrimp now accounts for about half of the total value of aquaculture production (>USD 250 billion). The global demand for prawns and shrimps seems insatiable, and much of the rapid expansion of intensive culture development has occurred in Asia (especially China, Thailand, Vietnam, and Indonesia) and Central America (especially Ecuador). Initial lack of government regulation led to severe problems of habitat destruction and eutrophication of coastal waters. With a better understanding of these problems and improved regulation of location and operation of prawn and shrimp farms, infectious diseases are now considered one of the major limiting factors of further development of the industry, alongside understanding the nutritional needs of animals and improving sustainable supply of feeds. For most crustacean species, culture still depends on the use of larvae produced in hatcheries, mainly from sexually mature females caught in the wild. Thus, genetic selection of disease-resistant animals is not possible, although it is likely that closure of the life cycle for some species will be achieved soon. This will decrease the pressure on natural stocks and enable selection of disease-resistant brood stock. Some crustacean species are cannibalistic or are indiscriminate feeders that will consume feces and molted exoskeletons, leading to rapid disease transmission through ponds and holding tanks. Until recently, knowledge of the ecology of pathogens and application of diagnostic methods has been poor. The development of gene probes and Q-PCR techniques is assisting in control methods such as selection of larvae and restrictions on movement of stocks unless certified as disease-free. Viruses also remain infectious in frozen seafood products, and molecular methods are used for screening imports and exports to prevent disease transmission. However, many of these techniques are expensive and require highly trained personnel, and it is necessary to develop cheaper and easier biosecurity methods for use by rural farmers in developing countries.

The expansion of intensive culture of shrimps and prawns, in both volume and geographic distribution, has resulted in the number of well-characterized viruses increasing from six in 1988 to nearly twenty today. One of the most devastating diseases in Asia has been white spot syndrome of penaeid prawns. This was first recognized in eastern Asia and has spread world-wide, with current losses estimated at over USD 1 billion. High mortalities (80–100%) result within 2–3 days for juveniles and 7–10 days for adults. The agent (WSSV) is a large, enveloped, rod-shaped, double-stranded DNA virus, recently named as *Whispovirus* and assigned as the sole member of the group Nimaviridae. Analysis of the genome sequence of the virus indicates that it differs from all known viruses, although some genes are weakly homologous to herpesvirus genes.

Whispovirus has a broad host range among crustaceans and infects many tissues, multiplying in the nucleus of the target cell. It is not known how the virus enters the shrimp and spreads within the body, but recent studies suggest that a viral envelope protein may interact with a host chitin-binding protein. There have been many studies on the immune responses of the host to infection. *Whispovirus* is a lytic virus and causes destruction of host tissues. Analysis of crustacean genomes is revealing the evolutionary history of the virus, showing that extensive acquisition of host–virus interaction-related genes was a key event in the origin of the extreme virulence of *Whispovirus*. Disease transmission has been controlled in some regions by use of integrated ELISA- and PCR-based screening of larvae to ensure that they are free of the disease (specific pathogen-free, SPF) before transfer to ponds; however, there is considerable sequence variation, which may limit the effectiveness of screening techniques. Animals acquire the infection from the water and show white spots on the inner surface of the shell (*Figure 11.11*). Interestingly, the white spots appear to be caused by a chitinase, and this enzyme may have been transferred to the viral genome by phage conversion.

Crustaceans may also be infected by parvoviruses, including infectious hypodermal and hematopoietic necrosis virus (IHHNV), hepatopancreatic parvovirus (HPV), and spawner mortality virus (SMV). SMV emerged in the 1990s in Queensland, Australia, and has been responsible for mortalities of 25–50% in black tiger prawns (*Penaeus monodon*). The virus originates in wild brood stock, as 25% of female spawners carry SMV. Again, integrated PCR and ELISA tests are being developed for high-throughput screening in an attempt to produce SPF stocks. Sequence analysis shows that there is considerable homology between SMV and some insect viruses. The four shrimp parvoviruses fall into two different clades that group with different insect parvoviruses.

Figure 11.11 Head of penaeid prawn infected with *Whispovirus*, showing characteristic white spots on the exoskeleton. Credit: Donald Lightner, University of Arizona.

Taura syndrome virus (TSV) is a small, non-enveloped, icosahedral, ssRNA virus belonging to the family Dicistroviridae; it has caused heavy mortalities in central America and the USA since the 1990s and several variants are known. Other viruses causing significant diseases in shrimp and other crustaceans include baculovirus penaei virus (BPV), monodon baculovirus (MBV), and midgut gland necrosis virus (MGNV). Animals acquire these viral infections from water, although it has been suggested that seabirds could act as a vector of some shrimp viruses by transmitting infectious virions in their feces. Larvae of aquatic insects may also be reservoirs of infection.

Despite the importance of wild crustaceans such as crabs and lobsters in natural populations and fisheries, knowledge of the importance of viruses in these species is very limited. Baculoviruses, a parvo-like virus, and reoviruses have been reported in the hepatopancreas of *Carcinus* spp., and early reports of viruses resembling herpesviruses, picornaviruses, rhabdoviruses, and bunyaviruses in other crabs have also largely not been pursued since their original descriptions in the 1970s. Application of microscopic and molecular methods is needed to demonstrate the importance of viruses in the ecology of wild decapod crustaceans.

Parasitic dinoflagellates also cause disease in crustaceans

Dinoflagellates in the genus *Hematodinium* and related genera can be very significant pathogens of commercially important crustaceans in many parts of the world, with devastating effects on fisheries for crabs and lobsters in North America and langoustines in Europe. Infection causes pathological alterations to the organs, tissues, and hemolymph of the animal. Juveniles and females often show high prevalence of infection, resulting in sudden crashes in populations. At lower levels of infection, biochemical changes in the tissues result in the production of bitter-tasting compounds, making the crabmeat unmarketable.

DISEASES OF FISH

Microbial diseases of fish cause major losses in aquaculture and natural populations

The first scientific description of microbial diseases in fish can be traced to the description of "red pest" in eels in Italy by Canestrini in 1893. This disease, which we now know as vibriosis caused by the bacterium *V. anguillarum*, led to mass mortalities in migrating eels during the eighteenth and nineteenth centuries. Such large-scale fish kills in the wild occur occasionally, particularly in estuaries and on coral reefs, and may be caused by a wide range of bacterial, viral, or protistan infections, or by harmful algal bloom (HAB) toxins. Viral hemorrhagic septicemia (VHS) virus (genus *Novirhabdovirus*) is thought to be responsible for large periodic fluctuations in the populations of wild shoaling fish in the North Sea, and major epizootics in Australian pilchards have occurred because of a herpesvirus. Apart from these mass mortality effects, it is hard to estimate the normal impact of disease on fish populations in the wild, since sick fish do not last long in the natural environment and are quickly removed by predators. One of the few pieces of evidence that infection plays a significant role in controlling natural fish populations comes from observations made in the 1970s that hatchery-reared salmon immunized against vibriosis before transfer to the ocean had a 50% greater survival than nonimmunized fish, as shown by the return rate of tagged fish. Fish such as salmon and eels may be particularly susceptible to acute infections owing to the pronounced physiological changes that occur during their migration from fresh- to saltwater, or *vice versa*. Some pathogens are found in a high proportion of wild fish; these tend to be those that cause slow-developing chronic diseases and may affect populations by impairing growth and reproduction rather than sudden acute mortality.

The development of intensive mariculture in the 1970s led to a rapid growth in the science of fish pathology. Early attempts to farm salmonid fish in intensive offshore pens and cages were frustrated by large-scale mortality and heavy economic losses. This experience has recurred with many different fish species in all parts of the world and at times has threatened the survival of the industry in some countries. The impact of disease should come as no surprise, since it is common in all forms of intensive culture in which single species of animals are

DO VIRUSES INFECT CRUSTACEAN ZOOPLANKTON?

As the most abundant members of the zooplankton, the small crustaceans known as copepods have great significance in global marine ecology and nutrient cycling. They are the main food source for larger zooplankton, many fish, whales, and seabirds but there is also evidence of non-predatory mortality (Elliott and Tang, 2011). Could their population density be controlled by viral infection, as occurs with many phytoplankton (see *Chapter 7*)? Drake and Dobbs (2005) investigated this in a model system by exposing *Arcartia tonsa* copepods to natural concentrations of viruses in seawater. They concluded that exposure to viruses had no effect on copepod fecundity, or survival of larvae or adults. The authors suggested that the effect of viruses on higher trophic levels may be much lower than that on smaller hosts such as protists and that the effects of viral infection in crustaceans may be below the limits of detection (although they did not determine whether infection actually occurred). Further studies are needed, as Dunlap et al. (2013) found molecular and microscopic evidence of abundant and actively replicating novel ssDNA circo-like viruses in the tissues of two copepod species.

WHAT ARE EPIZOOTICS?

The term "epizootic" is used to describe a situation where cases of disease within an animal population rise sharply or suddenly above the normal levels, which are referred to as "enzootic." It is a subjective term and is not quantitative. A small number of cases of an unexpected rare disease affecting a small number of fish farms might be described as an epizootic, while the term would also be used to describe cyclical increases in disease causing mass mortalities across a large geographic region. These terms are the equivalent of the familiar terms "epidemic" and "endemic" used when describing human diseases. The origin and development of epizootics depends on particular circumstances affecting the host and the pathogen, with host population density and environmental factors often playing a critical role. The ecological impact of epizootics depends on the general status of the population before the disease outbreak and the resilience of the ecosystem in recovering from mortalities.

reared at high-population densities. Economic factors demand highly intensive systems, and this can lead to stress of the cultured fish, which then succumb to disease transmitted rapidly through the dense populations.

Today, marine fish culture is a worldwide industry, with production approaching that of extensive fish capture. Atlantic salmon (*Salmo salar*) in Norway, Chile, Scotland, Canada, and Ireland is the most economically important species, with fish such as gilthead sea bream (*Sparus aurata*) and sea bass (*Dicentrarchus labrax*) dominating culture in the Mediterranean Sea. Japan has a long history of intensive mariculture, dominated by yellowtail (*Seriola quinqueradiata*), ayu (*Plecoglossus altivelis*), flounder *(Paralichthys olivaceus)*, and sea bream *(Pagrus major)*. A wide range of species are cultured in China and southeast Asia, often in small-scale operations for local consumption. In recent years, there has been expansion in the farming or ranching of high-value fish such as: Atlantic cod (*Gadus morhua*), mainly in Norway; turbot (*Scophthalamus maximus*), mainly in Spain and Portugal; or tuna *(Thunnus thynnus, T. orientalis,* and *T. maccoyi)*, in the Mediterranean, Japan, and Australia. While there has been criticism of the sustainability and environmental impact of these ventures—including disease problems—the benefits of closed life-cycle aquaculture of such commercially important fish species relieves the pressure on natural populations, whose stocks are in rapid decline. In view of the need to feed the world's growing population and recognition of the environmental impacts of land-based agriculture (especially meat production), expansion of open-ocean aquaculture is likely in the coming years, but careful regulation of environmental impacts will be essential.

Microbial infections of fish cause a variety of disease signs

The need to implement effective control measures following a disease outbreak puts pressure on investigators to determine the causes of mortality quickly. Experience, careful observation of the stock by the fish farmer, and good record keeping play a large part in identifying diseases. Microbial infection is usually characterized by a rising level of mortality accompanied by characteristic disease signs. These are very varied but may be broken down into three broad categories: (1) systemic bacteremia or viremia, in which there is rapid growth of the pathogen, often with few external disease signs other than hemorrhaging; (2) skin, muscle, and gill lesions; and (3) chronic proliferative lesions. Some highly lethal systemic infections such as those caused by *Aeromonas, Piscirickettsia*, and infectious salmon anemia virus (ISAV) may cause high mortality rates even when no external signs are present. In contrast, other agents causing diseases with relatively low mortality rates—such as *Tenacibaculum maritimum, Moritella viscosa*, and *Lymphocystis disease virus*—can result in external lesions such as ulcers, necrosis, and tumor-like growths that make fish unmarketable.

Postmortem changes are very rapid as a result of overgrowth by the normal microbiota of fish, so it is important to examine fish showing signs of infection before they succumb completely. External examination will often reveal the presence of gill and tissue erosion, eye damage, hemorrhages, abscesses, ulcers, or a distended abdomen. Internal inspection may reveal organ damage and fluids in the body cavity. To the experienced eye, these signs will often indicate a particular disease agent, but the diagnosis must be confirmed by identification of the pathogen.

For bacteria, traditional culture methods are still widely used, involving plating tissue samples onto various selective media and performing biochemical tests using diagnostic keys. Not all bacterial pathogens are amenable to this approach, as some grow very slowly in culture (notably *Renibacterium* and some mycobacteria). An alternative approach is the use of methods such as ELISA or fluorescent antibody techniques (FAT) to detect bacterial antigens in the blood or tissues, or to detect a high titer of host antibodies against the pathogen. The diagnosis of viral infections is more difficult and time-consuming because it relies on propagation of the virus in a suitable cell culture. FAT and ELISA are therefore the main methods used for rapid identification of viral infection. There has been some success with "dipstick"-type kits for rapid diagnosis, based on modifications of the ELISA technique. Monoclonal antibodies have been produced against specific pathogens, allowing rapid identification of disease; they can also be used in fish health programs for screening brood stock for previous exposure to pathogens. For many bacterial and viral pathogens, there are now accurate molecular diagnostic tests based on PCR amplification and gene probes. Genomic fingerprinting methods are often used for strain characterization, which is especially important in studies of epizootics.

Fish-pathogenic bacteria possess a range of virulence mechanisms

All bacterial infections involve a number of stages, with different bacterial products important as virulence factors. The possession and expression of genes encoding these factors varies greatly among the broad range of bacteria in various taxonomic groups that have been associated with disease in marine fish. The initial infection and colonization of the host often depends on the pathogen sensing the presence of exudates and excretion products from the host. Chemotaxis using flagellar or gliding motility may result in association with specific surfaces, such as the gut, skin, or gills. Bacteria often produce pili or some other surface structures that enable firm attachment to epithelial surfaces.

The most critical stage for bacteria during colonization is the ability to overcome the innate host defense mechanisms and adaptive immune system. Fish mucus provides a protective inhibitory function and infection of undamaged skin is rare. Bacteria that penetrate the tissues are subject to the antibacterial collectins and complement proteins found in serum, and resistance to these is often due to modifications of the cell envelope. Virulent bacteria may also possess mechanisms to resist phagocytic cells of the host, usually by possession of hydrophobic surface layers or by producing enzymes that cause lysis of phagocytic cells. Some bacteria can survive within phagocytic cells of the host; these possess a range of mechanisms to overcome the hostile low pH and antibacterial compounds found in the intracellular environment.

As noted on p.237, bacteria require iron for the activity of essential cellular functions. Vertebrate animals possess highly efficient systems for the transport and sequestration of iron. The serum protein transferrin binds iron with extreme avidity, reducing the concentration of free iron in tissues to 10^{-18} M, about 10^8 times lower than the concentration that bacteria require for growth. During infection, an even more efficient iron-binding protein (lactoferrin) is released and sequestered iron is removed to storage in the liver and other organs. Thus, successful pathogens must compete with the host's iron-binding system to obtain sufficient iron for growth, and many pathogens achieve this via the production of siderophores and an iron-uptake system.

The final stage in disease is the damage to host tissues and body systems. Often this occurs as a result of the host response "over-reacting" to the presence of the pathogen, producing damaging cytokines and inflammatory products. Many pathogens produce toxins that can cause death by affecting major organs or extracellular enzymes that cause destruction of cells or tissues.

Vibrios are responsible for some of the main infections of marine fish

As seen in the sections on invertebrate diseases, members of the family Vibrionaceae are again prominent as pathogens. As noted above, *V. anguillarum* (also classified as *Listonella anguillarum*) was the first species to be identified as a fish pathogen causing vibriosis in eels (from which it derives its specific name), producing ulcers, external and internal hemorrhages, and anemia. This "classical" vibriosis causes heavy losses in eel culture, especially in Japan, and in a very wide range of marine species; it has caused particular problems in the culture of salmon in North America, sea bass and sea bream in the Mediterranean (*Figure 11.12*), and yellowtail and ayu in Japan. In all species, the disease is characterized by a very rapid generalized septicemia usually accompanied by external hemorrhages. A number of serotypes of the pathogen exist, distinguished by differences in the O-antigen (lipopolysaccharide of the outer membrane), but only the O1 and O2 serotypes are widespread. There are some genetic differences between strains isolated from different geographic regions and fish species; such knowledge is important in the development of vaccines for particular applications

(see *Chapter 14*). Vibriosis was one of the first fish diseases for which vaccines were developed and they have generally been very effective.

One of the major factors in virulence, which explains the extremely rapid growth *in vivo*, is the possession of a 65-Mb plasmid (pJM1) that confers ability to acquire iron in the tissues of the host. Proof of the key role of the plasmid in virulence of *V. anguillarum* was obtained by inducing loss of the plasmid by curing and reintroduction by conjugation and by directed mutagenesis of key genes. Mutation of any of the genes can result in a reduction of virulence by as much as 10^5 times. In experimental infections in which iron uptake by transferrin is swamped by injection of excess iron, possession of the plasmid does not confer an advantage. The plasmid pJM1 contains genes for synthesis of the siderophore anguibactin, an unusual hydroxamate–catechol compound, and outer-membrane transport proteins and regulators of transcription involved in iron uptake (*Figure 11.13*). Expression of the multiple components occurs only under low-iron conditions, owing to negative control at the transcriptional level by the chromosomally encoded Fur protein and at least three plasmid-encoded regulators, the best studied of which is AngR. Outer-membrane proteins only expressed under iron-restricted conditions are known as IROMPs. Recently, additional siderophores and iron-uptake mechanisms have been found.

Figure 11.13 Iron uptake via the siderophore anguibactin in *Vibrio anguillarum*. Anguibactin (Ang) is secreted to the external environment, where it competes with the host protein transferrin (T) for circulating iron. Chelated iron (Fe) is reabsorbed into the cell via an outer-membrane receptor and membrane transport proteins (TonB, ExbB, ExbD) before transport across the cell membrane via the ATP-dependent fat system. Biosynthesis of anguibactin, the outer-membrane receptor and transport proteins are subject to complex regulation at the transcriptional level (not shown). Credit: based on a drawing by Jorge Crosa.

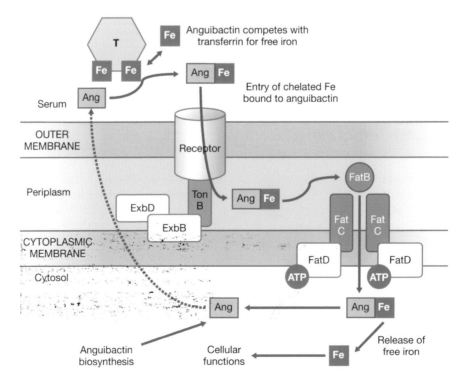

Quorum sensing (QS; p.102) via acyl homoserine lactone (AHL) signaling molecules is also an important factor in the growth and survival of *V. anguillarum* in its host and in the aquatic environment. There are three separate pathways involved in gene regulation of biofilm formation and production of protease, pigment, and exopolysaccharide. The hierarchical QS system consists of regulatory elements homologous to those found in both *Aliivibrio fischeri* (the LuxRI homologs VanRI,) and *V. harveyi* (the LuxMN homologs, VanMN). It is possible that some of the AHLs have effects on eukaryotic host cells, including an immunomodulatory function.

Other virulence factors are also involved in *V. anguillarum* pathogenicity, including a metalloprotease and a powerful hemolysin causing lysis of erythrocytes. This may contribute to the acquisition of iron by release of hemoglobin and probably accounts for the pale gills and anemia typical of infection. The structure of the LPS confers resistance to the bactericidal effects of the complement system present in normal serum. Also, the flagellum is essential for motility across the fish integument as directed mutations in the *flaA* gene result in reduced virulence. LPS antigens on the sheath of the flagellum (genes *virA* and *virB*) are also essential for virulence, probably by contributing to biofilm formation *in vivo*.

Vibrio ordalii was originally thought to be a biovar of *V. anguillarum* but was designated as a separate species based on a number of biochemical and genomic differences. It has caused major losses in the cage culture of salmon in Pacific coastal waters off Oregon, Washington State, and British Columbia. Infections with *V. ordalii* tend to be localized in muscle tissue rather than the generalized infections seen with *V. anguillarum*. Virulence has not been so well studied as that of *V. anguillarum* but complement resistance and a leucocytolytic toxin have been described.

Aliivibrio salmonicida (formerly *Vibrio salmonicida*) causes cold-water vibriosis (Hitra disease) in salmon and Atlantic cod farmed in northern regions such as Canada, Norway, and Scotland during the winter, when water temperatures drop below 8°C. Disease signs are broadly similar to those of *V. anguillarum*, but the two bacteria are serologically and genetically distinct. The pathogen is excreted in the feces of infected fish and seems to have good powers of survival in marine sediments, thus causing reinfection even if farm sites are "fallowed" for a season. Epidemiological studies suggest that there is interchange of the bacterium between populations of cod and salmon. Various strains can be distinguished by different plasmid profiles, but there does not seem to be a close association between plasmid possession and virulence. The bacterium produces various extracellular enzymes, a hydroxamate siderophore, and a *fur*-regulated iron-uptake system, all of which are implicated in virulence. Interestingly, significant amounts of the siderophore and the IROMPs required for transport are only expressed at temperatures below 10°C, which may explain the increased incidence of disease in the winter. This fact has been important in vaccine manufacture, since temperature-regulated surface proteins are important antigens. Another species that has caused salmon infections in cold northern waters was originally designated as *V. viscosus* but was reclassified as *Moritella viscosa*. This causes a condition known as winter ulcer disease, in which ulcerous lesions progress from the skin to the underlying muscle. Mortality rates are relatively low, but the appearance of the fish lowers its economic value considerably.

V. vulnificus causes infection of eels in Japan. The fish isolate is closely related to clinical isolates associated with human disease (p.337) but has been designated a separate taxon (biogroup 2) based on phenotypic, cultural, and serological properties. Iron acquisition and production of a capsule, hemolysin, protease, and other toxins have all been implicated in virulence.

Pasteurellosis affects warm-water marine fish

The expansion in the 1990s of sea bass and sea bream aquaculture in the Mediterranean and yellowtail culture in Japan was accompanied by outbreaks of a new disease that has caused heavy losses. The pathogen responsible was originally identified as *Pasteurella piscicida*, and the disease is still known as pasteurellosis, although the causative agent has been reclassified as *Photobacterium damselae* subsp. *piscicida*. When water temperatures exceed 20°C, acute mortalities up to 70% can occur, especially in the larval and juvenile stages. A more chronic

GENOMIC EVIDENCE SHOWS HOW FISH PATHOGENS ADAPT TO SPECIFIC HOSTS

The genome sequence of a strain of *Aeromonas salmonicida* subsp. *salmonicida* revealed that acquisition of mobile genetic elements, genome rearrangements, and gene loss appear responsible for adaptation of the bacterium to its specific host, salmonid fish (Reith et al., 2008). Many genome rearrangements have occurred, affecting regulation of gene expression. Analysis of further strains has revealed evolution of a diversity of constrained mesophilic and psychrophilic lifestyles (Vincent et al., 2016). The *Aer. salmonicida* genome contains numerous pseudogenes—defunct relatives of known genes that no longer encode proteins. The genome sequence of *Aliivibrio salmonicida* also reveals evidence of extensive gene rearrangements through incorporation of insertion sequences and the degeneration of many genes (Hjerde et al., 2008). Of particular interest is the loss of genes involved in the utilization of chitin, which are highly important in enabling vibrios to colonize surfaces in the marine environment (p.332). Gene loss or decay is typical of recently evolved genomes, often reflecting adaptation to a specific host (Pallen and Wren, 2007).

condition occurs in older fish. Interest in virulence mechanisms has focused largely on extracellular proteases, adhesive mechanisms, and the presence of a polysaccharide capsule. Iron concentration seems to be important in the regulation of expression of superoxide dismutase and catalase, which protect bacteria against reactive oxygen species (p.108) and may be important in intracellular survival. An iron-uptake system is associated with virulence, but as well as uptake by siderophores, the bacterium seems able to acquire iron by direct interaction between hemin molecules and outer-membrane proteins. Introduction of rapid PCR-based diagnostic methods and vaccines are proving effective in controlling the disease.

Aeromonas salmonicida has a broad geographic range affecting fish in fresh and marine waters

Aeromonas salmonicida was first described as a pathogen of trout in Europe in 1890 but is now known to have a broad host and geographic range. The taxonomy of *Aer. salmonicida* has been the subject of much debate over the years and it is now generally recognized that acute "typical" furunculosis in salmonids is caused by the subspecies *salmonicida*. In mariculture, this usually presents as a severe septicemia with acute mortalities. At the peak of the furunculosis outbreaks in Scotland and Norway in the 1980s, total industry losses neared 50% of stock. Externally, the fish show darkening and hemorrhages around the fins and mouth and internally there is extensive hemorrhaging and destruction of the organs. Other subspecies (*achromogenes*, *masoucida*, *pectinolytica*, and *smithia*) can be distinguished by differences in standard biochemical tests, pigment production, and molecular techniques such as gene probes and DNA–DNA hybridization. These subspecies are generally associated with dermal ulcerations in other marine species such as turbot (*Scophthalmus maximus*) and halibut (*Hippoglossus hippoglossus*). The name furunculosis derives from a boil-like necrotic lesion seen in a chronic form of the infection in older or more resistant fish (*Figure 11.14A*). Most strains can be easily isolated on laboratory media, although some are fastidious and have a specific requirement for heme.

The virulence mechanisms of *Aeromonas salmonicida* are extremely complex, and this pathogen is an excellent example of multifactorial virulence that has attracted the attention of numerous investigators for over half a century. There are many different interacting components of the factors responsible for entry, colonization, growth *in vivo*, and tissue damage. Genes for many of these virulence factors have been cloned and their properties studied in detail, with full genome sequences also now available. Virulent and avirulent strains are usually distinguished by the presence or absence of a regularly structured surface S-layer (p.69) known here as the "A-protein," the hydrophobic nature of which confers auto-agglutinating properties in culture (*Figure 11.14B*). Isolates possessing the 49-kDa

Figure 11.14 Infection of fish by *Aeromonas salmonicida*. A. Large boil-like lesion (furuncle) on the surface of infected salmon. B. The swollen skin lesion is filled with pink fluid containing blood and necrotic tissue. C. Growth of A⁺ (left) and A⁻ (right) cultures in broth: A⁺ cultures autoagglutinate owing to the hydrophobic S-layer (A-protein). D. Transmission electron micrograph (TEM, negatively stained) of an A⁺ strain showing regular structure of the surface A-protein. E. TEM section of a virulent A⁺ strain showing the additional layer (arrowed) external to the outer membrane. Note outer-membrane vesicles (V). Credits: A–B. Reprinted from Dallaire-Dufresne et al. (2014) with permission of Elsevier. C–E. Nigel Parker and Colin Munn, University of Plymouth.

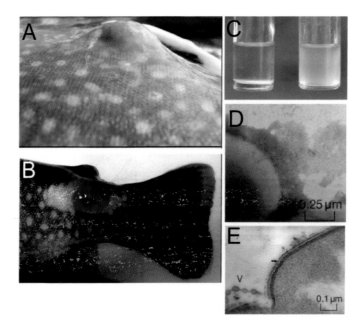

A-protein (A$^+$) can also be distinguished from A$^-$ strains by their growth on agar media containing Coomassie Blue or Congo Red dyes, which the A-protein absorbs. Electron microscopy shows the A-protein to be present as a structured array of tetrahedral subunits (*Figure 11.14C*) external to the typical Gram-negative outer membrane (*Figure 11.14D*). It is linked to the cell surface via the O polysaccharide sidechain of LPS. The main function of the A-protein is as a protective layer, which contributes to the bacterium's resistance to the bactericidal effects of complement in the serum of the host fish. Because of its hydrophobic nature, it also plays a role in adhesion to host tissue and survival within macrophages. *Aer. salmonicida* also produces a range of toxins, and many studies have been carried out on the activity of purified components, although it is now clear that these have synergistic interactions. The enzyme glycerophospholipid:cholesterol acyltransferase (GCAT) and a serine protease are particularly implicated as key virulence factors. GCAT forms a complex with LPS, which is hemolytic, leukocytolytic, cytotoxic, and lethal. Serine protease expression is regulated by an AHL-mediated QS mechanism, and this enzyme (in synergy with GCAT) is responsible for hemolysis. As in *V. anguillarum*, growth *in vivo* is dependent on the production of a siderophore (2,3 diphenol catechol) and the uptake of sequestered iron via IROMPs. A type III secretion system (T3SS, see p.339) also has a major role in pathogenesis by delivering a range of effector molecules that switch off the host's ability to recognize infection and prevent onset of disease. Elucidation of the various components in virulence and their immunological properties was a critical step in the formulation of modern vaccines, which have been largely successful in control of the disease in salmon mariculture since the mid-1990s.

The ecology of *Aer. salmonicida* has been the subject of much controversy, with some investigators suggesting that it is an obligate fish pathogen and others suggesting that it survives in the environment. Historical evidence suggests that it spread throughout Europe more than a century ago and spread into the wild salmon population, where it caused major losses in the 1930s. It is very likely that some wild populations are latently infected with *Aer. salmonicida*. The bacterium may enter the VBNC state (p.95), during which the cells undergo various morphological changes. One certainty is that the development of mariculture in enclosed bodies of water such as lochs in Scotland and fjords in Norway or Chile has led to a shift in the normal microbiota. Whereas *Aer. salmonicida* seems to have been primarily a freshwater organism, it has now adapted to life in seawater, and recent isolates may have a sodium requirement lacking in earlier strains. Farmed fish can transmit the disease to wild fish around sea cages and these can spread the pathogen to other sites. Wrasse introduced to net pens to remove sea lice parasites from salmon can become infected and thus constitute a reservoir of infection. Atlantic salmon have been shown to harbor latent *Aer. salmonicida* infections when they return from the ocean to spawn, so spread of the disease may be contributing to decline of wild populations.

Marine flexibacteriosis is caused by a weakly virulent opportunist pathogen

Members of the CFB group (p.305), many of which are pigmented and show gliding motility, are responsible for infections in a wide range of fish. Mostly, these bacteria are rather weak opportunist pathogens that colonize damaged tissue in fish weakened by stress, especially due to increased water temperature and nutritional deficiency. Most diseases caused by this group occur in fresh water, but *Tenacibacter maritimus* (previously classified as *Flexibacter marinus*) causes a disease called marine flexibacteriosis, sometimes referred to as "marine columnaris" or "black patch necrosis," which is widely distributed in numerous species of wild and farmed fish in Europe, Japan, North America, and Australia. It is characterized by excess mucus production, damage to the gills, tissue necrosis around the mouth and fins, skin lesions, and eventual death. It is most severe in juvenile fish at temperatures above 15°C. The clinical signs and the observation of long Gram-negative gliding bacteria in microscopic preparations are very distinctive. The bacteria are not easy to grow in the laboratory, as they require special low-nutrient media. There appear to be several O-antigen serotypes, which may be related to specific hosts. The disease responds to antibiotics administered by bath and some vaccines have been developed, with variable degrees of success.

Piscirickettsia and *Francisella* are intracellular proteobacteria infecting salmon and cod

Piscirickettsia salmonis was identified in 1989 as the etiological agent of a septicemic disease that caused severe economic losses in farmed salmon in Chile. Here, it causes predictable annual epizootics, but it has also caused sporadic outbreaks of disease in Norway, Iceland, and Canada. Infected fish stop feeding, become lethargic, and may show small white lesions or ulcers on the skin. The disease progresses rapidly, and death often occurs with few or no external symptoms. Internally, the most distinctive sign is off-white to yellow nodules in the liver. The disease seems to be transmitted both horizontally by blood-sucking ectoparasites and vertically via the eggs, making the testing and elimination of infected brood stock with ELISA and PCR assays an important element of disease control. *Piscirickettsia* was originally considered to be an obligate intracellular parasite and was formerly propagated only in suitable cell lines, but it can now be grown in broth or agar cultures using appropriate media. A range of serological tests has been developed, and genetic heterogeneity among strains can be studied by analysis of the 23S rRNA operon. Antibiotic treatment is ineffective, since there are very few agents capable of attacking intracellular bacteria without causing damage to host cells. A recombinant vaccine (p.402) based on an outer-membrane lipoprotein, OspA, shows promise for disease control.

Epizootics caused by *Francisella* have emerged as a particular problem in culture of Atlantic cod in Norway. This is a granulomatous inflammatory disease, resulting in extensive internal lesions, with nodules in the spleen, heart, kidney, and liver. The *Francisella* isolated from cod has recently been characterized as closely related to *Francisella philomiragia*, which is widely distributed in aquatic environments and may cause disease in humans with impaired immunity. Other *Francisella* spp. also cause infections in a range of other fish.

Intracellular Gram-positive bacteria cause chronic infections of fish

Renibacterium salmoninarum causes bacterial kidney disease (BKD) and is widely distributed in both wild and cultured salmon in many countries including Canada, the USA, Chile, Japan, Scotland, and Iceland. The expansion of salmon culture through international movement of eggs has assisted the spread of BKD and it causes significant losses in both Pacific and Atlantic salmon. BKD pathology is characterized by chronic, systemic tissue infiltration, causing granular lesions in the internal organs, especially the kidney. External signs include darkening of the skin, distended abdomen, exophthalmia, and skin ulcers. Significant changes in blood parameters are consistent with damage to the hematopoietic and lymphopoietic tissues of the kidney, liver, and spleen. The pathological signs are the result of the interactions between the host's cellular immune response and the pathogen's virulence mechanisms. Tissue destruction forms a focus of necrosis, owing to release of hydrolytic and catabolic enzymes and liberation of lytic agents from the bacteria.

Our understanding of the mechanisms of virulence of *R. salmoninarum* is hampered by the fact that the bacterium takes several weeks to grow on culture plates and does not form discrete colonies. Reproducible infection is also difficult to achieve in aquarium experiments. One important virulence factor is a 57-kDa surface protein, whose hydrophobic properties facilitate attachment to host cells. The key feature of *R. salmoninarum* is its ability to enter, survive, and multiply within host phagocyte cells. Binding of the complement component C3 to the bacterial surface enhances internalization of the bacterium because phagocytic cells possess a receptor for C3. Why does *R. salmoninarum* encourage uptake by cells that normally kill invading pathogens? The answer seems to lie in the pathogen's ability to survive (at least in part) the intracellular killing mechanisms of the macrophage and to replicate (albeit slowly) within the cells. As well as being resistant to reactive oxygen species, *R. salmoninarum* lyses the phagosome membrane in order to escape its strongly antibacterial environment (*Figure 11.15*). In the past, BKD has been difficult to diagnose, but disease management and control are helped by serological techniques (ELISA and FAT), together with recently developed gene probes and techniques for accurate differentiation of clinical isolates (based on PCR amplification of length polymorphisms in the tRNA intergenic spacer regions). These

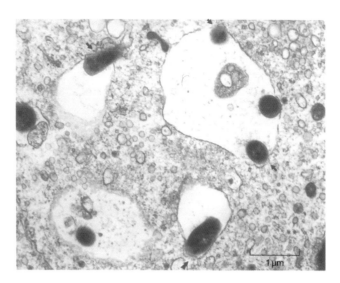

Figure 11.15 Intracellular growth of *Renibacterium salmoninarum*. Transmission electron micrograph showing lysis of the phagosome membrane *(arrows)*, prior to entrance of the bacterium into the cytoplasm. Image courtesy of S. K. Gutenberger and J. R. Duimstra, Oregon State University.

methods are used for certification of brood stock and eggs as disease-free and for implementing quarantine procedures to contain disease outbreaks. There are no effective antibiotic treatments and there have, therefore, been many attempts to develop a vaccine. These are thwarted by the slow growth of the pathogen, and recent work has focused on the use of recombinant DNA technology to produce fusion proteins and DNA vaccines (see p.402), as well as the use of a live preparation of a non-pathogenic *Arthrobacter* sp., which has some antigenic similarity to *R. salmoninarum*. Genes for several virulence factors, including the p57 protein and two hemolysins, have been cloned and expressed in *Escherichia coli* and the nature of the immune response investigated.

Species of the genera *Mycobacterium* and *Nocardia* also cause chronic, persistent infections with long periods of intracellular survival. Many different species of mycobacteria occur; these are widely distributed in seawater and sediments and colonize many species of fish. *Mycobacterium marinum* and *Mycobacterium fortuitum* are the best-known agents of disease. Disease develops slowly and usually affects mature fish; for this reason, mycobacteriosis is a particular problem in marine aquaria, but it has also emerged as a serious problem in the culture of sea bass and turbot. Most species of fish develop few external disease signs other than emaciation, although histopathological investigation reveals extensive granulomatous lesions with caseous necrotic centers. For this reason, the disease is often called fish tuberculosis, since there is some resemblance to the basic pathological mechanisms seen in the human condition caused by *Mycobacterium tuberculosis*. The delayed hypersensitivity reactions and involvement of cell-mediated immunity certainly show some parallels in fish and humans. *Nocardia* spp. cause similar chronic granulomatous conditions to mycobacteria, but the organisms can be distinguished in the laboratory. Again, because of their intracellular nature, there are few antimicrobial agents effective against these pathogens. Valuable aquarium fish are sometimes treated with isoniazid and rifampicin, but this is unwise given the danger of encouraging resistance to these drugs, valuable for the treatment of human tuberculosis. It should also be noted that *M. marinum* (and possibly other species) can cause a zoonotic infection in humans known as fish tank or aquarist's granuloma.

Some Gram-positive cocci affect the central nervous system of fish

The Gram-positive bacteria *Lactococcus garvieae* (formerly *Enterococcus seriolicida*) and *Streptococcus iniae* have been especially problematic as the cause of disease, often termed "streptococcosis" in both fresh and marine warm-water culture, especially in Japan. Several related species have also been described. They cause central nervous damage characterized by abnormal swimming behavior, often with exophthalmia and meningitis. Extracellular cytolytic toxins and an antiphagocytic capsule have been implicated in pathogenicity. *Streptococcus iniae* has caused epizootics in barramundi (*Lates calcarifer*) culture in

Australia and has also been implicated in extensive fish kills in the Caribbean Sea. It can also cause zoonotic infection by infection of wounds in workers handling infected fish, leading to severe cellulolytic infection.

Viruses cause numerous diseases of marine fish

Viruses have long been suspected as causative agents of disease in fish, but it was not until the successful development of fish cell culture methods in the 1960s that progress was made in diagnosing viral fish diseases. As with bacterial diseases, intensive aquaculture has exacerbated viral disease problems in fish. The mechanisms by which viruses produce disease in their fish hosts are less well defined than those of bacteria. Viruses must possess a mechanism to enter a susceptible host cell and grow within it; this is usually very specific for a particular type of cell and is determined initially by interactions between surface molecules on the virion and receptors on the host cell. Viruses possess many mechanisms to resist innate host defense mechanisms such as enzymes and antiviral substances in fish mucus. Following local replication at the point of entry, viruses are spread to adjacent cells or carried to other target tissues via the blood, lymph, or nervous systems. The outcome of infection depends on the balance between viral replication and its control by innate defense, interferons, antibodies, and cell-mediated immunity. Host damage is caused by various mechanisms, including killing infected cells because of production of toxic factors during viral replication, cytokines, inflammatory modulators, and necrosis. It is often the intensity of the host response to viral infection that produces the most damage.

Infectious salmon anemia (ISA) is one of the most serious diseases in salmon culture

ISA emerged as a new disease in farmed Atlantic salmon in Norway in the 1980s and has been a continuing problem since that time. Subsequent outbreaks causing major economic losses occurred in Canada, Scotland, the Faroe Islands, the USA, and Chile and has brought the industry to the brink of collapse on several occasions. There is significant variation in the severity of ISA, with mortalities in sea pens depending on the virus strain and susceptibility of fish stock. It is likely that the *Salmon isavirus* (ISAV) originated in wild salmonids in lakes of Norway, and that transmission probably evolved in the freshwater phase of trout. The development of salmon farming is thought to have changed the balance between ISAV and wild fish, leading to the rapid evolution of a highly pathogenic type. Wild fish may serve as reservoirs of infection. The virus infects the erythrocytes, leading to their lysis. ISA-infected fish are lethargic, with anemia (shown by very pale gills), exophthalmia, and hemorrhages in the eye chamber, skin, and muscle. ISAV is an enveloped negative-strand RNA virus of the Orthomyxoviridae family, which includes influenza virus. Influenza virus is known for its property of antigenic variation by mutation ("drift") and recombination ("shift") events in the segmented genome, leading to the emergence of new strains. This may also be occurring in ISAV, as isolates from Norway and North America now seem to be genetically distinct. Inactivated virus vaccines have been used in Chile and Norway and recombinant DNA vaccines have been developed.

Viral hemorrhagic septicemia (VHS) virus infects many species of wild fish

VHS virus (VHSV) is a rhabdovirus most commonly known as a causative agent of disease in the freshwater stage of trout and salmon culture, but which can also cause disease in marine net pens. VHSV has now been isolated from nearly 50 species of marine fish and is responsible for severe epizootics in wild populations, especially of shoaling species such as herring (*Clupea harengus*), mackerel (*Scomber* spp.), and sprat (*Sprattus sprattus*). As well as the impact of overfishing, fluctuations in natural populations due to viral infection may be driving some commercially important species close to extinction. Detailed phylogenetic studies based on nucleoprotein gene sequences of marine VHSV show that there are distinct populations of the virus and that aquaculture isolates probably originated in wild marine fish.

Lymphocystis virus causes chronic skin infection of fish

Infection with lymphocystis virus (a member of the iridovirus group) results in hypertrophy of skin cells, causing characteristic papilloma-like lesions, usually on the skin, fins, and tail. The virus seems to have worldwide distribution, causing particular problems in cultured and wild fish in the Mediterranean Sea and in many species of tropical coral reef fish. It is also one of the most common infections of long-lived aquarium fish. Its role in the ecology of natural populations is unclear; although the disease is not usually lethal, it does affect growth rates and probably makes infected fish more susceptible to predation. A large number of other iridoviruses have been reported from more than 140 different species of fish worldwide; there is considerable heterogeneity in genome sequence in this group.

Birnaviruses appear to be widespread in marine fish and invertebrates

Several isolates of birnavirus have been identified in various wild fish in waters around the British Isles, Japan, and the northwest Atlantic. These were originally considered to be closely related to the *Infectious pancreatic necrosis virus* (IPNV), which is best known as a cause of heavy mortalities in small fry in freshwater aquaculture hatcheries. Phylogenetic analysis of the marine birnaviruses from wild fish, based on amino acid sequence of the polyprotein gene, indicates that they form a group distinct from IPNV. Little is known about entry of these viruses and replication in the host, and their impact on wild populations is unclear.

Viral nervous necrosis (VNN) is an emerging disease with major impact

VNN is characterized by erratic behavior in infected fish, due to spongiform lesions in the brain (encephalopathy). The causative agent, *Betanodavirus* (a member of the Nodaviridae), has been reported in cultured marine fish worldwide and has emerged since the 1990s to have a major impact on the culture of several marine species. PCR analysis shows that wild and cultured fish are subclinically infected with betanodaviruses and constitute a reservoir of infection. The spread of these infections seems to be a direct consequence of the intensity of aquaculture and the movement of stock both within aquaculture areas and between distant geographic regions.

Protists cause disease in fish via infections, toxins, and direct physical effects

A large number of protists can infect wild, farmed, and aquarium fish. Often these are freeliving or benign parasites and the environmental and nutritional conditions that promote disease are largely unknown. For example, *Paramoeba perurans* causes sporadic, severe outbreaks in salmon culture in Tasmania. Diplomonad flagellates are commonly found in the gut of fish, in which they seem to have little effect, but under some circumstances they can cause systemic infection with high mortalities. Myxosporeans such as *Kudoa* sp. commonly cause muscle infections, causing cysts and softening of the tissue that impairs marketability of the fish. *Loma salmonae* is an obligate intracellular microsporidean parasite infecting the gills of many economically important fish, including both wild and cultured salmon and cod.

Excessive growth of phytoplankton leading to HABs in coastal waters can be responsible for mortalities in fish. Some of these blooms are caused by toxin-producing species that can also affect marine mammals (see below) or humans (see *Chapter 12*). Toxic genera such as *Alexandrium*, *Gymnodinium*, *Karenia*, and *Pseudo-nitzschia* have all been responsible for mass mortality in many parts of the world, both in wild and farmed fish. The increased use of coastal waters for fish and shellfish farming has undoubtedly contributed to stimulation of blooms by input of excess nutrients from the fish excreta and waste food. Caged fish cannot escape the effects.

Nontoxic algae can also kill fish. For example, the diatom *Chaetoceros convolutus* produces long barbs, which clog fish gill tissue causing excess mucus production, leading to death from reduction in oxygen exchange. Incidents involving the loss of more than 250,000 farmed

DID IMPORT OF TUNA FEED LEAD TO THE WORLD'S BIGGEST FISH KILL?

Two major epizootics occurred in the 1990s in the waters of South and Western Australia, resulting in the biggest ever single-cause fish kills. Mass mortalities of pilchards (*Sardinops sagax*) occurred, spreading at over 30 km a day and covering 6000 km of coastline.

There was a reduction in the spawning biomass of more than 75% (Ward et al., 2001), resulting in severe ecological consequences for other marine life and seabird predators, as well as major economic losses to an important commercial fishery. The causative agent was identified as a previously unknown herpesvirus. These epizootics coincided with the development of tuna ranching in South Australia; tuna were kept in large offshore sea cages and fed with frozen pilchards imported from California, Peru, Chile, and Japan to fatten them for the sushi and sashimi market. Although conclusive proof is hard to obtain, it seems likely that inadvertent introduction of the virus with feed from a different geographic region may have resulted in a disease outbreak because the Australian pilchard population had not previously encountered this variant of the virus and therefore had no immunity. Alternative explanations might be transmission by ballast water or by seabirds. Since 1998, there have been no more mass mortalities and the virus may now be endemic in Australia; because of the development of host "herd" immunity, it may now cause only low levels of mortality or exist as a latent infection (Whittington et al., 2008).

salmon at a time have occurred. Blooms of the flagellate *Heterosigma carterae* also cause heavy mortalities in Pacific coast salmon farms. Nontoxic microalgae can also have indirect effects on fish, as large blooms reduce light penetration and decrease the growth of sea-grass beds, which are often important nursery grounds for the young stages of commercially important fish. Also, clumping, sinking, and decay of phytoplankton can generate anoxic conditions.

DISEASES OF MAMMALS

Dinoflagellate and diatom toxins affect marine mammals

Marine mammals are susceptible to a variety of infectious disease conditions, as well as the effects of toxic HABs, as shown in *Figure 11.16*. The significance of HAB toxins has been extensively investigated because of their effects on human health, and the origin of blooms and the nature of the toxins is discussed in *Chapter 13*. Most HAB toxins have effects on the nervous system. Ingestion of HAB toxins by humans results in symptoms such as nausea, vomiting, diarrhea, temperature reversal effects, and paralysis, so it is reasonable to conclude that similar effects will also occur in marine mammals. If so, this would clearly affect feeding, buoyancy, heat conservation, breathing, and swimming. Neuropathological symptoms and mass mortality events in marine mammals are increasingly linked to toxins from HABs. One explanation for some mass strandings or beaching of social mammals such as whales and dolphins is that they become disoriented because of accumulation of toxins after feeding on contaminated fish or shellfish (*Figure 11.17*). In some studies, high levels of saxitoxins have been found in tissues of stranded killer whales (*Orcinus orca*) known to have been feeding on mussels in areas with a toxic *Alexandrium* bloom. Although the levels of toxin recovered from most tissues at postmortem often do not seem high enough to account for symptoms, it is likely that the toxins become concentrated in the brain. Brevetoxin shows a very high affinity for nerve tissue from the manatee (*Trichechus manatus*), which are often killed off the Florida coast during blooms of *Karenia brevis*, both by ingestion and inhalation. Saxitoxins also bind strongly to nerve tissue from several species of whale. Since the blood flow diverts to the brain during diving, this could deliver toxins absorbed from the gut to the brain, where they could accumulate in high enough concentrations to cause disorientation and other neurological effects, leading to stranding. Large-scale mass mortalities of baleen whales on the coast of Chile have been attributed to HAB toxins through association with El Niño events. However, it is important to recognize that stranding appears to have multiple environmental, physiological, and behavioral causes.

There is strong evidence for the role of the toxin domoic acid (DA), produced by the diatom *Pseudo-nitzschia* spp., in outbreaks of neuropathological illness in marine mammals and birds. The toxin was first associated with amnesic shellfish poisoning in humans (see p.347)

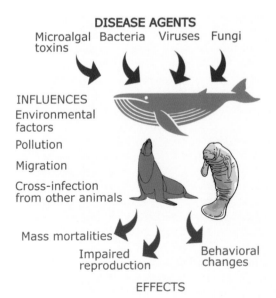

Figure 11.16 Microbes as agents of diseases in marine mammals. Various factors influence the susceptibility of the host and transmission of diseases.

and subsequently linked to periodic mass mortalities of sea lions, pelicans, and cormorants along the coast of northern California. Animals are poisoned by feeding on planktivorous fish and shellfish (such as anchovies, sardines, mussels, and clams) that have ingested large amounts of toxic *Pseudo-nitzschia*, which bloom as a result of climatic conditions and upwelling of nutrients. A pulse of domoic acid enters the food chain and the affected animals assimilate the toxin, which accumulates in the brain. Pregnant sea lion females are regularly exposed to DA in the diet, and this is especially harmful to development of the fetus brain, leading to abortion and neonatal death. Recent studies have also found that krill—the principal component of the diet of squid, baleen whales, and seabirds—can accumulate DA, which is then passed to higher trophic levels of the food chain.

Virus diseases cause mass mortalities in cetaceans and pinnipeds

In 1988, there were mass mortalities of more than 18000 common or harbor seals (*Phoca vitulina*) in Northern Europe, with over 60% population loss in some areas. At about the same time, an epizootic with similar mortality rates occurred in bottlenose dolphins (*Tursiops truncatus*) on the east coast of the USA. In 1992, there were mass mortalities of porpoises (*Phocoena phocoena*) in the Irish Sea and of striped dolphins (*Stenella coeruleoalba*) in the Mediterranean Sea, which reduced the population to 30% of its previous level. These disease outbreaks attracted considerable public concern; under pressure from environmentalist lobby groups, several government sponsored research projects were set up to investigate the problem. As a result of this work and subsequent studies, we now know that virus diseases are a significant cause of mortality in cetaceans (whales, dolphins, and porpoises) and pinnipeds (seals and sea lions). In the wild, it is difficult to study diseases of marine mammals except when mass mortalities occur, when animals are stranded, or if they become caught in fishing nets. Networks of marine conservation volunteers have helped in the collection of data, and it is now common practice to test stranded cetaceans and pinnipeds for viral diseases using immunological and molecular-based diagnostic tests, so that a worldwide picture of their importance is beginning to emerge. Obtaining blood or tissue samples from live animals is difficult and postmortem deterioration happens very quickly, so data are often sparse. Some knowledge about these diseases has also come from the study of captive animals in zoos and marine parks; the study of diseases of marine mammals has become a specialized branch of veterinary medicine.

Viruses from nine different families have been linked to diseases of marine mammals

The most important viruses infecting cetaceans and pinnipeds are the morbilliviruses; these are RNA viruses of the paramyxovirus group causing several mammalian diseases, of which the best-known examples are measles in humans and distemper in dogs. The first clues to

HOW DID A MICROBIAL TOXIN INSPIRE A CLASSIC HOLLYWOOD MOVIE?

Occasional mass mortality and erratic behavior of seabirds has been reported along the coast of northern California for many years. One such incident is thought to have been the inspiration—with some imaginative additions!—for the classic 1963 Alfred Hitchcock thriller movie *The Birds*, in which residents of a small town are terrified by the erratic behavior of birds. While on vacation in 1961, Hitchcock heard of an incident in the town of Capitola, California, in which seabirds crashed into cars and houses and died in large numbers. This inspired the idea for the film, adapted from a story set in Cornwall, England, by Daphne du Maurier. In 1991, another incident in Monterey Bay provided the first evidence to suggest that the birds had eaten fish containing the toxin domoic acid, originating from a diatom bloom. Extensive investigation involving scientists from many disciplines of another incident on this coast in 1998 provided conclusive evidence of domoic acid poisoning in birds and sea mammals and explained why this area is a hotspot for such events (Scholin et al., 2000).

Figure 11.17 Mass stranding of false killer whales in W. Australia, possibly caused by a toxic algal bloom. Image credit: Bahnfrend, CC BY 3.0 via Wikimedia Commons.

ℹ️ **DOGS WITH A NOSE FOR WHALE POOP**

Doucette et al. (2006) investigated whether the transfer of marine algal toxins through the food chain could explain poor health and impaired reproduction of highly endangered North Atlantic right whales (*Eubalaena glacialis*) in the Bay of Fundy. This required maneuvering a small boat into the vicinity of feeding whales, spotting a whale in the act of defecation, and scooping up the sample (scat) in a net. Doucette used this method to show high levels of saxitoxins in the feces of the whales and in the whales' primary food source (the copepod *Calanus finmarchius*), associated with a bloom of toxic dinoflagellates (*Alexandrium* spp.) Although right whale scat is bright orange and floats for about 30 min, sample collection at sea is a very hit-and-miss affair! A novel solution was pioneered by researchers at the New England Aquarium, who trained dogs to detect the scent of fresh scat; the dogs point to the direction of the source up to 1500 m away so that the helmsman can steer the boat toward it (Rolland et al., 2007).

the identity of the virus infecting seals in the 1988 epizootic came when serum was found to neutralize canine distemper virus (CDV). At first, it was thought that the disease might be linked to an outbreak of distemper in sled dogs in Greenland. However, using sequencing of the viral capsid proteins, the seal virus (now called phocine distemper virus, PDV) was shown to be a new species, although it shares some antigenic cross-reactions with CDV. One possible explanation for the sudden epizootic is transfer from another species (probably the harp seal, *Phoca groenlandica*), which migrates from a different geographic region in which the disease is enzootic. Serological studies have shown that PDV-like viruses are present in several species of marine mammals. By contrast, the virus responsible for outbreaks in seals in the Caspian Sea and Lake Baikal (*P. caspica* and *P. siberica* respectively) seems to be identical to CDV, and it almost certainly did come from dogs. Infection causes respiratory, gastrointestinal, and neurological symptoms, often with secondary bacterial infections such as pneumonia.

The morbilliviruses isolated from diseased porpoises, dolphins, and several species of whale appear to be closely related antigenically and genetically; they are now recognized as different strains of cetacean morbillivirus (CeMV). Phylogenetic studies show that CeMV is close to the ancestor of the morbillivirus group, so these viruses probably have a terrestrial origin and may have infected cetaceans for the several millions of years since they have populated the oceans. Infected animals show pneumonia-like symptoms and disturbance of diving, swimming, and navigation ability. In the enzootic state, the virus probably has long-term effects on population dynamics, causing mortalities mainly in young animals in which no immunity exists. Morbilliviruses typically lead to either rapid death or recovery with life-long immunity, with no persistent carrier state. They therefore require large populations to sustain themselves through the input of new susceptible hosts. As with seal distemper, epizootics in cetaceans probably occur as a result of cross-infection from different species or animals from other geographic regions once a sufficiently large population of susceptible individuals has built up to allow spread (this is similar to the decline in "herd immunity" that results in periodic measles epidemics in humans). The nutritional status of the host population and environmental factors are also important in determining the onset of epizootics. After the 1988 outbreak of phocine distemper in northern Europe, the harbor seal population slowly recovered. In 2002, a new epizootic of phocine distemper in harbor seals emerged, resulting in death of over 30000 seals, owing to the buildup of a threshold number of nonimmune animals (*Figure 11.18*).

Poxviruses have been implicated as the cause of tattoo-like skin lesions in several species of small cetaceans, although they do not usually cause significant mortalities. Papillomaviruses and herpesviruses cause genital warts (proliferation of the squamous epithelium) in porpoises,

Figure 11.18 Transmission of morbilliviruses in marine mammals. Major epizootics occur in the seal and dolphin host species shown. Black and gray silhouettes of host species indicate status as a possible reservoir or dead-end host, respectively. Phocine distemper virus (PDV) circulates mainly in seals with occasional spillover to sea otters and walruses. Canine distemper virus (CDV) circulates in terrestrial carnivores, causing occasional outbreaks in seals. Cetacean morbillivirus (CeMV) circulates mainly among cetaceans, with occasional spillover to pinnipeds. It is unknown if whale species act as disease carriers or spillover hosts. Redrawn from Jo et al. (2018) with permission of Elsevier.

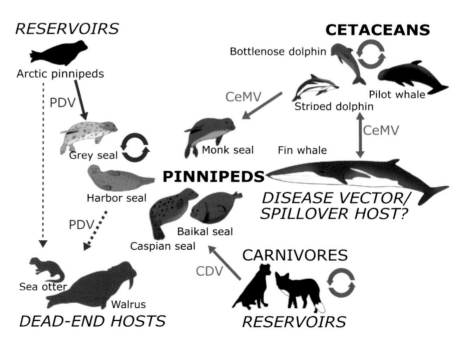

dolphins, and whales. As with their human equivalent, they are sexually transmitted and may affect population dynamics by disturbing reproduction and social behavior, since sexual "play" is an important part of cetacean social group interactions and genital warts can sometimes be large enough to interfere with copulation.

Various types of caliciviruses occur in a wide range of marine mammals, having been first described as San Miguel sea lion virus. From time to time, influenza A virus has been associated with mortalities in cetaceans and pinnipeds. As occurs in humans, animals infected with influenza become very weak and often succumb to secondary bacterial infections. The isolates are closely related to avian influenza in birds and are highly virulent. Seabirds are often associated with marine mammals during feeding at the surface and this favors transmission.

Several species of bacteria and fungi infect marine mammals

One of the most important globally distributed diseases of cetaceans and pinnipeds is brucellosis, caused by the bacterium *Brucella*, a highly contagious intracellular pathogen that also occurs in cattle, sheep, goats, and pigs. It infects the reproductive tract, especially the placenta and amniotic sac, leading to abortion. Frequent abortions have been observed to occur in closely monitored dolphin pods, and the bacterium could therefore have a significant effect on fertility and population dynamics. Several strains have been isolated from many different species, and serological studies show that up to 30% of small marine mammals surveyed have evidence of exposure. Phylogenetic analysis of the marine isolates of *Brucella* show that these may constitute two distinct new species named *Brucella ceti* and *Brucella pinnipedialis* with cetaceans and seals, respectively, as the preferred hosts.

Leptospirosis, caused by several species of *Leptospira*, occurs in many populations of seals and sea lions. Populations of Californian sea lions suffer a cyclical pattern of infection every 4–5 years; the mode of transmission is unclear, but probably occurs when the animals haul out on land. As in the human equivalent—Weil's disease, transmitted by rats—the main symptom is renal failure.

Tuberculosis is a chronic multiorgan disease caused by *Mycobacterium tuberculosis* and *M. bovis*, while other mycobacteria such as *M. marinum* and *M. fortuitum* can cause lesions in the skin and lungs. In seals, sea lions, and cetaceans, there have been several cases in zoos and marine parks. Other bacterial pathogens known to cause sudden mortality in captive cetaceans include *Burkholderia pseudomallei*, *Bordetella bronchiseptica*, and species of *Erysipelothrix*, *Streptococcus*, *Salmonella*, and *Pasteurella*. The significance of these pathogens in wild animals is not known.

In addition to primary pathogens, marine mammals can succumb to a host of opportunistic skin and respiratory infections caused by bacteria and fungi, such as pneumonia caused by *Aspergillus* spp., which is a particular problem in stressed captive animals. Direct acquisition of microbes of human origin, especially opportunist bacterial infections of wounds, may also occur when mammals swim in sewage-contaminated waters.

DISEASES OF SEA TURTLES

Sea turtles are affected by a virus promoting growth of tumors

Some novel herpesviruses are associated with infection of sea turtles (*Figure 11.19*). The chelonid fibropapilloma-associated herpesvirus has emerged as a major cause of debilitating tumors of the skin and internal organs in green (*Chelonia mydas*), loggerhead (*Caretta caretta*), and olive ridley (*Lepidochelys olivacea*) turtles. The large tumors interfere with feeding, behavior, and reproduction and are frequently observed in stranded animals, so there is concern that this disease may cause additional problems for the survival of these endangered species. The disease has been reported in Florida, Hawaii, and Australia and appears to have increased dramatically since its first description in 1938. Genomic analysis shows

DOES POLLUTION MAKE MARINE MAMMALS MORE PRONE TO DISEASE?

Environmental pollutants such as mercury, polychlorinated biphenyls (PCBs) and organochlorine pesticides accumulate in lipid-rich tissues, especially blubber and milk, with a variety of deleterious effects on marine mammals, especially reproductive failure. These compounds also impair the function of the immune system, so they may increase susceptibility of marine mammals to bacterial or viral diseases. An experiment in which different groups of seals were fed on fish caught in either highly polluted or less polluted areas showed that pollutants impaired the immune system, resulting in reduction in the natural killer cells that are critical in defense against virus (Ross et al., 1996; Ross, 2002). Mortality of porpoises and dolphins due to infectious diseases was linked with elevated levels of PCBs and heavy metals (Jepson et al., 1999; Bennett et al., 2001). Despite restrictions on the use of PCBs, biomagnification of these toxins is still occurring and continues to be linked with infectious diseases and other effects (Jepson et al., 2015). Because of the practical and ethical problems of conducting controlled experiments, it is impossible to prove conclusively that pollution is directly responsible for increased infections of marine mammals, but the circumstantial evidence is very strong.

Figure 11.19 Hawaiian green sea turtle (Chelonia mydas) with a large tumor caused by the fibropapilloma-associated turtle herpesvirus. Credit: George Balazs, NOAA.

that viruses belong to the alphaherpesvirus group, but those isolated from different species of turtle from different geographic regions are genetically distinct. The increase in disease seems to be due to unknown environmental factors affecting host susceptibility, rather than emergence of a more virulent form of the virus. A leech, *Ozobranchus* sp., acts as a mechanical vector for the virus. Other herpesviruses and papillomaviruses have also been implicated as the cause of disease in turtles reared in mariculture.

DISEASES OF SEAGRASSES AND SEAWEEDS

Heterokont protists cause ecologically important mortality of seagrasses

Diseases also affect the productivity of seagrasses such as eelgrass (*Zostera marina*) and turtlegrass (*Thalassia testudinum*) in shallow coastal habitats. Seagrass meadows are major habitats for marine life such as juvenile fish, turtles, dugongs, and manatees, and also act as feeding grounds for ducks and geese. Seagrass beds are under intense pressure due to destruction accompanying coastal development, boat activity, pollution, and the impacts of climate change, resulting in considerable ecological impact through loss of habitats and economic effects through impairment of recruitment to fishery stocks. Since the 1920s, there have been many reports of "wasting disease" affecting large areas of seagrass meadows in Europe and North America, resulting in the total loss of *Z. marina* from some areas. The causative agent was isolated in the 1980s and originally identified as a fungus or slime mold. Subsequent phylogenetic studies using rRNA gene sequencing revealed the primary pathogen to be a new species, *Labyrinthula zosterae*, a member of a monophyletic clade within the heterokonta protist group. Members of this group have an unusual morphology; the cells secrete an ectoplasmic membrane leading to a slimy net-like structure of actin-rich filaments, through which the nuclei are transported. The pathogen moves rapidly through the plant tissue through enzymatic degradation of cell walls, leading to blackening and eventually complete destruction of the tissue (*Figure 11.20*). The photosynthetic ability of tissue further away from the focus of infection is also severely inhibited, so the ecological impact of reduced productivity in eelgrass showing even minor disease signs could be significant. Other pathogens include the fungus-like oomycete *Phytopthora gemini*, which kills dormant seeds and developing seedlings. Disease outbreaks are predicted to rise due to climate change and pollution, but there is currently limited knowledge of the factors affecting these diseases.

Bacteria, fungi, and viruses infect marine macroalgae

Marine macroalgae (seaweeds) have a major function in forming habitat structure of tidal and subtidal waters and temperate reefs, providing food and shelter to a wide range of invertebrates and fish. Descriptions of seaweed diseases in natural habitats are surprisingly rare. In part, this is due to this being a neglected area of research, but it also seems that many seaweeds possess very effective chemical defense mechanisms against microbial colonization and infection. Indeed, many of these antimicrobial properties have been investigated for their biotechnological potential (see *Chapter 14*). Intensive aquaculture of red, brown,

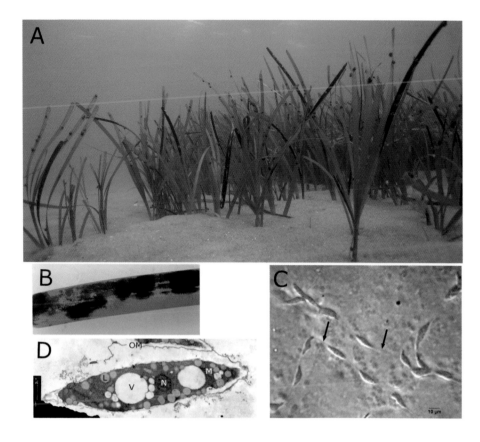

Figure 11.20 A. Eelgrass (*Zostera marina*) meadow in Peconic Estuary, Long Island, NY, showing characteristic black lesions indicating infection with *Labryrinthula* zosterae. B. Closeup of infected leaf. C. Light micrograph of *Labyrinthula* sp. on the surface of agar. Cells move through the outer matrix membrane of the trackway (arrows). D. Transmission electron micrograph of a single *Labyrinthula* sp. cell within infected tissue of the seagrass *Thalassia testudinum*. L, lipid vesicle; M, mitochondria; N, nucleus; OM, outer matrix membrane; V, vacuole. Credits: A. Kimberly Petersen Manzo and Chris Pickerell, Cornell Cooperative Extension Eelgrass Program; B–D. Tim Sherman, University of South Alabama.

and green algae as a source of foods, nutritional supplements, pharmaceuticals, and products such as carageenins and agar is a major industry in many countries, especially China, Japan, Indonesia, Korea, the Philippines, and Chile and some disease outbreaks have been described. The red alga *Porphyra*, harvested as the Japanese food crop nori, is susceptible to a wasting disease resulting in coalescing lesions caused by the oomycete *Pythium porphyrae*. Ascomycete fungi cause infections of *Laminaria*, *Sargassum*, and other kelps. Bacterial infections include "ice-ice disease" of the red seaweed *Kappachus alvarezii*, which is one of the most important cultured species for the production of carageenins. As we have seen in many other situations, *Vibrio* spp. are the most likely causative agents, colonizing the algal fronds if they are stressed by low salinity, high temperature, or poor light. Recent research on the microbiomes of kelps suggests that disease increases might be expected with ocean warming and acidification, due to shifts in the microbiome.

Reports of disease in natural ecosystems include those affecting the coralline algae associated with reefs, originally observed in the 1990s in the Cook Islands and now reported throughout the Indo-Pacific. Bacteria belonging to the genera *Planococcus*, *Bacillus*, and *Pseudomonas* have been identified in association with coralline lethal orange disease (CLOD), an infection that affects encrusting coralline algae. CLOD is identified by the appearance of bright orange coloration spreading to dead areas, although details of the pathogenic mechanisms are scant. Another disease is associated with a black fungal infection. As with other diseases, increased sea-surface temperatures seem to encourage the spread of infection. Other diseases lethal to coralline algae, with currently unknown etiology, have been reported from the Caribbean. Because coralline algae produce very hard deposits of calcium carbonate, they play an important role in cementing the structure of reefs and protecting them from wave damage. Disturbance of the microbiome also affects the settlement and metamorphosis of coral larvae and other invertebrates, so the emergence of these diseases is a cause for concern.

More than 50 species in most of the algal orders have been shown to contain VLPs, and about 20 algal viruses have been isolated and characterized. Most studies have been concerned with the role of viruses in infection of planktonic algae, especially prymnesiophytes, diatoms, and dinoflagellates (see *Chapter 7*) and there are surprisingly few reports of viruses in seaweeds. VLPs have been observed in the reproductive organs of several species of filamentous

> ### ⓘ SHIFTING PARADIGMS—FROM "PATHOGEN" TO "PATHOBIOME"
>
> Propelled by advances in high-throughput sequencings and bioinformatic analysis, our growing recognition of multicellular organisms and their associated microbes as a *holobiont* and knowledge of the importance of the complex interactions of the microbiome is changing our view of the nature of disease. Casadevall and Pirofski (2014) argue that "research on infectious diseases continues to be dominated by reductionist approaches ... what is needed is the simultaneous analysis of microbial and host variables using new analytical tools." We need to shift our focus onto the multipartite *interactions* between hosts and microbes. Pitlik and Koren (2017) hypothesize "that probably all diseases of holobionts, acute or chronic, infectious or non-infectious, and regional or systemic, are characterized by a perturbation of the healthy microbiome into a diseased pathobiome." We now have strong evidence that many complex disease of humans—including non-infectious conditions—are due to this imbalance in microbial communities and a huge number of variables shape its composition, This concept of *dysbiosis* can help to explain why we have found it so difficult to assign causes of disease in corals and other marine organisms.

brown algae in different geographic areas. Production of infectious virions is variable and host viability may not always be affected. The *Ectocarpus siliculosus* virus (EsV-1) has been sequenced, and study of the factors involved in viral replication is yielding valuable information about reproductive development in these algae, because virion release is synchronized with the release of spores or gametes, enabling interaction of viruses with their susceptible host cells. EsV-1 is designated as the founding member of the phaeoviruses, a group of the Phycodnaviridae. Host-specific phaeoviruses have been identified in several species of brown algae including kelps (Laminarinales) and are widely distributed in all temperate coastal waters. Their importance as diseases agents in the ecology and productivity of these algae is still unclear.

Conclusions

The ecological and economic impacts of many marine diseases caused by bacteria, protists, and viruses are increasing. Climate change, pollution, and other anthropogenic factors are undoubtedly influencing the incidence of disease in marine ecosystems. This chapter has provided many examples of evolution of new strains, changes in virulence, and transfer of pathogens between hosts and ecosystems—marine to freshwater, freshwater to marine, and terrestrial to marine. These shifts are often precipitated by movement of organisms between geographic areas, which may occur naturally through migration and inter-mingling of host species. However, we have also seen examples of major epizootics and the development of new reservoirs of infection emerging from the spread of intensive aquaculture and the international movement of fish, shellfish, feedstuffs, and seafood products. Adverse environmental factors may compromise the immune status or general physiology of the host and affect the virulence of pathogens. In many of the examples described, we have seen that it has proved impossible to reliably assign causality to a single pathogen and how difficult it is to unravel the complexity of host–environment–microbe interactions. In many cases, trying to distinguish organisms as "primary" or "opportunist" pathogens is no longer useful. Although interest in diseases of marine organisms has increased greatly in the past few years, the number of scientists directly involved in the detailed study of etiology and pathology of microbial infections of marine life remains very small and lags well behind those involved with study of disease in humans or in other natural ecosystems. Since climate change and other anthropological impacts are causing increased incidence and emergence of new diseases in some groups, we need much better baseline knowledge before we can implement strategies to prevent or limit the impact of diseases in the marine environment. This can only be achieved by increasing the number of researchers studying marine diseases, especially scientists with an integrated approach to the ecology of marine diseases, who employ the most effective technologies available to study organisms within a holobiont perspective. Finally, it is important to consider that the legacy of Koch, Pasteur, and the other founders of pathogenic microbiology in the 1880s may have unwittingly conditioned us to always seek a single microorganism as the causative agent of a specific disease. The reality is that many infectious diseases are multifactorial and depend on shifts in the normal microbial community and complex changes in the host, influenced by multiple environmental factors.

References and further reading

General articles

Baskin, Y. (2006) Sea sickness: the upsurge in marine diseases. *Bioscience* **56**: 464–469.

Burge, C.A., Friedman, C.S., Getchell, R. et al. (2016) Complementary approaches to diagnosing marine diseases: a union of the modern and the classic. *Philos. Trans. R. Soc. B Biol. Sci.* **371**: 20150207.

Burge, C.A., Mark Eakin, C., Friedman, C.S. et al. (2014) Climate change influences on marine infectious diseases: implications for management and society. *Ann. Rev. Mar. Sci.* **6**: 249–277.

Casadevall, A. & Pirofski, L. (2014) Microbiology: ditch the term pathogen. *Nature* **516**: 165–166.

Kasai, H., Nishikawa, S., & Watanabe, K. (2018) Viral diseases affecting aquaculture. In: *Seafood Safety and Quality*, Bari, Md.L. & Yamakzaki, K. (eds.), CRC Press, pp. 112–128.

Lafferty, K.D. (2009) The ecology of climate change and infectious diseases. *Ecology* **90**: 888–900.

Lafferty, K.D. (2017) Marine infectious disease ecology. *Annu. Rev. Ecol. Evol. Syst.* **48**: 473–496.

Lafferty, K.D., Porter, J.W., & Ford, S.E. (2004) Are diseases increasing in the ocean? *Ann. Rev. Ecol. Evol. System.* **35**: 31–54.

McCallum, H.I., Kuris, A., Harvell, C.D. et al. (2004) Does terrestrial epidemiology apply to marine systems? *Trends Ecol. Evol.* **19**: 585–591.

Munn, C.B. (2006) Viruses as pathogens of marine organisms—from bacteria to whales. *J. Mar. Biol. Assoc. U.K.* **86**: 1–15.

Pitlik, S.D. & Koren, O. (2017) How holobionts get sick—toward a unifying scheme of disease. *Microbiome* **5**: 64.

Starr, M., Lair, S., Michaud, S. et al. (2017) Multispecies mass mortality of marine fauna linked to a toxic dinoflagellate bloom. *PLoS One* **12**: e0176299.

Takemura, A.F., Chien, D.M., & Polz, M.F. (2014) Associations and dynamics of *Vibrionaceae* in the environment, from the genus to the population level. *Front. Microbiol.* **5**: 38.

Diseases of corals, sponges and echinoderms

Austin, B., Austin, D., Sutherland, R. et al. (2005) Pathogenicity of vibrios to rainbow trout (*Oncorhynchus mykiss*, Walbaum) and *Artemia nauplii*. *Environ. Microbiol.* **7**: 1488–1495.

Ben-Haim, Y. & Rosenberg, E. (2002) A novel *Vibrio* sp. pathogen of the coral *Pocillopora damicornis*. *Mar. Biol.* **141**: 47–55.

Ben-Haim, Y., Zicherman-Keren, M., & Rosenberg, E. (2003) Temperature-regulated bleaching and lysis of the coral Pocillopora damicornis by the novel pathogen Vibrio coralliilyticus. *Appl. Environ. Microbiol.* **69**: 4236–4242.

Bourne, D.G., Boyett, H. V, Henderson, M.E. et al. (2008) Identification of a ciliate (Oligohymenophorea: Scuticociliatia) associated with brown band disease on corals of the Great Barrier Reef. *Appl. Environ. Microbiol.* **74**: 883–888.

Cárdenas, A., Neave, M.J., Haroon, M.F. et al. (2018) Excess labile carbon promotes the expression of virulence factors in coral reef bacterioplankton. *ISME J.* **12**: 59–76.

Corinaldesi, C., Marcellini, F., Nepote, E. et al. (2018) Impact of inorganic UV filters contained in sunscreen products on tropical stony corals (Acropora spp.). *Sci. Total Environ.* **637–638**: 1279–1285.

Danovaro, R., Bongiorni, L., Corinaldesi, C. et al. (2008) Sunscreens cause coral bleaching by promoting viral infections. *Environ. Health Perspect.* **116**: 441–447.

Davy, S.K., Burchett, S.G., Dale, A.L. et al. (2006) Viruses: agents of coral disease? *Dis. Aquat. Organ.* **69**: 101–110.

Downs, C.A., Kramarsky-Winter, E., Segal, R. et al. (2016) Toxicopathological effects of the sunscreen UV filter, oxybenzone (benzophenone-3), on coral planulae and cultured primary cells and its environmental contamination in Hawaii and the US Virgin Islands. *Arch. Environ. Contam. Toxicol.* **70**: 265–288.

FFWC (2019) Florida Fish and Wildlife Conservation Commission. Stony Coral Tissue Loss Disease. Online at: https://myfwc.com/research/habitat/coral/disease/ (accessed 4 July 2019).

FKNMS (2018) Florida Keys National Marine Sanctuary. Case Definition: Stony Coral Tissue Loss Disease (SCTLD). Online at: https://nmsfloridakeys.blob.core.windows.net/floridakeys-prod/media/docs/20181002-stony-coral-tissue-loss-disease-case-definition.pdf (accessed 4 July 2019).

Garren, M., Son, K., Raina, J.-B. et al. (2014) A bacterial pathogen uses dimethylsulfoniopropionate as a cue to target heat-stressed corals. *ISME J.* **8**: 999–1007.

Garren, M., Son, K., Tout, J. et al. (2016) Temperature-induced behavioral switches in a bacterial coral pathogen. *ISME J.* **10**: 1363–1372.

Gavish, A.R., Shapiro, O.H., Kramarsky-Winter, E., & Vardi, A. (2018) Microscale tracking of coral disease reveals timeline of infection and heterogeneity of polyp fate. *bioRxiv*, 302778.

Gibbin, E., Gavish, A., Domart-Coulon, I. et al. (2018) Using NanoSIMS coupled with microfluidics to visualize the early stages of coral infection by *Vibrio coralliilyticus*. *BMC Microbiol.* **18**: 39.

Gibbin, E., Gavish, A., Krueger, T. et al. (2019) *Vibrio coralliilyticus* infection triggers a behavioural response and perturbs nutritional exchange and tissue integrity in a symbiotic coral. *ISME J.* **13**: 989–1003.

Hallegraeff, G., Coman, F., Davies, C. et al. (2014) Australian dust storm associated with extensive *Aspergillus sydowii* fungal "bloom" in coastal waters. *Appl. Environ. Microbiol.* **80**: 3315–3320.

Harvell, C.D., Montecino-Latorre, D., Caldwell, J.M. et al. (2019) Disease epidemic and a marine heat wave are associated with the continental-scale collapse of a pivotal predator (*Pycnopodia helianthoides*). *Sci. Adv.* **5**: eaau7042.

Hewson, I., Bistolas, K.S.I., Quijano Cardé et al. (2018) Investigating the complex association between viral ecology, environment, and Northeast Pacific sea star wasting. *Front. Mar. Sci.* **5**: 77.

Hewson, I., Button, J.B., Gudenkauf, B.M. et al. (2014) Densovirus associated with sea-star wasting disease and mass mortality. *Proc. Natl. Acad. Sci. USA* **111**: 17278–17283.

Hoff, M. (2007) What's behind the spread of white syndrome in Great Barrier Reef corals? PLoS Biol. **5**: e164.

Huggett, M.J. & Apprill, A. (2019) Coral microbiome database: integration of sequences reveals high diversity and relatedness of coral-associated microbes. *Environ. Microbiol. Rep.* **11**: 372–385.

Kimes, N.E., Grim, C.J., Johnson, W.R. et al. (2012) Temperature regulation of virulence factors in the pathogen *Vibrio coralliilyticus*. *ISME J.* **6**: 835–846.

Lawrence, S.A., Floge, S.A., Davy, J.E. et al. (2017) Exploratory analysis of *Symbiodinium* transcriptomes reveals potential latent infection by large dsDNA viruses. *Environ. Microbiol.* **19**: 3909–3919.

Lesser, M.P., Bythell, J.C., Gates, R.D. et al. (2007) Are infectious diseases really killing corals? Alternative interpretations of the experimental and ecological data. *J. Exp. Mar. Bio. Ecol.* **346**: 36–44.

Lohr, J., Munn, C.B., & Wilson, W.H. (2007) Characterization of a latent virus-like infection of symbiotic zooxanthellae. *Appl. Environ. Microbiol.* **73**: 2976–2981.

Luter, H.M. (2017) Sponge disease and climate change. In: *Climate Change, Ocean Acidification and Sponges*, Carballo, J.L. & Bell, J.J. (eds.), Springer, Cham, pp. 411–428.

Luter, H.M., Bannister, R.J., Whalan, S. et al. (2017) Microbiome analysis of a disease affecting the deep-sea sponge *Geodia barretti*. *FEMS Microbiol. Ecol.* **93**: fix074.

Mera, H. & Bourne, D.G. (2018) Disentangling causation: complex roles of coral-associated microorganisms in disease. *Environ. Microbiol.* **20**: 431–449.

Meron, D., Efrony, R., Johnson, W.R., et al. (2009) Role of flagella in virulence of the coral pathogen *Vibrio coralliilyticus*. *Appl. Environ. Microbiol.* **75**: 5704–5707.

Meyer, J.L., Castellanos-Gell, J., Aeby, G.S. et al. (2019) Microbial community shifts associated with the ongoing stony coral tissue loss disease outbreak on the Florida Reef Tract. *bioRxiv*, 626408.

Miller, M.W., Karazsia, J., Groves, C.E. et al. (2016) Detecting sedimentation impacts to coral reefs resulting from dredging the Port of Miami, Florida USA. *PeerJ* **4**: e2711.

Mills, E., Shechtman, K., Loya, Y., & Rosenbergg, I. (2013) Bacteria appear to play important roles in both causing and preventing the bleaching of the coral *Oculina patagonica*. *Mar. Eco. Prog. Ser.* **489**: 155–162.

Miner, C.M., Burnaford, J.L., Ambrose, R.F. et al. (2018) Large-scale impacts of sea star wasting disease (SSWD) on intertidal sea stars and implications for recovery. *PLoS One* **13**: e0192870.

Montalvo-Proaño, J., Buerger, P., Weynberg, K.D., & van Oppen, M.J.H. (2017) A PCR-Based assay targeting the major capsid protein gene of a dinorna-like ssRNA virus that infects coral photosymbionts. *Front. Microbiol.* **8**: 1665.

Munn, C.B. (2015) The role of vibrios in diseases of corals. *Microbiol. Spectr.* **3**: doi: 10.1128/microbiolspec.VE-0006-2014.

Mydlarz, L.D., Holthouse, S.F., Peters, E.C., & Harvell, C.D. (2008) Cellular responses in sea fan corals: granular amoebocytes react to pathogen and climate stressors. *PLoS One* **3**: e1811.

Nissimov, J., Rosenberg, E., & Munn, C.B. (2009) Antimicrobial properties of resident coral mucus bacteria of *Oculina patagonica*. *FEMS Microbiol. Lett.* **292**: 210–215.

Page, C. & Willis, B. Epidemiology of skeletal eroding band on the Great Barrier Reef and the role of injury in the initiation of this widespread coral disease. *Coral Reefs* **27**: 257–272.

Pascelli, C., Laffy, P.W., Kupresanin, M. et al. (2018) Morphological characterization of virus-like particles in coral reef sponges. *PeerJ* **6**: e5625.

Patterson, K.L., Porter, J.W., Ritchie, K.E. et al. (2002) The etiology of white pox, a lethal disease of the Caribbean elkhorn coral, *Acropora palmata*. *Proc. Natl. Acad. Sci. USA.* **99**: 8725–8730.

Pita, L., Rix, L., Slaby, B.M. et al. (2018) The sponge holobiont in a changing ocean: from microbes to ecosystems. *Microbiome* **6**: 46.

Pollock, F.J., Wada, N., Torda, G. et al. (2017) White syndrome-affected corals have a distinct microbiome at disease lesion fronts. *Appl. Environ. Microbiol.* **83**: e02799-16.

Precht, W.F., Gintert, B.E., Robbart, M.L. et al. (2016) Unprecedented disease-related coral mortality in Southeastern Florida. Sci. Rep. **6**: 31374.

Reshef, L., Koren, O., Loya, Y. et al. (2006) The Coral probiotic hypothesis. *Environ. Microbiol.* **8**: 2068–2073.

Reshef, L., Ron, E., & Rosenberg, E. (2008) Genome analysis of the coral bleaching pathogen *Vibrio shiloi*. *Arch. Microbiol.* **190**: 185–194.

Richardson, L.L. (1998) Coral diseases: what is really known? *Trends Ecol. Evol.* **13**: 438–443.

Rosenberg, E., Kushmaro, A., Kramarsky-Winter, E. et al. (2009) The role of microorganisms in coral bleaching. *ISME J.* **3**: 139–146.

Rypien, K. & Baker, D. (2009) Isotopic labeling and antifungal resistance as tracers of gut passage of the sea fan pathogen *Aspergillus sydowii*. *Dis. Aquat. Organ.* **86**: 1–7.

Rypien, K.L. (2008) African dust is an unlikely source of *Aspergillus sydowii*, the causative agent of sea fan disease. *Mar. Ecol. Ser.* **367**: 125–131.

Rypien, K.L., Andras, J.P., & Harvell, C.D. (2008) Globally panmictic population structure in the opportunistic fungal pathogen *Aspergillus sydowii*. *Mol. Ecol.* **17**: 4068–4078.

Santos, E. de O., Alves, N., Dias, G.M. et al. (2011) Genomic and proteomic analyses of the coral pathogen *Vibrio coralliilyticus* reveal a diverse virulence repertoire. *ISME J.* **5**: 1471–1483.

Sato, Y., Civiello, M., Bell, S.C. et al. (2016) Integrated approach to understanding the onset and pathogenesis of black band disease in corals. *Environ. Microbiol.* **18**: 752–765.

Sato, Y., Ling, E.Y.S., Turaev, D. et al. (2017) Unraveling the microbial processes of black band disease in corals through integrated genomics. *Sci. Rep.* **7**: 40455.

Shapiro, O.H., Kramarsky-Winter, E., Gavish, A.R. et al. (2016) A coral-on-a-chip microfluidic platform enabling live-imaging microscopy of reef-building corals. *Nat. Commun.* **7**: 10860.

Sharp, W. & Maxwell, K. (2018) Investigating the ongoing coral disease outbreak in the florida keys: collecting corals to diagnose the etiological agent(s) and establishing sentinel sites to monitor transmission rates and the spatial progression of the disease. Report, Florida Fish & Wildlife Research Institute. Online at https://floridadep.gov/sites/default/files/FWC-Sentinel-Site-Report-Final.pdf (accessed 4 July 2019).

Soffer, N., Brandt, M.E., Correa, A.M. et al. (2014) Potential role of viruses in white plague coral disease. *ISME J.* **8**: 271–283.

Soler-Hurtado, M.M., Sandoval-Sierra, J.V., Machordom, A., & Diéguez-Uribeondo, J. (2016) *Aspergillus sydowii* and other potential fungal pathogens in gorgonian octocorals of the Ecuadorian Pacific. *PLoS One* **11**: e0165992.

Stacy, B.A., Wellehan, J.F.X., Foley, A.M. et al. (2008) Two herpesviruses associated with disease in wild Atlantic loggerhead sea turtles (*Caretta caretta*). *Vet. Microbiol.* **126**: 63–73.

Sussman, M., Mieog, J.C., Doyle, J., et al. (2009) *Vibrio* zinc-metalloprotease causes photoinactivation of coral endosymbionts and coral tissue lesions. *PLoS One* **4**: e4511.

Sussman, M., Willis, B.L., Victor, S., & Bourne, D.G. (2008) Coral pathogens identified for White Syndrome (WS) epizootics in the Indo-Pacific. *PLoS One* **3**: e2393.

Sutherland, K.P., Berry, B., Park, A. et al. (2016) Shifting white pox aetiologies affecting *Acropora palmata* in the Florida Keys. *Phil. Trans. R. Soc. B* **371**: 20150205.

Sutherland, K.P., Shaban, S., Joyner, J.L. et al. (2011) Human pathogen shown to cause disease in the threatened eklhorn coral *Acropora palmata*. *PLoS One* **6**: e23468.

Sweet, M. & Bythell, J. (2012) Ciliate and bacterial communities associated with White Syndrome and Brown Band Disease in reef-building corals. *Environ. Microbiol.* **14**: 2184–2199.

Sweet, M. & Bythell, J. (2017) The role of viruses in coral health and disease. *J. Invertebr. Pathol.* **147**: 136–144.

Sweet, M., Bulling, M., & Cerrano, C. (2015) A novel sponge disease caused by a consortium of micro-organisms. *Coral Reefs* **34**: 871–883.

Sweet, M.J. & Bulling, M.T. (2017) On the importance of the microbiome and pathobiome in coral health and disease. *Front. Mar. Sci.* **4**: 9.

Sweet, M.J., Jones, R., & Bythell, J.C. (2011) Coral diseases in aquaria and in nature. *J. Mar. Biol. Assoc. UK* **92**: 791–801.

Thurber, R.V., Payet, J.P., Thurber, A.R., & Correa, A.M.S. (2017) Virus–host interactions and their roles in coral reef health and disease. *Nat. Rev. Microbiol.* **15**: 205–216.

Tout, J., Siboni, N., Messer, L.F. et al. (2015) Increased seawater temperature increases the abundance and alters the structure of natural *Vibrio* populations associated with the coral *Pocillopora damicornis*. *Front. Microbiol.* **6**: 432.

Ushijima, B., Videau, P., Burger, A.H. et al. (2014) *Vibrio coralliilyticus* strain OCN008 is an etiological agent of acute *Montipora* white syndrome. *Appl. Environ. Microbiol.* **80**: 2102–2109.

van de Water, J.A.J.M., Allemand, D., & Ferrier-Pagès, C. (2018) Host-microbe interactions in octocoral holobionts - recent advances and perspectives. *Microbiome* **6**: 64.

Walton, C., Hayes, N.K., & Gilliam, D.S. (2018) Impacts of a regional, multi-year, multi-species coral disease outbreak in Southeast Florida. *Front. Mar. Sci.* **5**: 323.

Ward, J.R. & Lafferty, K.D. (2004) The elusive baseline of marine disease: are diseases in ocean ecosystems increasing? *PLoS Biol.* **2**: e120.

Webster, N.S. Sponge disease: a global threat? *Environ. Microbiol.* **9**: 1363–1375.

Wegley Kelly, L., Haas, A.F., & Nelson, C.E. (2018) Ecosystem microbiology of coral reefs: linking genomic, metabolomic, and biogeochemical dynamics from animal symbioses to reefscape processes. *mSystems* **3**: e00162-17.

Weynberg, K.D. (2018) Ecosystems: from open waters to coral reefs. In: Malmstrom, C. (ed), *Environmental Virology and Virus Ecology*. Academic Press, pp. 2–38.

Wilson, B., Muirhead, A., Bazanella, M. et al. (2013) An improved detection and quantification method for the coral pathogen *Vibrio coralliilyticus*. *PLoS One* **8**: e81800.

Wilson, W.H. (2011) Coral viruses. In: Hurst, C.J. (ed), *Studies in Viral Ecology: Animal Host Systems*. Wiley-Blackwell, pp. 143–151.

Wood, E. (2018) Impacts of sunscreens on corals. Report by the International Coral Reef Initiative (ICRI). https://crm.gov.mp/wp-content/uploads/crm/ICRI_Sunscreen_0.pdf (accessed 4 March 2019).

Work, T. & Meteyer, C. (2014) To understand coral disease, look at coral cells. *Ecohealth.* **11**: 610–618.

Work, T.M., Richardson, L.L., Reynolds, T.L., & Willis, B.L. (2008) Biomedical and veterinary science can increase our understanding of coral disease. *J. Exp. Mar. Bio. Ecol.* **362**: 63–70.

Diseases of molluscs

Arzul, I., Corbeil, S., Morga, B., & Renault, T. (2017) Viruses infecting marine molluscs. *J. Invertebr. Pathol.* **147**: 118–135.

Bruto, M., Labreuche, Y., James, A. et al. (2018) Ancestral gene acquisition as the key to virulence potential in environmental *Vibrio* populations. *ISME J.* **12**: 2954–2966.

de Lorgeril, J., Escoubas, J.-M., Loubiere, V. et al. (2018) Inefficient immune response is associated with microbial permissiveness in juvenile oysters affected by mass mortalities on field. *Fish Shellfish Immunol.* **77**: 156–163.

Genard, B., Miner, P., Nicolas, J.-L. et al. (2013) Integrative study of physiological changes associated with bacterial infection in Pacific oyster larvae. *PLoS One* **8**: e64534.

Maloy, A.P., Ford, S.E., Karney, R.C., & Boettcher, K.J. (2007) *Roseovarius crassostreae*, the etiological agent of Juvenile Oyster Disease (now to be known as Roseovarius Oyster Disease) in *Crassostrea virginica. Aquaculture* **269**: 71–83.

Meyer, G.R., Lowe, G.J., Gilmore, S.R., & Bower, S.M. (2017) Disease and mortality among Yesso scallops *Patinopecten yessoensis* putatively caused by infection with *Francisella halioticida. Dis. Aquat. Organ.* **125**:79–84.

Mills, A.M., Ward, M.E., Heyl, T.P., & Van Dover, C.L. (2005) Parasitism as a potential contributor to massive clam mortality at the Blake Ridge Diapir methane-hydrate seep. *J. Mar. Biol. Assoc. United Kingdom* **85**: 1489–1497.

Romalde, J.L., Dieguez, A.L., Lasa, A., & Balboa, S. (2014) New *Vibrio* species associated to molluscan microbiota: a review. *Front. Microbiol.* **4**: 413.

Ward, M.E., Van Dover, C.L., & Shields, J.D. (2004) Parasitism in species of *Bathymodiolus* (Bivalvia: Mytilidae) mussels from deep-sea seep and hydrothermal vents. *Dis. Aquat. Organ.* **62**: 1–16.

Zielinski, F.U., Pernthaler, A., Duperron, S. et al. (2009) Widespread occurrence of an intranuclear bacterial parasite in vent and seep bathymodiolin mussels. *Environ. Microbiol.* **11**: 1150–1167.

Diseases of crustaceans

Bateman, K.S. & Stentiford, G.D. (2017) A taxonomic review of viruses infecting crustaceans with an emphasis on wild hosts. *J. Invertebr. Pathol.* **147**: 86–110.

Cano-Gõmez, A., Goulden, E.F., Owens, L., & Hõj, L. (2010) *Vibrio owensii* sp. nov., isolated from cultured crustaceans in Australia. *FEMS Microbiol. Lett.* **302**: 175–181.

Drake, L.A. & Dobbs, F.C. (2005) Do viruses affect fecundity and survival of the copepod *Acartia tonsa* Dana? *J. Plankton Res.* **27**: 167–174.

Dunlap, D.S., Ng, T.F.F., Rosario, K. et al. (2013) Molecular and microscopic evidence of viruses in marine copepods. *Proc. Natl. Acad. Sci. U. S. A.* **110**: 1375–1380.

Elliott, D. & Tang, K. (2011) Influence of carcass abundance on estimates of mortality and assessment of population dynamics in *Acartia tonsa. Mar. Ecol. Prog. Ser.* **427**: 1–12.

Flegel, T.W. (2009) Current status of viral diseases in Asian shrimp aquaculture. *Isr. J. Aquac.* **61**: 229–239.

Flegel, T.W. (2019) A future vision for disease control in shrimp aquaculture. *J. World Aquac. Soc.* doi.org/10.1111/jwas.12589

Flegel, T.W., Pasharawipas, T., Owens, L., & Oakey, H.J. (2005) Evidence for phage-induced virulence in the shrimp pathogen *Vibrio harveyi*. 329–338. *Dis. Asian Aqiuac.* V: 329–327.

Kawato, S., Shitara, A., Wang, Y. et al. (2018) Crustacean genome exploration reveals the evolutionary origin of white spot syndrome virus. *J. Virol.* **93**: e01144-18.

Labreuche, Y., Chenivesse, S., Jeudy, A. et al. (2017) Nigritoxin is a bacterial toxin for crustaceans and insects. *Nat. Commun.* **8**: 1248.

Munro, J., Oakey, J., Bromage, E., & Owens, L. (2003) Experimental bacteriophage-mediated virulence in strains of *Vibrio harveyi. Dis. Aquat. Organ.* **54**: 187–194.

Oakey, H.J., Cullen, B.R., & Owens, L. (2002) The complete nucleotide sequence of the *Vibrio harveyi* bacteriophage VHML. *J. Appl. Microbiol.* **93**: 1089–1098.

Wang, W. (2011) Bacterial diseases of crabs: a review. *J. Invertebr. Pathol.* **106**: 18–26.

Diseases of fish

Austin, B. & Newaj-Fyzul, A. eds. (2017) *Diagnosis and Control of Diseases of Fish and Shellfish*. John Wiley & Sons.

Austin, B., Pride, A.C., & Rhodie, G.A. (2003) Association of a bacteriophage with virulence in *Vibrio harveyi. J. Fish Dis.* **26**: 55–58.

Austin, B. & Zhang, X.-H. (2006) *Vibrio harveyi*: a significant pathogen of marine vertebrates and invertebrates. *Lett. Appl. Microbiol.* **43**: 119–124.

Beaz-Hidalgo, R. & Figueras, M.J. (2013) *Aeromonas* spp. whole genomes and virulence factors implicated in fish disease. *J. Fish Dis.* **36**: 371–388.

Cárdenas, C., Ojeda, N., Labra, Á., & Marshall, S.H. (2019) Molecular features associated with the adaptive evolution of Infectious Salmon Anemia Virus (ISAV) in Chile. *Infect. Genet. Evol.* **68**: 203–211.

Dallaire-Dufresne, S., Tanaka, K.H., Trudel, M.V. et al. (2014) Virulence, genomic features, and plasticity of *Aeromonas salmonicida* subsp. *salmonicida*, the causative agent of fish furunculosis. *Vet. Microbiol.* **169**: 1–7.

Escobar, L.E., Escobar-Dodero, J., & Phelps, N.B.D. (2018) Infectious disease in fish: global risk of viral hemorrhagic septicemia virus. *Rev. Fish Biol. Fish.* **28**: 637–655.

Guanhua, Y., Wang, C., Wang, X. et al. (2018) Complete genome sequence of the marine fish pathogen *Vibrio anguillarum* and genome-wide transposon mutagenesis analysis of genes essential for in vivo infection. *Microbiol. Res.* **216**: 97–107.

Hjerde, E., Lorentzen, M.S., Holden, M.T.G. et al. (2008) The genome sequence of the fish pathogen *Aliivibrio salmonicida* strain LFI1238 shows extensive evidence of gene decay. 9:616.

Ina-Salwany, M.Y., Al-Saari, N., Mohamad, A. et al. (2019) Vibriosis in fish: a review on disease development and prevention. *J. Aquat. Anim. Health.* **31**: 3–22 .

Kalatzis, P., Castillo, D., Katharios, P. et al. (2018) Bacteriophage interactions with marine pathogenic vibrios: implications for phage therapy. *Antibiotics* **7**: 15.

Kashulin, A., Seredkina, N., & Sørum, H. (2017) Cold-water vibriosis. The current status of knowledge. *J. Fish Dis.* **40**: 119–126.

Krkošek, M. (2017) Population biology of infectious diseases shared by wild and farmed fish. *Can. J. Fish. Aquat. Sci.* **74**: 620–628.

Pallen, M.J. & Wren, B.W. (2007) Bacterial pathogenomics. *Nature* **449**: 835–842.

Reith, M.E., Singh, R.K., Curtis, B. et al. (2008) The genome of *Aeromonas salmonicida* subsp. *salmonicida* A449: insights into the evolution of a fish pathogen. *BMC Genomics* **9**: 427.

Rønneseth, A., Castillo, D., D'Alvise, P. et al. (2017) Comparative assessment of *Vibrio* virulence in marine fish larvae. *J. Fish Dis.* **40**: 1373–1385.

Vanden Bergh, P. & Frey, J. (2013) *Aeromonas salmonicida* subsp. *salmonicida* in the light of its type-three secretion system. *Microb. Biotechnol.* **7**: 381–400.

Vincent, A.T., Trudel, M.V., Freschi, L. et al. (2016) Increasing genomic diversity and evidence of constrained lifestyle evolution due to insertion sequences in *Aeromonas salmonicida*. *BMC Genomics* 17: 44.

Wallace, I.S., McKay, P., & Murray, A.G. (2017) A historical review of the key bacterial and viral pathogens of Scottish wild fish. *J. Fish Dis.* 40: 1741–1756.

Ward, T., Hoedt, F., McLeay, L. et al. (2001) Effects of the 1995 and 1998 mass mortality events on the spawning biomass of sardine, *Sardinops sagax*, in South Australian waters. *ICES J. Mar. Sci.* 58: 865–875.

Whittington, R.J., Crockford, M., Jordan, D., & Jones, B. (2008) Herpesvirus that caused epizootic mortality in 1995 and 1998 in pilchard, *Sardinops sagax neopilchardus* (Steindachner), in Australia is now endemic. *J. Fish Dis.* 31: 97–105.

Disease of marine mammals

Bennett, P.M., Jepson, P.D., Law, R.J. et al. (2001) Exposure to heavy metals and infectious disease mortality in harbour porpoises from England and Wales. *Environ. Pollut.* 112: 33–40.

Doucette, G.J., Cembella, A.D., Martin, J.L. et al. (2006) Paralytic shellfish poisoning (PSP) toxins in North Atlantic right whales *Eubalaena glacialis* and their zooplankton prey in the Bay of Fundy, Canada. *Mar. Ecol. Prog. Ser.* 306: 303–313.

Doucette, G.J., Medlin, L.K., McCarron, P., & Hess, P. (2018) Detection and Surveillance of Harmful Algal Bloom Species and Toxins. In: *Harmful Algal Blooms*. John Wiley & Sons, Ltd, Chichester, UK, pp. 39–114.

Gulland, F.M.D., Dierauf, L.A., & Whitman, K.L. (2018) *CRC Handbook of Marine Mammal Medicine*, 3rd edition. CRC Press.

Gulland, F.M.D. & Hall, A.J. (2007) Is marine mammal health deteriorating? Trends in the global reporting of marine mammal disease. 4: 135–150.

Häussermann, V., Gutstein, C.S., Beddington, M., Cassis, D., Olavarria, C., Dale, A.C. et al. (2017) Largest baleen whale mass mortality during strong El Niño event is likely related to harmful toxic algal bloom. *PeerJ* 5: e3123.

Jepson, P.D., Bennett, P.M., Allchin, C.R. et al. (1999) Investigating potential associations between chronic exposure to polychlorinated biphenyls and infectious disease mortality in harbour porpoises from England and Wales. *Sci. Total Environ.* 244: 339–348.

Jepson, P.D., Deaville, R., Barber, J.L. et al. (2015) PCB pollution continues to impact populations of orcas and other dolphins in European waters. *Sci. Rep.* 6: 18573.

Jo, W.K., Osterhaus, A.D., & Ludlow, M. (2018) Transmission of morbilliviruses within and among marine mammal species. *Curr. Opin. Virol.* 28: 133–141.

Lefebvre, K.A., Bargu, S., Kieckhefer, T., & Silver, M.W. (2002) From sanddabs to blue whales: the pervasiveness of domoic acid. *Toxicon* 40: 971–977.

Nymo, I.H., Tryland, M., & Godfroid, J. (2011) A review of *Brucella* infection in marine mammals, with special emphasis on *Brucella pinnipedialis* in the hooded seal (*Cystophora cristata*). *Vet. Res.* 42: 93.

Pyenson, N.D., Gutstein, C.S., Parham, J.F. et al. (2014) Repeated mass strandings of Miocene marine mammals from Atacama Region of Chile point to sudden death at sea. *Proc. R. Soc. B Biol. Sci.* 281: 20133316–20133316.

Rolland, R.M., Hamilton, P.K., Kraus, S.D. et al. (2007) Faecal sampling using detection dogs to study reproduction and health in North Atlantic right whales (*Euhalaena glacialis*). *J. Cetac. Res. Man.* 8: 121–125.

Ross, P.S. (2002) The role of immunotoxic environmental contaminants in facilitating the emergence of infectious diseases in marine mammals. *Hum. Ecol. Risk Assess.* 8: 277–292.

Ross, P.S., De Swart, R.L., Timmerman, H.H. et al. (1996) Suppression of natural killer cell activity in harbour seals (*Phoca vitulina*) fed Baltic Sea herring. *Aquat. Toxicol.* 34: 71–84.

Scholin, C.A., Gulland, F., Doucette, G.J. et al. (2000) Mortality of sea lions along the central California coast linked to a toxic diatom bloom. *Nature* 403: 80–84.

Smith, J., Connell, P., Evans, R.H. et al. (2018) A decade and a half of *Pseudo-nitzschia* spp. and domoic acid along the coast of southern California. *Harmful Algae* 79: 87–104.

Whatmore, A.M., Dawson, C., Muchowski, J. et al. (2017) Characterisation of North American *Brucella* isolates from marine mammals. *PLoS One* 12: e0184758.

Diseases of turtles

Chaloupka, M., Balazs, G.H., & Work, T.M. (2009) Rise and fall over 26 years of a marine epizootic in Hawaiian green sea turtles. *J. Wildl. Dis.* 45: 1138–1142.

Greenblatt, R.J., Work, T.M., Balazs, G.H. et al. (2004) The *Ozobranchus leech* is a candidate mechanical vector for the fibropapilloma-associated turtle herpesvirus found latently infecting skin tumors on Hawaiian green turtles (*Chelonia mydas*). *Virology* 321: 101–110.

Greenblatt, R.J., Work, T.M., Dutton, P. et al. (2005) Geographic variation in marine turtle fibropapillomatosis. *J. Zoo Wildl. Med.* 36: 527–530.

Diseases of seagrasses and algae

Hughes, R.G., Potouroglou, M., Ziauddin, Z., & Nicholls, J.C. (2018) Seagrass wasting disease: nitrate enrichment and exposure to a herbicide (Diuron) increases susceptibility of *Zostera marina* to infection. *Mar. Poll. Bull.* 134: 94–98.

Lachnit, T., Thomas, T., & Steinberg, P. (2016) Expanding our understanding of the seaweed holobiont: RNA viruses of the red alga *Delisea pulchra*. *Front. Microbiol.* 6: 1489.

McKeown, D., Schroeder, J., Stevens, K. et al. (2018) Phaeoviral infections are present in *Macrocystis*, *Ecklonia* and *Undaria* (Laminariales) and are influenced by wave exposure in Ectocarpales. *Viruses* 10: 410.

Qiu, Z., Coleman, M.A., Provost, E. et al. (2019) Future climate change is predicted to affect the microbiome and condition of habitat-forming kelp. *Proc. Roy. Soc. B* 286: 20181887.

Sullivan, B.K., Trevathan-Tackett, S.M., Neuhauser, S., & Govers, L.L. (2018) Host-pathogen dynamics of seagrass diseases under future global change. *Mar. Poll. Bull.* 134: 75–88.

Zozaya-Vald, E., Roth-Schulze, A.J., Egan, S., & Thomas, T. (2017) Microbial community function in the bleaching disease of the marine macroalgae *Delisea pulchra*. *Environ. Microbiol.* 19: 3012–3024.

Chapter 12

Marine Microbes as Agents of Human Disease

The overwhelming majority of marine microbes that are naturally present in the sea (indigenous or autochthonous organisms) are nonpathogenic for humans. However, there are several important diseases caused by infection with bacteria naturally associated with marine animals as part of their normal microbiota. Humans can become infected when marine bacteria colonize the intestinal tract through ingestion of contaminated seawater or seafood; or they may enter the body via damaged skin or wounds. Such diseases are not normally transmitted from person-to-person, although cholera provides a very important exception. Humans may also become ill following exposure to toxins produced by bacteria, dinoflagellates, and diatoms, usually via ingestion of fish or shellfish that have accumulated the toxins in their tissues. A small number of infectious diseases of fish and marine mammals (zoonoses) can be transmitted to humans. Infectious diseases are also caused by viruses and bacteria introduced into the sea via sewage pollution; these are discussed in *Chapter 13*.

Key Concepts

- Several species of *Vibrio* whose normal habitat is seawater, plankton, and sediments in coastal and marine environments are major human pathogens that are becoming increasingly important as a result of climate change.

- Interactions with phages, extensive acquisition and exchange of genetic information, and complex regulatory mechanisms are responsible for the population structure and evolution of *Vibrio* spp. in the aquatic environment and their virulence in human infections.

- Several syndromes of "seafood poisoning" occur following consumption of fish or shellfish containing certain marine bacteria, their toxins, or their metabolic products.

- Zoonotic infections can be acquired by contact with fish or marine mammals.

- Toxins produced by cyanobacteria and microalgae (dinoflagellates and diatoms) may cause illness after ingestion of fish or shellfish that have accumulated the toxins by feeding.

- The incidence, distribution, and effects of toxic harmful algal blooms are increasing due to climate change and other anthropogenic factors; detection and management of outbreaks is a major activity of public health and fisheries authorities.

MICROBIAL INFECTIONS

Pathogenic vibrios are common in marine and estuarine environments

Most members of the genus *Vibrio* can tolerate a range of salinities and they are therefore widely distributed in marine and estuarine environments throughout the world, especially in warmer waters above 17°C. They are associated with plankton and the surfaces of marine animals and occur in high densities in filter-feeding shellfish. In particular, vibrios have numerous adaptations for colonization and utilization of chitin, which means that they are particularly associated with the exoskeleton of crustaceans, such as copepods. As discussed in *Chapter 11*, several species are important pathogens of marine fish and invertebrates, but there are also about 12 species that have been associated with human diseases. Of these, the most important are *Vibrio cholerae*, *V. vulnificus*, and *V. parahaemolyticus*, discussed in detail in this chapter. Other species are occasionally associated with diarrhea or wound infections, usually following consumption or handling of seafood.

Vibrio cholerae is an autochthonous aquatic bacterium

The disease cholera (caused by *V. cholerae*) was originally confined to parts of India until the early nineteenth century, when six major pandemics of the classical form of the disease spread to Europe and North America. Throughout the twentieth century, cholera continued to spread to become a major cause of mortality throughout the world, aided by increased intercontinental transport, wars, and natural disasters. The 1960s saw the emergence of the El Tor biotype from the Celebes Islands of Indonesia, which has spread to become the seventh global pandemic. The El Tor biotype has reduced virulence compared with the previous "classical" *V. cholerae* strains but is better adapted for transmission. In recent years, another more virulent El Tor variant has emerged in several parts of Asia and Africa. There are >200 serotypes of the bacterium, distinguished by the lipopolysaccharide O-antigens on their cell surface, although they are biochemically similar. Until 1993, all epidemic cases of cholera were caused by *V. cholerae* possessing antigen O1 (including the El Tor biotype), but in 1993 a new serotype, O139, emerged as the cause of epidemics that began in the Bay of Bengal. To date, this serotype has been largely confined to Asia and its incidence has recently declined.

Cholera is characterized by profuse diarrhea caused by fluid and electrolyte loss from the small intestine and is therefore easily spread from person-to-person via contamination of water and food. This concept developed from John Snow's famous deduction in 1854 that a particular point source (the Broad Street water pump) was responsible for clusters of disease outbreaks in London. However, the history of cholera through the ages reveals a striking association with the sea, and epidemiologists have long suspected that there might be an aquatic environmental reservoir of *V. cholerae*. Research by Rita Colwell and colleagues, beginning in the 1970s, changed our thinking about the ecology of *V. cholerae* and we now realize that the bacterium possesses numerous signaling and regulatory adaptations that equip it for life both as an autochthonous inhabitant of marine and estuarine waters and for colonization of the human host.

The survival of the pathogen in water is greatly affected by environmental conditions, particularly salinity, temperature, and nutrient concentration. *V. cholerae* can survive for long periods in seawater, but the numbers that can be isolated from water are often very low and insufficient to initiate infection. Comparison of direct epifluorescence and viable counts shows that a large proportion of the cells enter the "viable but nonculturable" (VBNC) state (see p.95). During transition to the VBNC state, the cells initiate an active program of change, resulting in reduced size, alteration of mRNA, and changes to the cell surface. This adaptation allows bacteria to survive adverse changes in nutrient concentration, salinity, pH, and temperature. *V. cholerae* associates with a range of phytoplankton and zooplankton, especially copepods, and with egg masses of chironomid flies. Like other vibrios, *V. cholerae* possesses chitin-binding proteins and chitinase, which play key roles in colonization of surfaces and biofilm formation in the environment, under the control of a two-component sensor system (see p.103). Attachment to chitin has been shown to lead to upregulation of numerous genes in chitin degradation and protein synthesis, whilst genes for motility and chemotaxis are suppressed.

CHOLERA REMAINS A MAJOR WORLD HEALTH PROBLEM

Cholera has been endemic in the Indian subcontinent for millennia. Seven global pandemics have resulted in untold human misery. Today, major epidemics occur regularly, especially in Africa and Asia during floods, earthquakes, wars, and other disasters. The disease requires notification to the World Health Organization (WHO) under international health regulations, but most cases are not reported because countries fear damage to trade and tourism. It is estimated that there are ~2.9 million cases, with up to 95,000 deaths annually, mostly in young children (WHO, 2017). Attempts to develop long-lasting effective vaccines for use in endemic areas have had only partial success. Formalin-killed whole cells combined with purified B-subunits has proved the most successful and can provide up to 85% protection, although this is short-lived. A live attenuated vaccine—in which the active A-subunit of the toxin has been deleted by genetic modification—has also been used. Vaccination is regarded as only a complementary prevention and control measure, with improvement of sanitation and clean water of paramount importance (WHO, 2017). This remains a major challenge in many overpopulated areas of the world.

Complex regulatory networks control human colonization and virulence of *V. cholerae*

Major changes in gene expression occur in the transition from the aquatic environment to the human host, and *vice versa*. These involve hundreds of genes, in sets (regulons) under the control of master regulators that in turn respond to environmental factors via signaling mechanisms and quorum sensing (p.49). Following ingestion, the bacteria encounter dramatic environmental changes, particularly the shock of temperature increase in the body and low pH as it encounters the acid of the stomach. There is a complex process of gene regulation as the bacterium responds to chemotactic signals, swims toward the surface of the lumen of the small intestine and attaches to the epithelium via pili on the bacterial surface. The bacteria then produce a toxin that affects membrane permeability, stimulating loss of electrolytes and water from the small intestine (although it does not lead to permanent damage to the cells). Each molecule of the toxin is composed of two different protein subunits. There are five B-subunits that bind to membrane receptors (carbohydrate-containing lipids known as gangliosides) on the surface of the gut cells, and one A-subunit that is inserted into the host cell membrane. The A-subunit is an ADP-ribosylating enzyme that modifies a host regulatory protein. This results in continued activation of the enzyme adenyl cyclase in the gut cells, leading to increased levels of cyclic AMP responsible for excessive fluid and electrolyte loss by activating a transmembrane regulator chloride channel. Fluids in the gut cells are replaced from interstitial tissue, plasma, and deeper tissues, and cholera victims usually die from extreme dehydration as fluids flow uncontrolled from the body. The dehydration can be limited by oral rehydration therapy or intravenous drips; in these circumstances, a patient can pass up to 20 liters of fluid a day. However, despite the effectiveness of this simple treatment, it is often very difficult to deliver during epidemics in field hospitals in remote areas.

The cholera pathogen possesses an amazingly complex set of interconnected signaling and regulatory pathways to maximize its dissemination and survival, both in the host and in the aquatic reservoir. Infection results in the copious secretion of water and salts by cells of the small intestine, leading to the characteristic "rice-water stool" containing over 10^7 viable bacteria per milliliter. The excreted bacteria are hyperinfectious, with an infectious dose about 100–1000 times lower than laboratory-cultured bacteria. This transient phenotypic change drives the rapid person-to-person transfer during a cholera epidemic, but decays when the bacterium enters the aquatic environment. In the late stages of infection, *V. cholerae* expresses genes that prepare it for survival outside of the host. These include genes involved in motility, transport, and regulation of attachment to chitin and its subsequent utilization. The ecology of *V. cholerae* in the aquatic environment and human hosts is illustrated in *Figure 12.1*.

The ecology of *V. cholerae* in the natural environment, and the initiation and decline of cholera epidemics is greatly affected by dynamic interactions between the bacteria and vibriophages. Some filamentous vibriophages are lysogenic (p.201) including CTXΦ, which encodes the cholera toxin, plus a wide range of lytic tailed vibriophages. These regulate the density and population structure of *V. cholerae* during plankton blooms due to environmental factors. In the host, the density of lytic vibriophages increases concomitantly with the accumulation of large numbers of *V. cholerae* in the intestinal fluids during late stages of infection. The continued action of these lytic vibriophages when they return to the aquatic environment may explain the decline of cholera epidemics.

Mobile genetic elements play a major role in the biology of *Vibrio* spp.

There is evidence that the ancestors of the modern *Vibrio* spp. have undergone major shifts in their genetic makeup, which explains their transition from autochthonous aquatic bacteria to pathogens. Genetic information in vibrios is divided between two distinct chromosomes. The larger chromosome (Chr1) is of similar size in all vibrios (typically ~3.0 Mb) and contains most of the "housekeeping" genes for growth and viability, as well as the major pathogenicity factors, which are clustered in distinct regions known as pathogenicity islands that have a different G:C base pair ratio to other parts of the chromosomes. Many of these genes show

OLD CLOTHES CAN PREVENT CHOLERA

Knowledge of the close association of *V. cholerae* with zooplankton as a key factor in the ecology of cholera has led to a simple but effective method of control. In Bangladesh, efforts to provide clean water from wells have led to disastrous levels of poisoning by arsenic leached from the rocks, and many people still rely on the collection of untreated surface water from ponds and rivers. This is an obvious source of cholera and other diseases, especially as fuel to make water safe by boiling is in very short supply. Therefore, field workers have trained local people to use a simple cloth filter when collecting water. An old sari folded eight times provides an effective trap for the zooplankton and small particles, to which most of the *V. cholerae* are attached. Trials conducted by Colwell et al. (2003) over a number of years have shown that this simple method can prevent almost 50% of new cases in villages where it is used. Follow-up studies after several years showed that the method was sustainable and still provided protection (Huq et al., 2010). Although this has little effect on person-to-person transfer once the disease is established in the population, this method reduces the opportunities for the pathogen to transfer from the aquatic environment into humans.

Figure 12.1 Schematic diagram of the ecological interactions of *Vibrio cholerae*. (1) In the aquatic reservoir, *V. cholerae* strains of diverse genetic makeup (shown by different colors) are associated with biofilms on particles, copepods etc. Attachment to chitin promotes natural transformation by DNA released via lytic phages and T6SS action. (2) Under appropriate environmental conditions, enrichment of pathogenic *V. cholerae* (shown as red cells) occurs; these can infect humans when contaminated water is ingested. (3) Bacteria migrate through the mucus to colonize the epithelium of the small intestine (4), mediated by attachment via toxin co-regulated pili (TCP). (5) Production of cholera toxin (CT) leads to excessive secretion of body fluids and salts from the intestine, leading to watery diarrhea containing large numbers of hyperinfectious *V. cholerae* (shown by red cells with yellow marking). (6) These escape from the mucus and propagate widespread epidemic transmission and seeding of the environment. Note that changes in the abundance of lytic phages in the host and environment also have a major effect on initiation and decline of epidemics (not shown, see text).

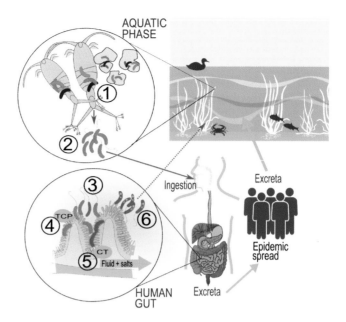

evidence of acquisition by horizontal transfer. The small chromosome (Chr2) is more variable in size (0.8–2.4 Mb) and gene composition. Although the chromosomes are independent, the replication of Chr2 depends on the replication of a specific gene in Chr1, so that termination of replication of both chromosomes and the subsequent cell division are coordinated. The genes on both chromosomes are expressed differently in the aquatic and human environments. This genome flexibility is thought to play a major role in the adaptation of vibrios to various niches during their evolution.

In *V. cholerae*, the *ctxA* and *ctxB* genes encoding the two subunits of the toxin are derived from integration into the chromosome of the genome of a phage (CTXΦ). Outside of areas contaminated by excreta from infected persons, most environmental isolates do not possess the *ctx* genes. Other genes, encoding accessory toxins, are involved in the morphogenesis of CTXΦ. The second major factor in pathogenicity is the presence of toxin coregulated pili (TCP) that coat the bacterial surface. The *tcp* genes encoding these pili are clustered in a 40 kb chromosomal pathogenicity island (VPI-1). This region may also be a phage genome (VPIΦ) and some authorities argue that the pili act as a receptor for CTXΦ. VPI-1 also encodes ToxT, which is a key regulatory element for the transcription of both *ctxAB* and the *tcp* operon; thus, the remarkable situation occurs in which a regulator on one mobile element controls the expression of a toxin on another.

ToxR and ToxS are transmembrane proteins that also regulate pili and toxin production in response to environmental signals. Intriguingly, homologs of both proteins have been found to be widely distributed in both pathogenic and nonpathogenic species in the family Vibrionaceae, so it appears that they evolved in the distant past as part of the ancestral genome. Their function is to control the synthesis of outer membrane porins that regulate transport of ions and small molecules; their relative abundance in the membrane alters in response to osmotic pressure. In *V. cholerae*, these signals have evolved to control toxin and pilus gene expression through a complex signal transduction cascade involving other membrane signaling proteins that are encoded by the CTX and TCP pathogenicity islands. The evolution of pandemic *V. cholerae* strains can be explained by these shifts in genetic composition. Two islands in the genome of seventh pandemic (El Tor) strains (VSP-1 and VSP-2) encode genes responsible for altered carbohydrate metabolism and functions thought to be essential for the characteristic environmental persistence and spread of the El Tor biotype. Another shift involved the horizontal gene transfer of genes encoding enzymes involved in synthesis of lipopolysaccharide, which led to the emergence of strain O139. The key to this extensive genome evolution is the large integron island (123.5 kb) found on the small chromosome; this is a cluster of genes that captures DNA sequences from a wide variety of sources, integrating them by site-specific recombination and rearranging them to form functional genes.

As noted above, attachment to chitin plays a key role in the life cycle of vibrios. Recently, attachment to chitin has been shown to induce production of a Type VI Secretion System (T6SS) whose primary function is the delivery of effector molecules that kill other bacteria in the environment (*Figure 12.2a*). This is thought to provide a competitive advantage to vibrios—by the liberation of nutrients from other bacteria—growing in biofilm communities on the surface of crustacean exoskeletons or on chitin rich particles in the water. Cells of closely related *V. cholerae* are protected from self-destruction by immunity proteins. Furthermore, this process releases DNA from other bacteria. Chitin also induces natural competence—the ability to acquire exogenous DNA by transformation—in which a type IV pilus produced on the cell surface can attach to DNA and retract into the cytoplasm (*Figure 12.2b*) before processing prior to integration of selected genes. This horizontal transfer of a large cluster of genes is believed to explain the continuous evolution of new strains of *V. cholerae*, such as the sudden emergence of the epidemic O139 Bengal serotype in 1992, and its rapid spread in India and Bangladesh. Chitin-induced regulatory programs and T6SS systems have also been demonstrated in other pathogenic and symbiotic vibrios—*V. vulnificus, V. parahaemolyticus, V. alginolyticus, V. coralliilyticus*, and *Aliivibrio fischeri*—and undoubtedly play a key role in their survival, population structure, and genetic adaptation in the natural aquatic environment as well as in their animal hosts.

Non-O1 and non-O139 serotypes of *Vibrio cholerae* are widely distributed in coastal and estuarine waters

There are numerous other strains of *V. cholerae* that do not possess either the O1 or O139 antigens characteristic of epidemic cholera strains. These non-O1 and non-O139 serotypes lack the major cholera toxin but they can cause localized mild to moderate gastroenteritis and

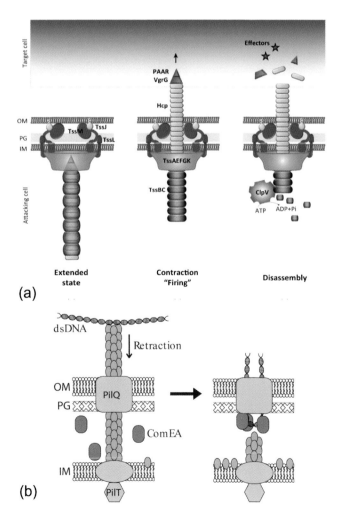

Figure 12.2 Processes in chitin-induced DNA acquisition in *Vibrio* spp. (a) Model of the structure and mechanism of the Type 6 Secretion System, showing the "aim, load, fire" mechanism of a contractile sheath-like structure and expulsion of a cell-puncturing device. Toxins and other effector molecules with various functions are delivered directly into target cells (other bacteria or animal host cells). (b) Model of the "catch and reel" mechanism of natural transformation, based on real-time fluorescence video microscopy of *Vibrio cholerae*. dsDNA released from other bacterial cells binds to the tip of a specialized pilus that is pulled through the PilQ pore via a molecular ratchet mechanism involving the periplasmic protein ComEA. Abbreviations: IM inner membrane, OM outer membrane, PG peptidoglycan. Credits: A. Reprinted from Cianfanelli et al. (2016) with permission of Elsevier. B. Reprinted from Ellison et al. (2018) with permission of Springer Nature.

Are human Vibrio infections a microbial barometer for climate change?

Sea surface temperatures (SST) are increasing. The well-known effects of high levels of atmospheric CO_2 (together with other gases such as methane and nitrous oxide) in promoting global warming via the "greenhouse effect" are a major topic of current socio-political concern. The most immediate effects in our oceans are the rise in sea surface temperature (SST) and ocean acidification (OA). Since 1950, the mean SST of the Indian, Atlantic, and Pacific Oceans has increased by 0.7°C, 0.5°C and 0.3°C and respectively (IPCC, 2018; *Figure 12.3*). Within this, there are great regional variations, with extended summer periods and extreme temperature anomalies ("heatwaves") more frequently observed at high latitudes and in enclosed bodies of water.

Seasonality of *Vibrio cholerae* epidemics. Rita Colwell and her numerous co-workers conducted many detailed studies of the environmental factors affecting the epidemiology of cholera outbreaks (reviewed by Lipp et al. (2002); Almagro-Moreno and Taylor (2013). Clear seasonal patterns linked to climate are apparent in areas where cholera is endemic. In India, retrospective analysis of epidemics since the nineteenth century shows strong linkage to monsoon events. In the Bengal region, regular screening of water for the abundance of *V. cholerae* shows strong correlation with increased SST, mean wave height, and blooms of phyto- and zooplankton. Baracchini et al. (2017) developed a model that shows how rainfall and temperature explain the seasonality of cholera and the dynamics of outbreaks in different years, depending on the fluctuations in the aquatic reservoir of the pathogen.

Other *Vibrio* spp. do not cause epidemics. However, *V. vulnificus*, *V. parahaemolyticus*, *V. alginolyticus*, non-O1/O139 *V. cholerae*, and other *Vibrio* spp. can all cause significant disease, either by wound infection or consumption of contaminated seafood. Baker-Austin et al. (2017) have dubbed vibrios "the microbial barometer of climate change," because *Vibrio* infections are increasing and being recorded in areas where they have not occurred previously.

Although the number of cases remains small, official reporting of non-cholera *Vibrio* infections is not mandatory in most countries, and the real public health impact is probably considerably greater than realized. Increasing SST and rising sea levels is leading to greater risk of infection from vibrios in estuarine and coastal waters in temperate regions, where they thrive in warm (>15°C) seawater with low to medium salinity (<25 ppt NaCl).

Long-term increase in the relative abundance of vibrios. Determining whether there have been changes in the numbers of particular microbes over long periods is difficult because of the lack of historical records. A study by Vezzulli et al. (2012) developed a pioneering approach by developing methods to apply modern DNA analysis techniques to archived samples from the Continuous Plankton Recorder (CPR) survey maintained by the SAHFOS/MBA Laboratory in Plymouth, UK. This sampler is towed by commercial ships and provides the longest running dataset of plankton samples collected over 60 years. It has a mesh size of 270 μm and was designed for collecting zooplankton, which are trapped on a moving band of silk. However, small zooplankton and phytoplankton and associated epibiotic bacteria are also trapped. Vezzulli and colleagues were able to extract DNA from formalin-preserved silk samples that had been collected from transects in two areas of the North Sea from 1961 to 2005. After, preliminary work to overcome the damaging effects of long-term formalin storage on DNA integrity, they used Q-PCR (p.47) with general and *Vibrio*-specific primers to provide estimates of the relative abundance of *Vibrio* spp. as a proportion of the total bacterial counts associated with the plankton. The authors found a long-term increase in relative *Vibrio* abundance, with a statistically significant correlation with the summer SST in samples taken around the Rhine estuary (but not the Humber estuary, which experiences lower peaks of summer temperature). Using high-throughput sequencing to compare samples taken in the 1961–1972 and 1998–2004 periods, Vezzulli et al. (2012) also showed that a major shift in bacterial community composition had occurred, with members of

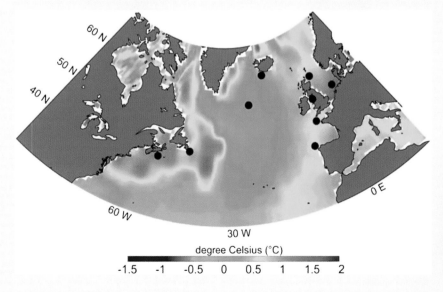

Figure 12.3 Changes in North Atlantic SST calculated as the difference between SST averages for the years 2000–2011 and 1890–1958. Reprinted from Vezzulli et al. (2016) with permission of National Academies of Science.

BOX 12.1   RESEARCH FOCUS

the Vibrionaceae increasing in dominance. This was explained by a "regime shift" in plankton composition, known to have occurred at this time due to changes in water circulation in the North Sea. Because of the strong association of vibrios with plankton, much of the observed increase observed in this study could be explained by a temperature-related increase in the abundance of plankton, especially chitin-rich copepods. It is not known if the number of free-living vibrios has also increased. In their follow-up study, Vezzulli et al. (2016) extended their analysis to 133 CPR samples collected at nine locations in the North Atlantic (marked as dots on *Figure 12.3*). In most of the areas, there was a strong positive correlation between decadal SST values and the relative abundance of vibrios. The increase was greatest in the last ten years surveyed, during which there was an abrupt rise in SST. Statistical modeling showed that the main driver of changes in *Vibrio* abundance is the general increase in the Northern Hemisphere Temperature, although the influence of other natural oscillations in chemical and physical processes are an additional contributory factor. As in their 2012 North Sea study, Vezzulli and colleagues found evidence from the CPR samples that long-term changes in phyto- and zooplankton abundance and species composition affected the relative abundance of vibrios and hypothesized that "a poleward transport of *Vibrio* species mediated by zooplankton will occur as a result of global warming."

Increased incidence of *Vibrio* infections. Baker-Austin et al. (2012) provided the first empirical evidence of a clear link between climate change and an unexpected increase in *Vibrio* infections observed in Northern Europe, especially around the Baltic Sea. This is one of the largest low-salinity seas, where recent increases in SST have been about seven times the global average, with particularly high summer temperatures. Statistical methods employed by Baker-Austin and colleagues showed that the number and

distribution of annual cases of *Vibrio* infections is strongly correlated with peaks in summer SST. Most infections are probably acquired through leisure activities such as swimming and watersports. Subsequently, Vezzulli et al. (2016) showed that the rise in *Vibrio* infections in Northern Europe matched the increase in abundance of plankton-associated vibrios in the archived CPR samples. This was especially notable in the heatwave summer of 2006, when an unprecedented 60 reported *Vibrio* infections in the North Sea and Baltic Sea areas coincided with one of the highest levels of the *Vibrio* abundance index from the CPR analysis. Similar trends of increasing *Vibrio* diseases were observed using data from the US Atlantic coast and northwest Spain.

A monitoring system to evaluate the risks of *Vibrio* infections. Models developed using the information from the Baker-Austin et al. (2012) study were used by the European Centre for Disease Prevention and Control to develop a web-based mapping system based on remote sensing data of SST and salinity of coastal waters from sampling stations throughout the world (Semenza et al., 2017). This provides daily prediction of areas that provide particularly amenable conditions for the growth of vibrios. In a study of the Swedish coast in the Baltic Sea, Semenza et al. found that areas predicted suitable for growth of vibrios showed a record increase in reported vibrio infections (mostly of the ear, wounds, or septicemia), during the heatwave summer of 2014. They suggest that the mapping tool could be used to issue risk-based public health warnings for recreational activities and harvesting shellfish at critical times. Baker-Austin et al. (2017) review the properties of vibrios that make them ideal candidates to assess the effect of climate change on the spread of marine waterborne diseases and discuss the further research and improved surveillance measures needed to provide reliable, quantifiable data for future risk-assessment models.

watery diarrhea, possibly as a result of the production of accessory toxins that damage the membranes or disrupt the cytoskeleton of gut cells. These strains are widely distributed as autochthonous components of coastal and estuarine waters and can survive and multiply in a wide range of seafoods; infections are most commonly associated with eating raw or undercooked seafood, especially oysters. Infections are seasonal and show a peak with increased water temperatures in late summer and early fall, coinciding with the warmest water temperatures. They are also capable of infecting wounds and causing redness and swelling at the site of infection. Septicemia can occur in people with reduced immunity or liver disease.

Vibrio vulnificus is a deadly opportunistic pathogen

V. vulnificus is part of the natural microbiota of estuarine and coastal water, sediment, plankton, and shellfish throughout the world, especially tropical or subtropical regions. Human infection occurs either by the infection of wounds or by primary septicemia following the consumption of contaminated shellfish. Increased awareness has recently led to recognition of cases in the Caribbean islands, Japan, and Taiwan, and there is growing evidence of the spread of infections to new areas, probably linked to climate change. In healthy individuals, consumption of contaminated raw shellfish usually results in an unpleasant, but not life-threatening, gastroenteritis. However, in individuals with chronic underlying diseases (especially liver cirrhosis, immunodeficiency, or diabetes) a septicemic form of the disease may result, with about 90% of cases requiring hospitalization and a mortality rate of about 50%. In the USA it has the highest economic impact of all food-related illnesses. The US Gulf states require mandatory health warnings to be displayed wherever raw oysters are served. The infective dose is believed to be very low, with estimates ranging from 100 to 300 bacteria needed to infect susceptible persons. Males are much more susceptible to infection, possibly owing to hormonal differences. A key feature of this pathogen is its extremely rapid growth in host tissues, and death can occur as little as 24 h after infection. Infected persons show fever, chills, shock, and large skin lesions filled with bloody fluid (*Figure 12.4A*). These can become necrotic and so extensive that surgical removal of tissue or amputation of limbs is often necessary. As well as causing food-borne disease, *V. vulnificus* can also cause a very serious infection following contamination of wounds. This can occur if existing open wounds are contaminated by seawater containing the pathogen. Wounds obtained from shucking oyster shells, fishing hooks, fish spines, and the like are a particularly likely source (*Figure 12.4B*).

Pathogenicity of *V. vulnificus* is due to the interaction of multiple gene products

V. vulnificus is clearly a very aggressive organism able to circumvent many of the body's defense mechanisms, especially when these are undermined by host susceptibility factors. An extracellular capsule is especially important in conferring resistance to phagocytosis and the bactericidal effect or complement in the serum. Strains of the pathogen that naturally lack the capsule, or mutants in which the genes encoding the capsule are disrupted, lose their virulence. A second factor contributing to the rapid growth of the pathogen is its ability to scavenge iron as a nutrient from the host. *V. vulnificus* also produces powerful cytolytic and proteolytic enzymes that are responsible for dissemination from the intestine and the extensive tissue damage. In fatal cases, death is primarily due to the effect of endotoxin (LPS) which stimulates overproduction of cytokines with consequent host damage. Many attempts have been made to identify genotypic or phenotypic markers of pathogenicity using DNA analysis—including recent full genome comparisons—but this is complicated by the extensive horizontal gene transfer prevalent in the vibrios.

Environmental factors affect the pathogenicity of *V. vulnificus*

Three biotypes of *V. vulnificus* have been described: Biotype 1 contains human clinical and related environmental strains; Biotype 2 is a pathogen of farmed fish; and Biotype 3 appears to be a hybrid strain causing wound infections associated with handling fish. Two distinct genotypes of *V. vulnificus* Biotype 1 have been described, based on allelic variation of a randomly amplified gene marker vcg (virulence correlated gene). Almost all human clinical isolates from septicemic cases were thought to belong to the C-type containing this marker, whilst most environmental strains (E-type) isolated from shellfish and water do not, although these do cause wound infections. Genome sequencing has not yet provided a satisfactory explanation for this. Updated phylogenetic classification systems are being developed,

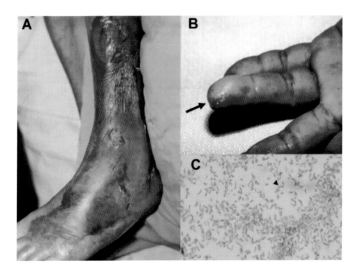

Fig. 12.4 Consequences of *Vibrio vulnificus* infection. A. Gangrene and hemorrhage in a patient with septicemia following infection by ingestion. B. Infection of a fish bone injury from which septicemia developed. C. Gram-negative curved bacilli isolated from a blood sample. Reprinted from Hsueh et al. (2004) with permission of CDC.

incorporating more detailed consideration of the divergence of strains based on the host and geographic sources of isolation. It is likely that some complex regulation of gene expression under different conditions—either in water, shellfish, or the human host—will explain this enigma and this may be revealed by transcriptomic studies.

V. vulnificus can be detected at low levels in seawater from many parts of the world, but it is most common in warm coastal waters of medium salinity, where it can reach densities of up to 10^4 CFU mL^{-1}. Filter-feeding shellfish such as oysters concentrate high levels of the bacterium, often exceeding 10^5 per gram of tissue. There is a strong seasonality to *V. vulnificus* infections, and this correlates with an inability to detect the pathogen in the winter, when the temperature drops below about 15°C. Like *V. cholerae* and many other pathogens, the bacterium enters a VBNC state as a stress response (in this case, reduced temperature rather than starvation). VBNC *V. vulnificus* cells have been shown to be virulent and can be resuscitated and cultured if protected from oxidative stress (see p.95).

Vibrio parahaemolyticus is the leading cause of seafood-associated gastroenteritis

V. parahaemolyticus occurs in marine and coastal waters throughout the world, colonizing the surface of many types of marine animals and plankton. It is the leading cause of gastroenteritis associated with the consumption of seafood and is the commonest cause of all types of food poisoning in Asia (especially Japan), because of the popularity of raw seafood. *V. parahaemolyticus* has recently become more important in the USA, Australasia, and Europe, largely due to the increasing international trade in seafood products, but possibly also due to rising seawater temperatures. After an incubation period of 12–24 h, infection results in watery diarrhea, abdominal cramps, nausea, vomiting, headache, fever, and chills. Most infected persons recover within a few days, but antibiotic treatment may be necessary for more serious infections. Mild cases are usually not reported, so the true incidence of infection is unknown. As with other vibrios, *V. parahaemolyticus* can also cause infection of wounds—occasionally leading to necrotizing fasciitis or septicemia—after exposure to contaminated seawater.

The disease is associated with raw or incompletely cooked seafood (particularly shrimp, crabs, and bivalve mollusks). It is thought that a large infective dose is required for sufficient numbers to survive the acidity of the stomach—use of antacid medicines may reduce this—and outbreaks often follow the cross-contamination of seafood that has been kept in a warm kitchen. The doubling time of the bacterium at 37°C can be as low as 10 min, permitting a massive expansion of the population within a few hours.

Although high numbers of *V. parahaemolyticus* can usually be recovered from harvested fish and shellfish, especially those from estuarine and coastal environments, only a small proportion of isolates appear to carry factors that make the bacterium pathogenic. Isolates from environmental samples are usually much less virulent than those isolated from clinical samples, so it is likely that there are either multiple types of the organism, or that it undergoes genetic changes and selective enrichment in the gut as seen with *V. cholerae*. The exact mechanisms by which the bacterium produces disease remain unclear, despite extensive research and the identification of numerous virulence factors, including adhesins, toxins, iron-sequestering mechanisms, and Type III and IV secretion systems (T3SS, T6SS).

Most strains isolated from clinical samples are "Kanagawa positive"—this is a reaction on a specialized type of blood agar—because they possess a thermostable direct hemolysin (TDH) and/or a thermolabile related toxin (TRH). Although defined by their pore-forming hemolytic activity on blood cells in the laboratory, the natural target of these toxins *in vivo* is the intestinal epithelial cells. Binding to epithelial cell membranes in the colon leads to disturbance of osmotic pressure, fluid loss and cell death. By contrast, most environmental isolates produce neither TDH nor TRH—even if they contain the *tdh* or *trh* genes detected using gene probes. About 10% of strains isolated from clinical cases are also non-toxigenic.

Further insight into the mechanisms of pathogenicity of *V. parahaemolyticus* came with the recognition of genes for two Type III secretion systems (T3SS-1 and T3SS-2) in genome

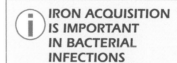

IRON ACQUISITION IS IMPORTANT IN BACTERIAL INFECTIONS

All bacteria require iron for the activity of essential cellular functions. Vertebrates possess highly efficient systems for transporting and storing iron. The blood serum protein transferrin binds iron with extreme avidity, reducing the concentration of free iron in tissues to 10^{-18} M, about 10^8 times lower than the concentration that bacteria require for growth. In response to infection, an even more efficient iron-binding protein (lactoferrin) is released, and this removes iron from the blood to storage in the liver and other organs. To overcome this host defense mechanism, invading pathogens must obtain enough iron for growth by competing with the host's iron-withdrawal system. One of the reasons that *Vibrio vulnificus* is able to cause rapid septicemic infection in people who have underlying liver damage is because the iron-storage and transport systems are disturbed, leading to excess levels of iron in the serum. This allows the bacteria to multiply so rapidly that they overwhelm the body's defenses (Baker-Austin and Oliver, 2018).

sequences. Like other vibrios, *V. parahaemolyticus* has two chromosomes. T3SS-1 is encoded on the large chromosome and is present in environmental and clinical isolates, whereas T3SS-2 is encoded on the small chromosome in a pathogenicity island that also encodes TDH and is present only in clinical isolates. The T3SS can be likened to a molecular syringe which pathogens use to inject proteins into host cells; this is known to be a major factor in the pathogenicity of other enteric pathogens such as *Shigella, Salmonella, Yersinia,* and enteropathogenic *Escherichia coli*. The T3SS-1 of *V parahaemolyticus* injects a number of proteins that induce autophagy in the host cell (a process of self-digestion of cell contents), followed by rounding and release of nutrients for the pathogen on cell death, whereas the T3SS-2 is thought to be responsible for cytotoxicity and enterotoxicity due to disruption of the cytoskeleton of intestinal cells. Recently, new Type VI Secretion Systems (T6SS) have been described in enteric pathogens (*Figure 12.2A*). *V. parahaemolyticus* possesses two such systems—T6SS-1 and T6SS-2—encoded on chromosomes 1 and 2 respectively. These have been shown to contribute to adhesion and cellular disruption of cell cultures, but their role *in vivo* is unclear.

There are numerous serotypes of *V. parahaemolyticus,* characterized by the O (LPS) and K (capsular) antigens. Until the 1990s, outbreaks of *V. parahaemolyticus* showed no clear association with particular serotypes, but since the mid-1990s, a clonal pandemic strain O3:K6 has spread throughout the world and is now the cause of most seafood-associated outbreaks. Further investigation reveals that this serotype designation conceals variations of genetic diversity of *V. parahaemolyticus* that can only be revealed by multi-locus sequencing typing (MLST, p.45) or full genome comparison. This shows that there are several genetic regions characteristic of pathogenicity islands, mostly present in the O3:K6 pandemic strains. All isolates of the O3:K6 pandemic serotype contain the genes for T3SS-2 and TDH. It seems that these strains have evolved within a short time frame by acquiring large regions of new DNA—possibly due to phage infection—and this may explain their prevalence and spread as human pathogens.

Research to develop reliable methods—combining classical culture methods with modern molecular approaches—for laboratory testing of *V. parahaemolyticus* and identification of pathogenic strains in the face of this continuing evolution of genetic variants is especially important due to the growing global trade in seafood products and the need to adopt internationally accepted standards to ensure public safety.

Microbes associated with fish and marine mammals can be transmitted to humans

Diseases of animals that are transmitted to humans are known as zoonoses. Some fish pathogenic bacteria can infect humans, usually via skin abrasions or wounds. Infection by *Mycobacterium marinarum* ("fish tank granuloma") is an important hazard for aquarium workers and hobby aquarists. It usually causes lesions on the hands but can cause more serious infections of the bones and joints in people with impaired immunity. *Streptococcus iniae* and *Edwardsiella tarda* have been responsible for small outbreaks of serious invasive infections in workers handling infected warm-water fish from aquaculture facilities. Other bacteria such as *Vibrio* spp., *Aeromonas* species, *Lactococcus garviae*, and *Erysipelothrix rhusiopathiae* from the skin or intestines of fish can cause range of skin and gastrointestinal infections, including some chronic conditions.

There is also a definite, albeit small, risk of acquiring diseases from marine mammals. There have been cases of infection by *Mycobacterium* sp. (causing respiratory and skin lesions) and *Brucella* infection (causing severe lethargy, fever, and headaches) in aquarium and research staff working with mammals. Marine conservation organizations highlight the need for caution when engaging in the rescue of stranded cetaceans. "Seal finger" is an extremely painful infection of the hands caused by *Staphylococcus* or *Erysipelothrix*, which occurs in seal hunters and research workers, especially in the Arctic regions of Norway and Canada. Seal bites can also cause serious infections, caused by a variety of poorly characterized bacteria, and are notoriously difficult to treat unless antibiotics are used promptly. The popularity of tourist activities such as swimming with dolphins, could also lead to a rise in transmission

WORMS AND MOTHS AS MODELS FOR PATHOGENICITY STUDIES

Many advances in the understanding the genetic and biochemical mechanisms by which bacteria produce human diseases come from the use of cell culture, but proof of *in vivo* effects depends on the use of animal models. For enteric diseases such as those caused by *Vibrio* spp., this often requires administration of pathogens directly into the peritoneum of mice or the exposed intestine (ileal loop) of rabbits, to measure inflammation or accumulation of fluid due to toxins. Such experiments require expensive specialized facilities and there is pressure to develop more ethically acceptable alternatives that can be used to conduct large-scale screening programs of multiple bacterial strains. In the case of *Vibrio parahaemolyticus*, it has proved very difficult to unravel the complexities of pathogenicity. To tackle this problem, researchers have used the nematode worm *Caenorhabditis elegans* (Durai et al., 2011) and larvae of the wax moth *Galleria mellonella* (Wagley et al., 2018) —these are both widely used in different branches of biomedical research—to determine the pathogenic potential of the highly divergent strains isolated from different courses. Wagley and colleagues found the wax moth model particularly useful, as the larvae were susceptible to both toxigenic and non-toxigenic strains of *V. parahaemolyticus,* and this is leading to the development of new genetic markers for the effective screening of clinical and environmental isolates of *V. parahaemolyticus.*

of zoonoses. Such activities also increase the likelihood of transmission *from* humans *to* marine mammals, although there is limited evidence of this to date. Of the viruses, influenza poses the most important risk and several instances of direct transmission from persons in close contact with infected seals have occurred. Evolution of new strains of influenza virus occurs regularly through antigenic changes, owing to recombination events in the fragmented genome. It is therefore possible that marine mammals could provide an opportunity for transmission of new variants to humans. Protozoan parasites such as *Giardia* may also have a reservoir in marine mammals.

DISEASES CAUSED BY MARINE MICROBIAL TOXINS

Scombroid fish poisoning results from bacterial enzyme activity

This type of food-borne intoxication is associated with eating fish of the family Scombridae, which include tuna, mackerel, marlin, and bonito. The tissue of these fish contains high levels of the amino acid histidine. If there is a delay or breakdown at any stage in the refrigeration process between catching the fish and consumption, bacteria from the normal microbiota can multiply and convert histidine to histamine (*Figure 12.5*). Bacteria isolated from fish associated with scombroid fish poisoning include *Raoultella planticola*, *Morganella morganii*, *Hafnia alvei*, and *Photobacterium phosphoreum*. The histamine is heat stable and will withstand normal processing such as canning. The spoilage may not be enough to alter the taste or smell of the fish, but levels of histamine above 1 mg g^{-1} can be enough to induce a rapid allergic-type response. Reddening of the face and neck, shortness of breath, and, in severe cases, respiratory failure can result within a few minutes of eating contaminated fish. Treatment with antihistamine drugs is beneficial and the person usually recovers fully. The incidence of scombroid poisoning is increasing because of the rising popularity of sushi and the import of fresh tuna and swordfish by airfreight over long distances, with opportunities for breaks in the cold chain. Some outbreaks involving groups of affected people may necessitate intervention by health authorities and destruction of suspected fish products.

Botulism is a rare lethal intoxication from seafood

Botulism is one of the most feared food-borne diseases. It is caused by *Clostridium botulinum*, a Gram-positive bacterium that forms endospores that survive high temperatures. The bacterium is a strict anaerobe and under appropriate conditions it produces a powerful neurotoxin in a range of different foodstuffs. Botulinum toxin is the most lethal toxin known; if ingested, tiny amounts of the toxin—estimated to be only about 200 ng—cause a severe flaccid paralysis. The toxin is a protease that cleaves proteins involved in the docking of neurotransmitter vesicles, blocking release of the neurotransmitters and therefore preventing muscle contraction. *C. botulinum* is normally incapable of growth within the adult gut, so disease is dependent on ingestion of preformed toxin, with symptoms usually occurring 12–36 h after ingestion. Unlike other fast-acting neurotoxins, the delay is caused by absorption of the toxin and transport via the central nervous system. (Note that the bacterium *can* grow in the gut of babies, and infant botulism is typically associated with spore-contaminated honey.) *C. botulinum* is classified into types A to G based on the serological properties of the toxin produced. Type E is most commonly associated with seafood and can be isolated from marine sediments and the intestinal contents of fish and crustaceans. Fortunately, food-borne

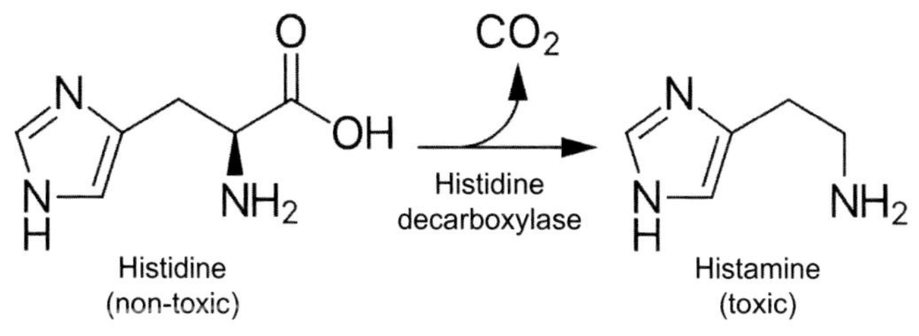

Figure 12.5 Enzymatic conversion of histidine to histamine.

botulism—especially associated with seafoods—is very rare because of strict regulations concerning commercial canning and other methods of preservation. Most salting, drying, and smoking methods of fish preservation are safe, but some cases have occurred following consumption of smoked fish in modified atmosphere packaging (p.360). Since smoking kills most, but not all, of the spores, such products should always be refrigerated to inhibit growth of any survivors and sold with restricted shelf life. Most cases occur from home-prepared ethnic foods. Botulism is a persistent serious health problem in indigenous Arctic communities in northern Canada and Alaska, where some traditional foods such as fish, whale, or seal meat are wrapped, buried in the permafrost and left to rot. Wound botulism can occur in workers handling fish, causing local paralysis.

Fugu poisoning is caused by a neurotoxin of bacterial origin

This intoxication is caused by the ingestion of tetrodotoxin (TTX; *Figure 12.6a*), which is found in the intestines, liver, and gonads of pufferfish, especially *Takifugu rubripes* and related species. The toxin is probably synthesized by environmental bacteria and accumulates in the organs of the fish. TTX is one of the most active neurotoxins known, and acts by blocking the flow of sodium ions in the nerves. Within a few minutes or hours of eating fish contaminated with the toxin, victims feel tingling sensations in the mouth and a sense of lightness, followed quickly by the onset of total paralysis. The person remains conscious but totally immobilized until the moment of death, which occurs in up to 50% of cases. In Japan, fugu is a prized delicacy for which diners in specialist restaurants (*Figure 12.6b*) are prepared to pay large amounts for the thrill of eating tiny portions of this risky food—usually served as elegant plates of translucent sashimi—with the added frisson of possibly ingesting a miniscule dose of the toxin in order to give a "buzz." Fishing for *T. rubipres* and the sale of fugu are now tightly regulated in Japan and fugu chefs must be specially licensed, so most of the 50 or so cases each year occur among fishermen and amateur cooks who prepare the dish. There are a few licensed fugu restaurants in the USA, but it is banned in Europe. Fish farmers in some parts of Japan have developed systems for production of nonpoisonous fugu by rearing fish on special diets and centralized preparation facilities have improved safety. Consumption of fugu has a long history and it is deeply embedded in Japanese culture, with associated economic significance. The development of safe fugu has met with problems

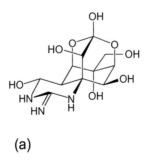

(a)

(b)

Figure 12.6 (a) Representative structure of tetrodotoxin (there are numerous analogues of the basic structure). (b) A typical fugu restaurant in Tokyo, where live puffer fish are displayed in tanks to be prepared by licensed chefs. Credit: Richard Tucker, CC-BY-SA 2.0.

of consumer acceptance and controversy about whether rules prohibiting the sale of fugu liver—"kimo," the most expensive and prized delicacy—should be relaxed.

TTX is widespread amongst marine animals

Besides pufferfishes, a wide variety of animals including triggerfishes, sunfishes, xanthid crabs, seastars, molluscs, and worms are known to contain TTX, in which its main function seems to be in defense against predators. Some marine animals also use TTX as a mechanism of attack; among the best-known examples are the blue-ringed octopus (*Hapalochlaena maculosa*) of northern Australia, which uses TTX in its venom to paralyze its prey of fish and crustaceans, and the tropical flatworm *Planocaeta multicentaculata*, which paralyzes its prey of cowrie mollusks. A few species of terrestrial salamanders, newts, and toads also produce the toxin. Animals that contain TTX appear to have single-point mutations in the amino acid sequence of a sodium channel protein, which makes them resistant to the toxin's effects. Because of the wide distribution of TTX in phylogenetically diverse groups it seems unlikely that the toxin is synthesized by the animals themselves—it is a very complex molecule with no known critical metabolic role—so the most likely explanation is that it is acquired through the food web. This is supported by the fact that farmed pufferfish that have been fed special diets are non-toxic. Various isolates of *Vibrio*, *Pseudomonas*, *Alteromonas*, *Shewanella*, *Flavobacterium*, and other bacteria have been isolated from the skin, intestinal tract, gonads, salivary glands, and elsewhere of TTX-containing animals. These bacterial isolates have also been reported to produce the toxin in culture, although there is much controversy about these results due to variations in growth conditions and detection methods. Some investigators consider the TTX-producing bacteria to be symbionts, but the functions of TTX for these bacteria and the evolution of this possibly symbiotic relationship remain unknown.

There is some concern that populations of TTX-containing fish have spread to other regions, where the dangers are less well known. For example, populations of the toadfish *Lagocephalus scleratus* have increased in the Eastern Mediterranean and there have been cases of TTX poisoning from this source in Turkey. Recently, there have been some concerns for public health following the detection of low levels of TTX in marine bivalves and gastropods in European waters. Although there are very few known cases of TTX poisoning associated with this source, studies are in progress to assess potential risks.

Some dinoflagellates and diatoms produce harmful toxins

The growth of marine plankton is affected by a wide range of factors that influence their spatial and temporal distribution. Seasonal periodic increases in plankton density (blooms) obviously have great ecological importance in ocean food webs. In addition, some planktonic microbes produce toxins that affect human health, and many of these can form exceptional blooms under certain conditions. These are frequently referred to as "red tides," although this colloquial term is something of a misnomer since not all toxic blooms are red in color—they may not even reach high enough densities to discolor the water. A more generally accepted term is "harmful algal blooms" (HABs), although this term is also not entirely accurate because health effects are not always associated with a distinct "bloom." Among the many thousands of species of phytoplankton, only about 150 are known to produce toxins, and only a few of these actually cause problems for human health. Examples of some of the main toxic dinoflagellates and diatoms are illustrated in *Figure 12.7*. The main health hazard for humans comes from eating fish or shellfish that have accumulated the toxin in their tissues through feeding in water containing high levels of toxin-producing plankton. Some toxins may also produce disease symptoms as a result of direct contact with contaminated water or inhalation of aerosols. Representative structures of the major algal toxins are shown in *Figure 12.8*. All of these phycotoxins are non-proteinaceous substances that remain active after cooking. This, together with the usually very rapid onset and neurological symptoms, is an important distinction of HAB intoxications from most bacterial and viral infections associated with consumption of shellfish. Most of the toxins cause paralysis by binding to the voltage-gated sodium channel of nerve cells, inhibiting influx of Na$^+$ ions and therefore blocking the generation of action potentials.

Figure 12.7 Examples of toxin-producing diatoms and dinoflagellates. Upper: light micrographs of A. *Dinophysis acuti* and B. *Pseudonitzschia pseudodelicatissima*. Lower: scanning electron micrographs (SEM) of C. *Karenia brevis* and D. *Gambierdiscus toxicus*. Credits: A. D. Cassis, T. Ivanochko, J. Shiller, B. Moore-Maley, J. Kim, S. Huang, A. Sheikh, G. Oka. Phytopedia: The Phytoplankton Encyclopedia Project. Published at www.eoas.ubc.ca/research/phytoplankton/B. Susanne Busch, IOW. C, D. Paula Scott, Florida Fish and Wildlife Commission.

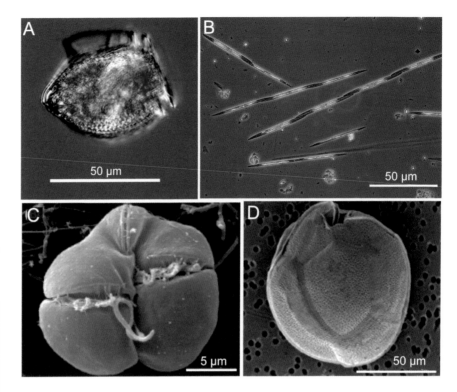

> **ⓘ TOXIC HABS ARE NOT NEW, BUT THEY ARE GROWING IN IMPORTANCE**
>
> Sudden changes in the appearance of coastal waters have been described for many centuries. For example, it is often surmised that the reference in the Bible to the "first plague … and the river [Nile] turned to blood" is a description of such a bloom, thought to have occurred about 1290 BCE). Ancient cultures such as Pacific Islanders and coastal tribes of North America had scouts who "read the sea" for changes in color, in order to warn communities about the dangers of harvesting fish and shellfish. From the fifteenth century onwards, seafaring explorers recorded outbreaks of fish and shellfish poisoning that affected them and their crews during their journeys. Fossil records show that blooms of toxic microalgae have occurred for millennia. Despite their long history, HABs have assumed growing importance since the late twentieth century because they are occurring more frequently and in regions where they have not previously occurred. This is due to stimulation of blooms by nutrient runoff, climate change, increased aquaculture in coastal waters, and transport of algae in ships' ballast water. Besides affecting human health, HABs damage other marine life (see Chapter 11) and have significant economic and social impacts on fisheries, aquaculture, and tourism.

Paralytic shellfish poisoning is caused by saxitoxins produced by dinoflagellates

Paralytic shellfish poisoning (PSP) has a worldwide distribution and is the best-known human illness associated with microalgae. PSP is caused by a group of over 50 closely related water-soluble toxins known as saxitoxins (STXs) and gonyautoxins. The main organisms responsible for production of PSP toxins are *Alexandrium* spp. and *Gymnodinium catenatum*, which have been described for many years in temperate waters of North America, Northern Europe, Scandinavia, and southeast Asia. In the past few decades these taxa have been increasingly reported from countries in the southern hemisphere, including Australia, New Zealand, and Chile. In tropical waters, PSP is largely caused by *Pyrodinium bahamense*. PSP results from eating shellfish, especially clams, oysters, mussels, and certain species of crabs, in which toxins produced by the dinoflagellates have built up through filter feeding. Crabs and lobsters can also accumulate the toxin by feeding on contaminated bivalve mollusks.

PSP has a dramatic onset, often within minutes of consuming contaminated shellfish. Tingling around the lips is quickly followed by numbness of the face and neck, nausea, headache, and difficulty in speech. In severe cases, muscular and respiratory paralysis can occur, and mortality can be more than 10% unless there is rapid access to medical services. Treatment consists of stomach pumping and administration of charcoal to absorb the toxins. In severe cases, artificial respiration may be required, but there are usually no long-lasting effects.

Many countries operate PSP-management programs, which include monitoring seawater for the density of toxin-producing dinoflagellates and the assay of shellfish samples for saxitoxins. Careful analysis has shown that there is no straightforward relationship between the presence of toxin-producing dinoflagellates and the level of particular toxins in shellfish tissue. Shellfish may modify the toxins or excrete them differentially according to climatic and physiological conditions, making management of outbreaks more difficult. Commercial collection of shellfish is prohibited when saxitoxin levels in the tissue exceed a certain threshold concentration (typically, 800 µg kg^{-1}). The traditional internationally accepted assay method involves intraperitoneal injection of an extract of shellfish meat into mice. The toxin is quantified as "PSP equivalents" by the time taken for the mice to die, but this assay does not give any information about the precise amounts of individual toxin components or their breakdown products. There have been many attempts to develop alternative methods that do not depend on the use

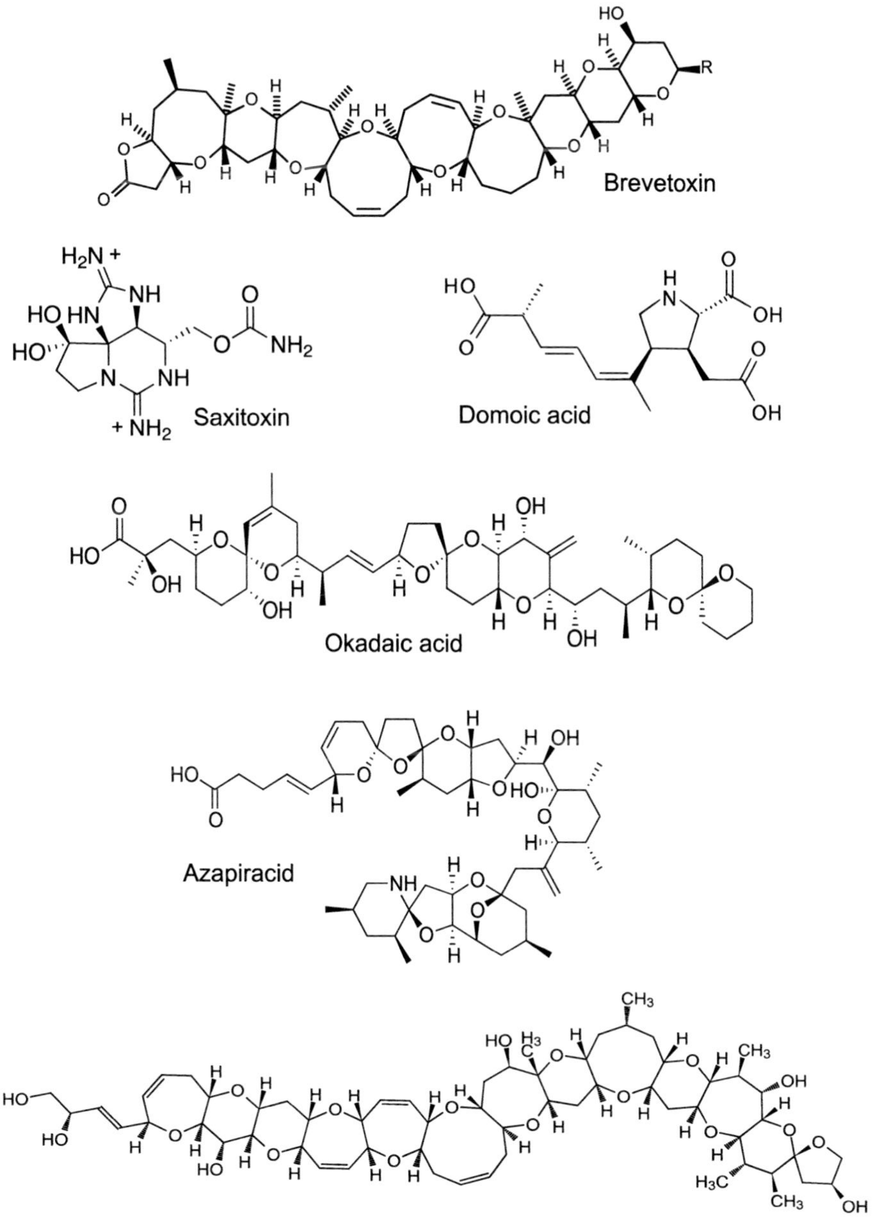

Figure 12.8 Representative structures of major dinoflagellate and diatom structures. Note that there is considerable variation in the structure of brevetoxins, ciguatoxins, and saxitoxins.

of animals and that give more reliable and informative results. This is complicated by the fact that the profile of different toxin analogues and their toxicity can vary greatly between different regions. High-pressure liquid chromatography fluorescence detection (HPLC-FLD) and liquid chromatography/mass spectrometry (LSMS) are now widely used as official methods by regulatory authorities. A receptor-binding assay based on the ability of native toxin in a sample to compete with radioactively-labeled saxitoxin for binding sites in rat cell membrane preparations also provides quantitative results. Commercially available immunological rapid "dipstick" test kits produced by Scotia™ and Neogene™ have been approved for regulatory use and allow shellfish producers to conduct simple qualitative tests on a boat or dockside before harvesting. A sample of shellfish tissue is ground and applied to a lateral flow strip, which provides a positive result (negative if below the 800 µg kg^{-1} saxitoxin equivalent level) within a few minutes. Quantitative results can be obtained using enzyme-linked immunosorbent assay (ELISA) methods including Abraxis™ and Europroxima™ kits. This screening is very cost-effective for use by shellfish producers to determine whether full testing by regulatory authorities is needed. The closure of commercial fisheries can have devastating economic and social effects, affecting the livelihood of whole communities for many years, so validation of methods appropriate to a particular fishery and area is essential. Authorities will also impose bans on collection of shellfish for personal consumption, and notices on beaches are a common sight in affected areas (*Figure 12.9*).

Figure 12.9 A typical beach notice prohibiting the collection of shellfish due to contamination by PSP or other toxins. Credit: Washington State Department of Health.

Brevetoxin causes illness via ingestion or inhalation during red tides

Neurotoxic shellfish poisoning (NSP) is caused by the toxin brevetoxin, produced by the dinoflagellate *Karenia brevis* (*Figure 12.*7C) and a few other species. The lipophilic toxin binds to voltage-gated sodium channels in nerve cells, leading to disruption of normal neurological function. Typical signs of intoxication include dilated pupils, paresthesia (abnormal sensations on the skin, such as tingling or burning), a reversal of hot–cold temperature sensation, vertigo, muscle pains, diarrhea, and nausea. Blooms of *K. brevis* are highly distinctive—these are the archetypal red tides—and are usually seasonal, starting in late summer and lasting for many months. They have been known in the Gulf of Mexico, especially the Florida coast, for several centuries; however, in recent years blooms have occurred more frequently and have been of longer duration. Since the late 1980s, blooms have also occurred on the eastern US coast and in New Zealand, probably involving other *Karenia* spp. Like PSP, NSP is caused by consumption of shellfish that have accumulated the toxin, although symptoms are usually milder. However, an additional threat comes from inhalation of the toxin. During extensive blooms, lysis of *K. brevis* cells and disruption by surf action leads to aerosols containing the toxin—these can be carried by wind up to 100 km inland—causing shortness of breath and eye irritation. In Florida, beach closures are necessary during red tides and health advisory notices are issued warning asthmatics and other susceptible people to stay indoors. The toxin also kills marine mammals, birds, and turtles (see p.320) and the blooms also cause massive fish kills, leading to further problems as coastal waters become anoxic. Thus, red tides have a major impact on fisheries and tourism, and government agencies are involved in close monitoring using satellite imagery and other techniques, so that precautionary management procedures can be put in place at the first signs of a bloom.

Dinophysiotoxins and azaspiracid toxins from shellfish result in gastrointestinal symptoms

Diarrhetic shellfish poisoning (DSP) is caused by eating shellfish (usually bivalve mollusks) that contain the toxin okadaic acid and other dinophysiotoxins produced by dinoflagellates of the genera *Dinophysis* (*Figure 12.7A*) and *Prorocentrum*. Okadaic acid is an inhibitor of protein phosphatases. Ingestion causes the intestinal cells to become highly permeable to water, leading to profuse diarrhea and vomiting which usually lasts for a day or two without serious effects. The symptoms resemble those caused by infection by enteric viruses from shellfish grown in sewage-polluted waters (p.362), but a growing body of evidence since the 1980s has shown that DSP can result from consuming shellfish from high-quality waters that are free of sewage contamination. Many outbreaks have been described in Japan and in European countries, particularly on the Atlantic coast of Spain, France, and the British Isles. A European Union Directive requires the monitoring of shellfish for the toxin—by LCMS and immunoassay kits like those used for STX—and this has resulted in closure of many traditional shellfish harvesting and aquaculture operations in areas where toxins have been detected. Unlike the other shellfish poisonings, a DSP risk can occur even in the absence of a bloom; *Dinophysis* densities as low as 200 cells mL^{-1} can lead to accumulation of unacceptable levels of toxin in shellfish. Okadaic acid has been shown to be carcinogenic in rats, and long-term exposure has been suggested as a possible cause of cancer in the digestive tract, although there seems to be insufficient epidemiological evidence to raise human health concerns.

In the 1990s, outbreaks of DSP-like disease associated with mussels harvested from Irish waters was linked to the consumption of mussels (*Mytilus edulis*). DSP toxins could not be identified, and a new group of related toxins called the azaspiracids (AZAs) was shown to be responsible. Since then, several outbreaks of AZA poisoning have been reported in several countries. The symptoms are severe diarrhea and vomiting, nausea, and headache, which may persist for several days. In a mouse bioassay, AZAs cause death from serious tissue and organ damage, with neurotoxic effects at high doses. However, their mode of action seems quite different from the other shellfish toxins. Development of improved chromatographic assays and immunoassays means that routine monitoring for AZAs is now being introduced, with the result that they are much more widespread than previously thought. The toxins seem to be associated particularly with *M. edulis* and there are no obvious links with bloom events. Rather, it seems as if the toxin accumulates gradually in the tissue of the mussel as a result of long-term exposure to low levels of the dinoflagellate *Azadinium spinosum*. The toxin can persist in the shellfish tissue for many months.

Amnesic shellfish poisoning is caused by toxic diatoms

Amnesic shellfish poisoning (ASP) was discovered in 1987, following an unusual outbreak of seafood poisoning in Prince Edward Island, Canada, in which more than 100 people were affected and three died. The symptoms included rapid onset of vomiting, disorientation, and dizziness that did not match those of other known diseases. As is usual in such large outbreaks, health professionals interviewed those affected and were surprised that many could not answer the usual questions about what they had eaten before feeling ill; they showed other signs of amnesia, which persisted for many months. This suggested that a new type of neurotoxin was involved, and this was identified as domoic acid produced by the diatom *Pseudo-nitzschia multiseries*. We now know that several other species of *Pseudo-nitzschia* (e.g. *P. pseudodelicatissima*, *Figure 12.7B*) can produce domoic acid and these have been isolated from many parts of the world. Human cases of ASP are rare, but they are almost certainly underreported, and in areas where *Pseudo-nitzschia* blooms occur, they may be associated with eating planktivorous fish such as anchovies. As discussed on p.320, domoic acid from *Pseudo-nitzschia* blooms also causes disease in marine birds and mammals and is increasingly recognized in the marine food chain.

Ciguatera fish poisoning has a major impact on the health of tropical islanders

In terms of public health, ciguatera fish poisoning (CFP) is undoubtedly the most important of the diseases caused by marine toxins, especially for inhabitants of tropical islands who rely on fishing. It is also a well-known hazard among sailors and travelers in the tropics (35°N–35°S) and was first documented by explorers of the Caribbean and Pacific in the 15th and 16th centuries. In 1774, Captain James Cook gave a detailed account in his ship's log of poisoning that affected him and his entire crew after eating fish in the New Hebrides, which

BENEFICIAL APPLICATIONS OF MARINE ALGAL TOXINS

Tetrodotoxin (TTX) and saxitoxins both cause paralysis by inhibiting transmission of the nerve impulse by interfering with the passage of sodium ions through the membrane of nerve cells. Indeed, studying the interaction of these toxins with membrane proteins was vital for our understanding of how sodium channels work. Ion channels are composed of proteins in the membrane which form a pore, enabling the passage of specific ions from the aqueous external environment, through the hydrophobic membrane to the interior of the cell. These toxins interact with the external surface of the channel and prevent the flux of ions. Besides their use as biological research probes, these molecules also have medical applications as anesthetics and therapeutic agents (Assunção et al., 2017). Modified TTX and saxitoxins have been used for prolonged anesthesia and relief of severe pain, prevention of brain damage after strokes, and as a treatment for heroin addiction. Sadly, their intense toxicity has meant that they have also been developed for use in biological warfare, terrorism, and espionage.

? DID CIGUATERA CAUSE THE MIGRATION OF PACIFIC ISLANDERS?

In the years between 1000 and 1450, multiple waves of migration occurred between the islands of Polynesia and distant lands, especially New Zealand. The reasons why islanders undertook such perilous voyages into the unknown across vast expanses of ocean have always been something of a mystery. By studying past climate conditions and examination of archaeological artifacts, a native of the Cook Islands has suggested that the migrations may have been due to increased chronic incidence of CFP affecting the population, who were heavily dependent on ciguatera-susceptible large carnivorous fish at the time (Rongo et al., 2009). Reliance on fishing in the islands declined after this time and the remaining population seem to have moved to smaller, safer fish species. However, the highest incidence of CFP still occurs today in the Cook Islands, and modern outbreaks have led to a decline in fishing and reliance on processed foods rather than fish, with a new wave of mass emigration to New Zealand and Australia in the 1990s.

was almost certainly a description of CFP. The causative agent of CFP was not discovered until 1977, when the toxin-producing dinoflagellate *Gambierdiscus toxicus* (*Figure 12.7D*) was isolated in the Gambier Islands in French Polynesia. Related species occur worldwide, and isolates vary greatly in the amount and types of toxin produced. Cyanobacteria have also recently been implicated in some outbreaks. The fat-soluble toxins become concentrated by biomagnification as they pass up the food chain, and disease is almost always associated with eating large predatory fish, as illustrated in *Figure 12.10*. More than 400 species of fish have been implicated in CFP, with moray eel and barracuda being the most common. CFP is characterized by a distinctive sequence of diarrhea, vomiting, abdominal pain, and neurological effects within a few hours of eating contaminated fish. Unusual symptoms include numbness and weakness in the extremities, aching teeth, and reversal of temperature sensation (cold things feel hot and hot things feel cold). In severe cases, this can progress rapidly to low blood pressure, coma, and death. However, the type and severity of symptoms depends on the particular "cocktail" of toxins consumed, and this varies with species of fish and geographic origin. There seem to be important differences in structure between the main CFP toxins implicated in Pacific and Caribbean cases. Doses that cause severe symptoms in one person may be harmless in another. It is also likely that other ancillary toxins are involved. In some people, neurological problems can persist for many years, and a repeat attack can be triggered by eating any fish (whether or not it contains toxin) or by drinking alcohol. The toxin has also been shown to cross the placenta to affect the fetus, and there are even reports of transmission to another person via sexual intercourse. In countries where the disease is endemic, there are a range of local remedies, but these are of unproven value. In severe cases, it is necessary to administer intravenous mannitol, which reverses the effect on sodium transport.

The incidence of CFP can be quite localized but unpredictable. Local fishermen will often believe that fish from a particular island or reef will be hazardous, whilst others nearby are not. Many travelers have learnt—to their cost—that such claims are often not reliable. *Gambierdiscus* adheres to macroalgae on the surface of dead coral and on the seabed, but little is known about the factors promoting colonization of the macroalgae by the dinoflagellates and its subsequent proliferation and toxicity. Damage to the reef by dynamite fishing,

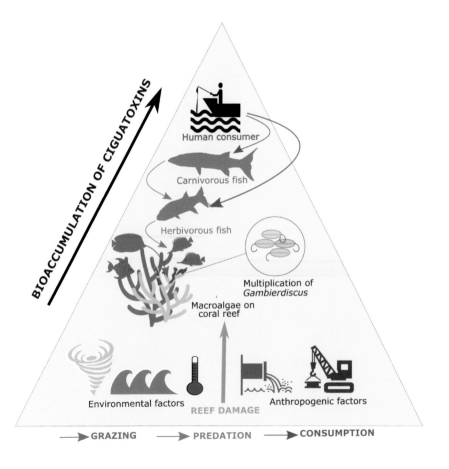

Figure 12.10 Schematic diagram showing the accumulation of ciguatoxin through the food chain on coral reefs.

diving, boat anchors, military action, or construction work promotes the initial colonization by the macroalgae. Increased frequency of bleaching and diseases of corals may also be leading to higher occurrence of CFP. However, this is only part of the story, as the incidence of toxic fish is very dependent on environmental factors such as rainfall, pollution, and nutrient runoff from the land. Also, it is likely that there are genetic differences among different strains of *Gambierdiscus* and factors that affect dinoflagellate growth may have a different effect on toxin production. Recently, reports of ciguatera-like illness associated with dinoflagellates in the genus *Ostreopsis* have been reported in Italy and Puerto Rico. The complex ecological, toxicological, and physiological factors seen with CFP illustrate dramatically how careful we must be when looking for a simple "cause and effect" in the etiology of a disease.

Bacteria influence the production of HAB toxins

Early in the study of the origin of PSP, microbiologists observed bacteria-like structures in electron micrographs of dinoflagellate cells and asked if these could be symbionts and whether they are involved in toxin production. Besides the intrinsic interest in the possible role of symbiosis in providing their eukaryotic host with a possible defensive mechanism, answering these questions is of importance in the monitoring and mitigation of fish and shellfish toxicity. Cyanobacterial toxins, especially microcystin, are an important health hazard in freshwater lakes and it is now known that certain cyanobacteria possess gene clusters involved in saxitoxin synthesis, although phylogenetic analysis suggests that these have evolved separately rather than by gene transfer. Recently, there have been accounts of PSP linked to cyanobacterial blooms in coastal lagoons.

Dinoflagellates form tight associations with numerous bacteria and many experiments have been conducted to compare levels of toxin production in dinoflagellate cultures before and after curing of contaminating bacteria, by treatment with antibiotics or enzymes. Although there is little evidence for autonomous production of dinoflagellate toxins by associated bacteria, they do influence the biosynthesis of the toxins—probably by supplying cofactors or precursors necessary for toxin production. Bacterial involvement also seems to be implicated in the production of domoic acid by the diatom *Pseudo-nitzschia*. Axenic cultures have greatly reduced levels of domoic acid, whereas levels are increased when certain species of bacteria isolated from the original culture are reintroduced.

Although the precise role of bacteria in synthesis of HAB toxins remains uncertain, there seems little doubt that the presence of bacteria has a considerable influence on the levels and nature of toxins released by the algae. An additional complication is that some of the bacteria naturally associated with dinoflagellates and diatoms can cause lysis of the cells, leading to sudden release of high levels of toxin, whilst other bacteria may metabolize algal products. The importance of such algicidal bacteria as a possible method of control of HABs is discussed in *Chapter 14*.

Dinoflagellate and diatom toxins may be antipredator defense mechanisms

The toxins produced by dinoflagellates and diatoms are usually complex secondary metabolites, and diversion of energy to their synthesis implies some ecological benefit. The extreme sensitivity of humans and other animals to these toxins and the fact that most act on similar mechanisms of nerve action is coincidental. It is usually presumed that toxins have evolved to deter predation by zooplankton. If this is correct, once physical and chemical conditions are favorable for the initial bloom of a toxic species, the production of toxins will deter predation and prolong the maintenance of high densities of the species. It seems that many zooplankton species, such as copepods, can discriminate dinoflagellates with low toxin content and feed selectively, and some flagellates and ciliates appear to be killed by dinoflagellate toxins. However, toxins do not give universal protection against grazing by all species of predator, and it may be that the toxins offer a selective advantage to organisms that produce them under low-nutrient conditions, by directing predation pressure onto competitors. It is notable that the production of toxin, and its cellular concentration, is very dependent on the supply of nutrients (especially phosphorus). This has led to the suggestion that the toxins may act as a reserve for storage when the nutrient supply is unbalanced. As discussed in *Chapter* 6, many dinoflagellates are mixotrophic, with the capacity for both phototrophic and heterotrophic nutrition, and the production of such secondary metabolites may be associated with major switches in metabolic pathways.

Complex factors affect the incidence of HABs and toxin-associated diseases

The expansion of HABs and associated diseases is due to complex interactions between a variety of climatic, environmental, and biological factors. Blooms occur due to particular combinations of physical conditions (temperature, sunlight, water stratification, circulation) and chemical conditions (levels of oxygen and specific nutrients). Species dispersal due to large-scale water movements is important, and unusual currents and storms may account for appearance of atypical blooms. For example, in some recent years, the *K. brevis* red tide off Florida has moved much further north because of formation of unusual circulation patterns in the Gulf Stream. The occurrence of some HABs has been closely correlated with the El Niño Southern Oscillation phenomenon. The increased incidence of HABs is often cited as evidence of disturbance to our oceans and atmosphere due to increased carbon dioxide levels.

One explanation for the increased geographic distribution of some HABs is transport of the causative organisms via ship movements. Modern large vessels pick up many thousands of liters of water as ballast, which can be transported to completely different geographic regions. Several studies have shown that the cysts of toxic dinoflagellates and other harmful "alien" organisms can be transported in this way. For example, the introduction of PSP into southern Australia in the 1980s is suspected to be due to transport from Japan and Korea. As a consequence, regulations of the International Maritime Organization require ships to have programs for management of ballast water—such as offshore exchange of ballast water, disinfection of the water before it is discharged, or transfer to a shore-based disinfection facility.

Nutrient enrichment of coastal waters is often linked to the increased incidence of HABs. Upwelling of nutrients into cold oligotrophic waters is responsible for many natural blooms, such as those that occur regularly off the California coast. Such natural phenomena may now be overlaid with eutrophication from anthropogenic sources of excess nutrients, and some studies have provided clear evidence for this. Although there are few long-term studies and it is difficult to compare the results from different areas, it seems likely that a significant increase in plankton growth can result from nutrient input—from sewage and runoff of agricultural fertilizers—into coastal waters and estuaries with limited exchange with the open ocean. For example, regular blooms of non-toxic microalgae such as *Phaeocystis* and *Emiliania* have become a regular occurrence in the North Sea and English Channel (see *Figure 6.8A*). Could such eutrophication also lead to an increase in toxic species? One widely held idea is that nutrient inputs from sewage or agricultural land runoff alter the ratio of particular nutrients, as well as the total loading, and that this may change the balance of the different plankton groups. Sewage is rich in nitrogen and phosphorus but has a low silicon content. Since diatoms specifically require silicon, this could favor the selective growth of dinoflagellates. The emergence of dinoflagellate blooms as a problem in the North Carolina and Maryland estuaries was linked to increased nutrient levels due to an expansion in poultry and pig farming on the surrounding land. The dinoflagellate *Pfiesteria piscicida* was implicated as the cause of fish mortalities and was also linked to neuropsychological effects in humans (although this was not proven, and the true nature of the problems remains something of a mystery). The expansion of mariculture, such as high-density salmon or shrimp culture, has been directly implicated in the increased frequency of HABs due to nutrient enrichment from uneaten food and excreta. For example, the increased incidence of *Alexandrium* blooms and high levels of PSP toxins in shellfish in Scotland during the 1990s was due to poorly sited salmon pens in lochs, bays, and inlets with limited water exchange.

Another obvious explanation for the apparent increase in HABs and associated diseases is increased awareness of toxic species and more extensive and effective monitoring, which is exacerbated by the increased use of coastal waters for aquaculture of shellfish and finfish. These activities act as sensitive indicators of potential problems and undoubtedly account for at least part of the apparent spread of PSP and DSP. Mass mortalities in fish farms can occur due to HABs, although these do not always involve toxic species.

Increasing incidence of HABs—both toxic and non-toxic—also pose considerable threats to seawater desalination facilities. These use a process of reverse osmosis which is effective at removing most toxins but can be overwhelmed if toxin concentrations are very high. Blooms also clog filters or cause biofouling of osmosis membranes.

Coastal waters must be regularly monitored to assess the development of HABs

Regular surveys by microscopic examination of the dynamics of phytoplankton populations within an area may give advance warning of an increase in particular species, which may sometimes precede a toxic bloom. Such surveys are time-consuming, and it is often difficult to distinguish toxic species or strains using morphological criteria. Some improvements can be achieved using flow cytometry or epifluorescence microscopy after labeling the target species with specific fluorescent antibodies. Unfortunately, background fluorescence is often a problem. Microphotography and automated image analysis of algal cells that are identified by artificial intelligence systems are also being used to identify particular types of algae. Remote sensing can be employed by equipping satellites with spectral scanners that detect pigments in surface waters, and it is becoming possible to identify specific types of algae based on their spectral signature. When coupled with physical measurements such as sea surface temperatures and current flows, satellite images are especially useful in tracking the development and movement of blooms. As genetic sequence data for toxic dinoflagellates and diatoms become available, gene probes are being increasingly used. A common method is to amplify microalgal DNA encoding 18S rRNA using eukaryote-specific primers. The resulting PCR products can be cloned and sequenced, leading to a specific oligonucleotide probe that will hybridize with DNA of the organism in water samples to provide quantitative estimates of density of particular HAB species using microarray or qPCR technology.

Surveys can be used to limit the economic and health impacts of HABs, but whether we can use such information to control their development or spread is questionable, since HABs often occur over huge areas. The effects of weather, ocean currents, and the many physical and biological factors that determine the development and eventual demise of a bloom are unpredictable. Chemical treatments, such as adding agents that promote agglutination and sinking of microalgal cells, can be applied locally to control blooms, but the ecological impact of such intervention on a large scale needs careful assessment. Flocculating clays are used widely in Japan and China to control blooms around fish farms. Algicidal bacteria and viruses capable of initiating lysis of dinoflagellates and diatoms have been isolated from seawater and investigated as possible biological control agents. Also, some microalgae have been shown to contain lysogenic viruses, which can be induced into the lytic cycle by particular conditions. Some success in initiating the collapse of bloom populations has been achieved in microcosm experiments, but much more research and evaluation of the ecological and "scale-up" issues is needed before biological control becomes a practical proposition. Such biotechnological approaches to the control of HABs are considered further in *Chapter 14*.

? IS IT POSSIBLE TO DEVELOP REAL-TIME HAB FORECASTS?

The ability to provide early warning of the development of specific HABs would be of great value to managers of fisheries and conservation programs, and for residents and tourists in coastal zones. Computer modeling based on environmental factors, such as water temperature and salinity, river flow, winds, and tides, and biological factors such as abundance and behavior of toxic dinoflagellates have been developed by NOAA state authorities in the USA to provide weekly prediction of *Karenia* and *Pseudo-nitzschia* HABs in California and Florida. In the near future, it is likely that automatic sensing devices using gene probe technology could be placed on offshore buoys, and information could be beamed to satellites and integrated with improved remote sensing signal-detection systems linked to powerful computer models leading to reliable real-time forecasts.

Conclusions

The most important human diseases associated with autochthonous marine bacteria are infections caused by bacteria in the genus *Vibrio*. These are adapted to a life in the aquatic environment but possess complex regulatory systems coordinating the expression of virulence factors when they transfer to the human host. *Vibrio* spp. show evidence of extensive gene transfer that explains the emergence of new pathogenic variants. Numerous intoxications are caused by dinoflagellates and diatoms. Factors that affect plankton blooms and the production of toxins harmful to humans are numerous, complex, and poorly understood. A combination of natural and anthropogenic effects is undoubtedly responsible for a worldwide increase in HABs, a wider range of toxic species and the emergence of "new" human diseases. Climate change is increasing the threat of disease in the human population. Bacterial pathogens show increased abundance in seawater at higher temperatures and people are more likely to become infected following coastal flooding storms or hurricanes, whilst the increased intensity of HABs and their spread to new regions can be linked to climate change.

References and further reading

Diseases caused by *Vibrio* spp.

Almagro-Moreno, S. & Taylor, R.K. (2013) Cholera: Environmental reservoirs and impact on disease transmission. *Microbiol. Spectrum* 1: 1–19.

Baker-Austin, C. & Oliver, J.D. (2018) *Vibrio vulnificus*: New insights into a deadly opportunistic pathogen. *Environ. Microbiol.* 20: 423–430.

Baker-Austin, C., Trinanes, J., Gonzalez-Escalona, N., & Martinez-Urtaza, J. (2017) Non-cholera vibrios: The microbial barometer of climate change. *Trends Microbiol.* 25: 76–84.

Baker-Austin, C., Trinanes, J.A., Taylor, N.G.H. et al. (2012) Emerging *Vibrio* risk at high latitudes in response to ocean warming. *Nat. Clim. Change* 3: 73–77.

Baracchini, T., King, A.A., Bouma, M.J. et al. (2017) Seasonality in cholera dynamics: A rainfall-driven model explains the wide range of patterns in endemic areas. *Adv. Water Resour.* 108: 357–366.

Borgeaud, S., Metzger, L.C., Scrignari, T., & Blokesch, M. (2015) The type VI secretion system of *Vibrio cholerae* fosters horizontal gene transfer. *Science* 347: 63–67.

Chourashi, R., Das, S., Dhar, D. et al. (2018) Chitin-induced T6SS in Vibrio cholerae is dependent on ChiS activation. *Microbiology* 164: 751–763.

Church, S.R., Lux, T., Baker-Austin, C. et al. (2016) *Vibrio vulnificus* type 6 secretion system 1 contains anti-bacterial properties. *PLoS One* 11: e0165500.

Cianfanelli, F.R., Monlezun, L., & Coulthurst, S.J. (2016) Aim, load, fire: The type VI secretion system, a bacterial nanoweapon. *Trends Microbiol.* 24: 51–62.

Colwell, R.R., Huq, A., Islam, M.S. et al. (2003) Reduction of cholera in Bangladeshi villages by simple filtration. *Proc. Natl. Acad. Sci.* 100: 1051–1055.

Durai, S., Pandian, S.K., & Balamurugan, K. (2011) Changes in *Caenorhabditis elegans* exposed to *Vibrio parahaemolyticus*. *J. Microbiol. Biotechnol.* 21: 1026–1035.

Ellison, C.K., Dalia, T.N., Ceballos, A.V. et al. (2018) Retraction of DNA-bound type IV competence pili initiates DNA uptake during natural transformation in *Vibrio cholerae*. *Nat. Microbiol.* 3: 773–780.

Faruque, S.M. & Mekanolos, J.M. (2012) Phage-bacterial interactions in the evolution of *Vibrio cholerae*. *Virulence* 3: 556–565.

Ghenem, L., Elhadi, N., Alzahrani, F., & Nishibuchi, M. (2017) *Vibrio parahaemolyticus*: A review on distribution, pathogenesis, virulence determinants and epidemiology. *Saudi J. Med. Med. Sci.* 93: 93–103.

Hartnell, R.E., Stockley, L., Keay, W. et al. (2019) A pan-European ring trial to validate an International Standard for detection of *Vibrio cholerae*, *Vibrio parahaemolyticus* and Vibrio vulnificus in seafoods. *Int. J. Food Microbiol.* 288: 58–65.

Heng, S.-P., Letchumanan, V., Deng, C.-Y. et al. (2017) *Vibrio vulnificus*: An environmental and clinical burden. *Front. Microbiol.* 8: 997.

Hsueh, P.-R., Lin, C.-Y., Tang, H.-J. et al. (2004) *Vibrio vulnificus* in Taiwan. *Emerg. Infect. Dis.* 10: 1363–1368.

Huq, A., Yunus, M., Sohel, S.S. et al. (2010) Simple sari cloth filtration of water is sustainable and continues to protect villagers from cholera in Matlab, Bangladesh. *MBio* 1: e00034-10.

IPCC (2018) Special Report: Global Warming of 1.5°C above pre-industrial levels. www.ipcc.ch/sr15/. Accessed 10 Jan 2019.

Le Roux, F. & Blokesch, M. (2018) Eco-evolutionary dynamics linked to horizontal gene transfer in vibrios. *Annu. Rev. Microbiol.* 72: 89–110.

Lekshmi, N., Joseph, I., Ramamurthy, T., & Thomas, S. (2018) Changing facades of *Vibrio cholerae*: An enigma in the epidemiology of cholera. *Indian J. Med. Res.* 147: 133–141.

Lipp, E.K., Huq, A., & Colwell, R.R. (2002) Effects of global climate on infectious disease: The cholera model. *Clin. Microbiol. Rev.* 15: 757–770.

Nelson, E.J., Harris, J.B. Morrisa Jr, J.G. et al. (2009) Cholera transmission: The host, pathogen and bacteriophage dynamic. *Nat. Rev. Microbiol.* 7: 693–702.

Okada, K., Iida, T., Kita-Tsukamoto, K., & Honda, T. (2005) Vibrios commonly possess two chromosomes. *J. Bacteriol.* 187: 752–757.

Phillips, K.E. & Satchell, K.J.F. (2017) *Vibrio vulnificus*: From oyster colonist to human pathogen. *PLOS Pathog.* 13: e1006053.

Roig, F.J., González-Candelas, F., Sanjuán, E. et al. (2018) Phylogeny of *Vibrio vulnificus* from the analysis of the core-genome: Implications for intra-species taxonomy. *Front. Microbiol.* 8: 2613.

Sakib, S.N., Reddi, G., & Almagro-Moreno, S. (2018) Environmental role of pathogenic traits in *Vibrio cholerae*. *J. Bacteriol.* 200: e00795-17.

Semenza, J.C., Trinanes, J., Lohr, W. et al. (2017) Environmental suitability of *Vibrio* infections in a warming climate: An early warning system. *Environ. Health Perspect.* 125: 107004.

Shapiro, B.J., Levade, I., Kovacikova, G. et al. (2016) Origins of pandemic *Vibrio cholerae* from environmental gene pools. *Nat. Microbiol.* 2: 16240.

Vezzulli, L., Brettar, I., Pezzati, E. et al. (2012) Long-term effects of ocean warming on the prokaryotic community: Evidence from the vibrios. *ISME J.* 6: 21–30.

Vezzulli, L., Grande, C., Reid, P.C. et al. (2016) Climate influence on *Vibrio* and associated human diseases during the past half-century in the coastal North Atlantic. *Proc. Natl. Acad. Sci. USA* 113: E5062–71.

Wagley, S., Borne, R., Harrison, J. et al. (2018) *Galleria mellonella* as an infection model to investigate virulence of *Vibrio parahaemolyticus*. *Virulence* 9: 197–207.

WHO (2017) Cholera vaccines: WHO position paper – August 2017. *Wkly. Epidemiol. Rec.* 92: 477–500.

Zoonoses from marine animals

Gauthier, D.T. (2014) Bacterial zoonoses of fishes: A review and appraisal of evidence for linkages between fish and human infections. *Vet. Rec.* 203: 27–35.

Haenen, O.L., Evans, J.J., & Berthe, F. (2013) Bacterial infections from aquatic species: Potential for and prevention of contact zoonoses. *Rev. Sci. Tech. (Int. Off. Epizoot.)* 32: 497–507.

Hunt, T.D., Ziccardi, M.H., Gulland, F.M. et al. (2008) Health risks for marine mammal workers. *Dis. Aquat. Org.* 81: 81–92.

Waltzek, T.B., Cortés-Hinojosa, G., Wellehan Jr, J.F.X., & Gray, G.C. (2012) Marine mammal zoonoses: A review of disease manifestations. *Zoonoses Pub. Health* **59**: 521–535.

Bacterial and algal toxins

Assunção, J., Guedes, A.C., & Malcata, F.X. (2017) Biotechnological and pharmacological applications of biotoxins and other bioactive molecules from dinoflagellates. *Mar. Drugs* **15**:.393.

Dorantes-Aranda, J.J., Campbell, K., Bradbury, A. et al. (2017) Comparative performance of four immunological test kits for the detection of Paralytic Shellfish Toxins in Tasmanian shellfish. *Toxicon* **125**: 110–119.

EFSA Panel on Contaminants in the Food Chain, Knutsen, H.K., Alexander, J., Barregard, L. et al. (2017) Risks for public health related to the presence of tetrodotoxin (TTX) and TTX analogues in marine bivalves and gastropods. *EFSA J.* **15**: e4752.

Feng, C., Teuber, S., & Gershwin, M.E. (2016) Histamine (scombroid) fish poisoning: A comprehensive review. *Clin. Rev. Allergy Immunol.* **50**: 64–69.

Jal, S. & Khora, S.S. (2015) An overview on the origin and production of tetrodotoxin, a potent neurotoxin. *J. Appl. Microbiol.* **119**: 907–916.

Magarlamov, T.Y., Melnikova, D.I., & Chernyshev, A.V. (2017) Tetrodotoxin-producing bacteria: Detection, distribution and migration of the toxin in aquatic systems. *Toxins* **9**: 166.

McLauchlin, J., Little, C.L., Grant, K.A., & Mithani, V. (2006) Scombrotoxic fish poisoning. *J. Pub. Health* **28**: 61–62.

Rongo, T., Bush, M., & van Woesik, R. (2009) Did ciguatera prompt the late Holocene Polynesian voyages of discovery? *J. Biogeogr.* **36**: 1423–1432.

Ecology and Detection of Harmful Algal Blooms

Gilbert, P.M., Berdalet, E., Burford, M.A. et al. (2018) *Global Ecology and Oceanography of Harmful Algal Blooms.* Springer Nature.

McPartlin, D.A., Loftus, J.H., Crawley, A.S., Silke, J. et al. (2017) Biosensors for the monitoring of harmful algal blooms. *Curr. Opin. Biotechnol.* **45**: 164–169.

Ottesen, E.A. (2016) Probing the living ocean with ecogenomic sensors. *Curr. Opin. Microbiol.* **31**: 132–139.

Chapter 13

Microbial Aspects of Marine Biofouling, Biodeterioration, and Pollution

The first section of this chapter discusses the economically important detrimental effects of marine microbes through biofouling of marine surfaces and structures, the biodeterioration of materials, and the spoilage of seafood. The second section considers the microbial diseases arising from the pollution of the sea by sewage and wastewater, together with methods of monitoring water and shellfish using microbial indicators and new approaches for detection of pathogens. In the final section, the role of marine microbes in biodegradation and bioremediation of oil and other chemical pollutants is discussed. The chapter concludes with discussion of the roles that microbes play in two areas of growing concern, namely the mobilization of toxic mercury and other pollutants into marine food webs, and plastic pollution of the oceans.

Key Concepts

- Microbial colonization often provides the first stage in biofouling of surfaces in the sea, leading to economic losses through interference with the efficient operation of ships and damage to structures such as piers and aquaculture facilities.

- Corrosion and deterioration of metal and wooden structures is initiated by microbes.

- Microbial activity results in harmful spoilage of fish and shellfish; it can also be used to produce seafood products with altered desirable properties.

- Pathogenic microbes introduced to the sea via sewage and wastewater constitute a health hazard through recreational use of coastal environments and contamination of shellfish.

- Bacterial indicators are used for monitoring coastal waters for fecal pollution but may be unreliable predictors of health risks, which are mainly associated with viruses.

- Many microbes degrade oil naturally and this may be enhanced using bioremediation.

- Microbial activity in sediments and pelagic aggregates leads to production of toxic mercury compounds that can contaminate fish.

- Interactions of microbes with polluting microplastics may affect their transport and persistence, with potentially major implications for ocean processes such as carbon cycling.

BIOFOULING AND BIODETERIORATION

Microbial biofilms initiate the process of biofouling

As previously discussed in *Chapter 3*, the surfaces of inanimate objects and living organisms in the sea are colonized by mixed microbial communities that form biofilms showing complex physical structures and chemical interactions. As shown in *Figure 13.1*, the process of biofouling usually begins with the formation of a molecular conditioning film formed by the deposition of organic matter on the surface within a few minutes of immersing a material into seawater. The conditioning film is composed of many compounds including amino acids, proteins, lipids, nucleic acids, and polysaccharides. This is followed over the next few hours by the attachment of pioneer motile bacteria as primary colonizers, followed within a few days by a diverse association of bacteria and benthic diatoms. This leads to a slimy biofilm up to 500 μm thick and consolidated by the production of sticky exopolymers. Compounds produced in the microbial biofilm act as cues for the settlement of other microbes, planktonic algal spores, and invertebrate larvae. Colonization by some macrofouling organisms such as barnacles can occur very rapidly.

The complex dense community that develops leads to the biofouling of all types of marine surfaces, including coastal plants, macroalgae, animals, piers and jetties, oil-drilling rigs, boat hulls, fishing gear, aquaculture cages, engineering materials, concrete and metal structures. *(Figure 13.2)*. Other effects include blockage of pipes and filters, reduced efficiency of heating and cooling plants, and interference with the efficient operation of boats and ships.

Sailing enthusiasts know the inconvenience and cost—in lost weekends—of scraping the bottom of yachts, while the economic effects of biofouling of larger ships are immense, leading to extreme increases in fuel usage due to frictional drag. Formation of a microbial biofilm alone can cause a 1–2% increase in drag. Subsequent colonization by macroalgae increases this to about 10% and, if unchecked, extensive colonization by hard-shelled invertebrates such as barnacles, tubeworms, bryozoans, or mussels can lead to 30–40% drag. This costs shipping companies and navies throughout the world over $200 billion every year because of losses in fuel efficiency and the cost of antifouling measures. Increased fuel usage also leads to increased emissions of CO_2 and sulfurous pollutants. In addition, ships and other floating debris can transport non-indigenous species around the world.

Various methods for the prevention of macrofouling of ships' hulls, such as the use of copper sheeting, have been in use for hundreds of years. Self-polishing copolymer paints containing copper or tin compounds have been used extensively in the second half of the twentieth century. Tributyl tin (TBT) is particularly effective at controlling the problem, but its use has been banned since 2008 by the International Maritime Organization, following the recognition that it has serious environmental consequences, with particular effects on the ecology and reproductive behavior of marine invertebrates and the immunity of marine mammals (see p.323). There is, therefore, an active search for effective "environment-friendly" alternatives. Ideally, products isolated from marine organisms should have low toxicity, be effective at low concentrations, and break down quickly if released into the environment. Hundreds of

Figure 13.1 Schematic representation of stages in the biofouling process, showing the critical role of microbial activities. Soft biofouling organisms include algae, bryozoans, tunicates, anemones, soft corals, and sponges. Hard biofouling organisms include barnacles, mussels, and tubeworms. Modified from Martín-Rodríguez et al., 2015, CC BY 4.0.

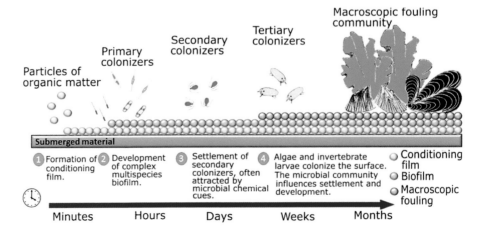

Figure 13.2 Examples of biofouling. (a) Polythene mesh immersed in North Atlantic water shows extensive colonization by hydroids (background), whereas copper alloy mesh (foreground) shows no biofouling. (b) Biofouling on a cast-iron statue in Antony Gormley's "Another Place" installation on Crosby Beach, Merseyside, England—shown four years after erection in the intertidal zone. Credits: A. Environment, B. Rept0n1x; CC BY-SA 3.0 via Wikimedia Commons.

molecules with antifouling properties have been investigated, with the richest sources being sessile marine organisms—especially sponges, corals, seaweeds, and ascidians—as these often possess mechanisms to avoid overgrowth by other organisms. As shown in *Figure 13.1*, there are various critical points at which the process of biofouling may be inhibited. From a microbiological perspective, our interest is in the prevention of the initial formation of a biofilm. One example of a successful development in this field is the discovery of brominated furanone compounds produced by the seaweed *Delisea pulchra*, which prevent biofilm formation by interfering with quorum sensing. A range of other such "quorum quenching" compounds are now being investigated,

The surfaces of some species of algae are colonized by bacteria that seem to interfere with cues for settlement of invertebrate larvae or spores of other algae. Bacteria belonging to the genera *Pseudoalteromonas* and *Phaeobacter*, isolated from bacterial biofilms on the surface of algae, produce a range of inhibitory compounds that have potential applications in antifouling treatments. A number of coatings for plastic surfaces have also been developed by mixing extracts of marine isolates of *Bacillus*, *Pseudomonas*, and *Streptomyces* with water-based resins. These "living paints" have shown promise in experimental trials, but problems with formulation and delivery of active compounds means that laboratory assays are often not confirmed in field trials.

Biofouling is also a major problem in desalination plants, in which seawater is treated by reverse osmosis. Microbial biofilms on filters decreases the flow of water and efficiency of salt extraction. They also cause reduced efficiency of heat exchangers in coastal powerplants.

It is important to note that while the negative aspects of these processes are considered here because of their deleterious consequences, microbial biofilms also play a very important, positive role. Microbial products may stimulate the settlement and metamorphosis of algae and invertebrate larvae, which is very important in marine ecology (e.g. in reef formation) and aquaculture (e.g. in settlement of mussels in suspended rope culture).

Microbes induce corrosion of metals, alloys, and composite materials

A range of aerobic and anaerobic bacteria are responsible for microbially induced corrosion (MIC), causing pitting and fracture stresses of metal structures, vessels, equipment, and instruments in the marine environment. Corrosion occurs due to attached bacteria in biofilms at the interface of the metal and seawater. Anaerobic sulfate-reducing bacteria (SRB) have long been known to be the main cause of marine corrosion, but we now recognize the importance of complex multi-species mixed biofilms containing microbes that carry out a range of biochemical and electrochemical processes synergistically. Where surfaces are open to oxygenated seawater, SRB can occupy the lower levels of a biofilm because oxygen is depleted by aerobic microbes in the upper layers. Many other species of bacteria and fungi can produce organic acids by fermentation that corrode steel, manganee and zinc alloys. MIC is a particular problem for the offshore oil and gas industry because it causes corrosion

BACTERIA CAN PRODUCE DEADLY GAS IN OIL PRODUCTION

The activity of sulfate-reducing bacteria (SRB) in oil and gas reserves can lead to high levels of sulfide that cause "souring" of the oil that cause problems in refining. More importantly, there have been numerous examples of leaks of toxic levels of H_2S gas on offshore drilling platforms that have resulted in fatalities to oil-rig workers. H_2S is toxic at very low levels. Trace amounts of the gas are easily detected by its characteristic "rotten eggs" smell, but it paralyzes the olfactory nerve at levels above 30 ppm—so a worker may smell the gas, investigate its source, and then be unable to smell higher concentrations. At low levels, H_2S causes headaches, blurred vision, and conjunctivitis. It can be instantly fatal at high levels, and even a few minutes exposure to 500–1000 ppm can cause long-term illness. Modern rigs employ a variety of sensors to detect escaping gas and workers receive special training in dealing with leaks, but in some oil-well "blowouts" drilling companies have been forced to set fire to rigs to burn off the escaping H_2S.

Teredinibacter turnerae is the only member of the endosymbiotic consortium of microbes in the shipworm *Bankia* spp. that has been cultured. The genome sequence reveals that the bacterium contains genes for a range of carbohydrate-degrading enzymes that specialize in the breakdown of wood as well as genes for nitrogen fixation (Yang et al., 2009). Stable isotope analysis shows that almost all of the carbon and up to half of the nitrogen in the shipworm is derived from the symbiont metabolism (Charles et al., 2018). O'Connor et al. (2014) made the surprising discovery that the digestive bacteria live in bacteriocytes in the gills, rather than the gut. Wood-degrading enzymes are secreted and transported—by mechanisms that are currently unknown—to the caecum, a part of the gut which is free of a resident microbiota. O'Connor and colleagues conclude that separation of the symbionts from the site of wood digestion eliminates competition with the host for soluble nutrients and enables the host to have greater control over the process. Identification of the small number of key bacterial enzymes transported to the gut opens up the possibility of developing efficient biotechnological digestion of lignocellulose—a major goal for the production of biofuels.

of carbon steel pipelines, drilling platforms and equipment and also leads to the "souring" of crude oil by the production of H_2S, with significant health and safety implications. Other bacteria can attack corrosion-resistant manganese or iron oxide films on the surface of iron and steel. Thermophilic sulfate-reducing archaea also occur in marine hydrothermal systems and oil reservoirs.

Control of biofilms is particularly difficult, because they are protected from the effects of many chemical treatments and repeated high concentrations of biocides—such as glutaraldehyde, quaternary ammonium, and phosphonium compounds—are needed. Corrosion of steel structures can be limited using cathodic protection, but this can lead to structural weakness unless the electrical potential is carefully applied and monitored; formation of excess hydrogen can cause metal fatigue. Routine monitoring by measuring levels of culturable acid-producing bacteria together with the application of biocides to fluids injected into the well can help to control the problem. Testing for SRB is more problematic because of the need for anaerobic culture, and development of molecular-based test kits will prove useful to the industry. Problems with the use of cathodic protection and environmental concerns about large-scale use of biocides at sea favor a biotechnological solution to the control of corrosive biofilms. Nitrate is sometimes added to injection fluids pumped into oil reservoirs during extraction, because nitrate-reducing bacteria may compete with SRB. However, this process needs careful monitoring because if nitrate enters the pipelines, nitrate reduction can be coupled with iron oxidation, leading to corrosion.

A special type of MIC occurs in the case of wrought iron or steel structures in deep water. Formations of rust aggregate into bioconcretious structures similar to a stalactite or icicle. They were first observed and called "rusticles" during exploration of the wreck of the *Titanic* 76 years after it sank in 1912 to a depth of 3.8 km off the coast of Newfoundland (*Figure 13.3*). Rusticles consist of iron oxides, carbonates, and hydroxides colonized by a complex syntrophic community of bacteria and fungi that corrode the metal to produce the iron compounds. This includes a novel species named *Halomonas titanicae*. The meter-length fragile structures are permeated by water channels that allow water to flow through the structure and are expected to result in complete destruction of the wreck by 2030. Such formations have subsequently been described on other deep-water wrecks and oil platform mooring chains.

Microbes cause biodeterioration of timber and marine wooden structures

Microbial colonization and decomposition of timber transported or stored at sea, or of wooden structures—such as wharves, jetties, piers, and boats—causes an immense amount of damage costing many billions of dollars. The main cause of damage is penetration by wood-boring invertebrates, of which the most important are the shipworms, so called because of the

Figure 13.3 Microbially induced rusticles on the steel hull of the Titanic. The rusticles pass through a cycle of growth, maturation and then fall away, over a probable 5–10-year cycle. Credit: Lori Johnston, RMS *Titanic* Expedition 2003, NOAA-OE.

damage caused to wooden boats throughout history. These are not worms, but small bivalves with long stretched bodies—sometimes up to 60 cm long and ~ 1 cm wide—in the family *Teredinidae*, which contains over 70 species. They tunnel into wooden structures using their serrated shell valves and secrete a calcium carbonate rich coating that protects the delicate tissue from crushing (*Figure 13.4*). As previously mentioned in *Chapter 11*, shipworms and other marine invertebrates that digest wood rely on an obligate symbiotic association with wood-degrading microbes. In one species of shipworm, *Bankia setacea*, the gammaproteo-bacterium *Teredinibacter turnerae* has been identified as a symbiont of a specialized region of the host gill. Wood is also degraded by small isopod crustaceans known as gribbles, such as *Limnoria lignorum* and related species that break down the surface of wood with cutting and grinding mouthparts. (Although notorious for their destructive role, shipworms and gribbles have an important ecological role due to the decay of fallen wood in coastal areas such as mangroves.)

Most types of wood used throughout history are susceptible to damage by wood borers, although shipbuilders have traditionally used the denser heartwood as it has some resistance. A few kinds of tropical hardwoods have high resistance to wood borers, but their use is unsustainable. The use of various preservative wood treatments delays deterioration, but these are expensive and cannot be used on new timber during transport. Treatments such as creosote and chromated copper arsenate cause considerable environmental damage and their use is now tightly regulated in many countries. Furfurylation is an environmentally-friendly process in which wood is modified by impregnation with furfuryl alcohol ($C_5H_6O_2$)—made from plant waste such as sugar cane or corn cobs—combined with heat and pressure treatment. This improves mechanical strength and resistance to biodeterioration. Identification of the symbionts and their mode of transmission provide a potential weak link in the process of biodeterioration, as it may be possible to identify compounds that inhibit bacterial colonization or metabolism.

Microbial growth and metabolism cause spoilage of seafood products

Fish and shellfish deteriorate very rapidly as a result of microbial activities, resulting in the rapid production of discoloration, slime, and unpleasant odors and flavors. The composition of microbial communities in freshly caught seafood varies considerably according to the animal species and the methods of fishing, handling, processing, and storage. Natural transformation of compounds within the fish tissue begins soon after death owing to autolytic processes, making catabolites available for decomposition by members of the bacteria that are normal inhabitants of the gut and skin of fish. Specific spoilage organisms include *Pseudomonas*, *Shewanella*, *Alteromonas*, *Moraxella*, *Acinetobacter*, *Cytophaga*, and *Flavobacteriumpolio*. Psychrophilic strains of *Pseudomonas* and *Shewanella* become the dominant spoilage organisms when fish is chilled on ice. Seafood is rich in poly-unsaturated fatty acids, which are prone to lipid oxidation leading to rapid loss of flavor and nutritional quality. Specific spoilage organisms are most often associated with the production of ammonia, amines, ketones, organic acids, and sulfur compounds that characterize "off" seafood. Trimethylamine is the most important of these metabolites, produced by reduction of trimethylamine oxide, naturally present in the tissue of many fish.

SHIPWRECKED SHIPS WRECKED

The preservation of submerged archaeological timbers is a specialized process. Waterlogged wood is protected when buried in anoxic sediments, but fungi and bacteria decay it rapidly once it is brought to the surface. The Swedish navy ship *Vasa*, which sank in Stockholm harbor on its maiden voyage in 1628, was discovered 333 years later in remarkably well-preserved condition because the hypoxic, low salinity conditions in the Baltic Sea prevented colonization by shipworms and degradative microbes. When the vessel was raised, impregnation with polyethylene glycol was used to fill voids in the wood structure. Despite this treatment, it was realized after ~20 years that unexpected decay was occurring after all. During its long period in anoxic conditions on the seabed—favored by sewage dumping in the harbor—sulfate reduction by bacteria led to accumulation of > 5 tonnes of sulfur in the timbers. Iron nails in the structure catalyzed an oxidative reaction forming sulfuric acid that is now hydrolyzing cellulose in the timber, reducing its stability (Sandström et al., 2002). The same problem has occurred with other recovered shipwrecks such as the *Mary Rose* and *Batavia*.

Figure 13.4 (a) Driftwood showing extensive boring damage by shipworm. (b) Teredo shipworm extracted from mangrove wood, Brazil. The body is ~0.5 m long. Credits: A. Michael C. Rygel. B. Depleswsk. CC-BY-SA-3.0 via Wikimedia Commons.

Bacteria grow rapidly to high levels (over 10^8 CFU per gram) in the nutrient-rich environment of fish tissue, which contains high concentrations of readily utilizable substrates such as free amino acids. Metabolic consortia develop in the mixed microbial community; for example, lactic acid bacteria (e.g. *Lactobacillus* and *Carnobacterium*) degrade the amino acid arginine to ornithine, which is further degraded to putrescine by enterobacteria. The production of siderophores is important for the acquisition of iron, since fish tissue contains limiting concentrations (see p.312). Lactic acid bacteria characteristically produce bacteriocins, which are highly specific membrane-active peptide antibiotics that inhibit certain other bacteria.

Most fish spoilage organisms are easily cultured and identified using standard microbiological methods, although there has been some recent use of molecular methods for characterization and early detection of spoilage organisms (e.g. gene probes for *Shewanella putrefaciens*) and chemical assays for total volatile basic nitrogen after extraction with trichloroacetic or perchloric acid. The growth kinetics of specific spoilage organisms can be determined by absorbance measurements in culture and used to construct mathematical models to predict rates of spoilage and shelf-life of fish products.

Processing, packaging, and inhibitors of spoilage are used to extend shelf-life

The food industry has devised various methods of extending shelf-life in "value-added" products by inhibiting or delaying spoilage, but many processes used with other foods are unsuitable because the texture and flavor of fish products is easily destroyed by processing, leading to reduced consumer acceptance. Many processes like salting, pickling, and smoking are based on traditional methods, while recent innovations include modified atmosphere packaging (*Figure 13.5*) in which products are packaged in a gas mixture containing low levels of oxygen and high levels of carbon dioxide and nitrogen. The composition of the gas mixture and the permeability of the plastic film used to wrap the product is optimized for specific food types. This can shift the ecology of the contaminating microbiota; for example, while CO_2 packaging of fresh fish in ice is highly effective, it can suppress the growth of respiratory bacteria so that fermentative *Photobacterium*, lactic acid bacteria, and enterobacteria become dominant. Spore-forming bacteria (*Bacillus* and *Clostridium*) can survive mild heat treatment (pasteurization) of vacuum-packed products. As well as spoilage, some fish products are the source of human pathogens. Fish- and shellfish-associated bacterial, viral, and toxic diseases were considered in *Chapter 12*, but other pathogens such as *Salmonella* spp., *Staphylococcus aureus*, and *Listeria monocytogenes* may be introduced from human sources during handling and processing. *Listeria* is of particular concern in lightly preserved ready-to-eat products such as cold-smoked fish and shellfish because of its ability to grow over a very wide temperature range and salinity. Control of these pathogens is an important requirement for commercial processors and modified atmosphere packaging is widely used to limit health risks.

Besides improvements in packaging methods, antioxidants and chemical inhibitors of microbial growth may be employed. Antibiotics such as tetracyclines were once added to ice on board fishing vessels and in markets to prevent spoilage of fresh fish. This practice is now prohibited because of concerns over antibiotic residues and resistance. However, bacteriocins from lactic acid bacteria are generally regarded as safe and are permitted in

Figure 13.5 Fish preservation, old and new. (a) Fishwives packing herring into barrels for salting, circa 1930. (b) A factory in Vietnam for processing fish reared by aquaculture, showing preparation under high levels of hygiene. Credits: (a) North Shields Library Services; (b) AccuDB. com.

(a)   (b)

foods (e.g. nisin is widely used in dairy products). With fish, addition of purified bacteriocins has produced some promising results, as has the addition of non-spoilage lactic acid bacteria as competitors of pathogens or spoilage organisms. It has been shown that quorum-sensing signaling molecules (see p.102) are produced during development of the microbial spoilage community, and interference with the quorum-sensing mechanism might offer a potential new method of control. Some natural preservatives such as plant extracts, essential oils, and bioactive peptides may be effective because of this quorum quenching effect.

Some seafood products are made by deliberate manipulation of microbial activities

The spoilage activities of microbes in fish are generally regarded as detrimental but use of fermented fish sauces have a long history, for example, the garum of the ancient Roman Empire. In some parts of the world, there are numerous ethnic food products that depend on microbial activities for preservation and flavors. Most of these are encountered in Asian countries, especially Indonesia, Thailand, the Philippines, and Japan. Many processes are conducted according to traditional recipes, but the microbiology of some has been investigated during commercialization of production for export, and pure starter cultures may sometimes now be used. Perhaps the best-known product is nam-pla (fish sauce), which is now widely used in the West due to the popularity of Thai cuisine. Nam-pla is made by fermentation of fish hydrolyzed by a high-salt concentration (15–20%), which encourages growth of the extreme halophile *Halobacterium salinarum*, leading to characteristic flavors and aromas. This appears to be the only food product that relies on activities of a member of the Archaea. Other high-salt products include som-fak, burong-isda, and jeikal, traditional Korean foods made from fermented shrimp. Other fermented fish products such as plaa-som use lower salt concentrations; in these, a microbial flora dominated by lactic acid bacteria and yeasts leads to a characteristic aroma. There must be enough salt present (2–8%) and a final pH of less than 4.5 to inhibit the growth of pathogens; nevertheless, contamination by *Staphylococcus* can be a problem. Garlic is a major ingredient of some recipes; not only does this serve as a carbohydrate source, but it also has antibacterial activity (some constituents of garlic inhibit quorum sensing). Ika-shiokara is made from squid and fish guts pickled in 2–30% salt in a process that depends on growth of the yeast *Rhodotorula*; if this sounds appealing, you will find it as a delicacy in Hokkaido, Japan. Nordic countries also have several traditional fermented fish dishes, of which the most famous is hákarl from Iceland. This is made from Greenland shark, traditionally buried on the seashore for several weeks or months, during which urea is converted to ammonia by *Moraxella* and *Acinetobacter* spp. before drying. Skate can also be processed in the same way. These foods were once an essential winter staple source of protein and energy for Icelanders but have recently gained a reputation as an item to be sampled as a dare by curious tourists, accompanied by shots of Brennivín liquor.

MARINE POLLUTION BY SEWAGE AND WASTEWATER

Coastal pollution by wastewater is a source of human disease

A large proportion of the world's population lives near the coast. Many of the largest urban settlements have grown up around river estuaries and natural harbors because of their importance for trade. As these towns and cities developed, it was an easy option to dispose of untreated sewage directly into the rivers and sea; the adage "the solution for pollution is dilution" was applied. In many developed countries, awareness of the problems arising from disposal in close proximity to the population led to long pipelines off the coast, but the grounds for this were usually esthetic rather than health-related. In many countries, vast quantities of untreated sewage are still disposed directly to sea. Even in highly developed countries, where most medium-sized communities will have sewage treatment works, human waste containing potential pathogens still finds its way into coastal waters from isolated dwellings, houseboats, and marinas. Mixture of untreated sewage with storm-water runoff is a regular occurrence, even in large cities with well-developed sewage systems.

HOW FRESH IS THAT FISH?

Wise consumers who purchase fresh fish know how to tell if fish is fresh and good to eat, based on appearance and smell. However, it is more difficult to assess prepacked fish bought from a supermarket, and "best before" and "use by" dates can be very unreliable, leading to excessive wastage for retailers as they necessarily err on the side of quality and safety. One novel solution to this problem is the development of "smart packaging" incorporating real-time sensors. For fish, the most promising approach is to detect the production of total volatile basic nitrogen (ammonia and volatile amines, TVB-N) produced by microbial activity during spoilage. For example, pH-sensitive dyes encapsulated within a polymer and covered by a gas-permeable film can change color in response to accumulation of TVB-N (Pacquit et al., 2006; Wells et al., 2019). Other approaches include electronic sensors and indicator labels that are activated during packaging; these measure temperature and time elapsed to indicate if the product is still fresh. Such sensors will have an important future in the retail market if their use proves to be a reliable indicator of freshness in various fish products under different conditions.

REDUCING DANGERS FROM SEWAGE POLLUTION

Every day, each human being produces between 100 and 500 g of feces and 1–1.5 L of urine, which are disposed via sewers, together with water used for flushing and washing. Sewage treatment involves three stages, which successively reduce the load of biological and chemical contaminants. Primary treatment consists of screening and separating solid material, before a second stage in which microbial processes occur in biofilms or flocs in trickling filters, activated sludge systems, or other types of bioreactors. Before discharge, sewage is usually subject to tertiary treatment such as filtration through sand or activated charcoal, or coagulation with iron or aluminum salts. Tertiary treatment results in approximately 10^7–fold reductions in fecal bacteria and up to 10^4–fold reductions in the number of viruses (Wang et al., 2018). Disinfection via ultraviolet light, chlorination, or ozone reduces pathogen levels further still, but these are expensive options requiring complex engineering works. Raw or partially treated sewage is still discharged in huge quantities into coastal waters in many parts of the world.

Since the mid-twentieth century, increased wealth and leisure time has led to greater use of the sea for recreational use, and a trip to the seaside has become an important feature of many peoples' lives. In the 1950s, awareness began to be focused on the potential hazards of swimming in sewage-polluted water. Public awareness of the problem became acute because of the growing popularity of seawater bathing and sports such as surfing, sail boarding, and diving. As discussed in *Box 13.1*, there are few well-documented large outbreaks of serious disease associated with recreational use of marine waters, but there are many epidemiological studies showing health risks for swimmers in waters polluted by wastewater. In most adults, these diseases are troublesome but not usually life-threatening; however, young children, old people, and those with an impaired immune system are at risk of acquiring more serious infections. Coastal waters used for recreation contain a mixture of pathogenic, opportunistic and non-pathogenic microorganisms derived from sewage effluents, bathers themselves, seabirds, and runoff from agricultural land contaminated by waste from livestock, as well as autochthonous marine bacterial pathogens such as the vibrios discussed in *Chapter 12*. Most of these pathogens are transmitted via the fecal–oral route and cause disease when swimmers unwittingly ingest seawater. Infection via the ears, eyes, nose, and upper respiratory tract and via open wounds may also occur. Aerosols may be a significant route of infection for surfers, causing mild respiratory illness. The number of organisms required to initiate infection will depend on the specific pathogen, the conditions of exposure, and the susceptibility and immune status of the host. Viruses and parasitic protozoa may require only a few viable units (perhaps 100 or less) to initiate infection, whereas most bacteria require large doses of thousands or millions. The types and numbers of various pathogens in sewage vary significantly according to the distribution of disease in the population from which sewage is derived, as well as geographic and climatic factors. As a general rule, the more serious infections transmitted via the fecal–oral route—diseases such as cholera, hepatitis, typhoid, and poliomyelitis—only become significant for swimming-associated illness if there is an epidemic in the local population. In this case, direct person-to-person transmission is likely to be far more important; however, recall the special case of cholera transmission discussed in *Chapter 12*.

Although bacteria are used as indicators of sewage pollution in seawater, as discussed below, they are much less important than viruses as a source of infection. Pathogenic enteric bacteria include *Salmonella*, *Shigella*, and pathogenic strains of *Escherichia coli*. Dermatitis ("swimmers' itch") and infection of cuts and grazes can be caused by bacteria such as *Staphylococcus*, *Pseudomonas*, and *Aeromonas* and the yeast *Candida* (as well as autochthonous *Vibrio* spp.). These organisms can also cause ear and eye infections.

Many of the pathogens in sewage effluent or land runoff also become associated with intertidal sediments, which provide a long-term reservoir and health risk. Viruses are protected by adsorption to sediment particles and bacteria may persist for long periods in a biofilm matrix or enter the VBNC state (see p.95). Levels of pathogens in the water column are correlated with resuspension of sediment particles, which increases water turbidity and prolongs survival.

Human viral pathogens occur in sewage-polluted seawater

More than 100 human enteric viruses belonging to three main families are transmitted via the fecal–oral route. The Adenoviridae are nonenveloped, icosahedral, double-stranded DNA viruses most commonly associated with infections of the upper respiratory tract, but there are several adenovirus types that cause gastroenteritis, leading to diarrhea and vomiting, primarily in children. The family Caliciviridae includes caliciviruses, astroviruses, and noroviruses. These are single-stranded RNA viruses, with small round or hexagonal virions. *Norovirus*, previously known as the Norwalk agent, is the most significant enteric virus in developed countries, probably causing about half of all cases of gastroenteritis. Millions of cases of infection by noroviruses occur each year through contamination of drinking water and foods by infected persons; the disease is a major problem as a cause of local epidemics in hospitals, hotels, holiday camps, and cruise ships. The infectious dose is thought to be very low (perhaps 1–10 virions), as stomach acid does not inactivate the virus and immunity to the virus is often incomplete —noroviruses also mutate rapidly leading to antigenic variation— so they are a major cause of disease associated with swimming or consumption of seafood.

The disease causes severe nausea, vomiting, and diarrhea, often accompanied by painful abdominal cramps, headaches, and chills.

The Picornaviridae are single-stranded RNA viruses with small icosahedral virions. This group includes polio virus (*Enterovirus C*), which infects the central nervous system, causing paralysis and death. Infections were at their peak in the 1940s and 1950s and prompted the development of effective vaccines and a World Health Organization global eradication program; there are now only a few thousand cases a year in a few regions of Africa and the Indian subcontinent. Coxsackieviruses also belong to this group and are probably the most abundant type in sewage; besides gastroenteritis, they are associated with meningitis and can sometimes cause paralysis.

Infection with hepatitis A virus is endemic in many low and middle-income countries and over 90% of children may be infected and excreting the virus. It is most commonly acquired by travelers to such countries, although there have been occasional outbreaks in Europe and North America associated with consumption of shellfish and other foods. Infection damages the function of the liver but is not usually serious.

Finally, the *Reoviridae* family includes the rotaviruses, which are the most serious causes of infant mortality in Africa and large parts of Asia. Even in high-income countries, a large proportion of young children are infected, meaning that sewage contains high numbers of these viruses.

Fecal indicator organisms (FIOs) are used to assess public health risks

The concept of an indicator organism stems from the idea that pathogens may be present in such low numbers in water that their detection is difficult, time-consuming, or expensive. An indicator is an easily cultivable organism present in sewage, whose presence indicates the possibility that pathogens may be present. Ideally, the indicator should be present in greater numbers and survive longer in the environment than the pathogens, providing a built-in safety margin. Early microbiologists recognized that *E. coli* and other coliform bacteria are present in large numbers (over 10^8 per gram) in the gut of warm-blooded animals and their feces—they are also very easy to culture—so they became the standard indicator for fecal pollution of water. The coliform bacteria are facultative anaerobic Gram-negative bacilli characterized by sensitivity to bile salts (or similar detergent-like substances) and the ability to ferment lactose at 35°C. The definition of the coliform group is operational rather than taxonomic, with 80 species in 19 genera; the best known of these are *Escherichia*, *Citrobacter*, *Enterobacter*, *Erwinia*, *Hafnia*, *Klebsiella*, *Serratia*, and *Yersinia*. The fecal coliforms—they may also be termed "thermotolerant coliforms," depending on the methodology used—are a subset of the coliforms originally distinguished by their ability to grow and ferment lactose at 44°C; the main member of this group is *E. coli*. Growth on selective media and fermentative properties are used in the traditional methods of identification and enumeration in water (*Figures 13.6, 13.9*). A defining feature of the coliforms is the enzyme β-galactosidase, which can be detected using o-nitrophenol-β-galactopyranoside (ONPG) as a substrate, resulting in a yellow color. *E. coli* can be distinguished by the production of β-glucuronidase, which can cleave either methylumbelliferyl-β-glucuronide (resulting in fluorescence under UV light) or proprietary chromogenic compounds (resulting in distinctive colored colonies on agar plates). These reactions form the basis of several commercial testing methods, which have gained widespread acceptance because they produce results more rapidly.

Coliforms and *E. coli* are unreliable FIOs for seawater monitoring

It is now generally recognized that, while appropriate for testing drinking water, coliform and *E. coli* counts are poor indicators of marine water quality. The total coliforms can be derived from a wide variety of sources, including plants and soil, so they often reflect runoff from the land as well as fecal pollution. It is also very difficult to distinguish whether *E. coli* in coastal waters originates from human or animal fecal contamination. Most significant of

Figure 13.6 Outline of membrane filtration method for enumeration of *E. coli* and intestinal enterococci in marine water samples (based on the USEPA method). Appropriate volumes of seawater are filtered through 0.45 μm membrane filters, which trap bacteria on the surface. Filters are rolled onto selective chromogenic agar. After incubation for 24 h, *E. coli* at 44°C produces red-magenta colonies on mTec agar (contains indoxyl-β-D-glucoside). Enterococci at 41°C produce blue colonies with a halo on MEI agar (contains indolyl-β-D-glucuronide).

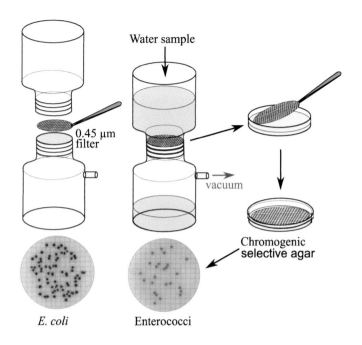

all, the survival of indicator coliforms and *E. coli* in the environment does not seem to reflect the survival of viral pathogens. Thus, they fail to meet one of the major requirements of a reliable indicator organism. Many studies have been carried out on the survival of bacteria and viruses in water. The effects of temperature, sunlight, pH, water turbidity, salinity, and presence of organic matter are highly complex interacting variables. Of these, the bactericidal effects of ultraviolet irradiation are the most important. Bacterial indicators have different survival characteristics in marine and fresh waters, while human viruses are inactivated at similar rates in both (*Figure 13.7*). Although many different studies have produced conflicting results, depending on the environmental conditions, a general conclusion is that levels of total coliforms are virtually useless as an indicator of health risks associated with marine waters, while levels of fecal coliforms (including *E. coli*) do not necessarily indicate good or poor water quality and are unreliable as predictors of health risks (see *Box 13.1*). In the USA, *E. coli* is used as an indicator in inland recreational waters but is no longer used in marine waters for regulatory purposes, although it is still in use in Europe and other countries.

Enterococci are more reliable FIOs for seawater monitoring

As early as 1980, it was proposed that the bacterial group known as the enterococci might be a better indicator of human pathogens for monitoring environmental water quality. These bacteria are consistently associated with the intestines of humans and warm-blooded animals and, as shown in *Figure 13.7*, generally have longer survival times in water than coliforms and *E. coli*. It was previously thought that different species were more consistently associated with humans or animals, and a high ratio of fecal coliforms to fecal enterococci has been interpreted as indicative of primarily human fecal contamination, whereas a low ratio was thought to indicate animal sources. These conclusions are now thought to be unreliable because of variable survival times of different species in seawater. Previously known as fecal streptococci, reclassification led to the recognition of a subgroup including the species *Enterococcus avium*, *E. faecalis*, *E. faecium*, *E. durans*, and *E. gallinarum*. These are differentiated from other streptococci by their ability to grow in 6.5% sodium chloride, at pH 9.6, and in a wide temperature range of 10–45°C; they are also resistant to 60°C for 30 min. Since the most common types found in the environment fulfil these criteria, the terms "fecal streptococci," "intestinal enterococci," and "*Enterococcus*" are used interchangeably. Although the number of enterococci excreted in feces is lower than *E. coli*, they are still much higher than the numbers of detectable pathogens. Methods of enumeration follow the same principles as those for *E. coli* as shown in *Figure 13.6* and a variety of specialized media incorporating chromogenic substrates, antibiotics, and tests for specific biochemical reactions can be used. Some semi-automated alternatives to membrane filtration have the advantage of producing same-day results. For example, the Enterolert® test kit quantifies enterococci by the

---

? COULD BUILDING SANDCASTLES LEAD TO DEADLY INFECTION?

We usually think of *E. coli* as a harmless member of the gut microbiota used, but several strains are pathogenic—of which the most serious is serotype O157:H7. This causes severe kidney damage and can be fatal, especially in children. Cattle are the main reservoir of *E. coli* O157:H7 and large numbers of the bacteria may be excreted in their feces. Serious epidemics in humans are usually associated with consumption of meat or salad crops contaminated by infected animals, but there have also been links between infection and environmental sources (Muniesa et al., 2006; Bintsis, 2018). In southwest England, the death of a child in 1999 and several other cases of O157:H7 infections linked to playing on the beach since that date are probably due to pollution from streams draining from land where cattle have grazed. This can be a problem on apparently pristine, rural beaches far from human habitation. Williams et al. (2007) and others have shown that *E. coli* O157:H7 can survive in beach sand and soil for various time periods, depending on the conditions. Several other studies have also linked digging in beach sand with enteric illness (Heaney et al. 2009).

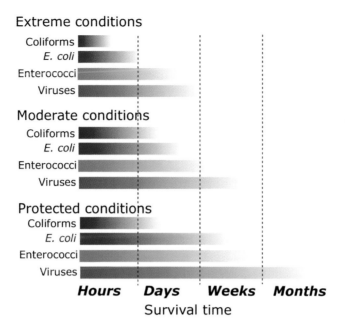

Figure 13.7 Simplified guide to typical survival times of fecal indicator organisms and enteric viruses in temperate coastal waters. Conditions that affect microbial survival in water are complex and interdependent. Extreme conditions would be typified by clear water, bright sunlight, high temperatures; protected conditions would include turbid water, cloud cover, low temperatures.

most probable number (MPN) method using growth in media with a fluorogenic substrate. The XplOrer64™ system measures growth of enterococci in a selective medium via changes in impedance. Under some conditions, autochthonous marine vibrios and aeromonads can interfere with the reactions and lead to false positive results. Also, runoff from farmland after heavy rainfall can significantly distort the counts of enterococci, and some studies have attempted to distinguish bacteria of human and animal origin.

Molecular-based methods permit quicker analysis of indicator organisms and microbial source tracking

The conventional culture-based methods used to assess microbial quality of water are widely used because they require relatively simple laboratory equipment, but they suffer from the disadvantage that it usually takes 18–96 h to obtain results. While regular water monitoring gives information about the long-term trends at particular sites, it does not provide information to the public about the state of the water on a particular day or time. Knowing that the levels of indicator organisms were unacceptably high a couple of days ago and that the beach users should have been advised not to swim does not inspire public confidence and does not adequately protect public health. Conversely, some authorities close beaches for several days after heavy rains as a precaution because past experience has shown that heavy rains may be associated with increased levels of indicator organisms; the reality may be that any "spike" in the levels of pathogens may have been and gone more quickly. This can be of considerable importance to the economy of coastal communities relying on beach trading as a source of income. For these reasons, there have been many efforts to develop more rapid methods of testing that will produce near real-time results in 2–4 h. This would allow management decisions to be made the same day and also permit better tracking of the source of contamination to its origin.

A number of qPCR-based studies have been carried out using primers targeting sequences of 16S rRNA genes or functional genes of indicator bacteria. DNA is extracted from water and amplified by multiplex PCR using a range of primers. PCR based methods suffer from the requirement for trained laboratory personnel in centralized laboratories, which could limit their usefulness if samples have to be transported long distances, although some trials of devices to carry out PCR assays in the field have shown some promise. Microarray techniques (see p.59) have the advantage that they can simultaneously assay for hundreds of different gene sequences. Beads can be coated with probes designed to hybridize with these amplicons and are also labeled with spectrally specific fluorescence markers so that quantification of specific targets can be measured in a fluorimeter. Generally, DNA-based methods detect genetic sequences from both dead and viable bacteria, which may overestimate concentrations when compared with culture-based methods. One promising alternative involves the immunomagnetic separation of *E. coli* or enterococci, using magnetic beads coated with antibodies directed against these bacteria. The beads are mixed with a water sample and selectively capture the bacteria; the beads are then collected from the sample using a magnet. The concentration of bacteria present is assessed by removing them from the beads and measuring the amount of ATP present by measuring bioluminescence in a luminometer, using a luciferin-luciferinase assay (see p.377). Because of considerable variations in composition of water samples and the possible low number of target FIOs, standardization of new methods and improvements in reliability and reproducibility of results are required before universal acceptance for regulatory purposes.

One of the main applications of new methods has been for microbial source tracking. The aim here is to identify populations of bacteria that are specific to particular animals, because recreational waters are often contaminated by nonpoint sources of fecal pollution of animal origin through runoff of farm waste via streams, or from wildlife including seabirds. Microbial source tracking usually involves culturing isolates of indicator bacteria from a wide range of sources and building up a library of profiling data, using phenotypic characteristics (such as antibiotic sensitivity or carbohydrate utilization) or genetic fingerprinting methods (see *Table 2.2*). These methods are very labor-intensive and time-consuming but reveal important information about temporal and geographic changes in communities associated with different animals.

Various alternative indicator species have been investigated

The limitations of the usual FIOs—coliforms, *E. coli*, and (to a lesser degree) enterococci—as reliable predictors of health risks in marine waters has led to investigations of alternative indicators. *Clostridium perfringens* has been used because it forms resistant endospores that survive for long periods in the environment. This is particularly valuable in monitoring long-term effects of sewage disposal, for example, in sediments and offshore sewage sludge dumps. The order Bacteroidales is a group of Gram-negative anaerobic bacteria including *Bacteroides*, *Prevotella*, and *Porphyromonas*. *Bacteroides fragilis* is one of the most abundant organisms in human feces and early studies showed that it dies rapidly on transfer to seawater, making it a good indicator of recent human pollution. Despite some promising research results, the requirement for anaerobic cultivation makes these methods unsuitable for many laboratories undertaking routine monitoring. However, detection and quantification of *Bacteroides* and *Prevotella* by qPCR has proved very useful in tracking sources of pollution. A variety of primer sets have been developed that distinguish between strains from humans and those of other animals such as cattle, horses, pigs, and dogs.

Methanogenic archaea have also been investigated as potential indicator species. Methanogens are widely distributed in marine sediments and activated sludge in sewage plants, but most members of the genus *Methanobrevibacter* seem to be restricted to the mouth and intestines of animals and the species *M. smithii* appears to be a human-specific species, present in human feces at densities between 10^7 and 10^{10} per gram. PCR assays targeting sequences of the *nifH* gene specific to *M. smithii* have been developed.

Phages are especially useful as indicator organisms, using the rationale that these viruses could be expected to have similar survival characteristics to viruses pathogenic for humans. Direct testing for enteric viral pathogens in seawater is technically very difficult and unreliable whereas phages are easily measured using plaque assays (plating concentrated water samples on lawns of susceptible host bacteria; see *Figure 7.2*) or with group-specific oligonucleotide probes. The most promising phages are the F+ ("male-specific") RNA coliphages that infect *E. coli* expressing sex pili on their surface. They are morphologically similar to several groups of enteric viruses and are present in very high numbers in sewage. Some success has been achieved in distinguishing animal and human sources and in establishing that they may have survival times similar to the human viruses when exposed to UV light under various conditions. Epidemiological studies have shown correlation between numbers of coliphages and gastrointestinal illness. Metagenomic analysis of viral sequences from the

SPICING UP POLLUTION CHECKS

Metagenomic analyses of viral sequences in the human gut has revealed a surprising result that may lead to new pollution indicators. Rosario et al. (2009) found that the *Pepper mild mottle virus* (PMMoV, genus *Tobamovirus*) is abundant in human feces, in treated wastewater and in coastal waters near discharge sites. Most animal feces do not contain PMMoV. Numerous studies have confirmed its strong potential as an indicator of human-specific pollution in environmental waters (Symonds et al., 2018). Why such high numbers of this plant virus are found in the human gut is unknown, although PMMoV RNA sequences are present in many spicy food products containing peppers. Colson et al. (2010) provided preliminary evidence of association with immune responses, abdominal pain and fever. They suggested that this might be the first example of a plant virus infecting humans. However, no causal relationship has been demonstrated (Balique et al., 2015) and epidemiological studies would be needed to rule out the effects of reactions to the spicy foods rather than viral infection.

human gut has revealed a previously unidentified phage (crAssphage), which probably infects *Bacteroides*. It is highly abundant in human fecal waste and qPCR assays to detect its concentration in water are being developed for use in pollution monitoring.

Countries have different quality standards for bathing waters

As discussed in *Box 13.1*, there have been many studies attempting to prove a link between the levels of FIOs and adverse health risks from recreational use of marine waters. The methodological or statistical basis of such studies has been the subject of much controversy. The difficulties of devising regulatory systems and quality measures that reliably reflect risk have led to variations in the microbiological monitoring programs and standards adopted by environmental agencies in different countries. In 2003, the World Health Organization (WHO) published guideline values for the microbiological quality of recreational waters. These used a statistical approach in which 95% of the counts must lie below a threshold value. Using this system, the estimated additional risk of bathing (compared with a control group) in waters with less than 40 colony-forming units (CFU) of enterococci per 100 mL is less than 1% (grade A water). Between 41 and 200 CFU per 100 mL (grade B water), this rises to a 1–5%, and between 201 and 500 CFU per 100 mL (grade C water), the increased risk is 5–10%. In grade D water, containing more than 500 enterococci per 100 ml, there is a >10% increased risk of contracting gastrointestinal illness. The risk of acquiring acute febrile respiratory infections is also significant in grade C and D waters. It should be noted that these risk factors apply to healthy adults, as children are usually excluded from this type of epidemiological study for ethical reasons. The risk of infection is probably much higher for lower age groups. It should also be noted that the data on which the standards are based are mainly obtained from Northern Europe and North America, so may not be applicable in all regions.

Despite these WHO guidelines, the nature of the indicators, the frequency of sampling, the methods of quantification, and the threshold values for compliance with standards all vary between countries. Because there can be considerable day-to-day variation in the numbers of indicator organisms detected, many systems are based on percentage compliance levels, such as the requirement in the European Union that 90 or 95% of measurements taken during the sampling period meet the prescribed standard. Other authorities such as the US Environment Protection Agency (EPA) use geometric mean values of data to even out unusually high or low results. The geometric mean is statistically more stable—although it can overlook the occasional high values at the top of the statistical distribution—whereas the 90/95% compliance system has the advantage of reflecting the occasional high levels of concern and is marginally easier for the public to interpret.

In Europe, the first definition of microbiological standards was published in 1976 as the European Community Bathing Water Directive. This defined testing protocols based on total and fecal coliform counts. The European legislation was used as the basis for improvements—at great expense—to coastal sewage disposal systems which were designed to comply with the standards for coliform and *E. coli* levels. Interpretation of epidemiological studies led to the realization that, although the motive of these schemes was laudable, the scientific basis for evaluating their effectiveness was questionable, since an increased risk of illness seemed to be associated with water-quality indicators well within the European limits. Increasing recognition that enterococci are a more reliable indicator than fecal coliforms led to the inclusion of a guide level for enterococci in amendments to the European legislation in 1998. After a long period of discussion and argument, revised microbiological standards were adopted in 2006, which shifted the emphasis from treatment and disposal of sewage to expanding the "Blue Flag" classification system for beaches and improvements in public information. On the basis of monitoring designated bathing beaches about 20 times from May to September, water quality is assessed based on samples from the previous four years using threshold values per 100 mL for *E. coli* (EC) and intestinal enterococci (IE). The categories are: Excellent (<100 IE, <250 EC); Good (<200 IE, <500 EC); Sufficient (185 IE, 500 EC); or Poor (>185 IE, >500 IE). The Excellent and Good standards are based on results from the 95th percentile, whereas the Sufficient standard—with an apparently lower limit than Good—is based on the 90th percentile. This has caused confusion and controversy. Health authorities will have to make decisions

BOX 13.1   RESEARCH FOCUS

Is it safe to swim? The health effects of recreational exposure to seawater

Correlation of FIO levels with bathers' perceptions of illness.
The first attempts to investigate this question were carried out in the 1950s in freshwater bathing sites. These showed that there was a significant increase in gastrointestinal, ear, nose, and throat symptoms in swimmers exposed to waters with high coliform counts. This was used by the US authorities to set the first standards for recreational waters. The epidemiological basis for defining these standards was questioned by many authors, most notably Cabelli et al. (1982), who undertook more detailed studies at US beaches. Rigorous attempts to ensure standardization of microbiological methods, reporting of symptoms by participants, and the use of suitable control groups were included. Cabelli's group found that levels of fecal coliforms were not a reliable indicator of health risks, but that there was a statistically significant correlation between the levels of enterococci and an increased incidence of gastrointestinal symptoms associated with swimming. As a result, the US Environment Protection Agency (USEPA) dropped the use of fecal coliforms as indicators in marine waters and set a new standard based on a geometric mean of 35 enterococci per 100 ml in five samples over a 30-day period. Many similar surveys have been undertaken, and meta-analyses of the data from numerous epidemiological studies have shown a consistent correlation between increased risk of gastrointestinal (GI) symptoms—especially in children —and the counts of FIOs, especially enterococci (Pruss, 1998; Wade et al., 2003; Arnold et al., 2016). Some studies have reported increased rates of skin symptoms among swimmers, but controlled studies have produced limited evidence of links with fecal contamination.

Controlled surveys. One of the major limitations of many studies is that they rely on *perception* of illness by those taking part, who are usually asked to respond to a questionnaire distributed among beachgoers. It is very difficult to distinguish the true effect of swimming from other effects; for example, illness might result from food consumed during a trip to the beach. In the 1990s, growing public concern and pressure from groups such as Surfers against Sewage and the Marine Conservation Society led the UK Government to sponsor new epidemiological research designed to overcome some of the criticisms of earlier studies. Carefully controlled beach surveys were conducted, in which volunteers were given a medical examination before randomized groups undertook supervised swims with water testing at frequent intervals. One week later, participants received a second medical examination plus testing of throat and ear swabs and fecal specimens. Randomization and detailed analysis of questionnaires allowed other risk factors such as food intake to be considered. The UK study (Fleisher et al., 1998; Kay et al., 2001) confirmed that there was close correlation between levels of enterococci in bathing water and a significant increase in risk of GI disease when the number exceeded 30–40 per 100 ml. A convincing dose-response curve was observed. In a new systematic meta-analysis of 40 studies, Leonard et al. (2018) employed rigorous selection criteria to determine the increased risk of becoming ill after seawater bathing and whether the level of increased risk depends on the nature of exposure. The results—based on over 120,000 people in the eligible studies—showed that bathing increased the odds of earache by 77% and the odds of GI symptoms by 29%, compared with not bathing. All the data in this analysis were obtained from high-income countries (US, UK, Australia, New Zealand, Denmark, and Norway) with a high level of pollution control, but the results show that further improvements would be beneficial.

Does surfing carry an increased risk? Participants in water sports such as body boarding, surfing, and scuba diving may have prolonged exposure and increased risks of ingestion of seawater or inhalation of aerosols. Despite many anecdotal reports of illnesses linked to water sports, there have been very few detailed studies of health risks from microbial infections. In a survey of surfers who spent a mean of 77 days a year surfing in Oregon, USA, Stone et al. (2008) concluded that surfers had a mean exposure of 11–86 CFU enterococci per day, with an epidemiological model predicting that they have a 23% probability of exceeding the exposure equivalent to the USEPA maximum acceptable gastrointestinal illness rates. Water sports activities may also take place outside of the official "bathing season," meaning that the authorities do not undertake checks of water quality. For example, in southwest England the best surfing conditions often occur in the winter, when potential pathogens may survive for longer or be resuspended from sediments by wave action following storms (Bradley and Hancock, 2003). In San Diego, California, a prospective cohort study of 654 adult surfers over two winters showed that surfing following stormy weather carried a three-fold increased incidence of earache/infections and a five-fold increase in infected open wounds (Arnold et al., 2017).

★ ★ ★ Excellent
★ ★ Good
★ Sufficient
— Poor

Figure 13.8 Example of beach signage showing EU Bathing Water Directive information. Upper sign indicates that bathing is not advised (or prohibited); lower sign indicates status of the classification of water quality at the beach according to the monitoring survey.

Despite the high incidence of head immersion and ingestion of water in surfers, their overall GI rate was similar to that previously reported for swimmers. Arnold and colleagues found that levels of FIOs were only a reliable indicator of risk within three days of heavy rain, which they attributed to urban runoff from storm drains. Other studies have also noted this lack of correlation between illness and FIOs when there is no well-defined point source of human fecal contamination (Fewtrell and Kay, 2015).

Exposure to antibiotic-resistant bacteria. Resistance to antibiotics has emerged as major threat to global health, and threatens to return us to the pre-antibiotic era, when many common infections will become untreatable and some surgical interventions and other medical procedures will become too dangerous to undertake (O'Neill, 2016). Antibiotic-resistant bacteria (ARB) are now widespread as a result of widespread use in medicine, agriculture, and aquaculture, which has encouraged the rapid evolution and spread of genes conferring antibiotic resistance. Resistance to β-lactam antibiotics (the widely used penicillin and cephalosporin families) is particularly concerning. Extended-spectrum β-lactamase (ESBL) enzymes encoded on mobile genetic elements are now prevalent in bacteria of the family Enterobacteriaceae and other pathogens associated with community and hospital-acquired infections, which are consequently very difficult to treat. There is increasing recognition of the transmission of ARB to humans from natural environments. Wastewater effluent has been shown to contribute to the dissemination of multidrug-resistant ESBL bacteria carrying the $bla_{CTX-M-15}$ gene (Amos et al., 2014). To determine whether there was a link between recreational use of coastal waters and acquisition of ARB, Leonard et al. (2018) devised an innovative epidemiological survey. A group of 150 frequent surfers was recruited, together with 150 matched controls with very little exposure to seawater. Dubbed the "Beach Bum Survey"—for obvious

reasons—the UK-based survey asked participants to collect their own rectal swabs, which were returned to the laboratory for isolation of E. coli resistant to cefotaxime (a third-generation cephalosporin, 3GC) and meropenem (a carbapenem). Resistant isolates were then screened for the mobile resistance genes using Q-PCR. Leonard and colleagues found 3GC-resistant E. coli in 15% of coastal sites. They isolated nearly 115000 E. coli, of which 140 (0.12%) were 3GC-resistant and 83 (0.7%) possessed $bla_{CTX-M-15}$. Based on the distribution of the resistant E. coli and using data on the number of water sports events, Leonard et al. (2018) concluded that surfers are at a particularly high risk of being exposed. Further analysis showed that—despite the low overall prevalence of colonization—surfers are four times as likely to be colonized by $bla_{CTX-M-15}$-containing E. coli as non-surfers. Several of the surfers in the survey were shown to be carrying 3GC-resistant pathogenic strains of E. coli; although they were asymptomatic, it is possible that future health impairment could lead to proliferation of the bacteria and disease. Despite the relatively small frequency of these ARB and the need for additional objective epidemiological information about the effects of harboring them, the study nevertheless provides worrying evidence of the spread of ESBL genes in the environment.

Concluding remarks. It is important that the general health benefits—physical outdoor activity and sense of well-being—of swimming, surfing, and other coastal pursuits are balanced against the mostly small additional risks of acute illnesses, which are usually mild and self-limiting (although risks for children are higher). However, continuing improvements to reduce pollution are essential and better provision of information is needed to enable the public to fully understand the meaning of water-quality standards so that they can evaluate risks and make informed choices according to local conditions.

on what constitutes an "acceptable risk," but public information on the interpretation of the water-quality results posted at many seaside beaches (*Figure 13.8*) is woefully inadequate to enable people to make an informed personal choice about the risks of bathing or undertaking water sports. In view of the importance of international tourism, public confidence in monitoring procedures is essential; and the current diversity of approaches is confusing. Surprisingly, the current European level of 100 enterococci per 100 mL (indicating "Excellent" water quality) is considerably greater than the level predicted by epidemiological studies as carrying a significant increased probability of illness and is much higher than that in many other developed countries. For example, the USA and Canada use a threshold value of 35 enterococci per 100 mL (based on a geometric mean of five samples over a 30-day period). In Australia and New Zealand, the standards for primary contact activities such as swimming and surfing are 150 fecal coliforms and 35 enterococci per 100 mL (with maximum permissible levels of 600 and 100 respectively); the authorities also define less stringent standards for secondary contact activities such as boating and fishing. It is surprising that the rank of "Excellent" water under European standards is only the equivalent of grade B in the WHO guidelines.

Shellfish from sewage-polluted waters can cause human infection

Molluscan bivalve shellfish such as oysters, mussels, clams, and cockles concentrate pathogens from their environment by filter feeding. For example, a single oyster filters about 5 liters of water per day and each square meter of a dense bed of mussels filters more than

10^5 liters per day. Thus, even if there are low concentrations of pathogenic microbes in the water, there is the potential for very large numbers to accumulate in the bodies of the animals. The rate of accumulation of microbes and their subsequent survival within the tissue depends on the species involved and numerous environmental factors, especially temperature.

Shellfish are often subjected to a process of disinfection known as depuration before they are sent to market for human consumption. Depuration involves transferring the animals for 24–48 h to free-flowing or recirculating clean seawater, which is disinfected by UV light, chlorination, or ozone treatment. The natural filtering activity of the bivalve mollusks results in elimination of their intestinal contents. The processes of depuration were developed in the early twentieth century to eliminate the risk of typhoid fever caused by the bacterium *Salmonella typhi*; at the time, this caused serious epidemics associated with eating shellfish grown in polluted water. Depuration is effective in removing *S. typhi* and other fecal bacteria including FIOs from shellfish. However, it is much less effective in the removal of sewage-associated viruses, naturally occurring bacterial pathogens such as *V. parahaemolyticus* and *V. vulnificus* (see p.338), dinoflagellate, and diatom toxins (see p.343), or chemical pollutants. Hepatitis A acquired from shellfish is of particular concern; although rarely fatal, it can cause a long, severe debilitating illness and some large outbreaks involving hundreds of cases—one outbreak in China affected 291,000 people—have occurred, for example, following the serving of contaminated shellfish at banquets or receptions. Norovirus is the most common cause of seafood-associated gastroenteritis but usually only comes to the attention of health authorities when outbreaks involving large groups of people occur. It is likely that thousands of undocumented cases of seafood-associated gastroenteritis occur each year, since most victims do not visit their doctor and, in any case, accurate diagnosis is difficult. As occurs with illnesses caused by the autochthonous marine bacterial pathogens *V. parahaemolyticus* and *V. vulnificus*, the problem is exacerbated by the fact that shellfish are often eaten raw or only partially cooked. Many of the enteric viruses are resistant to moderate heating, low pH, freezing, and drying, so normal food-processing methods have little effect.

Microbiological standards are used for classification of shellfish production areas

Shellfish growing areas are classified to assess their suitability for production for human consumption. The standards used in Europe (shown in *Table 13.1*) are based on end-product analysis of replicate samples of shellfish flesh by a most probable number

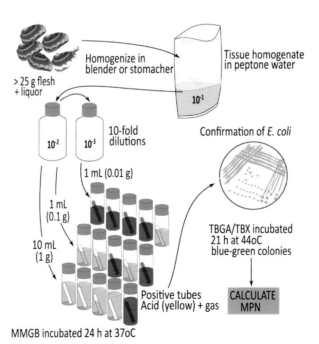

Figure 13.9 Outline of the most probable number (MPN) method for determining *E. coli* in shellfish. Media and conditions can vary; the version here is based on the EU Reference Laboratory protocol. Abbreviations: MMGB minerals modified glutamate broth; TBGA/TBX tryptone bile glucuronide agar (contains 5-bromo-4-chloro-3-indolyl-β-D glucuronide to detect β-glucuronidase activity). Calculation of MPN is based on the statistical probability that some tubes will be inoculated with a single organism from the dilutions used.

Table 13.1 *Classification of shellfish (bivalve molluscs) microbiological quality in the European Union according to Shellfish Hygiene Directive 2013*

Category	E. coli (MPN) per 100 g		Acceptability for human consumption
	Maximum level in 90% of samples	Overall maximum level	
A	230	700	May be harvested directly
B	4600	46000	Must be depurated in approved unit, relayed in an approved category A area, or heat treated
C	46000	–	Must be relayed for at least two months to meet category A or B requirements, or be heat treated

(MPN) method to quantify *E. coli* as an indicator of fecal pollution (*Figure 13.9*). In other countries, monitoring of FIOs in surface waters is used to classify shellfish production areas into one of five classes based on levels of total coliforms or fecal coliforms in surface waters. In the USA, approval for marketing of shellfish without prior treatment requires geometric mean levels in the water of *E. coli* $\leq$14 MPN per 100 mL (90th percentile $\leq$43). Shellfish from restricted areas with *E. coli* levels of $\leq$88 MPN per 100 mL (90th percentile $\leq$260) require depuration or relaying. Other areas may have periodic restrictions or be prohibited. Again, the use of *E. coli* as an indicator is a very unreliable predictor of health risk—no classification methods employ enterococci—and the standards have no epidemiological basis. Many outbreaks of viral illness—sometimes large and severe—have been associated with shellfish that appear to be of high standard as judged by the levels of bacterial FIOs. Shellfish for human consumption must also be tested for the presence of microbial toxins, for which there are separate standards (see p.344).

Direct testing for pathogens in shellfish is possible with molecular methods

The above discussion suggests that the indicator concept has major shortcomings when applied to marine waters and shellfish. Would it not be better to undertake direct testing for the pathogens themselves? The problem here is that most viruses, the pathogens that cause most concern, are difficult or impossible to culture. They are often also present at extremely low levels, yet nevertheless present a health hazard due to their low infectious dose. The usual method is to inoculate cell cultures of suitable human cells (such as enterocytes) with water samples (after concentration) and monitor for cell lysis or other cytopathic effects. Special safety precautions must be taken in laboratories using human cell culture, and the effects, if any, may take several weeks to appear. Attention has therefore turned to molecular biological techniques, in which the most promising approach has been detection of viral RNA sequences by the qRT-PCR method (p.47). Difficulties with the technique include selection of suitable primers that are representative of virus strains circulating in a particular region, and interference with the PCR amplification by inhibitors, which can be a particular problem in shellfish tissue. Careful optimization of techniques is required for extraction from the digestive tissue. Suitable primers that target the RNA virus group under study are added to samples of water or shellfish tissue extracts in the presence of the enzyme reverse transcriptase to make a cDNA copy, which is then amplified by *Taq* polymerase, usually using a nested PCR (p.47). The methods are still too time-consuming and expensive to be used in routine monitoring, but they are invaluable in confirming the sources of disease outbreaks and special studies such as environmental impact assessments of new sewage disposal schemes and classification of shellfish harvesting areas. Future advances will depend on linking infectious virus levels with epidemiological data on the incidence of disease.

OIL AND OTHER CHEMICAL POLLUTION

Oil pollution of the marine environment is a major problem

The world economy is totally dependent on petroleum products as a source of energy and raw materials for a vast range of products. Despite concerns about the impact of fossil fuels on climate change and the move towards sustainable sources, about 4 billion metric tons of crude oil are extracted annually and global demand is rising by about 1% every year. About 1.2 million metric tons of oil enters the oceans every year, of which nearly half comes from natural seepage from oil deposits under the seabed. As shown in *Figure 13.10*, pollution by oil tankers and spills from offshore oil rigs accounts for only a small proportion of oil entering the ocean, although it attracts the greatest public concern and has major impacts because of the sudden introduction of large amounts into the environment. Tankers take on water for ballast and, when this is discharged, considerable local contamination of the sea can occur if their oil separators are not functioning correctly. Occasionally, tankers collide or run aground, releasing large quantities of oil into the sea. Fortunately, in recent years, tighter regulations on ballast and greater use of double-hulled tankers has reduced the amount of oil spilt in this way, but major incidents still occur. The immediate damaging effect of this on marine wildlife and the economic impact on fisheries and tourism means that environmental agencies are under intense pressure to alleviate the problem quickly and effectively. Apart from immediate effects, there are concerns about toxic residues and long-term disruption of ecosystems. There are various strategies for dealing with spilled oil, including trapping with booms and skimming the surface to remove the oil, using absorbent materials to soak up oil, applying dispersant chemicals, or setting fire to oil slicks. The efficacy of these processes is very dependent on the location of the spill and weather conditions, and the use of dispersants or burning can have highly damaging effects on marine ecology.

Microbes naturally degrade oil in the sea

Fortunately, naturally occurring microbes break down most of the components of petroleum, which is a complex mixture of thousands of compounds, predominantly hydrocarbons. Oil is a natural product formed from algae buried in the sediments of the sea and lakes under high pressure and temperature over millions of years. Large quantities of crude oil have therefore been seeping naturally during this time, and many different microbes in more than 175 bacterial and archaeal genera have evolved efficient mechanisms to degrade the different hydrocarbon components—these have been isolated from seawater, shores, and sediments all over the world. A similar number of yeasts and filamentous fungi have also been described. Most hydrocarbon-degrading isolates are heterotrophs belonging to the Gammaproteobacteria, of which the most important are in the genera *Alcanivorax*, *Cycloclasticus*, *Fundibacter*, and *Oleispira*. Cyanobacteria (e.g. *Merismopedia*, *Microcoleus*, and *Phormidium*) and the alga *Ochromonas* have also been linked with hydrocarbon degradation. Some photoautotrophic species appear to accumulate hydrocarbons within vesicles, but not to degrade them and it is possible that they form syntrophic consortia with heterotrophic bacteria, leading to breakdown. Hydrocarbons are degraded by aerobic microbes in all parts of the water column and the upper surface of sediments, while anaerobes are active in deeper sediments and hydrocarbon seeps. In

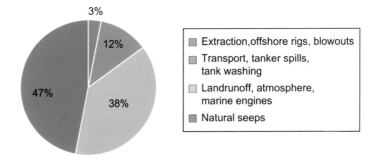

Figure 13.10 Sources of oil inputs to the sea (based on data from Transportation Research Board, National Research Council, Washington D.C., 2003.

anoxic environments, SRB and archaeal methanogens can degrade hydrocarbons as a sole source of carbon and energy. The breakdown of oil depends on the activities of a consortium of microorganisms, each responsible for transformation of a specific fraction. These degraders are always present, but normally constitute only a small proportion (<1%) of the microbiota of seawater, except in the location of natural seepage. However, following a pollution incident they multiply rapidly to 10% or more of the population. Immobilization of mixed microbial communities on biofilms may be particularly important in efficient biodegradation. *Box 13.2* describes how the use of dispersants at the wellhead enabled rapid degradation of oil in the *Deepwater Horizon* oil well blowout in 2010, one of the worst ever pollution incidents.

Physical and biological processes affect the fate of oil spills

When oil is released into the seas as a result of a spillage, it floats on the surface and the low-molecular-weight fractions evaporate quickly. Some components are water-soluble and photochemical oxidation occurs through the action of sunlight. The fate of an oil slick is very dependent on wave action and weather conditions. Sometimes, droplets of oil will become emulsified in the water and disperse quickly; in other cases, a water-in-oil emulsion will form to produce a thick viscous "chocolate mousse," which takes a very long time to disperse and gives rise to the familiar lumps of beach tar. Different crude oils vary, but usually over 70% of the hydrocarbons are biodegradable. Only the asphaltenes and resin components are recalcitrant to breakdown.

The biochemical processes involved in the biodegradation of oil have been studied extensively. Aerobic processes are responsible for the efficient biodegradation of oil. The initial step involves incorporation of molecular oxygen into *n*-alkanes by oxygenases, resulting in primary alcohols, which are then further oxidized to aldehydes and fatty acids. The fatty acids are then metabolized to acetyl-coenzyme A, which enters primary metabolism via the tricarboxylic acid (TCA) cycle. In this way, the *n*-alkanes are completely converted to carbon dioxide and water. There are many different routes for the degradation of the aromatic hydrocarbons, the key step being the cleavage of the aromatic ring; oxygen is again essential for this process. Polyaromatic hydrocarbons (PAHs) such as naphthalene, phenanthrene, and pyrene contain four or more aromatic rings and are only degraded slowly by a few types of bacteria. Consequently, PAHs that are not degraded in the water column accumulate in sediments and can persist for very long periods. Many harbors and estuaries that are used extensively by shipping have sediments that are chronically polluted with PAHs. In such anoxic sediments, anaerobic degradation occurs with the oxidative processes linked to the reduction of nitrate or sulfate, or the production of methane. Degradation probably begins with carboxylation reactions, followed by ring reduction and cleavage. The activity of burrowing benthic animals can enhance aerobic PAH degradation by microbes, owing to the mixing of sediments and introduction of oxygen through the elaborate ventilation systems that they construct; isolation of bacteria from such burrows may prove a rich source of new species.

Bioremediation of oil spills may be enhanced by emulsifiers and nutrients

Biodegradation proceeds most quickly when the oil is emulsified into small droplets and hydrocarbon uptake by microorganisms is stimulated by the production of biosurfactants (surface-active agents containing both hydrophilic and hydrophobic regions which reduce surface tension). Many microbes have hydrophobic surfaces and adhere to small droplets of oil, and many also produce extracellular compounds that disperse the oil. There has been great interest in developing natural biosurfactants as an alternative to chemical dispersants. For example, the bacterium *Acinetobacter calcoaceticus* is particularly effective in biodegradation of oil because it both adheres to hydrocarbons and produces an extracellular glycolipid biosurfactant called emulsan. Emulsan-deficient mutants grow very poorly on hydrocarbons. It may be possible to select naturally occurring strains or use genetic engineering to produce large amounts of biosurfactants, although optimizing the industrial-scale production of these compounds may be difficult. Emulsan is also used to reduce viscosity to aid in the extraction of crude oil.

BOX 13.2   RESEARCH FOCUS

How microbes cleaned up deep-sea oil after the *Deepwater Horizon* (DWH) disaster

Response to the DWH blowout. In April 2010, the DWH platform was drilling an exploratory well for oil at a depth of 1500 m, 322 miles off the coast in the Gulf of Mexico. A blowout led to a methane explosion and fire that killed 11 workers. The fire raged for 36 h, before the platform sank, leaving the wellhead seriously damaged and gushing huge quantities of crude oil into the deep sea. A massive operation involving hundreds of ships, aircraft, and shore-based operations was put in place to remove surface oil or trap it in containment booms in an attempt to protect the thousands of miles of beaches, estuaries, and wetlands on the Gulf Coast. Throughout the summer of 2010, nearly five million barrels of oil (7.8×10^5 m^3) of crude oil flowed unchecked into the sea, until a major engineering operation sealed the well with a concrete cap in mid-July. By the first week of August, oil could no longer be detected in the deep-water column or on the ocean surface, and in September, the US Coast Guard declared the well dead.

Use of chemical dispersants. Because of the scale of the release, more than 4×10^6 L of dispersant (Corexit 9500) were used to treat the oil as it escaped from the wellhead. Although it was reasoned that dispersing the oil might facilitate its breakdown by microbes, the main reason for using it at the well head was for safety reasons as the mixture of light crude oil and natural gas was very volatile

and would be highly flammable if it came to the surface. The injection of dispersant successfully forced some of the oil to rise to the surface at a distance from the wellhead as a result of creating small droplets that were less buoyant and moved more easily by ambient currents. Over 40% of the oil formed a plume—more than 200 m high and 2 km wide—of highly dispersed microdroplets of hydrocarbons and succinate compounds derived from the dispersants. The plume extended in the deep water over 35 km from the source (Atlas and Hazen, 2011) (*Figure 13.11*)

Effect on deep-sea microbial communities. From a microbiological perspective, one of the most interesting consequences of the spill was the fate of the massive plumes of oil droplets released in the deep water. This is a cold (2–5°C), high-pressure environment with very low levels of carbon, so it was not known how the microbial community would respond to such a sudden influx of hydrocarbons. An urgently coordinated study of the plume by a team from the Berkeley National Laboratory found evidence of rapid degradation of the oil (Hazen et al., 2010). The analysis was initially performed using the PhyloChip 16SrRNA microarray that was developed as a rapid method for the profiling of microbial populations in environmental samples. They found a significant stimulation of psychrophilic bacteria related to three families in

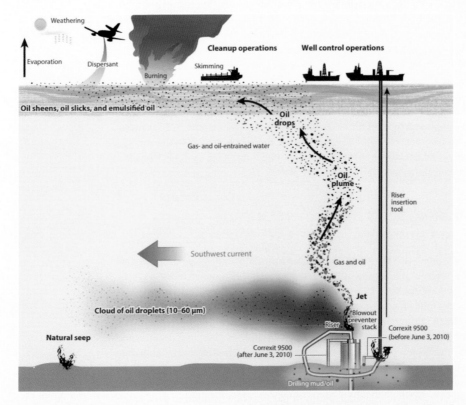

Figure 13.11 Schematic representation of the *Deepwater Horizon* oil spill. Not to scale (the well head well is 1500 m below the surface). Reprinted with permission from Atlas and Hazen (2011). Copyright 2011 American Chemical Society.

BOX 13.2   RESEARCH FOCUS

the order Oceanospirillales, dominated by a novel uncultured species closely related to *Oleispira antarctica* and *Oceaniserpentilla haliotis*. In later samples, the community composition shifted and was dominated by species of *Colwellia* and *Cycloclasticus* (Redmond and Valentine, 2012). Analysis of functional microbial genes in the cloud of dispersed oil correlated with changes in the concentration of various components of the oil and rapid biodegradation rates were demonstrated in laboratory studies, with the half-life of alkanes ranging from 1.2 to 6.1 days (Hazen et al., 2010). The authors concluded that the degradation proceeded so rapidly because the oil contained a high component of readily degradable volatile components, and because the oil was dispersed into small droplets, providing a large surface area amenable to attack by the bacteria. However, there did seem to be a constraint on the rate of degradation, possibly because of low levels of iron needed as a co-factor for degradative enzymes. This was reassuring in view of the fears that extensive oxygen depletion due to heterotrophic activity would create a massive anoxic "dead zone." In fact, the oxygen saturation inside the plume was only a few percent lower than outside the plume. Nowhere in the world has a higher concentration of natural seeps and offshore drilling than the Gulf of Mexico, so it is perhaps not surprising that the resident microbes were able to respond so rapidly to the influx of oil—resulting in a very different outcome to other catastrophic events such as the *Exxon Valdez* oil spill from a tanker on the Alaska shoreline (Atlas and Hazen, 2011). To obtain more information about the long-term fate of the oil, Hu et al. (2017) collected deep-sea water samples from the area in 2014 and conducted a laboratory simulation of the plume and reconstructed population genomes of the successive blooms of diverse bacteria observed, obtaining nearly complete genomes of the major hydrocarbon-degrading species. Hu and colleagues concluded that the rapidly degraded oil components were consumed by bacteria with substrate specialization; these included a new species belonging to the Oceanospirillaceae, which they named "*Ca.* Bermanella macondoprimitus." However, Delmont and Eren (2017) found no evidence that the 16S rRNA gene of the newly proposed species was present in the original community genome analysis of the plume and commented: "Although the simulation offers valuable insights into microbial succession patterns following the addition of oil, our analysis revealed that the dominant hydrocarbon-degrading bacteria in the simulation was not in the environment, and the one in the environment was not in the simulation."

Effects of dispersants on degradation. The use of such large quantities of chemical dispersants was highly controversial, because of their known toxicity and the fact that they had never been used in such deep water. Laboratory simulations to study the effects of the dispersant have also generated some controversy about their role in biodegradation. Experiments conducted by Kleindienst et al. (2015) found that the dispersants caused a major shift in microbial community composition through selection of dispersant-degrading *Colwellia*, which also bloomed during the DWH event. However, when oil was added to deep water samples in the absence of dispersants, hydrocarbon-degrading *Marinobacter* was selected. Kleindienst and colleagues concluded that dispersants can suppress rates of hydrocarbon degradation. However, Prince et al. (2016) argued that the laboratory methods used did not truly simulate the action of dispersants in the sea, where turbulence generates small droplets that diffuse apart and remain entrained in the water column. Kleindienst et al. (2016) rebutted these criticisms and maintained their argument that the efficacy of dispersants to reduce oil spill impacts is unclear, as undegraded chemically dispersed oil was shown to be present on Gulf coastlines several years after the spill.

Long-term impacts. Although almost all the suspended oil disappeared within a few months, deposits of undispersed oil have been transferred to the seabed and coastal sediments, while up to 30% of oil remains unaccounted for. Also, the dispersants used are known to be toxic to marine life and, as they had never been used in the deep sea previously, their long-term effects are not known. Whatever happens 1500 m down, oil that has contaminated the surface water and the shores, creeks, mangroves, and salt marshes of the Gulf Coast will behave very differently (reviewed by Kimes et al., 2014; King et al., 2015) . The legacy of the DWH disaster will undoubtedly affect the ecology and economy of the Gulf region for many decades to come and some sensitive areas may never fully recover.

Bioremediation is defined as a biological process to enhance the rate or extent of naturally occurring biodegradation of pollutants, although in a broader sense it can be used for deliberate use of any biological process that reverses environmental damage. The hydrocarbons in oil provide a carbon source for bacteria, but oil is deficient in other nutrients (especially nitrogen and phosphorus) and the supply of these is the main factor limiting the rate of degradation. Therefore, the most successful approach to bioremediation is the addition of inorganic or organic nutrients as fertilizers to accelerate natural processes. The process of seeding oil spills with exogenous microorganisms shown to have high degradative activity in the laboratory (bioaugmentation) has been less successful because they are rapidly outcompeted by the enrichment of naturally occurring microbes. Furthermore, the idea—popular in the 1980s—that genetic modification could be used to create a "superbug" suitable for use in the natural marine environment and capable of digesting all the different components of oil has proved to be misguided.

Laboratory studies provide little information about how well bioremediation treatments will work in the field. Some mesocosm and controlled release experiments in the field have been done, but these obviously must be limited in size and scope, as few authorities are willing

The oil tanker Exxon Valdez ran aground in Prince William Sound, Alaska in March 1989, releasing 11 million gallons of crude oil that contaminated 500 km of coastline in a pristine wilderness environment. Under the control of the US Environment Protection Agency, a range of approaches to bioremediation were tested, and the lessons learned have informed many other bioremediation applications. Nitrogen fertilizer was used in large amounts and its effectiveness was found to depend on the nature of the substrate to which the oil was bound. Enhancement of breakdown on pebbles, gravel, and large sand particles was much better than on fine sand particles. Bioremediation of exposed surfaces in Prince William Sound was successful—microbial activity was enhanced, and oil biodegradation was stimulated two- to five-fold. However, large amounts of oil remain in subsurface sediments trapped under boulders and pebbles, and it is unlikely that further bioremediation treatments would enhance degradation of the polyaromatic hydrocarbons, because few nutrients permeate down into the subsurface layers. Although most of the surface oil has gone, the ecological and economic legacy of the oil spill remains.

to allow the deliberate pollution of coastal waters. Therefore, most of our knowledge about the efficacy of bioremediation comes from studies of opportunity following large-scale spills from tanker accidents, in which investigators have little control over the prevailing conditions. Of such incidents, the cleanup after the *Exxon Valdez* spill in Alaska has been the most extensive study of bioremediation. Deliberate experimental contamination of shorelines in Norway and Canada had shown previously that, even under Arctic conditions, oil applied to shorelines would be degraded naturally within a few years, but that addition of agricultural fertilizers like ammonium phosphate, ammonium nitrate, or urea increased the initial rate of biodegradation up to ten times. Bioremediation has also had some success in other major oil spills, such as the *Prestige* tanker spill that affected the coast of Galicia in Spain in 2002 and in oil reserves deliberately released in Kuwait during the 1991 Gulf War. The best approach to fertilization seems to be to use oleophilic compounds, which stick to the oil and/or release nutrients slowly. One such slow-release fertilizer is a proprietary compound called Inipol™ EAP22; this is a microemulsion of urea in brine, encapsulated in an external phase of oleic acid and lauryl phosphate, co-solubilized by butoxy-ethanol. Bioremediation of sediment contamination is more problematic. In terrestrial situations, biodegradation is enhanced by tilling of contaminated soil, which introduces oxygen, but this is obviously impractical in marine situations. It may be possible to introduce oxygen to marine sediments by aeration pumps, chemical oxidants, or alternate electron acceptors, but these require careful evaluation of ecosystem effects. Bioremediation in sensitive habitats like mangroves and salt marshes is particularly difficult and the ecological impact is especially severe.

Bioremediation of petroleum products is applied extensively to clean up contaminated soil (e.g. to reclaim land polluted by spillage from oil tanks), and many commercial products developed for this purpose have been marketed for use in harbors and marinas. However, the scientific rigor with which these have been tested for marine bioremediation is questionable. A major problem in testing the effectiveness of bioremediation is monitoring the extent of degradation. Because breakdown proceeds in a progressive fashion, disappearance of compounds such as the alkanes and small aromatic compounds can be measured easily but monitoring removal of the more recalcitrant compounds is more problematic. One internal standard method is to measure the disappearance of biodegradable components in comparison with the concentration of hopanes, which are highly recalcitrant to breakdown.

In summary, bioremediation by the application of oleophilic or slow-release fertilizers is now generally accepted to have proved its worth as one component in the response to an oil spill. However, before it is used, careful attention must be paid to the nature of the substratum and the degree of penetration of oil into the sediments. Addition of exogenous organisms has not been successful, but further development of surfactant-producing strains and their formulation into products that allow them to compete with the stimulation of indigenous microorganisms may hold promise for the future. More research on changes in microbial community composition in response to introduction of oil containing different mixtures of hydrocarbons may also yield valuable information about the best ways to enhance natural processes of degradation.

Microbes can detoxify heavy metals from contaminated sediments

Pollution of the marine environment by heavy metals is a major threat to all forms of life due to their toxicity. The elements usually considered to be most important as environmental pollutants are chromium, arsenic, cadmium, tin, cobalt, nickel, antimony, lead, and mercury. These elements cause a range of biological effects, mainly through their interference with enzyme function. They can be carcinogenic, cause damage to the nervous and reproductive systems and are generally toxic at low levels to many aquatic organisms. Because they do not degrade, they accumulate and persist in the environment. The major sources in the marine environment are effluent from mining activity, discharge of industrial waste, atmospheric input from power stations, or terrestrial runoff from contaminated soils or landfill sites. Some marine bacteria and microalgae can gain resistance to high concentrations of metals in contaminated environments and are effective at immobilizing metals into a non-bioavailable form via biosorption or bioaccumulation.

Microbial systems can be used for ecotoxicological testing

The Microtox® system is an established biosensor for the rapid toxicity testing of water, sediment, and soil samples. It depends on inhibition of bioluminescence of *Aliivibrio fischeri* (p.103), which is supplied as a standardized freeze-dried culture. Light emission is measured using a photometer and is very sensitive to the presence of toxic chemicals at sublethal concentrations. Portable systems for field testing of chemical spills are now available. Bioassays based on the inhibitory effects on bioluminescence of dinoflagellates such as *Lingulodinium polyedra* or *Pyrocystis lunula* have also been developed. These methods are many times more sensitive and much easier to carry out than conventional ecotoxicological bioassays using fish or amphipods.

The study of natural marine microbial communities is an important aspect of monitoring the effects of pollution, temperature shifts, and other environmental disturbance. The DGGE technique for community analysis has been used widely in such studies, before its replacement by high-throughput sequencing (see p.49). Pollutant-degrading organisms can also be genetically modified to link the expression of degradative genes to reporter systems such as *lux* genes from *A.* fischeri or GFP genes from jellyfish (see p.35). These reporter systems have wide applications in cell biology as well as environmental investigations.

Microbial adsorption and metabolism affect accumulation of mercury

An indirect consequence of the activity of bacteria in soil and water is the transformation by microbial methylation reactions of trace elements such as mercury, arsenic, cadmium, and lead. These methylated elements are mobilized and can enter the food chain, causing deleterious health effects. The accumulation of mercury in marine fish is a particular health concern. Although it is a naturally occurring element released from volcanic eruptions and soil runoff, most mercury now enters the environment from anthropogenic sources, such as mining, coal-fired power stations, waste incinerators, and industrial processes. Mercury is also present in some pesticides and is an important component of discarded electronic equipment and batteries—erosion of coastal landfill sites due to rising sea levels and storms is a growing problem. It is estimated that up to 4.5×10^3 tonnes of mercury are deposited in the oceans each year, and levels are rising because of the increased emissions from rapidly developing industrialized countries in Asia.

Most mercury enters the sea in the form of the Hg^{2+} ion and adsorbs readily to particles in which it can be transformed by several types of anaerobic bacteria, especially sulfate- and iron-reducing bacteria and methanogens, to methylmercury (CH_3Hg^+ or MeHg) in a reaction mediated by the gene pair *hgcAB*. A second methyl group can also be attached to give dimethylmercury [$(CH_3)_2Hg$ or Me_2Hg]. Hg methylation mainly occurs in anoxic sediments and oxygen minimum zones (see p.235), but it also occurs in the mesopelagic zones of the open ocean, at great distances from industrial discharges. This explains why Pacific tuna—one of the major sources of mercury intake by humans—accumulate such high levels of MeHg even though they are living very distant from sources of mercury. It is likely that Hg from industrial sources is transported over long distances by atmospheric deposition and ocean currents and that dying diatoms and other algae scavenge Hg from the water column by adsorption to their cells as they sink into deeper waters. Anerobic bacteria containing the *hgcAB* genes convert the mercury during breakdown of organic matter within oxygen-depleted niches within marine snow aggregates. Thus, the mesopelagic habitat is a major entry point for mercury into marine food webs. The tissue levels of MeHg in different fish species depends on the depth at which they feed. In areas with high productivity and an efficient biological pump, MeHg adsorbed to diatom cells is rapidly transported to the seabed and accumulates in siliceous sediments (diatom oozes).

Microbial cycling is important in the distribution of persistent organic pollutants

Persistent organic pollutants include the organochlorine insecticides (e.g. DDT, aldrin, and chlordane), industrial chemicals (e.g. polychlorinated biphenyls, PCBs) and by-products (e.g.

THE HAZARDS OF MERCURY IN FISH

Mercury is a potent neurotoxin; it causes liver, kidney, and cardiovascular damage in adult humans and severely affects development of the fetal nervous system if consumed by pregnant women. Because MeHg is lipid soluble, it is concentrated in the food chain, especially in fish. One of the world's most serious environmental disasters occurred in Japan in the 1970s, when thousands of people who had eaten fish from the heavily polluted Minamata Bay were seriously affected by MeHg poisoning. In this case, hundreds of deaths and long-term health effects occurred. Because of its sequential concentration at each step of the food chain, high levels of MeHg occur in top predators such as tuna, shark, and marine mammals. Very high levels of mercury have been found in the tissues of Eskimo and Inuit communities who eat large amounts of fish, seal, and whale meat. The general health advice that fish is a good source of nutrition with significant health benefits has to be balanced against the possibility of consuming unsafe levels of MeHg. Health authorities in many countries advise pregnant women, nursing mothers, and young children to completely avoid fish with high levels of MeHg (e.g. swordfish, marlin, or shark) and to limit their intake of fish with moderate levels (e.g. tuna) to no more than twice a week.

dioxins and furans). They reach the sea via terrestrial runoff and atmospheric deposition. In numerous countries the manufacture and use of many of these chemicals has been prohibited, but they are highly persistent in sediments and dump sites and produced during incineration of waste. PCBs occur in many electronic products, and the disposal of unwanted equipment is a significant source of these chemicals. These chemicals are very resistant to photochemical, biological, and chemical degradation. Most persistent compounds are halogenated and highly soluble in lipids and they accumulate particularly in fatty tissues. Their semi-volatile nature allows them to vaporize or to be adsorbed onto atmospheric particles and they are therefore transported over great distances. For example, animals and humans living in remote polar regions have high levels of organic pollutants in their tissues despite these compounds not being used there in any significant amounts. Extensive evidence links these compounds to reproductive failure, impairment of the immune system, deformities, and other malfunctions in a wide range of marine life, and they are highly toxic to humans through ingestion of fish and other routes.

Microbial cycling of organic pollutants acquired by plankton in the upper parts of the ocean plays an important role in their distribution through ocean food webs. The bacterioplankton presents a large surface area for the adsorption of organic pollutants, and microbial loop processes release compounds during settlement of plankton debris and organic particles through the water column, but some particles with adsorbed pollutants will be buried in sediments. Disturbance of sediments by tides, currents, dredging, and the activity of benthic animals can release large quantities of the chemicals back into the water. PCBs are highly recalcitrant to degradation and there is an active search for microbes capable of breaking down the chlorine bonds, thus offering a potential use in bioremediation. Many aerobic bacteria can degrade the biphenyl ring in PCBs, but not the heavily chlorinated congeners. Anaerobic degradation of PCBs is known to occur, but isolation and identification of the organisms responsible has been elusive; different bacteria with distinct dehalogenases and congener specificities occur. Selective enrichment of PCB-contaminated sediments, accompanied by high throughput sequencing, can be used to identify new species of PCB degraders and the genes responsible for reductive dechlorination. Syntrophic consortia of members of the Dehalococcoidia class and sulfur-oxidizing Epsilonproteobacteria appear to be important. Bioelectrochemical systems (microbial fuel cells) are a promising technology for bioremediation of contaminated sediments—solid electrodes serve as an electron acceptor or donor to facilitate microbial oxidation or reduction.

Plastic pollution of the oceans is a major global problem

Different types of plastic have a variety of uses, replacing traditional materials and leading to many novel applications (*Table 13.2*). Every year, society produces over 350 million metric tons of plastic, and it is estimated that about 8 million MT of end-of-life plastics—equivalent to a garbage truckload every minute—finds its way into the marine environment. Because most plastics are resistant to degradation they are accumulating at an alarming rate and are currently the cause of great public concern, prompting government and international action to curb the use of single-use plastics, promote recycling and new approaches to reduce the current linear use of plastics, and move towards a more "circular economy." Some plastic polymers are now made using plant biomass (e.g. wood, starch, or cellulose), vegetable oils, or bacterial production. Manufacture of these bioplastics uses a renewable rather than fossil carbon source but changing the carbon source does not necessarily make the polymer any more degradable. A few biobased polymers, including polylactic acid, polyhydroxyalkonoates, and poly-3-hydroxy butyrate are fully biodegradable, but biologically derived PE or PA are not. Although they have some environmental benefits and some types biodegrade under controlled composting conditions, many so called biodegradable plastics (such as PE shopping bags incorporating corn starch) do not truly degrade; instead, they break up slowly and release many small particles of plastic. Beaches and coastlines throughout the world are littered with plastic debris and it is estimated that more than 10^5 tonnes of plastic waste circulates in the North Pacific and Atlantic gyres. Accumulations also occur in the South Atlantic and Pacific, as well as the Indian Ocean gyres, but these have not been so well sampled. Plastics can be found in waters from all over the world—from the surface to the deepest parts of the sea floor. Plastic waste is harmful to marine mammals, birds, and fishes. Ropes,

Table 13.2 *Examples of the most commonly used plastics*

Plastic	Abbreviation	Examples of uses
Polyethylene terephthalate	PET	Polyester fibers, packaging, bottles, clothing
Low-density polyethylene	LDPE	Shopping and food bags, agricultural mulches
High-density polyethylene	HDPE	Robust packaging containers
Polyvinyl chloride	PVC	Pipes, insulation, construction, clothing
Polypropylene	PP	Packaging, manufactured goods, textiles
Polystyrene	PS	Foam packaging, laboratory equipment
Polylactic acid	PLA	Fully biodegradable containers, 3D-printed parts
Polycarbonate	PC	Glasshouses, protective shields
Acrylic	PMMA	Transparent sheeting, optical devices
Polyamide (nylon)	PA	Clothing, tires, car parts, ropes, nets
Polyurethane	PU	Foam insulation and furnishings, moldings
Acrylonitrile butadiene styrene	ABS	Machined, molded, and 3D-printed parts; toys

discarded fishing nets, and plastic bags cause entanglement and suffocation, while syringes, cigarette lighters, straws, nurdles (pellets used in plastic manufacture), and other debris is mistaken for food by seabirds, fish, and turtles. But the readily visible pieces of plastic are just part of the problem. Most of the plastic that has entered the ocean cannot be accounted for. Some sinks to the bottom and will be buried, but much of it becomes smaller and smaller in size—eventually reaching nanometer size—due to fragmentation, abrasion, and weathering. Microplastics also enter the ocean in a primary form as beads used in cosmetic products and fibers from clothing. The importance of microbial interactions with these polluting particles is discussed in *Box 13.3*.

Conclusions

This chapter has shown that microbes have some serious deleterious effects due to colonization of marine surfaces, biodeterioration and corrosion of structures and materials, and spoilage of seafood products. These activities cause significant economic losses to maritime industries. Some control methods have been discussed here, while other biotechnological solutions are considered in the next chapter. We have seen that microbes play an important role in pollution of coastal waters by sewage, both as human pathogens introduced with fecal waste and also as indicators of pollution for environmental monitoring and evaluating risks to public health. There are severe limitations with conventional techniques, but considerable improvements are now being achieved through the use of new methodologies. By contrast, microbes play a highly beneficial role in the degradation of oil, most of which finds its way into the ocean from natural seepage. Augmentation or enhancement of these natural degradation processes for the bioremediation of coastal pollution from oil spills have been used with varying degrees of success. It remains to be seen whether microbes will be effective in the removal of other industrial pollutants of the modern world, especially persistent organic pollutants and plastics. One of the most worrying developments discussed is the realization that ocean microbes can convert mercury to a toxic form as part of the normal food web processes. This threatens the safety of pelagic fish as a food source and can only be controlled by limiting anthropogenic emissions from newly industrialized countries. We also need urgent solutions to prevent end-of-life plastics entering the ocean and better understanding of the fate and effects of the massive quantities of microplastics already in our oceans. Microbiologists will have a major role in investigating and ameliorating this global pollution problem.

BOX 13.3   **RESEARCH FOCUS**

Microorganisms and microplastics

How big (or small) is the problem? The term microplastics was first used by Thompson et al. (2004) to describe microscopic fragments of plastic. Since then, many studies have adopted a working definition of microplastics as <5 μm diameter. Recently, an international group of experts in this field have discussed the need for a unified terminology and system for categorizing different materials by size, shape, composition, and other parameters (Hartmann et al., 2019), so that researchers are "speaking the same language." They propose that the term microplastics is used for items 1 to <1000 μm and nanoplastics for 1 to <1000 nm. One of the main sources of primary microplastics are minute fibers released during the washing of garments made of synthetic fabrics. De Falco et al. (2018) estimated that a typical 5 kg wash load of woven polyester garments can release up to 6×10^6 microfibers. Other sources include microbeads (Napper et al., 2015) used in cosmetic exfoliants (now banned or restricted in many countries), air-blasting media, dust from vehicle tires, and road markings. Until quite recently, the fate of most of the microplastics was unknown—"Lost at sea – where is all the plastic?" (Thompson et al., 2004). We now know that microplastics can remain suspended for varying periods (depending on their density, size, and shape); they are transported via ocean currents and are found globally in the water column at all depths and in tidal, sub-tidal, and deep-sea sediments (Woodall et al., 2014). Microplastics are an appropriate size to be ingested by plankton-feeding animals, as demonstrated in a range of species, from zooplankton (Cole et al., 2013) to fish (Lusher et al., 2013), leading to concerns that they could interfere with natural feeding and negatively impact ecosystem function. Furthermore, plastics may also transfer toxic chemicals, either those present as a result of manufacture (e.g. phthalates and bisphenol A) or by adsorption of PCBs and other hydrophobic organic contaminants from the environment (Teuten et al., 2009), although Bakir et al. (2016) provide evidence that ingestion of microplastic is unlikely to provide a major pathway for the transfer of adsorbed chemicals from seawater to animals via the gut. Microbeads and microfibers in wastewater might also become colonized by pathogenic or ARB as they pass through water treatment plants. A report by the World Economic Forum (2016) suggests that, if current trends continue, the oceans could contain 1 tonne of plastic for every 3 tonnes of fish by 2025 and that plastics will exceed the weight of fish by 2050. Understanding the implications of this for ocean ecology and the consequences for human health (Wright and Kelly, 2017; Oliveira et al., 2019) is one of the most pressing scientific issues of our time.

Colonization of plastics by microbial biofilms. A number of studies have shown that plastics become rapidly colonized by microbes after introduction to seawater. Lobelle and Cunliffe, 2011) showed that LDPE food bags suspended in seawater became colonized within a few days by a firmly-attached bacterial biofilm, which decreased the hydrophobicity and buoyancy, causing the bags to sink below the surface after three weeks. Zettler et al. (2013) examined PP and PE plastic marine debris from the North Atlantic gyre using scanning electron microscopy (SEM) and high-throughput tag sequencing (p.54). They showed the presence of mixed biofilms containing diverse bacteria (over 1000 OTUs), diatoms, picoeukaryotes, and other protists covering up to 8% of the surface of the plastic—a community they termed the "plastisphere" (*Figure 13.12*). Sequence analysis showed that the microbial community was consistently distinct from that of the surrounding seawater. As discussed earlier in this chapter, such biofilms recruit a variety of other colonizers. Reisser et al. (2014) also observed bacteria-like cells, diatoms, coccolithophores, bryozoans, barnacles, and other organisms colonizing ocean microplastics around Australia. These plastic particles might therefore serve as a floating "microbial reef" that could disperse bacterial pathogens (such as *Vibrio* spp.), harmful algae, and invasive invertebrates. Significantly, both studies also observed bacterial cells embedded in pits and grooves on the surface of the plastic, suggesting that bacterial biodegradation of the plastics might be occurring. Zettler and colleagues noted that some OTUs matched bacteria known to be capable of degrading hydrocarbons. Harrison et al. (2014) conducted a mesocosm experiment in which they showed successional changes in community composition of bacteria on LDPE microplastics, detected over a two-week period using SEM and CARD-FISH (p.38). This study suggested that specific bacteria associated with hydrocarbon degradation might be selected. Oberbeckmann et al. (2016) also used SEM and tag sequencing to study PET bottles suspended in the North Sea. Bottles developed surface communities that were

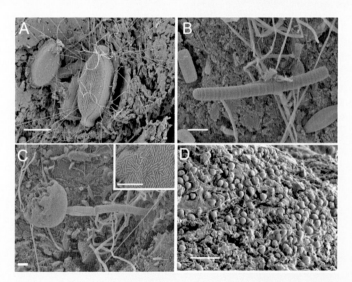

Figure 13.12 SEM images showing examples of the rich microbial community on plastic marine debris. A. Pennate diatom with possible filaments produced by *Hyphomonas*-like bacteria. B. Filamentous cyanobacteria. C. Stalked suctorian ciliate covered with ectosymbiotic bacteria (inset), along with diatoms, bacteria, and filamentous cells. D. Microbial cells pitting the surface. All scale bars are 10 μm. Reprinted with permission from Zettler et al. (2013). Copyright 2013 American Chemical Society.

BOX 13.3   **RESEARCH FOCUS**

significantly different from those in seawater, but not from glass surfaces, suggesting that the PET did not influence the nature of settling biofilm microbes. However, some OTUs were characteristic of the PET-associated biofilms, notably members of the Sphingobacteriales and Mycococcales and filamentous cyanobacteria in the genus *Phormidium*, as well as representatives of other bacterial families known to degrade hydrocarbons—as also observed by Zettler et al. (2013) in their North Atlantic gyre study.

How significant is microbial degradation of plastics? As noted in *Box 13.2*, many marine bacteria have evolved to degrade the hydrocarbons in crude oil, which has been seeping into the oceans for millions of years. However, plastics—mostly derived by polymerization and polycondensation of petroleum products—have only been accumulating in the natural environment for the past 50 or so years, and it might be expected that few organisms have evolved to degrade them efficiently. Yoshida et al. (2016) isolated a distinct consortium of PET-degrading bacteria, fungi and protists from sediments at a PET bottle recycling site. Through enrichment cultures, they isolated a novel bacterial species, given the proposed name *Ideonella sakaiensis*. Experimental treatments of PET film with this isolate resulted in complete degradation within six weeks at 30°C. The bacterium was shown to produce two key enzymes that hydrolyze PET and enable it to use it as a carbon and energy source. Through genome database searches, Yoshida postulated that the enzymes for PET metabolism may have evolved from hydrolases that target naturally occurring polymers such as cutin. However, Yang et al. (2016) commented that the results presented by Yoshida and colleagues exaggerated the rate of PET degradation because they used a low-crystallinity PET as substrate and did not demonstrate reduction in molecular weight of the polymer. Crystallinity of plastics such as PE, PP, and PET refers to the aggregation of segments of the long polymer chains in hard crystal-like domains, embedded in a soft amorphous polymer matrix. This has a great bearing on the weathering and degradation of plastics (Andrady, 2017). Yang et al. (2014, 2015) had previously demonstrated PE and PS degradation by bacteria in the guts of insect larvae. Such findings raise hopes that accelerated biodegradation of plastics might become feasible, under carefully controlled conditions, in industrial recycling plants through modification of such bacterial isolates or their enzymes, as shown for PETase by Austin et al. (2018). A better approach might be the artificial selection of natural microbial communities with improved biodegradation potential of recalcitrant polymers. Some success with this approach has been achieved by Wright et al. (2019) using chitin as a case study, although careful optimization of incubation times between generations is needed. Of course, such methods will be of no use to treat plastic waste already in the oceans, but the observations of apparent deformation of the surface of plastics in the studies of Zettler et al. (2013) and Reisser et al. (2014) hint that natural biodegradation does occur in the sea, albeit at a rate too slow to have any meaningful effect on plastic accumulation. Might it be possible that increased abundance and local concentration of microplastics in the environment could select for more efficient plastic-eating microbes? Might the usual efficient processes of adaptation and evolution through mutation and gene transfer then accelerate their spread? Perhaps, but this is likely to be a very slow process that can only be revealed by long-term mesocosm studies.

Could microplastics affect carbon cycling and food webs? There are various ways in which large numbers of microplastic particles or fibers could affect microbial processes in the food web and carbon pumps. Like all particles, they create a micro-environment that differs significantly from the surrounding water, creating "hotspots" of nutrients and microbial activity (see *Figure 1.11*). Zettler et al. (2013) suggested that the hydrophobic nature of plastic surfaces could lead to the surface concentration of micronutrients—an effect first described by Zobell (1943)—that could stimulate microbial activity in the upper layer of oligotrophic ocean gyres. In laboratory experiments designed to simulate the fate of microplastics in the surface ocean, Romera-Castillo et al. (2018) showed that suspension of PE and PP microplastics in sterile artificial seawater led to rapid leaching of dissolved organic carbon (DOC). This material probably includes truly dissolved substances such as additives used in plastic manufacture and nanoparticles released by fragmentation. After removal of the microplastics, the water was inoculated with natural bacterial community from surface sea water. Bacteria assimilated over half of the leached DOC and bacterial growth was stimulated by ~one order of magnitude. Although acknowledging the complexities of extrapolating the laboratory findings to ocean conditions, Romera-Castillo et al. (2018) show calculations that indicate the potential large impact of microplastics on carbon cycling. In another simulation, Galgani et al. (2018) showed that PS particles increased the production of CDOM (chromophoric or light-absorbing DOM, which is subject to photochemical transformations). This may increase the availability of low molecular weight compounds for microbial growth. Gewert et al. (2018) identified numerous degradation products (mainly dicarboxylic acids) from floating plastic fragments exposed to UV light. Galgani and colleagues extrapolated their findings to suggest that stimulation of CDOM release by microplastics acting as a hotspot for high microbial activity could alter carbon cycling at the sea surface. This could modify the penetration of solar radiation and change the air-sea flux of CO_2. Another way in which microplastics might affect ocean processes is via alteration of the rate of transport of carbon from the surface layers. Most plastics are buoyant, so how can the deep ocean be a major sink for microplastics? The answer seems to be that, after colonization by biofilms and association with TEPs, diatom cells, zooplankton feces, and other debris, they form an integral part of marine snow and sink through the water column. To simulate this process, Long et al. (2015) developed a flow-through roller tank to mimic the behavior of diatom aggregates as they settle through a layer of microplastic (PS) beads. Incorporation of beads into the aggregates increased the sinking rates of the beads from tens to hundreds of meters per day. However, different diatom aggregates varied in the amount of plastic incorporated and the effects this had on permeability, fragility, and sinking rates. The dense aggregates of diatom species that naturally sink rapidly showed decreased sinking rates after incorporating the low-density PS microbeads. In a similar roller bottle simulation, Porter et al. (2018) showed that laboratory generated marine snow can incorporate microplastics of different shapes, sizes and polymer types. Sinking rates of all tested microplastics (PA, PS, PE, PP fibers, and beads) increased when incorporated into marine snow. Buoyant polymers became negatively buoyant once incorporated into aggregates and sank rapidly. For example, PE microbeads floated on the surface, but sank at a relative rate of 659 m per day when incorporated into marine snow. The authors point out that these relative sinking rates may not be truly representative of real-world conditions, where a range of factors affects the rate of downward migration. Nevertheless, it seems that microplastics do have a major impact on the dynamics of marine snow transport and, hence, on microbially induced POM-DOM transition as the particles settle through the water column. It may be that changes in their surface properties and fluctuations in density might cause some particles to remain suspended. Recently, it was found that zooplankton copepods ingest PS microbeads and excrete them in their fecal pellets, which consequently have reduced density and structural integrity (Cole et al., 2016). This leads to greater fragmentation and reduction in sinking rates. Since zooplankton fecal pellets—containing large amounts of dense organic material—are responsible for the rapid transport of POM to the deep ocean (see p.230), natural processes of carbon flux could be altered considerably. Many further studies are needed to disentangle the complexities of the influence of microplastics and their interactions with microorganisms on their passage from source to sink. In particular, future studies need to recognize that the fate of plastics in the environment, and the organic pollutants that they absorb, depends on the physical and structural characteristics of the polymers, as well as their chemical composition (Andrady, 2017). Complex variables determine the effects of weathering, fragmentation, microbial colonization, and decay.

References and further reading

Biofouling, bideterioration, and corrosion of materials

Bixler, G.D. & Bhushan, B. (2012) Biofouling: Lessons from nature. *Trans. R. Soc. A* **370**: 2381–2417.

Charles, F., Sauriau, P.-G., Aubert, F., et al. (2018) Sources partitioning in the diet of the shipworm *Bankia carinata* (J.E. Gray, 1827): An experimental study based on stable isotopes. *Mar. Environ. Res.* **142**: 208–213.

de Carvalho, C.C.C.R. (2018) Marine biofilms: A successful microbial strategy with economic implications. *Front. Mar. Sci.* **5**: 126.

Enning, D. & Garrelfs, J. (2014) Corrosion of iron by sulfate-reducing bacteria: New views of an old problem. *Appl. Environ. Microbiol.* **80**: 1226–1236.

Jia, R., Unsal, T., Xu, D., et al. (2019) Microbiologically influenced corrosion and current mitigation strategies: A state of the art review. *Int. Biodeterior. Biodegrad.* **137**: 42–58.

Kip, N. & Van Veen, J.A. (2015) The dual role of microbes in corrosion. *ISME J.* **9**: 542–551.

Little, B.J., Lee, J.S., & Ray, R.I. (2008) The influence of marine biofilms on corrosion: A concise review. *Electrochim. Acta* **54**: 2–7.

Martín-Rodríguez, A.J., Babarro, J.M.F., Lahoz, F., et al. (2015) From broad-spectrum biocides to quorum sensing disruptors and mussel repellents: Antifouling profile of alkyl triphenylphosphonium salts. *PLoS One* **10**: e0123652.

O'Connor, R.M., Fung, J.M., et al. (2014) Gill bacteria enable a novel digestive strategy in a wood-feeding mollusk. *Proc. Natl. Acad. Sci. USA* **111**: E5096–104.

Pal, S., Qureshi, A., & Purohit, H.J. (2018) Intercepting signalling mechanism to control environmental biofouling. *3Biotech* **8**: 364.

Pandit, S., Sarode, S., Sargunaraj, F., & Chandrasekhar, K. (2018) Bacterial-mediated biofouling: Fundamentals and control techniques. In: *Biotechnological Applications of Quorum Sensing Inhibitors.* Springer Singapore, Singapore, pp. 263–284.

Sandström, M., Jalilehvand, F., Persson, I., et al. (2002) Deterioration of the seventeenth-century warship *Vasa* by internal formation of sulphuric acid. *Nature* **415**: 893.

Vigneron, A., Alsop, E.B., Chambers, B., et al. (2016) Complementary microorganisms in highly corrosive biofilms from an offshore oil production facility. *Appl. Environ. Microbiol.* **82**: 2545–2554.

Vigneron, A., Head, I.M., & Tsesmetzis, N. (2018) Damage to offshore production facilities by corrosive microbial biofilms. *Appl. Microbiol. Biotechnol.* **102**: 2525–2533.

Yang, J.C., Madupu, R., Durkin, A.S., et al. (2009) The complete genome of *Teredinibacter turnerae* T7901: An intracellular endosymbiont of marine wood-boring bivalves (shipworms). *PLoS One* **4**: e6085.

Sewage and wastewater pollution

Amos, G.C.A., Hawkey, P.M., Gaze, W.H., & Wellington, E.M. (2014) Waste water effluent contributes to the dissemination of CTX-M-15 in the natural environment. *J. Antimicrob. Chemother.* **69**: 1785–1791.

Arnold, B.F., Schiff, K.C., Ercumen, A., et al. (2017) Acute illness among surfers after exposure to seawater in dry- and wet-weather conditions. *Am. J. Epidemiol.* **186**: 866–875.

Arnold, B.F., Wade, T.J., Benjamin-Chung, J., et al. (2016) Acute gastroenteritis and recreational water: Highest burden among young US children. *Am. J. Public Health* **106**: 1690–1697.

Balique, F., Lecoq, H., Raoult, D., et al. (2015) Can plant viruses cross the kingdom border and be pathogenic to humans? *Viruses* **7**; 2074–2098.

Boehm, A.B., Graham, K.E., & Jennings, W.C. (2018) Can we swim yet? Systematic review, meta-analysis, and risk assessment of aging sewage in surface waters. *Environ. Sci. Technol.* **52**: 9634–9645.

Bradley, G., & Hancock, C. (2003) Increased risk of non-seasonal and body immersion recreational marine bathers contacting indicator microorganisms of sewage pollution. *Mar. Poll. Bull.* **46**: 791–794.

Cabelli, V.J., Dufour, A.P., McCabe, L.J., & Levin, M.A. (1982) Swimming-associated gastroenteritis and water quality. *Am. J. Epidemiol.* **115**: 606–616.

Colson, P., Richet, H., Desnues, C., et al. (2010) Pepper mild mottle virus, a plant virus associated with specific immune responses, fever, abdominal pains, and pruritus in humans. *PLoS One* **5**: e10041.

Dutilh, B.E., Cassman, N., McNair, K., et al. (2014) A highly abundant bacteriophage discovered in the unknown sequences of human faecal metagenomes. *Nat. Commun.* **5**: 4498.

Fewtrell, L. & Kay, D. (2015) Recreational water and infection: A review of recent findings. *Curr. Environ. Heal. Rep.* **2**: 85–94.

Fleisher, J.M., Kay, D., Wyer, M.D., & Godfree, A.F. (1998) Estimates of the severity of illnesses associated with bathing in marine recreational waters contaminated with domestic sewage. *Int. J. Epidemiol.* **27**: 722–726.

Haramoto, E., Kitajima, M., Hata, A.,et al. (2018) A review on recent progress in the detection methods and prevalence of human enteric viruses in water. *Water Res.* **135**: 168–186.

Hassard, F., Gwyther, C.L., Farkas, K., et al. (2016) Abundance and distribution of enteric bacteria and viruses in coastal and estuarine sediments—A review. *Front. Microbiol.* **7**: 1692.

Heaney, C.D., Sams, E., Wing, S., et al. (2009) Contact with beach sand among beachgoers and risk of illness. *Am. J. Epidemiol.* **170**: 164–172.

Jofre, J. & Blanch, A.R. (2010) Feasibility of methods based on nucleic acid amplification techniques to fulfil the requirements for microbiological analysis of water quality. *J. Appl. Microbiol.* **109**: 1853–1867.

Kay, D., Wyer, M., Fewtrell, L., et al. (2001) Health effects from recreational water contact.

Leonard, A.F.C., Singer, A., Ukoumunne, O.C., et al. (2018) Is it safe to go back into the water? A systematic review and meta-analysis of the risk of acquiring infections from recreational exposure to seawater. *Int. J. Epidemiol.* **47**: 572–586.

Leonard, A.F.C., Zhang, L., Balfour, A.J., et al. (2018) Exposure to and colonisation by antibiotic-resistant *E. coli* in UK coastal water users: Environmental surveillance, exposure assessment, and epidemiological study (Beach Bum Survey). *Environ. Int.* **114**: 326–333.

Mendes Silva, D. & Domingues, L. (2015) On the track for an efficient detection of *Escherichia coli* in water: A review on PCR-based methods. *Ecotoxicol. Environ. Saf.* **113**: 400–411.

Muniesa, M., Jofre, J., Garcia-Aljaro, C., & Blanch, A.R. (2006) Occurrence of *Escherichia coli* O157 : H7 and other enterohemorrhagic *Escherichia coli* in the environment. *Environ. Sci. Technol.* **40**: 7141–7149.

O'Neill, J. (2016) Tackling drug-resistant infections globally: Final report and recommendations. The review on antimicrobial resistance. Available online: https://amr-review.org/sites/default/files/160518_Final%20paper_with%20cover.pdf (accessed 2 February 2019).

Oliver, D.M., Hanley, N.D., van Niekerk, M., et al. (2016) Molecular tools for bathing water assessment in Europe: Balancing social science research with a rapidly developing environmental science evidence-base. *Ambio* **45**: 52–62.

Pruss, A. (1998) Review of epidemiological studies on health effects from exposure to recreational water. *Int. J. Epidemiol.* **27**: 1–9.

Quilliam, R.S., Taylor, J., & Oliver, D.M. (2019) The disparity between regulatory measurements of *E. coli* in public bathing waters and the public expectation of bathing water quality. *J. Environ. Manage.* **232**: 868–874.

Rosario, K., Symonds, E.M., Sinigalliano, C., et al. (2009) Pepper mild mottle virus as an indicator of fecal pollution. *Appl. Environ. Microbiol.* **75**: 7261–7267.

Stachler, E., Kelty, C., Sivaganesan, M., et al. (2017) Quantitative CrAssphage PCR assays for human fecal pollution measurement. *Environ. Sci. Technol.* **51**: 9146–9154.

Stone, D.L., Harding, A.K., Hope, B.K., & Slaughter-Mason, S. (2008) Exposure assessment and risk of gastrointestinal illness among surfers. *J. Toxicol. Environ. Heal. A-Curr. Issues* **71**: 1603–1615.

Symonds, E.M., Nguyen, K.H., Harwood, V.J., & Breitbart, M. (2018) Pepper mild mottle virus: A plant pathogen with a greater purpose in (waste)water treatment development and public health management. *Water Res.* **144**: 1–12.

USEPA Office of Science & Technology (2015) Review of coliphages as possible indicators of fecal contamination for ambient water quality. Available online: www.epa.gov/wqc/review-coliphages-possible-indicators-fecal-contamination-ambient-water-quality (accessed 2 Feb 2019).

Wade, T.J., Pai, N., Eisenberg, J.N.S., & Colford Jr, J.M. (2003) Do US Environmental Protection Agency water quality guidelines for recreational waters prevent gastrointestinal illness? A systematic review and meta-analysis. *Environ. Health Perspect.* **111**: 1102–.1109.

Wang, H., Sikora, P., Rutgersson, C., et al. (2018) Differential removal of human pathogenic viruses from sewage by conventional and ozone treatments. *Int. J. Hyg. Environ. Health* **221**: 479–488.

Williams, A.P., Avery, L.M., Killham, K., & Jones, D.L. (2007) Persistence, dissipation, and activity of *Escherichia coli* O157: H7 within sand and seawater environments. *FEMS Microbiol. Ecol.* **60**: 24–32.

Fish and shellfish spoilage and safety

Bellou, M., Kokkinos, P., & Vantarakis, A. (2013) Shellfish-borne viral outbreaks: A systematic review. *Food Environ. Virol.* **5**: 13–23.

Bintsis, T. (2018) Microbial pollution and food safety. *AIMS Microbiol.* **4**: 377–396.

Calci, K.R., Meade, G.K., Tezloff, R.C., & Kingsley, D.H. (2005) High-pressure inactivation of hepatitis A virus within oysters. *Appl. Environ. Microbiol.* **71**: 339–343.

Gram, L. & Dalgaard, P. (2002) Fish spoilage bacteria - problems and solutions. *Curr. Opin. Biotechnol.* **13**: 262–266.

Gyawali, P., Fletcher, G.C., McCoubrey, D.-J., & Hewitt, J. (2019) Norovirus in shellfish: An overview of post-harvest treatments and their challenges. *Food Control* **99**: 171–179.

Hassard, F., Sharp, J.H., Taft, H., et al. (2017) Critical review on the public health impact of norovirus contamination in shellfish and the environment: A UK perspective. *Food Environ. Virol.* **9**: 123–141.

Imamura, S., Kanezashi, H., Goshima, T., et al. (2017) Effect of high-pressure processing on human noroviruses in laboratory-contaminated oysters by bio-accumulation. *Foodborne Pathog. Dis.* **14**: 518–523.

Murchie, L.W., Cruz-Romero, M., Kerry, J.P., et al. (2005) High pressure processing of shellfish: A review of microbiological and other quality aspects. *Innov. Food Sci. Emerg. Technol.* **6**: 257–270.

Odeyemi, O.A., Burke, C.M., Bolch, C.C.J., & Stanley, R. (2018) Seafood spoilage microbiota and associated volatile organic compounds at different storage temperatures and packaging conditions. *Int. J. Food Microbiol.* **280**: 87–99.

Olatunde, O.O. & Benjakul, S. (2018) Natural preservatives for extending the shelf-life of seafood: A revisit. *Compr. Rev. Food Sci. Food Saf.* **17**: 1595–1612.

Pacquit, A., Lau, K.T., McLaughlin, H., et al.(2006) Development of a volatile amine sensor for the monitoring of fish spoilage. *Talanta* **69**: 515–20.

Potasman, I., Paz, A., & Odeh, M. (2002) Infectious outbreaks associated with bivalve shellfish consumption: A worldwide perspective. *Clin. Infect. Dis.* **35**: 921–928.

Tsironi, T.N. & Taoukis, P.S. (2018) Current practice and innovations in fish packaging. *J. Aquat. Food Prod. Technol.* **27**: 1024–1047.

Ventura De Souza, R., Campos, C., Hamilton, L., et al. (2017) A critical analysis of the international legal framework regulating the microbiological classification of bivalve shellfish production areas. *Rev. Aquacult.* **10**: 1025–1033.

Wells, N., Yusufu, D., & Mills, A. (2019) Colourimetric plastic film indicator for the detection of the volatile basic nitrogen compounds associated with fish spoilage. *Talanta* **194**: 830–836.

Wu, L., Pu, H., & Sun, D.-W. (2019) Novel techniques for evaluating freshness quality attributes of fish: A review of recent developments. *Trends Food Sci. Technol.* **83**: 259–273.

Oil pollution

Atlas, R. & Bragg, J. (2009) Bioremediation of marine oil spills: When and when not—The *Exxon Valdez* experience. *Microb. Biotechnol.* **2**: 213–221.

Atlas, R.M. & Hazen, T.C. (2011) Oil biodegradation and bioremediation: A tale of the two worst spills in U.S. history. *Environ. Sci. Technol.* **45**: 6709–6715.

Baelum, J., Borglin, S., Chakraborty, R., et al. (2012) Deep-sea bacteria enriched by oil and dispersant from the Deepwater Horizon spill. *Environ. Microbiol.* **14**: 2405–2416.

Bik, H.M., Halanych, K.M., Sharma, J., & Thomas, W.K. (2012) Dramatic shifts in benthic microbial eukaryote communities following the Deepwater Horizon Oil Spill. *PLoS ONE* **7**: e38550.

Delmont, T.O. & Eren, A.M. (2017) Simulations predict microbial responses in the environment? This environment disagrees retrospectively. *Proc. Natl. Acad. Sci. USA* **114**: E8947–E8949.

Hazen, T.C., Dubinsky, E.A., DeSantis, T.Z., et al. (2010) Deep-sea oil plume enriches indigenous oil-degrading bacteria. *Science* **330**: 204–208.

Hazen, T.C., Prince, R.C., & Mahmoudi, N. (2016) Marine oil biodegradation. *Environ. Sci. Technol.* **50**: 2121–2129.

Hu, P., Dubinsky, E.A., Probst, A.J., et al. (2017) Simulation of Deepwater Horizon oil plume reveals substrate specialization within a complex community of hydrocarbon degraders. *Proc. Natl. Acad. Sci. USA* **114**: 7432–7437.

Kimes, N.E., Callaghan, A.V., Suflita, J.M., & Morris, P.J. (2014) Microbial transformation of the Deepwater Horizon oil spill—Past, present, and future perspectives. *Front. Microbiol.* **5**: 603.

King, G.M., Kostka, J.E., Hazen, T.C., & Sobecky, P.A. (2015) Microbial responses to the *Deepwater Horizon* oil spill: From coastal wetlands to the deep sea. *Annu. Rev. Mar. Sci.* **7**: 377–401.

Kleindienst, S., Seidel, M., Ziervogel, K., et al. (2015) Chemical dispersants can suppress the activity of natural oil-degrading microorganisms. *Proc. Natl. Acad. Sci. USA* **112**: 14900–5.

Kleindienst, S., Seidel, M., Ziervogel, K., et al. (2016) Reply to Prince et al.: Ability of chemical dispersants to reduce oil spill impacts remains unclear. *Proc. Natl. Acad. Sci. USA* **113**: E1422–3.

Prince, R.C., Coolbaugh, T.S., & Parkerton, T.F. (2016) Oil dispersants do facilitate biodegradation of spilled oil. *Proc. Natl. Acad. Sci. USA* **113**: E1421.

Redmond, M.C. & Valentine, D.L. (2012) Natural gas and temperature structured a microbial community response to the Deepwater Horizon oil spill. *Proc. Natl. Acad. Sci. USA* **109**: 20292–7.

Ron, E.Z. & Rosenberg, E. (2014) Enhanced bioremediation of oil spills in the sea. *Curr. Opin. Biotechnol.* **27**: 191–194.

Transportation Research Board & National Research Council (2003) *Oil in the Sea III: Inputs, Fates, and Effects.* The National Academies Press, Washington, DC.

Organic pollutants and heavy metals

Ayangbenro, A.S. & Babalola, O.O. (2017) A new strategy for heavy metal polluted environments: A review of microbial biosorbents. *Int. J. Environ. Res. Public Health* **14**: 94.

Chakraborty, J. & Das, S. (2016) Molecular perspectives and recent advances in microbial remediation of persistent organic pollutants. *Environ. Sci. Pollut. Res.* **23**: 16883–16903.

Chen, F., Li, Z.-L., & Wang, A.-J. (2019) Acceleration of microbial dehalorespiration with electrical stimulation. In *Bioelectrochemistry Stimulated Environmental Remediation*. Springer Singapore, Singapore, pp. 73–92.

Cheng, K.Y., Karthikeyan, R., & Wong, J.W.C. (2019) Microbial electrochemical remediation of organic contaminants: Possibilities and perspective. In *Microbial Electrochemical Technology*, pp. 613–640, Elsevier.

Harding, G., Dalziel, J., & Vass, P. (2018) Bioaccumulation of methylmercury within the marine food web of the outer Bay of Fundy, Gulf of Maine. *PLoS One* **13**: e0197220.

Matturro, B., Frascadore, E., & Rossetti, S. (2017) High-throughput sequencing revealed novel *Dehalococcoidia* in dechlorinating microbial enrichments from PCB-contaminated marine sediments. *FEMS Microbiol. Ecol.* **93**: fix134.

Matturro, B., Ubaldi, C., & Rossetti, S. (2016) Microbiome dynamics of a polychlorobiphenyl (PCB) historically contaminated marine sediment under conditions promoting reductive dechlorination. *Front. Microbiol.* **7**: 1502.

Outridge, P.M., Mason, R.P., Wang, F., et al. (2018) Updated global and oceanic mercury budgets for the United Nations Global Mercury Assessment 2018. *Environ. Sci. Technol.* **52**: 11466–11477.

Qureshi, A.A., Bulich, A.A., & Isenberg, D.L. (2018) Microtox* toxicity test systems—Where they stand today. In Microscale Testing in Aquatic Toxicology, Wells, P.G. & Lee, K. (eds.), CRC Press, pp. 185–199.

Regnell, O. & Watras, C.J. (2019) Microbial mercury methylation in aquatic environments: A critical review of published field and laboratory studies. *Environ. Sci. Technol.* **53**: 4–19.

Sanchez-Ferandin, S. (2015) Assessing ecotoxicity in marine environment using luminescent microalgae: Where are we at? *Am. J. Plant Sci.* **6**: 2502–2509.

Senn, D.B., Chesney, E.J., Blum, J.D., et al. (2010) Stable isotope (N, C, Hg) study of methylmercury sources and trophic transfer in the northern Gulf of Mexico. *Environ. Sci. Technol.* **44**: 1630–1637.

Sunderland, E.M., Krabbenhoft, D.P., Moreau, J.W., et al. (2009) Mercury sources, distribution, and bioavailability in the North Pacific Ocean: Insights from data and models. *Global Biogeochem. Cycles* **23**: GB2010.

Wang, Y. & Tam, N.F.Y. (2019) Microbial remediation of organic pollutants. In *World Seas an Environmental Evaluation*, Sheppard, C. (ed.). Acdemic Press, pp. 283–303.

Zaferani, S., Pérez-Rodríguez, M., & Biester, H. (2018) Diatom ooze-A large marine mercury sink. *Science* **361**: 797–800.

Pollution by plastics

Andrady, A.L. (2017) The plastic in microplastics: A review. *Mar. Poll. Bull.* **119**: 12–22.

Austin, H.P., Allen, M.D., Donohoe, B.S., et al. (2018) Characterization and engineering of a plastic-degrading aromatic polyesterase. *Proc. Natl. Acad. Sci.* **115**: E4350–E4357.

Bakir, A., O'Connor, I.A., Rowland, S.J., et al. (2016) Relative importance of microplastics as a pathway for the transfer of hydrophobic organic chemicals to marine life. Environ. Pollut. **219**: 56–65.

Cole, M., Lindeque, P., Fileman, E., et al. (2013) Microplastic ingestion by zooplankton. *Environ. Sci. Technol.* **47**: 6646–6655.

Cole, M., Lindeque, P.K., Fileman, E., et al. (2016) microplastics alter the properties and sinking rates of zooplankton faecal pellets. *Environ. Sci. Technol.* **50**: 3239–3246.

De Falco, F., Gullo, M.P., Gentile, G., et al. (2018) Evaluation of microplastic release caused by textile washing processes of synthetic fabrics. *Environ. Pollut.* **236**: 916–925.

Galgani, L., Engel, A., Rossi, C., et al. (2018) Polystyrene microplastics increase microbial release of marine chromophoric dissolved organic matter in microcosm experiments. *Sci. Rep.* **8**: 14635.

Gewert, B., Plassmann, M., Sandblom, O., & MacLeod, M. (2018) Identification of chain scission products released to water by plastic exposed to ultraviolet light. *Env. Sci. Technol. Lett.* **5**: 272–276.

Geyer, R., Jambeck, J.R., & Law, K.L. (2017) Production, use, and fate of all plastics ever made. *Sci. Adv.* **3**: e1700782.

Gregory, M.R. (2009) Environmental implications of plastic debris in marine settings—Entanglement, ingestion, smothering, hangers-on, hitch-hiking and alien invasions. *Philos. Trans. R. Soc. Lond. B. Biol. Sci.* **364**: 2013–25.

Harrison, J.P., Schratzberger, M., Sapp, M., & Osborn, A.M. (2014) Rapid bacterial colonization of low-density polyethylene microplastics in coastal sediment microcosms. *BMC Microbiol.* **14**: 232.

Hartmann, N.B., Hüffer, T., Thompson R.C., et al. (2019) Are we speaking the same language? Recommendations for a definition and categorization framework for plastic debris. *Environ. Sci. Technol.* **53**: 1039–1047.

Lobelle, D. & Cunliffe, M. (2011) Early microbial biofilm formation on marine plastic debris. *Mar. Pollut. Bull.* **62**: 197–200.

Long, M., Moriceau, B., Gallinari, M., et al. (2015) Interactions between microplastics and phytoplankton aggregates: Impact on their respective fates. *Mar. Chem.* **175**: 39–46.

Lusher, A.L., McHugh, M., & Thompson, R.C. 2013. Occurrence of microplastics in the gastrointestinal tract of pelagic and demersal fish from the English Channel. *Mar. Poll. Bull.* **67**: 94–99.

Napper, I.E., Bakir, A., Rowland, S.J., & Thompson, R.C. (2015) Characterisation, quantity and sorptive properties of microplastics extracted from cosmetics. *Mar. Poll. Bull.* **99**: 178–185.

Oberbeckmann, S., Osborn, A.M., & Duhaime, M.B. (2016) Microbes on a bottle: Substrate, season and geography influence community composition of microbes colonizing marine plastic debris. *PLoS One* **11**: e0159289.

Oliveira, M., Almeida, M., & Miguel, I. (2019) A micro(nano)plastic boomerang tale: A never ending story? *Trends Anal. Chem.* **112**: 196–200.

Porter, A., Lyons, B.P., Galloway, T.S., & Lewis, C. (2018) Role of marine snows in microplastic fate and bioavailability. *Environ. Sci. Technol.* **52**:.7111–7119.

Reisser, J., Shaw, J., Hallegraeff, G., et al. (2014) Millimeter-sized marine plastics: A new pelagic habitat for microorganisms and invertebrates. *PLoS One* **9**: e100289.

Romera-Castillo, C., Pinto, M., Langer, T.M., et al. (2018) Dissolved organic carbon leaching from plastics stimulates microbial activity in the ocean. *Nat. Commun.* **9**: 1430.

Teuten, E.L., Saquing, J.M., Knappe, D.R.U., et al. (2009) Transport and release of chemicals from plastics to the environment and to wildlife. *Philos. Trans. R. Soc. B Biol. Sci.* **364**: 2027–2045.

Thompson, R.C., Olsen, Y., Mitchell, R.P., et al. (2004) Lost at sea: Where is all the plastic? *Science* **304**: 838.

Woodall, L.C., Sanchez-Vidal, A., Canals, M., et al. (2014) The deep sea is a major sink for microplastic debris. *R. Soc. Open Sci.* **1**: 140317.

World Economic Forum (2016) The New Plastics Economy: Rethinking the future of plastics. Available online: http://www3.weforum.org/docs/WEF_The_New:Plastics_Economy.pdf (accessed 2 February 2019).

Wright, R.J., Gibson, M.I., & Christie-Oleza, J.A. (2019) Understanding microbial community dynamics to improve optimal microbiome selection. *Microbiome* **7**: 85.

Wright, S.L. & Kelly, F.J. (2017) Plastic and human health: A micro issue? *Environ. Sci. Technol.* **51**: 6634–6647.

Yang, J., Yang, Y., Wu, W.-M., et al. (2014) Evidence of polyethylene biodegradation by bacterial strains from the guts of plastic-eating waxworms. *Environ. Sci. Technol.* **48**: 13776–13784.

Yang, Y., Yang, J., & Jiang, L. (2016) Comment on "A bacterium that degrades and assimilates poly(ethylene terephthalate)". *Science* **353**: 759.

Yang, Y., Yang, J., Wu, W.-M., et al. (2015) Biodegradation and mineralization of polystyrene by plastic-eating mealworms: Part 2. role of gut microorganisms. *Environ. Sci. Technol.* **49**: 12087–12093.

Yoshida, S., Hiraga, K., Takehana, T., et al. (2016) A bacterium that degrades and assimilates poly(ethylene terephthalate). *Science* **351**: 1196–1199.

Zettler, E.R., Mincer, T.J., & Amaral-Zettler, L.A. (2013) Life in the "plastisphere": Microbial communities on plastic marine debris. *Environ. Sci. Technol.* **47**: 7137–7146.

Zobell, C.E. (1943) The effect of solid surfaces upon bacterial activity. *J. Bacteriol.* **46**: 39–56.

Chapter 14

Marine Microbial Biotechnology

Biotechnology is defined broadly as the application of scientific and engineering principles to provide goods and services through mediation of biological agents. Biotechnological approaches used for monitoring and remediating pollution are considered in *Chapter 13*. The first part of this chapter considers examples of products from marine microorganisms, including enzymes, pharmaceuticals, antifouling agents, polymers, and biofuels. This is followed by discussion of biotechnological processes for the diagnosis and prevention of disease in marine systems, especially aquaculture. Major economic benefits derive from the industrial and biomedical exploitation of marine microbial processes—their enormous natural diversity and range of metabolic activities provide great opportunities for future exploitation. As our appreciation of the great variety of marine microbes grows with the study of different habitats, so the collection of microbes, their genes, and their products with useful properties continues to expand.

Key Concepts

- Marine microbes are the source of important enzymes and polymers used in many branches of industry.

- Microalgae may provide a significant future source of biofuels.

- Many pharmaceuticals and other health products have been obtained from marine microbes, but their potential has not yet been fully realized commercially.

- Study of bacterial colonization of marine surfaces has led to new approaches to antifouling and prevention of infection through interference with signaling mechanisms.

- Structural components of marine microbes are being exploited in nanotechnology, bioelectronics, and the development of new materials.

- Marine microbes and their products are increasingly used in vaccines, probiotics, and immunostimulants for aquaculture of finfish and shellfish, in order to overcome the problems of disease and resistance to antimicrobial compounds.

- Phages have potential uses in the biological control of bacterial diseases of fish, shellfish, and corals.

Enzymes From Marine Microbes Have Many Applications

Among the many beneficial activities of marine microbes shown in *Table 14.1*, enzymes feature prominently. Enzymes are widely used by industry and the global market in 2021 is estimated to be about USD 5 billion, growing by 4% a year. Many research institutes and commercial organizations have developed culture collections of marine bacteria, fungi, and microalgae, with initial attention being focused on culturable organisms that are easily collected from near-shore habitats. The full potential of the thousands of marine microbes already in culture collections has not yet been fully investigated. In recent years, the growing recognition of the importance of fungi as autochthonous inhabitants of diverse marine habitats has led to increased focus on their potential for biotechnological exploitation. The most commonly exploited enzymes are those that degrade polymers, especially proteins and carbohydrates. Examples of early successes include the production by *Vibrio* spp. of various types of extracellular proteases, some of which are tolerant of moderate salt concentrations and detergents. Another is the extraction of glucanases and other carbohydrate-degrading enzymes from *Bacillus* spp. isolated from mud samples.

Table 14.1 Some beneficial effects of activities or products from marine microbes

Application	Examples
Aquaculture	Disease diagnostics Nutritional supplements Pigments Probiotics Vaccines
Cosmetics	Liposomes Polymers Sunscreens
Environmental protection	Bioremediation of pollution Disease diagnostics Non-toxic anti-fouling agents Toxicology bioassays Waste processing
Food processing	Enzymes Flavors Preservatives Texture modifiers
Manufacturing industry	Bioelectronics Polymers Structural components
Fuels	Desulfurization of oil and coal Oil extraction Production of biofuels
Nutraceuticals	Anti-oxidative compounds Dietary supplements Health foods
Pharmaceuticals and biomedical devices	Antibacterial, antifungal, and antiviral agents Anti-tumor and immunosuppressive agents Biosensors Drug delivery Enzymes Neuroactive agents Self-cleaning implants
Textiles and papers	Enzymes Surfactants

The most successful developments have occurred with the isolation of enzymes from extremophilic microorganisms; some of the products that have resulted are shown in *Table 14.2*. Enzymes from thermophiles and hyperthermophiles have particular attractions for industrial processes, which often require high temperatures. Even at milder temperatures, thermophilic enzymes are beneficial because of their much greater stability. The structural features of thermophilic enzymes that confer stability and function at high temperatures (sometimes >100°C) are discussed in *Chapter 3*.

The use of proteases and lipases as stain removers in detergents is a particularly important application, since properties that confer high thermostability and activity are often combined with resistance to bleaching chemicals and surfactants used in these products. Some washing powders incorporate thermophilic cellulases and hemicellulases, which digest loose fibers and help to prevent "bobbles" on clothes after washing; these enzymes are also important in the manufacture of "stone-washed" denim. Enzyme production for biological detergents is a very large market accounting for approximately 30% of the total global production of enzymes. The first source of these enzymes was soil bacteria of the genus *Bacillus*, but enzymes from marine thermophiles have higher temperature optima and superior stability.

Modern food processing uses a wide range of enzymes. Almost all processed foods now rely on some form of modified starch product for the improvement of texture, control of moisture, and prolonged shelf life. Amylases hydrolyze α-1,4-glycosidic linkages in starch to produce a

Table 14.2 Some biotechnological applications of extremophilic microbes

Product	Applications
Thermophiles and hyperthermophiles	
Amylases, pullulanases, lipases, proteases	Baking, brewing, food processing
DNA polymerases	PCR amplification of DNA
Lipases, pullulanases, proteases	Detergents
S-layers	Ultrafiltration, electronics, polymers
Xylanases	Paper bleaching
Halophiles	
Bacteriorhodopsin	Bioelectronic devices, optical switches, photocurrent generators
Compatible solutes	Protein, DNA, and cell protectants
Lipids	Liposomes (drug delivery, cosmetics)
S-layers	Ultrafiltration, electronics, polymers
γ-linoleic acid, β-carotene, cell extracts	Health foods, dietary supplements, food colors, aquaculture feeds
Psychrophiles	
Ice nucleating proteins	Artificial snow, frozen food processing
Polyunsaturated fatty acids	Food additives, dietary supplements
Proteases, lipases, cellulases, amylases	Detergents
Alkaliphiles and acidophiles	
Acidophilic bacteria	Fine papers, waste treatment
Elastases, keratinases	Leather processing
Proteases, cellulases, lipases	Detergents
Sulfur-oxidizing acidophiles	Recovery of metals, sulphur removal from coal and oil

mixture of glucose, malto-oligosaccharides, and dextrins. All the remaining α-1,4-glycosidic branches in the products are hydrolyzed by pullulanase. When starch is treated with amylase and pullulanase simultaneously at high temperatures, it produces higher yields of desired end products. Pullulanases and other enzymes derived from the hyperthermophilic bacterium *Thermotoga maritima* have recently been introduced into food processing. A range of other carbohydrate-modifying enzymes have been isolated from marine thermophiles.

Agar-degrading enzymes can be isolated from many species of bacteria, including *Vibrio*, *Pseudomonas*, *Pseudoalteromonas*, *Alteromonas*, *Thalassomonas*, *Cytophaga*, and *Agarivorans*. Zones or craters of agar degradation are often seen surrounding colonies of bacteria isolated from seawater, sediments, algae, and marine invertebrates when grown on agar plates. Agar, obtained from red seaweeds such as *Gracilaria*, is added to many foods such as ice cream, glazes, and processed cheese to improve texture. Industrially, agar-degrading enzymes (agarases) are used to produce diverse oligosaccharides with a variety of properties and numerous applications in health foods and cosmetics. Agarases are also widely used in the molecular biology laboratory as a method of purifying DNA after separation on agarose gels. The enzymes are also useful for degradation of the cell wall of algae for extraction of labile substances with biological activities such as unsaturated fatty acids, vitamins, and carotenoids. As with many enzymes and other proteins isolated from bacteria, it is often more convenient for manufacturing to clone the genes responsible and express them in *E. coli* or another recombinant host. However, there can be problems with expression and correct folding of thermostable proteins in mesophilic hosts, and if post-translational modifications are required for enzyme function, the recombinant enzyme may not function as required.

Enzymes from psychrophilic microbes have many uses and isolates from the deep sea and polar regions have been exploited for use in food-processing applications in which low temperatures are required to prevent spoilage, destruction of key ingredients (e.g. vitamins), or loss of texture. For example, galactosidases from cold-adapted bacteria have been used for the removal of lactose from milk to improve digestibility, and xylanases are used in baking to improve crumb texture in bread. Proteases from psychrophiles can be used for tenderizing meat at low temperatures. Significant savings in energy costs result from the use of cold-water laundry detergents that incorporate proteases and lipases active at temperatures of 10°C or less.

DNA polymerases from hydrothermal vent organisms are widely used in molecular biology

The discovery of thermostable DNA polymerases used in the polymerase chain reaction (PCR) has been one of the most spectacular scientific advances since the discovery of DNA itself. The use of the PCR in research, disease diagnostics, and forensic investigations has led to a huge market; the market value of PCR enzymes, kits, and equipment was USD 7.4 billion in 2017 and is growing by ~6% each year. The original PCR enzyme, *Taq* polymerase, was isolated from *Thermus aquaticus* from a terrestrial hot spring. Although *Taq* is still the least expensive and most widely used enzyme for both research and diagnostic uses of the PCR, a number of alternative enzymes are now available that are superior in certain applications, and some of these were isolated from marine bacteria (*Table 14.3*). The choice of enzyme depends on the specific activity and sensitivity required with different amounts and lengths of template, nucleotide specificity, and various other factors, including cost. Enzymes from the hyperthermophilic vent bacteria have greater thermostability and activity at higher temperatures, particularly useful when amplifying GC-rich sequences. They may also have a higher fidelity of replication (due to integral proofreading ability), although careful optimization of the reaction is required; enzymes such as Vent™ or *Pfu* are often used in a mixture with *Taq*.

Metagenomics and bioinformatics lead to new biotechnological developments

Molecular biology methods have led to exciting new approaches in the search for new products from marine microbes. In an analogy with the gold rush, "bioprospectors" can use oligonucleotide hybridization probes to "pan" for genetic sequences of interest in the hope

SUNKEN WHALE CARCASSES MAY YIELD NOVEL ENZYMES

As described on p.269, when the carcass of a dead whale reaches the seafloor, it passes through sequential stages that support a diverse community of organisms for many years (Smith et al., 2015). Deep-sea whale carcasses thus present a unique high-pressure, low-temperature environment for the selection of microbes with potential benefits for industrial applications, but the difficulties of studying deep-sea whale falls means that only limited investigations have been possible. Metagenomic analysis of a whale fall at 4240 m in the South Atlantic Ocean revealed a number of microbial genes with biotechnological potential (de Freitas et al., 2019). Many had little sequence identity to previously described proteins and some were recognized as having potential uses in bioremediation of organic pollutants. Often, the skeleton is colonized by *Osedax* worms (*Figure 10.9*), which secrete acid to dissolve calcium phosphate in the bones. The root of the worm contains symbiotic Oceanospirillales bacteria that can metabolize complex carbon compounds (Goffredi et al., 2005), but the specific roles of host cells and symbionts are not clear (Goffredi et al., 2007; Miyamoto et al., 2017). Further information about these interactions could lead to more novel enzymes with interesting properties.

Table 14.3 *Some thermostable DNA polymerases and their sources*

DNA Polymerase	Organism	Source[a]	Half-life at 95°C	Proofreading
Taq	*Thermus aquaticus*	T (N or R)	40	–
Amplitaq®	*Thermus aquaticus*	T (R)	40	–
Vent™	*Thermococcus litoralis*	M (R)	400	+
Deep Vent™	*Pyrococcus GB-D*	M (R)	1380	+
Tth	*Thermus thermophilus*	T (R)	20	–
Pfu	*Pyrococcus furiosus*	M (N)	120	+
ULTma™	*Thermotoga maritima*	M (R)	50	+

[a]T = terrestrial hot spring; M = marine hydrothermal vent; (N) = natural; (R) = recombinant.

of reaping rich rewards. This enables the screening of communities without the need for culture. For example, comparison of sequence data for genes encoding proteins with a particular function can allow the identification of consensus sequences. These can be used to construct complementary oligonucleotide probes, which are then used to search for genes in DNA amplified from a community of interest. Bioinformatics data-mining tools can be used to identify gene homologs, and genes can then be cloned and expressed in suitable hosts. The disadvantage of this method is that it may fail to detect truly novel proteins with sequences unlike those previously described. This limitation is borne out by the results from metagenomic analyses and genome sequencing of cultured microbes. In both cases, large numbers of open reading frames (ORFs) in the genomes have no match with existing genes in databases. To overcome this, direct functional screening of metagenomic libraries can be employed to identify candidate genes associated with observed enzymatic activities. PCR primers for amplification of sequences from libraries can be rationally designed using sequence features characteristic of genes for particular enzymes and tailored for the possession of properties compatible with specific industrial applications. These approaches are leading to new enzymes such as esterases, lipases, chitinases, amidases, cellulases, alkane hydrolases, and proteases. Advances in sequencing technology and bioinformatics mean that we will undoubtedly be able to develop numerous new enzymes carrying out reactions that are completely unknown to us at present.

Polymers from marine bacteria have many applications

Microbial polymers are used in bioremediation, industrial processes, manufacturing industry, and food processing. The best-exploited compounds are extracellular polysaccharides that form the glycocalyx associated with biofilm formation and protection from phagocytosis. Applications in bioremediation were mentioned in *Chapter 13*. Other potential applications include underwater surface coatings, bioadhesives, drag-reducing coatings for ship hulls, dyes, and sunscreens. It has been suggested that the high production (up to 80% cell dry weight under appropriate conditions) of poly-hydroxy-β-hydroxyalkanoates by some marine bacteria as food reserves could be exploited to produce fully biodegradable plastics. Oil-derived plastics continue to dominate the market, but bacterially produced plastics are becoming more competitive because of increasing oil prices and the polluting effects of nondegradable plastics *(Box 13.1)*.

Microalgae can produce biofuels and edible oils

The global economy is dependent on oil for energy and as a source of chemicals. With increasing energy demands from a growing population, limited supplies of fossil fuels, concerns about CO_2 emissions and political instability in oil-producing countries, there has been considerable recent interest in using renewable biological materials to produce biofuels. Crops such as corn, sugar, soybeans, sunflowers, or palms can be fermented to alcohol for addition

AFTER THE GOLD RUSH—MICROBIAL NANOPARTICLES

Nanotechnology—the manipulation of atoms and molecules for the fabrication of products on microscopic scale—is a rapidly expanding field. Biosynthetic Au, Ag, Cd nanoparticles (typically 10–20 nm diameter) have many potential applications in electronics, new antimicrobial treatments, biological imaging, and as delivery agents for DNA vaccination or CRISPR gene-editing. Manufacture of nanoparticles by top-down size reduction involves environmentally damaging physical and chemical processes, so efforts to develop cost-effective eco-friendly methods are being developed. This can occur intracellularly due to reduction of metallic nanoparticles by enzymes on the cytoplasmic membrane or cell wall. Extracellular synthesis depends on extracellular microbial enzymes such as nitrate reductase—electrons are transferred to metal ions, which precipitate as nanoparticles (Patil and Kim, 2018). Examples of marine microbes employed for such synthesis include *Rhodococcus* spp., *Marinobacter pelagius*, and *Vibrio alginolyticus* (bacteria), *Euglena* spp. and *Nannochloropis* spp. (microalgae), and *Aspergillus sydowii* (fungus);(Vala, 2014).

to gasoline or used to produce biodiesel. However, the costs of production, competition with food production for use of high-quality land, and the ecological consequences of deforestation associated with such schemes make the use of land crops unsustainable. Therefore, there has been much interest in growing microalgae as a source of biodiesel and other biofuels including hydrogen, methane and ethanol.

Several species of microalgae may be suitable for large-scale culture. Marine examples include cyanobacteria (e.g. *Synechococcus*), diatoms (e.g. *Phaeodactylum tricornutum* and *Chaetoceros muelleri*), prymenesiophytes (e.g. *Emiliania huxleyi*), and chlorophytes (e.g. *Dunaliella tertiolecta* and *Botryococcus braunii*). Careful selection of species is needed depending on culture conditions and required end-products and the bacteria associated with the microalgae may influence growth and productivity. Autotrophic microalgae can be grown in saline or brackish ponds, lagoons or open circulating units with seawater and wastewater, requiring sunlight and some simple nutrients. Alternatively, they can be grown in photobioreactors; these are flow-through transparent glass or plastic columns, panels or tubes exposed to light (*Figure 14.1*). Although offering many advantages of better control and prevention of contamination, photobioreactors require large capital investment. Hybrid systems use photobioreactors to produce large volumes of single-species microalgae, which are then exposed to nutrient rich open ponds for production of biomass. Growth is very efficient because, unlike vascular plants, all cells in a microalgal culture are photosynthetic and carbon dioxide and nutrients are available directly to the cells. In addition, the concentration of carbon dioxide in an algal suspension is much higher than that in the atmosphere above a land plant. Algae therefore have the potential to produce about 30 times more energy per unit area than arable crops such as soybeans. Microalgae can also be grown heterotrophically, which can be applied for bioremediation of wastewater or effluent from food processing.

Microalgae can be harvested by flocculation or sedimentation, followed by filtration or centrifugation before drying. At present, downstream processing requires high energy inputs and accounts for 50–80% of the cost of production. Lysis of the cell wall to extract oils is problematic due to strengthening polymers. Various methods of lysis have been used, including mechanical shearing, solvent extraction, thermochemical treatment, microwave, and ultrasound disruption.

The starting points for oil production from microalgae are the lipids and fatty acids found in their membranes and stored in intracellular compartments as an energy reserve. Algal strains vary greatly in the types and quantity of lipids produced; selection of appropriate strains for biodiesel production depends on many factors. Careful control of growth conditions can lead to the cells containing over 75% of their mass as lipids. These lipids can be converted into a variety of compounds with similar properties to petroleum products, including diesel.

Figure 14.1 A serpentine photo-bioreactor Phyco-Flow™ installed at Cyanotech in Hawaii for culture of *Haematococcus pluvialis* for astaxanthin production. Image courtesy of Varicon Aqua, UK.

Whilst there have been considerable successes in small-scale production of microalgal bio-fuels, the costs of production of biodiesel remain much higher than other biofuels such as soybeans, jatropha, or palm oil to be a viable competitor. Much ingenuity will be needed to develop this process into a sustainable and cost-effective industry that will deliver the considerable environmental benefits. Supplies of carbon dioxide and sunlight are clearly not a problem, but organisms require other major nutrients, especially nitrogen and phosphorus. Microbial nitrogen fixation could perhaps be harnessed for supply to algal culture systems, but the only source of phosphorus is from the mining of natural deposits and supplies are running out quickly (this threatens all forms of agriculture). The challenge will be, therefore, to develop sustainable systems with almost perfect systems for recycling nutrients after harvesting.

Microalgal culture is also used for production of high-value human and animal foodstuffs and additives. Global pressures for the supply of vegetable oils—a major essential component of human and animal diets—leads to extensive land use, often resulting in destruction of natural habitats, such as felling of tropical forests for palm oil production. Microalgae are particularly important as a source of essential polyunsaturated fatty acids (PUFA) and carotenoids in fish and shellfish aquaculture and as a component of food supplements and nutraceuticals.

Genetic modification has been shown to enhance production of triacyl glycerides, omega 3 oils and triterpenoids and shows potential to increase productivity and lower production costs in microalgal culture. However, scale-up to industrial application will require careful risk assessment of environmental concerns.

Marine microbes are a rich source of biomedical products

Natural products provide many compounds used in medicine and health promotion. Throughout much of the twentieth century, the pharmaceutical industry was engaged in a continual search for new compounds with biological activities that can be exploited for treatment or prevention of disease. Many of our most successful drugs are secondary metabolites obtained from plants or bacteria and fungi isolated in the terrestrial environment, especially soil. However, in the last few decades, the effort necessary to isolate valuable new compounds has become increasingly disproportionate to the returns. Companies moved away from natural product-based drug discovery and came to rely more on the use of combinatorial chemistry to synthesize libraries of compounds, and structure–function analysis to chemically modify already existing compounds. Now, there is a resurgence of interest in natural products from marine habitats because of the development of new tools for underwater exploration and the discovery of the enormous diversity of marine organisms. Thousands of bioactive compounds isolated from marine microbes have been investigated for their biological activities, including anti-microbial, anti-cancer, anti-inflammatory, and antioxidant activities, and a selection of these is shown in *Table 14.4*. Cyanobacteria and Actinobacteria have been especially rich sources of potentially useful pharmaceutical agents, with many patents issued. Many of these are non-ribosomal peptides or complex polyketides, but there is enormous diversity in structure, so that different species and strains within species produce a variety of compounds with differences in pharmacological properties.

Many bioactive compounds from marine invertebrates are produced by microbes

For many years, marine invertebrates (especially sponges, bryozoans, and ascidians) have been investigated as a source of bioactive compounds, and many antibiotics and antitumor agents have been isolated. Many of these soft-bodied animals have evolved chemical defenses against predators or to prevent colonization or disease. Different species of these animals are host to a wide range of microbes, many of which are symbionts (see *Chapter 10*). It now seems likely that many of the compounds discovered are produced by the microbes inhabiting the animals, rather than by the host's own metabolism. However, although compounds isolated from sponges and bryozoans often share structural similarity with known microbial metabolites,

Table 14.4 Examples of pharmaceutical compounds from marine microbes

Agent	Producing organism	Activity
Abyssomicin	*Verrucispora* sp.	Inhibits para-amino-benzoic acid synthesis; broad-spectrum antibiotic.
Apratoxins	*Lyngbya* spp.	Interfere with signalling and transcription in tumor cells.
Bryostatins	"*Ca.* Endobugula sertula"	Inhibits protein kinase C; anti-tumor; treatment of leukemia and esophageal cancer; may be useful in treatment of Alzheimer's disease.
Coibamide A	*Leptolyngbya* sp.	Induces apotosis in tumor cells.
Cryptophycins	*Nostoc* spp.	Inhibits polymerization of tubulin in tumor cells.
Curacin	*Lyngbya majuscula*	Blocks mitosis in tumor cells.
Largazole	*Symploca* sp.	Inhibits histidine decarboxylase; anti-tumor; anti-epileptic; prevents mood swings.
Psymberin	Uncultivated sponge symbiont	Cytotoxic.
Salinosporamide A	*Salinospora* sp.	Inhibits protein breakdown in the proteasome; effective against a range of cancers; anti-malarial; may relieve some symptoms of Alzheimer's disease.

it is not easy to prove that they are indeed of microbial origin in the symbiosis. One notable example is the family of compounds known as bryostatins, isolated from the bryozoan *Bugula neritina (Figure 14.2A)*. Bryostatins are cytotoxic macrolides, which are very likely synthesized by an as-yet-uncultured member of the Gammaproteobacteria named "*Ca.* Endobugula sertula." Bryostatin-1 shows great promise for the treatment of certain types of leukemia and esophageal cancer in clinical trials and has also been found to have promising potential in the treatment of Alzheimer's disease. The compound ecteinascidin is a highly complex and potent antitumor agent that was isolated from the tunicate *Ecteinascidia turbinata (Figure 14.2B)* that proved extremely difficult to obtain in sufficient quantities. After many years, a metagenomic approach identified the biosynthetic gene cluster that is responsible for identifying the 25 enzymes needed for synthesis of ecteinascidin. Sequence analysis confirmed that these genes belong to an uncultured symbiotic bacterium "*Ca.* Endoacteinascidia frumentis." Eribulin is a synthetic analogue of halichondrin B originally isolated from the sponge

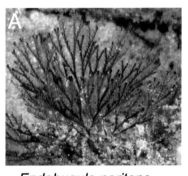
Endobugula neritans

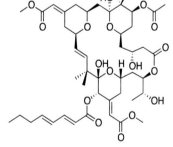
Bryostatin-1

Ecteinascidia turbinata

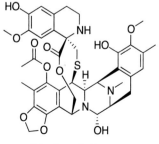
Trabectidin

Figure 14.2 Examples of successful pharmaceuticals derived from symbionts of invertebrates. A. *Bugula neritina* and bryostatin-1. B. *Ecteinascidia turbinata* and trabectedin. Image credits: A: Reprinted from Mans (2016), CC-BY-4.0. B: Sean Nash, CC-BY-2.0 via Flickr.

Halichondria okadai. It is proving very successful for treatment of breast cancer and liposarcoma. Obtaining sufficient supplies of the compound from its natural source became impossible due to shortage of supply, and laboratory synthesis of the complex molecule was eventually achieved using 67 separate synthetic steps. There is strong evidence that it is produced by a sponge-associated microbe, but the organism responsible has not yet been identified. If the phenomenon of production of useful metabolites by symbionts is found to be widespread, it overcomes the major problem of the need to harvest scarce marine animal life and the consequent disruption of ecosystems. Exploitation of microbial symbionts for pharmaceutical production could involve isolation or culture of the microbes, but since many cannot be cultured, it is likely to be more productive to identify and clone genes for recombinant expression.

Sponges are a particularly good source for new microbial compounds. As noted in *Chapter 10*, sponges contain highly diverse microbial communities comprising up to half of the sponge volume, and there appear to be sponge-specific microbiota that have coevolved for ~600 million years. Cultivation of sponge-associated microorganisms that produce bioactive compounds is being actively pursued by several research groups using a range of culture media and conditions; if successful, this approach has obvious advantages for large-scale inexpensive production of compounds. Several species of sponge-associated actinomycetes, vibrios, and fungi have been successfully isolated and shown to produce active compounds with antitumor, antibacterial, antifungal, anti-inflammatory, immunosuppressive, neuroactive, and other activities. However, despite employing various strategies to improve isolation, such as the addition of sponge tissue extracts to media, only a small proportion of the microbes present in sponges have been cultured. It is possible that many sponge microbes are obligate symbionts and will prove very difficult or impossible to culture as they may require specific metabolic intermediates from the host sponge.

Metagenomic approaches to identify and clone appropriate genes are increasingly used. Many of the most interesting secondary metabolites with biotechnological potential are produced by the action of polyketide synthases. These are giant modular enzymes composed of multiple protein units that contain a coordinated group of active sites. The complex polyketide molecules are synthesized in a stepwise manner from simple 2-, 3-, or 4-carbon building blocks. Bacterial polyketide synthases contain many catalytic domains organized into repeated sets of modules—each module incorporates one carbon unit into the growing polyketide chain. Several of these enzyme complexes have now been cloned, including that from "*Ca*. E. sertula," responsible for bryostatin synthesis, and a gene cluster responsible for synthesis of psymberin (a highly potent antitumor agent) from the metagenome of a complex microbial community in the sponge *Psammocinia* aff. *bulbosa*.

With so much potential from the sea, why are there so few new drugs?

Thousands of potential drugs from marine sources have shown great promise in initial trials, but only a handful have been fully developed by pharmaceutical companies for approved use in medicine. After demonstration of *in vitro* activity, agents must be rigorously tested for toxicity and potential adverse effects in cell cultures and laboratory animals. Probably only 1 in 1000 agents reaches the next stage of phase 1 clinical trials in human volunteers. The drug must then go through progressively larger phase 2 and phase 3 clinical trials involving thousands of patients before it is approved and licensed by government authorities for use as a drug. The process is very expensive and can take 10–15 years, so at each stage of development the company must evaluate the costs of continuing the process against the potential market returns — this cost–benefit analysis is especially important as patent protection on new drugs may last less than 20 years. The estimated cost of bringing a drug to market for human use is estimated at over USD 2 billion, which explains why so few potential agents reach the final stages of approval for general use.

Study of complex microbial communities may lead to new antibiotics

The spread of antimicrobial resistance (AMR) among pathogenic microbes is one of the greatest threats to human health. Many of the antimicrobial antibiotics used in the treatment of microbial infections were derived from terrestrial fungi (e.g. penicillins) and bacteria (e.g.

tetracyclines, aminoglycosides, and macrolides). Cephalosporin was obtained from the fungus *Cephalosporium*, isolated near a coastal sewage outfall. Drug-discovery companies have begun a renewed search for novel microbes in the marine environment. Large-scale rapid-throughput screening programs are underway in several parts of the world, investigating marine sediments, sponges, corals, and other habitats. It is likely that the intense competitive pressure found in dense mixed microbial communities, such as those on algal surfaces and in corals, will select for microbes producing antimicrobials. Biofilms are likely to be a rich source, and attention should be given to using new types of bioassays to detect metabolites that interfere with cell signaling or chemotaxis, as well as those that cause outright growth inhibition in the standard detection method.

Marine microbes provide various health-promoting products

A wide range of substances used for health promotion—often referred to as nutraceuticals—and includes functional foods, probiotics, and nutritional supplements. This is an expanding global market, valued at over USD 380 billion, growing at ~7% each year. New marine microbial products, especially those from microalgae, are likely to contribute to this area. For example, the polyunsaturated long-chain omega-3 fatty acid docosahexaenoic acid (DHA) promotes cardiovascular health and is especially important in fetal brain development; it is therefore taken by many consumers as a food supplement and added to infant formula feed and omega-3-enriched eggs and milk. With growing pressure on fish stocks and the possibility of contamination by methylmercury (see p.377), DHA is produced commercially by the marine heterotrophic dinoflagellate *Crypthecodinium cohnii*. A spin-off from this work was the development of DHA-enriched diets for larval stages in aquaculture. Newly isolated deep-sea psychrophilic bacteria and thraustochytrids (see p.183) may also be a good source of polyunsaturated fatty acids and other compounds with health benefits. These sources are also favored by vegetarians, and demand for non-animal sources of foods and food supplements is growing rapidly. It is necessary to prove that microorganisms with no history of use in food products are nontoxic and nonpathogenic and to develop suitable methods for large-scale cultivation.

A major part of the healthfood industry is concerned with antioxidative effects. Research with bacteria from coral reefs exposed to very high levels of visible light and ultraviolet irradiation indicates that they have novel mechanisms of reversing the resultant oxidative damage. This could lead to products that overcome some aspects of the aging process. Marine bacteria and fungi produce a diverse range of polysaccharides, some of which have been shown to have anti-inflammatory properties and ability to boost the immune system. These have potential applications as probiotics for humans and animals (see p.403).

Marine microbes have applications in biomimetics, nanotechnology, and bioelectronics

In materials science and technology, biomimetics is the term given to the process of "taking good designs from nature." Nanotechnology involves the construction of materials and functional objects assembled from the basic molecular building blocks, which offers the potential of new products ranging from new computer technology to microscopic machines. Marine microbes are proving to be a rich source for these new technologies. Bacterial and archaeal S-layers (see p.69) have applications in nanotechnology because of the ordered alignment of functional groups on the surface and in the pores, which allow chemical modifications and the binding of molecules in a very precise fashion. Isolated S-layer subunits can self-assemble as monolayers on solid supports such as lipid films, metals, polymers, and silicon wafers, which can be used to assemble different "building blocks" such as proteins, nucleic acids, lipids, and glycans. These have a wide range of applications in colloid and polymer science and the electronics industry. Examples include carriers for vaccine or drug delivery, "biochip" sensors and biocatalysts. The uniform size and alignment of pores in S-layers also makes them suitable for use as ultrafiltration membranes. Fusion proteins incorporating a range of specific properties such as antibodies or enzymes can be developed to allow highly precise construction and alignment of functional membranes for use in biosensors.

As described in *Chapter 6*, different species of diatoms construct their silica shells in a huge diversity of forms, with different shapes and surface architecture. Understanding the molecular basis by which diatoms achieve the construction of their frustules is leading to great advances in nano-assembly of materials into desired structures. Diatom silica has important applications in drug delivery, biosensors, tissue engineering, and energy storage devices. The calcite scales (coccoliths) of coccolithophores can also be modified by adsorption of proteins or incorporation of metal ions and genetic modification can lead to production of material with customized surface architecture.

The microscopic rotary motor of bacterial flagella has attracted the interest of engineers for some years, with suggestions that isolated basal bodies could form the basis of self-propelled micromachines (e.g. for targeted drug delivery systems). The recent discovery of chemotaxis and ultrafast swimming in marine bacteria (p.98) could lead to advances in this field, as further research leads to an understanding of the molecular basis of these processes.

Magnetotactic bacteria (p.43) produce intracellular magnetic crystals with a stable single-magnetic domain with uniform structure and purity that is difficult to achieve with chemical processes. These could have important applications in the electronics industry and for biomedical applications such as the production of magnetic antibodies, enzyme immobilization, cell separation, and in magnetic resonance imaging. Understanding the mechanisms by which bacteria construct the magnetosomes and the introduction of changes via genetic modification could lead to the ability to display particular proteins. These bacteria are very difficult to culture, but there is active research into optimization of culture media and incubation conditions to improve yields.

Biomolecular electronics relies on the use of native or genetically modified biological molecules such as proteins, chromophores, and DNA. One of the best examples to date is the use of bacteriorhodopsin isolated from *Halobacterium salinarum* (p.155). Bacteriorhodopsin changes its structure every few milliseconds to convert photons into energy. A chromophore embedded in the protein matrix absorbs light and induces a series of changes to the optical and electrical properties of the protein. Bacteriorhodopsin can store many gigabytes of information in three-dimensional films (holographic memories), and genetic modification produces proteins with various desirable properties. Bacteriorhodopsin has also been used to construct artificial retinas, and to make nano-sized solar cells and motion sensors.

Microbial biotechnology has many applications in aquaculture

Intensive aquaculture now supplies well over half of all fish and shellfish consumed and the industry continues to grow rapidly as natural stocks are depleted. As discussed in *Chapter 11*, the health of marine animals depends on interactions between the host, the pathogen, and the environment. In aquaculture, the most important practical measures to prevent or limit diseases are those that reduce stress and maintain good hygiene and overall health of the stock (*Table 14.5*). These factors are largely a matter of good husbandry and management practice. However, microbial biotechnology has played a key role in the massive expansion of intensive global aquaculture over the past few decades, especially in the control of water quality, development of fast and effective diagnostic procedures, control of disease with antibiotics and vaccines, and the development of nutrients and probiotics for stimulating growth and feed conversion.

Antimicrobial agents are widely used in aquaculture

Although many antimicrobial agents are effective against bacterial pathogens in the laboratory, the number of effective treatments for fish or shellfish disease is quite limited for several reasons. (Note: the term antimicrobials includes true antibiotics, which have a biological origin, as well as synthetic agents). Firstly, antimicrobials must be proven to be active against the pathogen but should produce minimal side effects in the host. The best chemotherapeutic antimicrobials work by targeting a process present in bacteria that is absent or different in their eukaryotic host. For example, penicillins (such as amoxicillin) target peptidoglycan

Table 14.5 *Important factors for optimizing the health of fish and shellfish in aquaculture. Items in bold type show practices dependent on microbial biotechnology*

Health factor	Practices
Design and operation of culture systems	Separation of hatchery and growing-on facilities Good management practices and record keeping
Hygiene	Disinfection of nets Protective clothing and equipment Prompt removal of moribund and dead fish **Bioremediation for improvement of water quality**
Nutrition	Careful monitoring of optimal growth rates at all stages of the life cycle **Immune stimulants as feed additives** **Probiotics for growth promotion and disease prevention**
Minimizing stress	Avoid netting, grading, overcrowding Maintain good water quality Avoid feeding before handling Use anasthesia during handling Breed "domesticated" lines of fish
Breaking the pathogen's life cycle	Disinfection of tanks and equipment Separate fish of different ages Fallowing sites for 6 months to 1 year
Diagnosis of disease	**Development of rapid methods (e.g. antibody tests and PCR assays)**
Eliminating vertical transmission	**Production of specific pathogen free brood stock, testing eggs and sperm for pathogen (e.g. PCR assays)**
Preventing geographic spread	Licensing system for egg and larval suppliers Notifiable disease legislation Movement restrictions from infected sites
Eradication	Slaughter policy for notifiable diseases Government compensation
Antimicrobial treatment	**Antibiotics and synthetic antimicrobial agents** **Sensitivity testing** Limit use to prevent evolution of resistance
Vaccination	**Immersion, oral, and injectable vaccines** **Ensure strains used for vaccines are appropriate for local disease experience** **Well-designed tests for evaluation of efficacy in appropriate species** **Assess the need for re-immunization (boosters)** **DNA vaccines**
Genetic improvement of stock	**Select for disease resistance traits** **Transgenics—incorporation of genes for disease resistance and growth promotion**

synthesis whilst oxytetracycline targets the 30S ribosomal subunit; both of these are unique to bacteria. Secondly, the agent must reach the site of infection in adequate concentrations to kill the pathogen or, more usually, inhibit its growth sufficiently to allow the host's immune system to eliminate the pathogen. The rate of uptake, absorption, transport to the tissues, and excretion rates vary greatly among different fish species. Very few agents are transported across eukaryotic membranes, which is why intracellular pathogens such as *Renibacterium salmoninarum* or *Piscirickettsia salmonis* are particularly difficult to control. Furthermore, because fish are poikilothermic, uptake and absorption of antimicrobials are very dependent on temperature. Therefore, the efficacy of a particular compound should be evaluated for each host–pathogen interaction under various environmental conditions. A third factor is the need to evaluate the rate of elimination of the drug to ensure that there are no unacceptable residues in the flesh of fish intended for human consumption. Again, because fish are poikilothermic, the rate of excretion and degradation depends on temperature, so it is necessary to calculate a "degree-day" withdrawal period between the last administration of the antimicrobial and the slaughter of the fish for the market. For example, the withdrawal period for oxytetracycline is 400 degree-days (e.g. 40 days at 10°C or 20 days at 20°C). With the growth of aquaculture, government agencies and large retailers in many countries now test farmed fish for antimicrobial residues, in the same way that they test meat or milk.

Antimicrobial agents are most commonly administered in medicated feed. These are often given to the whole population, but reduction in feeding is often one of the first signs of disease, so infected fish may not receive the appropriate dose of antimicrobial agents from medicated feed. Antimicrobials may sometimes be given by immersion of infected fish in a bath containing the agent, especially for gill and skin infections. Problems arise with both routes of administration because of wastage and contamination of the environment. Injection is rarely used, except for brood stock and aquarium fish.

Government regulatory authorities require a considerable amount of testing before licensing a drug for use, and the high costs of testing deter pharmaceutical companies from introducing new agents. The range of treatments is therefore very limited. Regulatory control is strict in the EU, USA, Canada, and Norway, with only a small number of agents officially approved. The routine use of antibiotics for disease prevention or growth promotion has been banned in many countries because of the problems of resistance and residues discussed below, but illegal use continues and in some countries with the most intensive aquaculture (especially in Asia and South America), there are no effective controls at all on antimicrobial usage.

Antimicrobial resistance (AMR) is a major problem in aquaculture

Bacteria possess three main strategies for AMR, as shown in *Table 14.6*. Individual bacterial isolates often possess more than one resistance mechanism, and individual antimicrobials may be affected by different resistance mechanisms in different bacteria. Bacteria possess *intrinsic* resistance to certain agents because of inherent structural or metabolic features of the bacterial species—this is almost always expressed by chromosomal genes. This type of resistance is relatively easy to deal with, but *acquired* resistance causes major problems in all branches of veterinary and human medicine. The use of almost every antimicrobial sooner or later leads to the selection of resistant strains from previously sensitive bacterial populations. This occurs via spontaneous mutations in chromosomal genes, which occur with a frequency of about 10^{-7}, or by the acquisition of plasmids or transposons. AMR genes carried on conjugative plasmids (R-factors) may spread rapidly within a bacterial population and may transfer to other species. Plasmid-borne AMR is frequently encountered in bacteria pathogenic to fish and shellfish. Emergence of a resistant strain at a fish site renders particular antimicrobials useless, and the resistance can easily spread until it is the norm for that species. Antimicrobials do not *cause* the genetic and biochemical changes that make a bacterium resistant, but they do select for strains carrying the genetic information that confers resistance. The more an antimicrobial is used, the greater the selection pressure for resistance to evolve. If an antimicrobial agent is withdrawn from use, the incidence of strains resistant to it sometimes declines, because the resistant bacteria now have no advantage and the additional

STEMMING THE TIDE OF ANTIMICROBIAL RESISTANCE (AMR) IN AQUACULTURE

Data for the quantity of antimicrobials used in aquaculture is scarce, because of unregulated use in many countries, with large quantities used prophylactically with little regard for sensitivity testing or proper withdrawal periods. This has been especially problematic during the rapid growth of shrimp and salmon aquaculture; for example, 700 g of antimicrobials are used to produce one tonne of shrimp in Vietnam and one tonne of salmon in Chile requires 1500 g. By contrast, only 4.8 g of antimicrobials are required to produce the same amount in Norway (Cabello et al., 2016). Aquaculture leads to dynamic hotspots for generation of resistance genes and it is essential that we analyze the transfer of AMR between farmed fish, the microbial community, and the environment as our reliance on aquaculture grows (Watts et al., 2017). Santos and Ramos (2018) argue that urgent global action is required through the "One Health" approach—integrating human, animal, and environmental health—to prevent a major crisis in the treatment of bacterial infectious diseases in both human and veterinary populations

Table 14.6 *Examples of the biochemical basis of acquired bacterial resistance to antimicrobials used in aquaculture*

Strategy	Example	Mechanism
Modification of the target binding site	Penicillins	Altered penicillin-binding membrane proteins
	Quinolones	Altered DNA gyrase
Enzymic degradation	Penicillins	β-lactamase production
Reduced uptake or accumulation	Tetracyclines	Altered membrane transport proteins → active efflux
Metabolic bypass	Sulfonamides	Hyperproduction of substrate (*p*-aminobenzoic acid)

Figure 14.3 Representation of pathways of selection and transfer of antimicrobial resistance (AMR) genes in an aquaculture system. Sediments provide an interface for microbial interactions, leading to transfer of AMR genes by horizontal gene transfer (HGT).

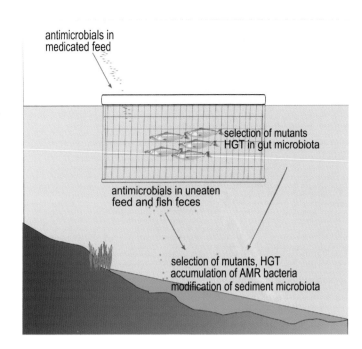

burden of extra genetic information makes them less competitive. However, this is not always the case, as plasmids frequently confer resistance to several antibiotics. AMR causes considerable problems in aquaculture. For example, at times between 20 and 30% of cases of furunculosis in Scottish salmon culture have been due to strains of *Aeromonas salmonicida* that are resistant to three or more antimicrobials.

Besides the obvious economic losses caused by inefficiency in disease control, there is great concern about the risks of antimicrobial usage in aquaculture to human health and environmental quality. Several studies have shown a buildup of resistant strains in sediments underneath sea cages in sites with poor water exchange, due largely to the accumulation of uneaten food and excretion of antimicrobials in the fish feces (*Figure 14.3*). Transfer of resistance genes to marine bacteria is known to occur, and AMR bacteria have been isolated from fish that have escaped from facilities where these agents are used excessively. Experimental studies have shown transfer to human pathogens and related bacteria, and this raises concerns about risks of transfer of resistance genes into the gut microbiota of consumers or fish-farm workers. Antibiotic residues could also prove toxic or cause allergies in some people who eat the fish. These concerns about aquaculture are part of a general awareness of the folly of indiscriminate use of antimicrobial agents in medicine, agriculture, and everyday products, with AMR emerging as a major global threat.

Vaccination of finfish is widely used in aquaculture

Teleost fish possess an efficient immune response and respond to the administration of microbial antigens by the production of antibodies (B-cell response) and cell-mediated immunity (T-cell response). The most common method of administering vaccines to small fish (up to about 15 g) is via brief immersion in a dilute suspension. Particulate antigens (such as bacteria) probably stimulate immunity after passage across the gills. Intraperitoneal injection is necessary for reliable protection with some vaccines, especially viral vaccines and bacterial vaccines that contain soluble components (e.g. most furunculosis vaccines). Injection vaccines are usually administered with an oil adjuvant, which ensures slow release of the antigen and a heightened immune response. Injection vaccines have high efficacy but have some drawbacks. Despite devices to convey fish from the water onto an injection table and the use of repeater syringes (*Figure 14.4*), injection vaccines incur high labor costs and they cannot be used on fish weighing less than about 15 g. Unless carefully managed, the stress associated with crowding, removal from the water, and injection causes mortalities and may even precipitate infection. The most desirable form of vaccine is one that can be administered orally. The main difficulty with this approach is

Figure 14.4 Vaccination of salmon. A. Fish are pumped from holding tanks and anesthetized before intraperitoneal (ip) injection with a bacterin suspension, using a repeating syringe. B. close up of ip injection. Credit: Kathy Taylor, Aqualife Services Ltd.

that microbial antigens are degraded in the fish's stomach and fore-gut before reaching the gut-associated lymphoid tissue in the hindgut, where the immune response occurs. This is overcome by microencapsulation of the vaccine in biodegradable polymers such as poly DL-lactide-co-glycolide. Many commercial vaccines using this, or similar, approaches are now available.

The simplest type of bacterial vaccine is a bacterin, which consists of a dense culture of bacterial cells killed by formalin treatment. Although technically simple, careful attention must be given to the quality control of media composition and incubation conditions, to ensure that the bacterin contains the appropriate protective antigens. One of the earliest successes in vaccine development was the *Vibrio anguillarum* vaccine effective against vibriosis in salmon. The protective antigen in this case seems to be lipopolysaccharide in the bacterial outer membrane. Commercial vaccines usually incorporate two or more serotypes to allow for antigenic variation and they often work well in a range of situations. Development of effective bacterins for *Vibrio* [now *Aliivibrio*] *salmonicida* and *Vibrio ordalii* also proved relatively straightforward, but the early success with the vibrios was not repeated with other diseases. For example, the breakthrough in development of an effective, long-lasting furunculous vaccine against *Aeromonas salmonicida* was only achieved after recognition of the crucial role of extracellular proteases and iron-regulated outer membrane proteins (see p.312), and manipulation of the culture and formulation conditions to ensure the correct blend of particulate and soluble antigens.

Inactivated vaccines can also be produced to protect against viral diseases. Viruses can only be propagated in cell culture and the cost of production is very high, although some commercial vaccines have been developed against infectious salmon anemia virus (ISAV), infectious pancreatic necrosis virus (IPNV), and salmon pancreas disease virus (SPDV). To ensure good protection, careful attention must be paid to the presence of specific antigens and it may be necessary to administer at a specific life stage of the fish.

Recombinant DNA technology is used to produce vaccines for diseases caused by viruses and some bacteria

Viruses can only be propagated in cell culture and some bacterial pathogens, notably *Renibacterium salmoninarum* and *Piscirickettsia salmonis*, are slow-growing and difficult to culture. The cost of production of inactivated vaccines using cell culture is prohibitively expensive, so attention turned to the use of recombinant DNA technology. Genes important in virulence—such as viral capsid proteins, bacterial toxins, or surface antigens—can be cloned and expressed in a recombinant host to produce a subunit vaccine. The most common method of achieving this is to fuse the gene of interest to a gene for β-galactosidase or maltose-binding protein, leading to production of a fusion or hybrid protein in a high-expression system such as *E. coli*. The fusion protein is produced in large amounts, sometimes up to 30% of total protein. If desired, the carrier protein can be removed and the cloned microbial antigen can be purified using chromatography, but in practice this is often not necessary as crude cell lysates make efficient vaccines, thus reducing processing costs. Sometimes, yeast or insect cell cultures (using a baculovirus vector) are used, especially if glycosylation of the protein is required. Additional genes for T-cell epitopes (e.g. from tetanus toxin or measles virus) may also be introduced to enhance the immunostimulatory potential of a vaccine by stimulating function of the T-cells in the immune system. This approach has been used to generate vaccines against several bacterial diseases, including furunculosis (based on the *A. salmonicida* serine protease), piscirickettsiosis (*P. salmonis* OspA membrane protein), and bacterial kidney disease (BKD; *R. salmoninarum* hemolysin and metallo-protease). Subunit viral vaccines against ISAV and IPNV are based on capsid proteins.

Live attenuated vaccines are effective but not widely used

Live vaccines, in which the virulence of the pathogen is attenuated, are very effective because the bacterium or virus replicates within the host and delivers antigens over a prolonged period. They are also better at stimulating mucosal and cell-mediated immunity without the need for adjuvants, and more suitable for oral delivery. Many human viral vaccines are based on this principle, although live bacterial vaccines have been less favored because their more complex genomes lead to the possibility of incomplete attenuation or subsequent reversion to virulence. Recombinant DNA technology allows the deletion and replacement of specific genes necessary for virulence and survival *in vivo* in a more controlled and targeted approach to attenuation. Such an approach has been used to construct a live *A. salmonicida* vaccine by deletion of the *aroA* gene, which encodes an essential amino acid biosynthesis pathway, not present in animals. Allelic replacement of this gene and subsequent further attenuation guards against the possibility of reversion. In trials, this vaccine was highly effective, and the vaccine strain could be engineered to deliver other antigens. A similar approach was used for *P. damselae* subsp. *piscicida* by removing a siderophore gene and for *S. iniae* by removing the gene for a surface virulence protein. Live vaccines for IPN have been successful in freshwater salmonid culture, although they have not proved effective against marine birnaviruses. Even with good evidence of protection and no evidence of reversion, live vaccines have not found favor with licensing authorities for use in fish, largely prompted by concerns about deliberate release into the environment. One exception is the use in Chile of a commercial live vaccine against BKD in Atlantic salmon, based on administration of *Arthrobacter davidanieli*. This bacterium is closely related to *R. salmoninarum* but is non-pathogenic. Good cross-protection has been reported.

DNA vaccination depends on fish cells expressing a protective antigen

Unlike conventional vaccination, which depends on the administration of an antigen, DNA vaccination, also known as genetic immunization, is based on the delivery of naked DNA containing the sequence for a region (epitope) of the protective antigen. The DNA containing this sequence is incorporated into a plasmid with appropriate promoters; if this is done correctly, the gene will be expressed in the fish host cells. Thus, the fish makes foreign antigens internally and then mounts an immune response to them. An alternative strategy is based on the expression of an antibody fragment inside the cell that can bind to and inactivate the pathogen. Fish cells efficiently express foreign proteins encoded by eukaryotic expression vectors. The first experiments using this technology were carried out using glycoprotein genes of the rhabdoviruses causing viral hemorrhagic septicemia (VHS) and infectious hematopoietic necrosis (IHN). These were highly successful, and large-scale trials have confirmed their efficacy in some, but not all, fish species tested. A DNA vaccine against salmon pancreas disease virus (SPDV, alphavirus) has also recently been approved for use in aquaculture. However, DNA vaccines for other fish viruses such IPNV, ISAV, and halibut nodavirus) have produced mixed results, and refinement of the vector plasmids and delivery system are required. In experimental trials, many of these vaccines have been effective after intramuscular injection. An alternative method of gene delivery into the skin or muscle is to use a "gene gun," which fires tiny gold particles coated with the DNA into the tissue. Issues of longevity of protection, stimulation of the correct immune responses, and safety need further investigation.

ARE DNA VACCINES THE KEY TO FUTURE SUCCESS?

Despite hundreds of experimental DNA vaccines, only four have been licensed for veterinary use to date. This is partly because some have given limited protection, but also because public opinion and licensing authorities are wary of the safety of genetically modified organisms (GMOs) in the environment or in food. However, the DNA is not introduced into the germ line and the plasmids and the vaccinated host are not considered to be GMOs (Collins et al., 2019). DNA vaccines are thought by most experts to have very low risks and with better understanding and acceptance they hold great potential in aquaculture, especially for control of diseases caused by viruses, eukaryotic parasites, and hard-to-culture bacteria. With advances in genomics, transcriptomics, and proteomics, it will become increasingly possible to better identify the key correlates of protective host responses for targeted vaccine design (Dalmo, 2018).

Probiotics, prebiotics, and immunostimulants are widely used in aquaculture

Interest in this area has occurred largely because of the need to reduce antibiotic usage due to concerns over antimicrobial resistance and residues. Probiotics have been very successful in other branches of agriculture such as poultry and pig farming and commercial production is now of considerable economic importance to the animal feed industry. Numerous studies in various species of finfish have shown that addition of certain bacteria to feeds leads to disease resistance or other benefits. In general, probiotic strains have been isolated from naturally occurring microbiota of fish. They are distinguished from vaccines because they do not require participation of the host immune system and may act directly by inhibition of pathogens via production of antimicrobial metabolites or bacteriocins, or by competition for nutrients or sites for colonization of the gut mucosa. Besides direct effects on pathogens or enhancement of nutrition, some probiotic products (in the broadest sense of the term) have immunostimulatory effects. A range of preparations incorporating lipopolysaccharides, glucans, and other microbial components have been shown to enhance host defenses. Whilst many studies lack scientific rigor and proof of the mechanisms of disease prevention by specific challenge, evidence of their efficacy is provided by major reductions in antibiotic usage. Some statistically rigorous studies have shown that probiotics improve the survival and growth of larvae and lessen the survival of pathogenic bacteria in the gut.

Many probiotic agents show antagonistic activity against pathogens *in vitro*, but this is no guarantee that they have this effect *in vivo*. Most antagonistic compounds are secondary metabolites produced during the stationary phase in culture, and it is unknown whether they are produced in significant amounts *in vivo*. Electron microscopy has been used to show how probiotic bacteria may compete with pathogens for colonization sites. Some probiotics may produce beneficial dietary compounds such as vitamins, antioxidants, and lipids or may provide their host with digestive enzymes, enabling the degradation of complex nutrients such as chitin, starch, or cellulose. These factors are becoming very important in the development of sustainable aquaculture, as there is a need to use alternative nutrient sources to replace the industry's reliance on fishmeal.

The gastrointestinal tract of fish and shellfish larvae becomes colonized by bacteria present in the water; therefore, control of water quality in the early larval stages is very important to prevent colonization by pathogens. Agents may be added to the water, incorporated into pelleted feeds, or used to enrich live feed such as rotifers or *Artemia*. A variety of commercial products incorporating dried endospores of *Bacillus* spp. have proved effective in limiting disease. These are cheap to manufacture, stable, and easy to administer. Other agents, such as lactic acid bacteria, *Roseobacter* spp., *Pseudoalteromonas*, *Pseudomonas*, nonpathogenic *Vibrio* spp., yeasts, and microalgae are effective, but these are more difficult to distribute commercially as live cultures. There does seem to be a specific indigenous intestinal microbiome in marine fish, composed mainly of Gram-negative, aerobic, anaerobic, and facultatively anaerobic bacteria. The most common genera isolated by culture from the gut of marine fish are *Vibrio*, *Photobacterium*, *Pseudomonas*, and *Acinetobacter*. Recent application of molecular methods shows that various other organisms are important and that the microbiome is very variable dependent on genetic, nutritional, and environmental factors.

There is a highly developed industry supplying a great range of products promoted as immunostimulants for use in shrimp, prawn, and lobster culture. Some products are promoted as vaccines, although this is usually considered an incorrect use of the term since it should only be applied to an agent that induces protection through long-lasting immune memory that is dependent on primary challenge with an antigen and the stimulation of clones of specific lymphocytes. Invertebrates lack this system, although crustaceans do possess active cells such as hemocytes that are responsible for inflammatory-type reactions, phagocytosis, and the killing of microbes via oxidative burst or microbicidal proteins. Immune stimulation can occur by enhancing nonspecific complement activation, phagocytosis, or cytokine production. A range of treatments, including live or killed bacteria, glucans, lipopolysaccharide, extracts of yeast, and algal cell walls, have been used, and promising effects have been reported.

WHAT ARE PROBIOTICS AND PREBIOTICS?

The word "probiotic" is derived from the Latin "for life." In its most common usage, probiotics are defined as live microbial feed supplements that stimulate health by inducing beneficial changes in the intestinal microbiota. Merrifield et al. (2010) propose a wider definition in aquaculture, stating that a probiotic effect may be based on direct benefits to the animal host (e.g. immunostimulation, production of inhibitors against pathogens, reduced stress response, and improved gastrointestinal morphology) as well as benefits to the fish farmer or consumer (such as improved fish appetite, growth, or quality). Prebiotics are dietary components that stimulate the growth of beneficial bacteria. A range of carbohydrates, such as inulin, and complex oligosaccharides are used and many studies show health benefits (Ringo et al., 2010).

BOX 14.1   RESEARCH FOCUS

"The enemy of my enemy is my friend"—phage therapy for marine diseases

Rediscovery of an old idea. Soon after phages were discovered in 1915 (p.212), it was proposed that they might be useful for treating bacterial infections of humans (reviewed by Wittebole et al., 2014). Over many years, phage therapy (PT) research institutes in Tbilisi and Wroclaw developed methods for isolating and propagating phages and using them to treat a variety of human infections. Phages were also employed for the treatment of cholera in India between 1925 and 1934, when vibriophages were administered at the start of cholera outbreaks and released into community drinking water for prophylaxis, resulting in significant reduction in death rates (Nelson et al., 2009) and there is now interest in reviving this approach (Bhandare et al., 2018). In the 1930s, pharmaceutical companies in the USA began to produce phages commercially, but there were problems with standardization and poor design of clinical trials. Furthermore, the discovery and rapid success of antibiotics meant that the idea of PT was quickly forgotten in Western countries in the 1940s, although it continued to be used successfully in Georgia, Russia, and Poland. In the 1980s, the problems of antimicrobial resistance (AMR) prompted scientists to reassess the potential of PT and carefully designed trials have proved that it can be very successful in treatment of bacterial infections of mammals. Although numerous US companies are developing PT and there is increased openness towards PT among Western medical professionals, much more research and robust clinical trials are needed before it will be widely accepted for human use (Górski et al., 2018). In contrast, there are many encouraging developments in the use of PT in agriculture and aquaculture (Gon Choudhury et al., 2017), although the lack of standardized research approaches and inconsistent reporting make it difficult to compare results from various studies (Richards, 2014).

Advantages and disadvantages of phages as therapeutic and biocontrol agents. Phages usually have a narrow host range directed against specific pathogens, and they should therefore not affect the normal host microbiota. They are self-perpetuating in the presence of susceptible bacteria, so repeated administration is not necessary. Also, phages coevolve with their host bacteria, so even if a bacterial strain acquires resistance, it should be easy to isolate new phages that are infective. There is also some evidence that accumulated phage resistance may lead to dominance of less virulent phenotypes (Kalatzis et al., 2018). However, this variability means that if the exact strain of bacterium causing an infection is unknown, it is necessary to inoculate with a "cocktail" of phages. Additional problems can occur because animals may mount an immune response to the phage proteins and the phage virions may not reach all parts of the body. Precautions are also needed to avoid the use of temperate phages, which could raise the potential of transfer of virulence or antibiotic resistance genes (Flegel et al., 2005).

PT in aquaculture of finfish. The first documented use of PT was by Nakai et al. (1999) to control a serious opportunistic disease problem caused by *Lactococcus garvieae* affecting yellowtail (*Seriola quinqueradiata*). They found that over 90% of *L. garviae*

strains were sensitive to two phages. In an experimental infection, 100% of fish survived intraperitoneal injection with *L. garviae* when phage was injected at the same time, whereas the survival rate was only 10% in fish injected with the bacteria alone. Partial (50%) protection was observed if 24 h elapsed between bacterial infection and injection of the phage. Imbeault et al. (2006) showed that addition of phages to the water delayed the onset of furunculosis caused by *Aer. salmonicida* in brook trout (*Salvelinus fontinalis*). However, Verner-Jeffreys et al., 2007) showed that Atlantic salmon (*Salmo salar*) injected with phage immediately after injection of *Aer. salmonicida* died more slowly, but eventual mortality was the same as the control treatment and there was some evidence of evolution of resistance to specific phage strains. These authors concluded that PT may not be suitable for such a highly virulent pathogen, although the injection challenge used in this study is not representative of the natural infection processes. There have been several promising studies showing the potential of phages for treatment or prevention of *Vibrio* diseases (Kalatzis et al., 2018). Higuera et al. (2013) showed that juvenile Atlantic salmon were protected from experimental infection with *V. anguillarum* in the laboratory and in fish farm conditions when phages were added to the tank water. The most promising application is probably the addition of the phages to the water to prevent infection during larval rearing, when vaccines would be ineffective because the immune system is not fully developed. Phages can also exclude the pathogenic bacteria from the water without harming the beneficial microbiota and remain in the water for a week or more, providing a prophylactic measure (Higuera et al., 2013). Rørbo et al. (2018) showed that a broad-host-range vibriophage reduced mortality of cod and turbot larvae challenged experimentally with four different *V. anguillarum* strains. For fry and juvenile fish, continuous delivery of phages via coated feed pellets would provide an easy method of routine administration. Christiansen et al. (2014) showed effective uptake of *Flavobacterium psychrophila* phages after oral administration to rainbow trout (*Oncorhynchus mykiss*). Huang and Nitin (2019) recently developed an edible protein-based biopolymer coating for feed pellets, which proved to be a cost-effective delivery method for phages.

PT in aquaculture of crustaceans. The need to prevent catastrophic infections by *Vibrio* spp., which are the primary opportunistic pathogens of shrimps and prawns, could make the effort of maintaining collections of different phage cocktails commercially worthwhile. Although there are a number of anecdotal reports since the 1990s, there are few published reports of rigorously conducted field trials in scientific journals. In India, Vinod et al. (2006) isolated several lytic phages of *V. harveyi* from farm or hatchery water used to rear *Penaeus monodon* shrimp. In laboratory microcosm experiments, addition of phages to water produced a reduction in the bacterial counts from 10^6 to 10^3 CFU mL^{-1} and 80% of larvae survived, compared with 25% in the control. In the same laboratory, Karunasagar et al. (2007) showed a dose-dependent effect on the formation of *V. harveyi* in tanks. In Australia,

BOX 14.1   RESEARCH FOCUS

Crothers-Stomps et al. (2010) isolated a collection of phages from two virus families with lytic activity against *V. harveyi* and showed their ability to control vibriosis in culture of larvae from the tropical rock lobster *Panulirus ornatus*. Lomelí-Ortega and Martínez-Díaz (2014) achieved successful control of experimental *V. parahaemolyticus* infection of whiteleg shrimp *Litopenaeus vannamei* larvae.

Could phages be used to control coral disease? The threat to coral reefs from infectious diseases, especially those caused by bacteria, was discussed in *Chapter 11*. Control of disease by large-scale use of antibiotics is impractical and the lack of an antibody response in corals means that vaccination is not possible—although Teplitski and Ritchie (2009) suggest that it may be possible to prime the immune defenses of certain corals. Eugene Rosenberg and coworkers at the University of Tel Aviv, Israel, have pioneered the concept of PT for control of bacterial coral diseases. In aquarium experiments, Efrony et al. (2007) isolated phages of the pathogen *Thallassomonas loyaena* from the Red Sea and showed that addition of one type of phage at a density of 10^4–10^6 PFU mL^{-1} protected the coral *Favia favus* against infection caused by incubation in water taken from an aquarium containing diseased coral. The phages increased in number, and levels remained high even after water in the aquarium was changed, indicating that the phage replicates and remains associated with coral tissue. The *T. loyaena* BA3 phage was characterized and shown to prevent transmission of the disease to healthy corals (Efrony et al., 2009). Similarly, the coral *Pocillopora damicornis* was protected against infection and reinfection by *V. coralliilyticus* using phages specific for this bacterium (Efrony et al., 2007). Some coral biologists argue that the proposal to treat reefs in this way is impractical and environmentally unsound. Thus, although additional laboratory experiments have shown that phage infection have the potential to control bacterial infections in laboratory-reared corals (Cohen et al., 2013; Jacquemot et al., 2018), there appears to be only one example of environmental application of PT for corals. A small-scale pilot field study on a threatened reef infected in the Red Sea gave promising results (Atad et al., 2012). Treatment of diseased *F. fava* corals with phage BA3 inhibited transmission of *T. loyaena* infection to nearby corals—only 5% of the healthy corals became infected when placed near phage-treated diseased corals, whereas 61% of healthy corals were infected in the no-phage control.

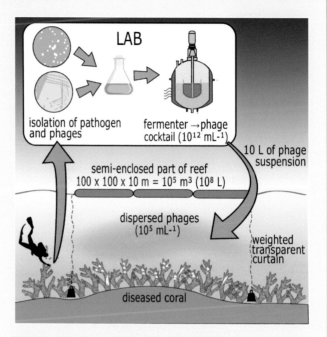

Figure 14.5 Representation of hypothetical phage treatment of diseased reef (see text for details).

Would it really be technically feasible to treat a disease outbreak to stop it spreading to surrounding parts of the reef? *Figure 14.5* shows that to treat an area of diseased coral of 10^4 m^2 in water 10 m deep (a water volume of 10^5 m^3) with an effective level of phage of ~10^5 mL^{-1}—this is 100 times the protective level in the Atad et al. trial—would require only 10 L of phage lysate, an amount easily obtainable from a laboratory fermenter. In the USA, Kellogg (2007) suggested that prospects are good for using phages to treat diseased corals in the seriously impacted reefs of the Florida Keys, although this idea has not been implemented. Whilst generally supportive of the concept, Teplitski and Ritchie (2009) argue that a better understanding of the role of vibrios as commensal bacteria in corals is needed before PT can be used in the field, especially the effect of phages on other members of the coral holobiont. In a situation in which some reefs are in critical danger of being lost forever (see *Chapter 11*) because of bleaching and disease, further field research is urgently needed to decide whether PT in corals is a feasible and worthwhile approach.

Conclusions

The untapped hidden treasure available from microbes in the sea is immense. Advances in molecular biology, especially high-throughput sequencing, powerful bioinformatics tools, and functional screening methods mean that many new potential products and processes for industry and medicine will be discovered in the coming years. Despite the great opportunities for biotechnology, it is disappointing that only a tiny fraction of the discoveries has been successfully exploited. Promising research leads from laboratory studies are often not fully exploited because of the high costs of developing new products and processes and the risk-averse nature of many commercial enterprises and government authorities. Furthermore, most of the successful developments have involved isolation of marine microbes and exploitation of their products and processes in land-based industries. Large-scale direct application of marine microbial biotechnology in the ocean setting should play a major role in ensuring new sources of energy and sustainable aquaculture to feed the world's growing population, and it could help to mitigate some of the environmental damage caused by human activities. We will face important economic. ethical and environmental issues to ensure that the exploitation of these discoveries can be properly harnessed for the benefit of humanity and our planet.

References and further reading

Enzymes

Beygmoradi, A. & Homaei, A. (2017) Marine microbes as a valuable resource for brand new industrial biocatalysts. *Biocatal. Agric. Biotechnol.* **11**: 131–152.

Bonugli-Santos, R.C., dos Santos Vasconcelos, M.R., Passarini, M.R.Z., et al. (2015) Marine-derived fungi: Diversity of enzymes and biotechnological applications. *Front. Microbiol.* **6**: 269.

de Freitas, R.C., Marques, H.I.F., da Silva, M.A.C., et al. (2019) Evidence of selective pressure in whale fall microbiome proteins and its potential application to industry. Mar. Genom. **45**: 201–227.

Goffredi, S.K., Johnson, S.B., & Vrijenhoek, R.C. (2007) Genetic diversity and potential function of microbial symbionts associated with newly discovered species of *Osedax* polychaete worms. *Appl. Environ. Microbiol.* **73**: 2314–2323.

Goffredi, S.K., Orphan, V.J., Rouse, G.W., et al. (2005) Evolutionary innovation: A bone-eating marine symbiosis. *Environ. Microbiol.* **7**: 1369–1378.

Miyamoto, N., Yoshida, M., Koga, H., & Fujiwara, Y. (2017) Genetic mechanisms of bone digestion and nutrient absorption in the bone-eating worm *Osedax japonicus* inferred from transcriptome and gene expression analyses. *BMC Evol. Biol.* **17**: 17.

Parte, S., Sirisha, V.L., & D'Souza, J.S. (2017) Biotechnological applications of marine enzymes from algae, bacteria, fungi, and sponges. *Adv. Food Nutr. Res.* **80**: 75–106.

Popovic, A., Tchigvintsev, A., Tran, H., et al. (2015) Metagenomics as a tool for enzyme discovery: Hydrolytic enzymes from marine-related metagenomes. In: *Prokaryotic Systems Biology*. Springer, Cham, pp. 1–20.

Smith, C.R., Glover, A.G., Treude, T., et al. (2015) Whale-fall ecosystems: Recent insights into ecology, paleoecology, and evolution. *Ann. Rev. Mar. Sci.* **7**: 571–596.

Biofuels and edible oils

Beacham, T.A., Sweet, J.B., & Allen, M.J. (2017) Large scale cultivation of genetically modified microalgae: A new era for environmental risk assessment. *Algal Res.* **25**: 90–100.

Gupta, A., Barrow, C.J., & Puri, M. (2012) Omega-3 biotechnology: Thraustochytrids as a novel source of omega-3 oils. *Biotechnol. Adv.* **30**: 1733–1745.

Lian, J., Wijffels, R.H., Smidt, H., & Sipkema, D. (2018) The effect of the algal microbiome on industrial production of microalgae. *Microb. Biotechnol.* **11**: 806–818.

Milano, J., Ong, C., Masjuki, H.H., et al. (2016) Microalgae biofuels as an alternative to fossil fuel for power generation. *Renew. Sust. Energy Rev.* **58**: 180–197.

Shuba, E.S. & Kifle, D. (2018) Microalgae to biofuels: 'promising' alternative and renewable energy, review. *Renew. Sustain. Energy Rev.* **81**: 743–755.

Vinayak, V., Manoylov, K.M., Gateau, H., et al. (2015) Diatom milking: A review and new approaches. *Mar. Drugs* **13**: 2629–2665.

Xue, Z., Yu, W., Liu, J., et al. (2018) Edible oil production from microalgae: A review. *Eur. J. Lipid Sci. Technol.* **120**: 1700428.

Bioactive compounds and pharmaceuticals

Bibi, F., Faheem, M., Azhar, E.I., et al. (2017) Bacteria from marine sponges: A source of new drugs. *Curr. Drug Metab.* **18**: 11–15.

Dias, D.A., Urban, S., & Roessner, U. (2012) A historical overview of natural products in drug discovery. *Metabolites* **2**: 303–336.

Gerwick, W.H. & Fenner, A.M. (2013) Drug discovery from marine microbes. *Microb. Ecol.* **65**: 800–806.

ul Hassan, S.S., Anjum, K., Abbas, S.Q., et al. (2017) Emerging biopharmaceuticals from marine actinobacteria. *Environ. Toxicol. Pharmacol.* **49**: 34–47.

Mans, D.R. (2016) Exploring the global animal biodiversity in the search for new drugs–marine invertebrates. *J. Transl. Sci.* **2**: 170–179.

Nanotechnology and bioelectronics

Ali, I., Peng, C., Khan, Z.M., & Naz, I. (2017) Yield cultivation of magnetotactic bacteria and magnetosomes: A review. *J. Basic Microbiol.* **57**: 643–652.

Charrier, M., Li, D., Mann, V.R., et al. (2019) Engineering the S-Layer of *Caulobacter crescentus* as a foundation for stable, high-density, 2D living materials. *ACS Synth. Biol.* **8**: 181–190.

Farjadian, F., Moghoofei, M., Mirkiani, S., et al. (2018) Bacterial components as naturally inspired nano-carriers for drug/gene delivery and immunization: Set the bugs to work? *Biotechnol. Adv.* **36**: 968–985.

Gordon, R., Losic, D., Tiffany, M.A., et al. (2009) The glass menagerie: Diatoms for novel applications in nanotechnology. *Trends Biotechnol.* **27**: 116–127.

Li, Y.-T., Tian, Y., Tian, H., et al. (2018) A review on bacteriorhodopsin-based bioelectronic devices. *Sensors* **18**: 1368.

Mishra, M., Arukha, A.P., Bashir, T., et al. (2017) All new faces of diatoms: Potential source of nanomaterials and beyond. *Front. Microbiol.* **8**: 1239.

Patil, M.P. & Kim, G.-D. (2018) Marine microorganisms for synthesis of metallic nanoparticles and their biomedical applications. *Colloids Surf. B: Biointerfaces* **172**: 487–495.

Ragni, R., Cicco, S.R., Vona, D., & Farinola, G.M. (2018) Multiple routes to smart nanostructured materials from diatom microalgae: A chemical perspective. *Adv. Mater.* **30**: 1704289.

Saeedi, P., Moosaabadi, J.M., Sebtahmadi, S.S., et al. (2012) Potential applications of bacteriorhodopsin mutants. *Bioengineered* **3**: 326–328.

Skeffington, A.W. & Scheffel, A. (2018) Exploiting algal mineralization for nanotechnology: Bringing coccoliths to the fore. *Curr. Opin. Biotechnol.* **49**: 57–63.

Sleytr, U.B., Schuster, B., Egelseer, E.-M., & Pum, D. (2014) S-layers: Principles and applications. *FEMS Microbiol. Rev.* **38**: 823–864.

Vala, A.K. (2014) Exploration on green synthesis of gold nanoparticles by a marine-derived fungus *Aspergillus sydowii*. **34**: 194–197.

Vargas, G., Cypriano, J., Correa, T., et al. (2018) Applications of magneto-tactic bacteria, magnetosomes & magnetosome crystals in biotechnology and nanotechnology: Mini-review. *Molecules* **23**: 2438.

Aquaculture vaccines, probiotics and antimicrobial resistance

Adams, A. & Thompson, K.D. Biotechnology offers revolution to fish health management. *Trends Biotechnol.* **24**: 201–205.

Cabello, F.C., Godfrey, H.P., Buschmann, A.H., & Dölz, H.J. (2016) Aquaculture as yet another environmental gateway to the development and globalisation of antimicrobial resistance. *Lancet Infect. Dis.* **16**: e127–133.

Charoonnart, P., Purton, S., Saksmerprome, V., et al. (2018) Applications of microalgal biotechnology for disease control in aquaculture. *Biology (Basel)* **7**: 24.

Collins, C., Lorenzen, N., & Collet, B. (2019) DNA vaccination for finfish aquaculture. *Fish Shellfish Immunol.* **85**: 106–125.

Dadar, M., Dhama, K., Vakharia, V.N., et al. (2017) Advances in aquaculture vaccines against fish pathogens: Global status and current trends. *Rev. Fish. Sci. Aquac.* **25**: 184–217.

Dalmo, R.A. (2018) DNA vaccines for fish: Review and perspectives on correlates of protection. *J. Fish Dis.* **41**: 1–9.

Dittmann, K.K., Rasmussen, B.B., Castex, M., et al. (2017) The aquaculture microbiome at the centre of business creation. *Microb. Biotechnol.* **10**: 1279–1282.

Hølvold, L.B., Myhr, A.I., & Dalmo, R.A. (2014) Strategies and hurdles using DNA vaccines to fish. *Vet. Res.* **45**: 21.

Merrifield, D.L. & Carnevali, O. (2014) Probiotic modulation of the gut microbiota of fish. In: *Aquaculture Nutrition*. John Wiley & Sons, Ltd., Chichester, UK, pp. 185–222.

Merrifield, D.L., Dimitroglou, A., Foey, A., et al. (2010) The current status and future focus of probiotic and prebiotic applications for salmonids. *Aquaculture* **302**: 1–18.

Munang'andu, H. (2018) Intracellular bacterial infections: A challenge for developing cellular mediated immunity vaccines for farmed fish. *Microorganisms* **6**: 33.

Nayak, S.K. (2010) Probiotics and immunity: A fish perspective. *Fish Shellfish Immunol.* **29**: 2–14.

Pérez-Sánchez, T., Mora-Sánchez, B., & Balcázar, J.L. (2018) Biological approaches for disease control in aquaculture: Advantages, limitations and challenges. *Trends Microbiol.* **26**: 896–903.

Ringø, E., Olsen, R.E., Gifstad, T.Ï., et al. (2010) Prebiotics in aquaculture: A review. *Aquac. Nutr.* **16**: 117–136.

Rodiles, A., Rawling, M.D., Peggs, D.L., et al. (2018) Probiotic applications for finfish aquaculture. In: *Probiotics and Prebiotics in Animal Health and Food Safety*. Springer, Cham, pp. 197–217.

Salonius, K., Siderakis, C., MacKinnon, A.M., & Griffiths, S.G. (2005) Use of *Arthrobacter davidanieli* as a live vaccine against *Renibacterium salmoninarum* and *Piscirickettsia salmonis* in salmonids. *Dev. Biol.* **121**: 189–197.

Santos, L. & Ramos, F. (2018) Antimicrobial resistance in aquaculture: Current knowledge and alternatives to tackle the problem. *Int. J. Antimicrob. Agents* **52**: 135–143.

Sayes, C., Leyton, Y., & Riquelme, C. (2018) Probiotic bacteria as an healthy alternative for fish aquaculture. *Antibiotic Use in Animals*. doi: 10.5772/intechopen.71206.

Watts, J.E.M., Schreier, H.J., Lanska, L., & Hale, M.S. (2017) The rising tide of antimicrobial resistance in aquaculture: Sources, sinks and solutions. *Mar. Drugs* **15**: e158.

Phage therapy

Atad, I., Zvuloni, A., Loya, Y., & Rosenberg, E. (2012) Phage therapy of the white plague-like disease of *Favia favus* in the Red Sea. *Coral Reefs* **31**: 665–670.

Bhandare, S., Colom, J., Baig, A., et al. (2018) Reviving phage therapy for the treatment of cholera. *J. Infect. Dis.* **219**: 786–794.

Christiansen, R.H., Dalsgaard, I., Middelboe, M., et al. (2014) Detection and quantification of *Flavobacterium psychrophilum*-specific bacteriophages *in vivo* in rainbow trout upon oral administration: Implications for disease control in aquaculture. *Appl. Environ. Microbiol.* **80**: 7683–7693.

Cohen, Y., Pollock, F.J., Rosenberg, E., & Bourne, D.G. (2013) Phage therapy treatment of the coral pathogen *Vibrio coralliilyticus*. *Microbiologyopen* **2**: 64–74.

Crothers-Stomps, C., Høj L, Bourne, D.G., et al. (2010) Isolation of lytic bacteriophage against *Vibrio harveyi*. *J. Appl. Microbiol.* **108**: 1744–1750.

Efrony, R., Atad, I., & Rosenberg, E. (2009) Phage therapy of coral white plague disease: Properties of phage BA3. *Curr. Microbiol.* **58**: 139–145.

Efrony, R., Loya, Y., Bacharach, E., & Rosenberg, E. (2007) Phage therapy of coral disease. *Coral Reefs* **26**: 7–13.

Flegel, T.W., Pasharawipas, T., Owens, L., & Oakey, H.J. (2005) Evidence for phage-induced virulence in the shrimp pathogen *Vibrio harveyi*. *Dis. Asian Aquac.* **V**: 329–337.

Gon Choudhury, T., Tharabenahalli Nagaraju, V., et al. (2017) Advances in bacteriophage research for bacterial disease control in aquaculture. *Rev. Fish. Sci. Aquac.* **25**: 113–125.

Górski, A., Międzybrodzki, R., Łobocka, M., et al. (2018) Phage therapy: What have we learned? *Viruses* **10**: 288.

Higuera, G., Bastías, R., Tsertsvadze, G., et al. (2013) Recently discovered *Vibrio anguillarum* phages can protect against experimentally induced vibriosis in Atlantic salmon, *Salmo salar*. *Aquaculture* **392–395**: 128–133.

Huang, K. & Nitin, N. (2019) Edible bacteriophage based antimicrobial coating on fish feed for enhanced treatment of bacterial infections in aquaculture industry. *Aquaculture* **502**: 18–25.

Imbeault, S., Parent, S., Lagacé, M., et al. (2006) Using bacteriophages to prevent furunculosis caused by *Aeromonas salmonicida* in farmed brook trout. *J. Aquat. Anim. Health* **18**: 203–214.

Jacquemot, L., Bettarel, Y., Monjol, J., et al. (2018) Therapeutic potential of a new jumbo phage that infects *Vibrio coralliilyticus*, a widespread coral pathogen. *Front. Microbiol.* **9**: 2501.

Kalatzis, P., Castillo, D., Katharios, P., et al. (2018) Bacteriophage interactions with marine pathogenic vibrios: Implications for phage therapy. *Antibiotics* **7**: 15.

Karunasagar, I., Shivu, M.M., Girisha, S.K., et al. (2007) Biocontrol of pathogens in shrimp hatcheries using bacteriophages. *Aquaculture* **268**: 288–292.

Kellogg, C.A. (2007) Phage therapy for Florida corals? *U.S. Geological Surv. Fact Sheet* 2007–3065.

Lomelí-Ortega, C.O. & Martínez-Díaz, S.F. (2014) Phage therapy against *Vibrio parahaemolyticus* infection in the whiteleg shrimp (*Litopenaeus vannamei*) larvae. *Aquaculture* **434**: 208–211.

Nakai, T., Sugimoto, R., Park, K.H., et al. (1999) Protective effects of bacteriophage on experimental *Lactococcus garvieae* infection in yellowtail. *Dis. Aquat. Organ.* **37**: 33–41.

Nelson, E.J., Harris, J.B., Morris, J.G., et al. (2009) Cholera transmission: The host, pathogen and bacteriophage dynamic. *Nat. Rev. Microbiol.* **7**: 693–702.

Richards, G.P. (2014) Bacteriophage remediation of bacterial pathogens in aquaculture: A review of the technology. *Bacteriophage* **4**: e975540.

Rørbo, N., Rønneseth, A., Kalatzis, P.G., et al. (2018) Exploring the effect of phage therapy in preventing *Vibrio anguillarum* infections in cod and turbot larvae. *Antibiotics* **7**: 42.

Teplitski, M. & Ritchie, K. (2009) How feasible is the biological control of coral diseases? *Trends Ecol. Evol.* **24**: 378–385.

Verner-Jeffreys, D.W., Algoet, M., Pond, M.J., et al. (2007) Furunculosis in Atlantic salmon (*Salmo salar* L.) is not readily controllable by bacteriophage therapy. *Aquaculture* **270**: 475–484.

Vinod, M.G., Shivu, M.M., Umesha, K.R., et al. (2006) Isolation of *Vibrio harveyi* bacteriophage with a potential for biocontrol of luminous vibriosis in hatchery environments. *Aquaculture* **255**: 117–124.

Wittebole, X., De Roock, S., & Opal, S.M. (2014) A historical overview of bacteriophage therapy as an alternative to antibiotics for the treatment of bacterial pathogens. *Virulence* **5**: 226–235.

Other articles

Kumar, J., Singh, D., Tyagi, M.B., & Kumar, A. (2019) Cyanobacteria: Applications in biotechnology. In: *Cyanobacteria*, ed. by Mishra, A.K., Tiwari, D.N. and Rai, A.N., Academic Press, pp. 327–346,

Luna, G.M. (2015) Biotechnological potential of marine microbes. In: *Springer Handbook of Marine Biotechnology*. Springer, Berlin, Heidelberg, pp. 651–661.

Martín-Rodríguez, A.J., Babarro, J.M.F., Lahoz, F., et al. (2015) From broad-spectrum biocides to quorum sensing disruptors and mussel repellents: Antifouling profile of alkyl triphenylphosphonium salts. *PLoS One* **10**: e0123652.

Pessôa, M.G., Vespermann, K.A.C., Paulino, B.N., et al. (2019) Newly isolated microorganisms with potential application in biotechnology. *Biotechnol. Adv.* **37**: 319–339.

Rahman, P.K.S.M., Mayat, A., Harvey, J.G.H., et al. (2019) Biosurfactants and bioemulsifiers from marine algae. In, *The Role of Microalgae in Wastewater Treatment*. Springer, Singapore, pp. 169–188.

Romano, S. (2018) Ecology and biotechnological potential of bacteria belonging to the genus *Pseudovibrio*. *Appl. Environ. Microbiol.* **84**: e02516-17.

Satheesh, S., Ba-akdah, M.A., & Al-Sofyani, A.A. (2016) Natural antifouling compound production by microbes associated with marine macroorganisms—A review. *Electron. J. Biotechnol.* **21**: 26–35.

Chapter 15
Concluding Remarks

The overarching theme of the book has been our ever-expanding realization of the diversity of microbes and their adaptations to the different physical, chemical, and biological conditions in their world. We have explored exciting new discoveries about how microbial activities underpin the biogeochemical processes that shape the function of our oceans and our planet. Novel types of microbes, biochemical pathways, ecological interactions, and biogeochemical effects are being discovered at ever-increasing rates and we are beginning to understand how the microscale effects of microbial processes can affect global processes in the oceans. The oceans undoubtedly hold many more surprises yet to be revealed.

Technological advances have driven most of the advances in our knowledge—we can now explore the properties and activities of marine microbes at every level from single cells to global assemblages. The next few years will undoubtedly see further technological progress and will be dominated by increasing advances and application of metagenomics, transcriptomics, and proteomics, plus major advances in single-cell biology. In particular, we can expect to see improved methods for exploring the function of the archaea, protists, and fungi in ocean processes, knowledge of which has lagged behind that of bacteria. However, sequence data alone is not enough to gain a true understanding of microbial processes and community interactions. We have seen several examples of how the renaissance of culture-based microbiology—guided by 'omics and advanced imaging—is dispelling the dogma that most marine microbes are uncultivable. It is difficult, but definitely worth the effort. This will be especially important to gain a proper understanding of viral ecology—we now have a great wealth of metagenomic data on ocean viromes, but only know the hosts of these viruses in a fraction of cases. Advances in this approach will allow us to devise hypothesis-driven approaches to investigate microbial communities. Soon, further progress towards a real understanding of the ecology of marine microbes and their role in diverse activities will be enhanced by automated sampling, robotic processing in the field, and real-time remote sensing, coupled with artificial intelligence approaches to the analysis and coordination of data.

Marine microbiology research has also made tremendous contributions to general biological knowledge, especially to our understanding of evolution. The discovery of giant viruses and the ancient archaeal lineages is bringing us ever closer to understanding the origins of eukaryotes during the early evolution of life on our planet. The study of marine viruses has revolutionized our understanding of the importance of viral infection of cells on the creation and dissemination of new genetic information—a major driver in the generation of diversity and selection processes of evolution. The many wondrous examples of symbiotic partnerships between microbes and marine animals have revealed the mechanisms of coevolution over hundreds of millions of years. This field has also led to discoveries of cellular and molecular interactions between microbes and their hosts that have direct applications for the advancement of human health.

Our planet is in in trouble. Climate change, pollution, and overexploitation of finite resources threaten the future of humanity and many of the spcies with which we share our planet, but we need to remember that *microbial* life has adapted to enormous changes in atmospheric and ocean chemistry that have occurred periodically over the past 3 billion years. We can only make informed guesses about *how* microbes will adapt to the extreme changes over the short timescale of human influence on the planet. But we can be sure they *will* adapt—whatever we do, or don't do, in response to these changes—and we ignore their influence at our peril. The atmospheric concentration of CO_2 has now passed 400 ppm—the highest it has been since the mid Pliocene, ~3 MYA. Even if it were possible for humanity to succeed in limiting future CO_2 emissions immediately and to develop technological solutions to reduce the atmospheric levels, the most optimistic prediction is that it would take 100–200 years to reduce atmospheric CO_2 below 400 ppm and hold the temperature increase below 1.5°C in the long term. We know that global warming caused by increasing atmospheric concentrations of CO_2, N_2O, CH_4, and other anthropogenic pollutants is causing shifts in ocean chemistry, stratification, and currents—altering microbial community composition and the balance of biogeochemical cycles. Additional anthropogenic factors such as eutrophication and the effect of huge quantities of microplastics on the sinking rates of particles are also deeply troubling and make predictions of future trends even more difficult. Better understanding will come from improved knowledge of the rates of primary production in different size classes and taxonomic groups of phytoplankton, and from better quantification of the activities of heterotrophic microbes in the remineralization of organic matter and its sequestration in the deep ocean. We also need better knowledge of the decay rates of marine viruses, and how they will be affected by climate change and altered climatic conditions. Furthermore, climate change and pollution are undoubtedly major contributors to the rise in diseases of marine organisms—we are witnessing complete shifts in marine ecosystem functioning due to dysbiosis and microbial infections induced by environmental change. Direct impacts on human health are increasing.

Surely human ingenuity can find methods to mitigate some of the environmental damage caused by our activities? Large-scale direct application of marine microbial biotechnology in the ocean setting could play a major role in ensuring new sources of energy and sustainable aquaculture to feed the world's growing population. New initiatives in bioremediation, geoengineering, and methods to control marine diseases and conserve organisms and ecosystems. need urgent investigation, and—after proper evaluation and international approval—we must be brave enough to attempt them. Microbiologists have a major role to play in investigating these developments and in educating the public and politicians about the need for urgent and large-scale international action to ameliorate the problems we face. I hope this book will play some part in helping the next generation of students and scientists to exert this influence.

Index